FROM RANDOM WALKS TO RANDOM MATRICES

FROM RANDOM WALKS TO RANDOM MATRICES

Selected Topics in Modern Theoretical Physics

Jean Zinn-Justin

IRFU/CEA, Paris-Saclay University
and
French Academy of Sciences

OXFORD
UNIVERSITY PRESS

Great Clarendon Street, Oxford, OX2 6DP,
United Kingdom

Oxford University Press is a department of the University of Oxford.
It furthers the University's objective of excellence in research, scholarship,
and education by publishing worldwide. Oxford is a registered trade mark of
Oxford University Press in the UK and in certain other countries

First published 2019
First published in paperback 2021

Impression: 1

Published in the United States of America by Oxford University Press
198 Madison Avenue, New York, NY 10016, United States of America

British Library Cataloguing in Publication Data
Data available

Library of Congress Cataloging in Publication Data
Data available

ISBN 978–0–19–878775–4 (Hbk.)
ISBN 978–0–19–285696–8 (Pbk.)

Printed and bound by
CPI Group (UK) Ltd, Croydon, CR0 4YY

Links to third party websites are provided by Oxford in good faith and
for information only. Oxford disclaims any responsibility for the materials
contained in any third party website referenced in this work.

Preface

This work gathers a selection of seminars and mini-courses delivered during the last decade in a number of countries. In addition, four articles from Scholarpedia have been included, in newly edited form.

As a consequence, the various chapters try to be as self-contained as possible, even though the topics have been selected to involve some common themes, which give a general consistency to the whole work.

This implies some redundancy, which we have tried to minimize.

The main themes are the renormalization group (RG), fixed points, universality and the continuum limit, which open and conclude the work. Other important and related themes are path and field integrals, the notion of effective quantum or statistical field theories, gauge theories, which is the mathematical structure at the basis of the interactions in fundamental particle physics, including quantization problems and anomalies, stochastic dynamical equations and summation of perturbative series.

On purpose, in the chapters derived from seminars, we have tried to keep the technical level as low as reasonably possible, considering the topics.

Some basic historical background concerning the construction of quantum field theory (QFT), the construction of the Standard Model of interactions at the microscopic scale and the emergence of RG is provided.

Chapter 1 discusses the asymptotic properties at large time and space of the familiar example of the random walk. The emphasis here is on *universality, the continuum limit, Gaussian distribution and scaling.* These properties are first derived from an exact solution and then recovered by *RG* methods. This makes the introduction of the whole RG terminology possible.

Chapter 2 is largely descriptive and introduces the notion of functional (path and field) integrals, for boson as well as fermion systems (based on the formalism of Grassmann integration), as they are used in physics.

Chapter 3 illustrates, using a number of examples, the essential role of functional integrals in physics: the path integral representation of quantum mechanics explains why many basic equations of classical physics satisfy a variational principle, the relation between QFT and the theory of critical phenomena in macroscopic phase transitions and so on.

Chapter 4 describes a few important steps which have led from the discovery of infinities in quantum electrodynamics to the concept of renormalization and RG. RG today plays an essential role in our understanding of the properties of QFT and of continuous macroscopic phase transitions.

In Chapter 5, we first recall the importance of the concept of scale decoupling in physics. We then emphasize that QFT and the theory of critical phenomena have provided two examples where this concept fails.

To deal with such a situation, new tools have been invented based on the concept of RG.

Chapter 6 describes how the perturbative RG of QFT has made it possible to derive universal properties of continuous macroscopic phase transitions and to calculate universal quantities using dimensional continuation and Wilson–Fisher's $\varepsilon = 4 - d$ expansion. Series summation methods are then required to determine precise values of critical exponents.

Chapter 7 is devoted to a discussion of the RG flow when the effective ϕ^4 field theory that describes the universal properties of critical phenomena depends on several coupling constants. Some examples illustrate the notion of *emergent symmetry*.

Chapter 8 is more technical: we try to explain the notion of effective field theory, illustrating the topic with examples, and stress that all QFTs applied to particle or statistical physics are effective (*i.e.*, not fundamental) low energy or large distance theories.

Many RG results are derived within the framework of the perturbative RG. However, this RG is the asymptotic form in some neighbourhood of a Gaussian fixed point of the more general and exact RG, as introduced by Wilson and Wegner. In Chapter 9, we describe the corresponding RG equations and give some indications about their derivation.

In the case of the spontaneous breaking of a continuous symmetry, the correlation length diverges at all temperatures in the ordered phase. Physical systems then display universal properties for all temperatures below T_c. This phenomenon is a consequence of the existence of Goldstone modes (massless bosons, in the QFT terminology). In Chapter 10, we consider the example of the $O(N)$ symmetry and show that an RG can be constructed and universal properties derived at and near dimension 2 in the framework of an $\varepsilon = d - 2$ expansion.

Chapter 11 is the first of four chapters that discuss various issues connected with the Standard Model of fundamental interactions at the microscopic scale.

In Chapter 11, we discuss the notion of gauge invariance, first Abelian and then non-Abelian, the basic geometric structure that generates interactions.

In Chapter 12, we describe the main steps in the construction of the electroweak component of the Standard Model. The recent discovery of the last predicted particle of the Standard Model, the Higgs particle (which was searched for a number of years), has been an additional and major confirmation of the validity of the model.

Chapter 13 is devoted to some aspects of quantum chromodynamics (QCD), the part of the Standard Model responsible for strong interactions. As an introduction, the geometry of non-Abelian gauge theories, based on parallel transport, is recalled. This geometric point of view leads naturally to the construction of lattice gauge theories, which can be studied by numerical simulations.

Chapter 14 describes the important BRST symmetry of quantized gauge theories. The explicit realization of the symmetry is not stable under renormalization. By contrast, some quadratic equations satisfied by the generating functional of vertex functions (the 1PI (one-particle-irreducible) correlation functions), the Zinn-Justin equation, is stable and is at the basis of a general proof of the renormalizability of non-Abelian gauge theories.

Chapter 15 is devoted to the following question: even though QFTs are only effective low energy or large distance theories, under which conditions is it possible to define renormalized QFTs consistent on all scales? On the basis of examples, it has been suggested that a necessary and, perhaps, sufficient condition is the existence of ultraviolet fixed points, a property called 'asymptotic safety'.

Chapter 16 deals with the important problem of quantization with symmetries, that is, how to implement symmetries of the classical action in the corresponding quantum theory. The proposed solutions are based on regularization methods. Obstructions encountered in the case of chiral theories are emphasized.

In Chapter 17, we exhibit various explicit examples, where classical symmetries cannot be transferred to quantum theories. The obstructions are characterized by anomalies. The examples involve chiral symmetries combined with currents or gauge symmetries. In particular, anomalies lead to obstructions in the construction of theories and have, therefore, constrained the structure of the Standard Model. Other applications are also described.

Chapter 18 deals with a few examples where the classical action has an infinite number of degenerate minima but, in quantum theory, this degeneracy is lifted by barrier penetration effects. Technically, this corresponds to the existence of instantons, solutions to classical equations in imaginary time. In the case of QCD, this leads to the famous strong CP violation problem.

Computer simulations of critical statistical systems or QFT models are performed with systems where sizes are finite. In transfer matrix calculations, all sizes but one are also finite. In systems where the correlation length is large, it is thus important to understand how the infinite size limit is reached. This problem is investigated in Chapter 19.

After the discovery of the predicted Bose–Einstein condensation, which is a property of free bosons, an interesting issue was the effects of weak repulsive interactions. In Chapter 20, it is shown that the Bose–Einstein condensation is then replaced by a superfluid phase transition. These theoretical considerations are illustrated by an evaluation of the variation of the transition temperature at weak coupling.

Chapter 21 is devoted to a study of QFT at finite temperature, a topic that is relevant to heavy ion high energy collisions. An interesting issue concerns the conditions for dimensional reduction, that is, when can the initial field theory be replaced by an effective field theory without time dimension?

In Chapter 22, we study stochastic dynamical equations, generalized Langevin equations, which describe a wide range of phenomena from Brownian motion to critical dynamics in phase transitions. In the latter case, dynamic RG equations can be derived. Dynamic scaling follows, with a scaling time that exhibits critical slowing down.

In QFT, all perturbative expansions are divergent series, in the mathematical sense. This leads to a difficulty when the expansion parameter is not small. In the case of Borel summable series, a number of techniques have been developed to derive convergent sequences from divergent series. This is the central topic of Chapter 23.

In a number of quantum mechanics and QFT examples exhibiting degenerate classical minima related by quantum tunnelling, the perturbative expansion is not Borel summable and the perturbation series do not define unique functions. An important issue is then what kind of additional information is required to determine the exact expanded functions.

While the QFT examples are complicated and their study is still at the preliminary stage, in quantum mechanics, in the case of some analytic potentials that have degenerate minima (like the quartic double-well potential), the problem has been completely solved. Some examples are described in Chapter 24.

While the study of the statistical properties of random matrices of large size has a long history going back to Wigner, in the mid-1980s it was realized that some ensembles of random matrices in the large size and so-called double scaling limit could be used as toy models for two-dimensional quantum gravity coupled to conformal matter and string theory, or as examples of critical statistical models on some type of random surfaces.

A tremendous development of random matrix theory followed, using increasingly sophisticated mathematical methods, and number of matrix models have been solved exactly. However, the somewhat paradoxical situation is that either the models can be solved exactly or little can be said about them. Therefore, since the solved models exhibit critical points and universal properties, it is tempting to use RG ideas to determine universal properties, without solving models explicitly.

Some non-trivial progress has been achieved along these lines, which are reported in Chapter 25, but no systematic method has been discovered to go beyond the simplest approximation schemes.

This situation again illustrates the fact that RG is a fundamental idea but not a method. In order for it to be efficient, it is necessary to find a proper way to implement it.

Acknowledgements. Many thanks to Elisabeth Farrell for her careful proofreading of the manuscript.

Saclay, 18 April 2019

Contents

1 The random walk: Universality and continuum limit 1
1.1 Random walk invariant under space and discrete time translations . . 1
1.2 Fourier representation 2
1.3 Random walk: Asymptotic behaviour from a direct calculation 3
1.4 Corrections to continuum limit 5
1.5 Random walk: Fixed points of transformations and universality 6
1.6 Local and global stability of fixed points 8
1.7 Brownian motion and path integral 11

2 Functional integration: From path to field integrals 13
2.1 Random walk, Brownian motion and path integral 14
2.2 The Wiener measure and statistical physics 18
2.3 Generalization 19
2.4 Gaussian path integrals: The quantum harmonic oscillator 23
2.5 Path integrals: Perturbation theory 24
2.6 Path integral: Quantum time evolution 26
2.7 Barrier penetration in the semi-classical limit 27
2.8 Path integrals: A few generalizations 28
2.9 Path integrals for bosons and fermions 30
2.10 Field integrals: New issues 32

3 The essential role of functional integrals in modern physics 35
3.1 Classical physics: The mystery of the variational principle 35
3.2 Quantum evolution: From Hamiltonian to Lagrangian formalism . . . 38
3.3 From quantum evolution to statistical physics 41
3.4 Statistical models at criticality and quantum field theory 43
3.5 Barrier penetration, vacuum instability: Instanton calculus 44
3.6 Large order behaviour and Borel summability: Critical exponents . . . 45
3.7 Quantization of gauge theories 47
3.8 Numerical simulations in quantum field theory 50
3.9 Quantization of the non-linear σ-model 51
3.10 N-component fields: Large N techniques 52

4　From infinities in quantum electrodynamics to the general renormalization group 53
4.1　QFT, RG: Some major steps 54
4.2　QED and the problem of infinities 55
4.3　The renormalization strategy 58
4.4　The nature of divergences and the meaning of renormalization 59
4.5　QFT and RG 60
4.6　Critical phenomena: Other infinities 61
4.7　The failure of scale decoupling: The RG idea 62
4.8　Phase transitions: Exact RG in the continuum 63
4.9　Effective field theory: From critical phenomena to particle physics . . . 66

5　Renormalization Group: From a general concept to numbers 69
5.1　Scale decoupling in physics: A basic paradigm 70
5.2　Fundamental microscopic interactions 71
5.3　Macroscopic phase transitions 72
5.4　Fixed points: The QFT framework 75
5.5　RG, correlation functions and scaling relations 77
5.6　Exponents: Practical QFT calculations 78
5.7　Results for three-dimensional critical exponents 79

6　Critical phenomena: The field theory approach 81
6.1　Universality and RG 82
6.2　RG in the continuum: Abstract formulation 84
6.3　Effective field theory 85
6.4　The Gaussian field theory 87
6.5　Gaussian fixed point and Gaussian renormalization 89
6.6　Statistical scalar field theory: Perturbation theory 90
6.7　Dimensional continuation and regularization 91
6.8　Perturbative RG 92
6.9　RG equations: Solutions 96
6.10　Wilson–Fisher's fixed point: ε-Expansion 97
6.11　Critical exponents as ε-expansions 99
6.12　Three-dimensional exponents: Summation of the ε-expansion 100

7　Stability of renormalization group fixed points and decay of correlations 101
7.1　Models with only one correlation length 101
7.2　Cubic anisotropy, a model with two couplings 103
7.3　General quartic Hamiltonian: RG functions 106
7.4　Running coupling constants and gradient flows 107
7.5　Fixed point stability and value of the potential 108
7.6　Fixed point stability and field dimension 110

8　Quantum field theory: An effective theory 111
8.1　Effective local field theory: The scalar field 112
8.2　Perturbative assumption and Gaussian renormalization 113

8.3   Fundamental interactions at the microscopic scale 120

8.4   Field theory with a large mass: An explicit toy model 121

8.5   An effective field theory: The Gross–Neveu model 124

8.6   Non-linear σ-model: Another effective field theory 130

9   The non-perturbative renormalization group 137

9.1   Intuitive RG formulation 137

9.2   Non-perturbative RG equations 139

9.3   Partial field integration: Some identities 142

9.4   Partial field integration in differential form 143

10   O(N) vector model in the ordered phase: Goldstone modes 145

10.1   Classical lattice spin model and regularized non-linear σ-model . . . 146

10.2   Perturbative or low temperature expansion 149

10.3   Zero momentum or IR divergences 153

10.4   Formal continuum limit: The non-linear σ-model 154

10.5   The continuum theory: Regularization 157

10.6   Symmetry and renormalization 158

10.7   Correlation functions in dimension $d = 2 + \varepsilon$ at one loop 162

10.8   RG equations 165

10.9   Zeros of the RG β-function: Fixed points 166

10.10   Correlation functions: Scaling form below T_c 168

10.11   Linear formulation 170

10.12   Two dimensions 174

11   Gauge invariance and gauge fixing 177

11.1   Gauge invariance: A few historical remarks 177

11.2   Variational principle, charged particle and gauge invariance 178

11.3   Gauge invariance: A charged quantum particle 181

11.4   Evolution of a charged particle: Path integral representation 185

11.5   Classical electromagnetism and Maxwell's equations 185

11.6   Gauge fixing in classical gauge theories 187

11.7   QED 188

11.8   Non-Abelian gauge theories 190

11.9   Quantization of non-Abelian gauge theories: Gauge fixing 193

11.10   General Relativity 194

12   The Higgs boson: A major discovery and a problem 195

12.1   Perturbative quantum field theory: The construction 195

12.2   Spontaneous symmetry breaking 196

12.3   Non-Abelian gauge theories 198

12.4   The classical Abelian Landau–Ginzburg–Higgs mechanism 199

12.5   Abelian and non-Abelian Higgs mechanism 200

12.6   Non-Abelian gauge theories: Quantization and renormalization . . . 201

12.7   The self-coupled Higgs field: A simple RG analysis 202

12.8   The Gross–Neveu–Yukawa model: A Higgs–top toy model 204

12.9   GNY model: The general RG flow at one loop 207

12.10   The fine tuning issue 208

13 Quantum chromodynamics: A non-Abelian gauge theory 209
13.1   Geometry of gauge theories: Parallel transport 210
13.2   Gauge invariant action 212
13.3   Hamiltonian formalism. Quantization in the temporal gauge 214
13.4   Perturbation theory, regularization 219
13.5   QCD: Renormalization group 220
13.6   Anomalies: General remarks 221
13.7   QCD: The semi-classical vacuum and instantons 222
13.8   Lattice gauge theories: Generalities 225
13.9   Pure lattice gauge theory 226
13.10   Wilson loop and the confinement property 229
13.11   Fermions on the lattice. Chiral symmetry 235

14 From BRST symmetry to the Zinn-Justin equation 237
14.1   Non-Abelian gauge theories: Classical field theory 238
14.2   Non-Abelian gauge theories: The quantized action 239
14.3   BRST symmetry of the quantized action 241
14.4   The ZJ equation and remormalization 244
14.5   The ZJ equation: A few general properties 247
14.6   BRST symmetry: The algebraic origin 250

15 Quantum field theory: Asymptotic safety 253
15.1   RG and consistency 253
15.2   Super-renormalizable effective field theories: The $(\phi^2)^2$ example . . 255
15.3   A renormalizable field theory: The $(\phi^2)^2$ theory in dimension 4 . . . 258
15.4   The non-linear σ-model 260
15.5   The Gross–Neveu model 262
15.6   QCD 263
15.7   General interactions and summary 264

16 Symmetries: From classical to quantum field theories 265
16.1   Symmetries and regularization 265
16.2   Higher derivatives and momentum cut-off regularization 268
16.3   Regulator fields 271
16.4   Abelian gauge theory, the theoretical framework of QED 273
16.5   Non-Abelian gauge theories 276
16.6   Dimensional regularization and chiral symmetry 278
16.7   Lattice regularization 280

17 Quantum anomalies: A few physics applications 283
17.1   Electromagnetic decay of the neutral pion and Abelian anomaly . . 284
17.2   A two-dimensional illustration: The Schwinger model 291
17.3   Abelian axial current and non-Abelian gauge fields 298
17.4   Non-Abelian anomaly and chiral gauge theories 301
17.5   Weak and electromagnetic interactions: Anomaly cancellation . . . 303

17.6 Wess–Zumino consistency conditions 305
17.7 Lattice fermions: Ginsparg–Wilson relation 307
17.8 Supersymmetric quantum mechanics and domain wall fermions . . . 313

18 Periodic semi-classical vacuum, instantons and anomalies 319
18.1 The periodic cosine potential 319
18.2 Instantons, anomalies and θ-vacua: CP^{N-1} models 322
18.3 Non-Abelian gauge theories: Instantons and anomalies 328
18.4 The semi-classical vacuum and the strong CP violation 332
18.5 Fermions in an instanton background: The $U(1)$ problem 332

19 Field theory in a finite geometry: Finite size scaling 335
19.1 Periodic boundary conditions and the problem of the zero mode . . 335
19.2 Cylindrical geometry: Two-dimensional field theory 338
19.3 Effective $(\phi^2)^2$ field theory at criticality in finite geometries 343
19.4 Momentum quantization in finite geometries 346
19.5 The $(\phi^2)^2$ field theory in a periodic hypercube 347
19.6 The $(\phi^2)^2$ field theory: Cylindrical geometry 353
19.7 Continuous symmetries: Finite size effects at low temperature . . . 358

20 The weakly interacting Bose gas at the critical temperature 361
20.1 Bose gas: Field integral formulation 361
20.2 Independent bosons: Bose–Einstein condensation 363
20.3 The weakly interacting Bose gas and the Helium phase transition . . 364
20.4 RG and universality 365
20.5 The shift of the critical temperature for weak interaction 368

21 Quantum field theory at finite temperature 373
21.1 Finite temperature QFT: General considerations 374
21.2 Scalar field theory: Effective theory for the zero mode 376
21.3 The $(\phi^2)^2_{1,d}$ scalar QFT: Phase transitions 379
21.4 Temperature effects: The temperature-dependent mass 380
21.5 Phase structure at finite temperature at one loop 381
21.6 RG at finite temperature 384
21.7 Effective action: Perturbative calculation 386
21.8 Effective action: φ-Expansion 388
21.9 The $(\phi^2)^2$ field theory at finite temperature in the large N limit . . 389
21.10 The non-linear σ-model at finite temperature for large N 392
21.11 The GN model at finite temperature for large N 399
21.12 Abelian gauge theories: The QED example 408
A21 Appendix One-loop contributions 415
A21.1 Γ and ζ functions 415
A21.2 The one-loop two-point contribution at $T = 0$ 416
A21.3 The thermal corrections at one loop 416

22 From random walk to critical dynamics 421
22.1 Random walk with gradient driving force 422

22.2 An elementary example: The linear driving force 423
22.3 The Fokker–Planck formalism 427
22.4 Path integral representation 432
22.5 The dissipative Langevin equation: Supersymmetric formulation . . 435
22.6 Critical dynamics: The Langevin equation in field theory 439
22.7 Time-dependent correlation functions and dynamic action 442
22.8 The dissipative Langevin equation and supersymmetry 444
22.9 Renormalization of the dissipative Langevin equation 446
22.10 Dissipative Langevin equation: RG equations in $4 - \varepsilon$ dimensions . . 446

23 Field theory: Perturbative expansion and summation methods 451
23.1 Divergent series in quantum field theory 451
23.2 An example: The perturbative $(\phi^2)^2$ field theory 453
23.3 Renormalized perturbation theory: Callan–Symanzik equations . . . 455
23.4 Summation methods and critical exponents 457
23.5 ODM summation method 461
23.6 Application: The simple integral $d = 0$ 465
23.7 The quartic anharmonic oscillator: $d = 1$ 468
23.8 ϕ^4 field theory in $d = 3$ dimensions 469

24 Hyper-asymptotic expansions and instantons 471
24.1 Divergent series and Borel summability 472
24.2 Perturbative expansion and path integral 474
24.3 The quartic anharmonic oscillator: A Borel summable example . . . 475
24.4 The double-well potential: Generalized Bohr–Sommerfeld
      quantization formulae 478
24.5 Instantons and multi-instantons 484
24.6 Perturbative and exact WKB expansions 488
24.7 Other analytic potentials: A few examples 490

25 Renormalization group approach to matrix models 493
25.1 One-Hermitian matrix models and random surfaces: A summary . . 494
25.2 Continuum and double scaling limits 495
25.3 The RG approach 496

Bibliography 501

Index 521

1 The random walk: Universality and continuum limit

The *universality of a large scale behaviour* and, correspondingly, the existence of a macroscopic *continuum limit*, emerge as collective properties of systems involving a *large number of random variables* whose individual distribution is sufficiently localized.

These properties, as well as the appearance of an asymptotic *Gaussian distribution when the random variables are statistically independent*, are illustrated here with the simple example of the *random walk* with discrete time steps.

We first recall how the asymptotic large time, large space behaviour can be derived and emphasize its *universal properties*.

We then take a *renormalization group* (RG) viewpoint. Inspired by RG ideas, we introduce transformations, acting on the transition probability, which decrease the number of time steps [1]. We show that *Gaussian distributions are attractive fixed points* for these transformations. *The continuum asymptotic limit with universal scaling properties* is then recovered.

The properties of the *continuum limit* can then be described by a *path integral*.

1.1 Random walk invariant under space and discrete time translations

We consider a stochastic process, a random walk, in discrete times, first on the real axis and then, briefly, on the lattice of points with integer coordinates.

The random walk is specified by:

an initial probability distribution $P_0(q)$ (q being a position) at time $n = 0$,

a probability density $\rho(q, q') \geq 0$ for the transition from the point q' to the point q, which we assume *independent of the (integer) time n*.

These conditions define a Markov chain, a *Markovian process*, in the sense that the displacement at time n depends only on the position at time n, but not on the positions at prior times, homogeneous or stationary, that is, *invariant under time translation*, up to the boundary condition.

1.1.1 Translation invariant random walk in continuum space

Probability conservation implies

$$\int \mathrm{d}q \, \rho(q, q') = 1 \,. \tag{1.1}$$

The probability distribution $P_n(q)$ for a walker to be at point q at time n satisfies the evolution equation

$$P_{n+1}(q) = \int \mathrm{d}q' \, \rho(q, q') P_n(q'), \quad \int \mathrm{d}q \, P_0(q) = 1 \,.$$

Equation (1.1) then implies $\int \mathrm{d}q \, P_n(q) = 1$.

From Random Walks to Random Matrices. Jean Zinn-Justin, Oxford University Press (2019).
© Jean Zinn-Justin. DOI: 10.1093/oso/9780198787754.001.0001

Translation invariance. We have already assumed ρ independent of n and, thus, the transition probability is *invariant under time translation.*

In addition, we now assume that the transition probability is also *invariant under space translations* and, thus,

$$\rho(q, q') \equiv \rho(q - q').$$

As a consequence, the evolution equation takes the form of the *convolution equation,*

$$P_{n+1}(q) = \int \mathrm{d}q' \, \rho(q - q') P_n(q'),$$

which also appears in the discussion of the central limit theorem of probabilities.

Local random walk. We consider only transition functions piecewise differentiable and with bounded variation. We further assume that the transition probability $\rho(q)$ satisfies a bound of exponential form,

$$\rho(q) \leq M \, \mathrm{e}^{-A|q|}, \quad M, A > 0,$$

a property of *exponential decay* that we call *locality.* Qualitatively, *large displacements have a very small probability.*

Generalization to $\mathbb{R}^d$. The generalization to a translation invariant walk in $\mathbb{R}^d$ is simple. In particular, in the case of space rotation symmetry, the modification of the evolution equation is straightforward.

1.2 Fourier representation

The evolution equation simplifies after Fourier transformation. We thus introduce

$$\tilde{P}_n(k) = \int \mathrm{d}q \, \mathrm{e}^{-ikq} \, P_n(q) \,,$$

which is also a *generating function of the moments of the distribution $P_n(q)$.*

The reality of $P_n(q)$ and the normalization of the total probability imply

$$\tilde{P}_n^*(k) = \tilde{P}_n(-k), \quad \tilde{P}_n(k = 0) = 1 \,.$$

Similarly, we introduce

$$\tilde{\rho}(k) = \int \mathrm{d}q \, \mathrm{e}^{-ikq} \, \rho(q),$$

which is also a *generating function of the moments of the distribution $\rho(q)$:*

$$\langle q^r \rangle = \int \mathrm{d}q \, \rho(q) q^r.$$

Finally, the exponential decay condition implies that *the function $\tilde{\rho}(k)$ is analytic in the strip $|\operatorname{Im} k| < A$* and, thus, has a convergent expansion at $k = 0$.

The evolution equation then becomes

$$\tilde{P}_{n+1}(k) = \tilde{\rho}(k) \tilde{P}_n(k).$$

To slightly simplify the analysis, we take as an initial distribution $P_0(q) = \delta(q)$, where δ is Dirac's distribution (the walker at initial time is at $q = 0$ with probability 1). With this choice of initial conditions, $\tilde{P}_0(k) = 1$ and, thus,

$$\tilde{P}_n(k) = \tilde{\rho}^n(k).$$

1.2.1 Generating function of cumulants

We introduce the generating function of the cumulants of $\rho(q)$,

$$w(k) = \ln \tilde{\rho}(k) \Rightarrow w^*(k) = w(-k), \quad w(0) = 0. \tag{1.2}$$

Then,

$$\tilde{P}_n(k) = e^{nw(k)}, \quad P_n(q) = \frac{1}{2\pi} \int dk\, e^{ikq}\, \tilde{P}_n(k) = \frac{1}{2\pi} \int dk\, e^{ikq+nw(k)}.$$

The regularity of $\tilde{\rho}(k)$ and the condition $\tilde{\rho}(0) = 1$ imply that $w(k)$ has a regular expansion at $k = 0$ of the form

$$w(k) = -iw_1 k - \frac{1}{2}w_2 k^2 + \sum_{r=3} \frac{(-i)^r}{r!} w_r k^r,$$

where w_r is the rth cumulant, for example,

$$w_1 = \langle q \rangle, \quad w_2 = \langle q^2 \rangle - \langle q \rangle^2 = \langle (q - \langle q \rangle)^2 \rangle \geq 0.$$

1.3 Random walk: Asymptotic behaviour from a direct calculation

With the hypotheses satisfied by P_0 and ρ, the determination of the asymptotic behaviour for $n \to \infty$ follows from arguments identical to those leading to *the central limit theorem of probabilities*.

For $n \to \infty$ and $w_1 \neq 0$, $w(k)$ is dominated by the first term and, thus,

$$P_n(q) \underset{n\to\infty}{\sim} \frac{1}{2\pi} \int dk\, e^{ikq-inw_1 k} = \delta(q - nw_1).$$

The random variable q/n converges with probability 1 towards its expectation value w_1 (the mean velocity).

For $n \to \infty$ and $w_1 = 0$, $w(k)$ is dominated by the term of order k^2 and, thus,

$$P_n(q) \underset{n\to\infty}{\sim} \frac{1}{2\pi} \int dk\, e^{ikq-nw_2 k^2/2} = \frac{1}{\sqrt{2\pi n w_2}}\, e^{-q^2/2nw_2}.$$

Then it is the random variable $q/\sqrt{n}$ that has as its limiting distribution, a Gaussian distribution with width $\sqrt{w_2}$.

The random variable that characterizes the deviation with respect to the mean trajectory,

$$X = (s - w_1)\sqrt{n} = \frac{q}{\sqrt{n}} - w_1\sqrt{n}, \tag{1.3}$$

and, thus, $\langle X \rangle = 0$, has, as limiting distribution, the *universal Gaussian distribution*

$$L_n(X) = \sqrt{n} P_n(nw_1 + X\sqrt{n}) \sim \frac{1}{\sqrt{2\pi w_2}}\, e^{-X^2/2w_2},$$

which depends only on the parameter w_2.

The neglected terms are of two types, multiplicative corrections of order $1/\sqrt{n}$ and additive corrections decreasing exponentially with n.

The result implies that *the mean deviation from the mean trajectory increases as the square root of time*, a characteristic property of *Brownian motion*.

1.3.1 Continuum time limit

The asymptotic Gaussian distribution of the deviation $\bar{q} = q - nw_1$ from the mean trajectory is

$$P_n(\bar{q}) \sim \frac{1}{\sqrt{2\pi n w_2}} \, \mathrm{e}^{-\bar{q}^2/2nw_2} \, .$$

By changing the time scale and by a continuous interpolation, one can define a diffusion process or Brownian motion in continuous time.

Let t and ε be two real positive numbers and n the integer part of t/ε:

$$n = [t/\varepsilon] \, . \tag{1.4}$$

One then takes the limit $\varepsilon \to 0$ at t fixed and, thus, $n \to \infty$.

If the time t is measured with a finite precision Δt, as soon as $\Delta t \gg \varepsilon$, time can be considered as a continuous variable for what concerns all expectation values of continuous functions of time.

One then performs the change of distance scale

$$\bar{q} = x/\sqrt{\varepsilon} \, .$$

Since the Gaussian function is continuous, the limiting distribution takes the form

$$P_n(q)/\sqrt{\varepsilon} \underset{\varepsilon \to 0}{\sim} \Pi(t, x) = \frac{1}{\sqrt{2\pi t w_2}} \, \mathrm{e}^{-x^2/2tw_2} \, . \tag{1.5}$$

(The change of variables $q \mapsto x$ induces a change of normalization of the distribution.)

In the limit $n \to \infty$ and in suitable macroscopic variables, one obtains a diffusion process that can entirely be described in *continuum time and space*. The limiting distribution is a solution of the diffusion or heat equation,

$$\frac{\partial}{\partial t} \Pi(t, \mathbf{x}) = \tfrac{1}{2} w_2 \frac{\partial^2}{(\partial x)^2} \Pi(t, x) \, .$$

The distribution $\Pi(t, x)$ implies a *scaling property* characteristic of the Brownian motion. The moments of the distribution satisfy

$$\langle x^{2m} \rangle = \int \mathrm{d}x \, x^{2m} \Pi(t, x) \propto t^m \, . \tag{1.6}$$

The variable $x/\sqrt{t}$ has time-independent moments.

Dimensions. As the change $\bar{q} = x/\sqrt{\varepsilon}$ also indicates, *one can thus assign to position x a scaling dimension $1/2$ in time units* (this also corresponds to *assigning a Hausdorff dimension 2 to a Brownian trajectory* in higher dimensions).

1.4 Corrections to continuum limit

We now study how deviations from the limiting Gaussian distribution decay when $\varepsilon \to 0$.

We express the distribution of $\bar{q} = q - nw_1$ in terms of $w(k) = \ln \tilde{\rho}(k)$,

$$\rho(\bar{q}) = \frac{1}{2\pi} \int dk \; e^{ik\bar{q}+iknw_1} \, e^{nw(k)} = \frac{1}{2\pi} \int dk \; e^{ik\bar{q}+n\bar{w}(k)},$$

where we have introduced

$$\bar{w}(k) = w(k) + ikw_1 \,.$$

With our assumptions, the expansion of the regular function $\bar{w}(k)$ in powers of k takes the form

$$\bar{w}(k) = -\frac{1}{2}w_2 k^2 + \sum_{r=3} \frac{(-i)^r}{r!} w_r k^r \,.$$

After the introduction of macroscopic variables, which for the Fourier variables correspond to $k = \kappa\sqrt{\varepsilon}$, one finds

$$n\bar{w}(k) = t\omega(\kappa) \text{ with } \omega(\kappa) = -\frac{w_2}{2!}\kappa^2 + \sum_{r=3} \varepsilon^{r/2-1} \frac{(-i)^r}{r!} w_r \kappa^r \,.$$

One observes that, when $\varepsilon = t/n$ goes to zero, to each additional power of κ there corresponds an additional power of $\sqrt{\varepsilon}$.

In the continuum limit, the distribution becomes

$$\Pi(t,x) = \frac{1}{2\pi} \int d\kappa \, e^{-i\kappa x} \, e^{tw(\kappa)} \,.$$

Differentiating with respect to the time t, one obtains

$$\frac{\partial}{\partial t}\Pi(t,x) = \frac{1}{2\pi} \int d\kappa \, w(\kappa) \, e^{-i\kappa x} \, e^{tw(\kappa)}$$

and in $w(\kappa)$, κ can then be replaced by the differential operator $i\partial/\partial x$.

One thus finds that $\Pi(t,x)$ satisfies the linear generalized 'partial differential equation'

$$\frac{\partial}{\partial t}\Pi(t,x) = \left[\frac{w_2}{2!} \left(\frac{\partial}{\partial x} \right)^2 + \sum_{r=3} \varepsilon^{r/2-1} \frac{1}{r!} w_r \left(\frac{\partial}{\partial x} \right)^r \right] \Pi(t,x).$$

In the expansion, each additional derivative implies an additional factor $\sqrt{\varepsilon}$ and, thus, the *contributions that contain more derivatives decrease faster to zero.*

1.5 Random walk: Fixed points of transformations and universality

We now derive the *universal properties* of the asymptotic random walk, that is, the existence of a *limiting Gaussian distribution* independent of the initial distribution with its *scaling property*, by a quite different method that *does not involve calculating the asymptotic distribution explicitly.*

For simplicity, we assume that the initial number of time steps is of the form $n = 2^m$. The idea then is to *recursively combine the time steps two by two*, decreasing the number of steps by a factor 2 at each iteration. We then look for the *fixed points* of such a transformation.

This method provides a simple *application of RG ideas to the derivation of universal properties.*

It also allows us to introduce some basic *RG terminology.*

1.5.1 *Time scale transformation and renormalization*

At each iteration, one replaces $\rho(q - q')$ by

$$[\mathcal{T}\rho](q - q') \equiv \int \mathrm{d}q'' \rho(q - q'')\rho(q'' - q') = \int \mathrm{d}q'' \rho(q - q' - q'')\rho(q''),$$

rescaling the time scale by a factor 2.

The transformation of the distribution $\rho(q)$ is non-linear but, applied to the function $w(k) = \ln \tilde{\rho}(k)$, it becomes a linear transformation, since

$$[\mathcal{T}\tilde{\rho}](k) = \tilde{\rho}^2(k) \ \Rightarrow \ [\mathcal{T}w](k) \equiv 2w(k).$$

This transformation has an important property: it is independent of m or n. In the language of dynamical systems, its repeated application generates a *stationary, or invariant under time translation, Markovian dynamics.*

Large time behaviour and fixed points. The large time behaviour is obtained by iterating the transformation, studying $\mathcal{T}^m$ for $m \to \infty$.

A limiting distribution necessarily is a fixed point of the transformation.

It corresponds to a function $w_*(k)$ (the notation $_*$ is not related to complex conjugation) that satisfies

$$[\mathcal{T}w_*](k) \equiv 2w_*(k) = w_*(k). \tag{1.7}$$

For the class of fast decreasing distributions, the function $w_*(k)$ has an expansion in powers of k of the form ($w_*(0) = 0$)

$$w_*(k) = -iw_1 k - \tfrac{1}{2}w_2 k^2 + \sum_{\ell=3} \frac{(-i)^\ell}{\ell!} w_\ell k^\ell \quad w_2 \geq 0. \tag{1.8}$$

With this assumption, the fixed point equation (1.7) has only the trivial solution $w_*(k) \equiv 0$.

To the time rescaling must be associated a rescaling (a *renormalization*, in quantum field theory terminology) of the random space variable q.

Random variable: renormalization. Non-trivial fixed points can be reached if the transformation is combined with a *renormalization of the distance scale*, $q \mapsto zq$, with $z > 0$. We thus consider the transformation

$$[\mathcal{T}_z w](k) \equiv 2w(k/z).$$

The transformation $\mathcal{T}_z$ provides a simple example of an *RG transformation*, a concept that we describe thoroughly in the framework of phase transitions.

The fixed point equation then becomes

$$[\mathcal{T}_z w_*](k) \equiv 2w_*(k/z) = w_*(k),$$

which determines the possible values of z and the corresponding functions $w_*(k)$.

Dimension of the random variable. Comparing the rescaling of time and the random variable q, one can attach to q a *scaling dimension d_q*, in time units, defined by

$$d_q = \ln z / \ln 2. \tag{1.9}$$

1.5.2 Fixed points: generic situation: $w_1 \neq 0$

In the expansion (1.8) of the function $w_*(k)$, in the generic situation, the first term w_1 does not vanish. Expanding the RG equation, at order k, one finds

$$2w_1/z = w_1 \text{ and } w_1 \neq 0 \; \Rightarrow \; z = 2.$$

Then, identifying the terms of higher degree, one concludes

$$2^{1-\ell} w_\ell = w_\ell \; \Rightarrow \; w_\ell = 0 \text{ for } \ell > 1.$$

Therefore, a fixed point solution is

$$w_*(k) = -iw_1 k.$$

The fixed points form a one-parameter family, but the parameter w_1 can also be absorbed into a normalization of the random variable q.

Since

$$\rho_*(q) = \frac{1}{2\pi} \int dk \; e^{ikq - iw_1 k} = \delta(q - w_1),$$

fixed points correspond to the certain distribution $q = \langle q \rangle = w_1$.

Since *space and time are rescaled by the same factor 2*, q has a scaling dimension $d_q = \ln z / \ln 2 = 1$ in time units.

Consistently, the fixed point corresponds to $q(t) = w_1 t$, the *equation of the mean path*.

1.5.3 Centred distribution

For a centred distribution, $w_1 = 0$ and one has to expand to order k^2. One finds the equation

$$w_2 = 2w_2/z^2.$$

Since the variance w_2 is strictly positive, except for a certain distribution, a case that we now exclude, the equation implies $z = \sqrt{2}$.

Again, the coefficients w_ℓ vanish for $\ell > 2$ and the fixed points have the form

$$w_*(k) = -\tfrac{1}{2}w_2 k^2.$$

Therefore, one finds the Gaussian distribution

$$\rho_*(q) = \frac{1}{2\pi} \int \mathrm{d}k \; \mathrm{e}^{ikq - w_2 k^2/2} = \frac{1}{\sqrt{2\pi w_2}} \, \mathrm{e}^{-q^2/2w_2}\,.$$

The dimension $d_q = \ln z/\ln 2 = \tfrac{1}{2}$ *is consistent with the scaling property* $x \propto \sqrt{t}$ *of Brownian motion.*

The two essential asymptotic properties of the random walk, *convergence towards a Gaussian distribution, and scaling property*, are thus reproduced by this RG-type analysis.

1.6 Local and global stability of fixed points

For a non-linear RG transformation, a global stability analysis is, in general, impossible. One can only study the local stability of fixed points. Here, since the transformation is linear, local and global stabilities are equivalent.

1.6.1 General analysis and RG terminology

Setting

$$w(k) = w_*(k) + \delta w(k),$$

one finds,

$$[\mathcal{T}_z \delta w](k) \equiv 2\delta w(k/z)\,.$$

One then looks for the *eigenvectors and eigenvalues* of the transformation $\mathcal{T}_z$:

$$[\mathcal{T}_z \delta w](k) = \tau \delta w(k).$$

To the eigenvalue τ, one associates the *exponent*

$$\alpha = \ln \tau/\ln 2\,.$$

The perturbation δw has an expansion in powers of k of the form,

$$\delta w(k) = \sum_{\ell=1}^{} \frac{(-i)^\ell}{\ell!} \delta w_\ell k^\ell\,.$$

Then,

$$[\mathcal{T}_z \delta w](k) = 2\delta w(k/z) = 2 \sum_{\ell=1} \frac{(-ik)^\ell}{\ell!} z^{-\ell} \delta w_\ell \,.$$

The expression shows that the functions k^ℓ with $\ell > 0$ are the *eigenvectors* of the transformation $\mathcal{T}_z$ and the corresponding eigenvalues are

$$\tau_\ell = 2z^{-\ell} \;\Rightarrow\; \alpha_\ell = \ln \tau_\ell / \ln 2 = 1 - \ell \ln z / \ln 2 \,.$$

For $n = 2^m$ time steps, after m iterations, the component δw_ℓ is multiplied by n^{α_ℓ} since

$$\mathcal{T}_z^m k^\ell = \tau_\ell^m k^\ell = 2^{m\alpha_\ell} = n^{\alpha_\ell} k^\ell.$$

The behaviour, for $n \to \infty$, of the component of δw_ℓ on the eigenvector k^ℓ thus depends on the sign of the exponent α_ℓ for the various values of ℓ.

Definitions

$\alpha > 0$: The perturbations correspond to unstable directions; a component on the corresponding eigenvector diverges for $m \to \infty$. In RG terminology, a perturbation corresponding to a positive exponent α and which thus leads away from the fixed point, is called *relevant*.

$\alpha = 0$: Such perturbations are called *marginal*. In the general RG framework, a stability study requires going beyond the linear approximation.

$\alpha < 0$: The perturbations correspond to stable directions and are called *irrelevant*.

The notion of universality. Universality, in the RG formulation, is a consequence of the property that all eigenvectors, but a finite number, are irrelevant.

Line of fixed points. Quite generally, the existence of a one-parameter family of fixed points implies the existence of an eigenvalue $\tau = 1$ and, thus, an exponent $\alpha = 0$. Indeed, let us assume the existence of a one-parameter family of fixed points $w_*(s)$,

$$\mathcal{T} w_*(s) = w_*(s) \,,$$

where $w_*(s)$ is a differentiable function of the parameter s. Then,

$$\mathcal{T} \frac{\partial w_*}{\partial s} = \frac{\partial w_*}{\partial s} \,.$$

1.6.2   Fixed point stability: $w_1 \neq 0$

Having introduced some elements of *RG terminology*, we now discuss eigenvectors and eigenvalues or corresponding exponents.

(i) $\ell = 1 \Rightarrow \tau_1 = 1$, $\alpha_1 = 0$. If one adds a term δw proportional to the eigenvector k to $w_*(k)$, $\delta w(k) = -i\delta w_1 k$, then

$$w_1 \mapsto w_1 + \delta w_1 \,,$$

which corresponds to a new fixed point. This change can also be interpreted as a linear transformation on k or on the random variable q.

Since the exponent α_1 vanishes, the corresponding eigen-perturbation is *marginal*.

(ii) $\ell > 1 \Rightarrow \tau_\ell = 2^{1-\ell} < 1$, $\alpha_\ell < 0$. The eigen-pertubations are *irrelevant* since they correspond to *negative exponents*. The components of δw on such eigenvectors converge to zero for n or $m \to \infty$.

1.6.3 Fixed point stability: $w_1 = 0$.

We now study the stability of the fixed point corresponding to the transformation $\mathcal{T}_{\sqrt{2}}$. One sets

$$w(k) = w_*(k) + \delta w(k),$$

and looks for the eigenvectors and eigenvalues of the transformation

$$[\mathcal{T}_{\sqrt{2}} \delta w](k) \equiv 2\delta w(k/\sqrt{2}) = \tau \delta w(k).$$

The eigenvalues are

$$\tau_\ell = 2^{1-\ell/2}.$$

The corresponding exponents are

$$\alpha_\ell = \ln \tau_\ell / \ln 2 = 1 - \ell/2.$$

The values can be classified as:

(i) $\ell = 1 \Rightarrow \tau_1 = \sqrt{2}$, $\alpha_1 = \frac{1}{2}$. This corresponds to an unstable direction; a component on such a eigenvector diverges for $m \to \infty$.

In the RG terminology, the perturbation, which corresponds to a positive exponent α and which thus leads away from the fixed point, is called *relevant*.

Here, a perturbation linear in k violates the condition $w_1 = 0$. One is then brought back to the study of the more stable fixed points with $w_1 \neq 0$.

(ii) $\ell = 2 \Rightarrow \tau_2 = 1$, $\alpha_2 = 0$. A vanishing eigenvalue characterizes a *marginal* perturbation. Here, the perturbation only modifies the value of w_2 and, again, can be interpreted as a linear transformation on the random variable.

(iii) $\ell > 2 \Rightarrow \tau_\ell = 2^{1-\ell/2} < 1$, $\alpha_\ell = 1 - \ell/2 < 0$. Finally, all perturbations $\ell > 2$ correspond to stable directions in the sense that their amplitudes converge to zero for $m \to \infty$ and are *irrelevant*.

Redundant perturbations. In the examples examined here, the marginal perturbations correspond to simple changes in the normalization of the random variables. In many problems, this normalization plays no role. One can then consider that *fixed points corresponding to different normalizations should not be distinguished.*

From this viewpoint, in both cases one has found really only one fixed point. The perturbation corresponding to the vanishing eigenvalue is then no longer called marginal but *redundant*, in the sense that it changes only an arbitrary normalization.

Other universality classes. Other values of $z = 2^{1/\mu}$, correspond formally to new fixed points of the form $|k|^\mu$, $0 < \mu < 2$ ($\mu > 2$ is excluded because the coefficient of k^2 is strictly positive).

However, these fixed points are *no longer regular functions* of k. They correspond to distributions that have no second moment $\langle q^2 \rangle$ and thus no variance: they decay only algebraically for large values of q. In the RG terminology, they correspond to different *universality classes*, distributions with other decay properties.

1.6.4 Random walk on a lattice of points with integer coordinates

The analysis can be generalized to a random walk on the points of integer coordinates like $\mathbb{Z}^d$ (the simplest example being motion on the line by one lattice spacing left or right with probability $1/2$). Then, $w(\mathbf{k})$ is a periodic function of period 2π for each space direction.

However, at each RG transformation, the period is multiplied by a factor $z > 1$. Thus, asymptotically, the period diverges and, at least for continuous observables, the discrete character of the initial lattice disappears.

In the d-dimensional lattice $\mathbb{Z}^d$, if the random walk has *hypercubic symmetry*,

$$k_i \mapsto -k_i, \quad k_i \leftrightarrow k_j, \quad i, j = 1, \dots, d,$$

the leading term in the expansion of $w(\mathbf{k})$ for $\mathbf{k}$ small is $\frac{1}{2}w_2\mathbf{k}^2$, because it is the only quadratic hypercubic invariant. Therefore, asymptotically, the random walk is *Brownian motion with space rotation symmetry*.

The lattice structure is only apparent in the first irrelevant perturbation because there exists two independent cubic invariant monomials of degree 4:

$$\sum_{\mu=1}^{d} k_\mu^4, \quad \left(\mathbf{k}^2\right)^2.$$

1.7 Brownian motion and path integral

An iteration of the evolution equation of the translation invariant random walk

$$P_n(q) = \int \mathrm{d}q' \rho(q - q') P_{n-1}(q'),$$

in the case of a certain initial position $q = q_0 = 0$, yields

$$P_n(q) = \int \mathrm{d}q' \mathrm{d}q_1 \, \mathrm{d}q_2 \, \cdots \, \mathrm{d}q_{n-1} \, \rho(q - q_{n-1}) \cdots \rho(q_2 - q_1) \rho(q_1).$$

If one is interested only in the asymptotic properties of the distribution, which have been shown to be independent of the initial transition probability, one can derive them, in the continuum limit, starting directly from *Gaussian transition probabilities* of the form

$$\rho(q) = \frac{1}{(2\pi w_2)^{1/2}} \, \mathrm{e}^{-q^2/2w_2}.$$

The iterated evolution equation becomes

$$P_n(q) = \frac{1}{(2\pi w_2)^{n/2}} \int \mathrm{d}q_1 \, \mathrm{d}q_2 \, \cdots \, \mathrm{d}q_{n-1} \, \mathrm{e}^{-\mathcal{S}(q_0, q_2, \dots, q_n)} \tag{1.10}$$

with $q_n = q$ and

$$\mathcal{S}(q_0, q_2, \dots, q_n) = \sum_{\ell=1}^{n} \frac{(q_\ell - q_{\ell-1})^2}{2w_2}.$$

We then introduce macroscopic time and space variables (see Section 2.1),

$$\tau_\ell = \ell\varepsilon\,, \quad \tau_n = n\varepsilon = t\,, \quad x_k = \sqrt{\varepsilon}\,q_k\,, \quad \text{with } 0 \le k \le n\,,$$

and a continuous, piecewise linear path $x(\tau)$ (Fig. 1.1),

$$x(\tau) = \sqrt{\varepsilon}\left[q_{\ell-1} + \frac{\tau - \tau_{\ell-1}}{\tau_\ell - \tau_{\ell-1}}\left(q_\ell - q_{\ell-1}\right)\right] \quad \text{for } \tau_{\ell-1} \le \tau \le \tau_\ell\,.$$

The relative powers of ε between space and time again reflect the scaling properties of the random walk.

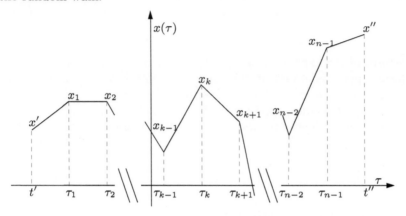

Fig. 1.1 A *piecewise linear path* contributing to the *time-discretized* path integral.

One verifies that $\mathcal{S}$ can be written as (with the notation $\dot{x}(\tau) \equiv dx/d\tau$)

$$\mathcal{S}\big(x(\tau)\big) = \frac{1}{2w_2}\int_0^t \big(\dot{x}(\tau)\big)^2 d\tau$$

with the boundary conditions

$$x(0) = 0\,, \quad x(t) = \sqrt{\varepsilon}q = \mathbf{x}\,.$$

Moreover,

$$P_n(q) = \frac{1}{(2\pi w_2)^{1/2}}\int\left(\prod_{\ell=1}^{n-1}\frac{dx(\tau_\ell)}{(2\pi w_2\varepsilon)^{1/2}}\right)e^{-\mathcal{S}(x)}\,.$$

In the continuum limit $\varepsilon \to 0$, $n \to \infty$ with t fixed, the expression becomes a representation of the distribution of the continuum limit,

$$\Pi(t,x) \sim \varepsilon^{-1/2}P_n(q),$$

in the form of a *path integral*, which we denote symbolically as

$$\Pi(t,x) = \int [dx(\tau)]\,e^{-\mathcal{S}(x(\tau))}\,,$$

where $\int[dx(\tau)]$ means sum over all continuous paths that start from the origin at time $\tau = 0$ and reach x at time t. The trajectories that contribute to the path integral correspond to *Brownian motion*, a random walk in *continuum time and space*. The representation of Brownian motion by path integrals, initially introduced by Wiener, is also called a *Wiener integral* [2] (see Section 2.1).

2 Functional integration: From path to field integrals

Prior to the second half of the twentieth century, the technical tools of theoretical physics were mainly differential or partial differential equations. However, in the twentieth century, systematic investigations of *large scale systems with quantum or statistical fluctuations* were started. New tools were then required. Among them, a major tool has been *functional integration*.

In this introductory chapter, which is an edited version of Ref. [3], we describe various forms of functional integrals as they are used in different branches of theoretical physics. A rigorous study of the mathematical properties of functional integrals is still an open subtopic of functional analysis and will not be considered here.

Path integrals involve weighted summation over all paths satisfying some boundary conditions and can be understood as extensions to an infinite number of integration variables of usual multidimensional integrals.

Path integrals are powerful tools for the study of *quantum mechanics*. Indeed, in quantum mechanics, physical quantities can be expressed as averages over all possible paths weighted by the exponential of a term proportional to the ratio of the classical action $\mathcal{S}$ associated to each path, divided by the Planck's constant $\hbar$ [4]. Thus, path integrals emphasize very explicitly the correspondence between classical and quantum mechanics and give an explicit meaning to the notion of quantum fluctuations. In particular, in the semi-classical limit $\mathcal{S}/\hbar \to \infty$, the leading contributions to the integral come from paths close to classical paths, which are stationary points of the action: path integrals lead to an intuitive understanding and simple calculations of physical quantities in the semi-classical limit.

The formulation of quantum mechanics based on path integrals is well adapted to systems with many degrees of freedom, where a formalism of Schrödinger type is much less useful. Therefore, it allows an easy transition from quantum mechanics to quantum field theory or statistical physics. In particular, generalized path integrals (known as field integrals) lead to an understanding of the deep relations between quantum field theory and the theory of critical phenomena in continuous phase transitions.

We first describe path integrals encountered in the study of *Brownian motion* [2] and of quantum statistical mechanics [5] (also *Euclidean time i.e.,* imaginary time quantum mechanics). This means that we consider the path integral representation of the matrix elements of the quantum statistical operator, or density matrix at *thermal equilibrium* $e^{-\beta \hat{H}}$, $\hat{H}$ being the quantum Hamiltonian and β the inverse temperature (measured in a unit where the Boltzmann constant k_B is 1).

This remarkable representation of quantum statistical systems also makes it possible, perhaps more surprisingly, to show a relationship between classical and quantum statistical mechanics. Indeed, for a whole class of quantum Hamiltonians, the integrand in the Euclidean time path integral defines a positive measure.

From Random Walks to Random Matrices. Jean Zinn-Justin, Oxford University Press (2019).
© Jean Zinn-Justin. DOI: 10.1093/oso/9780198787754.001.0001

We then define the *real time* (in relativistic field theory Minkowskian time) path integral [4], which describes the time evolution of quantum systems and corresponds, for time-translation invariant systems, to the evolution operator $\mathrm{e}^{-it\hat{H}/\hbar}$ (t being the real time).

Finally, we list a few generalizations: path integrals in the Hamiltonian formulation, path integrals in the holomorphic representation relevant for boson systems and, correspondingly, Grassmann path integrals for fermions [6].

Most of the applications of the path integral idea to physics involve, in fact, integrals over fields (see Chapter 3). In particular, field integrals are indispensable for the construction of quantized gauge invariant theories (see Chapter 13), which are at the basis of the Standard Model of fundamental interactions at the microscopic scale, as well as for describing *universal properties* of phase transitions (Chapter 6).

These applications rely on a pragmatic approach, focusing more on developing calculation tools (including large scale numerical simulations) than on establishing rigorous properties. Indeed, even though a number of interesting rigorous results have been proved, the construction of some realistic quantum field theories, like gauge theories in four space-time dimensions, remains a formidable mathematical challenge.

Path integrals: The origins. The first *path integral* seems to have been defined by *Wiener* [2], as a tool to describe the statistical properties of *Brownian motion*, inspired by the famous work of Einstein. If Wiener's work is rather well known, a less-known article of *Wentzel* [7] of about the same period introduces, in the framework of *quantum optics*, the notions of *sums over paths weighted by a phase factor*, of destructive interference between paths that do not satisfy classical equations of motion, and the *interpretation of the sum as a transition probability amplitude*.

Later, *Dirac* [8] calculated the matrix elements of the quantum evolution operator for infinitesimal time intervals. The generalization of the result to finite time intervals would have led to a path integral.

In physics, the modern history of path integrals really begins with the articles of *Feynman* [4], who formulated *quantum evolution in terms of sums over a set of trajectories weighted by* $\mathrm{e}^{i\mathcal{A}/\hbar}$, *where $\mathcal{A}$ is the value of the corresponding classical action (time integral of the Lagrangian) and $\hbar$ is Planck's constant*.

2.1 Random walk, Brownian motion and path integral

As a first example, we consider a random walk on the real line with discrete times $n = 0, 1, 2, \cdots$ (see Chapter 1). Such a stochastic process is specified by a probability distribution $P_0(q)$ for the position q at initial time $n = 0$ and a density ρ describing the probability of transition from the point q' to the point q. We assume that ρ is time independent, translation invariant and symmetric: $\rho(q - q') = \rho(q' - q)$.

The probability distribution $P_n(q)$ at time n satisfies the recursion relation or *master equation*

$$P_n(q) = \int \mathrm{d}q'\, \rho(q - q') P_{n-1}(q'), \quad \int \mathrm{d}q\, \rho(q, q') = 1\,. \tag{2.1}$$

Under rather general conditions (see Chapter 1), the most important being that $\rho(q)$ decreases fast enough for $|q|$ large (we call this a *local Markov process*), one can prove (a consequence of the central limit theorem of probabilities) that the distribution $P_n(q)$ converges asymptotically for large time and space towards a Gaussian distribution that *depends* only on the second moment of the transition probability.

Therefore, if one is interested only in large time and distance properties of the random walk, one can start directly from an *effective* Gaussian transition probability of the form

$$\rho(q) = \frac{1}{\sqrt{2\pi\xi}} \, \mathrm{e}^{-q^2/(2\xi)}, \tag{2.2}$$

where $\xi > 0$ characterizes the width of the distribution.

2.1.1   Continuum limit and path integral

In the example of a Gaussian transition probability, it is easy to calculate $P_n(q)$ explicitly by successive Gaussian integrations. However, for our purpose, it is more instructive to implement directly the recursion relation (2.1). If one assumes, for example, that the initial distribution is concentrated at the point $q = q_0$ (*i.e.*, $P_0(q) = \delta(q - q_0)$, where $\delta(q)$ is Dirac's generalized function (a distribution) also known as *Dirac function* or δ-function), one obtains at time n the probability distribution

$$P_n(q, q_0) = \int \mathrm{d}q_{n-1}\mathrm{d}q_{n-2} \cdots \mathrm{d}q_1 \, \rho(q - q_{n-1})\rho(q_{n-1} - q_{n-2}) \cdots \rho(q_1 - q_0). \tag{2.3}$$

In the Gaussian example (2.2), the expression takes the form

$$P_n(q, q_0) = (2\pi\xi)^{(1-n)/2} \int \mathrm{d}q_{n-1}\mathrm{d}q_{n-2} \cdots \mathrm{d}q_1 \, \mathrm{e}^{-\mathcal{S}(\mathbf{q})/\xi}, \tag{2.4}$$

where, defining $\mathbf{q} \equiv (q_0, q_1, \ldots, q_n)$ and $q \equiv q_n$,

$$\mathcal{S}(\mathbf{q}) = \frac{1}{2}\sum_{k=1}^{n}(q_k - q_{k-1})^2 \, .$$

We now introduce a time step $\varepsilon > 0$, the macroscopic time and space variables

$$\tau_k = t' + k\varepsilon, \quad x_k = \sqrt{\varepsilon}\, q_k, \quad \text{with } 0 \le k \le n,$$

(such that $\tau_0 = t'$, $\tau_n = t' + n\varepsilon \equiv t''$) and a continuous, piecewise linear path (see Fig. 1.1)

$$x(\tau) = \sqrt{\varepsilon}\left[q_{k-1} + \frac{\tau - \tau_{k-1}}{\tau_k - \tau_{k-1}}(q_k - q_{k-1}) \right] \text{ for } \tau_{k-1} \le \tau \le \tau_k \text{ and } k \ge 1, \tag{2.5}$$

with the boundary conditions

$$x(t') = \sqrt{\varepsilon}\, q_0 \equiv x', \quad x(t'') = \sqrt{\varepsilon}\, q \equiv x''. \tag{2.6}$$

One verifies that $\mathcal{S}(\mathbf{q})$ can then be rewritten as

$$\mathcal{S}(\mathbf{q}) = \mathcal{S}_\varepsilon(\mathbf{x}) \equiv \frac{1}{2} \int_{t'}^{t''} \dot{x}^2(\tau) \mathrm{d}\tau \,,$$

where $\dot{x}(\tau) \equiv \mathrm{d}x/\mathrm{d}\tau$.

The large time asymptotic behaviour is obtained by taking the large n limit at $(t'' - t')$ fixed and, thus, $\varepsilon = (t'' - t')/n \to 0$. This defines a *continuum limit* in time since the time step goes to zero.

In this limit, the normalized probability distribution $\Pi_0(t'', t'; x'', x')$ in the new variables is given by an Euclidean time *path integral* [2] that we denote by (see also Section 1.7)

$$\Pi_0(t'', t'; x'', x') = \lim_{n\to\infty} \frac{1}{\sqrt{\varepsilon}} P_n(q, q_0) = \int [\mathrm{d}x(\tau)] \, \mathrm{e}^{-\mathcal{S}_0(\mathbf{x})/\xi}, \qquad (2.7)$$

with

$$\mathcal{S}_0(\mathbf{x}) = \frac{1}{2} \int_{t'}^{t''} \dot{x}^2(\tau) \mathrm{d}\tau \,,$$

(the factor $1/\sqrt{\varepsilon}$ comes from the change of variables from q to x) where the symbol $[\mathrm{d}x(\tau)]$ (also denoted by $\mathcal{D}x(\tau)$ in the literature) means sum over all (trajectories) $x(\tau)$ satisfying the boundary conditions (2.6).

2.1.2 *Positive measure and correlation functions*

Since the integrand in the path integral is positive, $[\mathrm{d}x(\tau)] \, \mathrm{e}^{-\mathcal{S}_0(\mathbf{x})/\xi}$ defines a positive measure on paths, the so-called *Wiener measure*.

However, it is difficult to keep track of the absolute normalization in the continuum path integral limit. Therefore, one mostly uses path integrals to calculate expectation values.

In the case of a random walk, typical expectation values correspond to correlations between positions at different times, which can also be considered as generalized moments of the probability distribution.

For example, the n-point *correlation function* is defined by

$$\mathcal{F}^{(n)}(\tau_1, \tau_2, \ldots, \tau_n) = \langle x(\tau_1)x(\tau_2)\cdots x(\tau_{2n})\rangle_0$$

$$\equiv \mathcal{Z}^{-1} \int [\mathrm{d}x(\tau)] x(\tau_1)x(\tau_2)\cdots x(\tau_{2n}) \, \mathrm{e}^{-\mathcal{S}_0(\mathbf{x})/\xi}, \qquad (2.8)$$

with

$$\mathcal{Z} \equiv \int [\mathrm{d}x(\tau)] \, \mathrm{e}^{-\mathcal{S}_0(\mathbf{x})/\xi} \,.$$

In the ratio (2.8), the absolute normalization of the path integral then cancels.

2.1.3 Brownian paths

The form of $\mathcal{S}_0(\mathbf{x})$ determines the class of paths that contribute to the path integral, which are called in this case *Brownian paths*. As the factor $\sqrt{\varepsilon}$ in (2.5) suggests, Brownian paths satisfy a Hölder condition of order $1/2$, that is, for $\tau - \tau' \to 0$,

$$|x(\tau) - x(\tau')| = O(|\tau - \tau'|^{1/2}).$$

Therefore, generic Brownian paths are continuous but *not differentiable* and, thus, $\dot{x}(\tau)$ is not defined.

In this sense, the notation $\int \mathrm{d}\tau\, \dot{x}^2(\tau)$ has to be considered as a symbol and should not be taken literally. Nevertheless, it is a useful notation since, for $\xi \to 0$, the path integral is dominated by paths close to paths that leave $\mathcal{S}_0(\mathbf{x})$ stationary, that is, classical paths, which are differentiable (see the calculation in Section 2.1.4).

Finally, the continuity of the contributing paths allows understanding why it is possible that an integration over an increasingly dense set of points (or sum over all possible piecewise linear paths) eventually has a continuum limit.

2.1.4 Explicit calculation

Gaussian path integrals, like finite dimensional Gaussian integrals, are explicitly calculable. For example, the integral (2.7) of Brownian motion can be evaluated by a method that easily generalizes to other types of Gaussian integrals.

Varying the quantity $\mathcal{S}_0(\mathbf{x})$ with respect to the path $x(\tau)$, one obtains the classical equation of motion,

$$\ddot{x}(\tau) = 0\,.$$

The classical solution that satisfies the boundary conditions (2.6) of the path integral is

$$x_c(\tau) = x' + \frac{(\tau - t')}{(t'' - t')}(x'' - x').$$

One then changes variables in the path integral, $x(\tau) \mapsto r(\tau)$ with $x(\tau) = x_c(\tau) + r(\tau)$. At each time τ, the change is a translation and the associated Jacobian is 1. The boundary conditions on the new path $r(\tau)$ are $r(t'') = r(t') = 0$ and one finds

$$\mathcal{S}_0(\mathbf{x}) = S_{\mathrm{class.}} + \mathcal{S}_0(\mathbf{r}), \text{ with } S_{\mathrm{class.}} = \frac{1}{2}\frac{(x'' - x')^2}{(t'' - t')}.$$

The path integral (2.7) becomes

$$\Pi_0(t'', t'; x'', x') = \mathrm{e}^{-S_{\mathrm{class.}}/\xi} \int [\mathrm{d}r(\tau)]\, \mathrm{e}^{-\mathcal{S}_0(\mathbf{r})/\xi}\,.$$

The remaining path integral gives a normalization factor independent of x'', x'. Here, it can be determined by imposing the condition of probability conservation

$$\int \mathrm{d}x''\, \Pi_0(t'', t'; x'', x') = 1 \;\Rightarrow\; \Pi_0(t'', t'; x'', x') = \frac{1}{\sqrt{2\pi\xi(t'' - t')}}\, \mathrm{e}^{-S_{\mathrm{class.}}/\xi}, \quad (2.9)$$

although, as pointed out above, the determination of the normalization is not essential because the normalization cancels in all expectation values.

The result agrees with the asymptotic properties of *Brownian motion*.

2.2 The Wiener measure and statistical physics

The same path integral that describes Brownian motion has an interpretation in the framework of statistical physics.

2.2.1 Classical statistical physics

The expression (2.4) in the example (2.2) can also be considered as the classical partition function of $(n+1)$ particles on a one-dimensional lattice with spatial sites $k = 0, 1, \ldots, n$. The particle at site k deviates from its equilibrium position by the value x_k and particles have nearest-neighbour harmonic interactions:

$$\mathcal{Z}(x_n, x_0; \xi) = (2\pi\xi)^{(1-n)/2} \int \mathrm{d}x_{n-1}\mathrm{d}x_{n-2}\cdots\mathrm{d}x_1 \, \exp\left(-\frac{1}{2\xi}\sum_{k=1}^{n}(x_k - x_{k-1})^2\right).$$

The extremities of the chain are fixed at deviations x_0 and x_n, respectively.

Here, the parameter ξ has the interpretation of a temperature. The path integral (2.7) then corresponds to the continuum limit where the lattice spacing ε between two adjacent sites goes to zero at fixed total macroscopic length of the chain $L = n\varepsilon$.

2.2.2 Quantum statistical physics

Notation. From here on, we use Dirac's notation for vectors in a Hilbert space in the form of bras and kets $\langle \bullet |$, $| \bullet \rangle$. For example, in a generalized basis in which the position operator $\hat{q}$ is diagonal, the eigenvectors will be denoted by $|q\rangle$ and $\hat{q}|q\rangle = q|q\rangle$.

The density matrix $\hat{\rho}$ at thermal equilibrium at temperature $1/\beta$ (in units of the Boltzmann's constant k_B) is given by

$$\hat{\rho}(\beta) = \mathrm{e}^{-\beta\hat{H}}/\mathcal{Z},$$

where $\hat{H}$ is the quantum Hamiltonian (a linear operator acting on the Hilbert space of quantum states) and $\mathcal{Z}$ the partition function defined by

$$\mathcal{Z}(\beta) = \operatorname{tr} \mathrm{e}^{-\beta\hat{H}} \; \Rightarrow \; \operatorname{tr}\hat{\rho}(\beta) = 1.$$

For a free non-relativistic quantum particle of mass m, the quantum Hamiltonian $\hat{H}_0$ is simply

$$\hat{H}_0 = \hat{p}^2/(2m),$$

where in the position basis the momentum operator $\hat{p}$ is given by

$$\hat{p} = \frac{\hbar}{i}\frac{\partial}{\partial q} \; \Rightarrow \; [\hat{q}, \hat{p}] = i\hbar, \tag{2.10}$$

$\hbar$ being Planck's constant.

The matrix elements $\langle q | \, e^{-\beta \hat{H}_0} \, | q' \rangle$ then satisfy the partial differential equation

$$\frac{\partial}{\partial \beta} \langle q | \, e^{-\beta \hat{H}_0} \, | q' \rangle = \frac{\hbar^2}{2m} \frac{\partial^2}{(\partial q)^2} \langle q | \, e^{-\beta \hat{H}_0} \, | q' \rangle . \tag{2.11}$$

The continuum distribution $\Pi_0(t, t'; q, q')$ given by equation (2.9) satisfies the *diffusion equation*

$$\frac{\partial \Pi_0}{\partial t} = \frac{\xi}{2} \frac{\partial^2 \Pi_0}{(\partial q)^2} , \tag{2.12}$$

with the initial condition

$$\lim_{t \to t'} \Pi_0(t, t'; q, q') = \delta(q - q') .$$

Comparing equations (2.11) and (2.12), noting that $\Pi_0(t, t'; q, q')$ and $\langle q | \, e^{-\beta \hat{H}_0} \, | q' \rangle$ satisfy the same boundary conditions at initial time $t - t' \equiv \hbar\beta = 0$, we conclude that the matrix element of the density matrix at thermal equilibrium $\langle q | \, e^{-\beta \hat{H}_0} \, | q' \rangle$ is equal to $\Pi_0(t = \hbar\beta, 0; q, q')$ when $\xi = \hbar/m$ and, therefore, is given by the path integral

$$\langle q | \, e^{-\beta \hat{H}_0} \, | q' \rangle = \int [\mathrm{d}q(\tau)] \, e^{-\mathcal{S}_0(\mathbf{q})/\hbar} , \quad \text{with} \quad \mathcal{S}_0(\mathbf{q}) = \frac{1}{2} \int_0^{\hbar\beta} m \dot{q}^2(\tau) \mathrm{d}\tau , \tag{2.13}$$

with the boundary conditions $q(0) = q'$, $q(\hbar\beta) = q$.

This equation shows that, remarkably enough, the path integral of Brownian motion also yields the density matrix at thermal equilibrium of a free non-relativistic quantum particle.

2.3 Generalization

The path integral (2.13) has a generalization, directly relevant for quantum statistical physics, in the form

$$\Pi(t'', t'; q'', q') = \int [\mathrm{d}q(\tau)] \, e^{-\mathcal{S}(\mathbf{q})/\hbar} , \quad \text{with } q(t') = q', \; q(t'') = q'' , \tag{2.14}$$

where

$$\mathcal{S}(\mathbf{q}) = \int_{t'}^{t''} \mathrm{d}\tau \, \mathcal{L}_{\mathrm{e}}\big(\dot{q}(\tau), q(\tau)\big) \tag{2.15}$$

and the Euclidean Lagrangian is defined as

$$\mathcal{L}_{\mathrm{e}}(\dot{q}, q) = \tfrac{1}{2} m \dot{q}^2 + V(q) . \tag{2.16}$$

The parameter m can be identified with the mass of a non-relativistic quantum particle, and $V(q)$ is the potential. To be able to define the path integral, we assume that $V(q)$ is an analytic function of q and that

$$\int \mathrm{d}q \; e^{-\varepsilon V(q)} < \infty , \; \forall \varepsilon > 0 . \tag{2.17}$$

The Euclidean Lagrangian (2.16) is the imaginary time continuation of the usual classical Lagrangian (up to the sign). This explains why the potential $V(q)$ is *added* to the kinetic energy, while, in the Lagrangian of classical mechanics, the potential is *subtracted* from the kinetic energy.

One possible definition of this kind of path integrals refers to the Wiener measure:

$$\Pi(t'', t'; q'', q') = \Pi_0(t'', t'; q'', q') \left\langle \exp\left[-\frac{1}{\hbar} \int_{t'}^{t''} d\tau \, V\big(q(\tau)\big) \right] \right\rangle_0 , \qquad (2.18)$$

where the expectation value is defined in (2.8) with $\xi = \hbar/m$. With this normalization $\Pi(t', t'; q'', q') = \delta(q'' - q')$, which is the kernel associated with the identity operator.

2.3.1 Path integral and local Markov process

We introduce inside the path integral (2.14) the identity

$$1 = \int dq \, \delta\big(q - q(t)\big), \text{ with } t' < t < t'',$$

where $\delta(q)$ is Dirac's δ-function. Also, the expression (2.15) can be written as the sum

$$S(\mathbf{q}) = \int_{t'}^{t} d\tau \, \mathcal{L}_e(\dot{q}, q) + \int_{t}^{t''} d\tau \, \mathcal{L}_e(\dot{q}, q).$$

The path integral thus factorizes into the product of two path integrals with boundary conditions at t'' and t, and t and t', respectively, integrated over the intermediate point q. Therefore, the kernel $\Pi(t'', t'; q'', q')$, defined by the path integral (2.14), satisfies the Markov property in (Euclidean) time

$$\Pi(t'', t'; q'', q') = \int dq \, \Pi(t'', t; q'', q) \, \Pi(t, t'; q, q') \quad \text{for } t'' \geq t \geq t'. \qquad (2.19)$$

The multiplication rule also shows that $\Pi(t'', t'; q'', q')$ can be identified with the kernel or matrix element of an operator $\boldsymbol{\Pi}(t'', t')$ in an Hilbert space $\mathcal{H}$. In Dirac's bra-ket notation,

$$\langle q'' | \boldsymbol{\Pi}(t'', t') | q' \rangle \equiv \Pi(t'', t'; q'', q'). \qquad (2.20)$$

In operator notation, the relation (2.19) becomes

$$\boldsymbol{\Pi}(t'', t') = \boldsymbol{\Pi}(t'', t) \boldsymbol{\Pi}(t, t').$$

The Markov property (2.19) allows writing the path integral as a product of n path integrals integrated over intermediate points, corresponding to a time interval $\varepsilon = (t'' - t')/n$ that can be chosen arbitrarily small by increasing n. The path integral can thus be evaluated from its asymptotic behaviour for small time intervals. Then, with the boundary conditions $q(t) = q$, $q(t + \varepsilon) = q_\varepsilon$,

$$S(\mathbf{q}) = \int_{t}^{t+\varepsilon} d\tau \left[\tfrac{1}{2} m \dot{q}^2(\tau) + V\big(q(\tau)\big) \right] \sim m \frac{(q_\varepsilon - q)^2}{2\varepsilon} + \varepsilon V(q). \qquad (2.21)$$

For a given trajectory, the leading term in the expression when $\varepsilon \to 0$ is still the Brownian term (kinetic term in classical mechanics), which implies that the paths contributing to the path integral still satisfy the Hölder property $|q_\varepsilon - q| = O(\varepsilon^{1/2})$. In particular, this property implies that the argument q of V in (2.21) can be replaced by any value linearly interpolating between q and q_ε, the difference being of order $\varepsilon^{1/2}$ and hence negligible in this approximation.

The dominance of the Brownian term implies also a (spatial) locality property (and thus the denomination of *local Markov process*): for ε small, $\langle q_\varepsilon | \boldsymbol{\Pi}(t + \varepsilon, t) | q \rangle$ decreases exponentially when $|q - q_\varepsilon| \to \infty$. Since, at leading order in ε the normalization is provided by Brownian motion, one concludes that

$$\langle q_\varepsilon | \boldsymbol{\Pi}(t + \varepsilon, t) | q \rangle \sim \Pi_\varepsilon(t; q_\varepsilon, q)$$
$$= \sqrt{\frac{m}{2\pi\hbar\varepsilon}} \exp\left[-\left(m\frac{(q_\varepsilon - q)^2}{2\varepsilon} + \varepsilon V(q)\right)/\hbar\right]. \quad (2.22)$$

This leads to an alternative definition of the path integral (2.14) as the limit when $n \to \infty$ at $n\varepsilon = t'' - t'$ fixed of the n-dimensional integral

$$\langle q'' | \boldsymbol{\Pi}(t'', t') | q' \rangle = \lim_{n \to \infty} \int \prod_{k=1}^{n-1} \mathrm{d}q_k \prod_{k=1}^{n} \Pi_\varepsilon(\tau_{k-1}; q_k, q_{k-1}), \quad (2.23)$$

with the conventions $\tau_k = t' + k\varepsilon$, $q_0 = q'$, $q_n = q''$. At finite n, the right-hand side defines a Markov process in discrete times.

Time-dependent potentials. The path integral (2.14) has a straightforward generalization to time-dependent potentials $V(q, \tau)$. However, the Lagrangian (2.16) can no longer be identified with the continuation to imaginary time of a real time Lagrangian.

2.3.2 Path integrals and statistical physics

Classical statistical physics. If the interpretation of the path integral in terms of random walks (see (2.18)) and Markov processes is somewhat indirect, the interpretation in the framework of classical statistical physics is simple. In the form of the right-hand side of (2.23), the path integral appears as the continuum limit of the *classical* partition function of a one-dimensional lattice model. The configuration energy of the lattice model is (here we set $m = \hbar = 1$)

$$\mathcal{S}(\mathbf{q}) = \sum_{k=1}^{n} \left[\frac{1}{2\varepsilon}(q_k - q_{k-1})^2 + \varepsilon V(q_k)\right],$$

where we have assumed that V is time independent to enforce translation invariance. The continuum limit $\varepsilon \propto 1/n \to 0$ corresponds also to a kind of low temperature limit where the correlation length in the statistical model, which is proportional to $1/\varepsilon$, diverges.

Quantum statistical physics. From the evaluation of the path integral at short time intervals, one also infers that $\langle q''|\boldsymbol{\Pi}(t'',t')|q'\rangle$ satisfies the partial differential equation

$$\hbar\frac{\partial}{\partial t}\langle q|\boldsymbol{\Pi}(t,t')|q'\rangle = \left[\frac{\hbar^2}{2m}\frac{\partial^2}{(\partial q)^2} - V(q)\right]\langle q|\boldsymbol{\Pi}(t,t')|q'\rangle\,, \tag{2.24}$$

with the initial condition $\langle q|\boldsymbol{\Pi}(t',t')|q'\rangle = \delta(q-q')$ (*i.e.*, $\boldsymbol{\Pi}(t',t') = \mathbf{1}$).

Equation (2.24) is related by the formal substitution $t \mapsto it$ to the Schrödinger equation in real time,

$$i\hbar\frac{\partial}{\partial t}\langle q|\mathbf{U}(t,t')|q'\rangle = \left[-\frac{\hbar^2}{2m}\frac{\partial^2}{(\partial q)^2} + V(q)\right]\langle q|\mathbf{U}(t,t')|q'\rangle\,, \tag{2.25}$$

for the matrix elements of the quantum evolution operator $\mathbf{U}(t'',t')$. The operator solution of equation (2.25) with initial condition $\mathbf{U}(t',t') = \mathbf{1}$ is

$$\mathbf{U}(t,t') = \mathrm{e}^{-i(t-t')\hat{H}/\hbar}\,, \tag{2.26}$$

where

$$\hat{H} = \frac{\hat{p}^2}{2m} + V(\hat{q}) \tag{2.27}$$

is a time-independent quantum Hamiltonian and $\hat{q}$ and $\hat{p}$ are, respectively, the position and momentum operators with the commutation relation (2.10). Note also that with the condition (2.17), the Hamiltonian (2.27) has a discrete spectrum.

Equation (2.24), also called *imaginary time Schrödinger equation*, has then for its solution $\boldsymbol{\Pi}(t,t') = \mathrm{e}^{-(t-t')\hat{H}/\hbar}$, with the initial condition $\boldsymbol{\Pi}(t,t) = \mathbf{1}$. In particular,

$$\boldsymbol{\Pi}(t = \beta\hbar, 0) = \mathrm{e}^{-\beta\hat{H}}\,, \tag{2.28}$$

where the operator $\mathrm{e}^{-\beta\hat{H}}$ is proportional to the density matrix

$$\hat{\rho}(\beta) = \mathrm{e}^{-\beta\hat{H}}/\mathcal{Z}(\beta)\,,$$

which describes the thermal equilibrium of a quantum system with Hamiltonian $\hat{H}$ at temperature $1/\beta$, and $\mathcal{Z}(\beta) = \mathrm{tr}\,\mathrm{e}^{-\beta\hat{H}}$ is the partition function of the quantum system. It has the path integral representation

$$\begin{aligned}\mathcal{Z}(\beta) &= \mathrm{tr}\,\boldsymbol{\Pi}(\hbar\beta/2, -\hbar\beta/2)\\ &= \int \mathrm{d}q\,\Pi(\hbar\beta/2, -\hbar\beta/2; q, q) = \int [\mathrm{d}q(\tau)]\,\mathrm{e}^{-\mathcal{S}(\mathbf{q})/\hbar}\,,\end{aligned} \tag{2.29}$$

where $\mathcal{S}(\mathbf{q})$ is the imaginary time action (2.15) and one sums over all closed paths satisfying the *periodic boundary condition* $q(\hbar\beta/2) = q(-\hbar\beta/2)$.

2.4   Gaussian path integrals: The quantum harmonic oscillator

Gaussian path integrals, like finite dimensional Gaussian integrals, can be calculated explicitly. An example more general than the integral evaluated in Section 2.1.4 is provided by the path integral representation associated to the quantum harmonic oscillator defined by the Hamiltonian operator

$$\hat{H} = \frac{\hat{p}^2}{2m} + \frac{1}{2m}\,\omega^2\,\hat{q}^2\,, \tag{2.30}$$

where m is the mass of the particle, and $1/(2\pi\omega)$ is the period of the oscillations in the classical limit. Then, the expression (2.16) becomes

$$\mathcal{L}_{\mathrm{e}}(\dot{q}, q; \tau) = \tfrac{1}{2}m\left(\dot{q}^2 + \omega^2 q^2\right).$$

As in the Brownian motion example, in order to calculate the corresponding path integral, one first solves the classical equation of motion (the Euler–Lagrange equation corresponding to $\mathcal{L}_{\mathrm{e}}$)

$$-\ddot{q}(\tau) + \omega^2 q(\tau) = 0\,,$$

with the path integral boundary conditions $q(t/2) = q''$, $q(-t/2) = q'$. (Since the potential is time independent, we can choose $t' = -t/2$ and $t'' = t/2$ in the path integral (2.14)). The classical solution is

$$q_c(\tau) = \frac{1}{\sinh(\omega t)}\left[q'\sinh\big(\omega(t/2 - \tau)\big) + q''\sinh\big(\omega(\tau + t/2)\big)\right]$$

and the corresponding classical action is

$$\mathcal{S}(\mathbf{q_c}) = \frac{m\omega}{2\sinh\omega t}\left[(q'^2 + q''^2)\cosh\omega t - 2q'q''\right].$$

One then changes variables: $q(\tau) \mapsto r(\tau) = q(\tau) - q_c(\tau)$, such that $r(-t/2) = r(t/2) = 0$. Since

$$\mathcal{S}(\mathbf{q}) = \mathcal{S}(\mathbf{q_c}) + \mathcal{S}(\mathbf{r}),$$

the path integral becomes

$$\langle q''|\boldsymbol{\Pi}(t/2, -t/2)|q'\rangle = \mathrm{e}^{-\mathcal{S}(\mathbf{q_c})/\hbar}\int[dr(\tau)]\,\mathrm{e}^{-\mathcal{S}(\mathbf{r})/\hbar}\,.$$

The integral over $r(\tau)$ yields a normalization constant. Its dependence on ω can be determined by differentiating the path integral with respect to ω. The final normalization can be determined by comparing with the Brownian motion $\omega = 0$. One finds

$$\int[dr(\tau)]\,\mathrm{e}^{-\mathcal{S}(\mathbf{r})/\hbar} = \left(\frac{m\omega}{2\pi\hbar\,\sinh\omega t}\right)^{1/2}.$$

The quantum partition function $\mathcal{Z}(\beta) = \operatorname{tr} e^{-\beta \hat{H}}$ (see equation (2.29)) follows:

$$\mathcal{Z}(\beta) = \int dq \, \langle q | \boldsymbol{\Pi}(\hbar\beta/2, -\hbar\beta/2) | q \rangle = \frac{1}{2 \sinh(\beta\hbar\omega/2)}$$

$$= \frac{e^{-\beta\hbar\omega/2}}{1 - e^{-\beta\hbar\omega}} = \sum_{n=0}^{\infty} e^{-\beta\hbar\omega(n+1/2)} . \tag{2.31}$$

From equation (2.31) and the definition $\mathcal{Z}(\beta) = \operatorname{tr} e^{-\beta \hat{H}}$, one recovers the exact spectrum of the quantum Hamiltonian $\hat{H}$ whose eigenvalues are $E_n = \hbar\omega(n + \frac{1}{2})$.

2.5 Path integrals: Perturbation theory

Correlation functions. Correlation functions are (generalized) moments of the measure associated to the integrand in the path integral. They are defined by

$$\langle q(\tau_1)q(\tau_2) \cdots q(\tau_p) \rangle = \frac{1}{\mathcal{Z}} \int [dq(\tau)] q(\tau_1)q(\tau_2) \cdots q(\tau_p) \, e^{-\mathcal{S}(\mathbf{q})/\hbar}, \tag{2.32}$$

with

$$\mathcal{Z} = \int [dq(\tau)] \, e^{-\mathcal{S}(\mathbf{q})/\hbar},$$

a definition that extends to the interacting case the definition given for the free particle in Section 2.1.2.

In the example of periodic boundary conditions $q(-t/2) = q(t/2)$, $(q(t/2)$ integrated), which correspond to a summation over all closed trajectories, $\mathcal{Z}(\beta = t/\hbar)$ is the quantum partition function describing the thermodynamic equilibrium at temperature $1/\beta$.

2.5.1 Gaussian expectation values and Wick's theorem

We consider path integrals corresponding to *centred Gaussian measures*, for which the action is quadratic in the integration path $q(\tau)$, (simple examples being provided by Brownian motion and the quantum harmonic oscillator). Then, the symmetry $q \mapsto -q$ implies that correlation functions (2.32) odd in $q(\tau)$ vanish.

Moreover, for all centred Gaussian measures, correlation functions can be expressed in terms of the two-point function as stated by *Wick's theorem* (proved in a different context in [9]):

$$\langle q(\tau_1)q(\tau_2) \cdots q(\tau_{2\ell}) \rangle = \sum_{P\{1,2,\dots 2\ell\}} \langle q(\tau_{P_1})q(\tau_{P_2}) \rangle \cdots \langle q(\tau_{P_{2\ell-1}})q(\tau_{P_{2\ell}}) \rangle,$$

where $P\{1, 2, \dots, 2\ell\}$ are all possible (unordered) pairings of $\{1, 2, \dots, 2\ell\}$.

For example,

$$\langle q(\tau_1)q(\tau_2)q(\tau_3)q(\tau_4) \rangle = \langle q(\tau_1)q(\tau_2) \rangle \langle q(\tau_3)q(\tau_4) \rangle + \langle q(\tau_1)q(\tau_3) \rangle \langle q(\tau_2)q(\tau_4) \rangle$$
$$+ \langle q(\tau_1)q(\tau_4) \rangle \langle q(\tau_3)q(\tau_2) \rangle.$$

In the example of the quantum harmonic oscillator and periodic boundary conditions, in the limit $t \to \infty$, the two-point correlation function reduces to

$$\langle q(\tau_1)q(\tau_2) \rangle = \frac{\hbar}{2m\omega} e^{-\omega|\tau_1 - \tau_2|} .$$

2.5.2   Path integral: Perturbative calculation

We now assume that the potential $V(q)$ in the path integral is a polynomial in q, although perturbation theory can be generalized to analytic potentials.

In Section 2.5.1, it has been shown that Gaussian expectation values can be calculated explicitly: therefore, to evaluate a path integral, a possible method is to keep the quadratic part $(O(q^2))$ of the potential of $V(q)$ in the exponential as part of the Gaussian measure and to expand the remainder in a power series.

For illustration purpose, we consider the quartic anharmonic oscillator,

$$V(q) = \tfrac{1}{2}q^2 + \lambda q^4, \lambda > 0\,.$$

We set

$$\mathcal{S}(\mathbf{q}) = \mathcal{S}_0(\mathbf{q}) + \lambda \int \mathrm{d}\tau\, q^4(\tau), \quad \text{with} \quad \mathcal{S}_0(\mathbf{q}) = \tfrac{1}{2}\int \mathrm{d}\tau\, \big(\dot{q}^2(\tau) + q^2(\tau)\big).$$

The expansion of the corresponding path integral can then be written as (here we set $\hbar = 1$)

$$\mathcal{Z}(\lambda) = \int [\mathrm{d}q(\tau)]\, \mathrm{e}^{-\mathcal{S}_0(\mathbf{q})} \sum_{k=0}^{\infty} \frac{(-\lambda)^k}{k!} \left[\int \mathrm{d}\tau\, q^4(\tau)\right]^k$$

$$\sim \mathcal{Z}(0) \sum_{k=0}^{\infty} \frac{(-\lambda)^k}{k!} \left\langle \left[\int \mathrm{d}\tau\, q^4(\tau)\right]^k \right\rangle_0 ,$$

where $\langle \bullet \rangle_0$ means expectation value with respect to the Gaussian measure associated with $\mathcal{S}_0$, and the symbol $\sim$ emphasizes the property that the perturbative series is not convergent for $\lambda \neq 0$. Each term in the series can then be evaluated using Wick's theorem and the explicit form of the Gaussian two-point function. For example, at first order in λ,

$$\langle q^4(\tau) \rangle_0 = 3(\langle q^2(\tau) \rangle_0)^2.$$

The next order involves

$$\langle q^4(\tau) q^4(\tau') \rangle_0 = 9(\langle q^2(\tau) \rangle_0)^2 (\langle q^2(\tau') \rangle_0)^2 + 72\langle q^2(\tau) \rangle_0 \langle q^2(\tau') \rangle_0 (\langle q(\tau)q(\tau') \rangle_0)^2$$

$$+ 24(\langle q(\tau)q(\tau') \rangle_0)^4.$$

It is convenient to represent individual contributions graphically in terms of *Feynman diagrams* [10].

As stated above, such an expansion is divergent for all values of the parameter λ (Chapter 24). It is an asymptotic series, useful as such only if λ small enough. For larger values of the expansion parameter, series summation methods (like Borel summation), when applicable, are required.

2.6 Path integral: Quantum time evolution

Following Feynman [4], quantum time evolution (here we refer to *real physical time*) can be described in terms of (oscillatory) path integrals. In this formalism, considering a system described classically by the Cartesian coordinates $\mathbf{q} \equiv \{q^1, q^2, \ldots\}$, the matrix elements of the quantum evolution operator $\mathbf{U}(t'', t')$ between times t' and t'' are given by a sum over all possible trajectories (*paths*) $\mathbf{q}(\tau) \equiv \{q^1(\tau), q^2(\tau), \ldots\}$, which, in the simplest cases, can be written as

$$\langle \mathbf{q}'' \,|\mathbf{U}(t'', t')|\, \mathbf{q}' \rangle = \int [d\mathbf{q}(\tau)] \exp\left(\frac{i}{\hbar} \mathcal{A}(\mathbf{q})\right), \qquad (2.33)$$

with the boundary conditions

$$\mathbf{q}(t') = \mathbf{q}', \ \ \mathbf{q}(t'') = \mathbf{q}'', \qquad (2.34)$$

where the action $\mathcal{A}(\mathbf{q})$ is the time integral of the classical Lagrangian:

$$\mathcal{A}(\mathbf{q}) = \int_{t'}^{t''} d\tau \, \mathcal{L}\left(\mathbf{q}(\tau), \dot{\mathbf{q}}(\tau); \tau\right). \qquad (2.35)$$

Feynman's path integral (2.33) is valid when the term with two time derivatives in the Lagrangian (the kinetic term) has the form $\frac{1}{2}\sum_i m_i(\dot{q}^i)^2$: otherwise, the measure has to be modified and new problems arise. An example of the latter situation is provided when the coordinates q^i parametrize a Riemannian manifold, and the kinetic term is proportional to $\int dt \sum_{i,j} \dot{q}^i g_{ij}(\mathbf{q}) \dot{q}^j$, where g_{ij} is the metric tensor.

The formulation of quantum mechanics in terms of path integrals actually explains why equations of motion in classical mechanics can be derived from a variational principle (see also Section 3.1).

In the classical limit, that is, when the typical classical action is large with respect to $\hbar$, the path integral can be evaluated by using the *stationary phase method*. The sum over paths is dominated by the neighbourhood of paths that leave the action stationary: the *classical paths* that satisfy

$$\mathcal{A}\left(\mathbf{q} + \delta\mathbf{q}\right) - \mathcal{A}\left(\mathbf{q}\right) = O(\|\delta\mathbf{q}\|^2) \ \Rightarrow \ \frac{\delta\mathcal{A}}{\delta q^i} = 0 \ \Rightarrow \ \frac{\partial\mathcal{L}}{\partial q^i} - \frac{d}{dt}\frac{\partial\mathcal{L}}{\partial\dot{q}^i} = 0, \qquad (2.36)$$

with the boundary conditions (2.34). The leading order contribution is obtained by expanding the path around the classical path, keeping only the quadratic term in the deviation and performing the corresponding Gaussian integration.

This property generalizes to relativistic quantum field theory.

From the mathematical point of view, it is much more difficult to define rigorously the real time path integral than the imaginary time statistical path integral. A possible strategy involves, when applicable, the calculation of physical observables for imaginary time followed by an analytic continuation.

2.7   Barrier penetration in the semi-classical limit

We now illustrate with a simple example the evaluation of statistical (or imaginary time) path integrals in the semi-classical approximation. The reader is warned that the section is slightly *more technical* (see also Section 24.3).

The path integral associated with tr $e^{-t\hat{H}/\hbar}$,

$$\mathcal{Z}(t) = \int [\mathrm{d}q(\tau)]\, e^{-\mathcal{S}(\mathbf{q})/\hbar}, \quad \text{with } q(t/2) = q(-t/2), \tag{2.37}$$

is especially well suited to the evaluation, in the semi-classical limit $\hbar \to 0$, of specific quantum phenomena called *barrier penetration*, or tunnelling. Indeed, barrier penetration appears, in the semi-classical approximation, as formally related to classical motion in imaginary time in the classically forbidden region.

To explain the general idea, we consider the example

$$\mathcal{S}(\mathbf{q}) = \int_{-t/2}^{t/2} \mathrm{d}\tau \left[\tfrac{1}{2}\dot{q}^2(\tau) + V\big(q(\tau)\big) \right], \tag{2.38}$$

with the potential

$$V(q) = \tfrac{1}{2}q^2 - \tfrac{1}{4}\lambda q^4, \quad \lambda > 0. \tag{2.39}$$

The potential has one local minimum at $q = 0$ ($V = 0$) and two local maxima at $q = \pm 1/\sqrt{\lambda}$, and goes to $-\infty$ when $|q| \to \infty$.

As such, the integral (2.37) is not defined, because the potential V is not bounded from below. However, it can be defined by analytic continuation starting from $\lambda < 0$ and rotating the path in the complex q plane.

For $\lambda > 0$, in the classical limit, a particle at rest located at the minimum $q = 0$ stays there forever. However, due to quantum fluctuations (in the statistical interpretation of the path integral thermal fluctuations), the particle has a finite probability per unit time to leave the now metastable minimum, an effect called barrier penetration, or quantum tunnelling.

For $\lambda < 0$, the spectrum of the quantum Hamiltonian is real. After continuation, for $\lambda > 0$, one finds a complex energy spectrum (quantum resonances), the imaginary part of the energy eigenvalues being directly related to tunnelling.

In the path integral representation, it can be shown that the calculation of barrier penetration effects for $\hbar \to 0$ can be derived from an evaluation of the integral (2.37) by the steepest descent method, suitably generalized to path integrals.

One looks for non-trivial saddle points, here non-constant solutions of the classical equations of motion derived from the Euclidean action (2.38), which correspond formally to evolution in imaginary time. Moreover, if one is interested only in states with energies of order $\hbar$, then one has to take the limit $|t| \to \infty$.

Therefore, one looks for solutions that have a finite action on the real line. These solutions are called *instanton* solutions (see also Chapters 23 and 24), although initially instantons were called pseudoparticles [11, 12]. Here, the equation of motion obtained by varying $\mathcal{S}$ is

$$-\ddot{q}(\tau) + q(\tau) - \lambda q^3(\tau) = 0.$$

Due to time-translation invariance and symmetry $q \mapsto -q$, one finds two one-parameter family of instanton solutions (τ_0 is an integration constant),

$$q_c(\tau) = \pm \frac{\sqrt{2/\lambda}}{\cosh(\tau - \tau_0)} \quad \Rightarrow \quad \mathcal{S}(\mathbf{q}_c) = \int_{-\infty}^{+\infty} d\tau \, \mathcal{L}_e(\mathbf{q}_c) = \frac{4}{3\lambda}.$$

Completing the calculation of the saddle point contribution is a non-trivial exercise because it requires factorizing the path integral measure into an integration over τ_0 (a collective coordinate that parametrizes the solution and is related to the breaking of time-translation symmetry by the solution) before using a saddle point approximation for the other modes of the path.

One infers that, up to power-law corrections, the probability R per unit time of leaving the well is (equation (24.23))

$$R \underset{\lambda \to 0_+}{\sim} \left(\frac{8}{\pi \lambda} \right)^{1/2} e^{-4/(3\lambda)}.$$

2.8 Path integrals: A few generalizations

Up to now, we have described the simplest form of path integrals, which, for the point of view of quantum mechanics, involve only a classical Lagrangian with the general form (2.15). For more general Lagrangians or Hamiltonians, one encounters additional problems for defining path integrals.

2.8.1 The quantum particle in a static magnetic field

When the Lagrangian involves a term linear in the velocity, as in the example of a quantum particle in a magnetic field,

$$\mathcal{L}(\mathbf{q}, \dot{\mathbf{q}}) = \tfrac{1}{2} m \, \dot{\mathbf{q}}^2 - e \, \mathbf{A}(\mathbf{q}) \cdot \dot{\mathbf{q}}, \tag{2.40}$$

where $\mathbf{A}(\mathbf{q})$ is a given (smooth) vector potential, a new problem arises related to quantization. The classical Lagrangian together with the *correspondence principle* (replacing position and velocity by the corresponding quantum operators, which is not a principle but an educated guess) does not determine the quantum theory, because operators $\mathbf{A}(\hat{\mathbf{q}})$ and $\dot{\hat{\mathbf{q}}}$ no longer commute. Correspondingly, the naive continuum form of the path integral is not defined, because the continuum limit depends explicitly on the time-discretized form of the path integral and leads to a one-parameter family of different theories. This leads, for example, to the appearance of the undefined quantity sgn(0) in calculations.

From the path integral viewpoint, the problem is directly related to the property that the Brownian paths are not differentiable and, thus, $\dot{q}$ is not defined.

The underlying quantum Hamiltonian is *uniquely* determined by demanding either its Hermiticity or, equivalently, its gauge invariance.

To determine the corresponding path integral, one can either return to a time-discretized form consistent with the quantum Hamiltonian (which implies the mid-point rule in the argument of the vector potential), or, in the continuum, add a term with higher order time derivatives in the action, for example,

$$\mathcal{S} \mapsto \mathcal{S} + \varepsilon^2 \int_{t'}^{t''} \mathrm{d}\tau (\ddot{q}(\tau))^2,$$

and, then, study the $\varepsilon \to 0$ limit. This has the effect of restricting the integration to paths that satisfy a Hölder condition of order $3/2$ and are thus differentiable, in such a way that expectation values involving $\dot{\mathbf{q}}$ are then defined.

In the case of a magnetic field, this regularization does not violate gauge invariance and fixes the ambiguities ($\mathrm{sgn}(0) = 0$) in a way that is consistent with a gauge invariant Hermitian quantum Hamiltonian.

Gauge fixing. Note a peculiarity of the Lagrangian (2.40): the vector potential is defined only up the addition of a gradient term (a gauge transformation), since such an addition does not modify the equations of motion. It is thus possible to *fix the gauge*, that is, to determine the vector potential completely by an additional constraint. The choice $\nabla \cdot \mathbf{A} = 0$ (Coulomb gauge) renders the problem operator ordering irrelevant and eliminates the ambiguity (see also Chapter 11).

2.8.2 Hamiltonian formulation and phase space path integral

For a general classical Hamiltonian, the quantum evolution operator can *formally* be expressed in terms of a path integral involving an integration over phase space variables, position $\mathbf{q}$ and conjugate momentum $\mathbf{p}$ [13]:

$$\langle \mathbf{q}'' | \mathbf{U}(t'', t') | \mathbf{q}' \rangle = \int [\mathrm{d}\mathbf{p}(\tau)\mathrm{d}\mathbf{q}(\tau)] \exp\left(\frac{i}{\hbar}\mathcal{A}(\mathbf{p}, \mathbf{q})\right), \qquad (2.41)$$

with the boundary conditions

$$\mathbf{q}(t') = \mathbf{q}', \ \mathbf{q}(t'') = \mathbf{q}'', \qquad (2.42)$$

where the classical action $\mathcal{A}(\mathbf{p}, \mathbf{q})$ is now expressed in terms of the classical Hamiltonian $H(\mathbf{p}, \mathbf{q}; t)$:

$$\mathcal{A}(\mathbf{p}, \mathbf{q}) = \int_{t'}^{t''} \mathrm{d}\tau \left[\mathbf{p}(\tau) \cdot \dot{\mathbf{q}}(\tau) - H(\mathbf{p}(\tau), \mathbf{q}(\tau); \tau)\right]. \qquad (2.43)$$

This expression can heuristically be derived most conveniently by using mixed matrix elements $\langle \mathbf{p} | \mathbf{U}(t'', t') | \mathbf{q} \rangle$ where $\langle \mathbf{p} |$ is diagonal in the momentum basis.

When the Hamiltonian is quadratic in the conjugate momentum $\mathbf{p}$, the integral over $\mathbf{p}(\tau)$ is Gaussian, and the integration can be performed explicitly. One then recovers the classical Lagrangian (see Chapter 3). However, this may have the effect of modifying the $\mathbf{q}$-integration measure if the coefficient of the quadratic term in $\mathbf{p}(\tau)$ is not a constant.

In general, the definition of this path integral leads to difficulties reflecting the problems of associating quantum Hamiltonians to classical Hamiltonians involving products of p and q variables, because the corresponding operators $\hat{\mathbf{p}}$ and $\hat{\mathbf{q}}$ do not commute. It has mainly a heuristic value (see, however, Section 3.2.2), except in the semi-classical limit.

2.9 Path integrals for bosons and fermions

Up to now, we have described a path integral formalism relevant for distinct quantum particles. But quantum particles are either *bosons*, obeying the Bose–Einstein statistics or *fermions*, governed by Fermi–Dirac statistics. To describe the quantum evolution of several identical (and thus indiscernible) quantum particles, the path integral formulation has to be generalized.

2.9.1 Holomorphic formalism and bosons

In the case of bosons, it is based on the coherent state holomorphic formalism and the Hilbert space of analytic entire functions [14, 15].

For bosons occupying only a finite number of quantum states, the relevant path integral can formally be deduced from the phase space integral by a complex change of variables, up to boundary terms and boundary conditions. In the example of one quantum state, the change of variables is simply

$$z = (p + iq)/i\sqrt{2}, \quad \bar{z} = i(p - iq)/\sqrt{2}.$$

The complex variables z and $\bar{z}$ can be considered as classical variables associated with boson creation and annihilation operators.

The holomorphic path integral then takes the form [15]

$$\langle \mathbf{z}'' | \mathbf{U}(t'', t') | \bar{\mathbf{z}}' \rangle = \int [d\bar{\mathbf{z}}(\tau) d\mathbf{z}(\tau)] \, e^{\bar{\mathbf{z}}(t') \cdot \mathbf{z}(t')} \exp\left(\frac{i}{\hbar} \mathcal{A}(\mathbf{z}, \bar{\mathbf{z}}) \right), \qquad (2.44)$$

with the boundary conditions

$$\bar{\mathbf{z}}(t') = \bar{\mathbf{z}}', \; \mathbf{z}(t'') = \mathbf{z}'', \qquad (2.45)$$

where the classical action $\mathcal{A}(\mathbf{z}, \bar{\mathbf{z}})$ reads

$$\mathcal{A}(\mathbf{z}, \bar{\mathbf{z}}) = \int_{t'}^{t''} d\tau \left[-i\bar{\mathbf{z}}(\tau) \cdot \dot{\mathbf{z}}(\tau) \right.$$
$$\left. -H\left(i(\mathbf{z}(\tau) - \bar{\mathbf{z}}(\tau))/\sqrt{2}, (\mathbf{z}(\tau) + \bar{\mathbf{z}}(\tau))/\sqrt{2}; \tau \right) \right]. \qquad (2.46)$$

More generally, to N quantum states are associated N pairs of complex variables $(z_i, \bar{z}_i)$.

Even in the Gaussian example, this path integral in a straightforward interpretation suffers from the same ambiguities as in the example of a particle in a magnetic field. This leads also to the appearance of sgn(0) in calculations, which has to be defined by a regularization.

2.9.2 *Grassmann path integrals and fermions*

The understanding of Grassmann path integrals necessitates some prior knowledge of *Grassmann or exterior algebras*, including the definition and properties of *Grassmann differentiation and integration* [16].

For a Grassmann algebra with n generators θ_i, the algebraic rules can be summarized by

$$\theta_i \theta_j + \theta_j \theta_i = 0 \,,$$

$$\theta_i \frac{\partial}{\partial \theta_j} + \frac{\partial}{\partial \theta_j} \theta_i = \delta_{ij} \,,$$

$$\frac{\partial}{\partial \theta_i} \frac{\partial}{\partial \theta_j} + \frac{\partial}{\partial \theta_j} \frac{\partial}{\partial \theta_i} = 0 \,,$$

where the two last equations should be understood in the sense of multiplication and differentiation operators acting on elements of the Grassmann algebra.

The first equation implies that all elements of the Grassmann algebra can be written as linear combinations of elements of the form $\theta_{i_1} \theta_{i_2} \ldots \theta_{i_p}$ with $i_1 < i_2 \cdots < i_p$.

Finally, integration and differentiation are identical operations,

$$\int \mathrm{d}\theta_i \equiv \frac{\partial}{\partial \theta_i} \,.$$

Statistical or quantum physics. The description of the statistical properties or of the quantum evolution of fermion systems by path integrals requires the introduction of elements of an infinite dimensional Grassmann algebra and the integration over Grassmannian paths [16, 17].

For example, to describe a system with N available quantum states, one introduces the generators $\theta_i(\tau)$, $\bar{\theta}_i(\tau)$, $i = 1 \ldots N$, of a Grassmann algebra. They satisfy the commutation relations

$$\theta_i(\tau)\theta_j(\tau') + \theta_j(\tau')\theta_i(\tau) = 0 \,, \quad \theta_i(\tau)\bar{\theta}_j(\tau') + \bar{\theta}_j(\tau')\theta_i(\tau) = 0 \,,$$
$$\bar{\theta}_i(\tau)\bar{\theta}_j(\tau') + \bar{\theta}_j(\tau')\bar{\theta}_i(\tau) = 0 \,.$$

Then, rules of Grassmannian differentiation and integration can be formulated. It follows that the elements of the density matrix at thermal equilibrium, or the imaginary time path integral, take the form

$$\langle \boldsymbol{\theta}''|U(t'',t')|\bar{\boldsymbol{\theta}}'\rangle = \int_{\bar{\boldsymbol{\theta}}(t')=\bar{\boldsymbol{\theta}}'}^{\boldsymbol{\theta}(t'')=\boldsymbol{\theta}''} [\mathrm{d}\boldsymbol{\theta}(\tau)\mathrm{d}\bar{\boldsymbol{\theta}}(\tau)] \, \mathrm{e}^{-\bar{\boldsymbol{\theta}}(t') \cdot \boldsymbol{\theta}(t')} \exp\left[-\mathcal{S}(\boldsymbol{\theta},\bar{\boldsymbol{\theta}})\right] ,$$

with

$$\mathcal{S}(\boldsymbol{\theta},\bar{\boldsymbol{\theta}}) = \int_{t'}^{t''} \mathrm{d}\tau \left\{ \bar{\boldsymbol{\theta}}(\tau) \cdot \dot{\boldsymbol{\theta}}(\tau) + H\left[\boldsymbol{\theta}(\tau), \bar{\boldsymbol{\theta}}(\tau)\right] \right\} ,$$

where $H\left[\boldsymbol{\theta}(\tau), \bar{\boldsymbol{\theta}}(\tau)\right]$ represents the Hamiltonian acting on Grassmann functions.

2.10 Field integrals: New issues

While the path integral is an interesting topic for its own sake, the most useful physics applications are provided by a generalization: the field integral, where the integration over paths is replaced by an integration over real, complex or Grassmannian *fields*, (for a general reference see, *e.g.*, Ref. [18]).

For example, in a *local* field theory for a neutral scalar field $\phi(x)$, $x \in \mathbb{R}^d$, the partition function takes the form

$$\mathcal{Z} = \int [\mathrm{d}\phi(x)] \, \mathrm{e}^{-\mathcal{S}(\phi)/\hbar},$$

where the imaginary time (or Euclidean) action $\mathcal{S}(\phi)$ is a d-dimensional integral of an Euclidean Lagrangian density,

$$\mathcal{S}(\phi) = \int \mathrm{d}^d x \, \mathcal{L}_{\mathrm{e}}(\nabla\phi(x), \phi(x)).$$

Before regularization, a typical Lagrangian density is of the form,

$$\mathcal{L}_{\mathrm{e}}(\nabla\phi(x), \phi(x)) = \tfrac{1}{2}\big(\nabla_x\phi(x)\big)^2 + V\big(\phi(x)\big),$$

($\nabla \equiv \{\partial/\partial x_1, \dots, \partial/\partial x_d\}$). The Lagrangian density is a function of the field and its derivatives at the same point, a property called *locality*.

A simple, physical relevant example, is

$$\mathcal{L}_{\mathrm{e}}(\nabla\phi, \phi) = \tfrac{1}{2}\big(\nabla\phi(x)\big)^2 + \tfrac{1}{2}r\phi^2(x) + \frac{g}{4!}\phi^4(x), \tag{2.47}$$

where r and $g \geq 0$ are two parameters characterizing the model.

In a relativistic-covariant quantum field theory, the real time evolution (in $3+1$ space-time dimensions) is then given by

$$\int [\mathrm{d}\phi(x)] \exp\left(\frac{i}{\hbar}\mathcal{A}(\phi)\right),$$

where $\mathcal{A}$ now is the classical action, space-time integral of the classical Lagrangian density:

$$\mathcal{A}(\phi) = \int \mathrm{d}^4 x \, \mathcal{L}(\partial_\mu\phi(x), \phi(x)).$$

In the example of the Euclidean Lagrangian (2.47), the real time Lagrangian reads ($t \equiv x_0$, $\nabla \equiv \{\partial/\partial x_1, \partial/\partial x_2, \partial/\partial x_3\}$)

$$\mathcal{L}(\partial_\mu\phi(x), \phi(x)) = \tfrac{1}{2}\big(\partial_t\phi(x)\big)^2 - \tfrac{1}{2}\big(\nabla\phi(x)\big)^2 - \tfrac{1}{2}r\phi^2(x) - \frac{g}{4!}\phi^4(x).$$

While the algebraic properties of the path integral generalize easily, the field integral leads to new problems requiring new concepts like *regularization* and *renormalization*.

2.10.1 More general quantum field theories

Beside the scalar boson fields, other types of fields are also required. In the Standard Model, which describes fundamental interactions at the microscopic scale, in addition to the scalar boson Higgs field, one needs four component Dirac *Grassmann fields* with spin 1/2 for fermion matter and vector bosons in the form of *gauge fields*, associated with *gauge symmetries*, which generalize the vector potential in (2.40), to generate interactions (Sections 11.8). Moreover, after quantization of non-Abelian gauge theories, non-physical spinless fermions have to be introduced in order to be able to express the quantized action in local form (Section 11.9).

2.10.2 Regularization and effective field theories

A field theory with the Lagrangian (2.47) is called *effective* because its definition assumes implicitly the existence of an underlying microscopic model defined at a microscopic scale much smaller than the physical scale that the field theory intends to describe.

Technically, in dimensions $d > 1$, the gradient term $\frac{1}{2}(\nabla\phi(x))^2$ no longer selects fields regular enough, as a discrete or lattice approximation reveals and, as a consequence, field correlation functions are not defined at coinciding points (which require continuous fields). It is necessary to modify (in an non-physical way from the viewpoint of quantum physics, but not of classical statistical physics) the action at short distance, a procedure called *regularization*.

One possibility is to introduce quadratic terms in the field with higher order derivatives, which restrict the integration to fields satisfying a Hölder condition as in $d = 1$. The Pauli–Villars regularization amounts to the replacement

$$\left(\nabla_x\phi(x)\right)^2 \mapsto \nabla_x\phi(x)P(-\nabla_x^2/\Lambda^2)\nabla_x\phi(x),$$

where P is a polynomial of degree $n > (d-2)$, and Λ is the regularization momentum scale, a substitution for some more fundamental inverse microscopic scale. Another possibility is to introduce a lattice approximation with a lattice spacing $1/\Lambda$ (we do not discuss here the peculiarities of dimensional regularization).

2.10.3 Renormalization and renormalization group

It is then necessary to prove that, after regularization, with increasing Λ, physical observables become increasingly independent of the specific regularization and this involves the *perturbative renormalization theory* and a *renormalization group analysis*. This process finds a natural interpretation in the theory of continuous macroscopic phase transitions.

Note that the effective field theory viewpoint does not require taking the infinite cut-off limit, avoiding in this way the *triviality problem* (Section 8.2.5). By contrast, in the traditional presentation of quantum field theory as applied to particle physics, one requires the existence of an infinite Λ limit, called *renormalized theory*. This implies tuning the initial parameters of the model as a function of Λ, a procedure called *renormalization*, whose deeper physical meaning is not immediately clear.

3 The essential role of functional integrals in modern physics

A sizeable fraction of the theoretical developments in physics of the last sixty years would hardly be understandable without the use of path or, more generally, functional (*i.e.*, path and field) integration (see Chapter 2).

As examples, one can mention *thermal fluctuations*, with, as an important topic, *continuous phase transitions*, and *quantum fluctuations*, after Feynman had recognized that quantum evolution could derived from a weighted sum of classical trajectories. There, the main topic is *quantum field theory*, the basic formalism that describes fundamental interactions at the microscopic scale.

This chapter contains a brief review of what I feel are important theoretical insights and physics advances that functional integration has allowed.

The selection of topics is somewhat subjective (many other topics could have been added) but intends to convince the reader by explicit examples that, indeed, functional integration has become an essential mathematical tool of modern physics.

3.1 Classical physics: The mystery of the variational principle

The variational principle, or principle of least action, seems to have been proposed by Maupertuis and Euler (1744) and further developed by Euler. Starting in 1754, Lagrange made seminal contributions to the topic. In his treatise on analytical mechanics [19], Lagrange shows that the equations of motion of Newtonian mechanics can be derived from a variational principle.

3.1.1 Euler–Lagrange equations

The equations of Newtonian mechanics can be derived by defining a mathematical quantity, the action time integral of a Lagrangian,

$$\mathcal{A}(q) = \int_{t'}^{t''} dt\, \mathcal{L}\big(\mathbf{q}(t), \dot{\mathbf{q}}(t); t\big)$$

and expressing the stationarity of the action with respect to variations of the trajectory $\mathbf{q}(t)$ with fixed extremities $\mathbf{q}(t') = \mathbf{q}'$, $\mathbf{q}(t'') = \mathbf{q}''$. In Hamilton's form [20],

$$\delta \mathcal{A} = 0 \;\Rightarrow\; \frac{d}{dt}\frac{\partial \mathcal{L}}{\partial \dot{q}_i} = \frac{\partial \mathcal{L}}{\partial q_i}, \tag{3.1}$$

a form called Euler–Lagrange equations.

The simplest example is

$$\mathcal{L}(q, \dot{q}) = \tfrac{1}{2}m\dot{q}^2 - V(q) \;\Rightarrow\; m\ddot{q} = -V'(q).$$

From Random Walks to Random Matrices. Jean Zinn-Justin, Oxford University Press (2019).
© Jean Zinn-Justin. DOI: 10.1093/oso/9780198787754.001.0001

In this framework, *action and Lagrangian are only abstract mathematical quantities* that encode in a synthetic way the equations of motion and have been discovered to be quite useful for dealing, for example, with systems with constraints, or for exploiting conservation laws generated by continuous symmetries.

Note that the variational principle involves fixing both ends of the trajectory, while the equations of motion determine the trajectory from the boundary conditions at initial time, a feature that is somewhat intriguing and has led some scientists to call it a causality paradox.

3.1.2 *The particle in a static magnetic field*

Later, it was noticed that the equation of motion of a particle in a static magnetic field $\mathbf{B}$, which takes the form

$$m\ddot{\mathbf{q}} = e\dot{\mathbf{q}} \times \mathbf{B}(\mathbf{q}) \quad \text{where} \quad \nabla \cdot \mathbf{B}(\mathbf{q}) = 0\,,$$

quite remarkably, can also be derived from an action principle, *provided a new mathematical quantity is introduced,* the *vector potential,*

$$\mathbf{B}(\mathbf{q}) = -\nabla \times \mathbf{A}(\mathbf{q}).$$

The Lagrangian can then be written as [21]

$$\mathcal{L}(\mathbf{q}, \dot{\mathbf{q}}) = \tfrac{1}{2}m\dot{\mathbf{q}}^2 - e\mathbf{A}(\mathbf{q}) \cdot \dot{\mathbf{q}}\,. \tag{3.2}$$

In this classical framework, the *vector potential is not considered as a physical quantity* because it is defined only up to a gradient. Equivalent vector potentials are related by a *gauge transformation* (see also Chapter 11),

$$\mathbf{A}(\mathbf{q}) \mapsto \mathbf{A}(\mathbf{q}) + \nabla\Omega(\mathbf{q}).$$

Note that the Lagrangian is not invariant but changes only by a total derivative that does not affect the equations of motion.

3.1.3 *Electromagnetism and Maxwell's equations*

Maxwell's equations [22] (in the vacuum) can be written as (in Heaviside notation)

$$\nabla \cdot \mathbf{E} = \rho\,, \quad \nabla \times \mathbf{B} - \frac{\partial \mathbf{E}}{\partial t} = \mathbf{J}\,,$$

$$\nabla \cdot \mathbf{B} = 0\,, \quad \nabla \times \mathbf{E} + \frac{\partial \mathbf{B}}{\partial t} = 0\,,$$

where $\mathbf{E}$ and $\mathbf{B}$ are the electric and magnetic fields, respectively, ρ the charge, and $\mathbf{J}$ the current densities, respectively.

In quadri-covariant notation where $(i, j = 1, 2, 3)$

$$t \equiv x_0, \ F_{i0} = E_i, \ F_{ij} = -\sum_k \epsilon_{ijk} B_k, \ J_0 = \rho,$$

they take the covariant form $(\partial_\mu \equiv \partial/\partial x_\mu)$

$$\sum_{\mu=0}^{3} \partial_\mu F^{\mu\nu}(x) = J^\nu(x) \ \Rightarrow \ \sum_{\nu=0}^{3} \partial_\nu J^\nu(x) = 0.$$

In the covariant form, it becomes obvious that the electromagnetic tensor $F_{\mu\nu}(x)$ can be expressed in terms of a generalized vector potential, or *gauge field*, $A_\mu(x)$ under the form

$$F_{\mu\nu}(x) = \partial_\mu A_\nu(x) - \partial_\nu A_\mu(x). \tag{3.3}$$

The gauge field is defined only up to *an Abelian gauge transformation,*

$$A_\mu(x) \mapsto A_\mu(x) + \partial_\mu \Omega(x).$$

Then again, remarkably enough, with the introduction of this new mathematical quantity, Maxwell's equations can be derived by demanding the stationarity with respect to variations of the gauge field of a gauge invariant action,

$$\mathcal{A} = \int \mathrm{d}^4 x \, \mathcal{L}(\mathbf{A}, \dot{\mathbf{A}}),$$

where $\mathcal{L}(\mathbf{A}, \dot{\mathbf{A}})$ is the *Lagrangian density,*

$$\mathcal{L}(\mathbf{A}, \dot{\mathbf{A}}) = -\tfrac{1}{4} \sum_{\mu,\nu=0}^{3} F^{\mu\nu}(x) F_{\mu\nu}(x) - \sum_{\mu=0}^{3} J^\mu(x) A_\mu(x) \tag{3.4}$$

and $F_{\mu\nu}$ is expressed in terms of the gauge field A_μ by equation (3.3).

3.1.4 General Relativity

As a last example, we consider Einstein's relativistic theory of gravitation (or General Relativity) [23]. There again, the field equations can be derived from an action principle.

For example, in the absence of matter, in terms of the metric tensor $g_{\mu\nu}(x)$, the field equations read

$$R_{\mu\nu}(\mathbf{g}(x)) - \tfrac{1}{2} R(\mathbf{g}(x)) g_{\mu\nu} = 0, \tag{3.5}$$

where R is the scalar curvature and $R_{\mu\nu}$ the Ricci tensor.

These equations can be derived, by varying the metric, from Einstein–Hilbert's action,

$$\mathcal{A}(\mathbf{g}) = \int \mathrm{d}^4 x \, (-g(x))^{1/2} R(\mathbf{g}(x)), \tag{3.6}$$

where $g(x)$ is the determinant of the metric tensor. The action principle still holds in the presence of a cosmological constant and matter.

Note that, in the presence of *fermion matter*, the basic dynamical field is the *spin connection* instead of the metric tensor.

3.1.5 Quantum mechanics and the variational principle

The question then arises: why can all fundamental classical equations be derived by expressing the stationarity of a local action?

Quantum mechanics in its Hamiltonian formulation does not provide a straightforward answer. It can be considered as a *major success of quantum mechanics in the path integral formulation*, quantum field theory in the field integral formulation, that it provides a direct and simple explanation to this property.

According to Feynman [4], the matrix elements of the quantum evolution operator $\mathcal{U}(t'', t')$ is given by a path integral of the form

$$\mathcal{U}(t'', t'; q'', q') \propto \int_{q(t')=q'}^{q(t'')=q''} [\mathrm{d}q(t)]\, \mathrm{e}^{i\mathcal{A}(q)/\hbar}, \qquad (3.7)$$

where $\mathcal{A} = \int \mathrm{d}t\, \mathcal{L}$ is the classical action, time integral of the classical Lagrangian, and one sums over all possible trajectories $q(t)$ satisfying the boundary conditions at times t' and t'' ($\hbar$ is Planck's constant).

As a first consequence, Lagrangian and vector potential acquire a physical status and their appearance in classical physics is understood.

Moreover, in the classical $\hbar \to 0$ limit, the path integral can be calculated by the *stationary phase method* and thus is dominated by paths that leave the action stationary: these are precisely the *classical paths*. Therefore, the formulation of quantum mechanics based on path integrals explains the appearance of variational principles in classical physics. It also yields a convenient tool for calculating quantum corrections.

Finally, it leads to a more intuitive picture of quantum mechanics, giving a direct meaning to the notion of quantum fluctuations. It makes the question of the 'causality' of the variational principle irrelevant.

These properties generalize to the relativistic quantum field theory. In particular, the essential property of *locality*, that is, the action is the space-time integral of a Lagrangian density function of fields and their derivatives, implies that the classical field equations are partial differential equations.

Quantum gravity. The property that the *classical equations of General Relativity derive from a variational principle* suggests that the field integral over metrics of $\mathrm{e}^{i\mathcal{S}_{\mathrm{EH}}/\hbar}$, involving *Einstein–Hilbert's action* $\mathcal{S}_{\mathrm{EH}}$, properly regularized at short distance and with gauge fixing (of course, non-trivial issues), *should be directly relevant to quantum gravity.*

Einstein–Hilbert's action is most likely the leading term (with the cosmological constant) of a low energy or large distance expansion of an *effective quantum action*.

However, note that the classical action may have quantum corrections.

3.2 Quantum evolution: From Hamiltonian to Lagrangian formalism

In classical mechanics, the equations of motion can be derived from two alternative formalisms, one based on the Lagrangian $\mathcal{L}(q, \dot{q}; t)$ function of the position $q(t)$ and its time derivative $\dot{q}(t)$, and the other one on the Hamiltonian $H(p, q; t)$ function of *phase space variables*, the position q and the conjugate momentum p [20].

Hamiltonian and Lagrangian are related by a *Legendre transformation* ($\nabla_x \equiv (\partial/\partial x_1, \partial/\partial x_2, \ldots)$):

$$\mathcal{L}(\mathbf{q}, \dot{\mathbf{q}}; t) + H(\mathbf{p}, \mathbf{q}; t) = \mathbf{p} \cdot \dot{\mathbf{q}}, \quad \text{with} \quad \dot{\mathbf{q}} = \nabla_p H \Leftrightarrow \mathbf{p} = \nabla_q \mathcal{L}. \tag{3.8}$$

When the Hamiltonian has the simple form

$$H(\mathbf{p}, \mathbf{q}; t) = \mathbf{p}^2/2m + V(\mathbf{q}, t),$$

then $\dot{\mathbf{q}} = \nabla_p H = \mathbf{p}/m$ and

$$\mathcal{L}(\mathbf{q}, \dot{\mathbf{q}}; t) = \tfrac{1}{2} m(\dot{\mathbf{q}})^2 - V(\mathbf{q}, t).$$

To the variational principle based on the Lagrangian,

$$\mathcal{A}(q) = \int_{t'}^{t''} dt \, \mathcal{L}\big(\mathbf{q}(t), \dot{\mathbf{q}}(t); t\big),$$

corresponds an equivalent variational principle based on the Hamiltonian with the action

$$\mathcal{A}(p, q) = \int_{t'}^{t''} \big[\mathbf{p}(t) \cdot \dot{\mathbf{q}}(t) - H\big(\mathbf{p}(t), \mathbf{q}(t); t\big)\big] \, dt. \tag{3.9}$$

3.2.1 *Quantum evolution*

In quantum mechanics, evolution is governed by a unitary operator and this ensures *conservation of probabilities*. The evolution operator $\mathcal{U}(t'', t')$ between time t' and t'' for an infinitesimal time interval $t'' - t'$ is expressed in terms of a Hermitian operator $\mathbf{H}$ as,

$$\mathcal{U}(t'', t') = 1 - i\hbar(t'' - t')\mathbf{H}(t') + O(|t'' - t'|^2), \quad \mathbf{H} = \mathbf{H}^\dagger,$$

which can be identified with the quantum Hamiltonian. If the quantum Hamiltonian $\mathbf{H}$ is *time independent*, the evolution operator for an arbitrary time interval is given by

$$\mathcal{U}(t'', t') = \exp\left[-i(t'' - t')\mathbf{H}/\hbar\right]. \tag{3.10}$$

In the example of quantum particles with only positions as degrees of freedom, the *quantum Hamiltonian has the classical Hamiltonian $H(\mathbf{p}, \mathbf{q}; t)$ as a classical limit* ($\hbar \to 0$).

It is then possible to guess a path integral representation of the matrix elements of the evolution operator in terms of the classical Hamiltonian.

Path integral representation: The single particle on an axis. From the expression of the evolution operator in terms of the quantum Hamiltonian, one 'derives' a path integral representation of the matrix elements of the evolution operator $\mathcal{U}(t'', t')$ between times t' and t'' of the form [13],

$$\mathcal{U}(t'', t'; q'', q') = \int_{q(t')=q'}^{q(t'')=q''} [dp(t)dq(t)] \exp\left(i\mathcal{A}(p, q)/\hbar\right), \tag{3.11}$$

where $\mathcal{A}(p, q)$ *is the classical action in the Hamiltonian formalism* (equation (3.9)). With a proper interpretation, this form guaranties *unitary evolution*.

Special Hamiltonians. When the classical Hamiltonian H is a *quadratic form* in p, like

$$H(p, q; t) = p^2/2m + V(q, t),$$

the *integral over p is Gaussian and can be calculated explicitly:*

$$\int [\mathrm{d}p(t)] \exp \left[\frac{i}{\hbar} \int_{t'}^{t''} \mathrm{d}t \left(p(t)\dot{q}(t) - p^2(t)/2m \right) \right] \propto \exp \left[\frac{i}{\hbar} \int_{t'}^{t''} \mathrm{d}t \, \tfrac{1}{2} m \dot{q}^2(t) \right].$$

The integration amounts to replacing $p(t)$ by $m\dot{q}(t)$ as in the Legendre transformation, thus generating the Lagrangian, as anticipated in *Feynman's path integral*,

$$\mathcal{U}(t'', t'; q'', q') = \int_{q(t')=q'}^{q(t'')=q''} [\mathrm{d}q(t)] \exp \left(\frac{i}{\hbar} \int_{t'}^{t''} \mathrm{d}t \, \mathcal{L}(q, \dot{q}; t) \right)$$

with

$$\mathcal{L}(q, \dot{q}; t) = \tfrac{1}{2} m \dot{q}^2 - V(q, t).$$

General Hamiltonians. For more general Hamiltonians, the leading term in the exponential for $\hbar \to 0$ is still the Lagrangian but *quantum corrections* may have to be added.

3.2.2 *Relativistic quantum field theory*

In a relativistic quantum field theory, the *evolution is explicitly unitary in the Hamiltonian formulation* but *not explicitly relativistic covariant (i.e., with respect to the* Lorentz symmetry group) because *time and space play quite different roles.*

However, in the physically relevant theories, *the Hamiltonian is quadratic in the conjugate momenta,* and an analogous calculation allows expressing the quantum evolution in terms of an explicitly *covariant Lagrangian.*

This had been first noticed by Dirac [8] when calculating the evolution for an infinitesimal time interval in quantum electrodynamics (QED).

For example, for a scalar field $\phi(x)$, a simple Hamiltonian density takes the form

$$\mathcal{H}(\Pi, \phi) = \int \mathrm{d}^3 x \left[\tfrac{1}{2} \Pi^2(x) + \tfrac{1}{2} \left(\nabla_x \phi(x) \right)^2 + V\left(\phi(x) \right) \right], \tag{3.12}$$

where the field $\Pi(x)$ is the conjugate momentum.

The action in the Hamiltonian formulation then reads,

$$\mathcal{A}(\Pi, \phi) = \int \mathrm{d}t \, \mathrm{d}^3 x \left[\Pi(t, x)\dot{\phi}(t, x) - \mathcal{H}\left(\Pi(t, x), \phi(t, x) \right) \right]. \tag{3.13}$$

Again the Hamiltonian is quadratic in the conjugate momentum. The integration over $\Pi(t, x)$ can be performed and this amounts to substituting

$$\Pi(t, x) = \dot{\phi}(t, x).$$

The resulting action $\mathcal{A}(\phi)$ is the integral of the Lagrangian density,

$$\mathcal{A}(\phi) = \int dt \, d^3x \left[\tfrac{1}{2} \left(\dot{\phi}(t,x) \right)^2 - \tfrac{1}{2} \left(\nabla_x \phi(t,x) \right)^2 - V\left(\phi(t,x) \right) \right], \qquad (3.14)$$

and is relativistically covariant.

This example illustrates the *unifying power of the functional integral*, where analogous methods apply to the single quantum particle and to relativistic quantum field theory.

3.3 From quantum evolution to statistical physics

In quantum mechanics, the density matrix ρ at thermal equilibrium at a temperature $T = 1/\beta$ is given by (with $k_B = 1$)

$$\rho(\beta) = e^{-\beta \mathbf{H}} / \mathcal{Z}(\beta), \qquad (3.15)$$

where $\mathcal{Z}$ is the partition function,

$$\mathcal{Z}(\beta) = \operatorname{tr} e^{-\beta \mathbf{H}} \Rightarrow \operatorname{tr} \rho(\beta) = 1. \qquad (3.16)$$

(The thermal expectation value of an observable represented by a quantum Hermitian operator $\mathbf{A}$ is then $\operatorname{tr} \rho \mathbf{A}$.) Therefore, the density matrix is proportional to the analytic continuation for imaginary time of the evolution operator and obtained by setting $t = \hbar\beta/i$.

3.3.1 The single particle on an axis

For example, to the matrix elements of the evolution operator of a single particle in a potential on an axis,

$$\mathcal{U}(t'',t';q'',q') = \int_{q(t')=q'}^{q(t'')=q''} [dq(t)] \exp\left[\frac{i}{\hbar} \int_{t'}^{t''} dt \left[\tfrac{1}{2} m \dot{q}^2(t) - V\left(q(t) \right) \right] \right], \qquad (3.17)$$

correspond the elements of the density matrix,

$$\rho(\beta; q'', q') \propto \int [dq(t)] \, e^{-\mathcal{S}(q)}, \quad \mathcal{S}(q) = \int_{-\beta/2}^{\beta/2} d\tau \left[\frac{m}{2\hbar^2} \dot{q}^2(\tau) + V\left(q(\tau) \right) \right] \qquad (3.18)$$

with $q(-\beta/2) = q'$, $q(\beta/2) = q''$.

The partition function then is given by

$$\mathcal{Z}(\beta) = \int [\mathrm{d}q(t)] \, \mathrm{e}^{-\mathcal{S}(q)}$$

with the periodic boundary conditions,

$$q(-\beta/2) = q(\beta/2).$$

From quantum statistical to classical statistical physics. In the path integral representation of the density matrix, the integrand defines a positive measure on paths. Therefore, it can have another interpretation: $\mathcal{S}(q)$ can also be considered as the *configuration energy of a one-dimensional classical* system of length β with periodic boundary conditions.

3.3.2 Quantum field theory: Quantum and classical statistical physics

The correspondence between quantum and classical statistical physics, noticed in the example of the single particle, *generalizes to quantum field theory and field integrals* and becomes obvious through the functional integral formalism.

With analogous arguments, it can be shown that, for a scalar field $\phi(x)$ in *d space dimensions*, the *quantum partition function* at temperature $T = 1/\beta$ reads (in the simplest examples)

$$\mathcal{Z} = \int [\mathrm{d}\phi] \exp\left[- \int_{-\beta/2}^{\beta/2} \mathrm{d}t \int \mathrm{d}^d x \, \mathcal{L}_{\mathrm{E}}(\phi) \right], \qquad (3.19)$$

where $\mathcal{L}_{\mathrm{E}}$ is the Euclidean (imaginary time) Lagrangian density; t here is an inverse energy, and the (Bose) fields satisfy the periodic boundary conditions

$$\phi(-\beta/2, x) = \phi(\beta/2, x).$$

However, up to a reinterpretation of parameters, the same field integral representation yields a *classical partition function in $(d+1)$ space dimensions* for a system with finite size β and periodic boundary conditions in one space direction.

This observation is very relevant in the *theory of continuous phase transitions*, since it relates a class of *classical transitions in $(d+1)$ dimensions* to *quantum transitions at zero temperature ($\beta = \infty$) in d dimensions*.

More generally, the relation between classical and quantum statistical physics maps *finite temperature quantum* properties to *finite size properties in the classical theory*. The same *renormalization group* methods can then be applied to both situations (see Chapter 21).

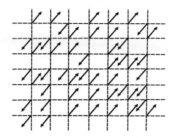

Fig. 3.1 Ising spins on a two-dimensional lattice: A configuration in the disordered phase.

3.4 Statistical models at criticality and quantum field theory

The field integral formulation emphasizes the relation between classical statistical physics and quantum field theory.

Following Wilson [24], it has been realized that large distance (or critical) universal properties near a continuous phase transition of a large class of statistical models with short range interactions could be described by an imaginary time local quantum field theory or, equivalently, a classical statistical field theory.

For example, the partition function of the d-dimensional Ising lattice model (Fig. 3.1), which has a $\mathbb{Z}_2$ reflection symmetry, is given by a sum over classical spins with nearest-neighbour (n.n.) interactions,

$$\mathcal{Z} = \sum_{\{S_i\}=\pm 1} \exp\left((J/T) \sum_{i,j \text{ n.n.}} S_i S_j \right), \quad J > 0,$$

where the multi-indices $i, j \in \mathbb{Z}^d$ denote lattice site positions.

For dimensions $d > 1$, at a critical temperature $T_c > 0$, the model has a continuous transition between an ordered magnetic and a disordered phase. At T_c, the correlation length diverges. This leads to a large distance physics with properties like algebraic decay of correlations and singularities at T_c, shared by a large class of models with a $\mathbb{Z}_2$ reflection symmetry and short range interactions. Many systems are non-magnetic, like liquid–vapour or binary mixtures.

Remarkably enough, these universal properties can then be inferred from an *effective ϕ^4* local quantum field theory (in imaginary time), which belongs to the same *universality class*.

The partition function reads,

$$\mathcal{Z}_{\phi^4} = \int [\mathrm{d}\phi] \exp\left[-\mathcal{S}(\phi) \right],$$

where one now *sums over all field configurations* $\phi(x)$ where $x \in \mathbb{R}^d$, and $\mathcal{S}(\phi)$ is the imaginary time action (or configuration energy),

$$\mathcal{S}(\phi) = \int \mathrm{d}^d x \left[\tfrac{1}{2} \left(\nabla_x \phi(x) \right)^2 + \tfrac{1}{2} r \phi^2(x) + \tfrac{1}{4!} g \phi^4(x) \right]. \tag{3.20}$$

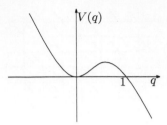

Fig. 3.2 The $V(q) = \frac{1}{2}(q^2 - q^3)$ potential.

The action has still to be regularized at short distance (see Section 2.10). This can be achieved by adding terms quadratic in the field with higher derivatives and which here reflect the initial lattice structure.

In the field theory, the phase transition occurs at a critical value r_c of the parameter r and, near r_c, $r - r_c$ is proportional to $T - T_c$.

The parameter $g > 0$ is an additional parameter required, at least, to be able to describe the low temperature (ordered) phase.

More generally, universal properties near a phase transition of N-component spin models with $O(N)$ symmetry whose partition function has the form,

$$\mathcal{Z} = \sum_{\{|\mathbf{S}_i|\}=1} \exp\left((J/T) \sum_{i,j \text{ n.n.}} \mathbf{S}_i \cdot \mathbf{S}_j\right)$$

can be inferred from an analogous $O(N)$ symmetric quantum field theory with the action

$$\mathcal{S}(\boldsymbol{\phi}) = \int \mathrm{d}^d x \left[\frac{1}{2}\left(\nabla_x \boldsymbol{\phi}(x)\right)^2 + \frac{1}{2}r\boldsymbol{\phi}^2(x) + \frac{1}{4!}g\left(\boldsymbol{\phi}^2(x)\right)^2\right]. \tag{3.21}$$

The field integral formulation of quantum field theory allows understanding directly the relation between the classical statistical physics of macroscopic phase transitions and quantum field theory.

Conversely, the relation between classical statistical physics and quantum field theory has suggested that *quantum field theory is only an effective low energy theory.* In particular, this remark applies to the *Standard Model of fundamental interactions at the microscopic scale.*

Polymers. Path integral techniques allow proving directly that the $N = 0$ limit of the $O(N)$ symmetric $(\boldsymbol{\phi}^2)^2$ field theory describes the *statistical properties of long polymers* [25] (mathematically, self-avoiding random walk).

3.5 Barrier penetration, vacuum instability: Instanton calculus

Quantum mechanics predicts for a non-relativistic theory with a finite number of particles the possibility to cross potential barriers. The probability per unit time of these *classically forbidden barrier penetration effects* can be evaluated in the semi-classical approximation ($\hbar$ small) by partial differential equation methods.

Alternatively, it can be determined by applying the *steepest descent method to the continuation of the path integral to imaginary time.* Saddle points then are *finite action solutions* of imaginary time equations of motion called *instanton* solutions. The interpretation is that, in the semi-classical approximation, barrier penetration (or tunnelling) is formally associated to classical motion in imaginary time in the classically forbidden region.

The *instanton method* imitates the evaluation of simple integrals and, unlike methods based on the Schrödinger equation, *its generalization to quantum field theory* is straightforward, showing again the *unifying power of functional integration.*

A simple example. The minimum of the one-dimensional potential $V(q) = \frac{1}{2}(q^2 - q^3)$ at $q = 0$ is classically stable but, in quantum mechanics, unstable by barrier penetration. The corresponding *imaginary time classical action* for a particle of mass 1 is

$$\mathcal{S}(q) = \int \mathrm{d}t \left[\tfrac{1}{2}\dot{q}^2(t) + \tfrac{1}{2}q^2(t) - \tfrac{1}{2}q^3(t) \right] . \tag{3.22}$$

The equation of motion is (imaginary time is equivalent to change $V \mapsto -V$)

$$\ddot{q}(t) = q(t) - \tfrac{3}{2}q^2(t).$$

Instanton solutions are

$$q(t) = \frac{1}{\cosh^2\big((t - t_0)/2\big)}, \tag{3.23}$$

(t_0 is an integration constant) which correspond to a trajectory that leaves $q = 0$ at $t = -\infty$, is reflected at $q = 1$ and returns to $q = 0$ at time $+\infty$. The corresponding contribution to the path integral is of order $\mathrm{e}^{-8/15\hbar}$ and is proportional to the probability per unit time of a quantum particle to leave the well.

Physics applications. Important phenomena in particle physics, like vacuum instability, lifting the *degeneracy of the periodic quantum chromodynamics (QCD) classical vacuum (c.f.,* the cosine potential), which leads to the notion of θ-vacuum and, thus, to the *strong CP problem,* or the *solution of the $U(1)$ problem,* are related to instantons [18] (see Chapter 18 for details).

3.6 Large order behaviour and Borel summability: Critical exponents

In quantum mechanics, for polynomial potentials, the spectrum can be calculated as a series expansion in the deviation of the potential from a quadratic harmonic potential. For example, if the potential reads,

$$V(q) = \tfrac{1}{2}\omega^2 q^2 + gq^N,$$

one can expand the spectrum in powers of the parameter g.

An important question is: what is the nature of the g expansion? The behaviour of the perturbative expansion at large orders is related to vacuum or ground state instabilities due barrier penetration.

Functional integral techniques and instantons [26] *have led to the determination of the large order behaviour of the perturbative expansion in quantum systems with analytic potentials and in a number of quantum field theories* [27].

An important outcome is that *perturbation series are always divergent.* Moreover, the knowledge of the large order behaviour has then direct implications on the problem of *Borel summability: divergent series that are Borel summable nevertheless define a unique analytic function* (see Chapter 23).

One condition is that *series can be Borel summable only if instantons correspond to non-physical values* of parameters of the action. For example, perturbative series corresponding to theories where instantons connect degenerate classical minima of the potential or where the minimum of the potential is only relative (Fig. 3.2), are not Borel summable [28] (see also Chapter 24).

Application to the evaluation of critical exponents. An important application has been the development of efficient numerical techniques to sum perturbative expansions in quantum field theory. This has allowed evaluating *critical exponents*, which characterize the singular behaviour of correlation functions near the critical temperature in *continuous phase transitions.*

For the $O(N)$ symmetric N vector model, critical exponents have been inferred from the imaginary time $O(N)$ symmetric $(\phi^2)^2$ quantum field theory [24]:

$$\mathcal{Z} = \int [\mathrm{d}\phi(x)] \, \mathrm{e}^{-\mathcal{S}(\phi)}$$

with

$$\mathcal{S}(\phi) = \int \mathrm{d}^d x \left[\tfrac{1}{2} \left(\nabla_x \phi(x) \right)^2 + \tfrac{1}{2} \phi^2(x) + \tfrac{1}{4!} g \left(\phi^2(x) \right)^2 \right]. \tag{3.24}$$

In the field theory formulation, one can derive *renormalization group* equations for correlation functions, which can then used to determine *universal critical properties* [29]. The corresponding *renormalization group functions* are obtained as power series in g, which have been shown to be *Borel summable for $d < 4$* and whose large order behaviour has been determined by instanton calculus (instanton solutions exist for non-physical values $g < 0$). The series have been summed with *summation methods based on a Borel transformation* [30] (see Chapter 23).

Critical exponents are given by the values of the renormalization group functions at the infrared fixed point value g^* of the coupling constant g. We display the value g^*, the correlation length exponent ν, the exponent β that characterizes the behaviour of the spontaneous expectation value of the field and the correction exponent $\theta = \omega \nu$ in Table 3.1 [31].

Instantons and the problem of non-Borel summability. In the case of *potentials with degenerate classical minima*, instanton calculus applied to the large order behaviour indicates that, due the tunnelling between the minima, the perturbative expansion is *non-Borel summable*, that is, does not determine unique functions. In simple quantum mechanics with analytic potentials, the problem can be studied systematically and it can be shown that *multi-instanton configurations* (generalized instantons) must be taken into account and new summation methods introduced.

Table 3.1

Estimates of critical exponents from $O(N)$ symmetric $(\phi^2)^2_3$ field theory
(Le Guillou and Zinn-Justin (1980), updated by Guida and Zinn-Justin (1998)).

N	0	1	2	3
g^*	26.63 ± 0.11	23.64 ± 0.07	21.16 ± 0.05	19.06 ± 0.05
ν	0.5882 ± 0.0011	0.6304 ± 0.0013	0.6703 ± 0.0015	0.7073 ± 0.0035
β	0.3024 ± 0.0008	0.3258 ± 0.0014	0.3470 ± 0.0016	0.3662 ± 0.0025
$\theta = \omega\nu$	0.478 ± 0.010	0.504 ± 0.008	0.529 ± 0.009	0.553 ± 0.012

These properties have deep relations with the *mathematical resurgence* theory (see Chapter 24) [38, 39].

3.7 Quantization of gauge theories

In *gauge theories*, like QED (the quantum version of Maxwell's equations; see Section 3.1.3), a straightforward application of Feynman's quantization rule (3.7) [4] leads to an undefined field integral representation of the evolution operator because the integrand does not depend on all degrees of freedom (integration variables) due to gauge invariance. However, field integral techniques have helped in solving the problem, as we now indicate.

3.7.1 QED

In QED, the classical action for the vector field (also called the gauge field) $A_\mu(x)$, coupled to a conserved current $J^\mu(x)$ representing charged matter, is (see Section 3.1.3)

$$\mathcal{A}(\mathbf{A}) = \int \mathrm{d}^4x \, \mathcal{L}(\mathbf{A}, \dot{\mathbf{A}}), \tag{3.25}$$

the integral of the Lagrangian density ($\mu, \nu = 0, 1, 2, 3$),

$$\mathcal{L}(\mathbf{A}, \dot{\mathbf{A}}) = -\tfrac{1}{4} \sum_{\mu,\nu} F^{\mu\nu}(x) F_{\mu\nu}(x) - \sum_\mu J^\mu(x) A_\mu(x), \quad \sum_\mu \partial_\mu J^\mu(x) = 0, \tag{3.26}$$

where $F_{\mu\nu}$ is the electromagnetic field related to the vector field by ($\partial_\mu \equiv \partial/\partial x_\mu$)

$$F_{\mu\nu}(x) = \partial_\mu A_\nu(x) - \partial_\nu A_\mu(x). \tag{3.27}$$

The action (3.25) is invariant under the shift or gauge transformation,

$$A_\mu(x) \mapsto A_\mu(x) + \partial_\mu \Omega(x), \tag{3.28}$$

a property called *gauge invariance*.

Naively, one would think that the quantum evolution operator is given by the field integral [4],

$$\mathcal{U} = \int [\mathrm{d}A_\mu(x)] \exp\left(\frac{i}{\hbar}\mathcal{A}(\mathbf{A})\right).$$

However, the *field integral is undefined because, due to the invariance under the gauge transformation (3.28), the integrand does not depend on four independent A_μ components.*

Gauge fixing. It is necessary to integrate only over a section of the gauge field space, intersecting all gauge orbits once. A section is defined by a gauge condition. As an example, we choose the covariant Landau's gauge

$$\nabla_x \cdot \mathbf{A}(x) = 0. \tag{3.29}$$

The problem then is to determine the *field integration measure on the gauge section* in such a way that *gauge invariant physical observables are independent of the choice of the section.*

One method is as follows [40]: one defines a *gauge invariant* function $\Delta(\mathbf{A})$ obtained from averaging the gauge condition over gauge transformations,

$$1 = \Delta(\mathbf{A}) \int [\mathrm{d}\Omega(x)] \prod_x \delta\big(\nabla_x^2\Omega(x) + \nabla_x \cdot \mathbf{A}(x)\big). \tag{3.30}$$

One then inserts the right-hand side into the gauge field integral,

$$\mathcal{U} \propto \int [\mathrm{d}A_\mu(x)\mathrm{d}\Omega(x)] \prod_x \delta\big(\nabla_x^2\Omega(x) + \nabla_x \cdot \mathbf{A}(x)\big)\Delta(\mathbf{A}) \exp\left(\frac{i}{\hbar}\mathcal{A}(\mathbf{A})\right).$$

After a change of variables in the form of a gauge transformation,

$$A_\mu(x) + \partial_\mu\Omega(x) = A'_\mu(x),$$

the integral over $\Omega(x)$ factorizes and yields an infinite constant factor $\int[\mathrm{d}\Omega(x)]$, which can be removed.

Finally, shifting $\Omega(x)$ in equation (3.30),

$$\Omega(x) + (\nabla_x^2)^{-1}\nabla_x \cdot \mathbf{A}(x) = \Omega'(x),$$

one finds that $\Delta(\mathbf{A})$ is a constant.

One concludes

$$\mathcal{U} \propto \int [\mathrm{d}A_\mu(x)] \prod_x \delta\big(\nabla_x \cdot \mathbf{A}(x)\big) \exp\left(\frac{i}{\hbar}\mathcal{A}(\mathbf{A})\right), \tag{3.31}$$

which is a defined field integral.

Because the result is so simple, the quantization of QED was guessed before methods based on field integration were proposed.

The $U(1)$ gauge group. The transformation (3.28) of the gauge field is associated with the Abelian group of $U(1)$ space-time dependent transformations (elements of the $U(1)$ group at each point). Indeed, the corresponding action on charged matter fields $\psi(x)$ has the form

$$\psi(x) \mapsto \mathrm{e}^{-ie\Omega(x)}\, \psi(x),$$

where e is the electric charge.

3.7.2 Quantization of non-Abelian gauge theories

The construction of the Standard Model of fundamental interactions at the microscopic scale (or particle physics) has required a generalization of the QED gauge invariance, the Abelian $U(1)$ group being replaced by the group $U(1) \times SU(2) \times SU(3)$, a group that has two non-Abelian factors.

All concepts required to quantize non-Abelian gauge theories in a covariant gauge, like the so-called *Faddeev–Popov method and 'ghosts'*, are based on field integrals.

Non-Abelian gauge theories [41]. We assume that the gauge group G is a semi-simple unitary Lie group. Matter fields transform like

$$\psi(x) \mapsto \mathbf{g}(x)\psi(x), \quad \mathbf{g} \in G. \tag{3.32}$$

For space-independent transformations, the gauge field $\mathbf{A}_\mu(x)$ transforms under the adjoint representation of the group G. When it is represented by an element of the Lie algebra $\mathfrak{L}(G)$ (an anti-Hermitian matrix), and $\mathbf{g}(x)$ is a space-time dependent group element, a gauge transformation takes the form

$$\mathbf{A}_\mu(x) \mapsto \mathbf{A}_\mu^{\mathbf{g}}(x) \equiv \mathbf{g}(x)\mathbf{A}_\mu(x)\mathbf{g}^{-1}(x) + \mathbf{g}(x)\partial_\mu\mathbf{g}^{-1}(x). \tag{3.33}$$

For matter fields, gauge invariance is enforced by replacing derivatives by covariant derivatives: $\mathbf{D}_\mu = \mathbf{1}\,\partial_\mu + \mathbf{A}_\mu$.

The associated curvature (generalization of the electromagnetic tensor)

$$\mathbf{F}_{\mu\nu}(x) = [\mathbf{D}_\mu, \mathbf{D}_\nu] = \partial_\mu\mathbf{A}_\nu(x) - \partial_\nu\mathbf{A}_\mu(x) + [\mathbf{A}_\mu(x), \mathbf{A}_\nu(x)]\,, \tag{3.34}$$

is a tensor for gauge transformations:

$$\mathbf{F}_{\mu\nu}(x) \mapsto \mathbf{g}(x)\mathbf{F}_{\mu\nu}(x)\mathbf{g}^{-1}(x).$$

The local gauge action

$$\mathcal{A}(\mathbf{A}) = -\frac{1}{4g^2}\int \mathrm{d}^4x\,\mathrm{tr}\sum_{\mu,\nu}\mathbf{F}_{\mu\nu}(x)\mathbf{F}^{\mu\nu}(x), \tag{3.35}$$

thus is gauge invariant.

By contrast with QED, the quantization of non-Abelian gauge theories, even without matter fields, does not follow from simple heuristic ideas (see Section 13.3) [42]. Due to gauge invariance, not all components of the gauge field are dynamical and a straightforward canonical quantization based on the operator formalism is impossible. A gauge fixing is required. A non-covariant quantization in the temporal (or Weyl) gauge $\mathbf{A}_0 = 0$ is possible but leads to a non-renormalizable and singular perturbative quantum field theory.

Again in the field integral formalism, the straightforward field integral representation of the evolution operator

$$U = \int [\mathrm{d}\mathbf{A}_\mu]\exp\left[i\mathcal{A}(\mathbf{A})/\hbar\right], \tag{3.36}$$

is not defined because the action $\mathcal{A}$, being gauge invariant, does not depend on all the integration fields.

It is necessary to integrate only once over each gauge copy, that is, to limit the integration to a section of gauge field space cutting each gauge-orbit once.

3.7.3 Covariant quantization: Faddeev–Popov's method

To implement, for example, the covariant gauge condition

$$\sum_\mu \partial_\mu \mathbf{A}_\mu(x) = 0 \tag{3.37}$$

(by inserting it in the field integral as a generalized δ-function) and still ensure the gauge independence of physical observables, a necessary condition for preserving unitarity, it is necessary to modify the integration measure over the gauge field.

The measure can be determined by averaging over all gauge transformations $\mathbf{g}(x)$ (a field integral) the condition $\sum_\mu \partial_\mu \mathbf{A}_\mu^g(x) = 0$. The resulting measure takes the form of a determinant that can be expressed in local form as a field integral over spinless (non-physical) fermion fields $\bar{\mathbf{C}}$ and $\mathbf{C}$ (the Faddeev–Popov 'ghost' fields), transforming under the adjoint representation) [40]. The δ-function imposing the constraint $\sum_\mu \partial_\mu \mathbf{A}_\mu(x) = 0$ can itself be written in a Fourier representation by introducing an auxiliary scalar field $\boldsymbol{\lambda}(x)$ (also transforming under the adjoint representation). If one slightly generalizes the method, expresses the constraint by a Fourier integral and integrates in the neighbourhood of the gauge section (3.37) with a Gaussian distribution of width ξ, one finally obtains the local action

$$\mathcal{S}(\mathbf{A}_\mu, \bar{\mathbf{C}}, \mathbf{C}, \boldsymbol{\lambda}) = \int \mathrm{d}^d x \, \mathrm{tr}\left[-\frac{1}{4e^2} \sum_{\mu,\nu} \mathbf{F}_{\mu\nu}^2(x) \right.$$
$$\left. + \frac{\xi e^2}{2} \boldsymbol{\lambda}^2(x) + \boldsymbol{\lambda}(x) \sum_\mu \partial_\mu \mathbf{A}_\mu(x) + \mathbf{C}(x) \sum_\mu \partial_\mu \mathbf{D}_\mu \bar{\mathbf{C}}(x) \right]. \tag{3.38}$$

It was eventually noticed that this quantized action has a *fermion-like symmetry, the BRST symmetry* [43]. *Supersymmetry* is a generalization of the BRST symmetry.

3.8 Numerical simulations in quantum field theory

In Sections 3.3.2 and 3.4, we described the relation between classical and quantum partition function in the continuum, and lattice models and quantum field theory. These relations have led to the *application of methods of statistical physics to non-perturbative studies of quantum field theories.*

The basic idea is to replace the continuum field integral by a lattice regularized form (Section 13.8). Then, *non-perturbative numerical techniques* become available, like strong coupling expansions or Monte Carlo-type simulations if the integrand defines a positive measure. This implies, in particular, that only quantum field theories in imaginary time, that is, quantum statistical models can be directly simulated.

An outstanding example is QCD, the origin of the strong nuclear force at the fundamental level. The physical spectrum of QCD cannot be calculated in the continuum by expanding around free field theory. Therefore, QCD has been investigated by large scale numerical simulations using a gauge invariant lattice approximation [44] called *lattice gauge theory*.

Gauge fields and gauge invariance. Below, we denote by $i, j, \ldots$ lattice sites belonging to a d-dimensional hypercubic lattice.

On the lattice, gauge fields are replaced by group elements $\mathbf{u}_{ij}$ associated to links joining neighbour sites i, j. Gauge transformations become

$$\mathbf{u}_{ij} \mapsto \mathbf{g}_i \mathbf{u}_{ij} \mathbf{g}_i^{-1}, \ \mathbf{g} \in G.$$

Wilson's plaquette action. In the absence of matter, the lattice action can be chosen to be Wilson's *plaquette action,* and the partition function reads

$$\mathcal{Z} = \int \prod_{\text{links}\{i,j\}} \mathrm{d}\mathbf{u}_{ij} \, \mathrm{e}^{-\beta_p \mathcal{S}_{\mathrm{W}}(\mathbf{u})}, \quad \mathcal{S}_{\mathrm{W}}(\mathbf{u}) = - \sum_{\text{plaquettes}\,(ijkl)} \mathrm{tr}\, \mathbf{u}_{ij} \mathbf{u}_{jk} \mathbf{u}_{kl} \mathbf{u}_{li}. \quad (3.39)$$

After addition of fermion matter, in the continuum limit $\beta_p \to \infty$, the lattice formulation yields a *non-perturbative definition of QCD.*

3.9   Quantization of the non-linear σ-model

The non-linear σ-model (see Chapter 10 and Section 21.10) is a quantum field theory that describes $O(N)$ symmetric ferromagnets in the low temperature phase dominated by Goldstone modes, which appears in the description of the chiral properties of low energy hadron physics and, because it is *asymptotically free in two dimensions,* gives some insight into the similar property of non-Abelian gauge theories in four dimensions.

The field $\boldsymbol{\phi}(x)$ is an N-component vector that lives on the sphere S_{N-1}:

$$\boldsymbol{\phi}^2(x) = 1. \quad (3.40)$$

In terms of the field $\boldsymbol{\phi}$, the action takes the form of a free action,

$$\mathcal{S}(\phi) = \frac{1}{2g} \int \mathrm{d}^d x \, [\nabla_x \boldsymbol{\phi}(x)]^2, \quad (3.41)$$

where the parameter g plays the role of the temperature for ferromagnets, but the constraint (3.40) generates interactions.

Within the perturbative expansion (an expansion in powers of g), the $O(N)$ symmetry is realized in the phase of spontaneous symmetry breaking, and the dynamical fields correspond to (massless) Goldstone modes (a situation that is realized, beyond perturbation theory, only at low temperature and for $d > 2$).

First perturbative calculations (based on the operator formalism) seemed to indicate that the $O(N)$ symmetry was explicitly broken by perturbative corrections. Within the operator formulation, a complicated calculation showed that the breaking term actually cancelled [45].

The field integral representation starting from the Hamiltonian formulation gives both the correct quantized form to all orders and the geometric explanation of the mistake as being equivalent to *omitting the necessary $O(N)$ invariant measure* [46],

$$\mathcal{Z} = \int [\mathrm{d}\phi] \prod_x \delta(\phi^2(x) - 1) \exp\left[-\mathcal{S}(\phi)\right]. \tag{3.42}$$

Moreover, to give a meaning to the model beyond perturbation theory, one can introduce a *lattice regularization*, and this yields an $O(N)$ lattice spin model. In this way, one can establish an indirect *relation between the non-linear σ-model and the $(\phi^2)^2$ statistical field theory*.

3.10 N-component fields: Large N techniques

In quantum field theories with $O(N)$ or $U(N)$ symmetries and fields in the vector representation, physical quantities can be evaluated in the large N limit, leading to interesting non-perturbative results, in particular when non-perturbative effects are essential [47].

At leading order, the large N results can be obtained by summing a subclass of Feynman diagrams, but functional integral techniques are much simpler and can be more easily extended to higher orders in $1/N$ (see Sections 21.9–21.11). The method does not depend on the number of space dimensions and applies to ordinary integrals as well as path and field integrals.

Applications include the study of the $(\phi^2)^2$ theory (and the calculation of critical exponents), the Gross–Neveu model, and so on.

For example, for a self-interacting scalar field $\phi(x)$, the basic idea is to introduce into the ϕ-field integral the identity

$$1 = \int [\mathrm{d}\lambda \mathrm{d}\rho] \exp\left\{ i \int \mathrm{d}^d x \, \lambda(x) \left[N\rho(x) - \phi^2(x)\right] \right\}, \tag{3.43}$$

where $\{\lambda(x), \rho(x)\}$ are two auxiliary fields. For the simple action,

$$\mathcal{S}(\phi) = \int \mathrm{d}^d x \left[\left(\nabla_x \phi(x)\right)^2 + V\left(\phi^2(x)\right)\right], \tag{3.44}$$

the substitution $\phi^2(x) \mapsto \rho(x)$ in V transforms the initial action into an action quadratic in ϕ, and the Gaussian integration over ϕ can be performed, making the N-dependence explicit. While the field ϕ may have large fluctuations, $\rho(x)$, which, for N large, is the average of a large number of degrees of freedom, is expected to have smaller fluctuations (central theorem of probabilities).

The resulting integral over the fields $\{\lambda(x), \rho(x)\}$ can then be evaluated, for N large, by the steepest descent method.

The large N method has been especially useful to give an insight into the structure of the $O(N)$ vector model in the critical domain, showing how the critical temperature goes to zero when the space dimension d goes to 2 and that the non-linear σ-model and the $(\phi^2)^2$ field theory with $O(N)$ symmetry are identical in the critical domain to all orders in $1/N$, establishing in this way a link between the $\varepsilon = 4 - d$-expansion of the $(\phi^2)^2$ field theory and the $\varepsilon = d - 2$-expansion of the non-linear σ-model.

4 From infinities in quantum electrodynamics to the general renormalization group

The constructions of *quantum (or statistical) field theories* (QFTs) and the deeply related *renormalization group* (RG), have been some of the major theoretical achievements in physics of the last century.

Indeed, QFT and RG play a crucial role in many, sometimes completely unrelated, domains of physics, most notably: the theory of *fundamental interactions at the microscopic scale* and the theory of *continuous macroscopic phase transitions*.

While QFT is the necessary framework to describe fundamental physics at the microscopic scale (gravity remains a separate issue), perhaps more surprisingly, it also provides a framework for discussing the universal singular properties of a large class of macroscopic continuous phase transitions (liquid–vapour, ferromagnetism, superfluidity, binary mixtures, *etc.*) at the transition, the properties of *dilute quantum gases* beyond the model of Bose–Einstein condensation, the statistical properties of *long polymer chains* (mathematically, self-avoiding random walk), percolation, and so on.

In fact, QFT offers, up to now, the *most powerful framework* in which physical systems characterized by *a large number of fluctuating degrees of freedom with local interactions* can be discussed.

However, initially, QFT was confronted with an unexpected problem: *infinities* appeared in the calculations beyond leading order of many physical quantities. Eventually, a method was discovered, *renormalization*, that made it possible to derive finite predictions from naively infinite expressions. The method would hardly have been convincing if the predictions had not been confirmed very soon after with increasing precision by experiments.

The renormalization process involves choosing a physical scale at which to define physical (or renormalized) parameters. A first form of *RG* emerged in this context, as a relation between renormalized parameters defined at different physical scales. However, its full meaning, as a perturbative form of a more general RG, was completely appreciated only later in the framework of macroscopic continuous phase transitions (a process in which *Wilson's contribution* was essential).

Beyond a number of direct physics applications, RG has made it possible to understand the origin and meaning of renormalizable QFT in particle physics, as well as universality in the theory of critical phenomena. It has led to a *satisfactory interpretation* of the origin and role of *renormalizable QFT* and of the renormalization process.

As a consequence, in the modern interpretation, *QFT's are only effective large distance theories*. Unlike earlier fundamental theories, they are *not necessarily consistent on all distance or energy scales*.

From Random Walks to Random Matrices. Jean Zinn-Justin, Oxford University Press (2019).
© Jean Zinn-Justin. DOI: 10.1093/oso/9780198787754.001.0001

4.1 QFT, RG: Some major steps

Often in science, because the focus is too much on specific breakthrough dates, one does not appreciate that many important advances result from long phases of exploration and development. Therefore, we feel it useful to recall a few important steps in the construction of QFT and the emergence and the first applications of the RG idea, to give an idea of the time scales involved.

1925: Heisenberg formulates the basis of *quantum mechanics* as a mechanics of matrices [48].

1926: Schrödinger publishes his famous equation [49], which bases quantum mechanics on the solution of a non-relativistic wave equation. Indeed, for accidental reasons, the spectrum of the hydrogen atom is better reproduced by the *non-relativistic Schrödinger equation* than by a *relativistic spinless** wave equation, the *Klein–Gordon equation* (1926).

1928: Dirac introduces another famous equation [50], a *relativistic wave equation* that takes into account the *spin 1/2 of the electron* and leads to a spectrum of the hydrogen atom in much better agreement with experiment, and this paves the way for a relativistic quantum theory.

1929–1930: Heisenberg and Pauli establish the general principles of *QFT* [51].

1930: Several unsuccessful attempts [52] are made to calculate the first quantum correction to the electron self-energy.

1934: Weisskopf publishes the first correct calculation of a quantum correction in *quantum electrodynamics (QED)*, a relativistic quantum theory describing the electromagnetic interactions between charged particles, and confirmation is made of the appearance of *infinities*, called ultraviolet (UV) divergences (since they are due to the contributions of very-short-wavelength photons) [53].

1937: *Landau* publishes his general *theory of phase transitions* [54].

1944: *Onsager reports the exact solution of the two-dimensional Ising model*, which exhibits non-mean-field critical exponents [55].

1947: The *Lamb shift* is measured [56] and there is a surprising agreement with a QED prediction obtained after *cancellation of infinities* between physical observables [57].

1946–1949: A general empirical strategy to eliminate divergences, called *renormalization*, is developed (Schwinger [58, 60], Tomanaga [59], Feynman [4, 10], Dyson [61]). Many articles concerning the construction of QED can be found in Ref. [62].

1950: Ginzburg and Landau introduce what is called nowadays the Abelian Higgs mechanism to describe the macroscopic properties of a superconductor in a magnetic field [63, 64].

1954–1956: A formal property of massless QED, called *RG*, is discovered whose deeper meaning is not fully understood (Peterman–Stückelberg [65], Gellman–Low [66], Bogoliubov–Shirkov [67]).

---

* The spin is the intrinsic angular momentum of particles, which takes integer (for bosons) or half-integer (for fermions) values in $\hbar$ units.

1954: Yang and Mills generalize classical electrodynamics to non-Abelian gauge groups. [41]

1964–1975: The *Standard Model* of fundamental interactions at the microscopic scale (Weinberg [68], Glashow *et al* [69], Salam), a *renormalizable QFT* based on the concepts of *non-Abelian gauge symmetry* and *spontaneous symmetry breaking* [70–72] (see Chapter 12) is formulated, quantized (Faddeev–Popov [40], DeWitt [73]) and shown to be perturbatively consistent ('t Hooft [74], 't Hooft–Veltman [75], Lee–Zinn-Justin [76]). With minor modifications (neutrino masses and oscillations), it still describes with a remarkable precision all interactions (except gravitation) between fundamental particles up to TeV energies.

1966–1972: Inspired by some premonitory ideas of Kadanoff [77], Wilson [78, 24] develops a more general *RG*, based on the iterative integration over short distance degrees of freedom, which includes the field theory RG in some limit, and which is able to explain *universal non-mean-field (or non-quasi-Gaussian)-like properties of continuous phase transitions* (like liquid–vapour, binary mixtures, superfluidity and ferromagnetism) or statistical properties of long polymers.

1972–1975: Following the work of Wilson and Fisher [79], Wilson [80, 81] and Wegner [82], several groups [83], including Brézin, Le Guillou and Zinn-Justin, develop efficient QFT techniques to *prove universal scaling relations* and *calculate universal quantities.*

1973: Politzer and Gross–Wilczek establish the *asymptotic freedom* of a class of non-Abelian gauge theories, which provides an RG explanation to the free particle behaviour of quarks at short distance inside nucleons [84].

1975–1976: Additional insight into the universal properties of phase transitions is provided by the study of the *non-linear σ-model* and the $(d-2)$ *expansion* (Polyakov [85], Brézin–Zinn-Justin [86]).

1977–1980: The calculation of the RG functions by field theory techniques, using a perturbative expansion within the Callan–Symanzik RG scheme [87], as suggested by Parisi [88], is initiated by Nickel [89], and the first precise estimates of critical exponents, based on Borel summation and conformal mapping, are published by Le Guillou and Zinn-Justin [30]. Using additional perturbative coefficients provided by Nickel, the results have were later confirmed and improved by Guida and Zinn-Justin (1998) [31].

4.2   QED and the problem of infinities

QED, a QFT. QED, which describes, in a quantum relativistic framework, electromagnetic interactions between charged particles, *is not a theory of individualized particles*, as in non-relativistic quantum mechanics, but a *QFT*. It is also a quantum extension of a classical relativistic field theory described by Maxwell's equations, where the dynamic variables are *fields*, the electric and magnetic fields.

Such a theory differs drastically from a theory of individual particles in the sense that fields have an *infinite number of fluctuating degrees of freedom*, the values of fields at each point in space. The non-conservation of particles in high energy collisions is a direct manifestation of this property.

Moreover, the *field theories* that describe microscopic physics have the essential property of *locality*, a generalization of the notion of point-like particles with contact interactions: *they lack a short distance structure, and this is the origin of infinities.* The infinities are directly generated by the high momentum photons in the first quantum correction of Fig. 4.1 and are since generally called *UV divergences.*

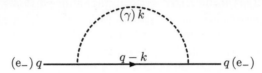

Fig. 4.1 Feynman diagram: An electron (solid line) of four-momentum q emits and reabsorbs a virtual photon (dashed line) of four-momentum k.

From a more general perspective, the fluctuations of the infinite number of degrees of freedom of fields combined with locality are the basic reasons why QFT's have somewhat unexpected properties.

4.2.1 First calculations: The problem of infinities

Shortly after the works of Dirac, Heisenberg and Pauli, the first calculations of the order $\alpha = e^2/4\pi\hbar c \approx 1/137$ (the fine structure constant) correction to the electron propagation in the photon field were reported. (The electric charge e is defined in terms of the Coulomb potential when written as e^2/R.)

Physicists wanted to investigate whether the newly formulated quantum theory could cure the disease of the *'classical relativistic model'* of the point-like electron: in a model where the electron is identified with a charged sphere of radius R, the self-energy generated by the Coulomb potential diverges as e^2/R when $R \to 0$. This leads to a problem because, in a relativistic theory, the electron self-energy contributes to its mass.

The QED action describing electromagnetic interactions between electrons, positrons and photons, associated with $\bar{\psi}, \psi$, four-component (spinor) fermion fields, and A_μ a vector field (the relativistic vector potential), respectively, has the form,

$$\mathcal{A}(\mathbf{A}, \bar{\psi}, \psi) = \int \mathrm{d}^4x \left[\bar{\psi}(x) \left(\not{\partial} + m_\mathrm{e} + ie\not{A}(x) \right) \psi(x) + \tfrac{1}{4}\textstyle\sum_{\mu,\nu} F_{\mu\nu}(x)F^{\mu\nu}(x) \right]$$

with $\not{A} \equiv \sum_\mu A_\mu \gamma^\mu$, γ^μ being the four 4×4 Dirac γ matrices and where

$$F_{\mu\nu}(x) = \partial_\mu A_\nu(x) - \partial_\nu A_\mu(x),$$

is the electromagnetic tensor.

The theory can be quantized, following Feynman, by introducing the field integral $\int e^{i\mathcal{A}/\hbar}$ suitably modified by some gauge fixing (see Section 3.7 and Chapter 11).

The first reported results (1930) [52] were incorrect. Indeed, the calculations were quite involved due to the use of a non-covariant perturbation theory, the problem of *gauge fixing* in the gauge invariant QED and, most importantly, the unclear role of Dirac holes (or negative energy states), which were predicted to be anti-electrons or positrons only in 1931 and were discovered experimentally in 1932 by Anderson [90].

The first correct calculation was published by Weisskopf (1934) [53] after a last mistake pointed out by Furry was corrected in an erratum.

In terms of Feynman diagrams (a representation imagined only more than a decade later [10]), the relevant physical processes consist in the emission and re-absorption by an electron of the energy-momentum q of a virtual photon of energy-momentum k, summed over all photon energy-momenta (see Fig. 4.1).

The contribution to the electron mass was still *infinite*, although the linear classical divergence was replaced by a softer logarithmic UV divergence (m_e is the electron mass in the absence of electromagnetic interactions),

$$\delta m_e = -3\frac{\alpha}{2\pi}m_e\ln(m_e c/\Lambda_\gamma) + O(\alpha^2),$$

where $\Lambda_\gamma \gg m_e c$ is the momentum at which the k integration involved in the calculation of the diagram of Fig. 4.1 is cut. This corresponds to a distance scale of order $R = \hbar/\Lambda_\gamma$.

It became slowly clear that the problem was very deep; these divergences seemed unavoidable consequences of *locality* (a relativistic field theory generalization of point-like particles with contact interactions) and *unitarity* (conservation of probabilities). Indeed,

(i) one must sum over the contribution of virtual photons with arbitrarily high energies because there is no short distance structure, and
(ii) due to the conservation of probabilities, all processes contribute additively.

These divergences suggested that QED was an incomplete theory, but it seemed very difficult to modify it without giving away some fundamental physics principle.

Dirac (1942) proposed *giving up unitarity*, but the physical consequences seemed hardly acceptable. A non-local relativistic extension (which would correspond to give an inner structure to all particles) was hard to imagine in a relativistic context, although Heisenberg (1938) proposed the introduction of a *fundamental length*. In fact, only in the 1980's were plausible candidates proposed in the form of *string theories*.

A more drastic solution was also put forward: Wheeler (1937) and Heisenberg (1943) suggested *abandoning QFT* completely in favour of a theory of physical observables (scattering data): the so-called *S-matrix theory*, an idea that became quite popular in particle physics in the 1960's in the theory of strong interactions (responsible for the strong nuclear forces).

4.2.2 Infinities and charged scalar bosons

In the meantime, more pragmatic physicists explored the nature and form of divergences in quantum corrections, calculating other physical quantities. Weisskopf investigated the divergences of the self-energy to all orders and concluded that QED with only charged fermions is affected by logarithmic divergences that *render the theory non-predictive but are numerically acceptable* if some new physics provides a momentum cut-off (like the range of nuclear forces, about 100 MeV) because α is small; by contrast, *charged scalar bosons lead to large quadratic divergences*, which are totally unacceptable because they would spoil the classical results [91].

Thus, can charged scalar bosons be fundamental particles? The question is still very relevant since the Standard Model of fundamental interactions at the microscopic scale (which describes particle physics) involves a scalar particle, called the *Higgs boson* [70–72], which was discovered in 2012 at the Large Hadron Collider (LHC) at CERN, with a mass of 125 GeV.

It becomes even more acute if one assumes that the Standard Model remains valid up to a possible *grand unification* ($\approx 10^{15}$ GeV) scale or to the gravitational *Planck mass* ($\approx 10^{19}$ GeV). It leads to the *fine tuning problem* and has been one motivation for introducing *supersymmetry* (a symmetry relating bosons and fermions). Supersymmetric particles are thus actively searched for at the LHC, with no success so far (2018).

4.3   The renormalization strategy

Although it was eventually noticed that, *in some suitable combinations* of physical observables at leading order, *divergences cancelled* (see, *e.g.*, Weisskopf [92]), the deeper meaning of this observation remained unclear.

An essential experimental input. In 1947 Lamb and Retheford [56] reported a precise measurement of the splitting between the levels $2s_{1/2}2p_{1/2}$ of the hydrogen atom (now called the Lamb shift); Rabi's group in Columbia measured the anomalous magnetic moment of the electron.

The first QED results. Remarkably enough, it was possible to organize the *calculation of the Lamb shift* in such a way that all infinities cancel (first approximate calculation by Bethe [57]) and the result agreed beautifully with experiment. Shortly after, Schwinger [58] obtained the leading contribution to the *anomalous magnetic moment of the electron*.

Soon, the idea of divergence subtraction was generalized to the concept of renormalization (building upon work of Kramers) and in 1949 Dyson [61], following work by Feynman, Schwinger and Tomonaga, gave the first proof that, after renormalization, divergences cancel to all orders of the perturbative expansion. The basic principles of *renormalization theory* were thus established.

The renormalization strategy. First, one modifies in a somewhat arbitrary and *non-physical* way the QFT by introducing a large energy-momentum cut-off Λ in the Fourier representation (equivalently, one modifies the QFT at a very short distance of order $\hbar/\Lambda$), to render all physical observables finite, a step called *regularization*.

One then calculates physical observables as functions of the parameters as they appear in the initial Lagrangian, called *bare parameters*, like the bare mass m_0 and the bare charge e_0 of the electron (mass and charge for vanishing interaction), or, equivalently, the bare fine structure constant $\alpha_0 = e_0^2/4\pi\hbar c$, in the form of perturbative expansions in α_0.

In particular, the physical mass m and the physical charge e have expansions like (β_2, γ_1 and C_1 are numerical constants)

$$\begin{cases} e^2/4\pi\hbar c \equiv \alpha = \alpha_0 - \beta_2\alpha_0^2 \ln(\Lambda/cm_0) + \cdots, \\ \qquad m = m_0\left[1 - \gamma_1\alpha_0 \ln(\Lambda C_1/cm_0) + \cdots\right]. \end{cases} \qquad (4.1)$$

One inverts the relations, expressing the parameters of the Lagrangian in terms of α and m as powers series in α,

$$\begin{cases} \alpha_0 = \alpha + \beta_2 \alpha^2 \ln(\Lambda/cm) + \cdots, \\ m_0 = m + \gamma_1 \, m\alpha \ln(\Lambda C_1/cm) + \cdots. \end{cases} \qquad (4.2)$$

One then inserts these expansions in all other physical observables, expressing them in terms of the *renormalized (or physical) parameters m* and α.

Most surprisingly, *all physical observables expressed in terms of renormalized fields and renormalized parameters then have an infinite cut-off Λ limit, independent of the specific regularization scheme (at least in a suitable large class).*

The renormalization method is *a priori* somewhat strange because it involves *tuning the initial parameters* of the Lagrangian as a function of physical parameters and the cut-off Λ, something that may not even be always possible beyond perturbation theory, especially if one insists on taking the infinite Λ limit.

This renormalization method would have been hard to accept if it had not led to QED predictions that *agreed, with unprecedented precision, with experiment.*

Moreover, from the success of the renormalization programme in QED emerged the very important concept of *renormalizable QFT*. Since *the renormalization procedure works only for a limited class of field theories*, this has severely constrained the structure of new possible theories.

4.4 The nature of divergences and the meaning of renormalization

Renormalized QED was obviously the right theory because its predictions agreed with experiment, but why? Several answers were proposed, for example:

(i) Divergences are a disease of the perturbative expansion in α_0 and a proper mathematical handling of the theory with non-perturbative input would free it from infinities. In the same spirit, *axiomatic QFT* tried to establish rigorous non-perturbative results from general principles.

(ii) More drastic, the problem was fundamental: *QFT was only defined by perturbation theory* but the procedure that generated the perturbative expansion had to be modified in order to generate automatically finite renormalized quantities. *The initial bare theory, based on a Lagrangian with divergent coefficients, was physically meaningless*; it provided a simple bookkeeping device to generate perturbative expansions.

This line of thought led, in particular, to the BPHZ formalism (Bogoliubov, Parasiuk [67], Hepp [93], Zimmerman [94]) and, eventually, to the Epstein–Glaser [95] work, where renormalized Feynman diagrams were generated directly and taken as the fundamental building blocks of the theory. While this approach significantly clarified the properties of renormalized perturbation theory, it had also the drawback of disguising the problem of infinities as if it had never existed in the first place.

(iii) The *cut-off* is an non-physical substitute for real physical cut-off effects that are generated by unknown or poorly understood additional interactions beyond QED (at the time the strong nuclear force was envisaged), but then the *origin of renormalizability*, which reflects some form of *short distance insensitivity*, has still to be understood.

This last viewpoint is the closest to modern thinking, except that the cut-off is no longer linked to the strong nuclear force but to some new physics still to be discovered, at a much higher energy scale.

4.5 QFT and RG

An intriguing consequence of the renormalization method in massless QED. In a QED with massless electrons, the renormalized (*i.e.*, physical) charge cannot be defined in terms of the interaction between non-existing static electrons, since massless particles propagate at the speed of light.

One must introduce some arbitrary mass or energy or momentum-scale μ to define the *renormalized charge e* in terms of the strength of the electromagnetic interaction at scale μ: it is the *effective charge* at scale μ.

But then the same physics can be parametrized by the effective charge e' at another scale, μ'. The set of transformations of physical parameters associated with this change of scale and required to keep physics constant was called *RG* (Peterman–Stückelberg [65], Gell-Mann–Low [66], Bogoliubov–Shirkov [67]).

One can then show that, in an *infinitesimal scale change*, the variation (or the flow) of the effective charge satisfies a differential equation of the form

$$\mu \frac{\mathrm{d}\alpha(\mu)}{\mathrm{d}\mu} = \beta\big(\alpha(\mu)\big), \quad \beta(\alpha) = \beta_2 \alpha^2 + O(\alpha^3), \tag{4.3}$$

where the function $\beta(\alpha)$ can be calculated as a series in α from perturbation theory.

In fact, even in a massive theory, the notion of the effective charge at some mass scale μ can be defined and used to parametrize physical observables, in particular, when one is probing energies much larger than the physical rest energies of particles.

Interpretation: At large distance, the strength of the electromagnetic interaction remains constant at the value measured through the Coulomb force. However, at distances much smaller than the wavelength $\hbar/m_e c$ associated with the electron (one explores in some sense the 'interior' of the particle), *screening* effects are observed. It is remarkable that this *short distance screening is related to renormalization.*

Gell-Mann and Low's initial hope, which was to determine the bare charge as the large momentum or high energy limit of the effective charge, failed due to the sign $\beta_2 > 0$, a sign of screening: the effective charge increases at large momentum (a phenomenon verified experimentally at the energy of the Z boson at CERN) until perturbation theory becomes useless.

A related issue: Landau's ghost. A leading log summation of high energy contributions to the electron propagator exhibits a non-physical (a 'ghost') pole (Landau and Pomeranchuk [96]) at a mass

$$M \propto m \, \mathrm{e}^{1/\beta_2 \alpha} \approx 10^{30} \mathrm{GeV}.$$

For Landau, this was a sign of *QED's inconsistency*, but Bogoliubov and Shirkov pointed out that this amounted to solving the RG flow equation for α small and using the solution for α large.

Still, we believe now that QED is, indeed, inconsistent but that this inconsistency shows up at an irrelevant high energy scale because α is so small.

4.5.1 The triumph of renormalizable QFT: The Standard Model

The principle of looking for renormalizable QFT's led, in the 1970s, to the construction of the Standard Model of particle physics which describes all interactions (but gravity) at the microscopic scale, based on *non-Abelian gauge symmetries*.

In particular, in the case of the strong nuclear force, the *negative sign* of the RG β-function has provided an explanation for the *weakness of interactions between quarks at short distance* as seen in deep inelastic experiments, also called *asymptotic freedom*, (Gross–Wilczek, Politzer 1973 [84]), and has led to quantum chromodynamics (QCD).

Over forty years of experiments, including the most recent experiments at the LHC at CERN, in particular the discovery of the Higgs boson [71], have confirmed the predictions of the Standard Model.

One outstanding problem remains: *how to incorporate gravitation into the framework of renormalizable theories*. The *failure* up to now of reaching this goal has led to the investigation of theories of non-field-theory type in the form of *string theories*.

However, a number of physicists still hope to be able to define a complete renormalized QFT consistent on all scales, beyond QCD, and this has led to the issue of *asymptotic safety* [97] (see also Chapter 15).

4.6 Critical phenomena: Other infinities

Second order or continuous macroscopic phase transitions, with *short range interactions*, are characterized, near the critical temperature by a collective behaviour leading to *correlations on large scales*. These correlations, generated dynamically, are characterized by a *correlation length*, which diverges at the critical temperature.

At the scale of the correlation length, non-trivial macroscopic phenomena are observed. The usual paradigm of *scale decoupling* leads one to expect that macroscopic physics could be described by a small number of well-chosen effective parameters, without explicit reference to the initial microscopic theory.

This idea leads to *mean field theory* and, in its more general form, to *Landau's theory* of critical phenomena [54].

Among the simplest predictions of such a theory, one finds the *universality* of the singular behaviour of thermodynamic quantities at the critical temperature T_c: for example, in a ferromagnetic system, the spontaneous magnetization M and the correlation length ξ behave like

$$M \propto \sqrt{T_c - T}, \quad \xi \propto 1/\sqrt{T - T_c},$$

these properties, and many others, being independent of the dimension of space, of symmetries, and of the form of the microscopic dynamics.

However, it slowly became apparent that these universal predictions disagreed with exact results like the solution of the Ising model [55], more precise experiments and lattice model calculations [98].

The results of mean field theory are reproduced by a weakly perturbed Gaussian field theory (which one can call *quasi-Gaussian*), a property that is, to some extent, related to the *central limit theorem of probabilities*.

However, for all space dimensions $d \leq 4$, an attempt to calculate corrections to the quasi-Gaussian theory leads to *infinities* when the correlation length diverges, invalidating the universal predictions of mean field theory.

Nevertheless, numerical investigations suggested that *some universality survived but in a more limited form*, the critical behaviour depending on the dimension of space as well as some general qualitative properties of models, like symmetries, but not the detailed form of interactions (provided they remain short range).

4.7 The failure of scale decoupling: The RG idea

The problem of infinities is deeply connected with the issue of *scale decoupling* in physics. Although scale decoupling is a property of most physical systems, it is not satisfied by QFT or continuous phase transitions near T_c. RG is the new tool devised to deal with situations where the property of scale decoupling fails.

4.7.1 Scale decoupling in physics

The decoupling of physical phenomena corresponding to very different length scales is a basic paradigm of physics: macroscopic objects can, in general, be described without explicit reference to the microscopic dynamics, except, perhaps, to determine a few macroscopic parameters. Provided one is able to discover the relevant degrees of freedom and parameters, one can construct models specifically adapted to the scale of phenomena.

This empirical property is essential for the predictivity of macroscopic models.

However, in the twentieth century, in two *a priori* very different domains of physics, fundamental particle physics and critical phenomena in macroscopic phase transitions, this commonly accepted idea has been challenged.

After a careful analysis, one observes that these systems are characterized by strong local (or short range) interactions between a very large (or infinite) number of dynamical degrees of freedom.

The *RG*, a completely new concept, has provided an explanation for the compatibility between a non-decoupling of scales and, nevertheless, the relative insensitivity of large distance physics to the microscopic structure, as being related to the existence of RG fixed points.

This rather abstract concept not only made it possible to understand such a property, but also inspired a number of precise practical calculations.

It has provided a new and natural interpretation to *renormalizable QFTs* as *effective large distance theories*.

4.7.2 The RG idea

We consider systems with *short range interactions* and at a continuous phase transition, where the correlation length is infinite.

The partition function is given by

$$\mathcal{Z} = \int \prod_x \mathrm{d}\phi(x)\, \mathrm{e}^{-\mathcal{H}_0(\phi)}, \tag{4.4}$$

where x is point in $\mathbb{R}^d$ or $\mathbb{Z}^d$, $\phi(x)$ represents a set of fluctuating variables taking continuous or discrete values, like the classical spins of the Ising lattice model, and $\mathcal{H}_0(\phi)$ is a rescaled configuration energy or Hamiltonian.

Kadanoff [77] proposed to calculate the partition function (4.4) via an iterative procedure based on a partial summation over short distance degrees of freedom, replacing in this way at each step the degrees of freedom by averages over a small volume (see also Fig. 5.1).

Starting from the partition expressed in terms of an initial Hamiltonian $\mathcal{H}_0(\phi)$, the partial summation yields a new Hamiltonian, a transformation called *RG* transformation and denoted symbolically $\mathcal{T}$. The iteration of the transformation generates a sequence of Hamiltonians,

$$\mathcal{H}_n(\phi) = \mathcal{T}\left[\mathcal{H}_{n-1}(\phi)\right], \quad n \geq 1.$$

Fixed points. One then looks for fixed points, the solution of

$$\mathcal{H}^*(\phi) = \mathcal{T}\left[\mathcal{H}^*(\phi)\right].$$

Note that the existence of fixed points requires combining the partial summation procedure with a ϕ renormalization (see also Section 1.5).

If an attractive fixed point can be found and the initial Hamiltonian belongs to the domain of attraction of the fixed point, then the sequence converges towards the fixed point,

$$\mathcal{H}_n(\phi) \underset{\Lambda \to 0}{\to} \mathcal{H}^*(\phi),$$

and the partial short distance insensitivity or universality is explained.

The existence of attractive fixed points of an RG made it possible to understand universality, that is, the relative insensitivity of large scale properties to the microscopic dynamics, within universality classes, when scales do not decouple.

4.8 Phase transitions: Exact RG in the continuum

Wilson [78, 24] reviewed critically and improved Kadanoff's proposal (RG is an idea, not a method). He argued that, even if initially the spin variables take only discrete values and space has a lattice structure, near the critical temperature, the spins could be replaced by fields in continuum space, leading to an effective local *statistical field theory* with arbitrary powers of the field.

Moreover, Wilson implemented the RG transformation using the Fourier representation of the field variables,

$$\phi(x) = \int \mathrm{d}^d k \, \mathrm{e}^{ikx} \, \tilde{\phi}(k). \tag{4.5}$$

He then defined an RG by averaging iteratively over the Fourier components $\tilde{\phi}(k)$ such that $|k|$ belongs to a momentum shell (see Fig. 4.2).

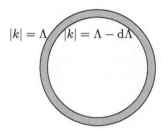

$|k| = \Lambda$ $|k| = \Lambda - \mathrm{d}\Lambda$

Fig. 4.2 Momentum-shell integration.

4.8.1 The exact RG

To construct an exact RG, one has to assume that the partition function, given by an integral of the form (4.4) over fields $\phi(x)$ with $x \in \mathbb{R}^d$, involves an effective Hamiltonian (or configuration energy) $\mathcal{H}(\phi)$ that is a general *local* (because the initial interactions are short range) Hamiltonian: a *local* Hamiltonian is a space integral of a linear combination of monomials involving products of the field and its derivatives at the same point.

To define an exact RG flow, one integrates over *infinitesimal momentum shells*: roughly, one integrates iteratively over the field components $\tilde{\phi}(k)$ (equation (4.5)) belonging to a momentum shell, (Fig. 4.2),

$$\Lambda - \mathrm{d}\Lambda \le |k| \le \Lambda, \quad \text{with } |\mathrm{d}\Lambda| \ll \Lambda. \tag{4.6}$$

However, to avoid introducing unwanted infrared singularities, it is necessary to replace the sharp cut-off implicit in equation (4.6) by a smooth cut-off function.

In the limit of an infinitesimal momentum shell, the recursion equation becomes a differential flow equation, the generalization of equation (4.3), of the general form

$$\Lambda \frac{\mathrm{d}}{\mathrm{d}\Lambda} \mathcal{H}_\Lambda(\phi) = \mathcal{T}\left[\mathcal{H}_\Lambda(\phi)\right]. \tag{4.7}$$

The operator $\mathcal{T}$ takes the explicit form of a quadratic functional (see equation (9.9)) of the Hamiltonian [100, 101].

In general, this equation is difficult to handle analytically. Even starting with a simple polynomial Hamiltonian, the RG equation generates more general local Hamiltonians with arbitrary powers of the field ϕ and higher order derivatives.

However, if the Hamiltonian is quadratic in the field (a free field theory in the particle physics interpretation), it is possible to identify a fixed point, the Gaussian fixed point.

The Gaussian fixed point. The partition function of the Gaussian fixed point has the form

$$\mathcal{Z} = \int [\mathrm{d}\phi(x)] \exp\left[-\mathcal{H}_{\mathrm{G}}(\phi)\right], \quad \mathcal{H}_{\mathrm{G}}(\phi) = \frac{1}{2}\int \mathrm{d}^d x \left(\nabla_x \phi(x)\right)^2,$$

equivalent, in particle physics, to a massless scalar field theory.

This fixed point is related to mean field theory. It is stable for dimensions $d > 4$ but unstable for $d < 4$ and this explains the failure of mean field theory. Near four dimensions, the perturbations responsible for the instability are

$$\int \mathrm{d}^d x \, \phi^2(x), \quad \int \mathrm{d}^d x \, \phi^4(x).$$

The first one induces a change of temperature, and the correlation length becomes finite. The second one is the leading perturbation on the critical surface (massless theory).

4.8.2 Asymptotic or perturbative RG equations

Dimension $d = 4$. For $d = 4$, the field theory including only these two perturbations,

$$\mathcal{H}(\phi) = \mathcal{H}_{\mathrm{G}}(\phi) + \int \mathrm{d}^4 x \left(\tfrac{1}{2}r\phi^2(x) + \tfrac{1}{4!}g\phi^4(x)\right), \tag{4.8}$$

where the parameter r plays the role of temperature near T_c and makes it possible to explore the neighbourhood of T_c (the critical domain), is exactly *renormalizable.*

It is then possible to write closed RG equations, of field theory type, *asymptotic* with respect to equation (4.7).

These equations are perturbative in the sense that they apply only in some neighbourhood of the Gaussian fixed point.

In the field theory framework, the RG transformation for the Hamiltonian reduces to a differential equation of the form

$$\Lambda\frac{\partial g}{\partial \Lambda} = \beta[g(\Lambda)]. \tag{4.9}$$

The fixed point equation then becomes

$$\beta(g^*) = 0.$$

In four dimensions, $\beta(g) = \beta_2 g^2 + O(g^3)$, with $\beta_2 > 0$, the equation has the unique solution $g^* = 0$, and the Gaussian fixed point is marginally stable, leading to logarithmic modifications of mean field predictions and, in particle physics, to the triviality issue.

Dimensional continuation and ε-expansion. To explore what happens near dimension 4, a trick has been invented, dimensional continuation, which extends the terms in the perturbative expansion of field theory to general real or complex values of the dimension d [102]. One can then study the field theory (4.8) in the form of a double series expansion in g and $\varepsilon = 4 - d$.

This theory is still renormalizable and satisfies a generalized form of the RG equation of the four-dimensional theory. The RG β-function in equation (4.9) then reads

$$\beta(g) = -\varepsilon g + \beta_2 g^2 + O(g^3), \ \beta_2 > 0 \,.$$

One discovers that the β-function has a zero $g^* \sim \varepsilon/\beta_2$ where the slope is positive and which has the interpretation of an infrared stable fixed point. This fixed point is at the origin of Wilson–Fisher's ε-expansion [79, 80] for critical exponents and other universal quantities [81, 82], whose properties can be further investigated by QFT methods [83].

General RG and QFT perturbative RG. Given a regularization, the general RG is exact but it acts in the space of all Hamiltonians or local field theory actions, a space depending on an infinite number of parameters, Therefore, the corresponding flow equations can be solved only by numerical methods, after drastic approximations.

The QFT RG is based on perturbation theory. It is exact for the sum to all orders of the leading terms in the critical domain. The RG of QFT appears, in the general framework, as an asymptotic RG for studying flows, and thus fixed points, that are close to the Gaussian fixed point. This is the situation of the ε-expansion.

One may then wonder whether the sum of the leading contributions is the leading contribution to the sum. Empirically, the successes in the calculation of critical exponents [89–31] or equation of state [99] yielding values in very good agreement with lattice models and experiments justifies the approach.

4.9 Effective field theory: From critical phenomena to particle physics

The theory of critical phenomena shows that *a dynamical generation of a large scale may generate a non-trivial large distance physics, which can be described by a universal renormalizable QFT* even when, initially, the statistical model involves a lattice, and the fluctuation degrees of freedom take only discrete values, as in the Ising model.

On the other hand, the condition that fundamental interactions at the microscopic scale should be describable by *renormalizable QFTs* has been one of the basic principles that have led to the Standard Model of microscopic physics. From the success of the programme, one might have concluded that *renormalizability is a new law of nature.*

The appearance of renormalizable QFTs in the theory of macroscopic phase transitions leads to a more natural explanation. One can thus speculate that fundamental interactions are described at some more microscopic scale (like the Planck length) by a *finite theory* that has no longer the nature of a local QFT.

Although such a theory should involve only some short microscopic scale, for reasons that can only be a matter of speculation, it generates strong correlations between a large number of degrees of freedom and a *large distance physics with almost massless particles.*

The example of critical phenomena then suggests that the low energy physics (in this framework, even the high energy colliders like the LHC at CERN still probe low energy physics) can automatically be described by a renormalizable QFT, endowed with a cut-off reflection of the initial short distance structure.

From this viewpoint, *the fine tuning of the Higgs bare mass,* equivalent to tuning the temperature to be close to the critical temperature in macroscopic phase transitions, becomes an important physics issue.

Moreover, additional non-renormalizable interactions should be expected (the irrelevant interactions of RG), suppressed by powers of the short distance scale. Non-renormalizable *quantum gravity* could be a first example of such an (RG) irrelevant interaction.

To summarize, *renormalizable quantum field theories* are only *effective large distance theories* (see Chapter 8) in which all non-renormalizable interactions have been neglected. They come endowed with a natural cut-off reflection of the existence of a more fundamental theory, but the specific implementation of the cut-off is not important because these theories are, to a large extent, *short distance insensitive, as seen in the renormalization process, a property that the corresponding RG helps to understand.* Moreover, they are *not necessarily consistent on all scales,* (like QED), having in general only a limited energy range of validity.

5 Renormalization group: From a general concept to numbers

Second order or continuous macroscopic phase transitions, with *short range inter-actions*, are characterized, near the critical temperature, by a collective behaviour leading to *correlations on large scales*. The scale of these dynamically generated correlations is characterized by the *correlation length*, which diverges at the critical temperature.

When the correlation length is large, at the scale of the correlation length non-trivial macroscopic phenomena are observed. The usual paradigm of *scale decoupling* leads one to expect that these phenomena could be described by macroscopic models without explicit reference to the initial microscopic theory.

This idea leads to *mean field theory* and, in its more general form, to *Landau's theory* of critical phenomena. Such a theory can also be called *quasi-Gaussian* (or perturbed Gaussian) with reference to the *central limit theorem of probabilities*.

Mean field theory leads to super-universal predictions for continuous phase transitions near the critical temperature, like the universal singular behaviour of thermodynamic quantities at the critical temperature, these properties being independent of dimension of space, of symmetries, and of the form of the microscopic dynamics.

However, these predictions are incorrect in the dimensions $d = 2$ and $d = 3$ relevant for statistical physics.

Moreover, *scale decoupling* is also inconsistent with the results of perturbative calculations in the theory of fundamental interactions at the microscopic scale, as the occurrence of infinities in a straightforward interpretation of four-dimensional quantum field theory (QFT) shows (see Chapter 4).

Nevertheless, empirical evidence in the study of critical phenomena indicates that a more limited form of short distance insensitivity (called *universality* in the statistical framework) survives. In the same way, the success of the renormalization theory implies also some partial short distance insensitivity in QFT.

To explain these unusual properties, which imply that, in some special situations, *scales do not decouple* but, nevertheless, some more limited form of short distance insensitivity survives, a new strategy has been invented based on the *renormalization group (RG)* idea. (Of course, this required first recognizing that the non-decoupling of scales was the central issue.)

In the RG approach, *universality* arises as the consequence of the existence of infrared stable *fixed points*. All initial Hamiltonians that lead to the same fixed point characterize a *universality class*.

Therefore, the implementation of the RG concept in QFT has clarified the meaning of the renormalization method, led to the concept of asymptotic freedom in microscopic physics, and allowed explicit calculations of universal critical quantities for macroscopic continuous phase transitions.

From Random Walks to Random Matrices. Jean Zinn-Justin, Oxford University Press (2019).
© Jean Zinn-Justin. DOI: 10.1093/oso/9780198787754.001.0001

5.1 Scale decoupling in physics: A basic paradigm

The decoupling of physical phenomena corresponding to very different scales of sizes is a basic paradigm in the modelling of most physics phenomena.

Example: The period of the pendulum

$$\tau \propto \sqrt{\ell/g},$$

where ℓ is the length of the pendulum, and $g \approx 9.8 \, \text{m/s}^2$ the gravity constant, can be determined, up to a numerical factor, by *simple dimensional analysis*. However, the argument relies on an implicit hypothesis: *sizes very different from the pendulum length ℓ*, like the size of the atoms or the radius of the earth, *do not affect the period.*

In the same way, orbits of planets can be determined with a very good precision by replacing planets and the sun by point-like objects, and omitting all other stars in the galaxy. Newtonian mechanics can deal with point-like objects, except in the peculiar case of head-on collision.

More generally, the description of macroscopic objects does not require the explicit knowledge of the microscopic theory, except perhaps to determine the numerical values of some effective parameters.

To summarize, provided one is able to discover the relevant degrees of freedom and parameters, one can devise physical models specifically adapted to the scale of phenomena.

This empirical observation, relevant for most physics phenomena, plays an essential role: otherwise, physics would be reduced to a simple parametrization of observations.

Indeed, if physics were sensitive to phenomena on all scales, predictive models would not exist, since any prediction would require a complete knowledge of all physical laws and all parameters of nature (not to mention the practical difficulty of determining systematically all macroscopic properties, starting from microscopic models).

Nevertheless, in the twentieth century, in two apparently quite distinct domains of physics, the physics of fundamental interactions at the microscopic scale, or particle physics, and the theory of continuous phase transitions (liquid–vapour, ferromagnetic, superfluid, *etc.*) this commonly accepted paradigm has failed.

In these examples, an infinite number of coupled degrees of freedom cannot be replaced by a finite number of effective macroscopic degrees of freedom.

To explain the compatibility between a non-decoupling of scales and, nevertheless, the relative insensitivity of large distance physics to the microscopic structure, a new concept has been invented: *the RG.*

Implementations of this rather abstract concept have been found that not only have made it possible to understand this relative insensitivity, but also have inspired a number of practical calculation methods.

The RG has also provided a new and natural interpretation of *renormalizable QFTs* as *effective large distance theories.*

5.2 Fundamental microscopic interactions

In a four-dimensional QFT like quantum electrodynamics (QED), a straightforward calculation of observables leads to *infinities* as the unavoidable consequence of *locality*, a relativistic generalization of *point-like particles with contact interactions*, and *conservation of probabilities*.

An empirical method, called *renormalization*, has eventually been found that yields finite results for measurable quantities: it is as follows

One temporarily modifies in a largely arbitrary, and necessarily non-physical way, the theory at a very short distance $1/\Lambda$ (a procedure called *regularization*). One then expresses all physical observables in terms of a small number of parameters related to the scale of observation, like the observed charge and mass of the electron, eliminating in this way the initial microscopic parameters (see also Section 4.3).

For a class of quantum field theories called *renormalizable* (like QED), one then shows that, at fixed physical parameters, all observables have a finite *perturbative* expansion in the limit $\Lambda \to \infty$ (removing the regularization); this implies that observables, once expressed in terms of *adapted parameters*, become *insensitive to the short distance structure*.

This strategy made it possible to construct the extremely successful Standard Model of particle physics, which describes all interactions at the microscopic scale up to TeV energies but gravitation.

Non-decoupling of scales. By itself, at first sight, the renormalization idea to introduce parameters adapted to the scale of the observed phenomena does not seem peculiar, since it is commonly used in physics. *However:*

(i) One seems unable to directly construct a model adapted to the scale of physics that one wants to describe. Moreover, to obtain finite results after removing the regularization, one has to tune the initial parameters of the Lagrangian (see, *e.g.*, equation (4.2)), something which may not always be possible beyond perturbation theory, and may not even be meaningful in the $\Lambda \to \infty$ limit.

(ii) *More strikingly: the physical (renormalized) parameters depend continuously on the scale of observation. This leads to the concept of the effective charge of the electron at a given scale.*

The latter property clearly indicates that *some very different scales of the relevant physics do not decouple*.

More precisely, as a direct consequence of the renormalization process, in a field theory with only one coupling α (like the ϕ^4 field theory or QED), for example, the effective charge $\alpha(\ell)$ at scale ℓ satisfies a flow equation, or *RG equation*, of the form

$$\ell \frac{\partial \alpha}{\partial \ell} = -\beta[\alpha(\ell)].$$

For example, at a scale corresponding to the Z vector boson mass ($\approx 2 \times 10^{-18}$ m), the value of the fine structure constant α is about $1/128$ instead of $1/137$, which is consistent with the flow equation.

The paradox of non-decoupling of scales and short distance insensitivity. The implications of renormalization theory are that physics is, to some extent, insensitive to the short distance structure, which is fortunate since the true physical short distance structure (because it is unknown or too complicated) has been replaced by an *ad hoc* non-physical short distance structure, an operation called *regularization* in QFT.

5.3 Macroscopic phase transitions

In continuous phase transitions for systems with short range interactions (like the Ising model), one observes near the transition temperature, a collective behaviour on large scales, which is characterized by a large correlation length.

The hypothesis of *scale decoupling* leads, as a consequence of the central limit theorem of probabilities, to a *mean field* or quasi-Gaussian theory and *super-universal properties* for large distance physics. However, such a conclusion disagrees both with experiment and with calculations in model systems. Again, different scales of physics do not decouple.

However, one observes a more limited *universal* behaviour within *universality classes*, suggesting that large scale phenomena are sensitive only to some qualitative properties of the microscopic interactions.

To explain this somewhat paradoxical situation, non-decoupling of scales and, still, relative insensitivity to microscopic interactions, a new strategy has been invented based on the RG idea.

5.3.1 The RG idea: Simple ferromagnetic systems

As an illustration, we consider a partition function corresponding to a classical model of spins ϕ interacting with nearest-neighbour interactions on a hypercubic lattice in d dimensions. In terms of the spins ϕ and the configuration energy (or Hamiltonian) $\mathcal{H}(\phi)$ (normalized by the temperature), the partition then reads (a is the lattice spacing)

$$\mathcal{Z} = \sum_{\{\phi(\mathbf{n}a)\},\ \mathbf{n}\in\mathbb{Z}^d} \exp\left[-\mathcal{H}(\phi)\right].$$

To calculate the partition function, one sums iteratively over short distance degrees of freedom, as suggested by Kadanoff [77] and completely reformulated by Wilson [78]; for example, one sums over the initial spins with the constraint that the average value in a each cell is fixed:

$$\phi(\mathbf{n}a) \mapsto \phi'(\mathbf{n}a) = \frac{\sqrt{Z}}{2d} \sum_{\substack{a\mathbf{n}'\ \text{neighbours} \\ \text{of } 2a\mathbf{n}}} \phi(a\mathbf{n}'),$$

where Z is a necessary spin renormalization factor.

This transformation leads to a new Hamiltonian, the function of average spins on a lattice with double lattice size (*cf.* Fig. 5.1 for a two-dimensional example). Iterating, one constructs an RG,

$$\mathcal{H}(\phi; 2^n a) = \mathcal{T}\left[\mathcal{H}(\phi; 2^{n-1}a)\right]. \qquad (5.1)$$

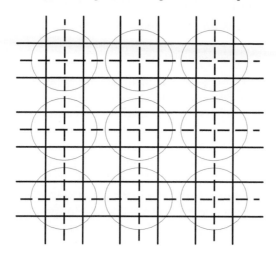

Fig. 5.1 Initial lattice with size a and lattice with size $2a$.

Note that, if the correlation length is infinite, after any number of iterations, there always remains a long distance physics at a scale much larger than the lattice spacing. By contrast, if the correlation length is finite, eventually the lattice spacing becomes of the size of the correlation length, the effective Hamiltonian corresponds to a collection of independent spins, and the partition function can be readily calculated.

5.3.2 Fixed points

If the correlation length is infinite, one can iterate the transformation (5.1) an infinite number of times without changing the large scale physics and then look for possible fixed points (this requires adjusting the renormalization factor Z, *cf.*, the example of the random walk, Section 1.5),

$$\mathcal{H}(\phi; 2^n a) \underset{n \to \infty}{\to} \mathcal{H}^*(\phi), \quad \mathcal{H}^*(\phi) = \mathcal{T}\left[\mathcal{H}^*(\phi)\right].$$

If the RG transformation has an attractive fixed point $\mathcal{H}^*$, for all initial Hamiltonians within some domain (the basin of attraction that defines a universality class), the iterated Hamiltonians converge towards the fixed point.

Thus, *RG allows understanding the universality of large scale or critical behaviour* but *within universality classes*, when very different scales do not decouple.

The fixed point may be attractive only for a subclass of Hamiltonians, and this requires a *stability analysis*.

The critical domain. One defines the *critical domain* as the domain where the correlation length is finite but very large compared to the microscopic scale. Then, a large number of iterations of the RG transformations is required before reaching the decoupling regime. The RG trajectories will first converge towards a fixed point, before moving away. One thus expects universal properties also in the critical domain.

5.3.3 *Scale non-decoupling and fixed points, a geometric analogue: Fractals*

To illustrate the notion of RG and fixed points, we propose a simple geometric analogy as displayed in Fig. 5.2.

In this slightly 'poetic' analogy, the straight line appears as an analogue of the Gaussian fixed point (see Section 5.4.1) and an example of scale decoupling. Indeed, if one starts from a smooth curve and enlarges one section repeatedly, eventually the section will converge towards a straight line that can be described without reference to a short distance structure (except if it is a physical line, in which case, eventually the pixel structure will become apparent but this happens at a completely different scale of phenomena).

By contrast, it is impossible to draw a mathematical fractal exactly. Indeed, each time one enlarges its image, new features appear. However, one can describe it by the rules that define how the picture changes in a change of scale (the equivalent of the RG). Moreover, from the way the geometric fractal is defined here, it is invariant under the scale transformation and, thus, the analogue of a non-Gaussian fixed point.

A Gaussian fixed point analogue: The straight line.

The RG transformation analogue.

A geometric fractal: The non-trivial fixed point analogue.

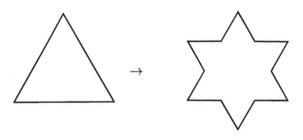

Fig. 5.2 RG and fixed points: A geometric analogue.

5.4 Fixed points: The QFT framework

Nature of fixed points. Even if initially space has a lattice structure and the spin variables take only discrete values, after a large number of iterations the lattice can be replaced by continuum space and the *effective* spins by fields $\phi(x)$ (like the electromagnetic field) defined at any point in space: a fixed point Hamiltonian $\mathcal{H}^*(\phi)$, if it exists, thus corresponds to a *local statistical field theory* (or QFT in imaginary time), and the partition function is given by a field integral of the form

$$\mathcal{Z} = \int [\mathrm{d}\phi(x)] \exp\left[-\mathcal{H}^*(\phi)\right], \qquad (5.2)$$

where the symbol $\int [\mathrm{d}\phi(x)]$ means sum over field configurations (see Section 2.10). The Hamiltonian $\mathcal{H}^*(\phi)$ is *local* because the initial interactions are short range (like nearest-neighbour interactions on a lattice): it has the form of a space integral over a function of the field and its space derivatives, taken at the same point.

We discuss below only reflection symmetric theories and thus $\mathcal{H}(\phi) = \mathcal{H}(-\phi)$.

5.4.1 The Gaussian fixed point

In d space dimensions, for $d > 2$, on the *critical surface* $T = T_c$ where the correlation length is infinite, the *Gaussian field theory*,

$$\mathcal{H}^*(\phi) = \mathcal{H}_{\mathrm{G}}^*(\phi) \equiv \tfrac{1}{2} \int \mathrm{d}^d x \left(\nabla_x \phi(x)\right)^2,$$

$(\nabla_x \equiv \{\partial/\partial x_1, \partial/\partial x_2, \ldots, \partial/\partial x_d\})$ *is a fixed point* (a result that is related to the central limit theorem of probabilities). The quadratic fixed point Hamiltonian is *local* since it is a space integral of the field $\phi(x)$ derivative at only the point x, and is thus relevant to the class of systems with *short range interactions*. The *Gaussian fixed point* is consistent with *Landau's* theory [54] or *mean field theory*.

Stability of the Gaussian fixed point. The more general local quadratic Hamiltonian,

$$\mathcal{H}_{\mathrm{G}}(\phi) = \tfrac{1}{2} \int \mathrm{d}^d x \left[(\nabla_x \phi(x))^2 + r\phi^2(x)\right],$$

where the parameter $r \geq 0$ characterizes the deviation from the critical temperature, makes it possible to describe physics in the disordered (high temperature) phase but makes sense only in the critical domain, that is, when the correlation $\xi \propto 1/\sqrt{r}$ is still very large compared to the microscopic scale.

Therefore, the *Gaussian fixed point* is always unstable with respect to a quadratic perturbation of the form $\int \mathrm{d}^d x\, \phi^2(x)$ (a *relevant* perturbation in the RG terminology, see Section 1.6) because it leads to a finite correlation length.

Moreover, even on the critical surface $T = T_c$, the Gaussian fixed point is *unstable* when the *dimension d of space is lower than 4*. Near dimension 4, the instability is induced by the perturbation

$$\int \mathrm{d}^d x\, \phi^4(x).$$

Note that such a perturbation is unavoidable. Indeed, to describe physics in the critical domain in the ordered phase, a double-well potential for constant fields is required ($r < 0$) and this implies generically an additional quartic term, as the example of mean field theory illustrates.

5.4.2 QFT perturbative RG

If another fixed point exists in a neighbourhood of the Gaussian fixed point, universal properties can be determined from a local statistical field theory (QFT in imaginary time) where the Hamiltonian in equation (5.2) reduces to

$$\mathcal{H}(\phi) = \mathcal{H}_{\mathrm{G}}(\phi) + \frac{1}{4!}g \int \mathrm{d}^d x\, \phi^4(x), \quad g > 0\,,$$

with

$$\mathcal{H}_{\mathrm{G}}(\phi) = \tfrac{1}{2} \int \mathrm{d}^d x \left[\nabla_x \phi(x) K(\nabla_x^2) \nabla_x \phi(x) + r\phi^2(x) \right]. \tag{5.3}$$

The parameter g is the strength of the perturbation, and the polynomial

$$K(z) = 1 + O(z) > 0\,, \quad \forall z\,,$$

of degree $\delta_K > d/2 - 1$ provides an *artificial* short distance structure that *removes* the non-physical short distance singularities (a *regularization* that ensures that the fields that contribute to the field integral (5.2) are at least continuous).

The critical theory is then obtained by tuning the parameter r in expression (5.3), which plays the role of the temperature, to a value $r_c(g)$ (a *mass renormalization* in QFT terminology).

The statistical field theory (5.3) is *renormalizable* at and near dimension 4 and, thus, admits a *QFT (perturbative) RG*.

The QFT RG appears, in the general RG framework, as an asymptotic RG, which makes it possible to look for fixed points that are close to the Gaussian fixed point in the sense that the contributions of all interactions *irrelevant* in the RG terminology (*i.e.*, corresponding to stable directions in Hamiltonian space, see Section 1.6) with respect to the Gaussian fixed point can be neglected.

In continuum space, it is possible (and convenient) to perform infinitesimal scale transformation $\ell \mapsto \ell(1 + \delta\ell/\ell)$. Then the transformation $g(\ell) \mapsto g(2\ell)$ corresponding to the transformation (5.1) is replaced by the differential equation

$$\ell \frac{\partial g}{\partial \ell} = -\beta[g(\ell)],$$

where the RG β-function can be calculated as a powers series in g.

The fixed point equation reduces to

$$\beta(g^*) = 0\,.$$

When the derivative $\beta'(g^*)$ is positive, the fixed point is attractive and governs the large distance behaviour.

Universal properties in the critical domain (near the transition temperature) are then derived by first determining the zeros g^* with positive slope of the β-function, and then calculating all other physical quantities for $g = g^*$.

5.5 RG, correlation functions and scaling relations

The β-function, as well as other RG functions, can be inferred from connected correlation functions,

$$W^{(n)}(x_1, x_2, \ldots, x_n) = \langle \phi(x_1) \cdots \phi(x_n) \rangle_{\text{connected}},$$

where $\langle \bullet \rangle$ means expectation value with respect to the measure $e^{-\mathcal{H}(\phi)}/\mathcal{Z}$. They satisfy RG equations of the form ($\rho = r - r_c \propto T - T_c$) [103]

$$\ell\frac{\partial}{\partial\ell}W^{(n)} = \left[\beta(g)\frac{\partial}{\partial g} - \tfrac{1}{2}n\big(d - 2 + \eta(g)\big) + \big(2 + \eta_2(g)\big)\rho\frac{\partial}{\partial\rho}\right]W^{(n)}. \qquad (5.4)$$

The infrared fixed point corresponds to a zero $g = g^*$ of the β-function with positive slope. For $g = g^*$, equation (5.4) reduces to

$$\ell\frac{\partial}{\partial\ell}W^{(n)} = \left[-\tfrac{1}{2}n\big(d - 2 + \eta(g^*)\big) + \big(2 + \eta_2(g^*)\big)\rho\frac{\partial}{\partial\rho}\right]W^{(n)}.$$

For $\rho = 0$, the two-point function thus has the scaling form,

$$W^{(2)}(x, T = T_c) \propto 1/|x|^{2d_\phi}, \quad \text{with} \quad d_\phi = \tfrac{1}{2}(d - 2 + \eta) \quad \text{and } \eta \equiv \eta(g^*),$$

where d_ϕ is the scaling dimension of the field.

For $T - T_c > 0$, in the critical domain, the two-point function has the more general scaling form

$$W^{(2)}(x, T) \propto \frac{1}{x^{2d_\phi}}F\big(x/\xi(T)\big),$$

where F is a universal function and

$$\xi(T) \underset{T \to T_c}{\propto} |T - T_c|^{-\nu}, \quad \nu = 1/\big(2 + \eta_2(g^*)\big).$$

The values $\eta(g^*) = \eta$ and $\eta_2(g^*) = 1/\nu - 2$ of the two RG functions $\eta(g)$ and $\eta_2(g)$ yield the only two independent critical exponents, η, which determines the scaling dimension of the field, and ν, which characterizes the singular behaviour of the correlation length. All other exponents follow from scaling relations resulting from the RG equations.

Note that the critical exponents, as well the ratios of critical amplitudes, are *pure numbers*. The theory has no adjustable parameters.

5.6 Exponents: Practical QFT calculations

The field theory RG has led to two calculation schemes:

(i) Following Wilson and Fisher [79, 80], generalizing the perturbative expansion to non-integer dimensions, one calculates *renormalized* correlation functions for the massless or critical $(T = T_c)$ field theory in $d = 4 - \varepsilon$ dimensions as ε-expansions [83], (the critical theory has no perturbative expansion for fixed dimensions $d < 4$).

(ii) Unlike the massless field theory, the massive theory exists for any dimension. One thus calculates *renormalized* correlation functions for the massive theory at fixed dimensions 2 and 3 by using the Callan–Symanzik form of RG equations [87], as suggested by Parisi [88].

All quantities are then obtained as perturbative expansions in powers of the renormalized coupling g_r, in particular the RG β-function.

The Wilson–Fisher ε-expansion. Near dimension 4, in the massless theory, the β-function takes the form $(\varepsilon = 4 - d)$

$$\beta(g_r) = -\varepsilon g_r + \beta_2 g_r^2 + \cdots, \quad \beta_2 > 0.$$

For $d < 4$, one finds one new stable fixed point: $g_r^* = \varepsilon/\beta_2 + O(\varepsilon^2)$ with $\beta'(g^*) = \varepsilon > 0$. It is then possible to solve the equation and calculate all universal quantities as power series in ε [83]. It remains to infer, from a small number of terms of a divergent series (the expansion is known up to order ε^5), information about the relevant physical dimensions, for example $\varepsilon = 1$ (see Chapter 23).

Fixed dimensions $d < 4$. One can also evaluate the β-function and other observables *directly in dimension 3* in the massive theory but, then, one has no longer a *small* expansion parameter. However, it has been noticed by Nickel [89] that Feynman diagrams in dimension 3 can be more easily evaluated than those near dimension 4.

For an N-component field $\boldsymbol{\phi}$ and the $O(N)$ symmetric Hamiltonian,

$$\mathcal{H}(\boldsymbol{\phi}) = \int \mathrm{d}^3x \left[\tfrac{1}{2}(\nabla_x \boldsymbol{\phi}(x))^2 + \tfrac{1}{2}r\phi^2(x) + \frac{g}{4!}(\phi^2(x))^2 \right],$$

Nickel has calculated all of the diagrams up to seven loops (in the terminology of Feynman diagrams), required for the calculation of the two independent critical exponents, but the diagrams contributing to the β-function, which are more difficult, have only been calculated up to six loops.

For example, to six-loop order, for $N = 1$, Nickel has obtained

$$\beta(\tilde{g}) = -\tilde{g} + \tilde{g}^2 - \tfrac{308}{729}\tilde{g}^3 + 0.3510695978\tilde{g}^4 - 0.3765268283\tilde{g}^5 + 0.49554751\tilde{g}^6$$
$$- 0.749689\tilde{g}^7 + O\left(\tilde{g}^8\right),$$

where, for all N, $\tilde{g} = (N+8)g_r/48\pi$.

One must then first determine the zero of the β-function, which is a number of order 1.

Large order estimates based on *instanton calculus* have shown that the coefficients β_k of the β-function at order k behave like (see, *e.g.*, Section 23.2.1)

$$\beta_k \underset{k \to \infty}{\propto} (-a_N)^k k^b k!$$

with $a_N = 9a/(N+8)$ and $a = 0.147774232\dots$. The perturbative expansion is thus *divergent* and since the relevant values of g are of order 1, the series is not useful directly.

To evaluate RG functions for a coupling of order 1, summation techniques are required (see Chapter 23). Le Guillou and Zinn-Justin [30] have developed specific techniques based on *Borel transformation and conformal mapping*. They make an efficient use of the known large order behaviour of perturbation theory.

Improved summation techniques and additional seven-loop contributions (but not for the RG β-function) have led to slightly more precise (but consistent) estimates of critical exponents (Guida and Zinn-Justin [31]).

Note that similar summation methods are required to derive some precise estimates from the ε-expansion for $\varepsilon = 1$, but the apparent errors seem to be larger, presumably because the series are shorter [32, 31] (see Table 6.1).

5.7   Results for three-dimensional critical exponents

Critical exponents have been evaluated for several values of N that correspond to the following transitions:

$N = 1$: liquid–vapour, binary mixtures, Ising systems,

$N = 2$: Helium superfluidity,

$N = 3$: ferromagnetic systems,

$N = 0$: in this formal limit, critical exponents describe some statistical properties of long polymers or of the self-avoiding random walk.

We display in Table 5.1 the results obtained by summing the three-dimensional perturbative expansions to the available order [31]. We give results for the five exponents $\gamma, \eta, \nu, \beta, \alpha$, although only two are independent, and the others can be inferred from the scaling relations:

$$\gamma = \nu(2 - \eta), \quad \beta = \tfrac{1}{2}(1 + \eta)\nu, \quad \alpha = 2 - 3\nu,$$

because the different series have been summed independently to check consistency. These exponents are defined by the critical behaviour for $T \to T_c$ of the different quantities,

$$\xi(T) \propto |T - T_c|^{-\nu}, \ \ \chi(T) \propto |T - T_c|^{-\gamma}, \ \ M(T) \propto (T_c - T)^{\beta}, \ \ C(T) \propto |T - T_c|^{\alpha},$$

$$W^{(2)}(x) \propto \frac{1}{|x|^{d-2+\eta}}, \ \ H(M, T) = M^{\delta} f[(T - T_c)M^{-1/\beta}],$$

where ξ is the correlation length, and $W^{(2)}$ is the two-point correlation function at T_c; in magnetic language, χ is the magnetic susceptibility, M is the magnetization, C is the singular part of the specific heat and H is the magnetic field expressed as a function of magnetization and temperature near T_c (the equation of state).

Finally, $\theta = \omega\nu$ is the correction exponent governing the leading correction to the singular behaviour for $T \to T_c$.

A comparison between these 'old' calculations based on RG field theory [30, 31] (see Table 5.1) and other theoretical methods shows quite good agreement (see, *e.g*, for Ising-like systems, more recent lattice high temperature and Monte Carlo results [33, 34], as well as recent results coming from the conformal bootstrap method [35]).

A comparison with available experimental results also shows very good agreement. Only the most precise experimental results coming from superfluid Helium transition (in low gravity experiments) [36],

$$\nu = 0.6705 \pm 0.0006\,, \quad \nu = 0.6708 \pm 0.0004\,,$$
$$\alpha = -0.01285 \pm 0.00038\,,$$

remain a challenge to field theory. Field theory RG yields:

$$\nu = 0.6703 \pm 0.0015\,, \quad \alpha = -0.011 \pm 0.004\,,$$

while lattice calculations claim [37]

$$\nu = 0.6717 \pm .0001\,, \quad \alpha = -0.0151 \pm 0.0003\,.$$

Future prospects. For the $d = 3$ series, a noticeable improvement can be expected from a seven-loop calculation of the β-function, since the value of g^* enters in the calculation of all other universal quantities.

Moreover, it would be useful if another group were to verify Nickel's calculations to check the precision of the evaluated diagrams.

However, to give an idea of the problem one faces, at *seven-loop order*, about *3500 diagrams* have to be evaluated, which are integrals of rather singular functions over up to *21 variables*.

The results produced by the ε-expansion (Table 6.1) are consistent but less precise, presumably because the series are shorter. Here again, longer series could improve the results.

Table 5.1

Critical exponents from the $O(N)$ symmetric $(\phi^2)^2_{d=3}$ field theory.

N	0	1	2	3
$\tilde{g}^*$	1.413 ± 0.006	1.411 ± 0.004	1.403 ± 0.003	1.390 ± 0.004
g_r^*	26.63 ± 0.11	23.64 ± 0.07	21.16 ± 0.05	19.06 ± 0.05
γ	1.1596 ± 0.0020	1.2396 ± 0.0013	1.3169 ± 0.0020	1.3895 ± 0.0050
ν	0.5882 ± 0.0011	0.6304 ± 0.0013	0.6703 ± 0.0015	0.7073 ± 0.0035
η	0.0284 ± 0.0025	0.0335 ± 0.0025	0.0354 ± 0.0025	0.0355 ± 0.0025
β	0.3024 ± 0.0008	0.3258 ± 0.0014	0.3470 ± 0.0016	0.3662 ± 0.0025
α	0.235 ± 0.003	0.109 ± 0.004	-0.011 ± 0.004	-0.122 ± 0.010
ω	0.812 ± 0.016	0.799 ± 0.011	0.789 ± 0.011	0.782 ± 0.0013
$\omega\nu$	0.478 ± 0.010	0.504 ± 0.008	0.529 ± 0.009	0.553 ± 0.012

6 Critical phenomena: The field theory approach

Continuous phase transitions, in physical systems with short range interactions, display, near the critical temperature, remarkable *universal* (*i.e.*, independent of the specific system within a wide class) macroscopic or large distance properties (like algebraic decay of correlations, or singular behaviour of thermodynamic quantities at the critical temperature).

Quantum field theory methods combined with renormalization group (RG) ideas have allowed proving universality as well as calculating universal quantities [105] (see also Chapter 5).

Universality of the large distance behaviour is related to infrared (IR) fixed points of the RG flow. Wilson and Fisher [79] succeeded in determining a set of fixed points (known as Wilson–Fisher fixed points) relevant for a large class of phase transitions (liquid–vapour, Helium, ferromagnets, *etc.*) by using a method that extends to complex (*i.e.*, non-integer) values of the space dimension the perturbative, Feynman diagram, expansion of quantum field theory, which is a standard approximation tool. In this way, they have generalized critical properties to non-integer values of the dimension.

They have discovered that near space dimension $d = 4$, more precisely $d = 4 - \varepsilon$ with $\varepsilon \to 0_+$, universal quantities can be calculated in the form of an ε-expansion. This approach has provided the first examples of analytic estimates of critical exponents in dimensions $d > 2$ that differ from their classical values (also known as *mean field* or *quasi-Gaussian values*).

In the framework of quantum field theories, the RG initially emerged as a consequence of the necessity to cancel infinities that appear in the perturbative expansion (the *renormalization* procedure) and the possibility of defining the parameters of the *renormalized theory* at different momentum scales (Chapter 4).

Although, the field theory RG is now understood to be an *asymptotic form*, in some neighbourhood of the Gaussian fixed point, of a more general RG, it has allowed confirming the Wilson–Fisher fixed point [83] and led to an understanding of universality for a large class of critical phenomena.

In particular, within the framework of the ε-expansion, quantum field theories methods have made it possible to derive general scaling properties [104] and calculate universal quantities [106, 32, 31], with the help of series summation methods, even if the most precise values of critical exponents derived by field theory methods are inferred from perturbation theory in fixed dimension 3 [31] (see also Section 5.5).

From Random Walks to Random Matrices. Jean Zinn-Justin, Oxford University Press (2019).
© Jean Zinn-Justin. DOI: 10.1093/oso/9780198787754.001.0001

6.1 Universality and RG

Gaussian fixed points. In Sections 4.8 and 5.4, we discussed the role of Gaussian distributions in critical phenomena. Gaussian distributions are found as fixed points in various RG implementations (see Section 6.5). Gaussian critical theories predict super-universal properties of critical phenomena (like mean field theory) but we have emphasized that, at least in dimensions $d \leq 4$, these predictions are incorrect because the Gaussian fixed point, in the RG sense, is unstable.

The study of the random walk clearly indicates that an asymptotic Gaussian distribution is the consequence of averaging a large number of statistically independent variables (the central limit theorem of probabilities). In the case of continuous phase transitions, as long as the correlation length is finite, the argument may be applicable but, at the critical temperature, the correlation length diverges. This suggests that all degrees of freedom are strongly correlated and that Gaussian distributions can no longer be expected to describe, in general, critical properties.

6.1.1 Quantum field theory: Renormalization and universality

Non-Gaussian examples of universality are provided by quantum field theories. In quantum electrodynamics (QED), the perturbative calculations exhibit (short distance or large momentum) infinities (see Chapter 4). It is then necessary to modify the short distance structure of the initial theory, a procedure called *regularization*, leading to a finite *regularized theory* but with some unavoidable *non-physical* short distance properties (*e.g.*, non-conservation of probabilities).

To such artificial modification is attached a short length scale known as a *short distance cut-off* and here denoted by $1/\Lambda$, (Λ having the dimension of an inverse distance and being known as an ultraviolet cut-off because, in QED, it cuts off high wavelengths of photons).

Nevertheless, and this was the surprise in the early stage of the construction of QED, a *universal* large distance theory, called *renormalized theory*, can be defined order by order in perturbation theory by parametrizing observables in terms of *physical (renormalized) parameters* rather than the initial (or *bare parameters*) of the regularized theory and then taking the infinite Λ limit. This procedure is called *renormalization*.

Universality of the renormalized theory refers to the independence of the renormalized theory from these short distance modifications in some large class.

The very existence of the renormalized theory in the 'ultraviolet limit' $\Lambda \to \infty$ (at fixed renormalized parameters) can be perturbatively, and, in simple (so-called super-renormalizable) cases, non-perturbatively, proved for a number of *local* quantum field theories, including non-Abelian gauge theories.

The renormalized theory resulting from this process is no longer related to a Gaussian field distribution, which in this framework corresponds to a simple free field theory.

However, note a puzzling feature of the renormalization theory: it requires varying the initial parameters of the theory as a function of the cut-off (at fixed physical parameters) while taking the infinite cut-off limit.

The perturbative RG idea: The origin. The definition of the renormalized parameters implicitly or explicitly requires the introduction of a new (large compared to $1/\Lambda$) physical distance scale known as the *renormalization scale*.

The relation between the renormalized (or physical) parameters in different parametrizations of the *same* renormalized theory corresponding to different values of the renormalization scale was historically called *RG* [65], and the linear partial differential equations for the correlation functions of the renormalized theory, resulting from variations of the parameters under an infinitesimal change of the renormalization scale, were called differential *RG equations*.

In quantum field theories, the existence of an RG is directly related to the existence of a renormalized theory but, conversely, an RG can be set up that gives a proper interpretation to the renormalized theory.

6.1.2 *Macroscopic continuous phase transitions: Universality*

Some statistical systems (random systems with a large number of degrees of freedom) with short range interactions exhibit continuous phase transitions with *universal* properties. The term *universal* is used in this context to emphasize the remarkable observation that many systems, including some that seem physically unrelated (like, *e.g*, the magnetic Ising model and liquid–vapour systems) share, sometimes unexpectedly, a set of non-trivial large scale properties at the phase transition. They belong to the same *universality class*.

Macroscopic continuous phase transitions (and thus with a correlation length diverging at the critical temperature) have been described for a long time by Landau's theory [54] of critical phenomena or *mean field theory*. Such a theory is interpreted in modern language as corresponding to a Gaussian distribution or perturbed Gaussian distribution (*quasi-Gaussian distribution*). One can call such a theory *super-universal* since it predicts large scale properties completely independent of the structure of the microscopic models.

However, for systems with short range interactions, in space dimensions smaller than or equal to 4, quasi-Gaussian models *do not* describe accurately the universal properties of phase transitions at large distance near the critical temperature.

Kadanoff's RG. To deal with this puzzling situation, Kadanoff introduced the idea of recursive summation over short degrees of freedom in lattice models [77]. This process, which generates an infinite sequence of Hamiltonians, was called RG. If, for a large class of initial Hamiltonians, this sequence has a common limit (an attractive RG *fixed point*), then *universality* of the large distance properties can be explained. The *basin of attraction* of a given fixed point in the space of Hamiltonians is called a *universality class*.

This idea were critically re-examined and the method improved by Wilson [78].

These articles are somewhat qualitative and show clearly that the RG is a general concept whose application, in addition, requires discovering a proper implementation.

6.1.3　From Wilson's momentum-shell integration to functional RG equations

Wilson [24] introduced a new implementation of the RG idea. First, he argues that, even if the initial model is defined on a lattice and the random variables take discrete values (like Ising spins), it can be replaced by a field theory in continuum space that has the same large scale physics.

He also replaces the iterative summation over short distance degrees of freedom by an iterative integration over Fourier components of the field with momenta belonging to momentum shells (see also Section 4.7.2).

This generates an RG that acts in the space of local Hamiltonians (local because the interactions are short range), a space that depends on an infinite number of parameters, generalizing thus the RG of quantum field theories.

A further elaboration of the idea, based on some form of infinitesimal momentum shell integration, then leads to the functional (or exact) RG equations (9.9) [100, 101].

Note that infinitesimal momentum shells can be defined in continuum space but *not* on the lattice.

If one starts from an initial Hamiltonian or configuration energy $\mathcal{H}$ of a statistical model with short range interactions characterized by a microscopic scale $1/\Lambda$, the RG transformations generate scale-dependent, effective Hamiltonians $\mathcal{H}_\lambda$ corresponding to increasing scales $1/(\lambda\Lambda)$, with $\lambda \in [0,1]$.

The (non-linear) flow equations that relate Hamiltonians associated to different scales are also called *RG equations*.

6.2　RG in the continuum: Abstract formulation

We assume that one has constructed an RG flow by integrating recursively in some way over short distance degrees of freedom in the sense of Kadanoff–Wilson and, correspondingly, determined an *effective Hamiltonian $\mathcal{H}_\lambda$* in the continuum, function of a scale parameter λ (such that $\mathcal{H}_1 = \mathcal{H}$). The scale parameter λ can then take any real positive value.

The denomination *RG* then refers to the property that $\ln \lambda$, where $\lambda > 0$ belongs to the dilatation (associated to the change of scale) semi-group, belongs to the additive group of real numbers.

RG flow and fixed points. If λ takes continuous values, one can construct an RG flow for infinitesimal variations of λ, of the form

$$\lambda\frac{\mathrm{d}}{\mathrm{d}\lambda}\mathcal{H}_\lambda = \mathcal{T}\left[\mathcal{H}_\lambda\right]. \tag{6.1}$$

The appearance of the derivative $\lambda\mathrm{d}/\mathrm{d}\lambda = \mathrm{d}/\mathrm{d}\ln\lambda$ reflects the multiplicative character of dilatations or scale changes. The RG equation (6.1) thus defines a dynamical process in 'time' $\ln\lambda$.

We require that the transformation $\mathcal{T}$, which acts in the 'space of Hamiltonians', is of *Markovian type*, that is, depends on $\mathcal{H}_\lambda$ but not on the trajectory that has led from $\mathcal{H}_{\lambda=1}$ to $\mathcal{H}_\lambda$.

We also require that the Markovian process is *stationary*, in such a way that the right-hand side depends on λ only through $\mathcal{H}_\lambda$ (but not on λ explicitly).

Finally, we require, and this is also an important condition, that the mapping $\mathcal{T}$ is sufficiently *differentiable* (*e.g*, infinitely differentiable). These conditions imposed to the RG flow (6.1) are at the origin of several important RG properties.

RG fixed points $\mathcal{H}^*$ are then solutions of the equation

$$\mathcal{T}(\mathcal{H}^*) = 0\,. \tag{6.2}$$

Fixed points: Local stability. Since the mapping $\mathcal{T}$ is *differentiable*, the local flow near a fixed point $\mathcal{H}^*$ can be studied by linearizing equation (6.1) at the fixed point:

$$\mathcal{T}(\mathcal{H}^* + \Delta\mathcal{H}_\lambda) \sim L^*\Delta\mathcal{H}_\lambda\,.$$

Formally, the solution of the linearized equations can be written as

$$\mathcal{H}_\lambda = \mathcal{H}^* + \lambda^{L^*}\left(\mathcal{H}_{\lambda=1} - \mathcal{H}^*\right).$$

The local RG flow near the fixed point is thus determined by the eigenvalues and eigenoperators of the linear operator L^*. Assuming, for simplicity, that L^* has a discrete real spectrum (a condition satisfied by all the simple models we study in this work), one can classify the eigenvalues according to their sign (see also Section 1.6).

In the RG terminology, if an eigenoperator is associated with a positive eigenvalue, it corresponds to a direction of instability and is called *relevant*.

Conversely, if an eigenoperator is associated with a negative eigenvalue, it corresponds to a direction of stability and is called *irrelevant*.

If an eigenvalue vanishes, the corresponding eigenoperator is called *marginal*, the linear approximation does not determine the property of RG flow and it is necessary to expand to higher orders. Generically, one expects flows with logarithmic behaviours.

However, one finds also operators with vanishing eigenvalue that correspond to a reparametrization of the field or order parameter space. Those are called *redundant* since they do not correspond to any physical observable.

6.3 Effective field theory

It is somewhat intuitive that, even if the initial statistical model is defined in terms of random variables associated to the sites of a space lattice, and taking only a finite set of values (like, *e.g.,* the classical spins of the Ising model), after repeated integrations over short distance degrees of freedom, the model converges towards a statistical field theory in continuum space [24].

We call such a theory *effective field theory* to emphasize that it does not represent a true microscopic model but is only designed to reproduce accurately its large distance properties.

We assume that the statistical field theory is defined in terms of a random real field $\phi(x)$ (the generalization to several fields is simple) in continuum space, $x \in \mathbb{R}^d$.

The partition function is then given by a field integral (*i.e.*, a sum over field configurations) of the form

$$\mathcal{Z} = \int [\mathrm{d}\phi(x)] \, \mathrm{e}^{-\mathcal{H}(\phi)} \,.$$

The important condition of *short range interactions* translates into the property of *locality* of the field theory: $\mathcal{H}(\phi)$ has the form of a space-integral over a linear combination of monomials in the field and its derivatives.

Moreover, we assume space translation invariance and, for simplicity, a $\mathbb{Z}_2$ reflection symmetry: $\mathcal{H}(\phi) = \mathcal{H}(-\phi)$. This $\mathbb{Z}_2$ symmetry characterizes the *universality class* of the Ising model.

When the effective field theory is inferred from a microscopic model, either by RG arguments or by mean field expansion, the coefficients of $\mathcal{H}(\phi)$ are regular functions of the temperature T near a critical temperature T_c where a continuous phase transition occurs and the $\mathbb{Z}_2$ symmetry is spontaneously broken.

Field correlation functions (generalized moments) are given by

$$\langle \phi(x_1)\phi(x_2) \cdots \phi(x_n) \rangle = \frac{1}{\mathcal{Z}} \int [\mathrm{d}\phi(x)] \phi(x_1)\phi(x_2) \cdots \phi(x_n) \, \mathrm{e}^{-\mathcal{H}(\phi)} \,. \qquad (6.3)$$

They can be derived from a generalized partition function in an external field $H(x)$,

$$\mathcal{Z}(H) = \int [\mathrm{d}\phi(x)] \, \exp \left[-\mathcal{H}(\phi) + \int \mathrm{d}^d x \, H(x)\phi(x) \right],$$

by functional differentiation as

$$\langle \phi(x_1)\phi(x_2) \cdots \phi(x_n) \rangle = \frac{1}{\mathcal{Z}(0)} \frac{\delta}{\delta H(x_1)} \frac{\delta}{\delta H(x_2)} \cdots \frac{\delta}{\delta H(x_n)} \mathcal{Z}(H) \bigg|_{H=0} \,.$$

We recall that functional differentiation satisfies the usual algebraic rules of differentiation and

$$\frac{\delta}{\delta \phi(x)} \phi(y) = \delta^{(d)}(x - y),$$

where $\delta^{(d)}$ is the d-dimensional Dirac δ-function (rather, distribution).

More directly relevant physical quantities are *connected* correlation functions (generalized cumulants). They can be derived by function differentiation from the free energy $\mathcal{W}(H) = \ln \mathcal{Z}(H)$ (omitting a temperature factor irrelevant here) in the external field H. The n-point function is given by

$$W^{(n)}(x_1, x_2, \ldots, x_n) = \frac{\delta}{\delta H(x_1)} \frac{\delta}{\delta H(x_2)} \cdots \frac{\delta}{\delta H(x_n)} \mathcal{W}(H) \bigg|_{H=0} \,.$$

Since we consider here only translational invariant theories, for any vector a,

$$W^{(n)}(x_1, x_2, \ldots, x_n) = W^{(n)}(x_1 + a, x_2 + a, \ldots, x_n + a). \qquad (6.4)$$

In particular, since $W^{(2)}(x, y) = W^{(2)}(x - y, 0)$, we denote it later by $W^{(2)}(x - y)$.

Connected correlation functions have the so-called *cluster* property: if one separates the points $x_1, \ldots, x_n$ in two non-empty sets, connected functions go to zero when the distance between the two sets goes to infinity. It is the large distance decay of connected correlation functions in the critical domain near the critical temperature T_c that may exhibit universal properties.

6.4   The Gaussian field theory

In the class of Gaussian theories (quadratic Hamiltonians), the simplest Gaussian even Hamiltonian is ($\nabla_x \equiv \{\partial/\partial x_1, \partial/\partial x_2, \ldots, \partial/\partial x_d\}$)

$$\mathcal{H}_{\mathrm{G}}(\phi) = \tfrac{1}{2}\int \mathrm{d}^d x \left[\left(\nabla_x \phi(x)\right)^2 + r\phi^2(x)\right], \quad r \geq 0. \tag{6.5}$$

The Hamiltonian for constant fields has a unique minimum at $\phi = 0$ and thus the expectation value $\langle \phi \rangle$ always vanishes: the Gaussian model can only describe a high temperature (disordered) phase $T \geq T_c$.

When derived from a microscopic model (either by a mean field approximation or by RG transformations), the parameter r is expected to be a regular function of the temperature and $r \propto T - T_c$ for $T \to T_c$.

In a Gaussian model, the only connected correlation function is the two-point function. All other, non-connected, correlation functions can be expressed in terms of the two-point function with the help of Wick's theorem [9].

6.4.1   The Gaussian critical theory

For $r = 0$, the two-point correlation function is the solution of the equation

$$-\nabla_x^2 W^{(2)}(x) = \delta^{(d)}(x), \tag{6.6}$$

which decreases at large distance (cluster property). Such a solution exists only for $d > 2$, where it is given by,

$$\langle \phi(x)\phi(0)\rangle \equiv W^{(2)}(x) = \frac{1}{(2\pi)^d}\int \frac{\mathrm{d}^d p \, \mathrm{e}^{ipx}}{p^2} = \frac{\Gamma(d/2 - 1)}{(4\pi)^{d/2}}\left(\frac{2}{|x|}\right)^{d-2}. \tag{6.7}$$

The two-point function has a power-law decay, showing that the correlation length is infinite, as expected for continuous phase transitions at the critical temperature.

Large distance singularity. For $d = 2$, the expression (6.7) is singular. In fact, the solution of equation (6.6) is proportional to $\ln|x|$. For $d < 2$, the solution is proportional to $|x|^{2-d}$. Thus, for $d \leq 2$, equation (6.6) has no solution obeying the cluster property (*i.e.*, no solution decreasing at large distance).

This pathology reflects the impossibility of continuous phase transitions with short range interactions in one dimension. For $d = 2$, a more detailed analysis shows that it reflects the impossibility of a spontaneous symmetry breaking with ordering for systems with continuous symmetries. By contrast, an example of a phase transition with continuous $O(2)$ symmetry but without ordering is known, the famous Kosterlitz–Thouless transition. In systems with discrete symmetries, like the Ising model, continuous phase transitions are also possible.

6.4.2 The non-critical Gaussian theory

The two-point function has the representation

$$W^{(2)}(x) = \frac{1}{(2\pi)^d} \int \frac{\mathrm{d}^d p \; \mathrm{e}^{ipx}}{p^2 + r} \, . \tag{6.8}$$

It has a so-called *Ornstein–Zernicke* form, a form also found in mean field theory. For $r > 0$, $W^{(2)}(x)$ decays exponentially at large distance as

$$W^{(2)}(x) \underset{|x|\to\infty}{\sim} \frac{1}{4\pi(2\pi\xi)^{(d-3)/2}} \frac{\mathrm{e}^{-|x|/\xi}}{|x|^{(d-1)/2}},$$

where, by definition, $\xi = 1/\sqrt{r}$ is the *correlation length*.

The critical domain. When r is of order 1 (the generic situation), the correlation length ξ is of order 1, which is the microscopic scale. Therefore, no large scale physics is generated. Large scale correlations implies *tuning* r to values such that $\xi \gg 1$ (r is a bare parameter, and $1/\xi$ the renormalized mass in quantum field theory terminology). This implies $r \ll 1$ and, then,

$$r \propto T - T_c \;\Rightarrow\; \xi = 1/\sqrt{r} \propto (T - T_c)^{-\nu}$$

with $\nu = 1/2$, where ν is the *correlation length exponent*. This is equivalent to tune the temperature T such that $T - T_c \ll 1$ and defines the *critical domain*.

Finally, note that the two-point function exists for $r \neq 0$, even in two dimensions.

6.4.3 Short distance singularities

With the Hamiltonian (6.5), the Gaussian model leads to a major difficulty: for $d \geq 2$, the two-point correlation function (and thus all other correlation functions) is not defined at coinciding points due to an ultraviolet or large momentum divergence of the Fourier integral (6.8).

The divergence is related to the singularity of the typical fields that contribute to the field integral, in particular, that they are not continuous.

This divergence is non-physical, since the field theory is assumed to be valid only at distances large with respect to the initial microscopic scale. Therefore, for $d \geq 2$, it is necessary to modify the Gaussian model at short distance in order to restrict the field integration to more regular fields, continuous to define expectation values of powers of the field at the same point, satisfying differentiability conditions to define expectation values of the field and its derivatives taken at the same point.

This can be achieved, for example, by adding to the Hamiltonian (6.5) terms with enough derivatives such that

$$\mathcal{H}_{\mathrm{G}}(\phi) \mapsto \mathcal{H}_{\mathrm{G}}(\phi) + \tfrac{1}{2} \sum_{k=2}^{k_{\max}} \alpha_k \int \mathrm{d}^d x \, \nabla_x \phi(x) (-\nabla_x^2)^{k-1} \nabla_x \phi(x), \tag{6.9}$$

where the coefficients α_k are constrained only by the positivity of the Hamiltonian. This modification, which leads to non-physical properties at short distance from the viewpoint of particle physics, is called *regularization*.

It is possible to verify by explicit calculation that, for $d > 2$, the modified Gaussian two-point function coincides for $|x| \gg 1$, $r \ll 1$, with the function (6.8) for all α_k. Moreover, for $2k_{\max} > d$, it has a $x = 0$ limit.

6.5 Gaussian fixed point and Gaussian renormalization

In $d > 2$ space dimensions, an RG flow can be constructed that has as a fixed point the critical Gaussian model (a free massless field theory in quantum field theory terminology), corresponding to the Hamiltonian

$$\mathcal{H}_{\mathrm{G}}^*(\phi) = \tfrac{1}{2} \int \mathrm{d}^d x \left(\nabla_x \phi(x) \right)^2. \tag{6.10}$$

The RG flow in the linear approximation at the fixed point can be implemented by the simple scaling,

$$\phi(x) \mapsto \lambda^{(2-d)/2} \phi(x/\lambda). \tag{6.11}$$

After the change of variables $x' = x/\lambda$, one verifies that $\mathcal{H}_{\mathrm{G}}^*(\phi)$ is, indeed, invariant.

6.5.1 Perturbing the Gaussian fixed point ($d > 2$)

The flow defined by the transformation (6.11) is the linearized RG flow at the Gaussian fixed point (Section 6.2). Eigenvectors of the linear flow (6.11) are monomials of the form

$$\mathcal{O}_{n,k}(\phi) = \int \mathrm{d}^d x \, O_{n,k}(\phi, x),$$

where $O_{n,k}(\phi, x)$ is a product of powers of the field and its derivatives at point x with $2n$ powers of the field (reflection $\mathbb{Z}_2$ symmetry) and $2k$ powers of $\partial_\mu \equiv \partial/\partial x_\mu$. Their RG behaviour under the transformation (6.11) is then given by a simple dimensional analysis. One defines the dimension of x as -1. The dimension of ∂_μ is then $+1$. The Gaussian dimension of the field is $[\phi] = (d-2)/2$. The dimension $[\mathcal{O}_{n,k}]$ of $\mathcal{O}_{n,k}$ is then

$$[\mathcal{O}_{n,k}] = -d + n(d-2) + 2k. \tag{6.12}$$

It can be verified that $\mathcal{O}_{n,k}$ scales like $\lambda^{-[\mathcal{O}_{n,k}]}$, and the corresponding eigenvalue of L^* thus is $\ell_{n,k} = -[\mathcal{O}_{n,k}]$.

When $\lambda \to +\infty$, for $\ell_{n,k} > 0$, the amplitude of $\mathcal{O}_{n,k}(\phi)$ increases; it is a direction of instability and, in the RG terminology, $\mathcal{O}_{n,k}(\phi)$ is a *relevant* perturbation.

For $\ell_{n,k} < 0$, the amplitude of $\mathcal{O}_{n,k}(\phi)$ decreases; it is a direction of stability and, in the RG terminology, $\mathcal{O}_{n,k}(\phi)$ is an *irrelevant* perturbation.

In the special case $\ell_{n,k} = 0$, the perturbation is *marginal*, and the linear approximation is no longer sufficient to discuss stability. Logarithmic behaviour in λ is then expected.

Finally, $\int \mathrm{d}^d x \, \phi^2(x)$ is always a direction of instability: it corresponds to introducing a deviation from the critical temperature and thus induces a finite correlation length.

The role of dimension 4. For $d > 4$, no other perturbation is relevant, and the Gaussian fixed point is stable on the critical surface (the surface of critical Hamiltonians).

At $d = 4$, one term becomes marginal: $\int \mathrm{d}^d x \, \phi^4(x)$, which below dimension 4 becomes relevant. The Gaussian fixed point for $\mathbb{Z}_2$ symmetric (Ising-like) systems or, more generally, for systems with an $O(N)$ symmetry, on the critical surface is unstable below dimension 4.

After dimensional continuation (a notion we define in Section 6.7.1), in dimension $d = 4 - \varepsilon$, ε positive and small, $\int \mathrm{d}^d x \, \phi^4(x)$ is the only relevant perturbation, and one expects to be able to describe critical properties with a Gaussian theory to which this unique term is added.

6.5.2 Gaussian renormalization

In what follows, we assume that, initially, the statistical system is very close to the Gaussian fixed point. The RG flow is then first governed by the linear flow.

Therefore, we implement the corresponding RG transformation: we introduce a parameter $\Lambda \gg 1$ and substitute $\phi(x) \mapsto \Lambda^{(2-d)/2}\phi(x/\Lambda)$. After the change of variables $x' = x/\Lambda$, the monomials $\mathcal{O}_{n,k}(\phi)$ are multiplied by $\Lambda^{-[\mathcal{O}_{n,k}]}$. In the quantum field theory terminology, this can be called a *Gaussian renormalization*.

The introduction of the parameter Λ has the effect of expressing the dimension (6.12) in terms of Λ: space coordinates x have dimension Λ^{-1}, derivatives have dimension Λ and the field dimension $\Lambda^{(d-2)/2}$. The Hamiltonian is dimensionless. The Hamiltonian (6.9) then becomes,

$$\mathcal{H}_\mathrm{G} = \tfrac{1}{2} \int \mathrm{d}^d x \left[\left(\nabla_x \phi(x)\right)^2 + r\Lambda^2 \phi^2(x) \right]$$

$$+ \tfrac{1}{2} \sum_{k=2}^{k_{\max}} \alpha_k \int \mathrm{d}^d x \, \nabla_x \phi(x) (-\nabla_x^2/\Lambda^2)^{k-1} \nabla_x \phi(x). \qquad (6.13)$$

6.6 Statistical scalar field theory: Perturbation theory

The standard analytic tool of quantum field theory is perturbation theory: in the field integral, one expands any non-quadratic contribution to the Hamiltonian in a power series, reducing its evaluation to an infinite sum of Gaussian expectation values, which can be expressed in terms of the Gaussian two-point function with the help of Wick's theorem.

6.6.1 The perturbed Gaussian or quasi-Gaussian model

To describe physics below T_c, terms have necessarily to be added to the Gaussian Hamiltonian (6.13) to generate a double-well potential for constant fields. The minimal addition, and the most relevant from the RG viewpoint, is of ϕ^4 type. This leads to the Hamiltonian

$$\mathcal{H}(\phi) = \mathcal{H}_\mathrm{G}(\phi) + \frac{g}{4!}\Lambda^{4-d} \int \mathrm{d}^d x \, \phi^4(x), \qquad (6.14)$$

where g is a dimensionless positive parameter, generically of order 1.

This addition induces a shift of the critical temperature. To recover a critical theory $(T = T_c)$, it is necessary to also add a ϕ^2 term with a specific g-dependent coefficient. One sets in the Hamiltonian (6.13),

$$r = r_c(g) + \rho/\Lambda^2, \qquad (6.15)$$

where $r_c(g) < 0$ is defined by the condition that $T = T_c$ for $\rho = 0$.

In quantum field theory terminology, the quantity $r_c(g)\Lambda^2$ is a *mass renormalization*. Moreover, ρ is much smaller than Λ^2 in the critical domain and this is at the origin of the *fine tuning problem*.

For dimensions $d > 4$, the ϕ^4 term is an irrelevant contribution that does not invalidate the universal predictions of the Gaussian model, and corrections to the Gaussian theory can be obtained by expanding in powers of the parameter g.

By contrast, for $d < 4$, the ϕ^4 contribution is relevant: the Gaussian fixed point is unstable and no longer governs the large distance behaviour. The perturbative expansion in powers of g of the critical theory $(T = T_c)$ contains so-called IR, that is, long distance, or zero momentum in Fourier space, divergences.

To determine the large distance behaviour of correlation functions, it becomes necessary to construct an RG in Hamiltonian space and this leads to equations (Chapter 9) that, in general, cannot be solved analytically. However, an ingenious method has been proposed: extend the definition of all terms of the perturbative expansion to arbitrary complex values of the dimension d, in the form of meromorphic functions. This allows studying the model analytically in dimension $d = 4 - \varepsilon$ as an expansion in powers of ε.

6.7 Dimensional continuation and regularization

To define dimensional continuation, we introduce the Fourier representation of the Gaussian two-point correlation function (or propagator) $\Delta(x)$, corresponding to the Hamiltonian, (6.9). Using translation invariance, we set

$$\Delta(x) \equiv \langle \phi(x)\phi(0) \rangle_{\mathrm{G}} = \frac{1}{(2\pi)^d} \int \mathrm{d}^d p \, \mathrm{e}^{-ipx} \, \tilde{\Delta}(p).$$

6.7.1 Dimensional continuation

A general representation useful for *dimensional continuation* of the Fourier transform of the Gaussian two-point function is

$$\tilde{\Delta}(p) = \int_0^\infty \mathrm{d}s \, \rho(s\Lambda^2) \, \mathrm{e}^{-sp^2}, \tag{6.16}$$

where $\rho(s) \to 1$ when $s \to \infty$.

To reduce the field integration to continuous fields and, thus, to render the perturbative expansion finite, one needs at least $\rho(s) = O(s^q)$ with $q > (d-2)/2$ for $s \to 0$. If one wants the expectation values of all local polynomials to be defined, one must impose to $\rho(s)$ to converge to zero faster than any power. In the context of quantum field theory, since the effect of the ρ-factor is to suppress $\tilde{\Delta}(p)$ for values of $|p| \gg \Lambda$, Λ is called the large momentum *cut-off*.

A contribution to perturbation theory (represented graphically by a Feynman diagram) takes, in Fourier representation, the form of a product of propagators integrated over a subset of momenta. With the representation (6.16), all momentum integrations become Gaussian and can be performed, resulting in explicit analytic meromorphic functions of the dimension parameter d.

For example, the contribution of order g to the two-point function (the first diagram of Fig. 6.1) is proportional to

$$\Omega_d = \int \frac{\mathrm{d}^d p}{(2\pi)^d}\, \tilde{\Delta}(p) = \int \frac{\mathrm{d}^d p}{(2\pi)^d} \int_0^\infty \mathrm{d}s\, \rho(s\Lambda^2)\, \mathrm{e}^{-sp^2} = \int_0^\infty \frac{\mathrm{d}s}{(4\pi)^{d/2}}\, s^{-d/2} \rho(s\Lambda^2),$$

which in the latter form is holomorphic for $2 < \operatorname{Re} d < 2(1+q)$.

6.7.2 *Dimensional regularization and ε-expansion*

To generate the ε-expansion and calculate universal quantities in the theory of critical phenomena, dimensional continuation is sufficient, since it makes it possible to explore the neighbourhood of dimension 4 and, following Wilson and Fisher [79], to calculate universal quantities as ε-expansions.

However, when one is interested only in the leading behaviour at large distance, an additional step is useful for practical calculations.

It can be verified that, if one takes $\operatorname{Re} d$ sufficiently small (including negative) so that, by naive power counting, all momentum integrals are convergent, one can then, after explicit dimensional continuation, take the infinite Λ limit. The resulting perturbative contributions become meromorphic functions with poles, in particular at dimensions at which large momentum, and low momentum in the critical theory, divergences appear [102].

This method of regularizing large momentum divergences is called *dimensional regularization* and is extensively used in quantum field theory [102]. It has also been used to calculate universal quantities in the theory of critical phenomena, like critical exponents, in the form of $\varepsilon = (4 - d)$-expansions [83, 106].

Note a peculiarity of dimensional regularization, which implies

$$\int \mathrm{d}^d k\, k^n = 0 \ \forall d > n \in \mathbb{Z}.$$

Therefore, dimensional regularization, a technique that does not seem to have a physical interpretation, performs an automatic *partial renormalization*, in the sense that it cancels what in momentum regularization would be power-law divergences and leaves only the equivalent of logarithmic divergences in the form of poles.

6.8 Perturbative RG

The perturbative RG, as it has been derived in the framework of the perturbative expansion of quantum field theory, relies on the *renormalization theory*. It takes the form of partial differential equations satisfied by correlation functions. For the ϕ^4 field theory, initially it has been formulated in space dimension $d = 4$. For critical phenomena, a minor extension is required that involves an additional expansion in powers of $\varepsilon = (4 - d)$, after dimensional continuation.

It is now understood that the field theory (or perturbative) RG and the more general functional RG [100, 101] are related. The field RG is the asymptotic form of the general RG valid in some neighbourhood of a Gaussian fixed point.

6.8.1   Critical theory: The renormalization theorem

It is convenient to express RG equations in terms of the Fourier representation of connected correlation functions. Taking into account translation invariance, one defines

$$(2\pi)^d \delta^{(d)}\left(\sum_{i=1}^{n} p_i\right) \tilde{W}^{(n)}(p_1, \ldots, p_n)$$

$$= \int \mathrm{d}^d x_1 \ldots \mathrm{d}^d x_n \, W^{(n)}(x_1, \ldots, x_n) \exp\left(i \sum_{j=1}^{n} x_j p_j\right), \qquad (6.17)$$

where, in analogy with quantum mechanics, the Fourier variables p_i are called momenta (and have dimension Λ).

First, we consider the critical theory ($\rho = 0$ in equation (6.15)) and denote by $\mathcal{H}_c$ the corresponding Hamiltonian (6.14).

To formulate the renormalization theorem, one introduces a momentum μ, the renormalization scale (which is a typical physical scale), and a parameter g_r characterizing the effective interaction at scale μ, called the renormalized interaction.

One can then find two dimensionless functions, $Z(\Lambda/\mu, g)$ and $Z_g(\Lambda/\mu, g)$, that satisfy (g and Λ/μ are the only two dimensionless combinations)

$$\Lambda^{4-d} g = \mu^{4-d} Z_g(\Lambda/\mu, g) g_r = \mu^{4-d} g_r + O(g^2), \quad Z(\Lambda/\mu, g) = 1 + O(g), \qquad (6.18)$$

calculable order by order in a double series expansion in powers of g and ε, such that all connected correlations functions

$$\tilde{W}_r^{(n)}(p_i; g_r, \mu, \Lambda) = Z^{-n/2}(g, \Lambda/\mu)\tilde{W}^{(n)}(p_i; g, \Lambda), \qquad (6.19)$$

called *renormalized*, have, order by order in g_r, finite limits $\tilde{W}_r^{(n)}(p_i; g_r, \mu)$ when $\Lambda \to \infty$ at p_i, μ, g_r fixed, independent of the details of regularization procedure, up to a multiplication of the renormalization constants Z and Z_g by finite functions of g_r.

The renormalization constant $Z^{1/2}(\Lambda/\mu, g)$ is the ratio between the Gaussian field renormalization $\Lambda^{(d-2)/2}$ and the renormalization in presence of the ϕ^4 potential.

Note that the applicability of this theorem *beyond perturbation theory*, implies that even for $\Lambda/\mu \to \infty$, for every value of g_r, one can find a corresponding value of $g(\Lambda)$. However, we will exhibit examples where, for any initial value g, the value of g_r goes to zero for $\Lambda \to \infty$ and the renormalized field theory in the infinite cut-off limit is trivial (a free field theory).

Perturbative universality. There is some arbitrariness in the choice of the *renormalization constants* Z and Z_g, since they can be multiplied by arbitrary finite functions of g_r. The constants can be completely determined by imposing three *renormalization conditions* to the renormalized correlation functions, which are then independent of the specific choice of the regularization. This a first very important result: since initial and renormalized correlation functions have the same large distance behaviour, this behaviour is, to a large extent, *universal*, since it can depend at most on one parameter, the parameter g in the ϕ^4 interaction.

6.8.2 RG equations for the critical theory

From equation (6.19) and the existence of a limit $\Lambda \to \infty$, a new equation follows, obtained by differentiating the equation with respect to Λ at μ, g_r fixed,

$$\Lambda \frac{\partial}{\partial \Lambda}\bigg|_{g_r, \mu \text{ fixed}} Z^{n/2}(g, \Lambda/\mu)\tilde{W}^{(n)}(p_i; g, \Lambda) \to 0. \tag{6.20}$$

In accordance with the perturbative philosophy, one then neglects all contributions that, order by order in g, decay as powers of Λ. One defines asymptotic functions $\tilde{W}_{\text{as.}}^{(n)}(p_i; g, \Lambda)$ and $Z_{\text{as.}}(g, \Lambda/\mu)$ as sums of the perturbative contributions to the functions $\tilde{W}^{(n)}(p_i; g, \Lambda)$ and $Z(g, \Lambda/\mu)$, respectively, that do not vanish for $\Lambda \to \infty$.

Using the chain rule, one then derives from equation (6.20),

$$\left[\Lambda \frac{\partial}{\partial \Lambda} + \beta(g, \Lambda/\mu)\frac{\partial}{\partial g} + \frac{n}{2}\eta(g, \Lambda/\mu)\right] \tilde{W}_{\text{as.}}^{(n)}(p_i; g, \Lambda) = 0,$$

where the functions β and η are defined by

$$\beta(g, \Lambda/\mu) = \Lambda \frac{\partial}{\partial \Lambda}\bigg|_{g_r, \mu} g, \quad \eta(g, \Lambda/\mu) = -\Lambda \frac{\partial}{\partial \Lambda}\bigg|_{g_r, \mu} \ln Z_{\text{as.}}(g, \Lambda/\mu).$$

Since the functions $\tilde{W}_{\text{as.}}^{(n)}$ do not depend on μ, the functions β and η cannot depend on Λ/μ and, finally, one obtains the RG equations [103]

$$\left[\Lambda \frac{\partial}{\partial \Lambda} + \beta(g)\frac{\partial}{\partial g} + \frac{n}{2}\eta(g)\right] \tilde{W}_{\text{as.}}^{(n)}(p_i; g, \Lambda) = 0. \tag{6.21}$$

From equation (6.18), one immediately infers that $\beta(g) = -\varepsilon g + O(g^2)$.

Remarks

(i) By replacing $\tilde{W}^{(n)}$ by $\tilde{W}_{\text{as.}}^{(n)}$, we have neglected in equation (6.21) contributions that vanish like $1/\Lambda^2$ up to logarithmic factors. These contributions can be cancelled by adding a suitable linear combination of local monomials of higher dimensions (dimension 6 in $d = 4$) like ϕ^6 or $(\nabla\phi)^2\phi^2$ to the action, and this procedure can be generalized recursively to cancel increasing smaller corrections.

(ii) Equation (6.21) remains valid even when the renormalized theory becomes trivial in the infinite cut-off limit.

6.8.3 RG equations in the critical domain above T_c

Correlation functions may also exhibit universal properties near T_c when the correlation length ξ is large with respect to the microscopic scale, here, $\xi\Lambda \gg 1$. To derive universal properties in the critical domain above T_c, one adds the ϕ^2 relevant term to the critical Hamiltonian (equation (6.15)):

$$\mathcal{H}(\phi) = \mathcal{H}_c(\phi) + \frac{\rho}{2} \int \mathrm{d}^d x \, \phi^2(x),$$

where ρ, the coefficient of ϕ^2, characterizes the (small in the microscopic scale) deviation from the critical temperature: $\rho \propto T - T_c$, $|\rho| \ll \Lambda^2$.

The renormalization theorem requires the introduction of a new renormalization factor $Z_2(\Lambda/\mu, g)$ associated with the parameter ρ. By a straightforward generalization of the arguments used for the critical theory, one derives an RG equation valid in the critical domain of the form [103],

$$\left[\Lambda\frac{\partial}{\partial\Lambda} + \beta(g)\frac{\partial}{\partial g} + \frac{n}{2}\eta(g) - \eta_2(g)\rho\frac{\partial}{\partial\rho}\right]\tilde{W}_{\mathrm{as.}}^{(n)}(p_i; \rho, g, \Lambda) = 0\,, \tag{6.22}$$

where a new RG function $\eta_2(g)$ related to $Z_2(\Lambda/\mu, g)$ appears.

The equations can be further generalized to deal with an external field (a magnetic field for magnetic systems) and the corresponding induced field expectation value (magnetization for magnetic systems). An RG equation for the equation of state follows.

6.8.4   Renormalized RG equations

For $d < 4$, if one is only interested in the leading scaling behaviour (and the first correction), it is technically simpler to use dimensional regularization and the renormalized theory in the so-called minimal (or modified) subtraction scheme. Equation (6.19) is asymptotically symmetric between initial and renormalized correlations, to the extent that, for a given value of the renormalized coupling g_{r}, one can indeed find an initial coupling g (the latter condition, which is non-perturbative, is not satisfied by an IR free theory).

One thus derives also (for the critical theory) [83],

$$\left[\mu\frac{\partial}{\partial\mu} + \tilde{\beta}(g_{\mathrm{r}})\frac{\partial}{\partial g_{\mathrm{r}}} + \frac{n}{2}\tilde{\eta}(g_{\mathrm{r}})\right]\tilde{W}_{\mathrm{r}}^{(n)}(p_i, g_{\mathrm{r}}, \mu) = 0$$

with the definitions

$$\tilde{\beta}(g_{\mathrm{r}}) = \mu\frac{\partial}{\partial\mu}\bigg|_g g_{\mathrm{r}}\,, \quad \tilde{\eta}(g_{\mathrm{r}}) = \mu\frac{\partial}{\partial\mu}\bigg|_g \ln Z(g_{\mathrm{r}}, \varepsilon)\,.$$

In this scheme, the renormalization constants (6.18) are obtained by going to low dimensions where the infinite Λ limit, at g_{r} fixed, can be taken. For example,

$$\lim_{\Lambda\to\infty} Z(\Lambda/\mu, g)\big|_{g_{\mathrm{r}} \text{ fixed}} = Z(g_{\mathrm{r}}, \varepsilon).$$

Then, order by order in powers of g_{r}, they have a Laurent expansion in powers of ε. In the minimal subtraction scheme, the freedom in the choice of the renormalization constants is used to reduce the Laurent expansion to the singular terms. For example, $Z(g_{\mathrm{r}}, \varepsilon)$ takes the form

$$Z(g_{\mathrm{r}}, \varepsilon) = 1 + \sum_{n=1}^{\infty} \frac{\sigma_n(g_{\mathrm{r}})}{\varepsilon^n}$$

with $\sigma_n(g_{\mathrm{r}}) = O(g_{\mathrm{r}}^{n+1})$.

A remarkable consequence is that the RG functions $\tilde{\eta}(g_{\mathrm{r}})$, and $\tilde{\eta}_2(g_{\mathrm{r}})$ when a ϕ^2 term is added, become independent of ε, and $\tilde{\beta}(g_{\mathrm{r}})$ has the simple dependence $\tilde{\beta}(g_{\mathrm{r}}) = -\varepsilon g_{\mathrm{r}} + \beta_2(g_{\mathrm{r}})$, where $\beta_2(g_{\mathrm{r}}) = O(g_{\mathrm{r}}^2)$ is also independent of ε.

6.9 RG equations: Solutions

RG equations can be solved by the method of characteristics. In the simple example of the critical theory and equation (6.21), one introduces a scale parameter λ and two functions of $g(\lambda)$ and $\zeta(\lambda)$ defined by

$$\lambda \frac{\mathrm{d}}{\mathrm{d}\lambda} g(\lambda) = -\beta\big(g(\lambda)\big), \ g(1) = g, \quad \lambda \frac{\mathrm{d}}{\mathrm{d}\lambda} \ln \zeta(\lambda) = -\eta\big(g(\lambda)\big), \ \zeta(1) = 1. \quad (6.23)$$

The function $g(\lambda)$ is the effective amplitude of the ϕ^4 term at the scale λ. Equation (6.21) is then equivalent to

$$\lambda \frac{\mathrm{d}}{\mathrm{d}\lambda} \left[\zeta^{n/2}(\lambda) \tilde{W}_{\mathrm{as.}}^{(n)}\big(p_i; g(\lambda), \Lambda/\lambda\big) \right] = 0,$$

which implies

$$\tilde{W}_{\mathrm{as.}}^{(n)}\big(p_i; g, \Lambda\big) = \zeta^{n/2}(\lambda) \tilde{W}_{\mathrm{as.}}^{(n)}\big(p_i; g(\lambda), \Lambda/\lambda\big)$$

and, thus,

$$\tilde{W}_{\mathrm{as.}}^{(n)}\big(p_i; g, \lambda\Lambda\big) = \zeta^{n/2}(\lambda) \tilde{W}_{\mathrm{as.}}^{(n)}\big(p_i; g(\lambda), \Lambda\big).$$

Here, the general flow equation (6.1) reduces to the first equation (6.23), and the large distance behaviour to the determination of the zeros of the function $\beta(g)$. When $\lambda \to \infty$, since $\beta(g) = -\varepsilon g + O(g^2)$, if $g > 0$ is initially very small, it moves away from the unstable Gaussian fixed point, as expected from the general RG analysis at the Gaussian fixed point. If one assumes the existence of another zero g^* with then $\beta'(g^*) > 0$ (an assumption that is confirmed by explicit calculations), then $g(\lambda)$ converges towards this fixed point.

From the definition (6.17), one infers that $\tilde{W}_{\mathrm{as.}}^{(n)}$ has dimension $[d - (d+2)n/2]$. Therefore,

$$\tilde{W}_{\mathrm{as.}}^{(n)}\big(p_i/\lambda; g, \Lambda\big) = \lambda^{(d+2)n/2-d} \tilde{W}_{\mathrm{as.}}^{(n)}\big(p_i; g, \lambda\Lambda\big)$$
$$= \lambda^{(d+2)n/2-d} \zeta^{n/2}(\lambda) \tilde{W}_{\mathrm{as.}}^{(n)}\big(p_i; g(\lambda), \Lambda\big). \quad (6.24)$$

If $g(\lambda)$ tends towards a fixed point value g^* and if $\eta(g^*) \equiv \eta$ is finite, equation (6.24) implies the universal behaviour

$$\tilde{W}_{\mathrm{as.}}^{(n)}\big(p_i/\lambda; g, \Lambda\big) \underset{\lambda\to\infty}{\propto} \lambda^{(d+2-\eta)n/2-d} \tilde{W}_{\mathrm{as.}}^{(n)}\big(p_i; g^*, \Lambda\big).$$

For the connected correlation functions in space, this result translates into

$$W_{\mathrm{as.}}^{(n)}\big(\lambda x_i; g, \Lambda\big) \underset{\lambda\to\infty}{\propto} \lambda^{-n(d-2+\eta)/2} W_{\mathrm{as.}}^{(n)}\big(x_i; g^*, \Lambda\big).$$

The exponent $d_\phi = (d - 2 + \eta)/2$ is the *scaling dimension of the field* ϕ, from the point of view of large distance properties.

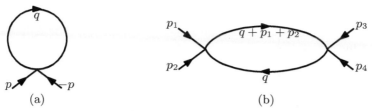

Fig. 6.1 One-loop divergent diagrams.

6.10 Wilson–Fisher's fixed point: ε-Expansion

For practical calculations, it is more convenient to deal with vertex functions than connected correlation functions. The generating functional of vertex functions is obtained from the generating functional of connected functions by a Legendre transformation that generalizes the relation between free energy and thermodynamic potential (or Hamiltonian and Lagrangian). From the viewpoint of Feynman diagrams, vertex functions are the sum of one-line irreducible diagrams.

For the two-point and four-point functions relevant here, the vertex functions $\tilde{\Gamma}^{(n)}$ in Fourier representation above T_c are given by

$$\tilde{\Gamma}^{(2)}(p) = 1/\tilde{W}^{(2)}(p), \quad \tilde{\Gamma}^{(4)}(p_1, p_2, p_3, p_4) = \tilde{W}^{(4)}(p_1, p_2, p_3, p_4)\Big/ \prod_{i=1}^{4} \tilde{W}^{(2)}(p_i).$$

The RG equations satisfied by the asymptotic vertex functions are obtained from equations (6.21) or (6.22) by simply changing $\eta(g)$ into $-\eta(g)$:

$$\left[\Lambda\frac{\partial}{\partial\Lambda} + \beta(g)\frac{\partial}{\partial g} - \frac{n}{2}\eta(g) - \eta_2(g)\rho\frac{\partial}{\partial\rho}\right]\tilde{\Gamma}^{(n)}_{\text{as.}}(p_i; \rho, g, \Lambda) = 0. \qquad (6.25)$$

6.10.1 The Ising class fixed point from the ϕ^4 field theory

At order g, in the critical theory ($\rho = 0$), one finds that $\tilde{\Gamma}^{(2)}$ is not modified, the order g contribution (diagram (a) of Fig. 6.1) being a constant that only shifts the critical value $r_c = r_c(g)$ (a mass renormalization in quantum field theory terminology). Thus,

$$\tilde{\Gamma}^{(2)}(p, g, \Lambda) = p^2 + O(g^2).$$

Inserting the expansion of $\tilde{\Gamma}^{(2)}$ into equation (6.25) for $n = 2$, one infers $\eta(g) = O(g^2)$.

The four-point vertex function at order g^2 is then given by

$$\tilde{\Gamma}^{(4)}(p_1, p_2, p_3, p_4, g, \Lambda) = \Lambda^\varepsilon g - \tfrac{1}{2}\Lambda^{2\varepsilon}g^2\left[B_d(p_1 + p_2) + B_d(p_1 + p_3) + B_d(p_1 + p_4)\right] + O(g^3),$$

where (diagram (b) of Fig. 6.1)

$$B_d(p) = \frac{1}{(2\pi)^d}\int \mathrm{d}^d q\, \tilde{\Delta}(q)\tilde{\Delta}(p - q).$$

In the logic of the ε-expansion, at leading order, one needs only B_4. Without the cut-off Λ, the integral would then diverge logarithmically. For Λ large, it is thus dominated by

$$B_4(p) \sim \frac{1}{(2\pi)^d} \int_{1<|q|<\Lambda} \frac{\mathrm{d}^4 q}{q^4} \sim N_4 \ln \Lambda$$

where N_d is the loop factor,

$$N_d = \frac{2}{(4\pi)^{d/2}\Gamma(d/2)}, \tag{6.26}$$

which, at higher orders, it is convenient not to expand in powers of ε. Then,

$$\tilde{\Gamma}^{(4)}(p_1, p_2, p_3, p_4, g, \Lambda) = g + g\varepsilon \ln \Lambda - \tfrac{3}{2}g^2 (\ln \Lambda + \text{ finite}) + O(g^3, g^2\varepsilon, g\varepsilon^2).$$

$\tilde{\Gamma}^{(4)}$ satisfies equation (6.25) for $n = 4$ and $\rho = 0$. Applying it to the explicit expansion, one concludes that

$$\beta(g) = -\varepsilon g + \frac{3}{16\pi^2}g^2 + O(g^3, \varepsilon g^2).$$

In the sense of an ε-expansion, the RG β-function has a zero g^* with a positive slope [83] (an IR fixed point equivalent to Wilson–Fisher's fixed point [79]),

$$g^* = \frac{3}{16\pi^2} + O(\varepsilon^2), \quad \omega \equiv \beta'(g^*) = \varepsilon + O(\varepsilon^2),$$

which governs the large momentum behaviour of correlation functions. Moreover, the exponent ω characterizes the leading correction to the critical behaviour.

Generalization. The results obtained for models with a $\mathbb{Z}_2$ reflection symmetry can easily be generalized to N-vector models with $O(N)$ orthogonal symmetry, which belong to different universality classes. Their universal properties can then be derived from an $O(N)$ symmetric field theory with an N-component field $\phi(x)$,

$$\mathcal{H}(\phi) = \int \mathrm{d}^d x \left[\tfrac{1}{2}\left(\nabla_x \phi(x)\right)^2 + \tfrac{1}{2}r\Lambda^2 \phi^2(x) + \frac{g}{4!}\Lambda^\varepsilon \left(\phi^2(x)\right)^2 \right] + \text{ regularization}.$$

Further generalizations involve theories with N-component fields but smaller symmetry groups, such that several independent quartic ϕ^4 terme are allowed. The structure of fixed points may then be more complicated (see Chapter 7).

6.10.2 ε-Expansion: A few general results

From the simple existence of the ε-expansion and the prediction of a fixed point, universal properties of a large class of critical phenomena (this includes scaling relations between critical exponents, and the scaling behaviour of correlation functions or the equation of state) can be derived to all orders in ε. Moreover, universal quantities can be calculated as ε-expansions (see, *e.g*, [104]).

Scaling relations between exponents. All the critical exponents that are usually defined can be related to two of them, for example,

$$\eta = \eta(g^*), \quad \nu = 1/(\eta_2(g^*) + 2),$$

where $d_\phi = \frac{1}{2}(d - 2 + \eta)$ is the dimension of the field, and ν characterizes the divergence of the correlation length ξ at T_c,

$$\xi \propto |T - T_c|^{-\nu}.$$

In magnetic language, the magnetic susceptibility exponent γ characterizing the divergence of the two-point correlation function at zero momentum at T_c, and the exponent α characterizing the singularity of the specific heat, are given by

$$\gamma = \nu(2 - \eta), \quad \alpha = 2 - \nu d.$$

Applied to the equation of state of magnetic systems, that is, the relation between applied magnetic field H, magnetization M and temperature T in the relevant limit $|H| \ll 1$, $|T - T_c| \ll 1$, the RG confirms Widom's scaling form

$$H = M^\delta f\big((T - T_c)/M^{1/\beta}\big),$$

where $f(z)$ is a universal (up to normalizations) calculable function. The two new exponents satisfy the relations

$$\delta = \frac{d + 2 - \eta}{d - 2 + \eta}, \quad \beta = \frac{1}{2}\nu(d - 2 + \eta).$$

Note that the relations involving the dimension d explicitly are not valid for the Gaussian fixed point.

6.11   Critical exponents as ε-expansions

The RG functions of the $(\phi^2)^2$ field theory are known up to five-loop order. Therefore, the ε-expansions of exponents are known, for generic N-vector models, up to ε^5 [106]. The results for the $\mathbb{Z}_2$ models are recovered by setting $N = 1$. For illustration, we give here only two successive terms in the expansion, referring to the literature for higher order results. In terms of the variable $v = N_d g$, where N_d is the loop factor (6.26), the RG functions $\beta(v)$ and $\eta_2(v)$ at two-loop order, $\eta(v)$ at three-loop order, are

$$\beta(v) = -\varepsilon v + \frac{(N + 8)}{6}v^2 - \frac{(3N + 14)}{12}v^3 + O(v^4)$$

and

$$\eta(v) = \frac{(N + 2)}{72}v^2 \left[1 - \frac{(N + 8)}{24}v\right] + O(v^4), \quad \eta_2(v) = -\frac{(N + 2)}{6}v\left(1 - \frac{5}{12}v\right) + O(v^3).$$

From the fixed point value, solution of $\beta(v^*) = 0$,

$$v^*(\varepsilon) = \frac{6\varepsilon}{(N+8)} \left[1 + \frac{3(3N+14)}{(N+8)^2}\varepsilon\right] + O(\varepsilon^3),$$

one infers the values of the critical exponents

$$\eta = \frac{\varepsilon^2(N+2)}{2(N+8)^2} \left[1 + \frac{(-N^2 + 56N + 272)}{4(N+8)^2}\varepsilon\right] + O(\varepsilon^4),$$

$$\gamma = 1 + \frac{(N+2)}{2(N+8)}\varepsilon + \frac{(N+2)}{4(N+8)^3}\left(N^2 + 22N + 52\right)\varepsilon^2 + O(\varepsilon^3).$$

Although this may not be obvious from these few terms, the ε-expansion is divergent for any $\varepsilon > 0$, as large order calculations based on instanton calculus have shown. For example, adding simply the known successive terms for $\varepsilon = 1$ and $N = 1$ yields

$$\gamma = 1.000\dots,\ 1.1666\dots,\ 1.2438\dots, 1.1948\dots,\ 1.3384\dots,\ 0.8918\dots.$$

Extracting precise numbers from the known terms of the series thus requires a summation method (see Chapter 23).

6.12 Three-dimensional exponents: Summation of the ε-expansion

We display below (Table 6.1) the values of several critical exponents of the $O(N)$ model obtained from Borel summation of the ε-expansion for $\varepsilon = 1$, that is, $d = 3$ (Le Guillou and Zinn-Justin [32], Guida and Zinn-Justin [31]). Since the ε-expansion is divergent, a summation procedure is required to extract precise numbers from the available expansions (see Chapter 23).

In Table 6.1, $N = 0$ corresponds to statistical properties of polymers (mathematically, the self-avoiding random walk), $N = 1$ corresponds to the Ising universality class, which also includes liquid–vapour and binary mixtures or anisotropic magnet phase transitions, $N = 2$ describes the superfluid Helium transition, and $N = 3$ corresponds to isotropic magnets.

Due to scaling relations like $\gamma = \nu(2-\eta)$, $\gamma + 2\beta = \nu d$, only two among the four are independent, but the series have been summed independently to check consistency. The results can be compared with the best available field theory results obtained from Borel summation of $d = 3$ perturbative series [31] (see Table 5.1) and show overall consistency but slightly larger apparent errors.

Table 6.1

Critical exponents in the $(\phi^2)_3^2$ field theory from the ε-expansion.

N	0	1	2	3
γ	1.1575 ± 0.0060	1.2355 ± 0.0050	1.3110 ± 0.0070	1.3820 ± 0.0090
ν	0.5875 ± 0.0025	0.6290 ± 0.0025	0.6680 ± 0.0035	0.7045 ± 0.0055
η	0.0300 ± 0.0050	0.0360 ± 0.0050	0.0380 ± 0.0050	0.0375 ± 0.0045
β	0.3025 ± 0.0025	0.3257 ± 0.0025	0.3465 ± 0.0035	0.3655 ± 0.0035

7 Stability of renormalization group fixed points and decay of correlations

The universal properties of a large class of macroscopic phase transitions with short range interactions can be described by statistical field theories involving scalar fields ϕ interacting through a ϕ^4-like quartic interaction.

The simplest critical systems have an $O(N)$ orthogonal symmetry and, therefore, the corresponding field theory has only one quartic interaction. The renormalization group (RG) flow involves only one ϕ^4-like coupling constant. Since, in addition, these field theories have many applications, they have been the most studied.

However, other systems do exist that are not $O(N)$ symmetric. A first class corresponds to Hamiltonians with several independent quadratic terms and, therefore, several correlation lengths. Then, generically, the correlation lengths diverge for different values of the temperature. When one diverges, only some components of the field are critical: the other components decouple and can be integrated out.

One can thus restrict the analysis to systems that are G symmetric, where G is a subgroup of the $O(N)$ group, and G has only one invariant quadratic polynomial but several independent invariant quartic polynomials.

We discuss here these more general physical systems from the RG viewpoint and within the framework of the $\varepsilon = 4 - d$ expansion. The flow of quartic interactions is then more complicated than in the $O(N)$ case and, in general, several fixed points are found. We discuss their relative stability and observe examples of RG-induced *emergent symmetries* of the critical theory, that is, symmetries larger than the symmetry of the initial Hamiltonian.

Moreover, we exhibit an intriguing relation between the stability of fixed points and the corresponding decay of the fixed point critical correlation functions that has been verified in many examples: *the stablest fixed point, which is unique, corresponds to the fastest decay of correlations* [104, 1, 107]. However, a rigorous general proof of the relation is still lacking.

7.1 Models with only one correlation length

The generic Hamiltonians that generate only one correlation length are invariant under a symmetry group G acting on the field, which is a subgroup of the $O(N)$ group and which admits only one quadratic invariant. Moreover, the field must transform under an irreducible representation of the group G.

Therefore, the two-point correlation function in the disordered phase, in component space, is proportional to the identity matrix. Denoting by ϕ_α, $\alpha = 1, \ldots, N$, the N components of the field, we can express this condition as ($\langle \bullet \rangle$ denotes expectation value)

$$\langle \phi_\alpha(x)\phi_\beta(y) \rangle = \frac{1}{N}\delta_{\alpha\beta} \sum_{\gamma=1}^{N} \langle \phi_\gamma(x)\phi_\gamma(y) \rangle . \tag{7.1}$$

From Random Walks to Random Matrices. Jean Zinn-Justin, Oxford University Press (2019).
© Jean Zinn-Justin. DOI: 10.1093/oso/9780198787754.001.0001

Moreover, we assume that the group G contains the reflection group $\mathbb{Z}_2$, $\phi \mapsto -\phi$ as a subgroup and *admits several, linearly independent, quartic invariant monomials* in ϕ, as the example of the cubic anisotropy will illustrate.

For this class of models, the effective Hamiltonians thus have the same quadratic terms as the $O(N)$ symmetric Hamiltonian, but differ by the quartic contributions: they contain several independent terms of $\int \mathrm{d}^d x \, \phi^4(x)$ type, one of them always being $O(N)$ symmetric:

$$\mathcal{H}(\phi) = \int \mathrm{d}^d x \left[\tfrac{1}{2}\left(\nabla_x \phi(x)\right)^2 + \tfrac{1}{2} r \phi^2(x) + \frac{1}{4!} \sum_a g_a V_a\left(\phi(x)\right) \right],$$

where the parameter r plays the role of the temperature near T_c and the monomials $V_a(\phi)$ are quartic in the field ϕ,

$$V_a(\lambda \phi) = \lambda^4 V_a(\phi),$$

derivative-free and linearly independent.

The $O(N)$ symmetric N-vector model. The critical properties of the simplest class of continuous phase transitions are described by an N-component field ϕ and the $O(N)$ symmetric $(\phi^2)^2$ statistical field theory, corresponding to the Hamiltonian (regularized at short distance or large momenta Λ)

$$\mathcal{H}(\phi) = \int \mathrm{d}^d x \left[\tfrac{1}{2}\left(\nabla_x \phi(x)\right)^2 + \tfrac{1}{2} r \phi^2(x) + \tfrac{1}{4!} u \left(\phi^2(x)\right)^2 \right].$$

For dimensions $d \geq 4$, the Gaussian fixed point $u = 0$ is stable and the critical behaviour can be described by mean field theory. By contrast, it is unstable for $d < 4$.

Following Wilson and Fisher [79], the RG equations can then be discussed analytically in $d = 4 - \varepsilon$ dimensions for ε infinitesimal. *Below four dimensions a non-Gaussian infrared fixed point $(u = u^* > 0)$ is found, which is stable against the Gaussian fixed point $(u = 0)$.* Universal quantities can then be calculated in the form of ε-expansions.

The field dimension. Parametrizing, as usual, the field dimension as $2d_\phi = d - 2 + \eta$, in such a way that the exponent η vanishes for the Gaussian fixed point, one derives from the spectral representation of the two-point function that η is *strictly positive for the non-Gaussian fixed point.* As a consequence, for a fixed point such that $u^* > 0$, *correlations decay at large distance faster than in the Gaussian theory.*

7.2   Cubic anisotropy, a model with two couplings

As an example, we first consider an N-component field ϕ_α, $\alpha = 1, \ldots, N$ and a Hamiltonian invariant under the hypercubic group, the finite group of transformations generated by

$$\phi_\alpha \mapsto -\phi_\alpha, \quad \phi_\alpha \leftrightarrow \phi_\beta \quad \text{for all } \alpha \text{ and } \beta.$$

The hypercubic symmetry group admits a *unique quadratic invariant, which ensures that all components of ϕ are critical simultaneously,* but *two independent quartic invariants.* The critical Hamiltonian $(T = T_c)$ of the symmetric ϕ^4-like theory has the general form [108],

$$\mathcal{H}_{\text{cubic}}(\phi)$$

$$= \int \mathrm{d}^d x \left\{ \tfrac{1}{2} \sum_\alpha \left[(\nabla_x \phi_\alpha(x))^2 + r_c \phi_\alpha^2(x) \right] + \frac{u}{24} \left(\sum_\alpha \phi_\alpha^2(x) \right)^2 + \frac{v}{24} \sum_\alpha \phi_\alpha^4(x) \right\}.$$

We parametrize the quartic interactions as $u = g\Lambda^{4-d}$, $v = h\Lambda^{4-d}$, where Λ is the cut-off and g, h are dimensionless.

Positivity. The two constants g, h must satisfy the two conditions $g + h \geq 0$, $Ng + h \geq 0$, to ensure that the Hamiltonian is positive for $\phi \to \infty$ and, thus, that the transition is second order. The first condition is obtained by choosing all ϕ_α zero but one, the second by taking them all equal.

7.2.1   RG and fixed points

In terms of a scale parameter λ, the two effective couplings at scale λ satisfy RG flow equations of the general form

$$\lambda \frac{\mathrm{d}g}{\mathrm{d}\lambda} = -\beta_g(g(\lambda), h(\lambda)), \quad \lambda \frac{\mathrm{d}h}{\mathrm{d}\lambda} = -\beta_h(g(\lambda), h(\lambda)), \ g(1) = g, \ h(1) = h. \quad (7.2)$$

A perturbative one-loop (in the sense of Feynman diagrams) calculation determines the two β-functions at leading order for $g, h = O(\varepsilon)$, $\varepsilon = 4 - d \to 0_+$. One obtains

$$\beta_g(g, h) = -\varepsilon g + \frac{1}{8\pi^2} \left(\frac{N+8}{6} g^2 + gh \right),$$

$$\beta_h(g, h) = -\varepsilon h + \frac{1}{8\pi^2} \left(2gh + \frac{3}{2} h^2 \right).$$

One looks for fixed points, solutions of $\beta_g = \beta_h = 0$ for $\varepsilon = 4 - d > 0$ and discusses their stability for $\lambda \to \infty$ as a function of the integer N. It is instructive to determine the symmetry of the Hamiltonians corresponding to the various fixed points, in particular, to the stable fixed point.

Solving the flow equations (7.2), one then studies the trajectories of the parameters g and h when the dilatation parameter λ varies.

One also tries to determine the nature of the phase transition as a function of the initial values of g and h.

It is simple to study such a two-dimensional flow because, at this order, all lines joining the origin to fixed points are RG invariant and cannot be crossed except at fixed points.

RG fixed points

One finds the four fixed points:

(i) The Gaussian fixed point: $g = h = 0$.

(ii) The decoupled fixed point: $g = 0$, $h = 16\varepsilon\pi^2/3$, which corresponds to N identical and decoupled field theories with a $\mathbb{Z}_2$ reflection symmetry (the Ising model class).

(iii) The isotropic or $O(N)$ symmetric fixed point: $h = 0$, $g = 48\varepsilon\pi^2/(N+8)$, (which is always present) and leads, when it is stable, to a *dynamically generated* $O(N)$ symmetry (an *emergent symmetry*), more extended than the hypercubic symmetry of the initial Hamiltonian.

(iv) Finally, a new fixed point, the *cubic fixed point*, is found with

$$g = \frac{16\pi^2\varepsilon}{N}, \qquad h = \frac{16\pi^2(N-4)\varepsilon}{3N}.$$

All fixed points belong to the half-plane $g \geq 0$. Only the cubic fixed point for $N < 4$ is such that $h < 0$. However, for $N \geq 1$, it satisfies the positivity condition $g + h \geq 0$ (and thus also $Ng + h \geq 0$).

7.2.2 Linearized flow and eigenvalues

The local stability of the four fixed points is determined by the eigenvalues of the matrix of partial derivatives, with respect to g and h, of the functions $-\beta_g, -\beta_h$, (with this convention, positive eigenvalue means instability):

$$
\begin{aligned}
\text{Gaussian fixed point:} \quad & \varepsilon\,, & \varepsilon\,, \\
\text{Decoupled (Ising-like) fixed point:} \quad & \tfrac{1}{3}\varepsilon\,, & -\varepsilon\,, \\
\text{Isotropic } O(N) \text{ symmetric fixed point:} \quad & \frac{N-4}{N+8}\varepsilon\,, & -\varepsilon\,, \\
\text{Cubic fixed point:} \quad & \frac{4-N}{3N}\varepsilon\,, & -\varepsilon\,.
\end{aligned}
$$

Both the Gaussian and the decoupled fixed points are always unstable.

The isotropic fixed point is stable for $N < N_c$, with $N_c = 4 + O(\varepsilon)$, a special example of a more general result concerning the $O(N)$ symmetric model, of *emergent symmetry* [104, 1]. Finally, the cubic fixed point is stable for $N > N_c$. At $N = N_c$ the two fixed points merge and then exchange roles. The general structure of the RG flow is displayed in Fig. 7.1.

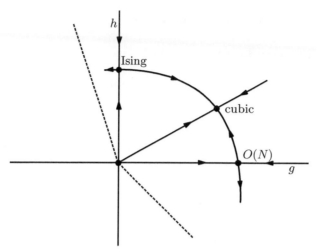

Fig. 7.1 Cubic anisotropy: RG flow for $N > 4$ in the (g, h) plane: The dotted lines correspond to stability limits.

7.2.3 Corresponding values of the exponent η

At order ε^2, the corresponding values of the exponent η are:

$$\text{Gaussian fixed point:} \quad \eta = 0 \,,$$

$$\text{Decoupled (Ising) fixed point:} \quad \eta_{\text{Is.}} = \frac{\varepsilon^2}{54} \,,$$

$$\text{Isotropic fixed point:} \quad \eta_{O(N)} = \frac{(N+2)\varepsilon^2}{2(N+8)^2} \,,$$

$$\text{Cubic fixed point:} \quad \eta_{\text{cub.}} = \frac{(N+2)(N-1)\varepsilon^2}{54N^2} \,.$$

It is then instructive to compare the values of η for the three non-trivial fixed points. The differences are:

$$\eta_{\text{cub.}} - \eta_{\text{Is.}} = \frac{N-2}{54N^2}\varepsilon^2,$$

$$\eta_{O(N)} - \eta_{\text{Is.}} = \frac{(N-1)(10-N)}{54(N+8)^2}\varepsilon^2,$$

$$\eta_{\text{cub.}} - \eta_{O(N)} = \frac{(N+2)(N-4)^3}{54N^2(N+8)^2}\varepsilon^2.$$

One notes that the largest value of η always corresponds to the stable fixed point.

Such a property was first noticed by Brézin, Le Guillou and Zinn-Justin. It was *discussed in the generic situation with four fixed points* in Ref. [104]. The issue was re-examined more systematically later in Refs. [1, 107].

7.3 General quartic Hamiltonian: RG functions

We now consider general Hamiltonians which depend on a field with N components ϕ_i and which have a symmetry such the *quadratic invariant in the field is unique*. A general critical Hamiltonian satisfying the assumption can be written as (up to regularization terms)

$$\mathcal{H}(\phi) = \int \mathrm{d}^d x \left\{ \frac{1}{2} \sum_{i=1}^{N} \left[(\nabla_x \phi_i(x))^2 + r_c \phi_i^2(x) \right] \right.$$
$$\left. + \frac{1}{4!} \Lambda^{4-d} \sum_{i,j,k,l=1}^{N} g_{ijkl}\, \phi_i(x)\phi_j(x)\phi_k(x)\phi_l(x) \right\}, \tag{7.3}$$

where g_{ijkl} is a tensor symmetric in its four indices. The condition that the Hamiltonian is bounded from below implies

$$\sum_{i,j,k,l=1}^{N} g_{ijkl}\, \phi_i\phi_j\phi_k\phi_l > 0 \quad \forall\, |\phi| = 1. \tag{7.4}$$

The condition (7.4) defines a structure of convex cone:
 (i) if g_{ijkl} satisfies the condition (7.4), sg_{ijkl} with $s > 0$ satisfies it;
 (ii) if g^1_{ijkl} and g^2_{ijkl} satisfy the condition (7.4), then $sg^1_{ijkl} + (1-s)g^2_{ijkl}$ satisfies condition (7.4) for $0 \le s \le 1$.

The two-point function. With the assumption that the quadratic invariant is unique, the two-point correlation function in the disordered phase is diagonal:

$$\langle \phi_i(x)\phi_j(y) \rangle_{\mathrm{conn.}} \equiv W^{(2)}_{ij}(x-y) = \delta_{ij} W^{(2)}(x-y). \tag{7.5}$$

This form ensures that the system has only one correlation length and that *all components ϕ_i of the field become critical simultaneously*. Otherwise, in a generic situation, the non-critical components would decouple at the transition and could thus be omitted.

The diagonal property (7.5) of the two-point function in the disordered phase implies that the tensor g_{ijkl} satisfies successive constraints in the perturbative expansion. For example, the first three orders yield

$$\begin{cases} \sum_k g_{ijkk} = \dfrac{\delta_{ij}}{N} \sum_{k,\ell} g_{kk\ell\ell}, \\[2mm] \sum_{k,l,m} g_{iklm}g_{jklm} = \dfrac{\delta_{ij}}{N} \sum_{k,l,m,n} g_{klmn}g_{klmn}, \\[2mm] \sum_{k,l,m,n,p} g_{iklm}g_{lmnp}g_{npkj} = \dfrac{\delta_{ij}}{N} \sum_{i,j,k,l,m,n} g_{qklm}g_{lmnp}g_{npkq}. \end{cases} \tag{7.6}$$

The RG β-functions. In the minimal subtraction scheme, the RG β-functions are given, up to order ε^3 (when g is assumed of order ε), by

$$\beta_{ijkl}(g) = -\varepsilon g_{ijkl} + \frac{N_d}{2} \sum_{m,n} (g_{ijmn}g_{mnkl} + g_{ikmn}g_{mnjl} + g_{ilmn}g_{mnkj}) \cdot$$

$$- \frac{N_d^2}{4} \sum_{m,n,p,q} (g_{ijmn}g_{mpqk}g_{npql} + 5 \text{ terms})$$

$$+ \frac{N_d^2}{48} \sum_{m,n,p,q} (g_{ijkm}g_{mnpq}g_{npql} + 3 \text{ terms}) + O(g^4), \tag{7.7}$$

where $N_d = 2/(4\pi)^{d/2}\Gamma(d/2)$ is the usual loop factor and where the additional terms are obtained by symmetrization over (i,j,k,l).

At order g^3, the field dimension can be inferred from the function

$$\eta(g) = \frac{N_d^2}{24N} \sum_{i,j,k,l} g_{ijkl}g_{ijkl} - \frac{N_d^3}{32N} \sum_{i,j,k,l,m,n} g_{ijkl}g_{klmn}g_{mnij} + O(g^4),$$

where the constraints (7.6) have been used.

For g small, the sign of the leading (quadratic) term is consistent with the general result $\eta(g) \geq 0$.

7.4 Running coupling constants and gradient flows

Gradient flow: Definition. We consider a flow or RG equation for a number of ϕ^4 like coupling constants g_a of the form,

$$\lambda \frac{dg_a}{d\lambda} = -\beta_a(g(\lambda)). \tag{7.8}$$

We assume that the β-functions can be written as

$$\beta_a(g) = \sum_b T_{ab}(g) \frac{\partial U(g)}{\partial g_b}, \tag{7.9}$$

where the matrix **T** with elements T_{ab} and the potential U are regular functions of the g_a, at least in some neighbourhood of vanishing couplings, and the matrix **T** is symmetric and positive.

Then, equation (7.8) defines a *gradient flow.*

Note that the general form (7.9) of the β-function is the only form of a gradient flow consistent with the transformation properties under reparametrization in the space of the coefficients g_a (diffeomorphisms). In fact, it is easy to verify that T_{ab} transforms under reparametrization as an inverse metric tensor.

7.4.1 The gradient property of the RG β-functions
It is simple to verify from the general expression (7.7) that the β-function, at this order, derives from a potential (Wallace and Zia [109]).

Explicitly, one finds

$$\beta_{ijkl}(g) = \frac{\partial U(g)}{\partial g_{ijkl}}$$

with

$$U(g) = -\frac{\varepsilon}{2} \sum_{i,j,k,l} g_{ijkl} g_{ijkl} + \frac{N_d}{2} \sum_{i,j,k,l,m,n} g_{ijkl} g_{klmn} g_{mnij}$$

$$-\frac{3N_d^2}{8} \sum_{i,j,k,l,m,n,p,q} g_{ijmn} g_{mpqk} g_{npql} g_{ijkl}$$

$$+\frac{N_d^2}{48} \sum_{i,j,kl,m,n,p,q} g_{ijkl} g_{ijkm} g_{mnpq} g_{npql} + O(g^5).$$

More generally, identifying the index a with the multi-index $\{ijkl\}$, it is possible to verify up to order g^5 (*i.e.*, for all known orders) that the β-functions of the general ϕ^4 models (7.3) can be written in the form (7.9).

7.4.2 *A few consequences*

The property of gradient flow has several consequences:

(i) Because T_{ab} is a positive matrix, *fixed points are extrema of the potential*:

$$\beta_a(g^*) = 0 \iff \frac{\partial U(g^*)}{\partial g_a} = 0.$$

Moreover, the potential decreases along an RG trajectory.

(ii) The *eigenvalues* of the matrix of first order partial derivatives of β at a fixed point are *real*.

(iii) Stable fixed points g^* are local minima of the potential, that is, the *matrix of second derivatives*

$$\frac{\partial^2 U(g^*)}{\partial g_a \partial g_b}$$

is *positive*.

7.5 Fixed point stability and value of the potential

Within the framework of the ε-expansion, one can prove two consequences of the property of gradient flow: *there exists at most one stable fixed point and the stable fixed point corresponds to the lowest value of the potential.*

Indeed, let us assume the existence of two fixed points corresponding to the parameters g^* and g'^*. We then consider the parameters g (admissible due to the convexity property) of the form

$$g(s) = sg^* + (1-s)g'^*, \quad 0 \le s \le 1,$$

and the corresponding potential

$$u(s) = U\big(g(s)\big).$$

As the explicit form shows, *at leading order, $u(s)$ is a third degree polynomial* in s.

7.5.1 First derivative

Due to the fixed point conditions at $s = 0$ and $s = 1$, the derivative

$$u'(s) = \sum_a g_a'(s) \frac{\partial U}{\partial g_a} = \sum_a (g_a^* - g_a'^*) \frac{\partial U}{\partial g_a}$$

vanishes at $s = 0$ and $s = 1$:

$$u'(0) = u'(1) = 0.$$

Therefore, since $u'(s)$ is a second degree polynomial, it has necessarily the form

$$u'(s) = As(1 - s).$$

7.5.2 Second derivative

The second derivative $u''(s)$ is given in terms of the matrix of second partial derivatives of U and, thus, the partial derivatives of the β-functions, by

$$u''(s) = \sum_{a,b} (g_a^* - g_a'^*) \frac{\partial^2 U(g(s))}{\partial g_a \partial g_b} (g_b^* - g_b'^*) = A(1 - 2s).$$

In particular, for $s = 0$ and $s = 1$,

$$A = \sum_{a,b} (g_a^* - g_a'^*) \frac{\partial^2 U(g'^*)}{\partial g_a \partial g_b} (g_b^* - g_b'^*), \quad -A = \sum_{a,b} (g_a^* - g_a'^*) \frac{\partial^2 U(g^*)}{\partial g_a \partial g_b} (g_b^* - g_b'^*).$$

At a stable fixed point, the matrix $\mathbf{U}''$ of partial second derivatives of U is positive. Thus, if g^* and g'^* are stable fixed points, A and $-A$ are both given by the *expectation value of a positive matrix* and are both positive, which is contradictory: *both fixed points cannot be simultaneously stable.*

More generally, the sign of A characterizes, in some sense, the relative stability of these two fixed points.

Let us assume, for example, $A < 0$, which is consistent with the assumption that g^* is stable. Then $u'(s) < 0$ in $[0, 1]$, and $U(g(s))$ is a decreasing function. Thus,

$$U(g^*) < U(g'^*).$$

In particular, *if g^* is a stable fixed point, it corresponds, among all fixed points, to the lowest value of the potential and thus it is unique* [1, 110] (for a recent analysis of fixed points in $d = 4 - \varepsilon$, $d = 3, 4, 6$, see Ref. [111]).

7.6 Fixed point stability and field dimension

For any fixed point g^*, the equation $\beta = 0$ implies, at leading order,

$$\left.\frac{\mathrm{d}U(\lambda g^*)}{\mathrm{d}\lambda}\right|_{\lambda=1} = 0 \;\Rightarrow\; \varepsilon \sum_{i,j,k,l} g^*_{ijkl} g^*_{ijkl} = \frac{3N_d}{2} \sum_{i,j,k,l,m,n} g^*_{ijkl} g^*_{klmn} g^*_{mnij}$$

and, thus,

$$U(g^*) = -\tfrac{1}{6}\varepsilon \sum_{i,j,k,l} g^*_{ijkl} g^*_{ijkl} + O(g^4),$$

a negative value and thus lower than the Gaussian fixed point value:

$$g^* \neq 0 \;\Rightarrow\; U(g^*) < U(0).$$

A relation between value of η and fixed point stability then follows from

$$\eta = \frac{N_d^2}{24N} \sum_{i,j,k,l} g^*_{ijkl} g^*_{ijkl} = -\frac{N_d^2}{4N\varepsilon} U(g^*).$$

The stable fixed point corresponds to the lowest value of U. It thus corresponds also to the largest value of the exponent η and, thus, the largest value of the field scaling dimension $d_\phi = \tfrac{1}{2}(d - 2 + \eta)$. Since, at large distance, at criticality, the two-point correlation functions decays like,

$$\langle \phi(x)\phi(0)\rangle \underset{|x|\to\infty}{\propto} 1/|x|^{2d_\phi},$$

the stable fixed point corresponds to the *fastest large distance decay of critical correlation functions*.

Beyond the ε-expansion. Vicari and Zinn-Justin [107] have investigated the results known beyond the ε-expansion. They have examined a number of numerical and exact results in three and two dimensions. All are all consistent with the conjecture, provided one adds the condition that the *fixed points that are compared must be related by RG flows.*

8 Quantum field theory: An effective theory

Effective field theory, a concept directly inspired from the theory of *critical phenomena* in statistical physics and based on *renormalization group* (RG) ideas (Section 6.3), is the modern framework of local quantum field theory (QFT). The terms appearing in the action are the first, most important terms of a large distance or low momentum expansion of an underlying more fundamental theory that, ultimately, will solve the unavoidable problem of short distance singularities.

The basic idea behind the notion of *effective field theory* is the following: one starts from a microscopic model involving an *infinite number of fluctuating degrees of freedom* whose interactions are characterized by a microscopic scale and in which, as a result of interactions, a length (called *correlation length* in statistical physics) much larger than the microscopic scale, or, equivalently, a mass much smaller than the characteristic mass scale of the initial model, is generated.

An outstanding example of such a situation is provided by statistical systems with *short range interactions* near the critical temperature of a continuous phase transition. These systems then exhibit a non-trivial *large distance* physics which has *universal properties*, properties independent, to a large extent, of the initial short distance structure.

To explain *universality*, Kadanoff [77] and Wilson [24]) have proposed constructing a sequence of *effective Hamiltonians* (or configuration energies or imaginary time actions) by summing iteratively over the degrees of freedom associated with the shortest distance. The transformation, an RG transformation, generates Hamiltonians describing all the same large distance physics.

If, for a whole class of initial Hamiltonians (a *universality class*), this sequence converges towards the same fixed point Hamiltonian, universality of the large scale physics can be explained.

For a large family of statistical models, the iterations eventually converge towards a *local statistical field theory* (analogous to a QFT in imaginary time).

Wilson [24] then proposed taking directly as an initial microscopic theory, an effective field theory that belongs to the same *universality class*, that is, that has the same large distance physics.

One calls such a field theory 'effective', to emphasize that it is not a microscopic model, but only a theory that reproduces accurately the *large distance properties*.

In contrast with the renormalizable field theories considered in particle physics, the effective theory contains all local monomials consistent with field content and symmetries: local monomials are space integral of powers of the field and its derivatives taken at the same point.

The Gaussian fixed point. An RG fixed point can be identified, the *Gaussian fixed point*, which has the form of a free massless scalar field theory, in the QFT framework.

From Random Walks to Random Matrices. Jean Zinn-Justin, Oxford University Press (2019).
© Jean Zinn-Justin. DOI: 10.1093/oso/9780198787754.001.0001

The RG flow near the fixed point is governed by a linear operator whose eigenvectors and eigenvalues characterize the local stability (Section 1.6).

Eigenvectors take the form of local monomials, which can be classified into relevant (direction of instabilities), marginal (which require a second order study and, in general, generate logarithmic flows) and irrelevant (corresponding to stability directions). It is simple to verrify that the irrelevant, marginal and relevant perturbations correspond to non-renormalizable, renormalizable and super-renormalizable interactions and mass terms, respectively, in the QFT terminology.

Therefore, in some neighbourhood of the Gaussian fixed point, when studying the RG evolution of the parameters of the action between the microscopic scale and the physical scale, when the ratio of these scales is small enough, one can omit the non-renormalizable interactions because they lead to very small corrections: the effective field theory can be approximated by a renormalizable, or super-renormalizable QFT.

The properties of correlation functions at the physical scale can then be inferred from RG arguments related to the *renormalization process*, which makes it possible to define renormalized functions that are asymptotically independent, at least perturbatively, from the microscopic structure (see also Chapter 4).

The QFT describing particle physics, called the Standard Model, is an incomplete theory due to the presence of ultraviolet (UV) infinities and requires a short distance modification (a regularization). It is renormalizable and makes it possible to define a finite theory at the physical scale by the renormalization process.

The example of statistical physics suggests then that the Standard Model is an effective theory, although the fundamental microscopic theory remains, up to now, a matter of speculations. Because the microscopic scale seems to be too small compared to the physical scale, the non-renormalizable interactions have not been observed yet.

Since all quantum or statistical field theories encountered in physics are effective in this sense, the denomination *effective field theory* is a kind of pleonasm and is sometimes restricted to field theories that are low energy approximations to other, more fundamental, known field theories when some masses are much larger than the energy or momentum scale one is probing. The large energy scale can also be provided by the temperature in the high temperature regime (see Chapter 21).

Conventions. In this chapter, compared to quantum evolution, we use an imaginary time formalism and, in applications to the theory of fundamental microscopic interactions (particle physics), we set the speed of light $c = 1$ and the Planck's constant $\hbar = 1$.

8.1 Effective local field theory: The scalar field

We consider now an effective local statistical (or quantum) field theory with a Hamiltonian or Euclidean action of QFT, which describes the large distance properties of some initial statistical model.

For simplicity, we consider below only systems that can be described by a one-component scalar field $\phi(x)$ and have a $\mathbb{Z}_2$ reflection (Ising-like) symmetry, although the generalization to, for example, $O(N)$ symmetric models is straightforward.

The Hamiltonian $\mathcal{S}(\phi)$ thus satisfies $\mathcal{S}(\phi) = \mathcal{S}(-\phi)$. We also assume space translation invariance.

The partition function can be expressed as an integral over fields $\phi(x)$ of the form,

$$\mathcal{Z} = \int [\mathrm{d}\phi]\, \mathrm{e}^{-\mathcal{S}(\phi)}, \tag{8.1}$$

where the action $\mathcal{S}(\phi)$ is *local* because the initial interactions are short range: a local theory is defined by the property that the action $\mathcal{S}(\phi)$ is a space integral of a *local* function of a field and its derivatives (*i.e*, all taken at the same point). Moreover, the action is a regular function of the parameters of the initial theory.

Physical observables can then be derived from correlation functions, expectation values of the form

$$\langle \phi(x_1)\phi_2(x_2)\cdots\phi_n(x_n)\rangle = \frac{1}{\mathcal{Z}} \int [\mathrm{d}\phi]\, \mathrm{e}^{-\mathcal{S}(\phi)}\, \phi(x_1)\phi_2(x_2)\cdots\phi_n(x_n). \tag{8.2}$$

General local field theory. A general local field theory in d Euclidean space dimensions can also be defined in terms of the Fourier components $\tilde{\phi}$ of the field

$$\phi(x) = \int \mathrm{d}^d k\, \mathrm{e}^{ikx}\, \tilde{\phi}(k).$$

The action can then be expanded in powers of $\tilde{\phi}(k)$ with contributions of the form

$$\frac{1}{2n!} \int \left(\prod_i \mathrm{d}^d k_i\, \tilde{\phi}(k_i) \right) \delta^{(d)}(k_1 + k_2 + \cdots + k_{2n}) \mathcal{S}_{2n}(k_1, k_2, \ldots, k_{2n}),$$

where the functions $\mathcal{S}_{2n}$ are analytic near zero momentum (all $k_i = 0$).

Functional RG equations. In this general set-up, the RG equations based on the integration over field components with large momentum soft shells, analogous to short distance summation in the spirit of Kadanoff and Wilson [77, 24], take the form of the quadratic functional equations (9.9) [100, 101]. However, these equations can only be solved, after rather drastic approximations, by numerical methods.

Fortunately, in some neighbourhood of the Gaussian fixed point, the equations can be simplified and lead to the perturbative RG of QFT.

8.2 Perturbative assumption and Gaussian renormalization

We now assume that the deviations from a free massless theory (the RG Gaussian fixed point), with action

$$\mathcal{S}^*(\phi) = \tfrac{1}{2} \int \mathrm{d}^d x \left(\nabla_x \phi(x) \right)^2, \tag{8.3}$$

are in some qualitative sense small (a notion that is defined empirically by the relevance of perturbation theory).

In this perturbative framework, one decomposes the local action into the sum of a quadratic part $\mathcal{S}(\phi)$ and interactions $\mathcal{V}_I(\phi)$,

$$\mathcal{S}(\phi) = \mathcal{S}_G(\phi) + \mathcal{V}_I(\phi). \tag{8.4}$$

The quadratic part $\mathcal{S}_G(\phi)$ is a general local quadratic action of the form,

$$\mathcal{S}_G(\phi) = \mathcal{S}^*(\phi) + \tfrac{1}{2}u_0 \int \mathrm{d}^d x\, \phi^2(x) + \tfrac{1}{2}\sum_{k=1} u_{k+1} \int \mathrm{d}^d x\, \left[\nabla_x \phi(x)(-\nabla_x^2)^k \nabla_x \phi(x)\right]$$

and $\mathcal{V}_I(\phi)$ is the space integral of a general, expandable, even function of the field and its derivatives, with terms at least of degree 4 in ϕ,

$$\mathcal{V}_I(\phi) = \int \mathrm{d}^d x\, V_I[\phi(x), \partial_\mu \phi(x), \ldots].$$

One then expands the field integrals (8.1, 8.2) in powers of the interaction $\mathcal{V}_I(\phi)$.

The action $\mathcal{S}_G(\phi)$ has been split into the sum of the action for a free field and a number of terms quadratic in the field with a suitable number of derivatives. These terms are required to render finite all terms in the perturbative expansion by removing all large momentum (or short distance) divergences, an operation called *regularization*.

The introduction of the regularization terms is an *ad hoc* procedure to replace the microscopic structure of the initial physical theory, either because it is too complicated or because it is unknown. The renormalization theory and the corresponding RG equations are then essential to prove that the large scale physics predictions are regularization independent, a form of short distance insensitivity.

Technically, the quadratic terms with higher order derivatives have to be chosen in such a way that the fields that contribute to the field integral are sufficiently regular for the Gaussian expectation values of all local monomials contributing to the action to be finite.

Finally, a generic local scalar field theory does not generate automatically a large correlation length, or a small mass. One parameter of the local action, which then plays the role of the temperature, has to be *tuned* to a critical value (analogous to the critical temperature), a problematic fine tuning from the viewpoint of particle physics. In the perturbative framework, one chooses the coefficient u_0 of ϕ^2.

8.2.1 *Gaussian renormalization and dimensional analysis*

The action (8.4) defines implicitly a microscopic scale. It is not the true microscopic scale; it is the shortest scale at which the local field theory can be used. However, from now on we call it the microscopic scale, and the theory with the action (8.4), the effective microscopic theory.

We also assume that, generically, in the absence of any additional information, the coefficients of the local expansion are numbers of order 1.

Large distance is then defined as a distance much larger than this microscopic scale. To describe only large distance physics, it is convenient to take the physical scale as a reference scale. We thus rescale distances $x \mapsto x'$ with

$$x = x'/a, \quad a \ll 1. \tag{8.5}$$

The initial microscopic scale is then characterized by the parameter a (related, for example, to the spacing of an initial lattice model) and, rather than studying physics at large distance or small momentum, one now studies the $a \to 0$ limit.

At this point, it is convenient to assume that the microscopic scale had initially no attached dimension (*i.e.,* for a lattice, the spacing was 1) and that a carries the dimension of a length. Then, 'large distance' means large compared to the microscopic scale a.

The quantity $\Lambda = 1/a$ has a momentum dimension (due to the convention $\hbar = 1$) and can be identified with the *cut-off* scale (in the QFT terminology), the scale at which the momentum integrals, in the Fourier representation of the perturbative expansion, must be cut. Indeed, the local expansion now breaks down below the scale a at which non-localities that render the initial theory finite may appear.

After the rescaling, the leading terms in the effective action are those with the smallest number of derivatives. For $\phi(x)$ small, the leading term is $\phi^2(x)$, which implies generically a correlation length of order a or, equivalently, a physical mass of order Λ. To generate a large correlation length or small mass, one must first tune the coefficient of ϕ^2 in the action in such a way that the kinetic $(\nabla_x \phi(x))^2$ term, which alone leads to a *critical or massless free theory*, becomes the leading term.

To characterize the relative size of the additional terms in the Hamiltonian (imaginary time or Euclidean action) with respect to the Gaussian fixed point, one then renormalizes the field, $\phi \mapsto \phi'$, to cancel the a dependence of the leading term corresponding to the Gaussian fixed point,

$$\int \mathrm{d}^d x \left(\nabla_x \phi(x)\right)^2 \mapsto a^{2-d} \int \mathrm{d}^d x \left(\nabla_x \phi(x)\right)^2 = \int \mathrm{d}^d x \left(\nabla_x \phi'(x)\right)^2$$

with

$$\phi'(x) = a^{(2-d)/2} \phi(x). \tag{8.6}$$

The combined rescaling and field renormalization can be called *Gaussian renormalization*. After this renormalization, the field has momentum or mass dimension $\frac{1}{2}(d-2)$.

In the new normalization, in the action (8.4) the quadratic part becomes (now omitting primes)

$$\mathcal{S}_\mathrm{G}(\phi) = \tfrac{1}{2} \int \mathrm{d}^d x \left(\nabla_x \phi(x)\right)^2 + \frac{1}{2a^2} u_0 \int \mathrm{d}^d x \, \phi^2(x)$$

$$+ \tfrac{1}{2} \int \mathrm{d}^d x \sum_{k=1} u_{k+1} a^{2k} \nabla_x \phi(x) (-\nabla_x^2)^k \nabla_x \phi(x). \tag{8.7}$$

More generally, a monomial $V_{n,k}$ contributing to $V(\phi)$, which involves n powers of ϕ and $2k$ derivatives acting in an unspecified way on the fields ϕ, is transformed into

$$V_{n,k}(\phi) \mapsto a^{2k+n(d-2)/2-d}V_{n,k}(\phi). \tag{8.8}$$

One recognizes in the power of the length a the dimension of the interaction vertex in the *power counting analysis of renormalization theory* [18].

8.2.2 Classification of interactions and the fine tuning problem

Quantities now have dimensions characterized by powers of the length a. Momenta acquire a dimension -1, position coordinates dimension 1, the field ϕ has dimension $\frac{1}{2}(2-d)$ and, more generally, all local monomials in the action have a dimension in units of a given by equation (8.8).

This dimensional analysis has direct implications for the role that the different terms in the action play, which can also be formulated in terms of the *stability with respect to local perturbations of the free massless theory*, which is also the mean field critical theory in the terminology of phase transitions and corresponds to the Gaussian fixed point from the point of view of *RG*.

A general observation is that the contributions to the action that are considered the most innocuous from the viewpoint of power counting in standard renormalization theory, now grow when $\Lambda = 1/a$ increases, while the 'dangerous ones', corresponding to *non-renormalizable interactions*, are the most suppressed.

The *marginal* situation corresponds to strictly *renormalizable interactions*, which are dimensionless, and whose RG flow with Λ is generally logarithmic with a behaviour which, even close to the free field theory, cannot be predicted by a leading order analysis but requires perturbative calculations and solving RG equations.

Mass term and fine tuning problem. The coefficient of ϕ^2 (in a free theory, the mass term) is multiplied by $1/a^2$. This indicates that, generically, no large distance scale is generated and the correlation length remains of order a or, equivalently, the scalar particle, if it exists, has a mass of the order of Λ. The local expansion is thus unjustified. Even in the absence of interactions, in the free field theory, a physical mass m implies $u_0 = m^2/\Lambda^2$ and thus requires a fine tuning of the coefficient u_0 of ϕ^2 of the order of $(m/\Lambda)^2 \ll 1$. This is equivalent to adjusting the temperature close to the critical temperature in statistical systems.

The tuning value is modified by interactions. We denote by u_{0c} the value of u_0 (negative for a ϕ^4 field theory) for which the theory is massless (or critical because the correlation length diverges) in the full theory. In terms of the inverse two-point function $\tilde{\Gamma}^{(2)}$ in Fourier representation,

$$u_0 = u_{0c} \ \Leftrightarrow \ \tilde{\Gamma}^{(2)}(p=0) = 0\,.$$

It is then convenient to set

$$u_0 = u_{0c} + a^2 r\,, \tag{8.9}$$

where r characterizes a small deviation from the massless theory because the factor a^2 cancels the Gaussian renormalization.

8.2.3 Renormalizable field theory

The scaling of local interactions depends on the space dimension. The physically relevant dimensions are four for particle physics, and two and three for statistical systems. We omit below dimension 2, which requires a specific discussion.

Dimension 3. In dimension 3, the field has dimension $\frac{1}{2}$, and the leading interaction, ϕ^4, has dimension 2. Its coefficient is proportional to $1/a$. If only the ϕ^2 coefficient is tuned, the interaction diverges with the cut-off $1/a$. This is the situation of critical phenomena when only the temperature can be adjusted.

From the viewpoint of renormalization theory, the ϕ^4 field theory is *super-renormalizable*. It is obtained by tuning not only the ϕ^2 contribution in the action but also the amplitude of the ϕ^4 interaction chosen to be proportional to a and the amplitude of the ϕ^6 interaction chosen to vanish. Then, the resulting theory is finite in the infinite cut-off ($a \to 0$) limit.

This has to be contrasted with the viewpoint of effective field theory. Generically, by tuning only the coefficient of ϕ^4 (in addition to the coefficient of ϕ^2) to a critical value, one obtains a theory in which the interaction is dominated by the ϕ^6 term and the field theory is renormalizable. An RG analysis shows that the effective ϕ^6 amplitude has a logarithmic behaviour. This field theory with two fine tuned parameters describes a tricritical behaviour.

Dimension 4. In four dimensions, the ϕ^4 coefficient has no a dependence in this leading order analysis. The same applies to the ϕ^3 interaction in six dimensions and to the ϕ^6 interaction in three dimensions.

A more detailed study, beyond leading order, based on RG equations [103], shows that the effective interaction strength has a logarithmic behaviour when the ratio between the length a and the correlation length decreases. A one-loop calculation determines whether the effective coupling increases or decreases for small coupling.

The renormalizable field theory. Neglecting all irrelevant interactions, one obtains the action of a renormalizable field theory, which can be written as

$$\mathcal{S}(\phi) = \mathcal{S}_{\mathrm{G}}(\phi) + \frac{1}{4!} g a^{d-4} \int \mathrm{d}^d x \, \phi^4(x). \tag{8.10}$$

The action still contains irrelevant quadratic contributions which provide the necessary large momentum cut-off.

8.2.4 Non-renormalizable interactions

Non-renormalizable interactions, like the ϕ^6 interactions in four dimensions, which cannot be dealt with by renormalization theory, appear in the effective field theory framework to be quite innocuous because they are multiplied by positive powers of a. They lead to very weak interactions. However, when added as a perturbation in the ϕ^4 field theory, they generate increasing divergences (this is the reason why they have not been considered in renormalizable QFTs). These divergences are partially cancelled by the powers of a (8.8). Moreover, a study of the *renormalization of local monomials of the fields* [104, 112], also called composite operators, shows that the contributions that do not vanish for $a \to 0$, or infinite cut-off, can be cancelled by shifts (as counter-terms) of the parameters of the renormalizable part of the action.

For example, in four dimensions, an $a^2\phi^6$ perturbation added to the renormalizable action (8.10) can be decomposed into the sum

$$a^2\phi^6(x) = C_1 a^{-2}\phi^2(x) + C_2\big(\nabla_x\phi(x)\big)^2 + C_3\phi^4(x) + a^2[\phi^6]_{\text{subtr.}},$$

where the $[\phi^6]_{\text{subtr.}}$ term generates only logarithmic divergences. The three first terms can be cancelled by a shift of the coefficients of terms already present in the action and the last term generates contributions fast decreasing at large distance.

Warning. The analysis is based on perturbation theory. Its validity beyond perturbation theory relies on the assumption that the perturbative effects do not modify qualitatively power counting, in the sense that the ranking of operators is not modified. A more general discussion would require a global study of functional *RG equations*.

8.2.5 *Renormalizable field theories and RG: The example of the ϕ_4^4 field theory*

In the general effective field theory, the RG implementation based on iterative integrations of short distance degrees of freedom or, in Fourier representation, on large momentum degrees of freedom leads to the functional RG equations (9.9) [100, 101]. However, near a Gaussian fixed point, these equations can be much simplified by omitting all non-renormalizable interactions. As an example, we first consider the ϕ^4 field theory in four dimensions.

We parametrize the action (8.10) as

$$\mathcal{S}(\phi) = \int \mathrm{d}^4 x \left[\tfrac{1}{2}\big(\nabla_\Lambda\phi(x)\big)^2 + \frac{u}{2a^2}\phi^2(x) + \frac{1}{4!}g\phi^4(x) \right], \quad g > 0, \tag{8.11}$$

with

$$\nabla_\Lambda \equiv \nabla\left(1 - \alpha_1 a^2\nabla^2 + \alpha_2 a^4\nabla^4 + \cdots\right), \quad \alpha_1, \alpha_2, \ldots > 0,$$

where $a = 1/\Lambda$ is the microscopic scale inverse of the momentum cut-off.

For $u = u_c$, the critical value at which the physical mass, or inverse correlation length, vanishes, using renormalization theory one proves that the connected correlation functions in Fourier representation for $|p_i| \ll \Lambda$ satisfy the *perturbative RG equations* [103] (perturbative because they can be proven to be valid only near the Gaussian fixed point),

$$\left(\Lambda\frac{\partial}{\partial\Lambda} + \beta(g)\frac{\partial}{\partial g} + \tfrac{1}{2}\eta(g)\right)\tilde{W}^{(n)}(p_i, g, \Lambda) = 0, \tag{8.12}$$

where perturbative contributions decreasing like $(\ln\Lambda)^\ell/\Lambda^2$ have been neglected.

Note that this equation is written in terms of the parameters of the microscopic theory considered as given: it is the renormalized parameters that evolve with scale.

Equation (8.12) can be solved by the method of characteristics. One introduces a scale parameter λ and the scale-dependent functions

$$\lambda\frac{\mathrm{d}}{\mathrm{d}\lambda}Z(\lambda) = \eta\big(g(\lambda)\big), \quad Z(1) = 1, \tag{8.13}$$

$$\lambda\frac{\mathrm{d}}{\mathrm{d}\lambda}g(\lambda) = \beta\big(g(\lambda)\big), \quad g(1) = g. \tag{8.14}$$

Then, equation (8.12) implies

$$\tilde{W}^{(n)}(p_i, g, \Lambda) = Z^{n/2}(\lambda)\tilde{W}^{(n)}(p_i, g(\lambda), \lambda\Lambda).$$

Rescaling $\lambda\Lambda \mapsto \Lambda$ and using dimensional analysis, one obtains

$$\tilde{W}^{(n)}(p_i, g, \Lambda/\lambda) = \lambda^{-n}\tilde{W}^{(n)}(\lambda p_i, g, \Lambda) = Z^{n/2}(\lambda)\tilde{W}^{(n)}(p_i, g(\lambda), \Lambda).$$

Therefore, the limit $|p_i|/\Lambda \to 0$ is equivalent to $\lambda \to 0$. The function $g(\lambda)$ is the effective interaction at distance scale a/λ. The integration of equation (8.14) yields

$$\ln \lambda = \int_g^{g(\lambda)} \frac{\mathrm{d}g'}{\beta(g')}.$$

The functions $\beta(g)$ and $\eta(g)$ have perturbative expansions that, at order g^2, are

$$\beta(g) = \frac{3}{16\pi^2}g^2 + O(g^3), \quad \eta(g) = O(g^2).$$

Near $g = 0$, $\beta(g)$ is positive. If it remains positive for larger values of g, $g(\lambda)$ is an increasing function of λ. For $\lambda \to 0$, $g(\lambda)$ goes to zero like

$$\ln \lambda \sim \frac{16\pi^2}{3}\int^{g(\lambda)} \frac{\mathrm{d}g'}{g'^2} \Rightarrow g(\lambda) \sim \frac{3}{16\pi^2}\frac{1}{|\ln \lambda|}.$$

The triviality issue. In renormalization theory and in the traditional presentation of QFT, one insists in taking the infinite cut-off limit, here, equivalently, the $\lambda = 0$ limit. This leads to the *triviality issue*, since $g(\lambda)$, which is equivalent to the renormalized coupling at the momentum scale $\mu = \lambda\Lambda$, then vanishes.

This problem is solved in the framework of *effective field theory*, since one assumes only $\mu \ll \Lambda$ but not Λ infinite. One simply expects in such a situation that the effective coupling is logarithmically small. This argument also applies to quantum electrodynamics (QED), but, since the fine structure constant is very small, the cut-off could be as large as the Planck mass (in units where speed of light $c = 1$), where local QFT is unlikely to be still applicable.

RG in $d = 4 - \varepsilon$ dimension. In the framework of the continuation of the Feynman diagrams to complex values of the dimension d, the precedent analysis can be generalized to the dimension $d = 4 - \varepsilon$ with $\varepsilon > 0$ infinitesimal.

The action becomes

$$S(\phi) = \int \mathrm{d}^4 x \left[\frac{1}{2}\left(\nabla_\Lambda \phi(x)\right)^2 + \frac{u}{2a^2}\phi^2(x) + \frac{g}{4!}a^{d-4}\phi^4(x)\right]. \tag{8.15}$$

The β-function has now an additional contribution of order g generated by the Gaussian dimension of the ϕ^4 interaction:

$$\beta(g) = -\varepsilon g + \frac{3}{16\pi^2}g^2 + O(g^3, g^2\varepsilon).$$

The β-function has the trivial zero $g = 0$ corresponding to the IR repulsive Gaussian fixed point and a non-trivial zero $g^* \sim 16\pi^2\varepsilon/3$, which is an infrared (IR) attractive fixed point that governs large distance properties and is at the origin of Wilson–Fisher's famous ε-expansion (see Chapter 6).

The super-renormalizable ϕ^4 field theory corresponds to a non-generic situation where one fixes $g(\lambda) \le g^*$ and tunes the initial value g accordingly.

8.3 Fundamental interactions at the microscopic scale

The local QFT that describes fundamental interactions at the microscopic scale is affected by unavoidable divergences, a consequence of locality and the conservation of probabilities. A recipe, called *renormalization*, has been found to deal with these divergences in a restricted class of QFTs, which are, therefore, called *renormalizable*.

The condition that microscopic physics should be describable by *renormalizable QFTs* has been one of the guiding principles that have led to the construction of the Standard Model of fundamental interactions at the microscopic scale.

From the spectacular success of the programme, one might have concluded that *renormalizability was a new law of nature.*

However, the construction of renormalizable field theories requires a momentum cut-off and a tuning of the parameters of the Lagrangian, a mysterious procedure that requires a physics explanation.

By contrast, *the theory of critical phenomena shows that, for a large class of models when a large scale and, correspondingly, non-trivial large distance physics is generated dynamically, large scale properties can be described at leading order by a renormalizable QFT.*

Such a scheme suggests a simpler and more natural explanation for the appearance of renormalizable QFTs in particle physics. One can speculate that fundamental interactions are described at some more microscopic scale (like the Planck length) by a finite theory that no longer has the nature of a local QFT. Although such a theory should involve only *some short microscopic scale*, for reasons that, at present, can only be a matter of speculation, it generates strong correlations between a large number of degrees of freedom and a non-trivial *large distance physics with very light particles* (all known particles in this context are very light).

This line of thought has considerably affected our view about the renormalization process.

In the traditional presentation of QFT, one introduces a temporary large momentum cut-off (Fourier representation of a short distance scale) to render the perturbative expansion finite (a process called *regularization*) and calculates physical observables as functions of the parameters of the Lagrangian and the cut-off, in particular, physical masses and coupling constants. One then eliminates the parameters of the initial Lagrangian in favour of direct relations between physical observables and takes the infinite cut-off limit.

When the *QFT is renormalizable*, the infinite cut-off limit exists *order by order* in perturbation theory and defines a regularization-independent, perturbative, *renormalized field theory.*

However, this process relies on *tuning all parameters of the initial Lagrangian as functions of the cut-off*, which then in the infinite cut-off limit generally diverge. The tuning is so difficult to justify that, at some point, it led to the claim that the *initial Lagrangian in QFT is non-physical*!

Moreover, the programme does not provide any rationale, other than technical, for eliminating *non-renormalizable interactions.*

By contrast, within the point of view of *effective field theory* [113], one assumes that a true cut-off exists, which is provided by a more fundamental theory. This theory could be another QFT or, ultimately, a fundamental *necessarily non-local* theory. Since this theory is unknown, the physical cut-off is replaced by an artificial non-physical cut-off that renders physical observables finite, a process called *regularization*.

It is then necessary to prove that physics results are largely independent of the specific regularization. In this framework, *renormalization theory* and the *RG* that follows play an essential role. The RG allows to evaluate the *effective parameters at the physical scale* and to show that physics at an energy or mass scale much lower than the cut-off is indeed asymptotically *independent of the regularization*.

In contrast with the traditional renormalization theory, here one assumes that the parameters of the Lagrangian are fixed and expected to be of order unity at the cut-off scale in the absence of any other knowledge (the *naturalness assumption*) and one wants to avoid tuning when it is not justified. From this viewpoint, *the fine tuning of the Higgs bare mass is a real issue that looks for an explanation.*

The success of this programme in macroscopic physics, in particular, initiated by Wilson in the context of macroscopic continuous phase transitions, gives confidence that it should also apply to particle physics.

Moreover, additional, *weak non-renormalizable interactions* should be expected (the irrelevant interactions of RG), suppressed by powers of the short distance scale. Non-renormalizable *quantum gravity* is a possible example of such an RG 'irrelevant' interaction.

Summary. In particle physics, *renormalizable QFTs* are only *effective large distance theories* in which all non-renormalizable interactions have been neglected because their contributions are assumed to be too small.

The naive classical action has to supplemented with a cut-off as an *ad hoc* substitution for the effect of a more fundamental theory, of unknown nature. The cut-off is provided through an regularization procedure and is assumed to be large compared to the physical scale but not necessarily infinite (avoiding in this way the *triviality issue*).

Fortunately, renormalizable QFTs are somewhat *short-distance insensitive, that is, regularization independent, due to their RG properties.*

They are *not necessarily consistent on all scales, (like QED)*, and may have only a limited energy or momentum range of validity.

8.4 Field theory with a large mass: An explicit toy model

We have mentioned that an effective field theory can describe not only the large scale properties of a non-field theory, like a statistical lattice model, but also the low energy properties of another field theory when one mass is very large.

We illustrate this mechanism with a simple example. We consider a four-dimensional model involving two scalar fields, ϕ and χ, and a *renormalizable* action sum of two terms,

$$\mathcal{S}(\phi, \chi) = \mathcal{S}_1(\phi) + \mathcal{S}_2(\chi, \phi).$$

The first term $\mathcal{S}_1(\phi)$ corresponds to the ϕ^4 field theory,

$$\mathcal{S}_1(\phi) = \int \mathrm{d}^4x \left[\tfrac{1}{2}\left(\nabla_x\phi(x)\right)^2 + \tfrac{1}{2}u_2\phi^2(x) + \frac{1}{4!}u_4\phi^4(x) \right] + \text{ regularization terms}\,,$$

and the second involves the χ field and the ϕ,χ interaction,

$$\mathcal{S}_2(\chi,\phi) = \tfrac{1}{2} \int \mathrm{d}^4x \left[\left(\nabla_x\chi(x)\right)^2 + M^2\chi^2(x) + g\chi^2(x)\phi^2(x) + g'\chi^4(x) \right].$$

An unspecified regularization with a large momentum cut-off Λ is implied.

We assume that the physical mass M of the field χ is much larger than the physical mass of the field ϕ and the energies one is probing. Therefore, at leading order, the χ^4 interaction can be neglected.

Since the particle associated to the field χ is not observed, one can integrate out the field χ. We set

$$\mathrm{e}^{-\mathcal{V}(\phi)} = \frac{1}{\mathcal{N}} \int [\mathrm{d}\chi]\, \mathrm{e}^{-\mathcal{S}_2(\chi,\phi)}, \quad \mathcal{N} = \int [\mathrm{d}\chi]\, \mathrm{e}^{-\mathcal{S}_2(\chi,0)}\,.$$

The integral over χ is Gaussian and yields the *non-local* addition to the action $\mathcal{S}_1(\phi)$ for the ϕ field ($\ln\det = \mathrm{tr}\ln$),

$$\mathcal{V}(\phi) = \tfrac{1}{2}\,\mathrm{tr}\ln\left[-\nabla_x^2 + M^2 + g\phi^2(x) \right]\left[-\nabla_x^2 + M^2 \right]^{-1}.$$

However, since the mass M of the χ field is large, one can expand $\mathcal{V}(\phi)$ in powers of ϕ^2 and then make a *local expansion* of each term.

8.4.1 Local expansion

The coefficient of $g\phi^2(x)$, in Fourier representation, is proportional to

$$\Omega_4(M) = \frac{1}{(2\pi)^4} \int \frac{\mathrm{d}^4p}{p^2+M^2} \sim \frac{1}{16\pi^2}\Lambda^2 - \frac{1}{8\pi^2}\ln(\Lambda/M) + O(1),$$

a divergent constant that renormalizes the ϕ mass term and thus shifts the critical value of u_2 where the ϕ mass vanishes.

The coefficient of the term of order g^2 in the Fourier representation is proportional to

$$g^2 \int \mathrm{d}^4x\, \mathrm{d}^4y\, \phi^2(x)\phi^2(y)\Delta^2(x-y),$$

where $\Delta(x)$ is the χ-field propagator. Then,

$$\Delta^2(x) = \frac{1}{(2\pi)^4} \int \mathrm{d}^4k\, \mathrm{e}^{ikx}\, B_4(k) \tag{8.16}$$

with

$$B_4(k) = \frac{1}{(2\pi)^4} \int \frac{\mathrm{d}^4p}{(p^2+M^2)((p+k)^2+M^2)}.$$

We rewrite the integral by using *Feynman's parametrization*, which here takes the form of the identity

$$\frac{1}{\alpha\beta} = \int_0^1 \frac{\mathrm{d}s}{[\alpha s + \beta(1-s)]^2}.$$

We apply it to $\alpha = (p+k)^2 + M^2$, $\beta = p^2 + M^2$, and, after the shift $p + sk \mapsto p$, the integral can be rewritten as

$$B_4(k) = \frac{1}{(2\pi)^4} \int_0^1 \mathrm{d}s \int \frac{\mathrm{d}^4p}{[p^2 + M^2 + s(1-s)k^2]^2}.$$

The integrand can be expanded in powers of k^2,

$$\frac{1}{[p^2 + M^2 + k^2 s(1-s)]^2} = \frac{1}{(p^2 + M^2)^2} - \frac{2k^2 s(1-s)}{(p^2 + M^2)^3} + \frac{3k^4 s^2 (1-s)^2}{(p^2 + M^2)^4} + \cdots . \quad (8.17)$$

The first term gives a divergent constant,

$$\frac{1}{(2\pi)^4} \int \frac{\mathrm{d}^4p}{(p^2 + M^2)^2} \sim \frac{1}{8\pi^2} \ln(\Lambda/M),$$

which, inserted in the integral (8.16), yields a local contribution renormalizing the coefficient of $\phi^4(x)$.

The second term is finite and yields

$$-\frac{2k^2}{(2\pi)^4} \int_0^1 \mathrm{d}s\, s(1-s) \int \frac{\mathrm{d}^4p}{(p^2 + M^2)^3} = -\frac{1}{192\pi^2} \frac{k^2}{M^2}.$$

It has the form of a local, non-renormalizable interaction proportional to

$$\frac{g^2}{M^2} \int \mathrm{d}^4x\, \phi^2(x) \nabla_x^2 \phi^2(x).$$

More generally, the higher order terms yield finite local, non-renormalizable interactions proportional to $\phi^2(x)(\nabla_x^2)^n \phi^2(x)/M^{2n}$.

The term of order ϕ^6 is finite and, at leading order, yields a contribution proportional to $\phi^6(x)/M^2$. The next term has two additional derivatives and a factor $1/M^4$. Quite generally, all contributions are local and their M dependence can be inferred from dimensional analysis: all contributions have mass dimension 4, and ϕ and ∇_x have mass dimension 1.

To summarize, the large M expansion generates a set of local interactions: two that do not vanish for M large but simply renormalize terms already present in the action $\mathcal{S}_1(\phi)$, and all others that correspond to non-renormalizable interactions and are suppressed by powers of M.

In an *effective field theory, small non-renormalizable interactions are the first observable remnants of a new massive particle or a new energy scale,* as the history of the theory of weak interactions in particle physics illustrates.

Successive perturbative contributions of these new interactions, which come endowed with a natural cut-off of order M, again renormalize the effective action and yield additional contributions that vanish for M large.

8.5 An effective field theory: The Gross–Neveu model

We now consider another field theory renormalizable in four dimensions, the Gross–Neveu–Yukawa (GNY) model, and derive from it an effective field theory adapted to the symmetric phase in the large distance, low energy regime, which we identify with the Gross–Neveu (GN) model [114–117].

In all dimensions, the GNY model has a discrete symmetry that prevents a fermion mass term and displays a phase transition with *spontaneous fermion mass generation*.

In four dimensions, the symmetry has the form of a discrete chiral symmetry. In the spontaneously broken phase, the model can be considered as a toy model, in the sense that it omits the important gauge interactions, to describe the interactions between top quarks and the Higgs boson.

Notation and conventions. In this section, we first consider Dirac fermions in two and four dimensions, in *imaginary time* before extending to dimension $d = 2 + \varepsilon$. This requires the introduction of Dirac γ matrices, which we choose to be Hermitian. In four dimensions, γ matrices are 4×4 matrices satisfying

$$\gamma_\mu \gamma_\nu + \gamma_\nu \gamma_\mu = 2\delta_{\mu\nu}\mathbf{1}\,, \quad \mu, \nu = 1, 2, 3, 4\,.$$

An additional matrix,

$$\gamma_5 = \gamma_1 \gamma_2 \gamma_3 \gamma_4\,,$$

which is associated with space reflections, is also defined.

The γ matrices act on four-component complex vectors $\psi, \bar{\psi}$ (called spinors) representing, in four dimensions, spin $1/2$ fermions.

In two dimensions, the two γ matrices can be identified with the Pauli matrices σ_1 and σ_2, and the analogue of γ_5 is σ_3:

$$\gamma_1 \equiv \sigma_1\,, \quad \gamma_2 \equiv \sigma_2\,, \quad \sigma_3 = -i\sigma_1\sigma_2\,.$$

The σ matrices act on two-component complex vectors.

Finally, for a d-dimensional vector with components Ω_μ, we use the standard notation

$$\slashed{\Omega} \equiv \sum_{\mu=1}^{d} \Omega_\mu \gamma_\mu\,.$$

8.5.1 The GNY model

The GNY model is described in terms of a $U(N)$ symmetric action for a set of $N > 1$ massless fermions $\{\psi^i, \bar{\psi}^i\}$, and a scalar field σ [118]. The action reads

$$\mathcal{S}(\bar{\psi}, \psi, \sigma) = \int \mathrm{d}^d x \left[-\bar{\psi}(x) \cdot \left(\slashed{\partial} + g\Lambda^{2-d/2}\sigma(x) \right) \psi(x) \right.$$

$$\left. + \tfrac{1}{2}\left(\nabla_x \sigma(x)\right)^2 + \tfrac{1}{2}r\Lambda^2\sigma^2(x) + \frac{\lambda}{4!}\Lambda^{4-d}\sigma^4(x) \right] + \text{ regularization terms}, \quad (8.18)$$

where Λ is the momentum cut-off, and r, g, λ dimensionless 'bare', that is, effective coupling constants at the large momentum scale Λ or microscopic scale $a = 1/\Lambda$.

In addition to the $U(N)$ symmetry, the GNY model has in all dimensions a $\mathbb{Z}_2$ reflection symmetry, which prevents the addition of a fermion mass term. In even dimensions, it can be identified with a discrete chiral symmetry. In four dimensions, it corresponds to the transformation

$$\psi(x) \mapsto \gamma_5 \psi(x), \quad \bar{\psi}(x) \mapsto -\bar{\psi}(x)\gamma_5, \quad \sigma(x) \mapsto -\sigma(x). \tag{8.19}$$

(In two dimensions, γ_5 is replaced by σ_3.)

Spontaneous $\mathbb{Z}_2$ symmetry breaking. In the tree approximation, the σ potential,

$$V(\sigma) = \tfrac{1}{2}r\Lambda^2\sigma^2 + \frac{\lambda}{4!}\Lambda^{4-d}\sigma^4,$$

leads to a phase transition at $r = 0$. For $r < 0$, it has two minima located at

$$|\sigma| = \Lambda^{(d-2)/2}\sqrt{-6r/\lambda},$$

corresponding to a spontaneous breaking of the $\mathbb{Z}_2$ symmetry and *fermion mass generation*. Then,

$$m_\psi = \Lambda|g|\sqrt{-6r/\lambda}, \quad \frac{m_\sigma}{m_\psi} = \sqrt{\frac{\lambda}{3g^2}}.$$

The masses are much smaller than the cut-off only if, at leading order, $|r| = O(1/\Lambda^2)$, the *standard fine tuning problem*.

The broken symmetry being discrete, no massless Goldstone boson is generated.

Beyond leading order: The transition point. In a pure σ^4 field theory, beyond leading order, the critical value r_c at which the transition occurs is always negative. This is understandable, since a double-well shaped potential is required to allow for a phase transition.

By contrast, the Yukawa $\bar{\psi}\psi\sigma$ interaction favours the transition and has the effect of increasing r_c. We assume below that the fermion effect is numerically dominant and that r_c is positive.

The m_σ/m_ψ ratio and RG. While, in the tree approximation, the ratio m_σ/m_ψ is arbitrary, an RG analysis indicates that the ratio is fixed in the large Λ limit.

8.5.2 Symmetric phase: The effective GN model

In the symmetric phase ($r > r_c$), the large distance, low energy physics is dominated by the massless fermions. In this limit, the massive scalar field σ can be integrated out. At leading order, one can keep only the quadratic σ contribution. The Gaussian integration leads to the replacement of $\sigma(x)$ by the solution of the σ field equation,

$$\sigma(x) = g\Lambda^{2-d/2} \int \mathrm{d}^d y \, \langle x| \, [-\nabla^2 + M^2]^{-1} \, |y\rangle \, \bar{\psi}(y)\cdot\psi(y) \underset{M\to\infty}{\sim} g\frac{\Lambda^{2-d/2}}{M^2}\bar{\psi}(x)\cdot\psi(x),$$

where, for convenience, we have set $M^2 = r\Lambda^2$.

The next term in the large M expansion, which we have neglected, is proportional to $\nabla^2 \bar{\psi}(x)\psi(x)/M^4$. For M large, the σ propagator is thus dominated by the mass term and, at leading order, one obtains the effective action,

$$S(\bar{\psi}, \psi) = -\int \mathrm{d}^d x \left[\bar{\psi}(x) \cdot \partial\!\!\!/\psi(x) + \tfrac{1}{2}\Lambda^{2-d}G\left(\bar{\psi}(x) \cdot \psi(x)\right)^2 \right]$$

with an attractive interaction since

$$G = g^2 \Lambda^2 / M^2 = g^2/r > 0\,.$$

This is the action of the GN model that we now study.

The GN model requires a symmetric UV cut-off, for example,

$$\partial\!\!\!/ \mapsto \partial\!\!\!/(1 - \nabla_x^2/\Lambda^2 + \cdots)\,.$$

8.5.3 The GN model: Four dimensions

In four dimensions, the GN model provides an interesting example of an *effective field theory*, reminiscent of the Fermi–Feynman–Gell-Mann model of fundamental weak interactions, but is simpler to analyse because, in the interaction term, currents are replaced by the scalar $\bar{\psi}\psi$ combination.

The action (15.6) then reads

$$S(\bar{\psi}, \psi) = -\int \mathrm{d}^4 x \left[\bar{\psi}(x) \cdot \partial\!\!\!/\psi(x) + \frac{G}{2\Lambda^2}\left(\bar{\psi}(x) \cdot \psi(x)\right)^2 \right]$$
$$+ \text{ regularization terms}\,. \tag{8.20}$$

The *perturbative GN model displays only a symmetric massless phase.*

The fermion field has mass dimension $3/2$. The model is *non-renormalizable*, and dimensional analysis indicates that the interaction is suppressed by a factor $1/\Lambda^2$. In the spirit of effective field theory, other local interactions could be added to the action but they are suppressed at least by a factor $1/\Lambda^4$.

Although the interaction is small and the tree approximation is expected to give the leading contribution, since the theory is not renormalizable divergences appear at higher orders in the perturbative expansion, which could cancel the $1/\Lambda^2$ factors. This problem requires a more detailed analysis.

Beyond leading order: One-loop contributions. At one-loop order, the contributions to the two-, four-, six- and eight-point functions are divergent.

The contribution to the two-point function is a constant that vanishes due to chiral symmetry.

To evaluate the four-point function, which has a quadratic divergence, one needs the fermion propagator,

$$\left\langle \bar{\psi}^i_\alpha(x)\psi^j_\beta(y) \right\rangle_0 \equiv \Delta^{ij}_{\alpha\beta}(y,x) = -i\frac{\delta_{ij}}{(2\pi)^4}\int \mathrm{d}^4 p\; \mathrm{e}^{-ip(x-y)} \frac{p\!\!\!/_{\alpha\beta}}{p^2}\,.$$

The first diagram of Fig. 8.1 is proportional to

$$N\frac{G^2}{\Lambda^4}\int \mathrm{d}^4q\,\frac{\mathrm{tr}\,\not{q}(\not{q}+\not{p})}{q^2(p+q)^2} = 4N\frac{G^2}{\Lambda^4}\int \mathrm{d}^4q\,\frac{(q^2+pq)}{q^2(p+q)^2}$$

$$= 2N\frac{G^2}{\Lambda^4}\left(2\int\frac{\mathrm{d}^4q}{q^2} - p^2\int\frac{\mathrm{d}^4q}{q^2(p+q)^2}\right).$$

The first integral is proportional to Λ^2 and acts like a renormalization, finite but dependent of the regularization, to the coupling constant G. The second integral contains a local logarithmically divergent contribution equivalent to an interaction of the form

$$\frac{\ln \Lambda}{\Lambda^4}\int \mathrm{d}^4x\,\bar{\psi}(x)\cdot\psi(x)\nabla_x^2\big(\bar{\psi}(x)\cdot\psi(x)\big)$$

and a cut-off independent finite, non-local, contribution of order G^2/Λ^4.

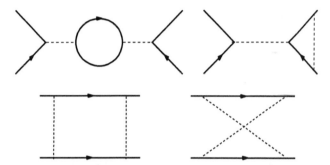

Fig. 8.1 Feynman diagrams, faithful representation: One-loop contributions to the four-point vertex function, full lines corresponding to Δ (fermions), dotted lines to faithful representations of four-fermion interactions.

The other diagrams again generate constant contributions of order $1/\Lambda^2$ that renormalize G, various contributions of the form of chiral invariant local interactions with two derivatives of order $\ln \Lambda/\Lambda^4$ and, finally, contributions that are short distance insensitive of order $1/\Lambda^4$.

The six-point function generates a $(\bar{\psi}\psi)^2\bar{\psi}\not{\partial}\psi$ local term of order $\ln \Lambda/\Lambda^6$ as well as finite contribution multiplied by $1/\Lambda^6$. The eight-point function generates a local $(\bar{\psi}\psi)^4$ term of order $\ln \Lambda/\Lambda^8$ and a finite contribution multiplied by $1/\Lambda^8$.

Higher orders. Quite generally, at higher orders, divergent contributions generate chiral invariant local interactions multiplied by powers of Λ determined by dimensional analysis ($\bar{\psi}\psi$ has dimension 3), multiplied by possible powers of $\ln \Lambda$. In addition, finite contributions are also generated, suppressed by increasing powers of Λ.

This analysis seems to indicate that physics is entirely governed by the first orders of perturbation theory. However, a problem could arise from the increasing powers of $\ln \Lambda$ that are generated at higher orders and which could sum to powers of Λ, invalidating the perturbative power counting. This problem can be tackled with RG methods.

8.5.4   The GN model in two dimensions

In two dimensions, the perturbative expansion does not exist, due to zero-momentum (IR) divergences, as the evaluation of the first diagram of Fig. 8.1 immediately shows. This indicates an instability of the symmetric phase. An IR regulator in the form of a fermion mass term, which breaks chiral symmetry, can be added to the action. Alternatively, finite temperature (see Section 21.11) or a finite box (Section 19.5), which do not break the symmetry, can be used.

In perturbation theory, the GN model is renormalizable in two dimensions. The $(\bar{\psi}\psi)^2$ interaction can be multiplicatively renormalized.

RG equations. One proves that, as a function of the cut-off Λ, the fermion mass m and the coupling G, vertex (or one-particle-irreducible (1PI)) functions in the Fourier representation satisfy the RG equations,

$$\left[\Lambda\frac{\partial}{\partial\Lambda} + \beta(G)\frac{\partial}{\partial G} - \frac{n}{2}\eta_\psi(G) - \eta_m(G)m\frac{\partial}{\partial m}\right]\tilde{\Gamma}^{(n)}\left(p_i; G, m, \Lambda\right) = 0\,. \qquad (8.21)$$

A two-loop calculation of the RG β-function yields

$$\beta(G) = -(N-1)\frac{G^2}{\pi} + (N-1)\frac{G^3}{2\pi^2} + O\left(G^4\right)\,. \qquad (8.22)$$

The field and mass renormalization RG functions $\eta_\psi(G)$ and $\eta_m(G)$ are

$$\eta_\psi(G) = \frac{2N-1}{8\pi^2}G^2 + O\left(G^3\right)\,, \quad \eta_m(G) = \frac{2N-1}{2\pi}G + O(G^2)\,. \qquad (8.23)$$

The *negative sign* of the leading term of the β-function shows that the model is *asymptotically (UV) free*: the Gaussian fixed point $G = 0$ is repulsive at low momentum and attractive for high momenta.

It shares this property with the non-linear σ-model in two dimensions but also with non-Abelian gauge theories in four dimensions.

The perturbative expansion is directly useful to evaluate large momentum properties but, in the chiral limit $m = 0$, the *spectrum is non-perturbative*. A number of arguments (in particular, the $1/N$ expansion that is studied in Section 21.11.2) lead to the conclusion that *chiral symmetry is always spontaneously broken* and that the spectrum includes *massive fermions and bosons*.

The mass scale $\mathcal{M}$ is RG invariant, and its dependence on the coupling constant is thus given by RG arguments. For $m = 0$, it satisfies

$$\left(\Lambda\frac{\partial}{\partial\Lambda} + \beta(G)\frac{\partial}{\partial G}\right)\mathcal{M}(\Lambda, G) = 0\,.$$

For dimensional reasons, $\mathcal{M} = \Lambda F(G)$. The RG equation reduces to

$$F(G) + \beta(G)F'(G) = 0\,. \qquad (8.24)$$

Thus,

$$\mathcal{M} = \Lambda F(G) \propto \Lambda \exp\left(-\int^G \frac{dg'}{\beta(g')}\right) \propto \Lambda G^{1/2(N-1)} \, e^{-\pi/(N-1)G},$$

the coefficient being non-perturbative.

The physical condition $\mathcal{M} \ll \Lambda$ then implies that the coupling constant G must be logarithmically small.

8.5.5 Beyond perturbation theory: $d > 2$

The analysis of the two-dimensional model suggests that, although the GN model has been specifically designed to describe the symmetry massless phase, it may also generate a spontaneously broken phase. This problem can be studied by different methods, including dimensional continuation.

$d = 2 + \varepsilon$ *dimensions.* By dimensional continuation, it is possible to define the perturbative expansion of the GN model for arbitrary real or complex values of the dimension d [118].

The perturbative expansion is no longer IR divergent. For $d = 2 + \varepsilon$ dimensions, $\varepsilon > 0$, in the framework of a double expansion in powers of G and ε, the β-function is given by

$$\beta(G) = \varepsilon G - (N-1)\frac{G^2}{\pi} + (N-1)\frac{G^3}{2\pi^2} + O\left(G^4\right).$$

The origin $G = 0$ is now an IR fixed point and, for G small, the physics is perturbative with chiral symmetry and massless fermions. However, for ε small, the β-function has another, IR repulsive, zero,

$$G_c = \frac{\pi\varepsilon}{N-1} + O(\varepsilon^2) \;\;\Rightarrow\;\; \beta'(G_c) = -\varepsilon + O(\varepsilon)^2.$$

Continuity with $d = 2$ indicates that G_c is a transition value between the small G symmetric, perturbative phase and a phase with spontaneous chiral symmetry breaking and massive particles. Since the broken symmetry is discrete, no Goldstone boson is generated.

$d > 2$: *Generic dimensions.* The analysis of the two-dimensional model and its formal extension to $d = 2 + \varepsilon$ dimensions indicate that, in $d > 2$ dimensions, a phase transition occurs at a value $G = G_c$ and, above G_c, a phase with spontaneous $\mathbf{Z}_2$ symmetry breaking, fermion mass generation and a scalar bound state, appears. The conjecture is comforted by the results of the large N expansion (Section 21.11).

However, although G_c is cut-off independent (it does not scale with the ratio between microscopic and physical scales), *its numerical value is not universal but depends on the regularization scheme.* Therefore, the existence of a transition may also depend on the regularization scheme.

From the viewpoint of an effective field theory, the transition value G_c of the dimensionless coupling G is a value where the concept of *perturbative effective field theory* loses its relevance. Even when one probes only a region of small momenta and small masses compared to the cut-off, although the interaction is formally very weak, *near and above G_c the physics is no longer perturbative.*

8.6 Non-linear σ-model: Another effective field theory

An important feature of low energy hadron physics is an approximate chiral symmetry, induced by the small mass of some quarks, realized in the phase of approximate spontaneous symmetry breaking (SSB). We focus here on the least explicitly broken symmetry, $SU(2) \times SU(2)$, due to the small mass of the **u** and **d** quarks, which in the scalar sector is equivalent to $O(4)$. To the almost SSB correspond three almost Goldstone bosons (the pions), which belong to a four-dimensional vector (the fundamental representation of $SO(4)$), the fourth component, a scalar field denoted by σ, being aligned with the direction of $O(4)$ breaking and being more massive (it is expected to lead to a broad resonance).

However, from the phenomenological viewpoint, chiral models are only predictive in the limit where the mass of the σ field is large. This suggests integrating out the massive component to generate an effective field theory. This leads to a difficulty, since the field with large mass is a member of a multiplet, and its elimination could induce an additional explicit breaking of the $O(4)$ symmetry.

8.6.1 The $O(N)$ symmetric $(\phi^2)^2$ field theory in the ordered phase

To study the problem of σ-integration, we concentrate on the boson sector. Moreover, it is convenient to generalize the $O(4)$ group to a generic $O(N)$ group, with the boson fields ϕ in the fundamental N-component representation.

We are led to consider the $O(N)$ symmetric action,

$$\mathcal{S}(\phi) = \int \mathrm{d}^d x \left[\tfrac{1}{2} \left(\nabla_x \phi(x) \right)^2 + \tfrac{1}{2} r \phi^2(x) + \tfrac{1}{4!} u \left(\phi^2(x) \right)^2 \right]$$
$$+ \text{ regularization terms} . \tag{8.25}$$

In the spontaneously broken (or ordered) phase $(r < r_c)$, the $O(N)$ symmetry is broken down to an $O(N-1)$ symmetry and this generates $(N-1)$ *Goldstone modes* associated with massless scalar particles [119]. At low momentum or large distance, the physics is entirely dominated by the interaction between Goldstone modes. It is thus natural to integrate out the last massive component of the N-component field. However, the problem is to avoid losing the information about the initial $O(N)$ symmetry of the model.

The problem can be solved by an appropriate parametrization of the field ϕ, and the integration over the massive mode then leads to the *non-linear σ-model* [120].

The integration. We consider the partition function

$$\mathcal{Z} = \int [\mathrm{d}\phi] \exp[-\mathcal{S}(\phi)].$$

We change variables in the field integral, setting

$$\phi(x) = \rho(x)\hat{\phi}(x) \text{ with } \hat{\phi}^2(x) = 1 .$$

This change is meaningful only in the broken phase where the expectation value $\langle \rho \rangle$ does not vanish and the ρ-field is *massive*. Otherwise, the change is singular for small fields.

The field integral becomes (assuming *a lattice or in the continuum a dimensional regularization* to deal with the non-trivial ρ measure):

$$\mathcal{Z} = \int \left[\rho^{N-1}(x)\mathrm{d}\rho(x) \right] \left[\mathrm{d}\Omega(\hat{\phi}(x)) \right] \exp\left[-\mathcal{S}(\rho, \hat{\phi}) \right],$$

where $\Omega(\hat{\phi})$ is the invariant measure on the sphere S_{N-1}, which *formally* can be written as

$$\left[\mathrm{d}\Omega(\hat{\phi}(x)) \right] = [\mathrm{d}\hat{\phi}(x)] \prod_x \delta\left(\hat{\phi}^2(x) - 1 \right),$$

and

$$\mathcal{S}(\rho, \hat{\phi}) = \int \mathrm{d}^d x \left\{ \tfrac{1}{2}\rho^2(x) \left(\nabla_x \hat{\phi}(x) \right)^2 + \tfrac{1}{2} [\nabla_x \rho(x)]^2 + \tfrac{1}{2} r \rho^2(x) \right.$$
$$\left. + \tfrac{1}{4!} u \rho^4(x) \right\}. \tag{8.26}$$

If one is interested only in momenta much smaller than the ρ-mass, or distances much larger than the corresponding correlation length, one can perturbatively integrate out the ρ-field.

The integration generates the effective action $\mathcal{S}_{\mathrm{eff.}}(\hat{\phi})$ for the field $\hat{\phi}$ defined by

$$\exp\left[-\mathcal{S}_{\mathrm{eff.}}(\hat{\phi}) \right] \propto \int \left[\rho^{N-1}(x)\mathrm{d}\rho(x) \right] \exp\left[-\mathcal{S}(\rho, \hat{\phi}) \right]. \tag{8.27}$$

To calculate the integral, we set

$$\mathcal{S}(\rho) = \int \mathrm{d}^d x \left[\tfrac{1}{2}\left(\nabla_x \rho(x) \right)^2 + \tfrac{1}{2} r \rho^2(x) + \tfrac{1}{4!} u \rho^4(x) \right].$$

We define

$$\mathcal{Z}(J) = \int \left[\rho^{N-1}(x)\mathrm{d}\rho(x) \right] \exp\left[-\mathcal{S}(\rho) - \tfrac{1}{2} \int \mathrm{d}^d x\, J(x)\rho^2(x) \right]$$

with $\mathcal{Z}(0) = 1$. Then,

$$\mathcal{S}_{\mathrm{eff.}}(\hat{\phi}) = -\ln \mathcal{Z}\left[J(x) \equiv \left(\nabla_x \hat{\phi}(x) \right)^2 \right].$$

The expansion of $\mathcal{Z}(J)$ in powers of J takes the form

$$\ln \mathcal{Z}(J) = -\tfrac{1}{2} \langle \rho^2 \rangle \int \mathrm{d}^d x\, J(x)$$
$$+ \tfrac{1}{8} \int \mathrm{d}^d x\, \mathrm{d}^d y \, \langle \rho^2(x)\rho^2(y) \rangle_{\mathrm{conn.}} J(x)J(y) + O(J^3),$$

where $\langle \bullet \rangle$ denotes an expectation value with respect to the normalized measure corresponding to $\mathrm{e}^{-\mathcal{S}(\rho)}$.

For large ρ mass M, all successive contributions become local. The first term,

$$S_{\text{eff.}}(\hat{\phi}) = \tfrac{1}{2} \langle \rho \rangle^2 \int \mathrm{d}^d x \left(\nabla_x \hat{\phi}(x) \right)^2, \tag{8.28}$$

has only two derivatives and thus governs the large distance behaviour. In the expression (8.28), one recognizes the action of the non-linear σ-model, the only action with two derivatives consistent with $O(N)$ symmetry.

The second term has four derivatives. It is suppressed by a factor $1/M^2$ and is subleading at large distance.

The perturbative ρ-integration. Each expectation value can be calculated perturbatively. We denote by v the expectation value of $\rho(x)$: $v = \langle \rho(x) \rangle$. We set

$$\rho(x) = v + \chi(x) \ \Rightarrow \ \langle \chi(x) \rangle = 0,$$

where v in the tree approximation is given by $v^2 = -6r/u$.

In terms of χ, the action (8.26) becomes

$$S(\chi) = \int \mathrm{d}^d x \left[\tfrac{1}{2} (\nabla_x \chi(x))^2 + \tfrac{1}{6} u v^2 \chi^2(x) + \tfrac{1}{6} u v \chi^3(x) + \tfrac{1}{24} u \chi^4(x) \right].$$

Neglecting all fluctuations of the field χ, one recovers the action (8.28) at leading order:

$$S_{\text{eff.}}^{(0)}(\hat{\phi}) = \tfrac{1}{2} v^2 \int \mathrm{d}^d x \left(\nabla_x \hat{\phi}(x) \right)^2. \tag{8.29}$$

Loop corrections are calculated with the χ propagator

$$\tilde{\Delta}_\chi(p) = \frac{1}{p^2 + M^2}, \quad M^2 = \tfrac{1}{3} u v^2,$$

and have then to be expanded for M large.

The contributions coming from the integration over χ first renormalize the coefficient v^2 of the *term with only two derivatives* in (8.29.)

The second term is proportional to the connected part of $\langle \rho^2(x) \rho^2(y) \rangle$,

$$
\begin{aligned}
\langle \rho^2(x) \rho^2(y) \rangle - \langle \rho^2(x) \rangle \langle \rho^2(y) \rangle &= \langle \rho^2(x) \rho^2(y) \rangle - \langle \rho^2(x) \rangle \langle \rho^2(y) \rangle \\
&= \langle \chi^2(x) \chi^2(y) \rangle - \langle \chi^2(x) \rangle \langle \chi^2(y) \rangle \\
&\quad + 4v^2 \langle \chi(x) \chi(y) \rangle \\
&= \Delta_\chi^2(x - y) + 4v^2 \Delta_\chi(x - y).
\end{aligned}
$$

The first term has been evaluated in Section 8.4 (equation (8.16)). It has a local expansion, derived from equation (8.17), where the leading contribution is proportional to $\ln(\Lambda/M)\delta(x - y)$, Λ being a cut-off, the next contribution proportional to $\nabla_x^2 \delta(x - y)/M^2$, and so on. The second term has a leading order contribution that is simply $4v^2 \delta(x - y)/M^2$.

Therefore, all loop contributions, except for a renormalization of the leading term, yield additional, *irrelevant in the RG terminology, ϕ-interactions with more derivatives, the leading one being proportional to $\int \mathrm{d}^d x ((\nabla_x \hat{\phi})^2)^2$, and counterterms for the non-linear σ-model, which in perturbation theory is more divergent that the initial $(\phi^2)^2$ field theory.*

Domain of validity. From the way the effective action has been derived, it is expected that its perturbative domain of validity is limited to the broken (or ordered) phase r fixed, $r < r_c$. In this regime, the non-linear σ-model (8.28) completely describes the long distance properties of the $(\phi^2)^2$ field theory.

For r close to r_c or $r > r_c$, the perturbative expansion of the non-linear σ is no longer useful. Physics becomes non-perturbative, and the configurations for which ρ vanishes become important. The interpretation of the non-linear σ-model as a *perturbative effective theory* is no longer valid.

8.6.2 The non-linear σ-model

The non-linear σ-model is defined in terms of the effective action resulting from the χ integration restricted to the leading term with two derivatives,

$$\mathcal{S}(\hat{\phi}) = \frac{\Lambda^{d-2}}{2g} \int \mathrm{d}^d x \left(\nabla_x \hat{\phi}(x)\right)^2, \quad \text{with} \quad \hat{\phi}^2(x) = 1. \tag{8.30}$$

The partition function is given by

$$\mathcal{Z} = \int [\mathrm{d}\Omega(\hat{\phi})] \, \mathrm{e}^{-\mathcal{S}(\hat{\phi})}. \tag{8.31}$$

The model requires a regularization, which, due to a non-trivial ϕ measure, can be a lattice regularization or, for practical calculations, dimensional regularization. Indeed, higher order derivatives do not take care of the measure and do not regularize one-loop diagrams.

At leading order, the $(N-1)$ independent field components are massless, and the model is in the broken phase. However, for $d = 2$, perturbative calculations show that the model has IR divergences due to would-be massless fields, indicating that the phase with SSB is unstable. It is necessary to add a mass term that breaks explicitly the $O(N)$ symmetry, which most conveniently can be generated by adding the linear coupling

$$\mathbf{c} \cdot \int \mathrm{d}^d x \, \phi(x).$$

The model is then renormalizable in the $d = 2$ dimension and non-renormalizable for $d > 2$.

It can be studied for $d = 2$ [85], for $d = 2 + \varepsilon$ and in the large N limit [86].

Two dimensions: RG equations. The model is renormalizable with two renormalization constants: a renormalization of the coupling and a renormalization of the radius of the sphere. RG equations follow. For the connected correlation functions, they can be written as (m is the physical mass)

$$\left[\Lambda \frac{\partial}{\partial \Lambda} + \beta(g)\frac{\partial}{\partial g} + \frac{n}{2}\zeta(g) + \left(\frac{\zeta(g)}{4} + \frac{\beta(g)}{2g}\right) m\frac{\partial}{\partial m}\right] \tilde{W}^{(n)}(p_i; g, m, \Lambda) = 0 . \quad (8.32)$$

The two RG functions at leading order are

$$\beta(g) = -\frac{(N-2)}{2\pi}g^2\left(1 + g/2\pi\right) + O(g^4), \quad (8.33a)$$

$$\zeta(g) = \frac{(N-1)}{2\pi}g + O(g^2). \quad (8.33b)$$

The sign for $N > 2$ of the first coefficient of the β-function shows that the model is *asymptotically free*. Moreover, in the limit $m \to 0$, the spectrum is non-perturbative, as is expected, since SSB with continuous symmetry is impossible [122].

In fact, from other considerations, it is known that the $O(N)$ symmetry is restored and, for $N > 2$, the spectrum contains N massive bosons.

Its physical mass M is an RG invariant and thus has the general form

$$M \propto \Lambda \exp\left[-\int^g \frac{dg'}{\beta(g')}\right] \propto \Lambda g^{-1/(N-2)}\, e^{-(2\pi)/(N-2)g} ,$$

where the multiplicative constant is non-perturbative.

The case $N = 2$ is peculiar because $SO(2)$ is an Abelian group. The content of the model *depends on the regularization*. In the simplest form, it is equivalent to a free field theory. In a lattice regularization, it leads to the famous Kosterlitz–Thouless phase transition [123].

Dimension $2 + \varepsilon$. Above dimension 2, the model can be studied in the form of a double series expansion in g and $\varepsilon = (d - 2)$. The β-function becomes [121]

$$\beta(g) = \varepsilon g - \frac{(N-2)}{2\pi}g^2 + O\left(g^3, g^2\varepsilon\right). \quad (8.34)$$

The slope at the Gaussian fixed point $g = 0$ is positive and, for g small, the large distance behaviour is governed by the Gaussian fixed point. The physics of the model is thus perturbative and describes weakly interacting massless Goldstone modes.

However, one finds an additional fixed point with a negative slope,

$$g^* = \frac{2\pi\varepsilon}{N-2} + O(\varepsilon^2), \quad \beta'(g^*) = -\varepsilon + O(\varepsilon^2),$$

a UV fixed point that governs the universal large momentum behaviour for momenta $|p| \ll \Lambda$.

Solving the RG equations, one finds that, at $g = g^*$, the field expectation value vanishes like

$$\langle \sigma \rangle \propto (g^* - g)^{-\zeta(g^*)/2\beta'(g^*)}, \quad -\frac{\zeta(g^*)}{2\beta'(g^*)} = \frac{N-1}{2(N-2)} + O(\varepsilon).$$

The result suggests that g^* corresponds to a phase transition between the spontaneously broken phase and a $g > g^*$ non-perturbative phase with symmetry restoration.

Conclusion. The interpretation of the non-linear σ-model as an effective perturbative theory is valid only for $0 < g < g^*$, except when the symmetry is explicitly broken.

This conclusion is comforted by the large N-expansion, which also indicates that, beyond perturbation theory, the non-linear σ-model with a proper regularization and the corresponding $(\phi^2)^2$ field theory have the same large distance physics (see also Chapter 10).

9 The non-perturbative renormalization group

The renormalization group (RG) of quantum field theory (QFT) is the asymptotic form of a more general RG that acts in the space of all Hamiltonians of *effective local theories* with the same field content and symmetries. For macroscopic phase transitions, the locality of the effective field theory follows from the assumption that the initial microscopic system has only short range interactions.

The simpler RG that appears in QFTs is valid in some neighbourhood of the Gaussian fixed point. It makes it possible to investigate the properties of renormalizable field theories as well as the existence of non-Gaussian fixed points close, in some sense, to the Gaussian fixed point. Examples are provided by infrared (IR) fixed points of N-vector models in $d = 4 - \varepsilon$ dimensions (see Chapter 6).

In this chapter, we describe the general non-perturbative (or *functional*) RG equations, using, for the purpose of illustration, the example of the Ising universality class in continuum space, that is, in the example of $\mathbb{Z}_2$ symmetric scalar field theories, as it was initially formulated by Wilson [24, 100] and Wegner [101]. We give a hint of their derivation.

Note that functional RG equations can also be used to give alternative proofs of perturbative renormalizability [124] within the framework of effective field theories.

9.1 Intuitive RG formulation

We consider a scalar field theory with a $\mathbb{Z}_2$ reflection symmetry, thus belonging to the Ising model universality class. In continuum space, the partition function is given by an integral over a field $\phi(x)$ (like a classical effective spin) of the form

$$\mathcal{Z} = \int [\mathrm{d}\phi] \exp\left[-\mathcal{H}(\phi)\right], \quad \mathcal{H}(\phi) = \mathcal{H}(-\phi), \tag{9.1}$$

where the Hamiltonian $\mathcal{H}(\phi)$ is translation invariant and local (see expression (8.4)), except for the quadratic terms in the field but whose non-locality is confined at very short distance $1/\Lambda_0$, which characterizes the scale of the initial short distance structure (the lattice spacing or the range of interactions).

To generate an RG flow, one then sums iteratively over short distance degrees of freedom, as suggested by Kadanoff [77] in the example of lattice models, but in Wilson's more useful formulation [24]. One considers *effective models* in continuum space, introduces the field Fourier representation,

$$\phi(x) = \int \mathrm{d}^d p \, \mathrm{e}^{ipx} \, \tilde{\phi}(p),$$

and sums over fields $\tilde{\phi}(p)$ whose arguments belong to a soft momentum shell.

From Random Walks to Random Matrices. Jean Zinn-Justin, Oxford University Press (2019).
© Jean Zinn-Justin. DOI: 10.1093/oso/9780198787754.001.0001

The assumption of short range interactions is implemented in momentum space in the form of a momentum cut-off Λ_0.

One limits the integration over $\tilde{\phi}(p)$ to momenta such that $|p| \leq \Lambda_0$, and denotes by $\mathcal{H}_1(\phi)$ the initial Hamiltonian. In the initial Wilson implementation, one then integrates over field components with momenta in a shell $\Lambda_0 - \mathrm{d}\Lambda_0 < |p| \leq \Lambda_0$, the equivalent of integrating in space over short distance degrees of freedom. This generates a new effective Hamiltonian $\bar{\mathcal{H}}_\lambda$ with $\lambda = \Lambda_0/(\Lambda_0 - \mathrm{d}\Lambda)$ given by

$$\exp\left[-\bar{\mathcal{H}}_\lambda(\phi)\right] = \int_{\Lambda_0 - \mathrm{d}\Lambda < |p| \leq \Lambda_0} [\mathrm{d}\tilde{\phi}] \exp\left[-\mathcal{H}_1(\phi)\right].$$

This transformation does not change the partition function and can be iterated. Since the goal is to determine possible fixed points of such a transformation, it is also necessary to rescale the field at each iteration (as the example of the random walk in Section 1.5 already shows). Introducing a renormalization factor Z that depends only on the running Hamiltonian $\mathcal{H}_\lambda$, one sets

$$\mathcal{H}_\lambda(\phi) = \bar{\mathcal{H}}_\lambda(\phi/Z^{1/2}).$$

Iterating, one constructs an RG, which in the limit of $\mathrm{d}\Lambda$ infinitesimal, takes the form

$$\lambda \frac{\mathrm{d}}{\mathrm{d}\lambda} \mathcal{H}_\lambda = \mathcal{T}\left[\mathcal{H}_\lambda\right], \tag{9.2}$$

$$\lambda \frac{\mathrm{d}}{\mathrm{d}\lambda} \ln Z(\lambda) = 2 - d - \eta\left[\mathcal{H}_\lambda\right], \tag{9.3}$$

where $\mathcal{T}$ is assumed to be a differentiable mapping (in some topology) of the space of Hamiltonians into itself and η to be a real continuous function which depends on λ only through $\mathcal{H}_\lambda$. As a function of the 'time' $\ln \lambda$, these equations define a stationary Markov's process.

For a continuous phase transition where the *correlation length is infinite* at the critical temperature, the RG transformations can be iterated without modifying the large distance properties. If they exist, IR *fixed points* then characterize large distance physics.

Fixed points. One looks for values of Z such that the IR fixed points solution of

$$\mathcal{T}\left[\mathcal{H}^*(\phi)\right] = 0, \tag{9.4}$$

can be found. *For critical systems ($T = T_c$) with divergent correlation length, the existence of attractive (IR) fixed points of an RG makes it possible to understand universality, within universality classes, of large distance properties of statistical systems.*

9.2   Non-perturbative RG equations

The non-perturbative RG, in continuum space and in the context of a local statistical field theories, determines a flow of Hamiltonians generated by a systematic and recursive integration over short distance, or large momenta in Fourier space, degrees of freedom. The simple momentum shell integration cannot be used as such because it would lead after iteration to artificial long range forces. One has to introduce a smooth cut-off in momentum space and preserve smoothness in the large momentum modes integration [125].

9.2.1   General local statistical field theory

The partition function is given by a field integral of the form (9.1). We decompose the Hamiltonian into the sum of two terms:

$$\mathcal{H}(\phi) = \frac{1}{2} \int \mathrm{d}^d x \, \mathrm{d}^d y \, \phi(x) K(x - y)\phi(y) + \mathcal{V}(\phi),$$

where K is a positive kernel and the interaction has the form

$$\mathcal{V}(\phi) = \int \mathrm{d}^d x \, V(\phi, x).$$

The functional $V(\phi, x)$ is a linear combination of monomials products of powers of the field $\phi(x)$ and its partial derivatives at the same point (*locality*).

     In a perturbative expansion Δ, the inverse of K,

$$\int \mathrm{d}^d z \, K(x - z)\Delta(z - y) = \delta^{(d)}(x - y),$$

is the propagator in Feynman diagrams.

     The propagator Δ has the Fourier representation

$$\Delta(x) = \frac{1}{(2\pi)^d} \int \mathrm{d}^d p \, \mathrm{e}^{ipx} \, \tilde{\Delta}(p).$$

We choose $\tilde{\Delta}(p)$ of the form

$$\tilde{\Delta}(p) = \frac{1}{p^2} C(p^2/\Lambda^2), \tag{9.5}$$

where $C(t)$ is an analytic function (or at least C^∞) such that $C(0) = 1$ and decreasing faster than any power for $t \to \infty$, for example,

$$C(t) = \mathrm{e}^{-t}.$$

The function C implements a short distance or large momentum *regularization*, which makes it possible to define Gaussian expectation values of all the products of the local monomials in the field (equivalent to the lattice spacing in lattice models).

The propagator (9.5) depends on a parameter Λ, usually called the *cut-off*, which is an *ad hoc* replacement for an initial short distance structure at scale $1/\Lambda$. One goal of the RG is then to demonstrate the existence of a set of properties, called *universal*, independent of the choice of the specific function C.

The pole of Δ indicates that we are interested in a critical theory where the *correlation length diverges*, that is, in the corresponding QFT terminology, a *massless theory*.

In this set-up, the RG idea is implemented by an *integration over field Fourier modes belonging to a smooth interval, approximately* $[\Lambda, \Lambda - d\Lambda]$, thus replacing the initial theory but a theory with a smaller cut-off and a modified interaction $\mathcal{V}$.

9.2.2 Functional RG equations

We define

$$\tilde{D}_\Lambda(k) = -\Lambda \frac{\partial \tilde{\Delta}_\Lambda(k)}{\partial \Lambda} = \frac{2}{\Lambda^2} C'(k^2/\Lambda^2). \tag{9.6}$$

The function $\tilde{D}_\Lambda$ has an essential property: *it has no pole at $k = 0$ and thus is not critical*. The function

$$D_\Lambda(x) = -\Lambda \frac{\partial \Delta_\Lambda(x)}{\partial \Lambda} = \int \frac{d^d k}{(2\pi)^d} e^{ikx} \tilde{D}_\Lambda(k) = \Lambda^{d-2} D_{\Lambda=1}(\Lambda x), \tag{9.7}$$

thus decays for $|x| \to \infty$ faster than any power if $C(t)$ is C^∞ and exponentially if $C(t)$ is analytical.

Moreover, we set

$$\tilde{L}_\Lambda(k) = \tilde{D}_\Lambda(k)/\tilde{\Delta}_\Lambda(k), \quad L_\Lambda(x) = \int \frac{d^d k}{(2\pi)^d} e^{ikx} \tilde{L}_\Lambda(k). \tag{9.8}$$

With the parametrization $\lambda = \Lambda_0/\Lambda$, where Λ_0 is the initial cut-off, the RG equations take the form of quadratic functional equations for a scale-dependent Hamiltonian $\mathcal{H}(\phi, \lambda)$:

$$\lambda \frac{d}{d\lambda} \mathcal{H}(\phi, \lambda) = -\frac{1}{2} \int d^d x \, d^d y \, D(x-y) \left[\frac{\delta^2 \mathcal{H}}{\delta\phi(x)\delta\phi(y)} - \frac{\delta\mathcal{H}}{\delta\phi(x)} \frac{\delta\mathcal{H}}{\delta\phi(y)} \right]$$
$$- \int d^d x \frac{\delta\mathcal{H}(\phi,\lambda)}{\delta\phi(x)} \left[\frac{1}{2}(d-2+\eta[\mathcal{H}]) + \sum_\mu x^\mu \frac{\partial}{\partial x^\mu} \right] \phi(x)$$
$$- \int d^d x \, d^d y \, L_\Lambda(x-y) \frac{\delta\mathcal{H}(\phi,\lambda)}{\delta\phi(x)} \phi(y), \tag{9.9}$$

where $\eta[\mathcal{H}]$ is an adjustable renormalization of the random field ϕ, necessary to allow for a convergence of the RG flow to fixed points.

Equation (9.9) has the stationary Markovian form (9.2) and leads to a fixed point equation of the form (9.4).

RG equation and locality. It is possible to write a number of different equations, but the equation derived here has one essential feature: the evolution preserves *locality*. Indeed, as we have already noted, the Fourier transform of $D_\Lambda(x)$ is

$$\tilde{D}_\Lambda(p) = -\Lambda \frac{\partial \tilde{\Delta}_\Lambda(p)}{\partial \Lambda} = \frac{2}{\Lambda^2} C'(p^2/\Lambda^2)$$

and thus is regular, having no pole at $p = 0$: the function $D_\Lambda(x)$ is a fast decreasing function.

Note that, even if at initial scale Λ_0, the interaction is a simple polynomial in ϕ, as the property is not preserved by the evolution. *Fixed points are, in general, complicated local functionals of the field ϕ*, since an infinite number of corrections to the scaling behaviour are cancelled. By contrast, in the perturbative RG, only the leading correction is eliminated at the fixed point.

Nevertheless, in the ϕ^4 example the fixed point equation can be solved near dimension 4 as a power series in the deviation from the Gaussian theory and thus the parameter g. One then recovers the results obtained more directly from the perturbative quantum field theory RG. *However, other approximation schemes may be considered, which are not of the perturbative type.*

9.2.3 Perturbative fixed points
The Gaussian fixed point. The *Gaussian field theory*,

$$\mathcal{Z} = \int [\mathrm{d}\phi(x)] \exp\left[-\mathcal{H}_{\mathrm{G}}(\phi)\right], \quad \mathcal{H}_{\mathrm{G}}(\phi) = \frac{1}{2}(2\pi)^d \int \mathrm{d}^d k \, \tilde{\phi}(-k)\tilde{\Delta}^{-1}(k)\phi(k),$$

is a fixed point. It differs from the free massless field theory only at the cut-off scale.

The *Gaussian fixed point* leads to *Landau* or mean field *theory*. The fixed point becomes *unstable* when the *dimension d of space is less than 4*.

Near dimension 4, the instability is induced by the perturbation

$$\int \mathrm{d}^d x \, \phi^4(x).$$

However, unlike what happens in the RG of QFT, even if the field theory at initial cut-off has the simple polynomial form

$$\mathcal{H}(\phi) = \mathcal{H}_{\mathrm{G}}(\phi) + \int \mathrm{d}^d x \left(\frac{1}{2}r\phi^2(x) + \frac{1}{4!}u\phi^4(x)\right),$$

where the parameter r plays the role of temperature near T_c, this property is not preserved by the RG flow described by equation (9.9).

If a non-Gaussian fixed point exists in a neighbourhood of the Gaussian fixed point, the fixed point equation can be solved in the form of a perturbative expansion in powers of u, and u is then determined by a close equation.

For the critical theory, in the framework of the $\varepsilon = 4 - d$-expansion, an (IR) fixed point has been identified, leading to calculation of universal quantities as ε-expansions.

At leading order, in the critical domain, the results of perturbative QFT can be recovered.

9.3 Partial field integration: Some identities

We now outline a derivation of equation (9.9). We decompose the derivation into several technical steps.

It is convenient to introduce the notation

$$(\phi K\phi) \equiv \int d^d x\, d^d y\, \phi(x) K(x-y)\phi(y), \quad K(x) = K(-x).$$

9.3.1 A basic identity

We define a first Hamiltonian,

$$\mathcal{H}_1(\phi) = \tfrac{1}{2}(\phi K_1 \phi) + \mathcal{V}_1(\phi),$$

and a second Hamiltonian that depends on the two fields ϕ_1, ϕ_2,

$$\mathcal{H}(\phi_1, \phi_2) = \tfrac{1}{2}\left[(\phi_1 K_2 \phi_1) + (\phi_2 \mathcal{K} \phi_2)\right] + \mathcal{V}_1(\phi_1 + \phi_2),$$

where the kernels $K_1, K_2, \mathcal{K}$ are strictly positive. We also define

$$\Delta_1 = K_1^{-1}, \quad \Delta_2 = K_2^{-1}, \quad \mathcal{D} = \mathcal{K}^{-1}.$$

Then, if

$$\Delta_1 = \Delta_2 + \mathcal{D} \quad \Rightarrow \quad K_1 = K_2(K_2 + \mathcal{K})^{-1}\mathcal{K},$$

a Gaussian integration on ϕ_1 at $\phi_1 + \phi_2$ fixed, leads to the identity

$$\int [d\phi_1\, d\phi_2]\, e^{-\mathcal{H}(\phi_1,\phi_2)} = \left(\frac{\det(\mathcal{D}\Delta_2)}{\det \Delta_1}\right)^{1/2} \int [d\phi]\, e^{-\mathcal{H}_1(\phi)}. \tag{9.10}$$

9.3.2 Other form of the identity: Partial integration

We now define

$$e^{-\mathcal{V}_2(\phi)} = (\det \mathcal{D})^{-1/2} \int [d\varphi]\, \exp\left[-\tfrac{1}{2}(\varphi \mathcal{K} \varphi) - \mathcal{V}_1(\phi + \varphi)\right],$$

as well as

$$\mathcal{H}_2(\phi) = \tfrac{1}{2}(\phi K_2 \phi) + \mathcal{V}_2(\phi).$$

Then, the identity (9.10) becomes

$$\int [d\phi]\, e^{-\mathcal{H}_2(\phi)} = \left(\frac{\det \Delta_2}{\det \Delta_1}\right)^{1/2} \int [d\phi]\, e^{-\mathcal{H}_1(\phi)}. \tag{9.11}$$

The identity has an interpretation resulting from the partial integration over the field ϕ. In the sense of kernels, the propagator $\mathcal{D}$ is positive and, thus, $\Delta_2 < \Delta_1$. Qualitatively, the propagator Δ_2 propagates less field modes than Δ_1 does.

9.4 Partial field integration in differential form

We now assume that Δ is a C^∞ function of a parameter s with

$$\Delta \equiv \Delta(s), \quad \frac{\mathrm{d}\Delta}{\mathrm{d}s} > 0.$$

For $s > s'$, we identify

$$\Delta_1 = \Delta(s), \quad \Delta_2 = \Delta(s'), \text{ and, thus, } \mathcal{D}(s, s') = \Delta(s) - \Delta(s') > 0.$$

Also,

$$K_1 = K(s) = \Delta^{-1}(s), \quad K_2 = K(s'), \quad \mathcal{K}(s, s') = [\mathcal{D}(s, s')]^{-1} > 0,$$

and

$$\mathcal{V}_1(\phi) = \mathcal{V}(\phi, s), \ \mathcal{V}_2(\phi) = \mathcal{V}(\phi, s'), \quad \mathcal{H}_1(\phi) = \mathcal{H}(\phi, s), \ \mathcal{H}_2(\phi) = \mathcal{H}(\phi, s').$$

The identity now reads

$$\int [\mathrm{d}\phi]\, e^{-\mathcal{H}(\phi, s')} = \left(\frac{\det \Delta(s')}{\det \Delta(s)}\right)^{1/2} \int [\mathrm{d}\phi]\, e^{-\mathcal{H}(\phi, s)}$$

with

$$e^{-\mathcal{V}(\phi, s')} = \left(\det \mathcal{D}(s, s')\right)^{-1/2} \int [\mathrm{d}\varphi]\, \exp\left[-\tfrac{1}{2}\left(\varphi \mathcal{K}(s, s')\varphi\right) - \mathcal{V}(\phi + \varphi, s)\right].$$

We set

$$s' = s - \sigma, \quad \sigma > 0,$$

and study the $\sigma \to 0$ limit. Then,

$$\mathcal{K}^{-1}(s, s - \sigma) = \mathcal{D}(s, s - \sigma) = -\sigma D(s) + O(\sigma^2) \tag{9.12}$$

with

$$D(s) = -\frac{\mathrm{d}\Delta(s)}{\mathrm{d}s} = -\lim_{\sigma \to 0} \frac{1}{\sigma} \mathcal{D}(s, s - \sigma).$$

In the limit $\sigma \to 0$, in the integral, the quadratic term in the field φ is multiplied by a factor $1/\sigma$ and, thus, the values of the field contributing to the integral are of order $\sqrt{\sigma}$. We thus expand $\mathcal{V}(\phi + \varphi, s)$ to order φ^2:

$$\mathcal{V}(\phi + \varphi, s) = \mathcal{V}(\phi, s) + \int \mathrm{d}^d x \frac{\delta \mathcal{V}(\phi, s)}{\delta \phi(x)} \varphi(x)$$

$$+ \frac{1}{2} \int \mathrm{d}^d x\, \mathrm{d}^d y \frac{\delta^2 \mathcal{V}(\phi, s)}{\delta \phi(x) \delta \phi(y)} \varphi(x)\varphi(y) + O\left(\varphi^3\right).$$

The expansion of $e^{-\mathcal{V}(\phi+\varphi,s)}$ follows. The expansion of the integral in powers of σ leads to a sum of Gaussian expectation values, which can be evaluated using Wick's theorem. Multiplying both members of the equation by $e^{\mathcal{V}(\phi,s)}$, one obtains

$$e^{\mathcal{V}(\phi,s)-\mathcal{V}(\phi,s-\sigma)}$$

$$= 1 + \frac{1}{2}\int \mathrm{d}^d x \,\mathrm{d}^d y \left[\frac{\delta\mathcal{V}}{\delta\phi(x)}\frac{\delta\mathcal{V}}{\delta\phi(y)} - \frac{\delta^2\mathcal{V}}{\delta\phi(x)\delta\phi(y)}\right]\langle\varphi(x)\varphi(y)\rangle + O(\sigma^{3/2}),$$

where, at leading order, $\langle\varphi(x)\varphi(y)\rangle$ is the Gaussian expectation value corresponding to the propagator (9.12),

$$\langle\varphi(x)\varphi(y)\rangle = -\sigma D(s; x-y) + O(\sigma^2).$$

Moreover,

$$e^{\mathcal{V}(\phi,s)-\mathcal{V}(\phi,s-\sigma)} = 1 + \sigma\frac{\partial}{\partial s}\mathcal{V}(\phi,s) + O(\sigma^2).$$

Collecting all terms of order σ, one finds the equation

$$\frac{\partial}{\partial s}\mathcal{V}(\phi,s) = \frac{1}{2}\int \mathrm{d}^d x \,\mathrm{d}^d y \, D(s; x-y)\left[\frac{\delta^2\mathcal{V}}{\delta\phi(x)\delta\phi(y)} - \frac{\delta\mathcal{V}}{\delta\phi(x)}\frac{\delta\mathcal{V}}{\delta\phi(y)}\right]. \qquad (9.13)$$

This equation yields a *sufficient condition* for the partition function

$$\mathcal{Z}(s) = (\det\Delta(s))^{-1/2}\int [\mathrm{d}\phi]\, e^{-\mathcal{H}(\phi,s)} \text{ with}$$

$$\mathcal{H}(\phi,s) = \tfrac{1}{2}(\phi K(s)\phi) + \mathcal{V}(\phi,s),$$

to be independent of the parameter s (sufficient since contributions with zero expectation value can be added to the equation).

Applying this result to the problem of momentum shell integration, we identify $s \equiv \ln\Lambda$.

The equation then relates a modification of the cut-off Λ, due to the integration over field Fourier modes close to Λ, to a modification of the interaction, generating in this way an RG.

To obtain an RG equation in standard form, it suffices to apply three simple modifications:

(i) replace in equation (9.13) $\mathcal{V}$ by the complete Hamiltonian $\mathcal{H}(\phi,s)$;

(ii) introduce a adjustable renormalization of the random field ϕ;

(iii) eliminate the explicit dependence on the parameter Λ in favour of a scale parameter in the right-hand side of the equation.

10 O(N) vector model in the ordered phase: Goldstone modes

In statistical systems in which a continuous symmetry is spontaneously broken, in the whole ordered phase, correlations decay algebraically and the large distance physics can be described in terms of weakly interacting *Goldstone modes* (massless Goldstone bosons, in the terminology of particle physics).

We consider here a simple example, the lattice $O(N)$ vector model, an $O(N)$ symmetric statistical model of classical spins (in the magnetic language) of unit length interacting on a lattice with nearest-neighbour interactions.

Wilson's renormalization group (RG) arguments [80], as well as a systematic expansion of corrections to the mean field approximation [28], indicate that, in the neighbourhood of the critical temperature, the large distance physics of the $O(N)$ model can be reproduced by a $(\phi^2)^2$ statistical or quantum field theory. However, such a field theory is not the most convenient tool to study the whole ordered low temperature phase.

By contrast, the formal continuum limit of the low temperature expansion leads to the $O(N)$ symmetric non-linear σ-model, a field theory characterized by a symmetry acting non-linearly on fields.

Conversely, the classical spin model can be considered as a non-perturbative *regularization (by a short distance modification)* of the non-linear σ-model to control the large momentum (ultraviolet (UV)) divergences of the field theory.

The non-linear σ-model can also be considered as an *effective field theory* with respect to the $(\phi^2)^2$ field theory, when one wants to describe only the ordered phase (see Section 8.6.1). Moreover, for N large, to all orders in an expansion in $1/N$, the $(\phi^2)^2$ and non-linear σ-models have identical large scale properties [86, 47] (see also Section 8.6.1).

The non-linear σ-model. Above two dimensions, the non-linear σ-model has a perturbative low temperature expansion that displays only the physics of weakly interacting Goldstone (massless) modes. In dimension 2, zero momentum (infrared (IR)) divergences (in the Fourier representation) appear and the perturbative expansion is no longer defined. This property is consistent with the Mermin–Wagner–Coleman theorem [122], which forbids spontaneous symmetry breaking with ordering of continuous symmetries in models with short range interactions.

After introduction of an explicit symmetry breaking that gives a small mass to Goldstone modes, a renormalized theory can be defined and RG equations follow.

A β-function can be calculated, which shows that, for $N > 2$, zero temperature is a UV stable fixed point [85, 86] (the property of *asymptotic freedom*), and indicates that, beyond perturbation theory, in the vanishing symmetric breaking limit and with a proper regularization (like the lattice model), the $O(N)$ symmetry remains unbroken and the spectrum exhibits N equal mass particles.

From Random Walks to Random Matrices. Jean Zinn-Justin, Oxford University Press (2019).
© Jean Zinn-Justin. DOI: 10.1093/oso/9780198787754.001.0001

These properties are reminiscent of what is observed in four-dimensional non-Abelian gauge theories, for example in *quantum chromodynamics* (QCD) (see Chapter 13). Notice also that both non-linear σ-models and non-Abelian gauge theories are geometric models in the sense that the form of interactions follows from a geometric principle.

To investigate the properties of the model in higher dimensions, a strategy reminiscent of what has been done for the $(\phi^2)^2$ field theory below four dimensions can be used: one defines a perturbative expansion both in the temperature and, after *dimensional continuation*, in the deviation $\varepsilon = d - 2$ from dimension 2.

One infers that, in dimension $d > 2$, a dynamic scale appears at which a *crossover* occurs between an asymptotic large distance behaviour dominated by Goldstone modes and a *universal short distance behaviour*, associated to the critical domain and still a property of the continuum limit [86].

The short distance behaviour is governed by a *UV stable fixed point: the critical temperature*, a property also necessary for *asymptotic safety* (Chapter 15). These properties suggest that it is possible to define a renormalized field theory consistent at all scales.

Let us point out that, unlike the $(\phi^2)^2$ field theory, the non-linear σ-model in the continuum cannot be regularized by a simple addition of terms with higher derivatives because such an addition is unable to regularize one-loop Feynman diagrams, a problem related to the choice of ordering of non-commuting quantum operators in products.

In the continuum, dimensional regularization is available because, in dimensions $d > 1$, it renormalizes to zero the commutator of field and conjugate momentum at the same point. However, it hides the role of the $O(N)$ invariant functional integration measure and is not meaningful beyond perturbation theory.

The regularized $(\phi^2)^2$ field theory can also be considered as a regularization of the non-linear σ-model in continuum space [126] (see Section 10.11).

Terminology. In the discussion that follows, we use the terms 'Hamiltonian' in the sense of the statistical theory of gases or configuration energy for lattice models and 'Euclidean (*i.e.*, imaginary time) action' in the sense of quantum field theory, equivalently.

10.1 Classical lattice spin model and regularized non-linear σ-model

We consider an $O(N)$ symmetric statistical model defined in terms of N-component classical spins of unit length interacting through nearest-neighbour interactions on a hypercubic lattice. The model can also be considered as a *short distance regularization* (and *non-perturbative definition*) of a quantum or, rather, statistical field theory in continuum space, the non-linear σ-model.

In a terminology adapted to the discussion of spontaneous symmetry breaking of continuous symmetries, the N-component classical spin or field $\phi(x)$ belongs to the quotient space $O(N)/O(N-1)$, which reduces to the sphere S_{N-1}:

$$\phi^2(x) = 1 \,. \tag{10.1}$$

To emphasize the correspondence between the spin and the non-linear σ-models, we introduce the finite differences

$$\nabla_\mu^{\text{lat.}} \phi(x) = \phi(x + a\epsilon_\mu) - \phi(x), \tag{10.2}$$

where x belongs to $(a\mathbb{Z})^d$, a d-dimensional cubic lattice, a is the lattice spacing and ϵ_μ is the unit vector in the μ direction.

The configuration energy is chosen to be of the form

$$\mathcal{E}(\phi) = \tfrac{1}{2} \sum_{\mu=1}^{d} \sum_{x \in (a\mathbb{Z})^d} \left(\nabla_\mu^{\text{lat.}} \phi(x) \right)^2 = \sum_{x, \mu} [1 - \phi(x + a\epsilon_\mu) \cdot \phi(x)]. \tag{10.3}$$

The partition function is then given by

$$\mathcal{Z} = \int \prod_{x \in (a\mathbb{Z})^d} \delta \left(\phi^2(x) - 1 \right) \mathrm{d}\phi(x) \exp\left[-\mathcal{E}(\phi)/T \right], \tag{10.4}$$

where T is the temperature of the classical statistical model. The measure and the integrand are explicitly invariant under the $O(N)$ group.

One recognizes the partition function of a lattice model with classical spins of unit length and nearest-neighbour ferromagnetic interactions.

10.1.1 Low temperature limit

Studying the perturbative expansion around the mean field approximation, one discovers that the large distance properties of the spin model in the critical domain near the transition temperature can be reproduced by the $(\phi^2)^2$ continuum field theory.

Here, by contrast, we are interested in properties of the model at low temperature in the ordered phase, or in a weak uniform external field, and this leads to a different kind of expansion.

At low temperature, in zero external field, the spin configurations that minimize the configuration energy satisfy

$$\left| \nabla_\mu^{\text{lat.}} \phi(x) \right| = 0 \quad \Rightarrow \quad \phi(x) = \mathbf{n}, \quad \mathbf{n}^2 = 1.$$

The leading configurations correspond to all spins aligned. Since, as a consequence of the $O(N)$ symmetry, all directions of the vector $\mathbf{n}$ are equivalent, the configuration energy admits a continuous set of equivalent leading configurations.

In the case of $O(N)$ invariant correlation functions, all minima give exactly the same contribution. The summation over all minima just gives a factor, the volume of the sphere S_{N-1}, which cancels with the normalization of the partition function.

However, in the case of non-$O(N)$-invariant correlation functions, a sum over all minima is equivalent to an $O(N)$ group average. Therefore, it would seem that only the $O(N)$ invariant correlation functions do not vanish. This issue is directly related to the existence of *spontaneous symmetry breaking*.

The problem can be analysed using, in particular, the cluster properties of connected correlation functions. One concludes:

in the disordered phase, the partition function involves a summation over all spin configurations and, thus, all leading configurations;

by contrast, *in the ordered phase, the statistical system is no longer ergodic and, therefore, the sum in the partition function is restricted to the subset of spin configurations that fluctuate around the direction of spontaneous magnetization.*

Postponing the question of whether one has eventually to average over all directions, one can choose one direction and integrate out the fluctuations around the corresponding configuration, generating in this way a *low temperature expansion.*

Finally, in the presence of a uniform external field, the degeneracy is lifted and the leading configuration corresponds to ϕ aligned along the (magnetic) field.

10.1.2 *Local parametrization*

We choose as the leading configuration the vector $\mathbf{n} = (1, \mathbf{0})$. To generate the low temperature expansion of the statistical model, it is then necessary to parametrize the spin vector in terms of independent variables. A parametrization of the unit sphere adapted to the low temperature expansion is, for example,

$$\phi(x) = (\sigma(x), \boldsymbol{\pi}(x)) , \tag{10.5}$$

where $\sigma(x)$ is the component of $\phi(x)$ along $\mathbf{n}$, and $\boldsymbol{\pi}(x)$ is an orthogonal $(N-1)$-component vector. At $\boldsymbol{\pi}$ fixed, $\sigma(x)$ is then determined by equation (10.1), which becomes

$$\sigma^2(x) + \boldsymbol{\pi}^2(x) = 1 .$$

Since $1 - \sigma \ll 1$, locally the solution, valid for a half sphere, is

$$\sigma(x) = \left(1 - \boldsymbol{\pi}^2(x)\right)^{1/2} . \tag{10.6}$$

Partition function. In terms of σ and $\boldsymbol{\pi}$, the configuration energy reads

$$\mathcal{E}(\phi) = \tfrac{1}{2} \sum_{x,\mu} \left(\nabla_\mu^{\mathrm{lat.}} \phi(x)\right)^2$$
$$= \mathcal{E}(\boldsymbol{\pi}, \sigma) \equiv \tfrac{1}{2} \sum_{x,\mu} \left[\left(\nabla_\mu^{\mathrm{lat.}} \boldsymbol{\pi}(x)\right)^2 + \left(\nabla_\mu^{\mathrm{lat.}} \sigma(x)\right)^2 \right] . \tag{10.7}$$

One then replaces σ by $\sqrt{1 - \boldsymbol{\pi}^2}$, and $\mathcal{E}$ becomes a function of $\boldsymbol{\pi}$ only.

The partition function becomes

$$\mathcal{Z} = \int \prod_{x \in (a\mathbb{Z})^d} \frac{\mathrm{d}\boldsymbol{\pi}(x)}{(1 - \boldsymbol{\pi}^2(x))^{1/2}} \exp\left[-\frac{1}{T} \mathcal{E}(\boldsymbol{\pi}, \sqrt{1 - \boldsymbol{\pi}^2}) \right] , \tag{10.8}$$

where $\mathrm{d}\boldsymbol{\pi}(1 - \boldsymbol{\pi}^2)^{-1/2}$ is the invariant measure on the sphere in the $\boldsymbol{\pi}$ parametrization.

Non-linear representation of the $SO(N)$ group. In what follows, we discuss transformations of the $O(N)$ group close to the identity and thus belonging to the subgroup $SO(N)$ of matrices with determinant 1.

The sub-group $SO(N-1)$ of $(N-1) \times (N-1)$ matrices that leaves σ invariant acts linearly on the vector $\boldsymbol{\pi}$ as

$$\boldsymbol{\pi} \mapsto \mathbf{O}\boldsymbol{\pi} \,, \text{ with } \mathbf{O}\mathbf{O}^T = 1 \,, \text{ and } \det \mathbf{O} = 1 \,.$$

One can decompose the set of generators of the Lie algebra of $SO(N)$ into the set of generators of the Lie algebra of $SO(N-1)$ and the complementary set.

This complementary set corresponds to infinitesimal group transformations that generate the variations

$$\boldsymbol{\pi} \mapsto \boldsymbol{\pi} + \mathcal{D}_\omega \boldsymbol{\pi} \text{ with } \mathcal{D}_\omega \boldsymbol{\pi} = \boldsymbol{\omega}\big(1 - \boldsymbol{\pi}^2\big)^{1/2} \,, \tag{10.9}$$

where the constants ω_α are the infinitesimal parameters of the transformation.

Since the transform of the vector $\boldsymbol{\pi}$ is a non-linear function of $\boldsymbol{\pi}$, one speaks here of a *non-linear representation* of the $SO(N)$ group.

The transformation of the field σ is then a consequence of the relation $\sigma = \sqrt{1 - \boldsymbol{\pi}^2}$ and of the transformation (10.9) of the field $\boldsymbol{\pi}$. One recovers the transformation of the linear representation, since

$$\mathcal{D}_\omega \sigma \equiv \mathcal{D}_\omega \big(1 - \boldsymbol{\pi}^2\big)^{1/2} = -\big(1 - \boldsymbol{\pi}^2\big)^{-1/2} \boldsymbol{\pi} \cdot [\mathcal{D}_\omega \boldsymbol{\pi}](x) = -\boldsymbol{\omega} \cdot \boldsymbol{\pi} \,.$$

10.2 Perturbative or low temperature expansion

We indicate now how a low temperature expansion, that is, an expansion in powers of the temperature T for $T \to 0$, can be generated.

10.2.1 The $\boldsymbol{\pi}$-integration

We have chosen to sum over configurations in the neighbourhood of the configuration $\boldsymbol{\pi}(x) = 0$. For $T \to 0$, the fields $\boldsymbol{\pi}(x)$ that contribute to the partition function are then such that

$$\big|\nabla_\mu^{\text{lat.}} \boldsymbol{\pi}(x)\big| = O\left(\sqrt{T}\right) \,.$$

Combined with the choice of expanding around $\boldsymbol{\pi}(x) = 0$, this implies

$$|\boldsymbol{\pi}(x)| = O\left(\sqrt{T}\right) \,.$$

The T dependence in the integrand is typical of integrals calculable by the steepest descent method. The values of $\boldsymbol{\pi}(x)$ of order 1 give exponentially small contributions to the field integral (of order $\exp(-\text{const.}/T)$), which are negligible to all orders in the perturbative expansion. As a consequence, to all orders in the expansion in powers of T, in the $\boldsymbol{\pi}$-integral *one can integrate freely over the components of $\boldsymbol{\pi}(x)$ from $+\infty$ to $-\infty$*, the constraints generated by

$$|\boldsymbol{\pi}(x)| \leq 1 \,,$$

again leading to exponentially small corrections.

The $\boldsymbol{\pi}$-integral at leading order is given by a Gaussian integral, and the perturbative expansion reduces, as usual, to the evaluation of Gaussian expectation values.

10.2.2 *The configuration energy and the measure*

Loop expansion. The dependence of the configuration energy on T shows that T orders the *loop expansion* in the sense of Feynman diagrams. Indeed, for a diagram contributing to a vertex function (the sum of one-particle or one-line irreducible Feynman diagrams), the following relation holds:

$$L = I - V + 1,$$

where L is the number of loops, I the number of internal lines and V the number of vertices. The same diagram is affected by the power $T^I \times T^{-V} = T^{L-1}$.

Interaction vertices. A problem is that the configuration energy $\mathcal{E}(\boldsymbol{\pi}, \sigma)$ is not polynomial. However, since $\boldsymbol{\pi}$ is of order $\sqrt{T}$,

$$\sigma = (1 - \boldsymbol{\pi}^2)^{1/2} = 1 - \tfrac{1}{2}\boldsymbol{\pi}^2 + O(\boldsymbol{\pi}^4),$$

$$\mathcal{E}(\boldsymbol{\pi}, \sigma) = \frac{1}{2}\sum_{x,\mu}\left(\nabla^{\text{lat.}}_\mu \boldsymbol{\pi}(x)\right)^2 + O(\boldsymbol{\pi}^4).$$

The expansion generates an infinite number of interaction vertices, even in $\boldsymbol{\pi}$. However, topological relations satisfied by Feynman diagrams show that, for a given connected correlation function and at a finite loop order (*i.e.*, a finite power of T), only a finite number of terms of the expansion contribute.

The functional measure. The measure term

$$\prod_x \left(1 - \boldsymbol{\pi}^2(x)\right)^{-1/2} = \exp\left[-\frac{1}{2}\sum_x \ln\left(1 - \boldsymbol{\pi}^2(x)\right)\right],$$

expanded in powers of $\boldsymbol{\pi}$, generates additional derivative-free, vertices. Since the measure term contains no factor $1/T$, it begins contributing only at one-loop order, confirming that the measure term is a quantum correction.

10.2.3 *The propagator*

In terms of the Fourier representation of the field,

$$\boldsymbol{\pi}(x) = \int \mathrm{d}^d p \, \mathrm{e}^{ipx} \, \tilde{\boldsymbol{\pi}}(p), \tag{10.10}$$

where p belongs to a Brillouin zone of linear size $2\pi/a$, the finite difference becomes

$$\nabla^{\text{lat.}}_\mu \boldsymbol{\pi}(x) = \int \mathrm{d}^d p \, \mathrm{e}^{ipx} \left(\mathrm{e}^{iap_\mu} - 1\right) \tilde{\boldsymbol{\pi}}(p).$$

The configuration energy can then be written as

$$\sum_{x,\mu}\left(\nabla^{\text{lat.}}_\mu \boldsymbol{\pi}(x)\right)^2 = 2(2\pi/a)^d \int \mathrm{d}^d p \, \tilde{\boldsymbol{\pi}}(p) \cdot \tilde{\boldsymbol{\pi}}(-p) \sum_\mu \left(1 - \cos(ap_\mu)\right).$$

The propagator, or two-point correlation function in the Gaussian approximation, of the π-field reads

$$\langle \pi_\alpha(x)\pi_\beta(0)\rangle|_{T\to 0} \sim \frac{\delta_{\alpha\beta}}{(2\pi)^d} \int \mathrm{d}^d p \, \mathrm{e}^{-ipx} \, \tilde{\Delta}(p)$$

with

$$\tilde{\Delta}(p) = \frac{2a^d T}{d - \sum_\mu \cos a p_\mu} = \frac{a^{d-2}T}{p^2 + O(a^2 p^4)}. \tag{10.11}$$

At leading order, the correlation length of the π-field thus is infinite (the field is massless in the particle terminology): the *low temperature expansion corresponds automatically to the situation where the $O(N)$ symmetry is spontaneously broken*. The π-field corresponds to the *Goldstone modes* associated with the spontaneous symmetry breaking of the $SO(N)$ symmetry that preserves the $SO(N-1)$ subgroup. Non-trivial large distance properties are generated.

The component σ, which is the massive partner of the π-field in the linear realization, has been eliminated by the constraint $\phi^2(x) = 1$.

Finally, one can verify that these properties are independent of the specific parametrization (10.5) of $\phi(x)$.

Other groups. The situation is a special case of a general result. If an initial continuous (Lie) symmetry group G is spontaneously broken down to a subgroup H, this generates a number of Goldstone bosons equal to the number of generators of the group G minus the number of generators of H.

In particle physics, at low energy the sector of strong interactions has an approximate $SU(2) \times SU(2)$ symmetry with six generators. The symmetry is broken down to $SU(2)$ with three generators. The three pions $\pi_\pm, \pi_0$ have a small mass compared to other particles and are the quasi-Goldstone particles associated to the symmetry breaking.

Still in particle physics, in the theory of weak and electromagnetic interactions, the group G is $SU(2) \times U(1)$, which has four generators and the subgroup H is $U(1)$, with one generator. One thus expects three Goldstone bosons. In fact, due to the gauge symmetry, the Goldstone bosons are replaced by the longitudinal parts of three massive vector fields, $Z, W_\pm$.

10.2.4 Gaussian fixed point and perturbations

For what follows, it is convenient to extend the definition of $\pi(x)$ by an infinitely differentiable interpolation to arbitrary real values of x. We can then expand the finite difference (10.2),

$$\nabla^{\text{lat.}}_\mu \pi(x) = \pi(x + a\epsilon_\mu) - \pi(x) = a\partial_\mu \pi(x) + \tfrac{1}{2}a^2 \partial^2_\mu \pi(x) + \cdots.$$

It is clear that, in the large distance, small lattice spacing limit, the leading contribution comes from the first derivative. The same argument applies for $\mathrm{D}_\mu \sigma(x)$.

Keeping only the leading contributions at large distance, one obtains the continuum partition function

$$\mathcal{Z} = \int [\mathrm{d}\Theta(\boldsymbol{\pi})] \exp\left[-\frac{a^{2-d}}{T}\mathcal{S}(\boldsymbol{\pi})\right], \tag{10.12}$$

where the measure $\mathrm{d}\Theta(\boldsymbol{\pi})$ is the product over all space points of $(1 - \boldsymbol{\pi}(x))^{-1/2}$ and the action reads

$$\mathcal{S}(\boldsymbol{\pi}) = \tfrac{1}{2}\int \mathrm{d}^d x \left[\left(\nabla_x\boldsymbol{\pi}(x)\right)^2 + \left(\nabla_x\sigma(x)\right)^2\right] \tag{10.13}$$

with

$$\sigma(x) = \sqrt{1 - \boldsymbol{\pi}^2(x)}.$$

The factor a^{-d} in front of the action in expression (10.12) is generated by the transformation of a sum on integers into a continuum integral.

Expanding in powers of $\boldsymbol{\pi}$, at leading order one finds

$$\frac{a^{2-d}}{T}\mathcal{S}(\boldsymbol{\pi}) = \frac{a^{2-d}}{2T}\int \mathrm{d}^d x \left(\nabla_x\boldsymbol{\pi}(x)\right)^2.$$

Therefore, for $d \geq 2$, $\boldsymbol{\pi}$ is of order $a^{d/2-1}$. Rescaling then the field by setting

$$\boldsymbol{\pi}(x) = a^{d/2-1}\boldsymbol{\pi}'(x),$$

one finds that $\boldsymbol{\pi}'(x)$ has a mass dimension $[\boldsymbol{\pi}'] = \tfrac{1}{2}(d-2)$ typical of a scalar field.

Moreover, a monomial contributing to the action with r derivatives and $2n$ powers of $\boldsymbol{\pi}'(x)$ is multiplied by a power $a^{\ell_{nr}}$ given by

$$\ell_{nr} = (n-1)(d-2) + r - 2.$$

The analysis of perturbations to the Gaussian fixed point

$$\int \mathrm{d}^d x \left(\nabla_x\boldsymbol{\pi}'(x)\right)^2$$

then reduces to dimensional analysis and the quantities ℓ_{nr} are also the eigenvalues associated with the eigenvectors of the linearized RG. As a consequence, in RG terminology:

(i) $\ell_{12} = 0$ corresponds to the Gaussian term and to a redundant operator;

(ii) for $d > 2$, for $n > 1$ or for $r > 2$, $\ell_{nr} > 0$ and thus all monomials are irrelevant. The leading large distance behaviour for $T < T_\mathrm{c}$, in the ordered phase, is given by the Gaussian theory.

(iii) For $d = 2$, the monomials with $r > 2$ are irrelevant, and the monomials with $r = 2$ are all marginal. It is necessary to study systematically the corrections to the Gaussian approximation.

(vi) For $d < 2$, the same monomials become relevant. This is expected, since it is known that a phase transition is then impossible.

The measure contribution. However, for $d > 2$, an additional problem arises: the measure generates derivative-free vertices, which thus seem relevant. In fact, these vertices maintain the symmetry $O(N)$ of the model and, in particular, cancel the mass corrections generated in the perturbative expansion by the other vertices. They play a role somewhat analogous to the $r_c \phi^2$ term in the $(\phi^2)^2$ Hamiltonian.

From this general analysis emerges the *special role of dimension* 2. We will thus study more systematically the model at and near two dimensions.

Lattice spacing and large momentum cut-off. In what follows, because we mainly work in Fourier space, *we introduce the momentum cut-off* $\Lambda = 1/a$ and discuss the small momentum, large cut-off limit rather than the large distance, small lattice spacing limit.

10.3 Zero momentum or IR divergences

Since the propagator behaves like $1/p^2$ for $|p| \to 0$, the perturbative expansion may exhibit zero momentum (IR) divergences. For example, the expectation value of the field σ at the one-loop order, in the initial normalization, is given by

$$\langle \sigma(x) \rangle = 1 - \tfrac{1}{2} \langle \pi^2(x) \rangle + O\left(\pi^4\right)$$

$$= 1 - \frac{(N-1)}{2(2\pi)^d} T \Lambda^{2-d} \int \frac{\mathrm{d}^d p}{p^2 + O(p^4)} + O\left(T^2\right).$$

The singularity at $p = 0$ is integrable for $d > 2$, but diverges at the critical dimension $d = 2$, the dimension at which the non-linear σ-model is renormalizable.

(i) The appearance of IR divergences in the perturbative expansion is consistent with the Mermin–Wagner–Coleman theorem. The theorem states that *spontaneous breaking with ordering ($\langle \phi \rangle \neq 0$) of continuous symmetries is impossible for $d \leq 2$ in models with short range interactions.* In the limit $d = 2$ (and except for $N = 2$, *c.f.* Section 10.12.3), the critical temperature T_c vanishes and the perturbative expansion is meaningful only in the presence of a IR regularization. As a consequence, the perturbative expansion gives no direct information about the long-distance properties of the model.

(ii) For $d > 2$, investigating the $(\phi^2)^2$ field theory (the N-vector model), one finds that the $O(N)$ symmetry is spontaneously broken at low temperature. Consistently, the perturbative expansion of the σ-model, which predicts also spontaneous symmetry breaking, has no zero momentum divergences. At low temperature, the large distance behaviour is dominated by massless Goldstone modes. In contrast with the $(\phi^2)^2$ field theory, the *perturbative expansion gives no indication about a possible critical temperature with symmetry restoration.*

10.3.1   IR regularization

To be able to define a perturbative expansion in *two dimensions*, one must first introduce an IR regularization. In an infinite volume, this necessitates giving a non-zero mass (*i.e.*, a finite correlation length) to the field π. Since the vanishing mass is a consequence of the spontaneous breaking of the $O(N)$ symmetry, it is necessary to *break the symmetry explicitly.*

One can, for example, add to the Hamiltonian an explicit mass term. However, a more convenient (technically and geometrically) method consists in adding a small uniform external field (a magnetic field in the magnetic language) linear in σ:

$$\mathcal{E}(\boldsymbol{\pi},\sigma) \mapsto \mathcal{E}(\boldsymbol{\pi},\sigma;H) = \mathcal{E}(\boldsymbol{\pi},\sigma) - H \sum_{x\in(a\mathbb{Z})^d} \sigma(x)\,, \quad H>0\,.$$

An immediate consequence of the modification is that the minimum of the configuration energy is no longer degenerate. Indeed, the minimum is now obtained by maximizing also the $H\sigma$ contribution and this implies $\boldsymbol{\pi}=0$ at the minimum.

Moreover, an expansion of σ in powers of $\boldsymbol{\pi}$,

$$\sigma = \left(1-\boldsymbol{\pi}^2\right)^{1/2} = 1 - \tfrac{1}{2}\boldsymbol{\pi}^2 + O\left((\boldsymbol{\pi}^2)^2\right),$$

shows that the quadratic terms in $\mathcal{E}(\boldsymbol{\pi},\sigma;H)$ lead to the new $\boldsymbol{\pi}$ propagator,

$$\tilde{\Delta}_{\alpha\beta}(p) = \delta_{\alpha\beta} \frac{T\Lambda^{2-d}}{p^2 + H + O(p^4)}.$$

The linear σ term has generated a mass for the field $\boldsymbol{\pi}$ proportional to $H^{1/2}$ for H small. It has also generated new, derivative-free, interactions.

In the case of the $(\boldsymbol{\phi}^2)^2$ field theory, the breaking term $\mathbf{H}\cdot\boldsymbol{\phi}$ is linear in the independent field component σ and, therefore, does not generate a new renormalization constant. The same result can be proved here, even though the component σ is a non-linear function of the field $\boldsymbol{\pi}$.

10.4 Formal continuum limit: The non-linear σ-model

The divergence of the correlation length corresponding to the field $\boldsymbol{\pi}$ implies the existence of a non-trivial large distance physics and, thus, of a continuum limit. One can try to describe the continuum limit by studying directly the formal continuum limit of the lattice model.

We have shown that the leading corrections to the Gaussian model come from the vertices with two derivatives. The most general configuration energy, or Euclidean action, invariant under the $O(N)$ group, a function of $\boldsymbol{\phi}$ and its derivatives that involves at most two derivatives (the leading terms at large distance), reads

$$\mathcal{S}(\boldsymbol{\phi}) = \frac{1}{2}\int \mathrm{d}^d x\, \nabla_x \boldsymbol{\phi}(x)\cdot\nabla_x\boldsymbol{\phi}(x)\,, \tag{10.14}$$

up to a multiplicative constant. (Translation invariance and isotropy in $\mathbb{R}^d$ have been used.) Indeed, each symmetric and derivative-free local term is a function of $\boldsymbol{\phi}^2(x)$ and thus reduces to a constant, and $\boldsymbol{\phi}(x)\cdot\nabla_x\boldsymbol{\phi}(x)$ vanishes.

The large distance physics in a model with $O(N)$ symmetry can thus be described, in dimension $d > 2$, below T_c, by the non-linear σ-model. The corresponding partition function can be formally written as (equation (10.12))

$$\mathcal{Z} = \int [\mathrm{d}\Theta(\phi)] \exp\left[-\frac{\Lambda^{d-2}}{T} \mathcal{S}(\phi) \right],$$

where the measure $\mathrm{d}\Theta(\phi)$ is the product over all space points of $\delta(\phi^2(x) - 1)\mathrm{d}\phi(x)$.

Parametrized in terms of the components $(\boldsymbol{\pi}, \sigma)$, a form adapted to a perturbative expansion for $T \to 0$, the action reads (*cf.* equation (10.13))

$$\mathcal{S}(\boldsymbol{\pi}) = \tfrac{1}{2} \int \mathrm{d}^d x \left[\left(\nabla_x \boldsymbol{\pi}(x)\right)^2 + \left(\nabla_x \sigma(x)\right)^2 \right] \tag{10.15}$$

with

$$\sigma(x) = \sqrt{1 - \boldsymbol{\pi}^2(x)}\,.$$

The euclidean action (10.15) can also be written in a coordinate independent way as

$$\mathcal{S}(\boldsymbol{\pi}) = \frac{1}{2} \int \mathrm{d}^d x \sum_{\alpha,\beta} G_{\alpha\beta}\left(\boldsymbol{\pi}(x)\right) \nabla_x \pi_\alpha(x) \nabla_x \pi_\beta(x), \tag{10.16}$$

where $G_{\alpha\beta}$ is the metric tensor on the sphere, a form then valid in any parametrization of the sphere. Here,

$$G_{\alpha\beta}(\boldsymbol{\pi}) = \delta_{\alpha\beta} + \frac{\pi_\alpha \pi_\beta}{1 - \boldsymbol{\pi}^2}\,.$$

The invariant integration measure over the $\boldsymbol{\pi}$ field, in a general parametrization, is then the covariant volume element on the sphere. Here,

$$\sqrt{\det G_{\alpha\beta}}\,\mathrm{d}\boldsymbol{\pi} = \left(1 - \boldsymbol{\pi}^2\right)^{-1/2} \mathrm{d}\boldsymbol{\pi}\,.$$

Finally, the temperature T characterizes the deviation of the statistical field theory from the Gaussian theory.

Perturbative expansion. Correlation functions are formally given by the expansion in powers of **J** of the functional

$$\mathcal{Z}(\mathbf{J}) = \int \left[\frac{\mathrm{d}\boldsymbol{\pi}(x)}{(1 - \boldsymbol{\pi}^2(x))^{1/2}} \right] \exp\left(-\frac{1}{T}\mathcal{S}(\boldsymbol{\pi}) + \int \mathrm{d}^d x\, \mathbf{J}(x) \cdot \boldsymbol{\pi}(x) \right). \tag{10.17}$$

The large distance properties of the perturbative expansion are the same as those of the lattice model in Section 10.2, except for the functional measure that requires a special discussion.

In the Fourier representation, the propagator (10.11) is replaced by the simpler expression,

$$\Delta_{\pi\pi}(p) = \frac{T}{p^2}\,. \tag{10.18}$$

Non-linear σ-model: The problem of the measure. A problem appears immediately: in continuum space, the functional integration measure is not defined. From a formal viewpoint, the measure can be interpreted as a determinant and thus, since $\ln \det = \operatorname{tr} \ln$, its logarithm as a trace:

$$\ln \prod_{x} \left(1 - \pi^2(x)\right)^{-1/2} = \exp\left[-\tfrac{1}{2}\operatorname{tr}\mathbf{G}\right]$$

with

$$\mathbf{G}(x,y) = \delta^d(x-y)\ln\left(1-\pi^2(x)\right) \;\;\Rightarrow\;\; \operatorname{tr}\mathbf{G} = {}^{\text{`}}\delta^{(d)}(0)'\int d^dx\,\ln\left(1-\pi^2(x)\right).$$

The undefined quantity $\delta^{(d)}(0)$ has the equivalently undefined Fourier representation,

$$\delta^{(d)}(0) = \int \frac{d^dk}{(2\pi)^d}.$$

This expression appears also elsewhere in perturbative calculations.

One must thus find a regularization for what appears formally as an additional, infinite, quantum correction of order T to the effective Hamiltonian.

10.4.1 Correlation functions with σ insertions

To express the implications of the complete $O(N)$ symmetry for correlation functions, it is convenient to consider also correlation functions with $\sigma(x)$ insertions. The generating functional of correlation functions of π and σ fields can be written as

$$\mathcal{Z}(J,K) = \int[d\pi]\exp\left[-\frac{1}{T}\mathcal{S}(\pi) + \int d^dx\left(\mathbf{J}(x)\cdot\pi(x) + K(x)\sigma(x)\right)\right] \quad (10.19)$$

and the correlation functions are obtained by expanding in powers of $\mathbf{J}$ and K. The generating functional of connected correlation functions is

$$\mathcal{W}(J,K) = \ln\mathcal{Z}(J,K).$$

The generating functional of vertex functions is defined by the Legendre transformation

$$\mathcal{W}(J,K) + \Gamma(\pi,K) = \int d^dx\,\pi(x)\cdot\mathbf{J}(x), \quad \text{with} \quad \pi_i(x) = \frac{\delta\mathcal{W}(J,K)}{\delta J_i(x)}, \quad (10.20)$$

where the symbol δ indicates a functional derivative, and π is here a classical field.

Only one-line irreducible Feynman diagrams contribute to vertex functions.

From these relations, one can derive

$$J_i(x) = \frac{\delta\Gamma(\pi,K)}{\delta\pi_i(x)}, \quad \frac{\delta\Gamma(\pi,K)}{\delta K(x)} + \frac{\delta\mathcal{W}(J,K)}{\delta K(x)} = 0. \quad (10.21)$$

With these definitions, in the classical limit, one finds

$$\Gamma(\pi,K) = \Gamma_0(\pi,K) = \frac{\mathcal{S}(\pi)}{T} - \int d^dx\,K(x)\sigma(x) \equiv \Sigma(\pi,K). \quad (10.22)$$

Finally, after the substitution $K(x) \mapsto K(x) + H/T$ constant, $\mathcal{Z}(J,K)$ generates correlation functions corresponding to an $O(N)$ symmetric action with a symmetry breaking term linear in σ.

10.5   The continuum theory: Regularization

Besides the contributions of the functional measure, the perturbative expansion of the continuum field theory suffers from other short distance, or large momentum, divergences. It is thus necessary to modify the theory at short distance to render all diagrams finite, an operation called *regularization*.

Lattice regularization. The initial lattice model defined in Section 10.1 clearly provides a suitable regularization and also yields a non-perturbative definition of the σ-model. Moreover, this is the only regularization that allows discussing the role of the functional measure in the perturbative expansion. On the other hand, perturbative calculations are complicated and a differential RG can be defined only asymptotically, since the dilatation parameter cannot be varied continuously.

10.5.1   Dimensional regularization

Dimensional regularization preserves the $O(N)$ symmetry of the configuration energy. Moreover, dimensional regularization performs a partial renormalization: as a consequence of the rule

$$\delta^d(0) = \frac{1}{(2\pi)^d} \int d^d k = 0 \,,$$

all divergences that, in a cut-off regularization appear as power-law divergences, are cancelled. Simultaneously, the contributions coming from the measure can be ignored and, therefore, the perturbative expansion has no large momentum divergences for $d < 2$. Due to its technical simplicity, this is, generally, the regularization one uses in explicit perturbative calculations.

A theoretical inconvenience is that the role of the functional measure is obscured. The possibility that the perturbative expansion can be defined without taking explicitly into account the measure term strongly suggests that the model is not completely defined by the perturbative expansion.

Therefore, for a deeper theoretical discussion of the non-linear σ-model, lattice regularization remains necessary.

Two dimensions. In two dimensions, one finds both UV and IR divergences and this is a source of another difficulty. To define the perturbative expansion by dimensional regularization, it is necessary to give a mass (a finite correlation length) to the field π and thus to break the $O(N)$ symmetry explicitly. The addition to the action of a term linear in σ (a magnetic field in ferromagnetic systems) provides the simplest symmetry breaking mechanism and is used below (see also Section 10.3.1).

10.5.2   Derivative or Pauli–Villars's regularizations

Any regularized version of the non-linear σ-model must remain invariant under the transformations of the $O(N)$ group. It is more complicated to implement this condition here than in the case of the $(\phi^2)^2$ field theory because the $O(N)$ symmetry relates the interaction terms in the Hamiltonian to the quadratic part.

A simple method consists in starting again from a description of the model in terms of the constrained $\phi(x)$ field, because the configuration energy (10.14) is formally identical to a Gaussian critical Hamiltonian.

Rescaling distances $x \mapsto \Lambda x$ (Λ is the cut-off), we thus replace $\mathcal{S}(\phi)$ by

$$\mathcal{S}_\Lambda(\phi) = \frac{\Lambda^{d-2}}{2} \int d^d x\, \phi(x) \cdot \left(-\nabla_x^2 + \frac{\alpha_2}{\Lambda^2}\nabla_x^4 - \frac{\alpha_3}{\Lambda^4}\nabla_x^6 + \cdots \right) \phi(x), \qquad (10.23)$$

where the α_i's are arbitrary dimensionless parameters, but chosen such that $\mathcal{S}_\Lambda$ remains bounded from below.

Expressing then $\phi(x)$ in terms of $\pi(x)$, one discovers that the large momentum behaviour of the propagator is improved (fields contributing to the field integral are more regular), but, simultaneously, new, more singular, interactions have been introduced. The result is that *all diagrams can be regularized except one-loop diagrams*, whose degree of divergence has not changed.

This property is related to another limitation of Pauli–Villars regularization: it does not solve the problem of the divergent contributions generated by the functional measure. Actually, it can be shown that the remaining one-loop divergences generated by the interactions are needed to formally cancel the divergent contributions coming from the measure (*e.g.,*, by using the phase space field integral).

In fact these divergences, which are present even for $d = 1$, are a consequence of the problem of *ordering of non-commuting quantum operators in products*. Indeed, the path integral representation of quantum mechanics, formally depends only on the classical Hamiltonian, which does not determine completely the quantum Hamiltonian when products of non-commuting operators appear.

One way to solve the one-loop problem, is to enforce the Ward–Takahashi (WT) identities, relations between correlation functions that result from the $O(N)$ symmetry (equation (10.27)). Once the one-loop diagrams have been defined, one inserts them into higher order diagrams when they appear as subdiagrams.

Finally, another regularization is provided by the $(\phi^2)^2$ field theory in the spontaneously broken phase, itself regularized by Pauli–Villars's regularization.

10.6 Symmetry and renormalization

In this section, we assume lattice regularization even though we use continuum notation. Dimensional regularization can also be used but then, as already explained, the role of the functional measure cannot be discussed.

We denote by Λ the momentum cut-off, representative of the scale of microscopic theory (like the inverse lattice spacing). We substitute in the generating functional (10.19) $T \mapsto \Lambda^{2-d}T$ in such a way that T becomes dimensionless:

$$\mathcal{Z}(J, K) = \int [d\pi] \exp\left[-\Sigma(\pi, K) + \int d^d x\, \mathbf{J}(x) \cdot \pi(x) \right] \qquad (10.24)$$

with

$$\Sigma(\pi, K) = \frac{\Lambda^{d-2}}{T}\mathcal{S}(\pi) - \int d^d x\, K(x)\sigma(x). \qquad (10.25)$$

Assuming a suitable regularization, one can discuss the renormalization of the non-linear σ-model, that is, how to modify the Hamiltonian to generate a finite perturbative expansion in the limit where the regularization is removed ($\Lambda \to \infty$).

A quantum field theory is called renormalizable by power counting if one can find suitable additions (called counter-terms) to the Hamiltonian to renormalize the perturbative expansion. When the theory has a symmetry, one has, in addition, to prove that counter-terms can be found that preserve the symmetry.

The non-linear σ-model is renormalizable by power counting in dimension 2 but, *a priori*, any even local monomial in $\boldsymbol{\pi}$ containing at most two derivatives can appear as a counter-term. Moreover, it is simple to prove that the counter-terms can preserve the residual symmetry $O(N-1)$ because it is linearly realized.

However, the non-linear part of the $O(N)$ transformations, as defined explicitly in equation (10.9), cannot survive renormalization because the radius of the sphere (10.1) is renormalized. To determine the general form of the renormalized action, it is necessary to express the non-linear part of the $O(N)$ symmetry in the form of functional WT identities.

10.6.1 *WT identities and master equation*

The $O(N)$ invariance of the action (10.16) implies the $O(N)$ invariance of $\mathcal{W}(J, K)$. In particular, the invariance under the infinitesimal transformation,

$$\mathbf{J}(x) \mapsto \mathbf{J}(x) + \boldsymbol{\omega} K(x), \quad K(x) \mapsto K(x) - \boldsymbol{\omega} \cdot \mathbf{J}(x), \quad \text{with } \boldsymbol{\omega} \text{ constant, } |\boldsymbol{\omega}| \ll 1,$$

implies

$$\int \mathrm{d}^d x \left(K(x) \frac{\delta \mathcal{W}(J, K)}{\delta J_i(x)} - J_i(x) \frac{\delta \mathcal{W}(J, K)}{\delta K(x)} \right) = 0. \qquad (10.26)$$

Using the relations (10.20) and (10.21), one can translate equation (10.26) into an equation for $\Gamma(\boldsymbol{\pi}, K)$, the generating functional of vertex functions. One obtains the master equation,

$$\int \mathrm{d}^d x \left(\frac{\delta \Gamma}{\delta \pi_i(x)} \frac{\delta \Gamma}{\delta K(x)} + K(x) \pi_i(x) \right) = 0. \qquad (10.27)$$

This functional equation expanded in powers of the π and K fields generates an infinite number of relations between vertex (and then correlation) functions, which express the non-linear part of the symmetry. In particular, it determines relations between the divergent parts of vertex functions, which have to be removed by adding counter-terms to the action.

It can be verified that the π-action (10.25) in an external space-dependent external field satisfies the same equation:

$$\int \mathrm{d}^d x \left(\frac{\delta \Sigma}{\delta \pi_i(x)} \frac{\delta \Sigma}{\delta K(x)} + K(x) \pi_i(x) \right) = 0. \qquad (10.28)$$

One then proves recursively on the number of loops (in the sense of Feynman diagrams) that the theory can be renormalized while preserving equations (10.27) and (10.28).

Solution of the master equation. The renormalized action is then the general local solution to equation (10.28), restricted by power counting: in two dimensions, the action density has dimension 2, the field has dimension 0, its derivative has dimension 1, and H has dimension 2. Therefore, the solution of equation (10.28) can have at most a term linear in K, since a term quadratic in K has at least dimension 4. We thus set

$$\Sigma_\mathrm{r}(\boldsymbol{\pi}, K) = \Sigma_\mathrm{r}(\boldsymbol{\pi}) - \int \mathrm{d}^d x \, \sigma_\mathrm{r}\big(\boldsymbol{\pi}(x)\big) K(x). \tag{10.29}$$

Moreover, the coefficient $\sigma_\mathrm{r}(\boldsymbol{\pi})$ is dimensionless and thus a derivative-free function of $\boldsymbol{\pi}^2(x)$ while $\Sigma_\mathrm{r}(\boldsymbol{\pi})$ has dimension 2 and contains terms with at most two derivatives.

From the coefficient of $K(x)$ in equation (10.28), one infers

$$\sigma_\mathrm{r}(\boldsymbol{\pi}) \frac{\delta \sigma_\mathrm{r}}{\delta \boldsymbol{\pi}(x)} + \boldsymbol{\pi}(x) = 0. \tag{10.30}$$

Since $\sigma_\mathrm{r}(\boldsymbol{\pi})$ is derivative-free and $O(N-1)$ invariant, the solution of equation (10.30) is simply

$$\sigma_\mathrm{r}^2\left(\boldsymbol{\pi}^2\right) + \boldsymbol{\pi}^2(x) = \boldsymbol{\phi}_\mathrm{r}^2(x) = Z^{-1}. \tag{10.31}$$

The equation shows that $\sigma_\mathrm{r}(\boldsymbol{\pi})$ is the renormalized σ-field, that Z is the field renormalization constant and that the radius of the sphere has been renormalized.

The equation taken for $K(x) = 0$ then yields

$$\int \mathrm{d}^d x \, \frac{\delta \Sigma_\mathrm{r}(\boldsymbol{\pi})}{\delta \pi_i(x)} \sigma_\mathrm{r}\left(\boldsymbol{\pi}(x)\right) = 0. \tag{10.32}$$

The equation expresses the property that $\Sigma_\mathrm{r}(\boldsymbol{\pi})$ is invariant under an infinitesimal transformation of the form

$$\delta \boldsymbol{\pi}(x) = \boldsymbol{\omega} \, \sigma_\mathrm{r}\big(\boldsymbol{\pi}(x)\big)$$

or, solving equation (10.31),

$$\delta \boldsymbol{\pi}(x) = \boldsymbol{\omega} \left(Z^{-1} - \boldsymbol{\pi}^2(x)\right)^{1/2}. \tag{10.33}$$

This is the renormalized form of the non-linear part of the $O(N)$ transformations. Equations (10.31) and (10.32) show that the renormalized functional $\Sigma_\mathrm{r}(\boldsymbol{\pi})$ is still $O(N)$ invariant: simply the radius of the sphere has been renormalized. From the power counting analysis, we know that Σ_r has at most two derivatives. Expressed in terms of renormalized fields, it is thus proportional to

$$\int \left[(\nabla \boldsymbol{\pi}_\mathrm{r}(x))^2 + (\nabla \sigma_\mathrm{r}(x))^2 \right] \mathrm{d}^d x \,,$$

where we have now introduced the notation $\boldsymbol{\pi}_r$ to indicate the fields of the unrenormalized action. Two renormalization constants thus are needed: the renormalization of the radius of the sphere, which is also a *field renormalization*, and a global renormalization of the action, corresponding to a *renormalization of the temperature T*.

10.6.2 *Renormalization constants and renormalized action*

In the initial action in a space-dependent field (10.25),

$$\Sigma(\boldsymbol{\pi}, K) = \int \mathrm{d}^d x \left\{ \frac{\Lambda^{d-2}}{2T} \left[(\nabla_x \boldsymbol{\pi}(x))^2 + (\nabla_x \sigma(x))^2 \right] - H(x)\sigma(x) \right\}, \quad \text{with}$$

$$\sigma(x) = \sqrt{1 - \boldsymbol{\pi}^2(x)}, \tag{10.34}$$

we set $K(x) = H\Lambda^{d-2}/T$, which amounts adding a symmetry breaking term linear in σ in the action.

Introducing a renormalization scale μ and the corresponding renormalized dimensionless coupling g, which is the effective coupling at scale μ, one can write the renormalized action in a space-dependent external field K as

$$\Sigma_{\mathrm{r}}(\boldsymbol{\pi}_{\mathrm{r}}, K) = \frac{\mu^{d-2}}{2g} \frac{Z}{Z_T} \int \left[(\nabla \boldsymbol{\pi}_{\mathrm{r}}(x))^2 + (\nabla \sigma_{\mathrm{r}}(x))^2 \right] \mathrm{d}^d x - \int K(x) \sigma_{\mathrm{r}}(x) \, \mathrm{d}^d x \tag{10.35}$$

with the relation

$$\sigma_{\mathrm{r}}(x) = \left[Z^{-1} - \boldsymbol{\pi}_{\mathrm{r}}^2(x) \right]^{1/2}. \tag{10.36}$$

In the renormalized action, we then set $K(x) = h\mu^{d-2}/g$, where h is the effective parameter at scale μ.

Comparing the expressions (10.35) and (10.34), one finds

$$\boldsymbol{\pi}(x) = Z^{1/2}\boldsymbol{\pi}_{\mathrm{r}}(x) \ \Rightarrow \ \sigma(x) = Z^{1/2}\sigma_{\mathrm{r}}(x),$$

the relations

$$\Lambda^{2-d}T = \mu^{2-d}gZ_T \tag{10.37}$$

and

$$h\frac{\mu^{d-2}}{g\sqrt{Z}} = H\frac{\Lambda^{d-2}}{T}. \tag{10.38}$$

The master equation (10.27) makes it also possible to discuss the effect of the symmetry breaking on correlation functions by expanding around $K(x) = \text{constant} \neq 0$.

Let us point out that μ plays the role of an intermediate scale characteristic of the critical domain around the Gaussian fixed point. The large distance behaviour corresponds to momenta small compared to μ, the universal short distance behaviour, when it exists, to momenta large compared to μ.

Other parametrizations. It is possible to discuss renormalization in the case of other parametrizations of the sphere. However, then the renormalization of the parametrization is more complicated and may involve an infinite number of renormalization constants.

10.7 Correlation functions in dimension $d = 2 + \varepsilon$ at one loop

The non-linear σ-model is *renormalizable in two dimensions*, and some of its properties can be inferred from RG arguments. To be able to discuss also physics for dimension $d > 2$, beyond the leading large distance behaviour of non-interacting Goldstone modes, we use dimensional continuation and set [86]

$$d = 2 + \varepsilon, \quad \varepsilon \geq 0,$$

in formal analogy to what has been done to explore the neighbourhood of dimension 4 in the $(\phi^2)^2$ field theory.

We then study the field theory within the framework of a double series expansion in powers of the temperature T and of ε.

In what follows, we assume *regularization by terms with higher derivatives*, despite the one-loop problem because it is more physical than dimensional regularization and calculations are easier than with lattice regularization. The one-loop problem is solved by imposing WT identities.

10.7.1 The field expectation value at one-loop order

At one-loop order, wee only need the $\boldsymbol{\pi}$ propagator,

$$\tilde{\Delta}_{\pi\pi}(p) = \frac{\Lambda^{2-d}T}{p^2 + H}. \tag{10.39}$$

The field expectation value $\bar{\sigma}$ of the composite field $\sigma(x)$ is given by

$$\bar{\sigma} \equiv \langle \sigma(x) \rangle = \left\langle \sqrt{1 - \boldsymbol{\pi}^2(x)} \right\rangle$$
$$= 1 - \tfrac{1}{2}\left\langle \boldsymbol{\pi}^2(x) \right\rangle + O(T^2) = 1 - \tfrac{1}{2}(N-1)T\Lambda^{2-d}\Omega_d(\sqrt{H}) + O(T^2), \tag{10.40}$$

where have defined (see Fig. 10.1)

$$\Omega_d(m) = \frac{1}{(2\pi)^d} \int \frac{\mathrm{d}^d q}{q^2 + m^2} \tag{10.41}$$

with the cut-off Λ implied. For $d = 2 + \varepsilon$,

$$\Omega_d(m) = \frac{1}{2\pi} \ln(\Lambda/m) + O(1) + O(\varepsilon).$$

Thus,

$$\bar{\sigma} = 1 - \frac{N-1}{4\pi}T\ln(\Lambda/\sqrt{H}) + O(T^2, T\varepsilon). \tag{10.42}$$

At this order, $\bar{\sigma}$ is not affected by the measure problem and can be regularized by a momentum cut-off provided by adding terms with higher derivatives to the action.

Fig. 10.1 The one-loop diagram Ω_d.

10.7.2 The two-point vertex function at one-loop order

WT identity. In momentum regularization, the two-point vertex function (or inverse two-point connected correlation function) is affected by divergences that cannot be regularized, even though formally they cancel. The problem can be solved by enforcing a WT identity obtained by differentiating equation (10.27) with respect to $\pi_j(y)$ and setting $\boldsymbol{\pi} = 0$, $K(x) = H\Lambda^{d-2}/T$. One finds

$$-\int \mathrm{d}^d x \, \langle \sigma(x) \rangle \, \Gamma_{ij}^{(2)}(x-y) + \delta_{ij} \frac{H\Lambda^{d-2}}{T} = 0 \,,$$

where the relation

$$\frac{\delta \Gamma}{\delta K(x)} \bigg|_{K=0,\pi=0} = - \langle \sigma(x) \rangle \,,$$

has been used. After Fourier transformation, the WT identity becomes

$$\bar{\sigma} \tilde{\Gamma}_{ij}^{(2)}(p=0) = \delta_{ij} \frac{H\Lambda^{d-2}}{T} \,. \tag{10.43}$$

Calculation. At one-loop order, the calculation requires the vertex $V^{(4)}$ of order π^4. One first uses the identity

$$\int \mathrm{d}^d x \, [\boldsymbol{\pi}(x) \cdot \nabla_x \boldsymbol{\pi}(x)]^2 = \frac{1}{4} \int \mathrm{d}^d x \, [\nabla_x(\boldsymbol{\pi}^2(x))]^2 = -\frac{1}{4} \int \mathrm{d}^d x \, \boldsymbol{\pi}^2(x) \nabla_x^2(\boldsymbol{\pi}^2(x)) .$$

In the Fourier representation,

$$-\frac{1}{4} \int \mathrm{d}^d x \, \boldsymbol{\pi}^2(x) \nabla_x^2(\boldsymbol{\pi}^2(x))$$

$$= \frac{1}{4} \int \mathrm{d}^d x \int \mathrm{d}^d p_1 \, \mathrm{d}^d p_2 \, \mathrm{e}^{ix(p_1+p_2)} \, \tilde{\boldsymbol{\pi}}(p_1) \cdot \tilde{\boldsymbol{\pi}}(p_2) \int \mathrm{d}^d p_3 \, \mathrm{d}^d p_4 \, (p_3 + p_4)^2$$

$$\times \mathrm{e}^{ix(p_3+p_4)} \, \tilde{\boldsymbol{\pi}}(p_3) \cdot \tilde{\boldsymbol{\pi}}(p_4)$$

$$= \frac{1}{4} (2\pi)^d \int \mathrm{d}^d p_1 \, \mathrm{d}^d p_2 \, \mathrm{d}^d p_3 \, \mathrm{d}^d p_4 \, \delta^{(d)} \left(\sum p_i \right) (p_3 + p_4)^2 \, \tilde{\boldsymbol{\pi}}(p_1) \cdot \tilde{\boldsymbol{\pi}}(p_2)$$

$$\times \tilde{\boldsymbol{\pi}}(p_3) \cdot \tilde{\boldsymbol{\pi}}(p_4).$$

One then verifies that the contribution of order $|\boldsymbol{\pi}|^4$ coming from the expansion of $H\sigma$ has the effect of replacing $(p_3 + p_4)^2$ by $(p_3 + p_4)^2 + H$ and leads to the vertex $\tilde{V}^{(4)}$ shown in Fig. 10.2.

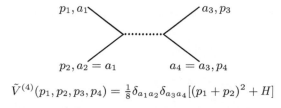

$$\tilde{V}^{(4)}(p_1, p_2, p_3, p_4) = \tfrac{1}{8} \delta_{a_1 a_2} \delta_{a_3 a_4} [(p_1 + p_2)^2 + H]$$

Fig. 10.2 Vertex coefficient of $\prod_{i=1,4} \tilde{\pi}_{a_i}(p_i)$. This is a faithful representation: the dotted line does not correspond to a propagator but makes it possible to represent faithfully the flow of group indices by full lines.

The calculation of the one-loop diagram shown in Fig. 10.3 follows. The diagram decomposes, in a faithful representation, into the sum of the two diagrams shown in Fig. 10.4, where the flow of group indices is explicit. It can be expressed in terms of the function Ω_d.

Fig. 10.3 Two-point function: One-loop contribution.

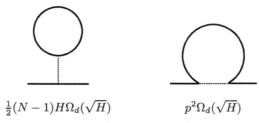

$$\tfrac{1}{2}(N-1)H\Omega_d(\sqrt{H}) \qquad\qquad p^2\Omega_d(\sqrt{H})$$

Fig. 10.4 Two-point function: faithful one-loop diagrams.

For example, the second diagram in Fig. 10.4 is given by

$$\frac{1}{(2\pi)^d}\int \mathrm{d}^d q \frac{(p+q)^2+H}{q^2+H} = \frac{p^2}{(2\pi)^d}\int \frac{\mathrm{d}^d q}{q^2+H} + \frac{1}{(2\pi)^d}\int \mathrm{d}^d q\,.$$

The contribution proportional to $\int \mathrm{d}^d q$ is formally cancelled by the π^2 contribution coming from the measure at leading order. Moreover, like the measure contribution, it vanishes in dimensional regularization.

If the two divergent contributions did not cancel exactly, an additional H independent constant, proportional to Λ^d, would be generated.

For the two-point vertex function, which has the form $\delta_{ij}\tilde{\Gamma}^{(2)}$, one then finds

$$\tilde{\Gamma}^{(2)}(p) = \frac{\Lambda^{d-2}}{T}\left(p^2+H\right) + \left[p^2+\tfrac{1}{2}(N-1)H\right]\Omega_d(\sqrt{H}) + O(T). \qquad (10.44)$$

Using the result (10.40), the WT identity (10.43) then takes the explicit form,

$$\left[1-\tfrac{1}{2}(N-1)T\Lambda^{2-d}\Omega_d(\sqrt{H})\right]\left[\frac{\Lambda^{d-2}H}{T}+\tfrac{1}{2}(N-1)H\Omega_d(\sqrt{H})\right] = \frac{\Lambda^{d-2}H}{T}+O(T).$$

The equation is exactly satisfied at one-loop order, leaving no room for an additional constant contribution to $\tilde{\Gamma}^{(2)}$.

Since the expansion of the measure gives constant divergent contributions to all π correlation functions, one can verify the consistency of the cancellation mechanism by using the WT identities obtained by successive differentiations of the master equation with respect to the π- field.

The one-loop two-point function (10.44) expanded in ε becomes

$$\tilde{\Gamma}^{(2)}(p) = \frac{\Lambda^\varepsilon}{T}\left(p^2 + H\right) + \frac{1}{2\pi}\left[p^2 + \tfrac{1}{2}(N-1)H\right]\ln(\Lambda/\sqrt{H}) + O(T,\varepsilon). \qquad (10.45)$$

10.8   RG equations

Since the perturbative expansion is renormalizable, RG equations follow.

In the case of the action (10.34), the relation between regularized and renormalized π-field vertex functions in the Fourier representation can be written as (in slightly symbolic notation because we omit the group indices)

$$Z^{n/2}(T, \Lambda/\mu)\tilde{\Gamma}^{(n)}(p_i; T, H, \Lambda) = \tilde{\Gamma}_{\mathrm{r}}^{(n)}(p_i; g, h, \mu), \qquad (10.46)$$

where the vertex functions have, for $H \propto h \neq 0$, a finite expansion in powers of T or g, and ε.

From this relation, it is possible to derive RG equations in two ways, either by differentiating with respect to μ, at fixed Λ, T, H, or by differentiating with respect to Λ, at fixed μ, g, h. In both cases, one uses the property that the perturbative renormalized functions have a finite $\Lambda \to \infty$ limit. In the case of the initial (bare) vertex or correlation functions, one expresses, in addition, that they are μ independent.

The RG equations satisfied by the initial vertex functions then are,

$$\left[\Lambda\frac{\partial}{\partial\Lambda} + \beta(T)\frac{\partial}{\partial T} - \frac{n}{2}\zeta(T) + \rho(T)H\frac{\partial}{\partial H}\right]\tilde{\Gamma}^{(n)}(p_i; T, H, \Lambda) = 0. \qquad (10.47)$$

In particular, the RG functions defined by

$$\Lambda\frac{\partial}{\partial\Lambda}\bigg|_{\mu,g,h \text{ fixed}} T = \beta(T),$$

$$\Lambda\frac{\partial}{\partial\Lambda}\bigg|_{\mu,g,h \text{ fixed}} -\ln Z = \zeta(T),$$

$$\Lambda\frac{\partial}{\partial\Lambda}\bigg|_{\mu,g,h \text{ fixed}} \ln H = \rho(T),$$

cannot depend on Λ/μ. Using the relation (10.37), one derives the explicit expression of the β-function of the non-linear σ-model,

$$\beta(T) = \varepsilon T\left(1 - T\frac{\mathrm{d}\ln Z_T}{\mathrm{d}T}\right)^{-1}. \qquad (10.48)$$

It follows that

$$\zeta(T) = -\beta(T)\frac{\mathrm{d}\ln Z}{\mathrm{d}T}.$$ (10.49)

Moreover, from equation (10.38) ($\mu^{d-2}h/g = Z^{1/2}(g)H/T$), one infers

$$\rho(T) = -\varepsilon + \frac{1}{2}\zeta(T) + \frac{\beta(T)}{T}.$$ (10.50)

To be able to discuss correlation functions involving the field σ, one also needs the RG equations satisfied by the connected correlation functions $W^{(n)}$:

$$\left[\Lambda\frac{\partial}{\partial\Lambda} + \beta(T)\frac{\partial}{\partial T} + \frac{n}{2}\zeta(T) + \left(\frac{1}{2}\zeta(T) + \frac{\beta(T)}{T} - \varepsilon\right)H\frac{\partial}{\partial H}\right]\widetilde{W}^{(n)}(p_i; T, H, \Lambda)$$
$$= 0.$$ (10.51)

10.8.1 RG functions: One-loop calculation

Equation (10.48) implies that $\beta(T) = \varepsilon T + O(T^2)$. Applying equation (10.51) for $n = 1$ to the one-loop expression (10.42) for $\bar{\sigma}$, one then infers

$$\zeta(T) = \frac{N-1}{2\pi}T + O(T^2, T\varepsilon).$$ (10.52)

Similarly, expressing that the one-loop expansion (10.45) of a two-point vertex function (the inverse π two-point connected correlation function) satisfies the RG equation (10.47) for $n = 2$, one recovers the RG functions $\zeta(T)$ and $\beta(T)$ at one-loop order. Using the relation (10.50), one obtains

$$\zeta(T) = \frac{(N-1)}{2\pi}T + O\left(T^2, T\varepsilon\right), \quad \beta(T) = \varepsilon T - \frac{(N-2)}{2\pi}T^2 + O\left(T^3, T^2\varepsilon\right).$$ (10.53)

The expansion of $\zeta(T)$ derived from $\tilde{\Gamma}^{(2)}(0)$ is identical to the result obtained from $\langle\sigma\rangle$, as also implied by the WT identity. The RG β-function is determined by the coefficient of p^2 in $\tilde{\Gamma}^{(2)}$.

The two RG functions have been, after some time, calculated up to four-loop order.

10.9 Zeros of the RG β-function: Fixed points

The large scale properties of the non-linear σ-model are determined by the behaviour of the effective temperature $T(\lambda)$ in a change of scale. The effective temperature is the solution of the equation

$$\lambda\frac{\partial T(\lambda)}{\partial\lambda} = -\beta(T(\lambda)), \quad T(1) = T,$$ (10.54)

where large scale corresponds to $\lambda \to \infty$.

The RG flow is governed by the sign of the β-function (equation (10.53)). The zeros of the β-function correspond to RG fixed points.

The origin $T = 0$ is always a fixed point, the Gaussian fixed point and $\beta'(0) = \varepsilon$. In $d = 2 + \varepsilon$ dimensions, $\varepsilon > 0$, for $N > 2$, another fixed point is found,

$$T_c = \frac{2\pi\varepsilon}{N - 2} + O(\varepsilon^2), \quad \beta'(T_c) = -\varepsilon + O(\varepsilon^2),$$

which is IR repulsive.

From the form of the β-function, one concludes:

(i) For $d = 2$, $N > 2$, due to the negative sign of the β-function, $T = 0$ is a unique and IR marginally unstable fixed point, the IR instability being a direct consequence of the IR divergences that massless Goldstone bosons would generate. The spectrum of the theory thus cannot be inferred from the perturbative expansion, and the implicit perturbative assumption of spontaneous symmetry breaking at low temperature is inconsistent. As we have already mentioned, this result is confirmed by a rigorous analysis, which indicates that the $O(N)$ symmetry is restored and the spectrum contains N equal mass states.

Still, for $N > 2$, $T = 0$ is also a *marginally stable UV fixed point* (a property called *asymptotic freedom* in the literature) and it thus governs the *universal large momentum behaviour* of renormalized correlation functions.

In the special case $d = 2$, $N = 2$, the β-function vanishes identically and the model requires a specific analysis.

(ii) For $d > 2$, that is, $\varepsilon > 0$, $T = 0$ is a stable IR fixed point, *the symmetry $O(N)$ is spontaneously broken at low temperature in zero field*. The effective interaction that determines the large distance behaviour, goes to zero for all effective temperatures for which the β-function remains positive. As a consequence, the large distance properties of the model can be inferred from the low temperature expansion and RG arguments, by replacing the parameters at scale Λ by the effective parameters obtained by solving the RG equations.

Critical temperature. Finally, one observes that, at least for $\varepsilon > 0$ and $N > 2$, in the sense of an ε-expansion, the RG function $\beta(T)$ has a non-trivial zero,

$$T_c = \frac{2\pi\varepsilon}{N - 2} + O\left(\varepsilon^2\right) \Rightarrow \beta(T_c) = 0, \quad \text{with} \quad \beta'(T_c) = -\varepsilon + O\left(\varepsilon^2\right),$$

which, since $\beta'(T_c) < 0$, corresponds to an *IR repulsive fixed point*. *Critical temperatures* have such a property and this suggests that T_c is the temperature where the $O(N)$ symmetry is restored. The consequences of this result will be discussed later.

Here, let us point out only that T_c is also a *UV fixed point, that is, it governs the large momentum behaviour of the renormalized theory*. Therefore, the perturbative analysis that indicates that the theory is not renormalizable for $d > 2$ cannot be trusted. Indeed, correlation functions have, for large momenta, a non-perturbative behaviour. The existence of this UV fixed point (asymptotic safety, see Chapter 15), opens the possibility that there could exist *a renormalized quantum field theory consistent for all distance scales in the continuum.*

10.10 Correlation functions: Scaling form below T_c

We now solve the RG equations for $d > 2$ at fixed temperature below T_c (the ordered phase) by a method different from the method of characteristics.

We introduce the three functions of the temperature,

$$\sigma_0(T) = \exp\left[-\frac{1}{2}\int_0^T \frac{\zeta(T')}{\beta(T')}dT'\right] = 1 - \frac{1}{4\pi\varepsilon}(N-1)T + O(T^2), \quad (10.55)$$

$$\xi(T) = \Lambda^{-1}T^{1/\varepsilon}\exp\left[\int_0^T\left(\frac{1}{\beta(T')} - \frac{1}{\varepsilon T'}\right)dT'\right]$$

$$= \Lambda^{-1}T^{1/\varepsilon}\left(1 + \frac{(N-2)}{2\pi\varepsilon^2}T + O(T^2)\right), \quad (10.56)$$

$$H_0^{-1}(T) = \frac{\Lambda^{d-2}}{T}\sigma(T)\left[\xi(T)\right]^d \sim -\frac{(N-1)}{\Lambda^2 4\pi\varepsilon}T^{d/(d-2)}. \quad (10.57)$$

Since the functions β and ζ are regular at T_c, one can also integrate the RG equations for $T > T_c$ but the boundary conditions in the definition of σ_0 and ξ have to be changed. The function ξ is then proportional to the finite correlation length in zero field.

Using also dimensional analysis, it is simple to verify that the solutions of the RG equations can be written as

$$\tilde{\Gamma}_{\mathrm{r}}^{(n)}(p_i; T, H, \Lambda) = \xi^{-d}(T)\sigma_0^{-n}(T)G^{(n)}\left(p_i\xi(T), H/H_0(T)\right). \quad (10.58)$$

The equation shows that $\xi(T)$ has, in zero external field, the nature of a *correlation length*.

For connected correlation functions, the same analysis leads to

$$\tilde{W}_{\mathrm{r}}^{(n)}(p_i; T, H, \Lambda) = \xi^{d(n-1)}(T)\sigma_0^n(T)F^{(n)}\left(p_i\xi(T), H/H_0(T)\right). \quad (10.59)$$

The field expectation value (the one-point function), which in magnetic systems is the magnetization, has the form

$$\langle\sigma(x)\rangle \equiv \bar{\sigma} = \sigma_0(T)F^{(0)}\left(H/H_0(T)\right).$$

Setting $H = 0$ in the equation, one concludes that $\sigma_0(T)$ is the *spontaneous magnetization*. By inverting the relation, one obtains the equation of state in a scaling form valid for any $T < T_c$,

$$H = H_0(T)f\left(\bar{\sigma}/\sigma_0(T)\right). \quad (10.60)$$

The vertex functions can also be expressed in terms of the field expectation value as

$$\tilde{\Gamma}_{\mathrm{r}}^{(n)}(p_i, T, \bar{\sigma}, \Lambda) = \xi^{-d}(T)\sigma_0^{-n}(T)G_M^{(n)}\left(p_i\xi(T), \bar{\sigma}/\sigma_0(T)\right). \quad (10.61)$$

These scaling forms are consistent with the solutions of the RG equations in the $(\phi^2)^2$ field theory below T_c: the two independent functions $\xi(T)$ and $\sigma_0(T)$ correspond to the two independent critical exponents ν, β, in the $(\phi^2)^2$ field theory.

They extend the large distance scaling form of correlation functions, valid in the critical region, to all temperatures below T_c.

The length $\xi(T)$: A crossover scale. Below T_c, in zero external field H, due to the presence of Goldstone modes, the correlation length is infinite and correlation functions decay algebraically at large distance. Still, we have introduced a length scale $\xi(T)$ to solve the RG equations.

In fact, $\xi(T)$ is a *crossover scale* between two different universal behaviours of correlation functions. For distances large with respect to $\xi(T)$, the behaviour of correlation functions is governed by Goldstone modes and can thus be inferred from the low temperature perturbative expansion. By contrast, *for distances much shorter than $\xi(T)$, but still large compared to the microscopic scale,* the behaviour of correlation functions is governed by the UV stable fixed point T_c.

Since T_c is also a critical temperature, correlation functions exhibit a critical behaviour. At T_c, $\xi(T)$ diverges and correlation functions have a critical scaling behaviour at all distances. When such a crossover scale exists, one can hope to define, in the continuum, a *renormalized quantum field theory consistent on all scales.*

10.10.1 *Critical exponents*

From the form of the RG functions, one infers the behaviour of thermodynamic quantities when T approaches T_c (for $N > 2$, $0 < \varepsilon \ll 1$). Below T_c the RG invariant scale is provided by the crossover length $\xi(T)$, while above T_c it is the correlation length. Both diverge like

$$\xi(T) \sim \Lambda^{-1} |T_c - T|^{1/\beta'(T_c)} .$$

The crossover or correlation length exponent is thus given by

$$\nu = -\frac{1}{\beta'(T_c)}.$$

For $d \to 2_+$, the exponent ν thus behaves like

$$\nu = 1/(d-2) + O(1) .$$

The spontaneous field expectation value $\sigma_0(T)$ vanishes at T_c like

$$\ln \sigma_0(T) \sim -\frac{1}{2}\frac{\zeta(T_c)}{\beta'(T_c)} \ln(T_c - T) \sim \nu d_\phi \ln(T_c - T) ,$$

This behaviour yields, in magnetic systems, the magnetization exponent $\beta = \nu d_\phi$ (not be confused with the β-function) and the anomalous dimension of the field

$$d_\phi = \tfrac{1}{2}\zeta(T_c).$$

At leading order, one finds

$$d_\phi = \frac{N-1}{2(N-2)}(d-2) + O\left((d-2)^2\right) . \tag{10.62}$$

10.10.2 Non-linear σ-model and $(\phi^2)^2$ field theory

Although we have found scaling behaviours consistent between the non-linear σ-model and the $(\phi^2)^2$ field theory, there is an important difference between the RG equations of both theories: the $(\phi^2)^2$ theory depends on two parameters: the coefficient of ϕ^2, which plays the role of the temperature, and the coefficient of $(\phi^2)^2$, which makes it possible to interpolate between the Gaussian and the IR stable fixed point, and has no apparent equivalent in the non-linear σ-model.

The correlation functions of the $(\phi^2)^2$ theory have an exact scaling form only at the IR fixed point, while, in the σ-model, it has been possible to eliminate all corrections to scaling, related to irrelevant operators, order by order in the perturbative expansion. One is thus led to the following remarkable conclusion: *the correlation functions of the non-linear σ-model with $O(N)$ symmetry are identical to the correlation functions of the $(\phi^2)^2$ field theory at the IR stable fixed point.* This conclusion is further supported by the comparison between both theories within the framework of the $1/N$ expansion.

10.11 Linear formulation

We now implement the constraint (10.1),

$$\phi^2(x) = 1\,,$$

by replacing the corresponding δ-function by its Fourier representation. Therefore, we introduce an auxiliary field $\lambda(x)$ and write the generating functional $\mathcal{Z}(\mathbf{J})$ of ϕ correlation functions as

$$\mathcal{Z}(\mathbf{J}) = \int [\mathrm{d}\lambda\,\mathrm{d}\phi]\exp\left[-\frac{\Lambda^{d-2}}{T}\left(\mathcal{S}(\phi,\lambda) - \int \mathrm{d}^d x\,\mathbf{J}(x)\cdot\phi(x)\right)\right], \qquad (10.63)$$

where the integration contour for λ is parallel to the imaginary axis, and

$$\mathcal{S}(\phi,\lambda) = \tfrac{1}{2}\int \mathrm{d}^d x\left\{[\nabla\phi(x)]^2 + \lambda(x)\left[\phi^2(x) - 1\right]\right\}. \qquad (10.64)$$

Note that, here, $\mathbf{J}(x)$ is an N-component source for the N-component field $\phi(x)$.

We choose an extremum of the potential as a starting point for perturbation theory:

$$\lambda(x) = 0\,,\quad \sigma(x) \equiv \phi_N(x) = 1\,,\quad \pi_i(x) \equiv \phi_i(x) = 0\,,\quad \text{for } i < N\,. \qquad (10.65)$$

The propagator of the $(N-1)$ remaining components $\boldsymbol{\pi}(x)$ is

$$\tilde{\Delta}\pi\pi(p) = \Lambda^{2-d}T/p^2.$$

In this formulation, the $O(N)$ symmetry is realized linearly but, nevertheless, the symmetry in the tree approximation is automatically spontaneously broken.

The consequences of such a situation have already been analysed in Section 10.2.3. Moreover, the problem of the interpretation of the measure is solved. On the other hand the generalization to other models on symmetric spaces is not easy.

Regularization. A first step is to introduce terms with higher derivatives, replacing ∇ by

$$\nabla_\Lambda \equiv \nabla \left(1 - \alpha_1 \nabla^2/\Lambda^2 + \alpha_2 \nabla^4/\Lambda^4 + \cdots\right).$$

However, the problem of one-loop divergences is not solved. A solution is to add a quadratic term in λ, which transforms the non-linear σ-model into a $(\phi^2)^2$ field theory,

$$\mathcal{S}(\phi, \lambda) = \tfrac{1}{2} \int \mathrm{d}^d x \left\{ [\nabla_\Lambda \phi(x)]^2 + \lambda(x) \left[\phi^2(x) - 1\right] - \lambda^2(x)/g\Lambda^2 \right\}$$

and, eventually, take the limit $g \to \infty$ [126].

We now set $\sigma(x) = 1 - \varsigma(x)$ and the action becomes

$$\mathcal{S}(\varsigma, \boldsymbol{\pi}, \lambda) = \tfrac{1}{2} \int \mathrm{d}^d x \left\{ [\nabla_\Lambda \varsigma(x)]^2 + [\nabla_\Lambda \boldsymbol{\pi}(x)]^2 + \lambda(x) \left[\boldsymbol{\pi}^2(x) + \varsigma^2(x) - 2\varsigma(x)\right] \right.$$
$$\left. - \lambda^2(x)/g\Lambda^2 \right\} \tag{10.66}$$

The action contains now a $\varsigma\lambda$ term. At leading order, in the Fourier representation, the connected two-point functions are

$$\begin{pmatrix} \tilde{\Delta}_{\varsigma\varsigma} & \tilde{\Delta}_{\varsigma\lambda} \\ \tilde{\Delta}_{\lambda\varsigma} & \tilde{\Delta}_{\lambda\lambda} \end{pmatrix}_c (p) = \frac{\Lambda^{2-d}T}{1 + K(p^2)/g\Lambda^2} \begin{pmatrix} \frac{1}{g\Lambda^2} & -1 \\ -1 & -K(p^2) \end{pmatrix}, \tag{10.67}$$

where we have denoted by $K(p^2)$ the regularized form of p^2, $K(p^2) = p^2 + O(p^4/\Lambda^2)$.

The interaction vertices are

$$-\frac{\Lambda^{d-2}}{2T} \int \mathrm{d}^d x \left(\lambda(x)\varsigma^2(x) + \lambda(x)\boldsymbol{\pi}^2(x)\right).$$

Linear symmetry breaking. The addition of the linear symmetry breaking term $-H \int \mathrm{d}^d x \, \sigma(x)$, after the shift $\lambda(x) \mapsto \lambda(x) + H$, leads to the addition of the mass term to the action (10.66),

$$\tfrac{1}{2} H \int \mathrm{d}^d x \left[\boldsymbol{\pi}^2(x) + \varsigma^2(x)\right],$$

provided one modifies also the regularization term $\lambda^2/\Lambda g$.

The propagator of the components $\boldsymbol{\pi}(x)$ then becomes

$$\tilde{\Delta}\pi\pi(p) = \frac{\Lambda^{2-d}T}{K(p^2) + H}.$$

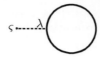

Fig. 10.5 Field expectation value: full lines for the $\pi\pi$ propagator, and a dotted line for the $\varsigma\lambda$ propagator.

The $\lambda\varsigma$ two-point matrix becomes

$$\begin{pmatrix} \tilde{\Delta}_{\varsigma\varsigma} & \tilde{\Delta}_{\varsigma\lambda} \\ \tilde{\Delta}_{\lambda\varsigma} & \tilde{\Delta}_{\lambda\lambda} \end{pmatrix}_c (p) = \frac{\Lambda^{2-d}T}{1 + \left(K(p^2) + H\right)/g\Lambda^2} \begin{pmatrix} \frac{1}{g\Lambda^2} & -1 \\ -1 & -K(p^2) - H \end{pmatrix}. \quad (10.68)$$

The field expectation value at one-loop. At one loop for $g \to \infty$, one finds (Fig. 10.5)

$$\langle \sigma \rangle = 1 - \langle \varsigma \rangle = 1 - \tfrac{1}{2}(N-1)T\Lambda^{2-d}\Omega_d(\sqrt{H}) + O(T^2),$$

in agreement with the result (10.40).

The π two-point vertex function. The calculation involves three diagrams as shown from left to right in Fig. 10.6), one with two $\lambda\pi^2$ vertices, one with one $\lambda\pi^2$ vertex and one $\lambda\sigma^2$ vertex, and one with two $\lambda\pi^2$ vertices but a different topology.

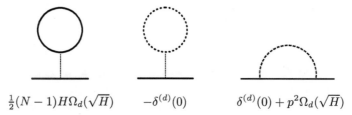

$$\tfrac{1}{2}(N-1)H\Omega_d(\sqrt{H}) \qquad -\delta^{(d)}(0) \qquad \delta^{(d)}(0) + p^2\Omega_d(\sqrt{H})$$

Fig. 10.6 Two-point function: full lines for the π propagator and dotted lines for the $\lambda\lambda$ and $\lambda\sigma$ propagators: $\delta^{(d)}(0)$ is a shorthand notation for $\frac{1}{(2\pi)^d}\int \mathrm{d}^d q$.

Summing the three contributions, one recovers the result (10.44) but one notices the explicit cancellation of the infinite factor $\delta^{(d)}(0)$ between the last diagrams. This verifies again the impossibility of using a momentum cut-off to regularize all diagrams, since, after the formal cancellation of $\delta^{(d)}(0)$, higher momentum terms added to the numerator and denominator cancel out and leave a logarithmic divergence.

Nevertheless, to prove exact cancellation, one can use WT identities. It is convenient here to use the form (10.26). Differentiating with respect to $J_j(y)$ and setting $\mathbf{J} = 0, K = H/T$, after Fourier transformation, one obtains

$$\frac{H\Lambda^{d-2}}{T}\widetilde{W}_{ij}^{(2)}(0) - \delta_{ij}\langle\sigma\rangle = 0.$$

Since $\tilde{W}_{ij}^{(2)}(p) = [\tilde{\Gamma}^{(2)}(p)]^{-1}$, one recovers equation (10.43), which, as we have already verified, implies exact cancellation.

Power counting and renormalization. Power counting tells us that the dimensions of the fields in dimension 2 are $[\phi] = 0$, $[\lambda] = 2$. Therefore, the renormalized action remains linear in λ and the term in $\lambda(x)$ takes the general form

$$\int \mathrm{d}^d x \, Z_\lambda \big(\phi^2(x) \big) \lambda(x).$$

The integral over $\lambda(x)$ then yields $Z_\lambda(\phi^2(x)) = 0$. This equation has at least a perturbative solution starting with $(\phi^2(x) = 1$. We denote by Z^{-1} the solution,

$$\phi^2(x) = Z^{-1}. \tag{10.69}$$

With this constraint, the most general action with not more than two derivatives is proportional to the initial action expressed in terms of the renormalized field.

Therefore, one recovers the results of the previous sections. The main disadvantage of this formulation is that it introduces an infinite number of renormalization constants as an intermediate step. Moreover, its generalization to other homogeneous spaces is complicated and aesthetically not especially appealing.

Finally, the non-linear formulation emphasizes the connection with other geometric models like gauge theories. On the other hand, the linear formulation clarifies the discussion of multiple insertions of general $O(N)$ invariant operators. In particular, one realizes that the insertions of operators of dimension larger than 2 will eventually generate terms of degree larger than 1 in $\lambda(x)$. The integration over λ then no longer implies the strict constraint (10.69).

10.11.1 *IR divergences and $O(N)$ symmetric functions*

Starting from the low temperature expansion, we have been able to describe not only the complete structure of the low temperature phase, as expected, but also in the non-Abelian case $N > 2$ (the rotation group $SO(N)$ is Abelian, that is, commutative for $N = 2$) to discover, in $2 + \varepsilon$ dimensions, the existence of a critical temperature and to determine the corresponding critical behaviour.

What is slightly puzzling in this result is that the perturbative expansion is sensitive only to the local structure of the sphere $\phi^2 = 1$ even though symmetry restoration seems to depend on the global structure of the sphere.

This is confirmed by the peculiarity of the Abelian case $N = 2$: locally the circle cannot be distinguished from the non-compact straight line and, thus, the σ-model becomes a free field theory.

For $N > 2$, by contrast, the sphere has a characteristic local curvature. However, one still faces the problem that different regular compact manifolds may have the same local metric and, thus, lead to the same perturbative expansion. They all lead to the same low temperature physics.

However, the preceding results concerning the critical behaviour are physically relevant only if they are still valid when ε is no longer infinitesimal and T is close to T_c, a condition that cannot be verified directly. In particular, the low temperature expansion may miss terms decreasing like $\exp(\mathrm{const.}/T)$, which may be qualitatively important and dependent on the regularization scheme.

For example, the physics for $N = 2$ is not well reproduced. The Kosterlitz–Thouless phase transition [123], which relies on effects invisible in the low temperature expansion and whose very existence depends on specific regularization schemes, like lattice regularization, is not found.

Conversely, a direct relation between the $(\phi^2)^2$ and σ models in continuum space, is provided for suitable regularizations by the large N expansion to all orders. The large N analysis suggests that the preceding considerations are valid for the N-component model, at least for large enough values of N.

10.12 Two dimensions

We now examine briefly the model in dimension 2, the dimension in which it is renormalizable, because it possesses some interesting properties.

10.12.1 Non-Abelian group: $N > 2$

For the $N > 2$, where the $O(N)$ group is non-Abelian, the non-linear σ-model shares an important property with QCD, the model describing strong interactions in the theory of fundamental microscopic interactions: the non-linear σ-model has the property of *asymptotic freedom* (for large momenta) because the first coefficient in the expansion of the β-function is negative,

$$\beta(T) = -\frac{(N-2)}{2\pi}T^2\left(1 + T/2\pi\right) + O(T^4), \tag{10.70}$$

unlike, for example, the $(\phi^2)^2$ field theory.

Therefore, the large momentum behaviour of universal correlation functions is entirely calculable from the perturbative expansion and RG arguments.

By contrast, since the Gaussian fixed point is IR unstable, in the limit of zero field H, the *spectrum of the σ-model is non-perturbative*. This is consistent with the expected absence of symmetry breaking with ordering, which implies a spectrum with N degenerate massive states.

Assuming symmetry restoration, one can obtain the temperature dependence at low temperature of the correlation length in zero external field or the lowest common mass. Adapting expression (10.56), one finds

$$\xi^{-1}(T) = m(T) = K\Lambda T^{-1/(N-2)}\,\mathrm{e}^{-2\pi/[(N-2)T]}$$

$$\times \exp\left[-\int_0^T \mathrm{d}T'\left(\frac{1}{\beta(T')} + \frac{2\pi}{(N-2)T'^2} - \frac{1}{(N-2)T'}\right)\right],$$

$$\underset{T\to 0}{\sim}\; K\Lambda T^{-1/(N-2)}\,\mathrm{e}^{-2\pi/[(N-2)T]}\;. \tag{10.71}$$

However, the integration constant K, which relates the physical mass to the RG scale, cannot be calculated by perturbative techniques.

Finally, the various scaling forms imply that the perturbative expansion at fixed external field is valid, at small momenta or large distances, for $H/H_0(T)$ large.

10.12.2 *O(N) invariant functions and IR singularities*

The configuration energy in zero field H with $O(N)$ symmetry has a sphere of classical degenerate minima. For $d = 2$, to define the perturbative expansion, one is forced to add to the energy a linear term that breaks the symmetry and selects a particular classical minimum. We have stated, and this can be easily verified, that for $d = 2$, in general, correlation functions have IR divergences when the parameter H goes to zero, a property that is consistent with the absence of spontaneous symmetry breaking.

However, perturbative calculations can be set up in a different way: one does not introduce a symmetry breaking term but, rather, a set of *collective coordinates* that parametrize the set of classical minima. In a finite volume with periodic boundary conditions, one then expands as usual, starting from a fixed minimum, but only the modes of the field that do not correspond to a global rotation (the non-zero momentum modes) are taken into account in the perturbative expansion. Eventually, one sums exactly over all classical minima.

Clearly, after this last summation, only $O(N)$ invariant correlation functions survive. Elitzur had conjectured, and David [127] has proved, that, *in two dimensions, $O(N)$ invariant correlation functions obtained by this procedure have a regular expansion at low temperature.* This implies that, if one calculates perturbatively $O(N)$ invariant correlation functions in a non-zero field and then one takes the limit $H = 0$, the limit is finite.

A simple example gives an idea about the cancellation mechanism that leads to an IR finite expansion. The $O(N)$ invariant two-point function at one-loop order reads,

$$\langle \sigma(x)\sigma(0) + \boldsymbol{\pi}(x) \cdot \boldsymbol{\pi}(0) \rangle = 1 - \tfrac{1}{2} \left\langle (\boldsymbol{\pi}(x) - \boldsymbol{\pi}(0))^2 \right\rangle + O(\pi^4)$$

$$= 1 - (N-1)T \int \frac{\mathrm{d}^2 p}{(2\pi)^2} \left(\frac{1 - \mathrm{e}^{ipx}}{p^2} \right) + O(T^2).$$

The numerator in the integrand vanishes at $p = 0$, cancelling the IR divergence.

10.12.3 *The Abelian SO(2) model*

In the simplest continuum limit, after the parametrization

$$\phi(x) = \{\cos\theta(x), \sin\theta(x)\},$$

the action becomes quadratic:

$$\mathcal{S}(\theta) = \tfrac{1}{2} \int \mathrm{d}^d x \, [\nabla_x \theta(x)]^2 .$$

The β-function thus vanishes to all orders in the low temperature expansion, a property that reflects that, locally, a circle and straight line are undistinguishable.

The continuum action has a larger symmetry than the initial lattice action, since it is invariant under any translation of the θ field by a constant.

Unlike the generic non-linear σ-model with $O(N)$ symmetry, $N > 2$, the $O(2)$ symmetric model is not asymptotically free but, by contrast, exhibits a line of fixed points.

The two-point function of the θ-field (a would-be massless scalar field) does not exist in two dimensions, since the integral representation of the two-point function,

$$\Delta(x) = \frac{1}{2\pi} \int \frac{\mathrm{d}^2 p \; \mathrm{e}^{ipx}}{p^2},$$

is IR divergent.

However, here the physical correlation functions are the periodic functions $\sin\theta$ or $\cos\theta$ or, equivalently, $\mathrm{e}^{\pm i\theta}$. These correlation functions exist, provided they have the symmetry of the action, that is, are invariant under constant translations of the θ field.

Even though the RG β-function vanishes identically, local functions of the fields must be renormalized and the $\mathrm{e}^{\pm i\theta}$ correlation functions have a non-Gaussian dimension for all temperatures T.

The expectation values of the non-invariant functions $\mathrm{e}^{\pm i\theta}$ and, thus, $\langle \phi \rangle$ vanish. An explicit calculation yields the invariant two-point function. One finds

$$\left\langle \mathrm{e}^{i\theta(x) - i\theta(0)} \right\rangle \propto x^{-T/2\pi}. \tag{10.72}$$

Despite the absence of symmetry breaking, the function has an algebraic behaviour at low temperature with a temperature-dependent exponent

$$\eta = T/2\pi \;\Rightarrow\; d_\phi = T/4\pi\,, \tag{10.73}$$

a behaviour consistent with the existence of a *line of fixed points*.

Similarly, all other invariant connected correlation functions decay algebraically for all temperatures.

By contrast, in a lattice regularization, the model exhibits the famous Kosterlitz–Thouless transition [123], which separates a low temperature phase with an infinite correlation length, but *without order*, from a phase with a finite correlation length.

11 Gauge invariance and gauge fixing

In classical electrodynamics, the structure of the field equations is such that the electric field $\mathbf{E}(t, \mathbf{x})$ and the magnetic field $\mathbf{B}(t, \mathbf{x})$ can be expressed in terms of a scalar field $A_0(t, \mathbf{x})$ (*scalar potential*) and a vector field $\mathbf{A}(t, \mathbf{x})$ (*vector potential*). The term *gauge invariance* refers to the property that a whole class of scalar and vector potentials, related by *gauge transformations*, describe the same electric and magnetic fields [128].

As a consequence, the dynamics of the electromagnetic field and of a charged system in an electromagnetic background do not depend on the choice of the representative $(A_0(t, \mathbf{x}), \mathbf{A}(t, \mathbf{x}))$ within the appropriate class.

The concept has been then extended to non-relativistic and relativistic electromagnetic quantum theories.

While electrodynamics is an example of an Abelian gauge theory, the concept of gauge invariance has later been generalized to other theories like, for example, non-Abelian gauge or Yang–Mills [41]) field theories or General Relativity [23].

11.1 Gauge invariance: A few historical remarks

Kirchhoff uses the components of the vector potential [129], where he extends the work of Weber [130] on electromagnetic induction. In this article, Kirchhoff also notices that the vector potential and the scalar potential satisfy a relation that in modern language can be called a gauge-fixing condition.

The component form of Maxwell's equations can be found in equations (54), (56), (112), (115) of Maxwell's 1861 article [22]. Maxwell, influenced by Thomson (later named Lord Kelvin) and Stokes, expresses, in equation (55) of the article, $\mu(\alpha, \beta, \gamma)$ (in modern language the magnetic field $\mathbf{B} = \mu\mathbf{H}$ as appropriate for linear materials) in terms of the curl of vector potential, whose components are denoted there (F, G, H). The components of the vector potential, are chosen to satisfy the constraint (57), which, in modern language, is the Coulomb gauge condition $\nabla \cdot \mathbf{A} = 0$. Maxwell also used the components of the vector potential also in an earlier (1855) article [131]. Helmholtz [132] exhibits a family of vector potentials dependent on a real parameter and makes the seminal observation that, while the differential form of the magnetic energy among current elements depends on the value of the parameter, the resulting integrated energy does not, or, in modern language, is gauge invariant.

Weyl introduces in three papers from [133] the concept of local scale invariance, which is called *Massstab Invarianz* in the first two articles, and *Eich Invarianz* in the third one. The *Eich Invarianz*, first translated as *calibration invariance*, was then translated as *gauge invariance*.

From Random Walks to Random Matrices. Jean Zinn-Justin, Oxford University Press (2019).
© Jean Zinn-Justin. DOI: 10.1093/oso/9780198787754.001.0001

Weyl's attempt to unify gravity and electromagnetism, which culminated in Refs. [137], was unsuccessful. It turned out from the works of Schrödinger [134], Fock [135], London [136] and Weyl [137] that quantum wave functions should vary by a phase factor under a gauge transformation and that Weyl's real scale should be replaced by a phase change involving the usual four-dimensional electromagnetic vector potential.

The years 1930-1950 saw the construction of quantum electrodynamics (QED), the quantum theory extending Maxwell's classical theory (see Section 4.1).

In 1954, Yang and Mills [41] introduced non-Abelian gauge fields.

More information can be found, for example, in Refs. [139–141] and Chapters 12–14.

11.2 Variational principle, charged particle and gauge invariance

The emphasis on the variational principle in a chapter devoted to gauge invariance has the following motivation. In the example of electromagnetism, with the introduction of a *vector potential*, classical equations can be derived from a variational principle. From the property that a whole class of vector potentials lead to the same classical equations, have emerged the concepts of gauge transformations and *gauge invariance*. The introduction of a vector potential has later played a major role in the quantization of theories with electromagnetic interactions.

In the following, the coordinates $\mathbf{q} = (q^1, q^2, q^3)$ refer to the standard Cartesian orthogonal coordinates, also denoted by $\mathbf{x} := (x^1, x^2, x^3)$. We also use the differential (vector) operator $\nabla \equiv (\partial/\partial x^1, \partial/\partial x^2, \partial/\partial x^3)$.

11.2.1 Euler–Lagrange equations

An important class of equations of motion in classical physics can be derived from a variational principle.

Given a dynamical system expressed in terms of some coordinates $\mathbf{q} \equiv (q^1, q^2, \ldots)$, one constructs a Lagrange function or *Lagrangian* $\mathcal{L}(\mathbf{q}(t), \dot{\mathbf{q}}(t); t)$, where the total derivative of a quantity X with respect to the time t has been denoted by $\dot{X}$.

One then defines the action as the time integral of the Lagrangian:

$$\mathcal{A}(\mathbf{q}) = \int_{t'}^{t''} dt \, \mathcal{L}\left(\mathbf{q}(t), \dot{\mathbf{q}}(t); t\right). \tag{11.1}$$

Imposing the stationarity of the action with respect to variations of the trajectory $\mathbf{q}(t)$, with fixed boundary conditions $\mathbf{q}(t') = \mathbf{q}'$ and $\mathbf{q}(t'') = \mathbf{q}''$, one recovers the equations governing the classical motion in a form called *Euler–Lagrange equations* ($\delta/\delta q(t)$ denotes functional differentiation):

$$\frac{\delta \mathcal{A}}{\delta q^i(t)} = 0 \quad \Rightarrow \quad \frac{\partial \mathcal{L}}{\partial q^i} - \frac{d}{dt}\frac{\partial \mathcal{L}}{\partial \dot{q}^i} = 0. \tag{11.2}$$

Classical gauge transformations. The addition of a total derivative to the Lagrangian,

$$\mathcal{L} \mapsto \mathcal{L}_\Omega = \mathcal{L} + \frac{\mathrm{d}\Omega(t, \mathbf{q}(t))}{\mathrm{d}t} = \mathcal{L} + \frac{\partial\Omega(t, \mathbf{q}(t))}{\partial t} + \nabla_q \Omega(t, \mathbf{q}(t)) \cdot \dot{\mathbf{q}}, \qquad (11.3)$$

changes the action only by boundary terms,

$$\mathcal{A} \mapsto \mathcal{A}_\Omega = \mathcal{A} + \Omega(t'', \mathbf{q}'') - \Omega(t', \mathbf{q}') \qquad (11.4)$$

and, thus, does not affect the classical equations of motion.

The Hamiltonian. The Hamiltonian, Legendre transform of the Lagrangian,

$$H(p, q; t) + \mathcal{L}(q, \dot{q}; t) = \sum_i \dot{q}^i p_i, \qquad p_i = \frac{\partial\mathcal{L}}{\partial\dot{q}^i}, \qquad (11.5)$$

is a function of a position $\mathbf{q}$ and a conjugate momentum $\mathbf{p}$, together called *phase space coordinates*. In the transformation (11.3), the conjugate momentum becomes

$$\mathbf{p} \mapsto \mathbf{p} + \nabla_q\Omega \quad \Rightarrow \quad H_\Omega(\mathbf{p}, \mathbf{q}; t) = H(\mathbf{p} - \nabla_q\Omega, \mathbf{q}; t) - \frac{\partial\Omega}{\partial t},$$

a confirmation that, in general, *the conjugate momentum is not a physical observable.*

The action in the phase space formalism reads

$$\mathcal{A}(\mathbf{p}, \mathbf{q}) = \int \mathrm{d}t\, [\mathbf{p}\dot{\mathbf{q}} - H(\mathbf{p}, \mathbf{q}; t)].$$

After the gauge transformation, it becomes

$$\mathcal{A}_\Omega(\mathbf{p}, \mathbf{q}) = \int \mathrm{d}t\, \left(\mathbf{p}\dot{\mathbf{q}} - H(\mathbf{p} - \nabla_q\Omega, \mathbf{q}; t) + \frac{\partial\Omega}{\partial t}\right)$$

and, after the change $\mathbf{p} - \nabla_q\Omega \mapsto \mathbf{p}$, one recovers the relation (11.4):

$$\mathcal{A}_\Omega(\mathbf{p}, \mathbf{q}) = \mathcal{A}(\mathbf{p}, \mathbf{q}) + \Omega(t'', \mathbf{q}'') - \Omega(t', \mathbf{q}').$$

11.2.2   The motion of the charged particle: The principle of gauge invariance

The equation of motion for a classical particle of mass m and charge e in a magnetic and an electric field can be written as

$$m\,\ddot{\mathbf{q}} = e\,\dot{\mathbf{q}} \times \mathbf{B}(t, \mathbf{q}) + e\,\mathbf{E}(t, \mathbf{q}), \qquad (11.6)$$

where $\mathbf{E}(t, \mathbf{q})$ is the electric field, and $\mathbf{B}(t, \mathbf{q})$ the magnetic field, which satisfies Gauss's law for magnetism in differential form,

$$\nabla_q \cdot \mathbf{B} = 0. \qquad (11.7)$$

Remarkably enough, these equations can be derived from a variational principle at the price of introducing a vector $\mathbf{A}(t, \mathbf{q})$ and a scalar potential $A_0(t, \mathbf{q})$ and considering the Lagrangian (see also Section 3.1)

$$\mathcal{L}(\mathbf{q}, \dot{\mathbf{q}}) = \tfrac{1}{2} m \, \dot{\mathbf{q}}^2 - e \, \mathbf{A}(t, \mathbf{q}) \cdot \dot{\mathbf{q}} - e \, A_0(t, \mathbf{q}). \tag{11.8}$$

Note that we have chosen for $\mathbf{A}$ a sign convention opposite to the more usual one, to ensure a consistency of conventions between Abelian and non-Abelian gauge transformations.

Then, in the combined transformations

$$A_0(t, \mathbf{q}) \mapsto A_0^{\Omega}(t, \mathbf{q}) = A_0(t, \mathbf{q}) + \frac{1}{e} \frac{\partial \Omega(t, \mathbf{q})}{\partial t}, \tag{11.9}$$

$$\mathbf{A}(t, \mathbf{q}) \mapsto \mathbf{A}^{\Omega}(t, \mathbf{q}) = \mathbf{A}(t, \mathbf{q}) + \frac{1}{e} \nabla_q \Omega(\mathbf{t}, \mathbf{q}), \tag{11.10}$$

the Lagrangian (11.8) transforms as

$$\mathcal{L} \mapsto \mathcal{L}_\Omega = \mathcal{L} - \frac{\mathrm{d}\Omega(t, \mathbf{q})}{\mathrm{d}t}, \tag{11.11}$$

cancelling the variations (11.3) of the Lagrangian in zero field. The Lagrangian (11.8) thus is *gauge invariant*.

The Euler–Lagrange equations (11.2) corresponding to the Lagrangian (11.8) take the form (11.6) with the identification

$$\mathbf{E}(t, \mathbf{q}) = \frac{\partial \mathbf{A}(t, \mathbf{q})}{\partial t} - \nabla_q A_0(t, \mathbf{q}), \tag{11.12}$$

$$\mathbf{B}(t, \mathbf{q}) = -\nabla_q \times \mathbf{A}(t, \mathbf{q}). \tag{11.13}$$

The latter equation and the identity $\nabla \cdot (\nabla \times \mathbf{A}) = 0$ imply Gauss's law (equation (11.7)).

Gauge invariant observables. Using the identity $\nabla \times (\nabla \Omega) = 0$, it is possible to verify directly that the fields $\mathbf{E}$ and $\mathbf{B}$ corresponding to the scalar and vector potentials $A_0^{\Omega}, \mathbf{A}^{\Omega}$, which were obtained by the *gauge transformations* (11.9) and (11.10), do not depend on Ω: $\mathbf{E}$ and $\mathbf{B}$ are *gauge invariant* quantities.

Scalar and vector potentials related by a gauge transformation belong, from the physics viewpoint, to an *equivalence class*.

11.2.3 Enforcing gauge invariance: A dynamic principle

We start from the Lagrangian of a free particle of mass m,

$$\mathcal{L} = \tfrac{1}{2} m \left[\dot{\mathbf{q}}(t) \right]^2, \tag{11.14}$$

and want to complete it in such a way that it becomes itself gauge invariant, that is, invariant under the transformations (11.3). This can be achieved by introducing a vector field $\mathbf{A}(t, \mathbf{q})$ and a scalar field $A_0(t, \mathbf{q})$ and considering the modified Lagrangian (11.8).

Demanding gauge invariance of the Lagrangian, that is, invariance of the Lagrangian in a transformation (11.3), is a *dynamic principle*: it requires introducing vector and scalar potentials and, thus, generates electromagnetic interactions.

11.2.4 The classical Hamiltonian in a magnetic and electric field

From the classical Lagrangian (11.8), after the Legendre transformation,

$$H(\mathbf{p}, \mathbf{q}) + \mathcal{L}(\mathbf{q}, \dot{\mathbf{q}}) = \mathbf{p} \cdot \dot{\mathbf{q}} \quad \text{with} \quad p_i = \frac{\partial \mathcal{L}}{\partial \dot{q}^i} = m\dot{q}^i - eA_i,$$

one derives the *classical Hamiltonian*, a function of the phase space coordinates $(\mathbf{p}, \mathbf{q})$:

$$H(\mathbf{p}, \mathbf{q}; t) = \frac{1}{2m} \left(\mathbf{p} + e\mathbf{A}(t, \mathbf{q}) \right)^2 + eA_0(t, \mathbf{q}). \tag{11.15}$$

Under a gauge transformation (equations (11.9) and (11.10)), the Lagrangian transforms by a total time derivative (equation (11.11)). In the Hamiltonian framework, this corresponds to a shift of the conjugated momentum and the scalar potential, since

$$H_\Omega(\mathbf{p}, \mathbf{q}; t) = \frac{1}{2m} \left(\mathbf{p} + e\mathbf{A}^\Omega(t, \mathbf{q}) \right)^2 + eA_0^\Omega(t, \mathbf{q}) = H(\mathbf{p} - e\nabla\Omega, \mathbf{q}; t) - e\frac{\partial\Omega}{\partial t}.$$

In particular, in presence of a magnetic field, the relation between velocity and momentum becomes

$$\dot{q}^i = \frac{\partial H}{\partial p_i} = \frac{1}{m} \left(p_i + eA_i \right),$$

showing that the conjugate momentum $\mathbf{p}$ is gauge dependent unlike the generalized conjugate momentum $\mathbf{p} + e\mathbf{A}(t, \mathbf{q})$, which is gauge invariant.

11.3 Gauge invariance: A charged quantum particle

Quantum mechanics: A few elements. A possible construction of quantum mechanics is based on vectors ψ of unit norm belonging to a complex Hilbert space $\mathcal{H}$ (pure states), which represent physical states, and operators $\hat{O}$ acting on $\mathcal{H}$. In particular, physical observables are described by self-adjoint (Hermitian) operators.

Physics predictions in quantum mechanics involve operator expectation values of the form $(\psi, \hat{O}\psi)$ (where $(\bullet, \bullet)$ denotes the scalar product in $\mathcal{H}$).

It follows that they are not affected by unitary transformations U ($UU^\dagger = U^\dagger U = 1$), which act both on state vectors and on operators as

$$\psi \mapsto U\psi, \quad \hat{O} \mapsto U\hat{O}U^\dagger. \tag{11.16}$$

As a consequence, the multiplication of state vectors by a phase (a $U(1)$ global group transformation) $\psi \mapsto e^{i\theta}\psi$ leaves operators and physics predictions unchanged.

The time evolution of state vectors $\psi(t) \in \mathcal{H}$ is governed by the *Schrödinger equation* [49]

$$i\hbar\frac{\partial\psi(t)}{\partial t} = \hat{H}\psi(t), \tag{11.17}$$

where the *Hamiltonian (quantum) operator* $\hat{H}$ is self-adjoint to guaranty *unitary evolution*, and $2\pi\hbar$ is Planck's constant.

To a spinless non-relativonic particle in continuum space, which classically can be described by phase space coordinates, position $\mathbf{q}$ and conjugate momentum $\mathbf{p}$, are associated now self-adjoint quantum operators $\hat{\mathbf{q}}$ and $\hat{\mathbf{p}}$, which satisfy the canonical commuting relations

$$[p_j, q^k] = -i\hbar\delta_j^k.$$

11.3.1 Quantum Hamiltonian in a static magnetic field and gauge invariance

We consider the Hamiltonian (11.15) first restricted to $\mathbf{A}(\mathbf{q})$ time independent and $A_0(t, \mathbf{q}) = 0$ (a static magnetic field).

To quantize the classical theory, the 'correspondence principle' suggests starting from the classical Hamiltonian (11.15), replacing phase space coordinates by the corresponding quantum operators and fixing the order of operators in products by the Hermiticity condition (unitary evolution). This leads to the Hamiltonian operator

$$\hat{H} = \frac{1}{2m}\left(\hat{\mathbf{p}} + e\mathbf{A}(\hat{\mathbf{q}})\right)^2. \tag{11.18}$$

Gauge transformation. It is possible to verify the identity

$$U\left(\hat{\mathbf{p}} + e\mathbf{A}(\hat{\mathbf{q}})\right)U^\dagger = \hat{\mathbf{p}} + e\mathbf{A}^\Omega(\hat{\mathbf{q}}), \tag{11.19}$$

where $\mathbf{A}^\Omega$ is a gauge transform of $\mathbf{A}$,

$$\mathbf{A}^\Omega(\mathbf{q}) = \mathbf{A}(\mathbf{q}) - \frac{1}{e}\nabla\Omega(\mathbf{q}) \tag{11.20}$$

and U is the unitary operator,

$$U = e^{i\Omega(\hat{\mathbf{q}})/\hbar}. \tag{11.21}$$

A gauge transformation of the vector potential, therefore induces a unitary transformation on the covariant conjugate momentum operator $\hat{\mathbf{p}} + e\mathbf{A}(\hat{\mathbf{q}})$ and, thus, on the Hamiltonian:

$$U\hat{H}(\mathbf{A})U^\dagger = \frac{1}{2m}\left[U\left(\hat{\mathbf{p}} + e\mathbf{A}(\hat{\mathbf{q}})\right)U^\dagger\right]^2 = \hat{H}(\mathbf{A}^\Omega),$$

where we have introduced the notation $\hat{H}(\mathbf{A})$ to emphasize the dependence of the Hamiltonian on the vector potential.

We now perform the corresponding unitary transformation on the vector $\psi(t)$ in the Schrödinger equation (11.17),

$$\psi_\Omega(t) = e^{ie\Omega(\hat{\mathbf{q}})/\hbar}\psi(t), \tag{11.22}$$

a transformation also called gauge transformation in this context, and obtain the unitary equivalent Schrödinger equation

$$i\hbar\frac{\partial\psi_\Omega(t)}{\partial t} = \hat{H}(\mathbf{A}^\Omega)\psi_\Omega(t).$$

We can now formulate the principle of gauge invariance in quantum mechanics: all physics observables should be invariant in a simultaneous gauge transformation on the vector field $\mathbf{A}$ and the state vector $\psi(t)$ (equations (11.20) and (11.22)). Gauge invariance ensures that physics results do not depend on the specific choice of the vector potential in the equivalence class.

The principle of gauge invariance requires that operators $\hat{O}$ corresponding to physical observables should undergo the same unitary transformation:

$$\hat{O}(\mathbf{A}^{\Omega}) = U\hat{O}(\mathbf{A})U^{\dagger}. \qquad (11.23)$$

For an operator function of $\hat{\mathbf{p}}$ and $\hat{\mathbf{q}}$, this implies that it can depend on $\hat{\mathbf{p}}$ only through the combination $\hat{\mathbf{p}} + e\mathbf{A}(\hat{\mathbf{q}})$.

Conversely, the condition of gauge invariance, in the sense of demanding that physics results should remain unchanged in a gauge transformation (11.22) without performing the corresponding unitary transformations (11.16) on operators, is a *dynamic principle*, in contrast with the condition of global invariance: it implies the introduction in the Hamiltonian of a vector potential and, thus, the presence of a magnetic field as well as a specific form of physics operators.

11.3.2 *The Schrödinger representation*

The Hilbert space can be chosen in such a way that state vectors are represented by functions $\psi(t, \mathbf{q})$ (*wave functions*) on which $\hat{\mathbf{q}}$, $\hat{\mathbf{p}}$ act like

$$\hat{\mathbf{q}}\psi(t, \mathbf{q}) = \mathbf{q}\psi(t, \mathbf{q}), \quad \hat{\mathbf{p}}\psi(t, \mathbf{q}) = \frac{\hbar}{i}\nabla\psi(t, \mathbf{q}).$$

The unitary transformation U in equation (11.21) then takes the form

$$\psi(t, \mathbf{q}) \mapsto \psi_{\Omega}(t, \mathbf{q}) \equiv e^{ie\Omega(\mathbf{q})/\hbar}\psi(t, \mathbf{q}) \equiv U(\mathbf{q})\,\psi(t, \mathbf{q}). \qquad (11.24)$$

Acting on wave functions $\psi(t, \mathbf{q})$, these transformations generate, at each point $\mathbf{q}$ in space, a representation of the multiplicative group $U(1)$ of complex numbers of modulus 1. Group transformations that vary in space are called *local group transformations*, in contrast to *global* (or *rigid*) group transformations, where the transformations act in the same way at all points (here, $\Omega(\mathbf{q})$ constant for all $\mathbf{q}$).

Gauge invariance implies here the invariance of physics results under *local $U(1)$ transformations*.

Covariant derivative and Schrödinger equation. The covariant conjugate momentum acts as a first order differential operator. One defines the *covariant derivative*

$$\frac{i}{\hbar}\left(\hat{\mathbf{p}} + e\mathbf{A}(\hat{\mathbf{q}})\right) \mapsto \mathbf{D} = \nabla_{q} + \frac{ie}{\hbar}\mathbf{A}(\mathbf{q}).$$

As a consequence of the transformation (11.19), the covariant derivative $\mathbf{D}$ satisfies

$$U(\mathbf{q})\,\mathbf{D_A}U^{\dagger}(\mathbf{q}) = \mathbf{D_{A^{\Omega}}} \quad \Leftrightarrow \quad U(\mathbf{q})\,\mathbf{D_A} = \mathbf{D_{A^{\Omega}}}\,U(\mathbf{q}), \qquad (11.25)$$

where we have used the notation $\mathbf{D_A}$ to emphasize its dependence on $\mathbf{A}$. Transformations (11.25) imply that $[\mathbf{D}\psi](x)$ and $[\mathbf{D}^2\psi](x)$ gauge transform as tensors.

Therefore, implementing gauge invariance in this context amounts to *replacing normal derivatives by covariant derivatives*. The quantum Hamiltonian is then a second order differential operator that can be written as

$$\hat{H} = -\frac{\hbar^2}{2m}\mathbf{D}^2.$$

The components D_i of $\mathbf{D}$ satisfy

$$[D_i, D_j] = \frac{ie}{\hbar}(\nabla_i\mathbf{A}_j - \nabla_j\mathbf{A}_i) = \frac{ie}{\hbar}F_{ij}, \quad \text{with} \quad F_{ij}(\mathbf{q}) = -\sum_k \epsilon_{ijk}B_k(\mathbf{q}),$$

$$(11.26)$$

where $\mathbf{B}$ is the magnetic field (11.13), and ϵ_{ijk} is the completely antisymmetric tensor, with $\epsilon_{123} = 1$.

Gauge transformations of covariant derivatives (11.25) imply

$$U(\mathbf{q})\,F_{ij}(\mathbf{A})U^\dagger(\mathbf{q}) = F_{ij}(\mathbf{A}^\Omega).$$

Thus F_{ij}, because it commutes with $U(\mathbf{q})$ (which is just a phase), is gauge invariant.

11.3.3 Time-dependent gauge transformations

One can generalize the formalism to time-dependent vector potentials and gauge transformations:

$$\psi_\Omega(t, \mathbf{q}) \mapsto \psi_\Omega(t, \mathbf{q}) = \mathrm{e}^{ie\Omega(t,\mathbf{q})/\hbar}\psi(t, \mathbf{q}) \equiv U(t, \mathbf{q})\,\psi(t, \mathbf{q}). \qquad (11.27)$$

This implies introducing a scalar potential A_0 and replacing in the Schrödinger equation (11.17) the time derivative by the *covariant time derivative*

$$D_0 = \frac{\mathrm{d}}{\mathrm{d}t} + \frac{ie}{\hbar}A_0$$

or, equivalently, to use the quantized version of the classical Hamiltonian (11.15)), which can be written as

$$i\hbar D_0\psi(t, \mathbf{q}) = -\frac{\hbar^2}{2m}\mathbf{D}^2\psi(t, \mathbf{q}). \qquad (11.28)$$

The Schrödinger equation (11.28) then describes the evolution of a spinless charged particle in magnetic and electric fields.

In analogy with equation (11.26), one also finds

$$[D_0, D_i] = \frac{ie}{\hbar}E_i, \qquad (11.29)$$

where E_i is the electric field (11.12).

11.4 Evolution of a charged particle: Path integral representation

Following Feynman [4], quantum mechanics can alternatively be formulated in terms of path integrals. In this formalism, the matrix elements of the quantum evolution operator $\mathcal{U}(t'', t')$ between times t' and t'' are given by a sum over all possible trajectories $\mathbf{q}(t)$ (*paths*), which, in the simplest cases, can be written as (see Chapter 2)

$$\langle \mathbf{q}'' | \mathcal{U}(t'', t') | \mathbf{q}' \rangle = \int [d\mathbf{q}(t)] \exp\left(\frac{i}{\hbar}\mathcal{A}(\mathbf{q})\right) \tag{11.30}$$

with boundary conditions $\mathbf{q}(t') = \mathbf{q}'$, $\mathbf{q}(t'') = \mathbf{q}''$, where $\mathcal{A}(\mathbf{q})$ is the classical action defined in equation (11.1).

In the case of a spinless non-relativistic particle in electric and magnetic fields, the Lagrangian (11.8), under a gauge transformation, changes by a total time derivative, equation (11.11). The action

$$\mathcal{A}(\mathbf{q}) = \int_{t'}^{t''} dt \left[\tfrac{1}{2} m \dot{\mathbf{q}}^2 - e\,\mathbf{A}(t, \mathbf{q}) \cdot \dot{\mathbf{q}} - e\,A_0(t, \mathbf{q})\right],$$

changes then by boundary terms and, correspondingly, the evolution operator transforms as

$$\langle \mathbf{q}'' | \mathcal{U}(t'', t') | \mathbf{q}' \rangle \mapsto U(t'', \mathbf{q}'') \langle \mathbf{q}'' | \mathcal{U}(t'', t') | \mathbf{q}' \rangle U^\dagger(t', \mathbf{q}'),$$

a property consistent with the transformation (11.27) of the wave function.

However, note that in the case of the Lagrangian (11.8), due to the appearance of the time derivative $\dot{\mathbf{q}}$ in the interaction, the path integral is not fully defined, because typical paths are not differentiable, but the commutation of time integration and path integration requires a specific definition, which formally amounts to choosing $\mathrm{sgn}(0) = 0$.

11.5 Classical electromagnetism and Maxwell's equations

Maxwell's equations [22] in the vacuum can be written (in local differential form) as:

$$\nabla \times \mathbf{E} + \frac{\partial \mathbf{B}}{\partial t} = 0 \quad \text{(Faraday's law)}, \tag{11.31}$$

$$\nabla \cdot \mathbf{B} = 0 \quad \text{(Gauss's law for magnetism)}, \tag{11.32}$$

$$\nabla \cdot \mathbf{E} = \rho \quad \text{(Gauss's law)}, \tag{11.33}$$

$$\nabla \times \mathbf{B} - \frac{\partial \mathbf{E}}{\partial t} = \mathbf{J} \quad \text{(Ampère–Maxwell's law)}, \tag{11.34}$$

where $\rho(t, \mathbf{x})$ is the charge density, $\mathbf{J}(t, \mathbf{x})$ is the current density and, again, $\mathbf{E}(t, \mathbf{x})$ is the electric field and $\mathbf{B}(t, \mathbf{x})$ is the magnetic field. (We use the international system of units extended with the conditions of unitary electric and magnetic constants $\epsilon_0 = \mu_0 = 1$, enforcing that the speed of light c in these units equals 1.)

Relativistic formalism. Maxwell's equations are consistent with special relativity: classical electromagnetism is a relativistic theory. The relativistic invariance of a theory is elegantly highlighted by expressing its observables in terms of quadri-vectors and quadri-tensors, whose components are labelled by Greek indices $\mu, \nu, \ldots$ running from 0 to 3.

Due to their different behaviour under Lorentz transformations, upper indices and lower indices are called contravariant and covariant, respectively. Two quadri-vectors of special interest are the coordinates x^μ, where $x^0 \equiv t$ is the time component, x^1, x^2, x^3 are the space components, and the corresponding derivatives are

$$\partial_\mu = \frac{\partial}{\partial x^\mu} = \left(\frac{\partial}{\partial t}, \frac{\partial}{\partial x^1}, \frac{\partial}{\partial x^2}, \frac{\partial}{\partial x^3} \right).$$

The Minkowski metric $\eta_{\mu\nu} = \mathbf{diag}(+1, -1, -1, -1)$ (respectively, its inverse $\eta^{\mu\nu}$) is used to lower contravariant indices (to raise covariant indices) like, for example,

$$V_\mu = \sum_{\nu=0}^{3} \eta_{\mu\nu} V^\nu.$$

To express Maxwell's equations in quadri-covariant notation, one introduces the antisymmetric electromagnetic tensor $F_{\mu\nu}$ defined by

$$F_{0i} = E_i, \quad F_{ij} = -\sum_k \epsilon_{ijk} B_k, \quad (i, j, k = 1, 2, 3),$$

as well as the quadri-current $J^\mu = (\rho, J^i)$ (E_i, B_i, J^i being, respectively, the i-th component of the 3-vectors $\mathbf{E}, \mathbf{B}, \mathbf{J}$).

In relativistic form, Faraday's and Gauss's laws (equations (11.31) and (11.32)) are combined into

$$\partial_\lambda F_{\mu\nu} + \partial_\mu F_{\nu\lambda} + \partial_\nu F_{\lambda\mu} = 0, \tag{11.35}$$

which are (known as Bianchi identities), while Gauss's law and Ampère–Maxwell's laws (equations (11.33) and (11.34)) give rise to

$$\sum_\mu \partial_\mu F^{\mu\nu} = J^\nu \implies 0 = \sum_{\nu,\mu} \partial_\nu \partial_\mu F^{\mu\nu} = \sum_\nu \partial_\nu J^\nu. \tag{11.36}$$

In a contractible manifold, following Poincaré's lemma, Bianchi identities (11.35) can be integrated by introducing a *gauge field* (a generalized vector potential) $A_\mu(x)$ such that

$$F_{\mu\nu} = \partial_\mu A_\nu - \partial_\nu A_\mu. \tag{11.37}$$

The gauge field $A_\mu(x)$ has for components the non-relativistic scalar potential $A_0(x)$ and the vector potential $A_i(x)$), $i = 1, 2, 3$.

One then verifies that two gauge fields $A_\mu(x)$ and $A_\mu^\Omega(x)$ related by the gauge transformation

$$A_\mu^\Omega(x) = A_\mu(x) - \partial_\mu \Omega(x), \tag{11.38}$$

correspond to the same electromagnetic tensor (11.37). In terms of non-relativistic scalar and vector potentials, equation (11.38) corresponds to equations (11.9) and (11.10).

As in the example of a particle in a magnetic field, Maxwell's equations (11.36) can be derived by demanding the stationarity of an action expressed in terms of the vector potential, after the identification (11.37). In presence of a conserved current J^μ, $\sum_\mu \partial_\mu J^\mu = 0$, the Lagrangian density reads

$$\mathcal{L}(A, \dot{A}; J) = -\tfrac{1}{4} \sum_{\mu,\nu} F_{\mu\nu} F^{\mu\nu} - \sum_\mu J^\mu A_\mu \tag{11.39}$$

($F_{\mu\nu}$ being expressed in terms of A_μ by equation (11.37)) and the action then is

$$\mathcal{A}(A; J) = \int \mathrm{d}^4 x \, \mathcal{L}(A, \dot{A}; J). \tag{11.40}$$

As in the example of the non-relativistic particle in a magnetic field, the action changes by boundary terms under the gauge transformation (11.38); if $\Omega(x)$ is a smooth function vanishing at space-time infinity, the action is invariant since then

$$\mathcal{A}^\Omega - \mathcal{A} = \int \mathrm{d}^4 x \sum_\mu J^\mu(x) \partial_\mu \Omega(x)$$

$$= \int \mathrm{d}^4 x \sum_\mu \left[\partial_\mu \left(J^\mu(x)\Omega(x) \right) - \left(\partial_\mu J^\mu(x) \right) \Omega(x) \right] = 0 \,.$$

11.6   Gauge fixing in classical gauge theories

In classical electromagnetism, the *gauge-fixing* problem is the problem of choosing a representative in the class of equivalent potentials, convenient for practical calculations or most suited to physical intuition (however, *quantization requires gauge fixing*).

Some usual gauges are:

- $\sum_{i=1}^3 \partial_i A_i(x) = 0$, known as *Coulomb's gauge*,
- $A_0(x) = 0$, known as the *temporal gauge* (or *Hamiltonian* or *Weyl's gauge*),
- $\sum_\mu n^\mu A_\mu(x) = 0$, its generalization, where n_μ can be a space-like quadri-vector, and known as the *axial gauge*,
- $\sum_\mu n^\mu A_\mu(x) = 0$, where n is a null-like quadri-vector, is known as the *light cone gauge*,

  and
- the relativistically covariant gauge $\sum_\mu \partial^\mu A_\mu(x) = 0$, known as the *Lorenz's gauge* or *Landau's gauge*.

Note that, some of these conditions do not fix the gauge field representative completely. The form and the meaning of the residual invariance depend on the choice of gauge fixing.

Finally, it is simple to generalize these gauges to theories with non-Abelian gauge invariance.

11.7 QED

In QED the evolution operator can be expressed as an integral over fields, a generalization of the path integrals of non-relativistic quantum mechanics.

11.7.1 Gauge field coupled to a conserved current

In analogy with non-relativistic quantum mechanics, one expects the time-evolution operator to be given by an integral over all classical fields,

$$\mathcal{U} = \int \prod_\mu [\mathrm{d} A_\mu(x)] \exp\left(\frac{i}{\hbar} \mathcal{A}(A; J)\right),$$

where the action is given by equations (11.40) and (11.39).

However, due to gauge invariance, the action depends only on three (out of four) degrees of freedom of the gauge fields and, thus, the integral over the full space of gauge fields *is not defined*. To solve the problem, it is necessary to *fix the gauge* (see also Sections 11.6 and 11.9). Gauge fixing can be achieved by restricting the integration over fields on a gauge section fixed by a constraint $\mathcal{E}(A, x) = 0$, or, more generally, by fixing $\mathcal{E}(A, x) = h(x)$ and by integrating over $h(x)$ with a Gaussian field distribution. The latter method leads to the addition to the action of the *non-gauge-invariant* contribution

$$\mathcal{A}_{\text{gauge}}(A) = \frac{1}{2\xi} \int \mathrm{d}^4 x\, \mathcal{E}^2(A, x). \tag{11.41}$$

The relativistic-covariant choice $\mathcal{E}(A, x) = \sum_\mu \partial_\mu A^\mu(x)$ gives in the limit $\xi \to 0$ *Landau's gauge* $\sum_\mu \partial_\mu A^\mu(x) = 0$. The special value $\xi = 1$ in equation (11.41) corresponds to *Feynman's gauge*.

When $\mathcal{E}(A, x)$ is linear in A, one then replaces in the field integral the initial gauge invariant action by

$$\mathcal{A}(A, J) + \mathcal{A}_{\text{gauge}}(A).$$

When $\mathcal{E}$ is not linear, the integration measure over gauge fields has to be modified.

11.7.2 Charged matter fields

One calls a field theory *local* when the action is a space-time integral of a Lagrangian density, which is a function of the fields and their derivatives taken at the same space-time point.

To construct a gauge invariant local action describing the interaction of matter with the gauge field $A_\mu(x)$, one can start from an action for charged matter fields that is local and invariant under global $U(1)$ transformations.

As an example of matter fields, we consider free spin $1/2$ Dirac fermions with charge e_χ and mass m. The fields are then four-component conjugate complex anticommuting (*i.e.*, belonging to a Grassmann algebra: see Section 2.9.2) vectors χ and $\bar\chi$ (called spinors). The Lagrangian density reads

$$\mathcal{L}_{\text{matter}} = \bar\chi(x)\left(\slashed\partial + im\right)\chi(x), \tag{11.42}$$

where γ^μ are the 4×4 Dirac matrices, satisfying $\gamma^\mu\gamma^\nu + \gamma^\nu\gamma^\mu = 2\eta^{\mu\nu}$, and we use, for any quadri-vector v_μ, the *standard notation* $\slashed v \equiv \sum_\mu \gamma^\mu v_\mu$. The action $\mathcal{A}_{\text{matter}} = \int \mathrm{d}^4 x\, \mathcal{L}_{\text{matter}}$ is invariant under the global $U(1)$ group transformations,

$$\chi(x) \mapsto U\chi(x), \quad \bar\chi(x) \mapsto U^*\bar\chi(x), \quad \text{with} \quad U = e^{ie_\chi \Omega/\hbar}, \tag{11.43}$$

where Ω is a space-time-independent constant. It follows from Noether's theorem [138] that this invariance implies (classically) the existence of a conserved current and of a conserved charge.

Gauge invariance requires invariance of the new action under the local group transformations obtained by replacing Ω, U in equation (11.43) by the two space-time functions $\Omega(x), U(x)$. This is achieved by replacing, in the matter Lagrangian density (11.42), ∂_μ by the *covariant derivative* $D_\mu = \partial_\mu + ie_\chi A_\mu(x)/\hbar$, which transforms under a four-dimensional generalization of equation (11.25) as

$$D_\mu(A^\Omega)[U(x)\chi(x)] = U(x)D_\mu(A)\chi(x).$$

Note that $[D_\mu, D_\nu] = ie_\chi F_{\mu\nu}/\hbar$.

Again, under the condition that $\mathcal{E}(A,x)$ is linear in A, quantum evolution of the total system (matter and gauge field) can be inferred from the field integral

$$\mathcal{U} = \int [\mathrm{d}A_\mu][\mathrm{d}\bar\chi][\mathrm{d}\chi]$$

$$\times \exp\frac{i}{\hbar}\left[\mathcal{A}(A) + \int \mathrm{d}^4 x\, \bar\chi(x)\left(\slashed{D} + im\right)\chi(x) + \mathcal{A}_{\text{gauge}}(A)\right].$$

Physical observables. The quantum field theory generated by this procedure does not explicitly satisfy the unitarity requirement (related to conservation of probabilities) and seems to depend on the gauge-fixing function and the parameter ξ. *Ward–Takahashi identities*, a set of relations among Green's functions, makes it possible to prove that the expectation values of *gauge invariant* quantities do not depend on these specific choices (a property called *gauge independence* in this context) and also satisfy unitarity.

For example, the two-point function $\langle\bar\chi(x)\chi(y)\rangle$, for $x \neq y$, is not physical. By contrast, expectation values of quantities involving products of operators of the form $\bar\chi(x)\chi(x)$ are physical. In addition, S-matrix elements are also physical observables.

11.7.3 Parallel transport

Gauge invariance is related to the geometric notion of parallel transport. Parallel transport makes it possible to define non-local gauge invariant quantities. One application is the construction of lattice gauge theories, which are useful to study non-perturbatively gauge theories by numerical simulations (see Chapter 13).

In the Abelian $U(1)$ example (QED), it is simple to define parallel transporters, which are curve-dependent elements of the $U(1)$ group. One considers the quantity,

$$U[C_{xy}] = \exp\left[-ie\oint_C \sum_\mu A_\mu(s)\mathrm{d}s^\mu\right],\tag{11.44}$$

where $\oint_C$ means integral along the continuous, piecewise differentiable oriented curve C with x as the origin and y as the extremity.

In a gauge transformation, a charged field $\bar{\psi}, \psi$ and the gauge field transform like

$$\bar{\psi}(x) \mapsto \mathrm{e}^{-i\Lambda(x)}\,\bar{\psi}(x),\ \psi(x) \mapsto \mathrm{e}^{i\Lambda(x)}\,\psi(x),\quad A_\mu(x) \mapsto A_\mu(x) - \frac{1}{e}\partial_\mu\Lambda(x),$$

and thus

$$e\oint_C \sum_\mu A_\mu(s)\mathrm{d}s^\mu \mapsto e\oint_C \sum_\mu A_\mu(s)\mathrm{d}s^\mu - \Lambda(y) + \Lambda(x).$$

The transformation of $U[C(x,y)]$ is then

$$U[C(x,y)] \mapsto \mathrm{e}^{i\Lambda(y)-i\Lambda(x)}\,U[C(x,y)]$$

and the product

$$\bar{\psi}(x)U[C_{xy}]\psi(y)$$

is gauge invariant. Similarly, the parallel transporter along a closed curve is gauge invariant (see also Section 13.1).

11.8 Non-Abelian gauge theories

As is usual in the domain of strongly quantum and relativistic physics, in this section we use SI units constrained by the additional requirement $\hbar = c = 1$.

11.8.1 Classical field theory

Yang and Mills [41] have generalized the structure of QED to a situation where the Abelian gauge group $U(1)$ is replaced by an non-Abelian Lie group G of $N \times N$ unitary matrices.

To construct a gauge invariant action, we consider, for example, complex fields $\phi(x)$ transforming globally (*i.e.*, the group elements are space-time independent) under the group G as

$$\phi(x) \mapsto \mathbf{g}\phi(x),\quad \mathbf{g} \in G\,,\tag{11.45}$$

and a group invariant Lagrangian density for matter fields $\mathcal{L}_{\mathrm{matter}}(\phi, \partial_\mu\phi)$:

$$\mathcal{L}_{\mathrm{matter}}(\mathbf{g}\phi, \partial_\mu\mathbf{g}\phi) = \mathcal{L}_{\mathrm{matter}}(\phi, \partial_\mu\phi)\,.\tag{11.46}$$

The goal is to promote the invariance of the action under the global transformations (11.45) to an invariance under local transformations (*i.e.*, space-time dependent):

$$\phi(x) \mapsto \mathbf{g}(x)\phi(x), \quad \mathbf{g}(x) \in G \;\; \forall x. \tag{11.47}$$

Again, a problem arises with field derivatives. As in the Abelian example, the solution is to replace derivatives by covariant derivatives. Here, *covariant derivatives* are $N \times N$ matrices of the form

$$\mathbf{D}_\mu \equiv \mathbf{D}_\mu(\mathbf{A}) = \mathbf{1}\,\partial_\mu + \mathbf{A}_\mu(x), \tag{11.48}$$

where the *gauge field* $\mathbf{A}_\mu(x)$ belongs to the Lie algebra of the group G. (The non-Abelian gauge field should not be confused with the 3-vector potential $\mathbf{A}(x)$ used in previous sections.) By definition, the covariant derivative $\mathbf{D}_\mu$ is a tensor under general gauge transformations, that is, transforms linearly as

$$\mathbf{g}(x)\left(\mathbf{1}\,\partial_\mu + \mathbf{A}_\mu(x)\right)\mathbf{g}^{-1}(x) = \mathbf{1}\partial_\mu + \mathbf{A}_\mu^{\mathbf{g}}(x)$$
$$\Leftrightarrow \; \mathbf{g}(x)\mathbf{D}_\mu(\mathbf{A}) = \mathbf{D}_\mu(\mathbf{A}^{\mathbf{g}})\,\mathbf{g}(x), \tag{11.49}$$

where, as a consequence, the gauge transform $\mathbf{A}_\mu^{\mathbf{g}}$ of the gauge field is given by

$$\mathbf{A}_\mu^{\mathbf{g}}(x) = \mathbf{g}(x)\mathbf{A}_\mu(x)\mathbf{g}^{-1}(x) + \mathbf{g}(x)\partial_\mu\mathbf{g}^{-1}(x), \quad \mathbf{g}(x) \in G \;\forall x. \tag{11.50}$$

The transformation is linear in the special case of a constant $\mathbf{g}(x) = \mathbf{g}_0$ (global transformation), but in general is affine. The property (11.49) of the covariant derivative then implies

$$\mathcal{L}_{\mathrm{matter}}(\mathbf{g}(x)\phi, D_\mu \mathbf{g}(x)\phi) = \mathcal{L}_{\mathrm{matter}}(\mathbf{g}(x)\phi, \mathbf{g}(x)D_\mu\phi) = \mathcal{L}_{\mathrm{matter}}(\phi, D_\mu\phi) \tag{11.51}$$

and, thus, the matter action is now gauge invariant.

In general, the gauge field $\mathbf{A}_\mu(x)$ has a mathematical interpretation as a Lie-valued *connection* and is used to construct covariant derivatives acting on fields, whose form depends on the representation of the group G under which the field transforms (for global transformations).

The commutator of two covariant derivatives (11.48),

$$\mathbf{F}_{\mu\nu}(x) = [\mathbf{D}_\mu, \mathbf{D}_\nu] = \partial_\mu \mathbf{A}_\nu(x) - \partial_\nu \mathbf{A}_\mu(x) + [\mathbf{A}_\mu(x), \mathbf{A}_\nu(x)], \tag{11.52}$$

is no longer a differential operator and corresponds to the *curvature* of the connection, which is a tensor for gauge transformations (*i.e.*, it transforms linearly):

$$\mathbf{F}_{\mu\nu}(x) \mapsto \mathbf{g}(x)\mathbf{F}_{\mu\nu}(x)\mathbf{g}^{-1}(x).$$

The *curvature tensor* (or field-strength) $\mathbf{F}_{\mu\nu}(x)$ is an element of the Lie algebra of G. Since $\mathbf{F}_{\mu\nu}$ is a tensor, the local action for the gauge field

$$\mathcal{A}(\mathbf{A}) = \frac{1}{4e^2} \int \mathrm{d}^4x \operatorname{tr} \sum_{\mu,\nu} \mathbf{F}_{\mu\nu}(x)\mathbf{F}^{\mu\nu}(x), \tag{11.53}$$

is gauge invariant.

When $\mathbf{g}(x)$ is close to the identity, that is, $\mathbf{g}(x) = \mathbf{1} + \omega(x) + O(\|\omega\|^2)$, the transformation (11.50) takes the form

$$\mathbf{A}_\mu^{\mathbf{g}}(x) - \mathbf{A}_\mu(x) = -\mathbf{D}_\mu\omega(x) + O(\|\omega\|^2),$$
$$\text{with } \mathbf{D}_\mu\omega(x) \equiv \partial_\mu\omega(x) + [\mathbf{A}_\mu(x), \omega(x)], \tag{11.54}$$

where $\mathbf{D}_\mu$ in equation (11.54) is the covariant derivative acting on fields belonging to the Lie algebra of the group G. This makes it possible to write the classical field equations corresponding to the action (11.53) in the form

$$\sum_\mu \mathbf{D}_\mu \mathbf{F}^{\mu\nu}(x) = 0. \tag{11.55}$$

Since the curvature tensor (11.52) is quadratic in the gauge field and the covariant derivative (11.54) linear, the field equations are cubic, showing that non-Abelian gauge fields are self-interacting, in contrast with the Abelian case.

In both the Abelian and the non-Abelian case, physical observables are related to expectation values of products of gauge invariant polynomials in the fields (called also *gauge invariant operators*).

Parallel transport. As all connections, the gauge connection generates *parallel transport*. Parallel transport becomes especially relevant in the framework of *lattice gauge theories*, where space-time is replaced by a discrete lattice, and gauge fields by group elements associated with links (Chapter 13).

11.8.2 Gauge fields and differential geometry

The concepts of gauge fields and covariant derivatives can be translated into the language of differential geometry, based on differential forms.

To the gauge field $\mathbf{A}_\nu$ is associated a Lie-valued *connection 1-form*,

$$\mathbf{A} = \sum_\mu \mathbf{A}_\mu \mathrm{d}x^\mu$$

and, to the curvature tensor $\mathbf{F}_{\mu\nu}$, a Lie-valued curvature 2-form,

$$\mathbf{F} = \sum_{\mu,\nu} \mathbf{F}_{\mu\nu} \mathrm{d}x^\mu \wedge \mathrm{d}x^\nu.$$

In terms of differential forms, the relation (11.52) then reads

$$\mathbf{F} = 2\left(\mathrm{d}\mathbf{A} + \mathbf{A} \wedge \mathbf{A}\right).$$

11.9 Quantization of non-Abelian gauge theories: Gauge fixing

In gauge theories, due to gauge invariance, not all components of the gauge field are dynamical, and the usual quantization method is not applicable: a gauge fixing is required. However, a straightforward extension of the ideas that worked for QED fails here. In contrast with QED, the quantized form of non-Abelian gauge theories cannot be easily guessed, even in the absence of matter fields.

A formal quantization in Weyl's or temporal gauge $\mathbf{A}_0 = 0$ is still possible, but leads to a theory that is not explicitly relativistic covariant and has a singular perturbative expansion. Transformations that can be easily explained only in the framework of field integrals (see also Section 3.7), make it possible to go over to explicitly covariant formulations.

11.9.1 *Gauge fixing in gauge field integrals*

In quantum gauge theories, the problem of gauge fixing is much more involved than in classical field theory. Indeed, since the integrand in field integrals giving physical observables is constant along gauge orbits (the set of all gauge fields obtained from one representative by gauge transformations), the naive field integrals (sums over all field configurations) are not defined. It becomes necessary to fix the gauge, that is, to integrate only over a section of gauge field space that—ideally—contains only one representative per orbit. Generalizing the formalism of QED, one defines the gauge section by a set of equations of the form $\mathcal{E}^a(A) = 0$, where the index a runs over the generators of the Lie algebra. The simplest, relativistically covariant gauges correspond to

$$\mathcal{E}^a(A) \equiv \sum_\mu \partial^\mu A_\mu^a(x) = 0. \tag{11.56}$$

Very often, one actually integrates not only over a gauge section, but over a whole neighbourhood of a section, by fixing $\mathcal{E}^a(A) = h^a(x)$ and by integrating over $h^a(x)$ with a (functional) Gaussian distribution. This procedure amounts to adding to the action a (non-gauge- invariant) gauge-fixing term, which is a generalization of the term (11.41).

Moreover, it is necessary to integrate over gauge fields with a measure ensuring that physical results are independent of the choice of the section. For covariant gauges and in contrast with QED, such a measure is non-trivial [40] (see also Ref. [73]). In the case of Landau's gauge (11.56), it takes the form of the absolute value of the determinant of a differential operator, $\det \sum_\mu \partial_\mu \mathbf{D}^\mu$.

For small gauge fields, the determinant can be chosen positive and can be written in local form by the introduction of non-physical spinless fermion fields, the so-called Faddeev–Popov 'ghost' fields [40].

BRST invariance (see Becchi, Rouet, Stora [43] and references therein) emerges in this context, as a substitute for the broken classical gauge invariance. It can then be proved that this construction yields a consistent theory in the sense of a perturbative expansion: the perturbative expansion is generated by keeping the quadratic part of the action and expanding all other terms in a power series, reducing the field integral to an infinite sum of Gaussian expectation values.

However, the problem of integration over suitable gauge sections is more subtle at a non-perturbative level. Indeed, one would like the section to cut all gauge orbits once or, at least, the same number of times, but this is not generally the case in non-Abelian gauge theories, as first pointed out by Gribov [142]. (One speaks then of Gribov copies.) In particular, the Faddeev–Popov determinant changes sign when two Gribov copies merge and, if the absolute value of the determinant is not taken into proper account, the integration measure is no longer positive.

One idea is then to restrict the integration to the region enclosing only one Gribov copy, but this is not easy to achieve in practice. The only known non-perturbative definition of gauge theories is obtained by replacing continuum space-time by a lattice leading to *lattice gauge theories*. Then, at least in a finite volume, gauge fixing is not necessary and Gribov's problem is avoided.

Gauge fixing and locality. In renormalizable *covariant gauges*, the theory is non-local, even if the introduction of non-physical fermions makes it possible to write the action in local form. In the non-covariant axial or temporal gauges, the action remains formally local but the theory is not renormalizable and the propagator then has non-physical zero-momentum singularities. Only theories with complete gauge symmetry breaking, in such a way that all gauge fields become massive, can be quantized in a physical unitary gauge, but the theory is then non-renormalizable, like massive QED.

Non-Abelian gauge theories and parallel transport. The notion of parallel transport generalizes immediately to non-Abelian gauge theories and is at the basis of lattice gauge theories. In the non-Abelian case, the explicit expression of the parallel transporter is more complicated than in the Abelian example because the gauge field $\mathbf{A}_\mu^\alpha(x)t_\alpha$ is an element of the Lie algebra of G, and the matrices representing the field at different points do not commute. It can be formally written as a path-ordered integral (see Section 13.1).

11.10 General Relativity

Einstein's relativistic theory of gravitation [23], also known as *General Relativity*, has properties somewhat analogous to gauge theories. Here, invariance under diffeomorphisms $x \mapsto x(y)$ (locally regular changes of coordinates) in a (pseudo-) Riemannian manifold $\mathcal{M}$ replaces gauge invariance. In this context, the Levi–Civita connection $\mathbf{\Gamma}_\mu$ (with components, the Christoffel symbol $\Gamma_{\nu\mu}^\lambda$), which generates parallel transport, plays the role of the gauge field.

The similarity with gauge theories is even more striking when the vielbein formalism is introduced, which is required in the case of matter fields with spin. A *vielbein* e_μ^α is a locally flat frame in the space tangent to the manifold. A gauge transformation corresponds to a change of local frame (a local Lorentz transformation). Gauge invariance corresponds to the independence of field equations from the choice of the local frame. The *spin connection* $\omega_\mu^{\alpha\beta}$, which can be expressed in terms of the vielbein, then plays the role of the gauge field.

12 The Higgs boson: A major discovery and a problem

The discovery of the Higgs boson, the only particle predicted by the Standard Model of fundamental interactions at the microscopic scale that was still missing, is an impressive, both experimental and theoretical, achievement.

The Higgs boson has been produced by the specifically designed proton–proton LHC at CERN (Geneva) but this discovery is the result of about 40 years of experimental efforts involving also several previous colliders, most notably LEP (the Large Electron–Positron collider at CERN) and the Tevatron (the proton–antiproton collider at Fermilab).

At the LHC, the first collisions were recorded in November 2009. A first phase run at about half the design energy (first 7 TeV and then 8 TeV) was completed by the end of 2012. In July 2012, the two multi-purpose detectors, Atlas and CMS could report the discovery of a new particle, a boson with a mass of about 125 GeV (130 times the proton mass). The new boson is the second-heaviest fundamental particle after the top quark (173 GeV).

The analysis of all 2011–2018 data taken at 7, 8 and 13 TeV is still consistent, at the precision of the experiments, with the conclusions that the discovered particle has the properties required to be the Higgs boson, a unique scalar boson predicted by the Standard Model, a model that describes all of the fundamental interactions except gravity.

The discovery of the Higgs boson is a final experimental confirmation of the validity of the model, at least up to the available collider energies.

Moreover, up to now (2018), no sign of new physics beyond the Standard Model has been detected.

The experiments have been guided by the predictions of the Standard Model, a theoretical construction that itself involved several decades of theoretical effort, which we now try to briefly describe.

12.1 Perturbative quantum field theory: The construction

The construction of a *quantum field theory* generally involves three main steps:

(i) construction of a classical theory,
(ii) quantization,
(iii) renormalization.

Moreover, in the example of *gauge theories*, due to the necessity of *gauge fixing*, an additional step is required, the proof of **unitarity**, which is related to the proof of *gauge independence*. The proof is especially involved in the non-Abelian case, owing to the presence of spinless non-physical fermions, the *Faddeev–Popov 'ghost' fields*. Moreover, in the case of *spontaneous symmetry breaking*, the *decoupling of the would-be massless Nambu–Goldstone bosons* has to be demonstrated.

From Random Walks to Random Matrices. Jean Zinn-Justin, Oxford University Press (2019).
© Jean Zinn-Justin. DOI: 10.1093/oso/9780198787754.001.0001

Only when this programme is completed, does one know that the theory has suitable physical properties and can perturbative calculations be safely performed.

12.1.1 Towards a model for weak and electromagnetic interactions

Specifically, the construction of the present model of weak and electromagnetic interactions has involved a number of intermediate steps. The most important are:

(i) construction of quantum electrodynamics, an Abelian gauge theory that is a quantum extension of Maxwell's electromagnetism,

(ii spontaneous breaking of continuous symmetries in scalar boson and fermion theories, with the correlated appearance of massless particles, called Goldstone bosons,

(iii) construction and quantization of Non-Abelian gauge theories,

(iv) discovery of the Abelian Landau–Ginzburg–Higgs mechanism, which combines Abelian gauge theories and spontaneous symmetry breaking,

(v) discovery of the non-Abelian Higgs mechanism, which combines non-Abelian gauge theories and spontaneous symmetry breaking.

Finally, the Higgs mechanism contains itself two ingredients: *mass given to all physical particles* and the *absence (or decoupling) of would-be massless Goldstone bosons.*

The well-documented history of the construction of *QED*, a lengthy process extending over about 20 years (1930–1950), will not be described here (but a few elements can be found in Chapter 4).

12.2 Spontaneous symmetry breaking

The notion of *spontaneous symmetry breaking* originates from *statistical physics and the theory of phase transitions.* A general framework was provided by Landau's theory (1937) [54].

As an illustration, we display in Fig. 12.1 an $O(2)$ symmetric potential, which, instead of a minimum located at the symmetry point, has a circle of minima. In the quantum framework, when the classical picture survives quantum fluctuations, this situation leads to a degenerate ground state. The choice of one specific minimum (resp., the choice of a specific ground state) then breaks spontaneously the $O(2)$ symmetry.

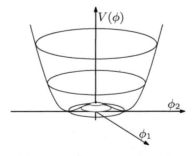

Fig. 12.1 An $O(2)$ symmetric potential $V(\phi)$ with degenerate minima.

12.2.1 Relativistic quantum field theory

In the context of the relativistic quantum theory in $d = 4$ dimensions (time + space), the notion of spontaneous symmetry breaking was initially introduced to describe the approximate $SU(2) \times SU(2)$ chiral symmetry of low energy hadron physics. In particular, *spontaneous symmetry breaking implies the existence of massless particles (Nambu–Goldstone bosons)* and this is consistent with the small pion mass [119].

The simplest relevant model is the $O(4) \sim SU(2) \times SU(2)$ symmetric $(\phi^2)^2$ theory for a four-component, self-interacting, scalar boson field $\phi \equiv (\sigma, \pi)$. The Euclidean action (before regularization) reads

$$ S(\phi) = \int d^4 x \left[\tfrac{1}{2} \left(\nabla_x \phi(x) \right)^2 + \tfrac{1}{2} r \phi^2(x) + \frac{g}{4!} \left(\phi^2(x) \right)^2 \right] , \quad g > 0 , \tag{12.1} $$

where ∇_x stands for the 4-vector $\partial / \partial x_1, \ldots, \partial / \partial x_4$.

In the classical limit, for $r < 0$, the extremum of the action is degenerate, corresponding to the sphere $\phi^2(x) \equiv v^2 = -6r/g$, and one says that the *symmetry is spontaneously broken*. A physical theory corresponds to a particular choice of an extremum $\phi = \mathbf{v}$.

Shifting the field by its expectation value, $\phi = \mathbf{v} + \chi$, one finds

$$ \tilde{S}(\chi) = S(v + \chi) - S(v) = \int d^4 x \left[\tfrac{1}{2} \left(\nabla_x \chi(x) \right)^2 + \frac{g}{6} \left(\mathbf{v} \cdot \chi(x) \right)^2 \right. $$
$$ \left. + \frac{g}{6} \mathbf{v} \cdot \chi(x) \chi^2(x) + \frac{g}{4!} \left(\chi^2(x) \right)^2 \right] . $$

Examining the quadratic terms in χ, one observes that, in the classical limit, the perturbative spectrum contains one massive particle, corresponding to the projection of χ on $\mathbf{v}$, with mass squared $m^2 = \tfrac{1}{3} g v^2$ and three massless particles, which are called *(Nambu–)Goldstone bosons* and correspond to the transverse directions. Still, in the classical limit the result survives when one takes into account the non-quadratic terms. This can be verified by the change of fields $\phi \mapsto (\rho, \mathbf{v})$, setting

$$ \phi(x) = (v + \rho(x)) \mathbf{n}(x) \text{ with } \mathbf{n}^2(x) = 1 , $$

a change that is legitimate only for $v \neq 0$. Then, the action becomes

$$ S(\rho, \mathbf{n}) = \int d^4 x \left\{ \tfrac{1}{2} \left[\left(\nabla_x \rho(x) \right)^2 + (v + \rho(x))^2 \left(\nabla_x \mathbf{n}(x) \right)^2 \right] + \tfrac{1}{2} r (v + \rho(x))^2 \right. $$
$$ \left. + \frac{g}{4!} (v + \rho(x))^4 \right\} . $$

One notices that the field $\mathbf{n}$ with three independent degrees of freedom has no mass term.

It remains then to show that the existence of Goldstone bosons survives quantum corrections to all orders in a perturbative expansion around the classical vacuum.

Spontaneous symmetry breaking and renormalization. As a first step, it is necessary to regularize the theory by introducing a cut-off without destroying the $O(4)$ symmetry of the model.

Then, while the *renormalization* of the theory in the symmetric phase is simple, it is more complicated in the broken phase because the Feynman rules are no longer explicitly symmetric. The proof that the renormalization constants of the symmetric theory renormalize also the theory in the broken phase, is based on the use of generalized *Ward–Takahashi (WT) identities* [143, 144]). WT identities also confirm the existence of massless Goldstone bosons.

12.3 Non-Abelian gauge theories

The classical non-Abelian generalization of Maxwell's electrodynamics has been proposed by Yang and Mills [41].

12.3.1 Classical theory

Global symmetry corresponds to invariance of the action under global (space-time-independent) group transformations acting on matter fields (see also Section 11.8.1). Gauge symmetry is obtained by further demanding invariance under *space-time dependent* transformations. This requires the introduction of a *gauge field* (or connection) $\mathbf{A}_\mu(x)$ associated to the adjoint representation of the Lie group. To the linear transformation of matter fields,

$$\phi(x) \mapsto \mathbf{g}(x)\phi(x),$$

where $\mathbf{g}(x)$ is a space-dependent group element, corresponds the affine transformation of the gauge field,

$$\mathbf{A}_\mu(x) \mapsto \mathbf{g}(x)\partial_\mu \mathbf{g}^{-1}(x) + \mathbf{g}(x)\mathbf{A}_\mu(x)\mathbf{g}^{-1}(x).$$

The coupling of matter fields to the gauge field is obtained by replacing ordinary partial derivatives by covariant derivatives,

$$\partial_\mu \mapsto \mathbf{D}_\mu \equiv \partial_\mu \mathbf{1} + \mathbf{A}_\mu(x).$$

The classical gauge action is defined in terms of the curvature tensor

$$\mathbf{F}_{\mu\nu}(x) = [\mathbf{D}_\mu, \mathbf{D}_\nu] = \partial_\mu \mathbf{A}_\nu(x) - \partial_\nu \mathbf{A}_\mu(x) + [\mathbf{A}_\mu(x), \mathbf{A}_\nu(x)].$$

In the absence of matter fields, in d space-time dimensions it reads

$$\mathcal{A}(\mathbf{A}_\mu) = -\frac{1}{4g^2} \int \mathrm{d}^d x \, \mathrm{tr}\, \mathbf{F}_{\mu\nu}(x)\mathbf{F}^{\mu\nu}(x).$$

12.3.2 The problem of quantization

According to the correspondence principle formulated in terms of Feynman's path integral, in simple examples the quantum evolution operator can be derived from a path or field integral of the form ($2\pi\hbar$ is Planck's constant)

$$\int [\mathrm{d}\mathbf{A}_\mu]\, e^{i\mathcal{A}(\mathbf{A})/\hbar},$$

where $\mathcal{A}(\mathbf{A})$ is the classical action.

In gauge theories, the field integration over the gauge field is not defined because the action is constant along a gauge orbit. The quantization necessitates a gauge section. Moreover, the trial and error method used in the Abelian case to quantize QED no longer works in the non-Abelian case. This was demonstrated by Feynman [42], who pointed out that, after the usual QED-like gauge fixing, unitarity at *one loop* implies the *addition of spinless fermion quanta*. After this work, the quantization of non-Abelian gauge theories remained for some time an open problem.

A solution to this very difficult problem was eventually proposed, using *field integral methods*, in 1967 by Faddeev and Popov [40], who obtained the (in general, non-trivial) integration measure on the gauge section and then found a local interpretation of the measure as being reproduced by an integration over spinless (thus non-physical) fermions (the Faddeev–Popov 'ghost' fields: see also Section 3.7.3).

12.4 The classical Abelian Landau–Ginzburg–Higgs mechanism

The first example of a classical Abelian Landau–Ginzburg–Higgs mechanism can be found in an effective macroscopic model for a *superconductor in a magnetic field*, proposed by Ginzburg and Landau [63].

Describing a superconductor in a magnetic field near the transition temperature T_c, the Ginzburg–Landau Hamiltonian (in the terminology of statistical physics) reads,

$$\mathcal{H}(A,\psi) = \int \mathrm{d}^3x \left[\frac{1}{4\mu_0} \sum_{\mu,\nu} (\partial_\mu A_\nu - \partial_\nu A_\mu)^2 + \sum_\mu \frac{1}{2m} \left| \left(\frac{\hbar}{i}\partial_\mu - 2eA_\mu \right) \psi \right|^2 \right.$$
$$\left. + \alpha(T - T_c)\,|\psi|^2 + \tfrac{1}{2}\beta\,|\psi|^4 \right],\quad \alpha, \beta > 0\,,$$

where A_μ is a vector potential that generates the magnetic field and ψ is a complex scalar field representing electron pairs. For $T < T_c$, the expectation value $\langle\psi\rangle \equiv \mathbf{v} \neq 0$. In the absence of the gauge field, the spectrum corresponds to a 'massive' particle and a 'massless' Goldstone boson in the particle physics terminology. It describes *the low temperature phase of a superfluid*.

In the presence of the gauge field, one can combine a change of field with a gauge transformation, setting

$$\psi(x) = \big(v + \rho(x)\big)\, e^{i\theta(x)/\hbar}\,,\quad A_\mu(x) = B_\mu(x) + \frac{1}{2e}\partial_\mu\theta(x).$$

The transformation completely eliminates $\theta(x)$ from the Hamiltonian and, therefore, massless Goldstone bosons are no longer generated. Only two massless particles remain. In the terminology of superconductivity, the inverses of the 'masses' of ψ and A, represent the *coherence length* and the *penetration length* of the magnetic field in the superconductor, respectively.

Replacing three-dimensional space by four-dimensional space-time, which, in the classical limit for the arguments given above, is irrelevant, and with some reinterpretation of the parameters, one obtains the relativistic Abelian Landau–Ginzburg–Higgs model.

A proper knowledge of this early work on superconductivity might have shown that the main remaining issues were the *quantization of the Landau–Ginzburg–Higgs mechanism* and the extension to *non-Abelian gauge theories*.

12.5 Abelian and non-Abelian Higgs mechanism

In quantum field theory, the idea that gauge fields could acquire a mass through interactions seems to have been proposed first by Schwinger [145]. In a first article, the proposal is not very explicit and no reference to spontaneous symmetry breaking is made. However, in the second article, it is illustrated by the exact solution of a $(1+1)$-*dimensional QED model with massless fermions* (now called the Schwinger model), although, in modern thinking, the model rather illustrates *confinement*, even if it also exhibits spontaneous chiral symmetry breaking.

Inspired by Schwinger, Anderson then published an article that was based on spontaneous symmetry breaking [146], and was more directly influential. On the basis of the plasmon theory and superconductivity, he conjectured that it should be possible to construct a model where gauge fields acquire a mass, and massless Goldstone bosons are avoided. However, an explicit model was missing.

Some discussion and confusion about the relevance of the *non-relativistic* examples to particle physics, specifically, the *avoidance of the massless Goldstone bosons*, followed (*e.g.*, Klein and Lee [147], Gilbert [148]: Gilbert's negative arguments are refuted by Higgs [149], who suggests that massless Goldstone bosons could be avoided in the presence of gauge fields).

Then a series of papers of 'optimistic' papers were published, where the authors obviously assume that gauge theories with spontaneously broken symmetries should be relevant.

Within the classical approximation, Englert and Brout [70] showed that, *in non-Abelian gauge theories, spontaneous symmetry breaking signalled by scalar field expectation values, can give masses to gauge bosons*. However, little is said about the fate of scalar particles and, in particular, about the absence of massless scalar (Goldstone) bosons.

In Ref. [71], Higgs discusses in the *classical limit* an explicit field theory example in which charged scalar particles are coupled to an Abelian gauge field. *The absence of Goldstone bosons and vector mass generation are shown for linearized field equations.*

The *extension to the non-Abelian situation* is discussed explicitly in the form of an $SU(3)$ theory and the spectrum of the gauge fields and scalars, after some specific symmetry breaking, described.

In Ref. [72], Guralnik, Hagen and Kibble consider a QED-like field theory model, an Abelian gauge field interacting with charged scalar particles. *They show, in the linear approximation and in a specific non-relativistic (radiation) gauge, that, after spontaneous symmetry breaking, the vector field becomes massive and the Goldstone boson decouples.*

In 1966, Higgs [150] reports a more thorough study of the Landau–Ginzburg–Abelian Higgs model, with QED coupled to a scalar charged field with spontaneous symmetry breaking, in the classical approximation and in the one-loop approximation in a specific gauge. Moreover, *in the classical approximation,* he shows that *a gauge transformation can be found that transforms the initial Lagrangian into a Lagrangian with only physical degrees of freedom,* a massive 'Higgs' scalar field and a massive vector field. However, this representation, often called *'unitary gauge',* is known to lead to non-renormalizable field theories.

Finally, the Higgs argument, that *a gauge transformation can be found that transforms the initial classical Lagrangian into a Lagrangian with only physical degrees of freedom,* is generalized to the *non-Abelian situation* by Kibble [151].

12.6 Non-Abelian gauge theories: Quantization and renormalization

Quantization. In sharp contrast with QED (the Abelian case) and as Feynman had already shown [152], *the quantization of non-Abelian gauge theories, which involves gauge fixing, could not be achieved following simple heuristic arguments.* This rather involved problem was eventually solved by *field integral methods* by Faddeev and Popov in 1967 in [40]. This opened the way for a systematic study of non-Abelian gauge theories as quantum field theories.

Towards renormalization. Two articles of 't Hooft [74], where it was argued that some models based on non-Abelian gauge theories, *both in the symmetric phase and in the case of spontaneous symmetry breaking,* should be *renormalizable,* then triggered an intensive theoretical activity.

Abelian Landau–Ginzburg–Higgs model: Full quantum theory and renormalization. Motivated by 't Hooft's work, Lee [153] first proved rigorously that the Abelian Landau–Ginzburg–Higgs model is *renormalizable, proving unitarity and gauge independence of the S-matrix to all orders,* in particular, that it involves no massless scalar boson but only massive scalar and vector fields. This article also seems to be at the origin of the denomination 'Higgs phenomenon' and, thus, presumably as a consequence, 'Higgs boson'.

12.6.1 Non-Abelian gauge theories: Renormalization

The first complete proofs of renormalizability relied on a set of *generalized WT identities* derived independently by Slavnov and Taylor [154].

They were used to prove renormalizability of non-Abelian gauge theories in the broken phase by Lee and Zinn-Justin [76].

This completed the theoretical proof of the validity of the model of weak and electromagnetic interactions [68].

The complexity of these proofs is a consequence of gauge fixing and spontaneous symmetry breaking, which completely *destroy the beautiful geometric structure of the initial Lagrangian*, leading to the appearance of a number of non-physical particles, would-be Goldstone bosons and Faddeev–Popov ghost fields.

Also relevant to this topics are articles by Fradkin and Tyutin, 't Hooft and Veltman, Fujikawa, Lee and Sanda and Ross and Taylor, among others.

12.6.2 BRST symmetry

An additional important technical advance was provided by the observation by Becchi, Rouet and Stora, and Tyutin, that the *quantized gauge action* has an unexpected fermion-like symmetry, which is now called the *BRST symmetry* [43], a property linked directly much more to the gauge-fixing procedure [155] than to gauge symmetry.

This observation was immediately used to derive a *completely general, much more transparent proof*, of renormalizability covering all semi-simple groups, renormalizable gauges and so on, based on the Zinn-Justin equation [156]. Gauge independence and unitarity in this general context was then dealt with in Ref. [157].

Finally, all technical details concerning the theoretical developments described in this chapter can be found in Zinn-Justin's work [18].

12.7 The self-coupled Higgs field: A simple RG analysis

With the Higgs boson, the last missing particle predicted by the Standard Model of all fundamental interactions at the microscopic scale except gravity has been discovered. Since the mass of the Higgs particle has now been determined, several properties of the renormalization group (RG) trajectories as well as the fine tuning problem can be re-examined. For example, the arguments based on the consistency limit of the Higgs model, and which led to an upper-bound on the possible Higgs mass, remain now only mildly relevant. In fact, due to the 'small' Higgs mass, the Higgs model could be consistent up to very high energies. On the other hand, if no new physics is discovered at higher energies, the amount of *fine tuning* of the Higgs mass parameter in the Lagrangian will rapidly increase, lending an additional piece of mystery to the Standard Model.

12.7.1 The self-coupling approximation

In the Standard Model, the Higgs field, through its various couplings, gives masses to all fundamental particles. The observed masses determine the corresponding couplings. Before the experimental discovery of the Higgs boson, the Higgs field expectation value was known, but not the Higgs self-interaction and, thus, the Higgs mass. However, the design of a new collider to detect the Higgs particle depended to some extent on a possible upper-bound on the Higgs mass. We summarize here some arguments that were used [158].

The Higgs field Lagrangian. The field theory reduced to the Higgs field, corresponds to a four-component real vector $\mathbf{H}$ interacting through a quartic $O(4)$ symmetric interaction. In Euclidean (*i.e.*, imaginary time) notation, the action can be written as ($\nabla \equiv (\partial/\partial x_1, \ldots, \partial/\partial x_4)$)

$$\mathcal{S}(\mathbf{H}) = \int \mathrm{d}^4 x \left[\tfrac{1}{2}(\nabla_\Lambda \mathbf{H}(x))^2 + \tfrac{1}{2} r \mathbf{H}^2(x) + \frac{1}{4!} \lambda (\mathbf{H}^2(x))^2 \right], \qquad (12.2)$$

where ∇_Λ is a regularized form of ∇, for example,

$$\nabla_\Lambda = \nabla \left[1 - \alpha_1 \nabla^2/\Lambda^2 + \alpha_2 \nabla^4/\Lambda^4 \right], \qquad \alpha_1, \alpha_2 > 0,$$

as required to cure the unavoidable problem of ultraviolet infinities. The cut-off Λ is an *ad hoc*, and non-physical, substitute for the scale of some new physics beyond the Standard Model, and r, λ are parameters relevant to the cut-off scale. The condition $m_H \ll \Lambda$, where m_H is the $\mathbf{H}$ physical mass, implies the tuning $|r - r_c| \ll \Lambda^2$, where r_c is the critical value where m_H vanishes.

In the perturbative regime, the Higgs mass increases with the self-coupling. For λ large enough, the Higgs mass is mostly determined by the Higgs self-coupling, justifying the analysis of a model restricted to the Higgs field (12.2).

In the pure $(\mathbf{H}^2)^2$ field theory and in the perturbative regime, simple RG arguments are applicable. The *effective coupling constant* $\lambda_\mathrm{r} = \lambda(\mu/\Lambda)$ at physical scale $\mu \ll \Lambda$ is given by

$$\int_\lambda^{\lambda(\mu/\Lambda)} \frac{\mathrm{d}\lambda'}{\beta(\lambda')} = \ln(\mu/\Lambda). \qquad (12.3)$$

The perturbative expansion of the RG β-function is

$$\beta(\lambda) = \beta_2 \lambda^2 + \beta_3 \lambda^3 + O\left(\lambda^4\right), \quad 8\pi^2 \beta_2 = 2, \quad \beta_3/\beta_2^2 = -13/24.$$

The positivity of β_2 indicates that the $(\mathbf{H}^2)^2$ field theory is infrared (IR) free: the effective coupling constant λ_r flows to zero for $\Lambda/\mu \to \infty$ (the famous *triviality* problem).

For $\lambda_\mathrm{r} = \lambda(\Lambda/\mu)$ small, one infers from equation (12.3),

$$\ln(\Lambda/\mu) = \frac{1}{\beta_2 \lambda_\mathrm{r}} + \frac{\beta_3}{\beta_2^2} \ln \lambda_\mathrm{r} + K(\lambda) + O(\lambda_\mathrm{r}), \qquad (12.4)$$

where $K(\lambda) = O(1)$ can only be determined by non-perturbative methods.

Still for λ_r small, perturbation theory relates the Higgs field expectation value, which is known from the Z vector boson mass ($|\langle \mathbf{H}\rangle| \equiv h \approx 250\,\mathrm{GeV}$), and the Higgs mass. At leading order, one finds

$$m_H^2 = \tfrac{1}{3} \lambda_\mathrm{r} h^2 + O\left(\lambda_\mathrm{r}^2\right).$$

Neglecting higher order corrections, one can substitute $\lambda_\mathrm{r} = 3 m_H^2/h^2$ into equation (12.4). Keeping only the two singular terms in the right-hand side of equation (12.4), one obtains a explicit relation between the two ratios Λ/μ and m_H/h:

$$\ln(\Lambda/\mu) = \frac{1}{\beta_2 \lambda_\mathrm{r}} + \frac{\beta_3}{\beta_2^2} \ln \lambda_\mathrm{r}, \quad \text{with } \lambda_\mathrm{r} = 3 m_H^2/h^2.$$

To minimize higher order corrections, one can choose for the renormalization scale $\mu = h$ or $\mu = m_H$.

If the Higgs field is really associated to a physical particle, the mass must be smaller than the cut-off. One then obtains an upper-bound for m_H ranging between 600 GeV and 700 GeV, in agreement with the then available computer simulation values. Moreover, the corresponding value of λ_r is such that renormalized perturbation theory at leading order should still be semi-quantitatively correct.

Before the Higgs boson discovery, these RG arguments indicated that a proton collider like the LHC), which would be able to explore physics in the TeV range, would discover the Higgs particle, or discover some new physics, or both.

Later, the direct search and the study of radiative corrections at LEP, even though they varied only like $\ln m_H$, pointed towards a lower mass for the Higgs boson, one lying between 114 and 200 GeV.

Conversely, since the physical coupling constant λ_r at scale μ is now known from the Higgs mass, one can infer from the equation an estimate of the cut-off scale (the maximal energy scale below which, in the absence of some new physics, the field theory can still be physically consistent). One finds a scale of the order of 10^{24} GeV.

The full Standard Model RG equations and the low physical value of the Higgs mass now suggest that the Standard Model could be consistent up to very high energies of the order of 10^{10} GeV, much beyond any foreseeable collider range.

Only the problem of the *fine tuning* of the mass parameter of the Higgs field renders this situation very unlikely.

12.8   The Gross–Neveu–Yukawa model: A Higgs–top toy model

Since the Higgs mass is 'small', the self-coupling of the Higgs field is indeed small and, even to get a rough understanding of the RG flow, coupling to other particles can no longer be neglected. Among them important ones are the coupling to the weak vector bosons $Z, W_\pm$. For what concerns matter fields, since the Higgs-particle coupling is proportional to the particle mass, the quark top coupling to the Higgs particle is by far the largest one.

Even neglecting, for simplicity, vector bosons, one can obtain a semi-quantitative picture from a model that concentrates on the Higgs–top physics, the Gross–Neveu–Yukawa (GNY) model [118].

In addition to the scalar field self-interaction, the model contains a Yukawa-like boson–fermion interaction. The model is renormalizable in four dimensions.

In the GNY model, as in the Standard Model, particles receive masses by spontaneous chiral symmetry breaking. However, no Goldstone bosons are generated because the chiral symmetry is discrete.

From the RG viewpoint, the model is IR free. In the model, with the assumption that *coupling constants have generic values at the cut-off scale*, for a cut-off large compared to the physical masses, the ratio of fermion and boson masses can be asymptotically predicted.

More generally, the RG flow can be studied as a function of the physical masses when the physical ratio differs from the prediction.

12.8.1 The GNY model

The GNY model involves a set of N massless fermions $\{\psi^i, \bar{\psi}^i\}$ and a scalar field H. It has a discrete Z_2 chiral symmetry and a $U(N)$ symmetry [118]. Under the chiral reflection, the fields transform like

$$\psi(x) \mapsto \gamma_5 \psi(x), \quad \bar{\psi}(x) \mapsto -\bar{\psi}(x)\gamma_5, \quad H \mapsto -H. \tag{12.5}$$

This symmetry prevents the addition of a fermion mass term to the action.

The $U(N)$ symmetry is implemented by the transformation

$$\psi(x) \mapsto U\psi(x), \quad \bar{\psi}(x) \mapsto U^\dagger \bar{\psi}(x), \quad U \in U(N).$$

A renormalizable action then reads

$$\mathcal{S}(\bar{\psi}, \psi, H) = \int \mathrm{d}^4 x \Bigg[-\bar{\psi}(x) \cdot (\slashed{\partial} + gH(x))\psi(x)$$

$$+ \tfrac{1}{2}\left(\nabla H(x)\right)^2 + \tfrac{1}{2} r \Lambda^2 H^2(x) + \frac{\lambda}{4!} H^4(x) \Bigg],$$

where the parameters g and λ are dimensionless and a momentum cut-off Λ, consistent with the symmetries, is implied, for example, implemented by adding higher derivatives,

$$\slashed{\partial} \mapsto \slashed{\partial} P(-\nabla^2/\Lambda^2), \quad \nabla^2 \mapsto \nabla^2 Q(-\nabla^2/\Lambda^2),$$

where the polynomials P, Q satisfy $P(0) = Q(0) = 1$ and $P(z), Q(z) > 0$ for $z > 0$.

The model describes a physics of *spontaneous chiral symmetry breaking* and *fermion mass generation*.

We set $r = r_c + \tau/\Lambda^2$, where for $r = r_c$ the physical masses of ψ and H vanish and a phase transition occurs. The physical domain, defined by the property that momenta and masses are much smaller than the cut-off Λ, corresponds to values $|\tau| \ll \Lambda^2$ and this leads to the usual *fine tuning* problem.

The parameter τ, in the terminology of macroscopic phase transitions, plays the role of the deviation from the critical temperature.

The renormalized action. Since we are interested in situations where masses and momenta are much smaller than the cut-off Λ, we introduce a characteristic physical scale $\mu \ll \Lambda$, technically called renormalization scale, and express the action, then called renormalized action, in terms of rescaled fields and effective parameters at the scale μ. Denoting by g_r, λ_r the renormalized couplings (the effective couplings at scale μ), one can write the renormalized action as

$$\mathcal{S}_r(H, \psi, \bar{\psi}) = \int \mathrm{d}^4 x \Bigg\{ -Z_\psi \left[\bar{\psi}(x) \cdot \left(\slashed{\partial} + g_r Z_g Z_H^{1/2} H(x) \right) \psi(x) \right] + \tfrac{1}{2} Z_H \left(\nabla H(x)\right)^2$$

$$+ \tfrac{1}{2}\left(Z_H r_c \Lambda^2 + \tau_r Z_m \right) H^2(x) + Z_\lambda \frac{\lambda_r}{4!} Z_H^2 H^4(x) \Bigg\}, \tag{12.6}$$

where $Z_\psi, Z_g, Z_H, Z_\lambda, Z_m$ are renormalization constants.

The phase transition in the tree approximation. In the tree approximation where the renormalized action reduces to

$$S_{\text{tree}}\left(\bar{\psi}, \psi, H\right) = \int \mathrm{d}^4 x \left[-\bar{\psi}(x) \cdot \partial \psi(x) - g_{\mathrm{r}} H(x) \bar{\psi}(x) \cdot \psi(x) \right.$$
$$\left. + \tfrac{1}{2}\left(\nabla H(x)\right)^2 + \tfrac{1}{2}\tau_{\mathrm{r}} H^2(x) + \frac{\lambda_{\mathrm{r}}}{4!} H^4(x) \right],$$

a phase transition occurs, for $\tau_{\mathrm{r}} = 0$, between a fermion massless symmetric phase for $\tau_{\mathrm{r}} > 0$ and a phase for $\tau_{\mathrm{r}} < 0$ where the chiral symmetry is spontaneously broken and a fermion mass is generated. The H expectation value

$$\langle H \rangle = \pm \sqrt{-6\tau_{\mathrm{r}}/\lambda_{\mathrm{r}}},$$

gives a mass to the fermions by a mechanism reminiscent of the Standard Model of weak-electromagnetic interactions.

The fermion and boson masses are then

$$m_\psi = g_{\mathrm{r}} |\langle H \rangle|, \quad m_H = \sqrt{\frac{\lambda_{\mathrm{r}}}{3}} |\langle H \rangle| \ \Rightarrow \ \frac{m_H}{m_\psi} = \frac{1}{g_{\mathrm{r}}}\sqrt{\frac{\lambda_{\mathrm{r}}}{3}}. \qquad (12.7)$$

12.8.2 RG equations and mass ratio

Beyond the tree approximation, the **GNY** model, like the $(\mathbf{H}^2)^2$ field theory, can be studied by RG techniques.

Most important here are the RG β-functions. In particular, they describe the variation of the coupling constants from the scale Λ, where they are assumed to be of order 1, to the physical mass scale $\mu \ll \Lambda$. At one-loop order, one finds

$$\beta_\lambda = \frac{1}{8\pi^2}\left(\tfrac{3}{2}\lambda^2 + 4N\lambda g^2 - 24Ng^4\right), \quad \beta_{g^2} = \frac{2N+3}{8\pi^2}g^4.$$

IR freedom and the triviality issue. One easily verifies that the origin $\lambda = g^2 = 0$ is IR stable. Assuming that the couplings $\lambda(\Lambda)$ and $g(\Lambda)$ are generic (*i.e.*, of order 1, numerically $8\pi^2$, which is the loop factor) at the cut-off scale Λ, and solving the RG equations, one infers that the coupling constants at a scale $\mu \ll \Lambda$ decrease like

$$g^2(\mu) \sim \frac{8\pi^2}{(2N+3)\ln(\Lambda/\mu)}, \quad \lambda(\mu) \sim \frac{8\pi^2 R_*(N)}{(2N+3)\ln(\Lambda/\mu)}$$

with

$$R_*(N) = \frac{1}{3}\left[-(2N-3) + \sqrt{4N^2 + 132N + 9}\right].$$

The infinite cut-off limit leads to the *triviality* issue since $\lambda(\mu) = g^2(\mu) = 0$.

For $\Lambda \gg \mu$ but finite, the ratio of H and fermion masses approaches the limit,

$$\lim_{\Lambda \to \infty} \frac{m_H^2}{m_\psi^2} = \frac{\lambda(\mu)}{3g^2(\mu)} = \tfrac{1}{3}R_*(N) = \frac{1}{9}\left(-(2N-3) + \sqrt{4N^2 + 132N + 9}\right).$$

As a function of N, the limiting ratio m_H/m_ψ thus varies from about 1.20 to 2, which corresponds to the $\bar{\psi}\psi$ threshold and the large N limit, when N varies from 1 to ∞.

IR freedom of the theory and the assumption that the couplings are generic at the cut-off scale imply that the mass ratio m_H/m_ψ approaches a fixed limit for $\Lambda/\mu \gg 1$.

12.9 GNY model: The general RG flow at one loop

We analyse now the general RG flow. Two-dimensional flows can be easily studied because, quite generally, RG trajectories can only meet at fixed points, here $g = \lambda = 0$. Setting,

$$t = \ln(\Lambda/\mu),$$

one can write the solutions of the one-loop differential equations explicitly as

$$g^2(t) = \frac{8\pi^2}{(2N+3)t}$$

$$\frac{\lambda(t)}{g^2(t)} = 1 - \frac{2N}{3} + \frac{\sqrt{4N^2+132N+9}}{3} \cdot \frac{\tanh\left(\dfrac{\sqrt{4N^2+132N+9}}{2(2N+3)}\ln(t)\right) + \theta}{1 + \theta\tanh\left(\dfrac{\sqrt{4N^2+132N+9}}{2(2N+3)}\ln(t)\right)},$$

where θ is an integration constant. The fixed lines correspond to the limits $\theta = \pm 1$.

One immediately verifies that the lines $g = 0$ and $\lambda = R_* g^2$ are fixed RG trajectories and thus cannot be crossed. By contrast, the line $\lambda = 0$ can be formally crossed and the RG trajectories then enter a non-physical region (see Fig. 12.2). Close to the line, new interactions, which have been neglected in the RG analysis, have to be taken into account and new physics is required.

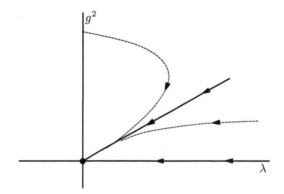

Fig. 12.2 RG flow: full lines correspond to fixed lines.

Comparison with physical values. Identifying the boson with the Higgs field and the fermion with the top field, it is interesting to put numbers on the vacuum expectation value and couplings (from data 2018):

$$\langle H \rangle = 246\,\text{Gev}\,,\quad m_\psi = 173.1 \pm 0.9\,\text{Gev}\,,\ m_H = 125.18 \pm 0.16\,\text{Gev}\,.$$

Then,

$$\lambda \approx 0.78\,,\quad g^2 \approx 0.50\,.$$

The coupling constants divided by the loop factor $8\pi^2$ are small and this justifies the use of perturbation theory and indicates that IR freedom is relevant since it predicts small renormalized couplings when the initial couplings at a large momentum cut-off scale are of order 1.

However, the physical ratio

$$R = \frac{\lambda}{3g^2} \approx 0.52$$

is smaller than the prediction of the model.

The main neglected contributions correspond to Higgs couplings to vector bosons and, therefore, the picture we find can only be qualitative but the analysis is here much simplified and, thus, more transparent.

However, more realistic calculations that include known particles [159], have been performed and seem to indicate that the physical Higgs mass lies very close to a fixed line. Depending on the precise top mass, deviations appear but at a very high energy scale, at least 10^{10} GeV.

This RG result is puzzling since it points towards the possibility that the Standard Model could be valid up to such a high energies. However, the problem of the fine tuning of the Higgs bare mass, which we have not discussed so far, then becomes extremely severe.

12.10 The fine tuning issue

More precise calculations confirm that, from the viewpoint of the RG trajectories, the Standard Model could be consistent up to about 10^{10} GeV, which corresponds to energies not accessible by any collider experiments. The main remaining serious issue is the fine tuning problem. The parameter to be fine tuned is the coefficient of the H^2 term in the Lagrangian, which is proportional to the cut-off squared. The one-loop order contribution to m_H^2 is of the form

$$\frac{1}{8\pi^2}\Lambda^2(\lambda - 2Ng^2).$$

Except for supersymmetric theories where the boson and fermion contributions automatically cancel, the amount of fine tuning f is thus of the order of

$$f = 8\pi^2 f_0 \frac{m_H^2}{\Lambda^2},$$

where f_0 in the GNY model is a number of order 1. While at presently available energies there is some tension, the fine tuning argument is not yet compelling. However, at higher energies, fine tuning will become an essential issue. For example, it is very hard to believe that the present Standard Model could still describe physics at the microscopic scale at 10^{10} GeV, which corresponds to a fine tuning of order 10^{-14}.

13 Quantum chromodynamics: A non-Abelian gauge theory

Quantum chromodynamics (QCD) is an approximation to the Standard Model of fundamental interactions at the microscopic scale that takes into account only the strong nuclear force. It describes the interactions between quarks mediated by particles (gluons) corresponding to non-Abelian gauge fields (Section 11.8). The gauge group (colour group) is $SU(3)$. Table 13.1 lists the quarks, organized in three generations, to which are attached flavour quantum numbers.

With the present matter content, QCD has the essential property of *asymptotic freedom*, a renormalization group (RG) property that explains why quarks behave like almost free particles at short distance, and are confined (as confirmed by numerical simulations), a property that explains why no freely propagating quarks are observed, only colour neutral bound states.

The form of the QCD Lagrangian leads to a number of unanswered questions, for example, concerning the extreme diversity of quark masses, which range from a few MeV to 173 GeV; the origin of the $SU(3)$ gauge group; the existence of three generations or families, which makes a weak CP (the product of C, the charge conjugation, and P, the parity) violation possible, and may thus have contributed, to some extent, to the matter–anti-matter unbalance in the universe; and the strong CP violation problem.

Table 13.1

Quarks: the three generations (2018).

Charge 2/3 quarks	Charge −1/3 quarks
u, $m = 2.2 \pm 0.5$ MeV	**d**, $m = 4.7 \pm 0.5$ MeV
c, $m = 1.27 \pm 0.03$ GeV	**s**, $m = 92$ to 104 MeV
t, $m = 173.1 \pm 0.9$ GeV	**b**, $m = 4.18 \pm 0.04$ GeV

In this chapter, we first recall some geometric aspects of the construction of non-Abelian gauge theories [41]. We then discuss the Lagrangian of QCD and a few physical consequences. Finally, we describe lattice gauge theory, which provides a non-perturbative regularization of gauge theories and an essential tool for numerical simulations.*

---

* In this chapter, *summation over repeated space or space-time indices* is always assumed, except when discussing lattice gauge theories. Imaginary time notation in four-dimensional space-time is used except where stated otherwise.

From Random Walks to Random Matrices. Jean Zinn-Justin, Oxford University Press (2019).
© Jean Zinn-Justin. DOI: 10.1093/oso/9780198787754.001.0001

13.1 Geometry of gauge theories: Parallel transport

The basic geometric structure of non-Abelian gauge theories is rooted in the notion of parallel transport. From this viewpoint, there is a continuity between gauge theories in continuum space and on a space (or space-time) lattice. We consider in this chapter only *compact Lie groups*.

13.1.1 *Gauge transformations, gauge invariance and parallel transport*

As a (physically relevant) example, we consider matter fields consisting of N Dirac fermions in four dimensions forming a vector multiplet transforming under the fundamental representation of the $SU(N)$ group (the group of complex unitary $N \times N$ matrices with determinant 1):

$$\psi^g(x) = \mathbf{g}\psi(x), \quad \mathbf{g} \in SU(N).$$

A gauge transformation is a *local group transformation*: fields transform differently under the group at each point of space-time (see also Chapter 11):

$$\psi_g(x) = \mathbf{g}(x)\psi(x), \quad \mathbf{g}(x) \in SU(N). \tag{13.1}$$

For products of fields taken at the same point, global invariance ($\mathbf{g}$ constant) implies local invariance. This is no longer true for invariant functions of products of fields taken at different points (or, as a limiting case, derivatives of fields).

The problem can be solved by introducing *parallel transport*.

Parallel transport. Let C_{xy} be an *oriented, piecewise differentiable curve* going from a point x, the origin, to a point y, the extremity. We denote by C^{-1} the curve with opposite orientation to C. If $[C_1]_{xy}$ and $[C_2]_{yz}$ are two curves such that the extremity y of C_1 coincides with the origin of C_2, we can consider the joined curve $C_1 \cup C_2$. We impose ($\bullet_x$ means the point x)

$$C_{xy} \cup C_{yx}^{-1} = \bullet_x .$$

The parallel transporter $\mathbf{U}(C)$ is an element of $SU(N)$ *depending on an oriented curve C*, which satisfies

$$\mathbf{U}(C \equiv \{\bullet\}) = \mathbf{1}, \tag{13.2a}$$

$$\mathbf{U}(C^{-1}) = \mathbf{U}^{-1}(C), \tag{13.2b}$$

$$\mathbf{U}\big((C_1 \cup C_2)_{xz}\big) = \mathbf{U}\big((C_1)_{xy}\big)\mathbf{U}\big((C_2)_{yz}\big). \tag{13.2c}$$

Consistently with the rules (13.2), in a gauge transformation, $\mathbf{U}(C_{xy})$ transforms like

$$\mathbf{U}^g(C_{xy}) = \mathbf{g}(x)\mathbf{U}(C_{xy})\mathbf{g}^{-1}(y). \tag{13.3}$$

It follows that the field

$$\psi_U(x) = \mathbf{U}(C_{xy})\psi(y), \tag{13.4}$$

transforms by $\mathbf{g}(x)$ instead of $\mathbf{g}(y)$ and a quantity like $\bar{\psi}(x)\mathbf{U}(C_{xy})\psi(y)$ is gauge invariant. Finally, *for any closed curve C*, $\mathrm{tr}\,\mathbf{U}(C_{xx})$ *is gauge invariant*.

Up to this point, the construction has been valid for both continuum and discretized space. We now concentrate on field theories in continuum space; gauge theories on a lattice are further discussed in Sections 13.8 and 13.9.

13.1.2 *Gauge theories in the continuum*

In the continuum, in the limit of an infinitesimal differentiable curve, one can parametrize $\mathbf{U}(C_{xy})$ with $y_\mu = x_\mu + \mathrm{d}x_\mu$ in terms of the *connection or gauge field* $\mathbf{A}_\mu(x)$, a space-vector and a matrix (anti-Hermitian) belonging to the representation of the Lie algebra of $SU(N)$:

$$\mathbf{U}(C_{xy}) = 1 + \mathbf{A}_\mu(x)\mathrm{d}x_\mu + o\left(\|\mathrm{d}x_\mu\|\right).$$

The transformation properties of $\mathbf{A}_\mu(x)$ are obtained by expanding equation (13.3) at first order in $\mathrm{d}x_\mu$. One infers ($\partial_\mu \equiv \partial/\partial_\mu$, $\mu = 0, 1, 2, 3$)

$$\mathbf{A}_\mu^g(x) = \mathbf{g}(x)\mathbf{A}_\mu(x)\mathbf{g}^{-1}(x) + \mathbf{g}(x)\partial_\mu\mathbf{g}^{-1}(x). \qquad (13.5)$$

From the point of view of global transformations ($\mathbf{g}$ constant), the field $\mathbf{A}_\mu(x)$ transforms by the adjoint representation of the group $SU(N)$. However, the transformation (13.5) being affine, $\mathbf{A}_\mu(x)$ is not a tensor for gauge transformations.

Covariant derivative. In the limit of an infinitesimal curve,

$$\psi_U = \left(1 + \mathbf{A}_\mu(x)\mathrm{d}x_\mu\right)\left(\psi(x) + \partial_\mu\psi(x)\mathrm{d}x_\mu\right) + o\left(\|\mathrm{d}x_\mu\|\right)$$
$$= \left(1 + \mathrm{d}x_\mu\mathbf{D}_\mu(\mathbf{A})\right)\psi(x) + o\left(\|\mathrm{d}x_\mu\|\right)$$

with $\mathbf{D}_\mu(\mathbf{A}) = 1\,\partial_\mu + \mathbf{A}_\mu$. The differential operator $\mathbf{D}_\mu(\mathbf{A})$ is called the *covariant derivative*. Its explicit form depends on the tensor on which it is acting.

The identity

$$\mathbf{g}(x)\left(1\,\partial_\mu + \mathbf{A}_\mu\right)\mathbf{g}^{-1}(x) = 1\,\partial_\mu + \mathbf{g}(x)\mathbf{A}_\mu(x)\mathbf{g}^{-1}(x) + \mathbf{g}(x)\partial_\mu\mathbf{g}^{-1}(x)$$

(the products have to be understood as products of differential and multiplicative operators), which, using equation (13.5), can be rewritten as

$$\mathbf{g}(x)\mathbf{D}_\mu(\mathbf{A})\mathbf{g}^{-1}(x) = \mathbf{D}_\mu(\mathbf{A}^g) \equiv \mathbf{D}_\mu^g, \qquad (13.6)$$

shows that $\mathbf{D}_\mu$ *is a tensor*

Infinitesimal gauge transformation. Setting,

$$\mathbf{g}(x) = 1 + \boldsymbol{\omega}(x) + o\left(\|\boldsymbol{\omega}\|\right),$$

in which $\boldsymbol{\omega}(x)$ belongs to the Lie algebra of $SU(N)$, one derives from equation (13.5) the form of the infinitesimal gauge transformation of the field $\mathbf{A}_\mu$,

$$\mathbf{A}_\mu(x) - \mathbf{A}_\mu^g(x) = \partial_\mu\boldsymbol{\omega} + [\mathbf{A}_\mu, \boldsymbol{\omega}] \equiv \mathbf{D}_\mu^{\mathrm{adj.}}\boldsymbol{\omega}. \qquad (13.7)$$

The equation yields the form of the covariant derivative $\mathbf{D}_\mu^{\mathrm{adj.}}$ acting on vectors transforming under the adjoint representation. It is simple to verify the relation

$$\partial_\mu\boldsymbol{\omega}^g + \left[\mathbf{A}_\mu^g, \boldsymbol{\omega}^g\right] = \mathbf{g}(x)\{\partial_\mu\boldsymbol{\omega} + [\mathbf{A}_\mu, \boldsymbol{\omega}]\}\mathbf{g}^{-1}(x),$$

in which $\mathbf{A}_\mu^g$ is given by equation (13.5), and $\boldsymbol{\omega}^g$ by

$$\boldsymbol{\omega}^g(x) = \mathbf{g}(x)\boldsymbol{\omega}(x)\mathbf{g}^{-1}(x).$$

Curvature tensor. The commutator of two covariant derivatives,

$$\mathbf{F}_{\mu\nu}(x) = [\mathbf{D}_\mu, \mathbf{D}_\nu] = \partial_\mu \mathbf{A}_\nu(x) - \partial_\nu \mathbf{A}_\mu(x) + [\mathbf{A}_\mu(x), \mathbf{A}_\nu(x)] \,,$$

is no longer a differential operator. It is again an element of the Lie algebra of the group and transforms, as a consequence of equation (13.6), as

$$\mathbf{F}^g_{\mu\nu}(x) = \mathbf{g}(x)\mathbf{F}_{\mu\nu}(x)\mathbf{g}^{-1}(x). \tag{13.8}$$

$\mathbf{F}_{\mu\nu}$ *is a tensor, the curvature tensor*, a generalization of the electromagnetic field of quantum electrodynamics (QED). The *curvature tensor is associated with parallel transport along an infinitesimal closed curve.*

13.1.3 Parallel transport: Explicit expressions in the continuum

In the Abelian $U(1)$ example (QED), the parallel transporter, which is an element of the $U(1)$ group, takes a simple, explicit form (see Section 11.7.3, equation (11.44)). It is given by the curvilinear integral

$$U[C_{xy}] = \exp\left[-ie \oint_{C_{xy}} A_\mu(s)\mathrm{d}s_\mu\right],$$

as a gauge transformation shows.

In the non-Abelian case, the explicit relation is more complicated because the gauge field $\mathbf{A}^\alpha_\mu(x)t_\alpha$ is an element of the Lie algebra of G, and the matrices representing the field at different points do not commute. It can be formally written as (P means path-ordered integral along the curve C.)

$$\mathbf{U}[C_{xy}] = \mathrm{P}\left\{\exp\left[\oint_{C_{xy}} \sum_\alpha A^\alpha_\mu(s)\mathbf{t}^\alpha \mathrm{d}s_\mu\right]\right\},$$

where $\mathbf{t}^\alpha$ belong to the Lie algebra of G.

In the continuum, the expression can be defined by an expansion in powers of $\mathbf{A}_\mu$.

13.2 Gauge invariant action

Matter fields. For fermions transforming by the fundamental representation of $SU(N)$, the action

$$\mathcal{S}_{\mathrm{F}}(\bar\psi, \psi) = -\int \mathrm{d}^4x \, \bar\psi(x) \left(\slashed{D} + M\right)\psi(x), \tag{13.9}$$

where M is a mass and $\slashed{\Delta} \equiv \Delta_\mu \gamma_\mu$ (γ_μ are the four Dirac γ-matrices), is gauge invariant.

Gauge fields. The simplest gauge invariant action $\mathcal{S}(\mathbf{A}_\mu)$, a function of the gauge field $\mathbf{A}_\mu$ only, has the form

$$\mathcal{S}(\mathbf{A}_\mu) = -\frac{1}{4g^2} \int \mathrm{d}^d x \ \mathrm{tr} \, \mathbf{F}_{\mu\nu}(x)\mathbf{F}_{\mu\nu}(x), \tag{13.10}$$

where g is the gauge coupling constant. The sign in front of the action takes into account that, with our convention, the matrix $\mathbf{F}_{\mu\nu}$ is anti-Hermitian.

This *action can be derived from the gauge invariant quantity* $\mathrm{tr}\, U(C)$ *in the limit of an infinitesimal closed contour* C (Fig. 13.1):

$$\mathrm{tr}\, \mathrm{P} \exp\left(\oint_C \mathbf{A}_\mu(x)\mathrm{d}x_\mu\right) - \mathrm{tr}\, \mathbf{1} \sim \tfrac{1}{2} \mathrm{tr}\, [u_\mu \mathbf{F}_{\mu\nu}(x)v_\nu]^2 \ .$$

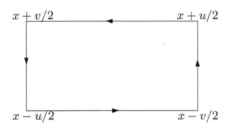

Fig. 13.1 The loop C.

The result is not rotation invariant at fixed vectors u, v but, after averaging over rotations acting on u, v, one obtains a result proportional to action (13.10).

Two remarks are immediately in order:

(i) In contrast with the Abelian case, because the gauge field transforms non-trivially under the group, (the gauge field is 'charged'), the *curvature tensor* $\mathbf{F}_{\mu\nu}$ *is not gauge invariant and, thus, not directly a physical observable.* The action (13.10) is no longer a free field action; the gauge field has self-interactions and even the spectrum of the pure gauge action is non-perturbative (some analytic results can be obtained only in dimension 2 $(1+1)$).

Lattice gauge theory provides a framework for non-perturbative investigations.

(ii) As in the Abelian case, the action, because it is gauge invariant, does not provide a dynamics to the degrees of freedom of the gauge field which correspond to gauge transformations and, therefore, some *gauge fixing is required.* From the field integral viewpoint, the integral

$$\mathcal{Z} = \int [\mathrm{d}\mathbf{A}_\mu] \exp\left[\frac{1}{4g^2} \int \mathrm{d}^d x \ \mathrm{tr}\, \mathbf{F}_{\mu\nu}(x)\mathbf{F}_{\mu\nu}(x)\right]$$

is not defined because the integrand is constant along gauge orbits.

13.2.1 Component form

A basis of generators of the Lie algebra of the $SU(N)$ matrix group can be chosen in the form of a set of $N \times N$ anti-Hermitian, traceless matrices $\mathbf{t}_a$. Both the gauge field and the curvature field tensor can be expanded on such a basis:

$$\mathbf{A}_\mu(x) = \sum_a A_\mu^a(x)\,\mathbf{t}_a\,, \quad \mathbf{F}_{\mu\nu}(x) = \sum_a F_{\mu\nu}^a(x)\,\mathbf{t}_a\,.$$

In terms of the structure constants of the Lie algebra defined by

$$[\mathbf{t}_a, \mathbf{t}_b] = \sum_c f_{abc}\,\mathbf{t}_c\,,$$

the components of the tensor can be written more explicitly as

$$F_{\mu\nu}^a(x) = \partial_\mu A_\nu^a(x) - \partial_\nu A_\mu^a(x) + \sum_{b,c} f_{bca} A_\mu^b(x) A_\nu^c(x). \tag{13.11}$$

The formulation of the Abelian gauge fields of the previous sections can be recovered (up to simple normalization factors) for $N = 1$, $\mathbf{t}_1 \mapsto ie_\phi$, and thus, $f_{111} = 0$.

13.3 Hamiltonian formalism. Quantization in the temporal gauge

Non-Abelian gauge theories can be quantized in a simple way in the temporal or Weyl gauge, using a Hamiltonian formalism. This leads to a field theory that, at least at a formal level, is unitary because it corresponds to a Hermitian Hamiltonian. In this form, the field theory can be easily discussed in the semi-classical approximation. However, it lacks relativistic covariance and this is the source of many difficulties. In particular, it is difficult to define a perturbative expansion.

13.3.1 Classical field equations

In real time field theory, we denote time as $t \equiv x_0 = ix_d$, and the corresponding field component as $\mathbf{A}_0 = -i\mathbf{A}_d$. We use the notation $\dot{Q}$ for the time derivative of Q. Space components will carry Roman indices $(\mathbf{A}_i, x_i)$.

The variation of the real time form of the action (13.10) leads to the classical gauge field equations

$$\mathbf{D}_\mu \mathbf{F}^{\mu\nu}(x) = 0\,. \tag{13.12}$$

Equation (13.12) does not lead to a standard quantization because the action does not depend on $\dot{\mathbf{A}}_0$, the time derivative of $\mathbf{A}_0$. Therefore, $\mathbf{A}_0$ is not a dynamical variable, the $\mathbf{A}_0$ *field equation is a constraint equation* that can be used to eliminate $\mathbf{A}_0$ from the action.

However, in the absence of a mass term, the reduced action does not depend on all space components of the gauge field. *Only the combination* $\left[\delta_{ij} - \mathrm{D}_i(\mathrm{D}_\perp^2)^{-1}\mathrm{D}_j\right]\dot{\mathbf{A}}_j$ *appears* ($\mathrm{D}_\perp^2$ is the covariant space Laplacian). But, *in contrast with the Abelian case, the projector acting on* $\mathbf{A}_i$ *depends on the field itself.* It follows that the procedure which leads to Coulomb's gauge does not work here, at least in its simplest form. Therefore, it is simpler to begin with a quantization in the temporal (or Weyl) gauge $\mathbf{A}_0 = 0$.

13.3.2 *Weyl's or temporal gauge: Classical theory*

Since the gauge transform of a solution to the field equations is again a solution, one can restrict fields by a gauge condition. One gauge condition especially well-suited to the construction of a Hamiltonian formalism (in particular, useful for finite temperature quantum field theory), is

$$\mathbf{A}_0(t, x) = 0, \tag{13.13}$$

which defines the *temporal or Weyl's gauge.*

In this gauge, the field equations simplify. We define,

$$\mathbf{E}_k(t, x) = -\dot{\mathbf{A}}_k(t, x)/g^2.$$

The field equations become (separating time and space components)

$$\dot{\mathbf{E}}_k(t, x) = \mathbf{D}_l \mathbf{F}_{lk}(t, x), \tag{13.14a}$$

$$\mathbf{D}_l \mathbf{E}_l(t, x) = 0. \tag{13.14b}$$

The first equation is a *dynamic equation* that can be directly derived from the initial Lagrangian in which the gauge condition (13.13) has been used,

$$\mathcal{L}(\mathbf{A}_k) = -\frac{1}{g^2} \operatorname{tr} \int \mathrm{d}^{d-1}x \left[\tfrac{1}{2}\dot{\mathbf{A}}_k^2(t, x) - \tfrac{1}{4}\mathbf{F}_{kl}^2(t, x) \right]. \tag{13.15}$$

The expression defines a conventional Lagrangian for the space components of the gauge field and shows that $\mathbf{E}_k$ is the momentum conjugate to $\mathbf{A}_k$.

The corresponding Hamiltonian is

$$\mathcal{H}(\mathbf{E}, \mathbf{A}) = -\operatorname{tr} \int \mathrm{d}^{d-1}x \left[\frac{g^2}{2}\mathbf{E}_k^2(x) + \frac{1}{4g^2}\mathbf{F}_{kl}^2(x) \right]. \tag{13.16}$$

By contrast, the second equation is a *constraint equation, a non-Abelian general-ization of Gauss's law.* The only relevant solutions of the field equations are those that satisfy the constraint.

Time independent gauge transformations. The gauge condition $\mathbf{A}_0 = 0$ is left in-variant by time-independent gauge transformations. In infinitesimal form (equation (13.7)),

$$\mathbf{A}_0(x) - \mathbf{A}_0^\omega = \dot{\boldsymbol{\omega}} + [\mathbf{A}_0, \boldsymbol{\omega}] = 0 \text{ if } \dot{\boldsymbol{\omega}} = 0.$$

Therefore, *time-independent gauge transformations form a symmetry group* of the Lagrangian (13.15) and thus of the Hamiltonian (13.16). The quantities $\mathbf{D}_l \mathbf{E}_l$ are the generators, in the sense of Poisson brackets, of the symmetry group. Therefore, the constraint (13.14b) is compatible with the classical motion: the only acceptable solutions are those that are invariant under time-independent gauge transforma-tions.

13.3.3 Quantum gauge theory in the temporal gauge

The classical considerations immediately generalize to the quantum theory, *the quantum operators* $\mathbf{D}_l\mathbf{E}_l$, *generators of a symmetry group, commute with the Hamiltonian.* The space of admissible physical states $\Psi(\mathbf{A})$ is thus restricted by the quantum generalization of Gauss's law:

$$\mathbf{D}_l\mathbf{E}_l\Psi(\mathbf{A}) \equiv \mathbf{D}_l\frac{1}{i}\frac{\delta}{\delta\mathbf{A}_l(x)}\Psi(\mathbf{A}) = 0\,.$$

The *equation implies that physical states are invariant under time-independent gauge transformations:* they belong to the invariant sector of the symmetry group, a subspace that is left invariant by quantum evolution.

Quantization in the temporal gauge then follows from standard arguments. Returning to the *Euclidean formalism* with $x \equiv \{x_1,\dots,x_d\}$, one concludes that the partition function can be written as

$$\mathcal{Z} = \int [\mathrm{d}\mathbf{A}_\mu] \prod_x \delta(\mathbf{A}_d(x)) \exp\left[\frac{1}{4g^2}\int \mathrm{d}^d x \,\mathrm{tr}\,\mathbf{F}^2_{\mu\nu}(x)\right]. \qquad (13.17)$$

Note that, at zero temperature, the perturbative vacuum is automatically gauge invariant and Gauss's law plays no role, *but this is no longer true at finite temperature.*

Perturbative expansion in the temporal gauge. The field theory in the temporal gauge is not explicitly space-time covariant and this leads to several difficulties in the perturbative expansion:

(i) The theory is not renormalizable in the sense of power counting. Indeed the propagator in this gauge,

$$\tilde{W}^{(2)}_{ij}(\mathbf{k}_\perp, k_d) = \frac{1}{k^2}\left(\delta_{ij} - \frac{k_i k_j}{\mathbf{k}_\perp^2}\right) + \frac{1}{k_d^2}\frac{k_i k_j}{\mathbf{k}_\perp^2},$$

in which $\mathbf{k}_\perp$ is the 'space' part of $\mathbf{k}$, does not decrease at k_d fixed for large spatial momenta $|\mathbf{k}_\perp|$.

(ii) The double pole at $k_d = 0$ leads to potential infrared (IR) divergences.

These problems can be solved by showing that *gauge invariant observables* can equivalently be derived from another quantum action which leads to a theory that is *explicitly covariant and renormalizable by power counting.*

13.3.4 Covariant generalized Landau's gauge

By a set of transformations on the field integral, which change generic correlation functions but not gauge invariant observables, one can pass to a covariant gauge constraining $\partial_\mu\mathbf{A}_\mu(x)$ a generalized form of Landau's gauge.

The partition function or vacuum functional $\mathcal{Z}$ then becomes

$$\mathcal{Z} = \int \left[\mathrm{d}\mathbf{A}_\mu \, \mathrm{d}\bar{\mathbf{C}} \, \mathrm{d}\mathbf{C} \, \mathrm{d}\boldsymbol{\lambda} \right] \exp\left[-\mathcal{S}(\mathbf{A}_\mu, \bar{\mathbf{C}}, \mathbf{C}, \boldsymbol{\lambda}) \right], \tag{13.18}$$

where $\mathcal{S}$ is the *local action*,

$$\mathcal{S}(\mathbf{A}_\mu, \bar{\mathbf{C}}, \mathbf{C}, \boldsymbol{\lambda}) = -\frac{1}{4g^2} \int \mathrm{d}^d x \, \mathrm{tr}\, \mathbf{F}_{\mu\nu}^2(x) + \mathcal{S}_{\mathrm{gauge}}(\mathbf{A}_\mu, \bar{\mathbf{C}}, \mathbf{C}, \boldsymbol{\lambda}), \tag{13.19}$$

$$\mathcal{S}_{\mathrm{gauge}} = \int \mathrm{d}^d x \, \mathrm{tr}\left[\frac{\xi g^2}{2} \boldsymbol{\lambda}^2(x) + \boldsymbol{\lambda}(x) \partial_\mu \mathbf{A}_\mu(x) + \mathbf{C}(x) \partial_\mu \mathbf{D}_\mu \bar{\mathbf{C}}(x) \right], \tag{13.20}$$

$\bar{\mathbf{C}}$ and $\mathbf{C}$ being non-physical spinless fermions, the *Faddeev–Popov 'ghost' fields*, and $\boldsymbol{\lambda}$ a scalar boson field, all transforming under the *adjoint representation* of the gauge group.

Except in the limit $\xi = 0$, it is also possible to integrate over $\boldsymbol{\lambda}(x)$ to find the new quantum action

$$\mathcal{S}(\mathbf{A}_\mu, \bar{\mathbf{C}}, \mathbf{C}) = \int \mathrm{d}^d x \, \mathrm{tr}\left\{ -\frac{1}{g^2}\left[\frac{1}{4}\mathbf{F}_{\mu\nu}^2(x) + \frac{1}{2\xi}\left(\partial_\mu \mathbf{A}_\mu(x) \right)^2 \right] \right.$$
$$\left. + \mathbf{C}(x) \partial_\mu \mathbf{D}_\mu \bar{\mathbf{C}}(x) \right\}. \tag{13.21}$$

The obvious drawback of the covariant gauge, which leads to a covariant, local and renormalizable theory, is the *lack of explicit positivity and thus unitarity*. In particular, Faddeev–Popov fermions, being spinless, are non-physical, because they do not obey to the spin–statistics connection.

Gribov's ambiguity. As pointed out by Gribov, *in contrast to the Abelian case, depending on the value of the gauge field* $\mathbf{A}_\mu(x)$, *the gauge condition*

$$\partial_\mu \mathbf{A}_\mu(x) = \boldsymbol{\nu}(x)$$

has not always a unique solution, a problem called *Gribov's ambiguity* [142].

When two solutions merge, the operator $\partial_\mu \mathbf{D}_\mu(\mathbf{A})$ *has zero eigenvalues*. The integral over the ghost fields $\mathbf{C}, \bar{\mathbf{C}}$ fields the factor

$$\int [\mathrm{d}\mathbf{C} \mathrm{d}\bar{C}] \exp\left[-\mathbf{C}(x) \partial_\mu \mathbf{D}_\mu \bar{\mathbf{C}}(x) \right] = \det \partial_\mu \mathbf{D}_\mu(\mathbf{A}).$$

(In fact, the ghost fields have been introduced to represent by a local action the determinant.)

This implies that the field integral representation of the gauge theory in the covariant gauge is *may not be meaningful beyond perturbation theory*. The same ambiguity has been shown to arise for a large class of gauge conditions.

13.3.5 BRST symmetry

One obvious consequence of the necessity of gauge fixing is that the quantized action is no longer gauge invariant. On the other hand the *quantized action*, in the covariant gauge, now has a BRS or *BRST (Becchi–Rouet–Stora–Tyutin) symmetry* [43], a consequence of the stochastic dynamics given to the degrees of freedom of the gauge group variables. It is invariant under transformations that exchange bosons and fermions (*cf.* supersymmetry). In the form (13.19) of the action, it can be written as [156]

$$\begin{cases} \delta \mathbf{A}_\mu(x) = -\varepsilon \mathbf{D}_\mu \bar{\mathbf{C}}(x), & \delta \bar{\mathbf{C}}(x) = \varepsilon \bar{\mathbf{C}}^2(x), \\ \delta \mathbf{C}(x) = \varepsilon \boldsymbol{\lambda}(x), & \delta \boldsymbol{\lambda}(x) = 0, \end{cases} \tag{13.22}$$

where ε is a Grassmann generator (it anticommutes with $\mathbf{C}$ and $\bar{\mathbf{C}}$ and $\varepsilon^2 = 0$).

One can also introduce the BRST functional differential operator

$$\mathcal{D} = \int \mathrm{d}^d x \ \mathrm{tr} \left[-\mathbf{D}_\mu \bar{\mathbf{C}}(x) \frac{\delta}{\delta \mathbf{A}_\mu(x)} + \bar{\mathbf{C}}^2(x) \frac{\delta}{\delta \bar{\mathbf{C}}(x)} + \boldsymbol{\lambda}(x) \frac{\delta}{\delta \mathbf{C}(x)} \right]. \tag{13.23}$$

Thanks to the introduction of the auxiliary field $\boldsymbol{\lambda}(x)$ [156], the BRST operator is nilpotent, with a vanishing square):

$$\mathcal{D}^2 = 0.$$

It can be called the BRST cohomology operator.

This relation reflects the property that two successive BRST transformations automatically yield zero.

The BRST symmetry of the quantized action can then be expressed by the equation

$$\mathcal{D}\mathcal{S}(\mathbf{A}_\mu, \bar{\mathbf{C}}, \mathbf{C}, \boldsymbol{\lambda}) = 0.$$

In the cohomological terminology, $\mathcal{S}$ is *BRST closed*.

Moreover, the additional contribution $\mathcal{S}_{\text{gauge}}$ generated by the quantization procedure, in the cohomological terminology *is BRST exact*, that is,

$$\mathcal{S}_{\text{gauge}} = \mathcal{D} \int \mathrm{d}^d x \ \mathrm{tr} \, \mathbf{C}(x) \left[\partial_\mu \mathbf{A}_\mu(x) + \tfrac{1}{2}\xi g^2 \boldsymbol{\lambda}(x) \right], \tag{13.24}$$

a crucial property that implies *gauge independence* of gauge invariant observables.

Ward–Takahashi (WT) identities associated with the BRST symmetry can be summarized by a *quadratic functional differential equation* satisfied by the unrenormalized action $\mathcal{S}$ and the generating functional Γ of *vertex or 1PI* (one-particle-irreducible) correlation functions,[2] which one can write symbolically as (see Chapter 14)

$$\mathcal{S} * \mathcal{S} = 0, \quad \Gamma * \Gamma = 0.$$

---

[2] WT identities are identities satisfied by vertex or correlation functions, consequences of the symmetry properties of the action.

These equations can be used to prove that counter-terms can be added to the action such that the renormalized action $\mathcal{S}_{\text{ren.}}$ and vertex functional still satisfy the same equations, (Zinn-Justin [156]):

$$\mathcal{S}_{\text{ren.}} * \mathcal{S}_{\text{ren.}} = 0 \,.$$

The solution of the equation (which relies on BRST cohomology techniques) implies *structural stability of the quantized action under renormalization.*

The strategy generalizes to the renormalization of all gauge invariant operators, but the derivation is less straightforward [160].

13.4 Perturbation theory, regularization

With respect to the Abelian case, the new features of the non-Abelian case are the presence of gauge field self-interactions and ghost terms. In four dimensions, as in the Abelian case, the gauge field has dimension 1. The ghost fields have a simple δ_{ab}/p^2 propagator and canonical dimension 1 in four dimensions. The interaction terms have all dimension 4 and, therefore, the *theory is renormalizable by power counting in four dimensions.*

The power counting for matter fields is the same as in the Abelian case, since the coupling to matter fields differs only by geometric factors corresponding to group indices. For example, the coupling to fermions generated by the covariant derivative is simply $\gamma_\mu t^a_{ij}$.

Infrared (IR) divergences. In the covariant gauge, and in the absence of a Landau–Ginzburg–Higgs mechanism that provides a mass to gauge fields, only the gauge $\xi = 1$, Feynman's gauge, leads to a theory that is obviously IR finite. In contrast to the Abelian case, it is impossible to give an explicit mass to the gauge field and to then construct a theory which is both unitary and renormalizable. On the other hand, one wants eventually to prove the gauge independence of the theory and, therefore, one must be able to define it for more than one gauge. One way to introduce an IR regulator is to consider the theory in a finite volume.

13.4.1 Regularization

Dimensional regularization is very convenient for practical calculations and works in a simple way, except in the presence of chiral fermions.

Lattice regularization, which is also relevant for non-perturbative calculations can be used generally since a method for handling chiral fermions has been discovered [166, 167] (related to the Ginsparg–Wilson relation [165]: see Section 17.7).

Finally, *momentum or Pauli–Villars's type regularizations* work even in the chiral case but they regularize all diagrams except one-loop diagrams [76, 161]. The regularized pure gauge action takes the form

$$\mathcal{S}(\mathbf{A}_\mu) = \int \mathrm{d}^d x \, \text{tr} \, \mathbf{F}_{\mu\nu}(x) P(\mathbf{D}^2/\Lambda^2) \mathbf{F}_{\mu\nu}(x),$$

the gauge function in Landau's gauge $\partial_\mu \mathbf{A}_\mu$ being changed into

$$\partial_\mu \mathbf{A}_\mu \longmapsto Q(\partial^2/\Lambda^2)\partial_\mu \mathbf{A}_\mu \,,$$

in which P, Q are polynomials.

Indeed, with this regularization, both the gauge field propagator and the ghost propagator can be made arbitrarily convergent. However, the *covariant derivatives generate new, more singular interactions.* One then verifies that the *power counting of one-loop diagrams remains unchanged while higher order diagrams can be made superficially convergent* by taking the degrees of P and Q large enough. Auxiliary fields are required to deal with the one-loop diagrams.

For matter fields, the situation is the same as in the Abelian case: for example, massive fermions contributions can be regularized by adding a set of regulator fields, massive fermions and bosons with spin.

Again in the case of chiral fermions, global chiral properties can be preserved, but problems arise with local chiral transformations (see Section 13.6). However, the problem of the compatibility between gauge symmetry and quantum corrections is reduced to an explicit verification of the WT identities for the one-loop diagrams. Note that *the preservation of gauge symmetry is necessary for the cancellation of non-physical states in physical amplitudes* and, thus, essential for the physical consistency of the quantum field theory.

13.4.2 WT identities and renormalization

From the BRST symmetry follow relations between correlation functions, which are called WT identities, and thus relations between divergences. They can be used to derive the form of the renormalized action (see Chapter 14). We give here the result only in the example of the pure gauge action in the covariant gauge, assuming that the gauge group G is simple. Counter-terms can be adjusted such that the renormalized form of the action (13.21) is given by the substitution:

$$\begin{cases} g^2 \longmapsto Z_g g^2\,, & \mathbf{A}_\mu \longmapsto Z_A^{1/2}\mathbf{A}_\mu\,, \\ \xi \longmapsto Z_A Z_g^{-1}\xi\,, & \mathbf{C}\bar{\mathbf{C}} \longmapsto Z_C\mathbf{C}\bar{\mathbf{C}}\,. \end{cases}$$

This result has a simple interpretation: the gauge structure is preserved and the coefficient of $(\partial_\mu \mathbf{A}_\mu)^2$ is unrenormalized, exactly as in the Abelian case. However, unlike the Abelian case, the gauge transformation of the gauge field and, more generally, the form of the covariant derivative, are modified by the gauge field renormalization.

13.5 QCD: Renormalization group

Quantum chromodynamics consists of a set of quarks characterized by a *flavour quantum number,*and which are also triplets of a gauged symmetry, the $SU(3)$ *colour*, realized in the symmetric phase. Their interactions are mediated by the corresponding gauge fields (*gluons*):

$$\mathcal{S}(\mathbf{A}_\mu, \bar{\mathbf{Q}}, \mathbf{Q}) = -\int \mathrm{d}^4x \left[\frac{1}{4g^2}\,\mathrm{tr}\,\mathbf{F}_{\mu\nu}^2(x) \right.$$
$$\left. + \sum_{\text{flavours}} \bar{\mathbf{Q}}_f(x)\,(\slashed{D} + m_f)\,\mathbf{Q}_f(x) \right]. \qquad (13.25)$$

The most important physical arguments in favour of such a model are

(i) Quarks behave almost like free particles at short distances, as indicated by deep inelastic scattering experiments or the spectrum of bound states of heavy quarks. This is consistent with the sign of the RG β-function in non-Abelian gauge theories with not too many fermions. The general form is [84]

$$\beta(g^2) = - \left[\tfrac{11}{3} C(G) - \tfrac{4}{3} T(R) \right] \frac{g^4}{8\pi^2} + O\left(g^6\right), \tag{13.26}$$

where $C(G)$ depends on the gauge group G, and R on the fermion representation. In the case of the $SU(N)$ group with N_F fermions in the fundamental representation, the values of $C(G)$ and $T(R)$ are

$$C(G) = N, \quad T(R) = \tfrac{1}{2} N_\mathrm{F}$$

and, therefore,

$$\beta(g^2) = - \left(\frac{11N}{3} - \frac{2N_\mathrm{F}}{3} \right) \frac{g^4}{8\pi^2} + O\left(g^6\right). \tag{13.27}$$

The β-function is negative for small coupling, provided

$$N_\mathrm{F} < 11N/2, \tag{13.28}$$

which, in the case of $SU(3)$, implies at most 16 flavours. If this condition is met, $g = 0$ is an ultraviolet (UV) fixed point that governs the large momentum behaviour.

Moreover, according to *Coleman–Gross's theorem* [162], in four dimensions only non-Abelian gauge symmetries share this property, which is called *asymptotic freedom*.

(ii) No free quarks have ever been observed at large distance (but they manifest themselves indirectly in jet production). This is consistent with the simplest picture in which the β-function (which, due to asymptotic freedom, is negative at small coupling) remains negative for all couplings in such a way that the effective coupling constant grows without bounds at large distances. The conjecture that no free quarks can be observed is called the *confinement* hypothesis. This confinement property is largely validated by numerical simulations of lattice QCD.

13.6 Anomalies: General remarks

Anomalies arise when semi-classical symmetries of the theory cannot be implemented in the full quantum theory because ordering in products of non-commuting quantum operators is involved. An elementary example showing the role of operator ordering in implementing symmetries is provided by a non-relativistic particle in a static magnetic field.

Starting from the classical Hamiltonian,

$$H = \frac{1}{2m} \left[\mathbf{p} + e\mathbf{A}(\mathbf{q}) \right]^2,$$

in the quantization when one wants to replace the classical phase space coordinates $\mathbf{p}$, $\mathbf{q}$ by the quantum operators $\hat{\mathbf{p}}$, $\hat{\mathbf{q}}$, a problem of operator order arises in the product $\hat{\mathbf{p}} \cdot \mathbf{A}(\hat{\mathbf{q}})$. It is fixed by two independent conditions: gauge invariance and Hermiticity:

$$\hat{H} = \frac{1}{2m} \left[\hat{\mathbf{p}} + e\mathbf{A}(\hat{\mathbf{q}}) \right]^2.$$

Fortunately, *these two symmetries can be implemented, are consistent and yield the same result.*

For the same reason, in gauge theories, the order in the products of quantum operators is important. When gauge symmetry is confronted with fermion chiral symmetry, a conflict in the implementation of both symmetries may appear that is called an *anomaly*.

Technically, an indication that a problem may arise comes from the impossibility of regularizing one-loop diagrams by higher derivatives methods in the case of chiral gauge field theories (a feature related to operator ordering).

Indeed, gauge theories with massless fermions and chiral symmetry can be found where the axial current is not conserved. The divergence of the axial current is then called an *anomaly*. This leads in particular to *obstructions to the construction of gauge theories when the gauge field couples differently to the two fermion chiral components*. Several consequences are physically important, such as the *constraint of anomaly cancellation in the theory of weak-electromagnetic interactions*, the *electromagnetic decay of the π_0 meson*, or the *U(1) problem*.

Since the only possible source of anomalies are one-loop fermion diagrams in gauge theories when chiral properties are involved, this reduces the problem to an analysis of the properties of fermions in the background of gauge fields, or equivalently to the *properties of the determinant of the gauge covariant Dirac operator* (see Chapter 17 for details).

13.7 QCD: The semi-classical vacuum and instantons

We first consider pure gauge theories. Moreover, for any gauge group that contains the Lie group $SU(2)$ as a subgroup, the issues of semi-classical vacuum and instantons can be discussed within the $SU(2)$ subgroup. Since $SU(2)$ is locally isomorphic to $SO(3)$, for pure gauge theories one can use a $SO(3)$ notation.

The gauge field $\mathbf{A}_\mu$ is a three-component vector and the gauge action reads

$$\mathcal{S}(\mathbf{A}_\mu) = \frac{1}{4g} \int \left[\mathbf{F}_{\mu\nu}(x) \right]^2 \mathrm{d}^4 x$$

with then

$$\mathbf{F}_{\mu\nu}(x) = \partial_\mu \mathbf{A}_\nu(x) - \partial_\nu \mathbf{A}_\mu(x) + \mathbf{A}_\mu(x) \times \mathbf{A}_\nu(x).$$

We define the dual of the tensor $\mathbf{F}_{\mu\nu}$ by

$$\tilde{\mathbf{F}}_{\mu\nu}(x) = \tfrac{1}{2} \sum_{\rho,\sigma} \epsilon_{\mu\nu\rho\sigma} \mathbf{F}_{\rho\sigma}(x),$$

where $\epsilon_{\mu\nu\rho\sigma}$ is the totally antisymmetric tensor with $\epsilon_{1234} = 1$.

13.7.1 *The θ-vacuum and instantons*

In the semi-classical approximation, the structure of the vacuum in non-Abelian gauge theories can be discussed most conveniently in the temporal gauge $\mathbf{A}_4 = 0$ (Section 13.3). The minima of the classical action then correspond to gauge field components $\mathbf{A}_i$, $i = 1, 2, 3$, which are pure gauge functions of the three space variables x_i:

$$-\tfrac{1}{2} i \mathbf{A}_m(x_i) \cdot \boldsymbol{\sigma} = \mathbf{g}(x_i) \partial_m \mathbf{g}^{-1}(x_i), \quad \mathbf{g}(x) \in SU(2).$$

The gauge functions $\mathbf{g}(x)$ define mappings of the group elements $\mathbf{g}$ into compactified $\mathbb{R}^3$ (because $\mathbf{g}(x)$ goes to a constant for $|x| \to \infty$). These mappings belong to different homotopy classes. Since $SU(2)$ is topologically equivalent to S_3, one is led to the study of the homotopy classes of *mappings from S_3 to S_3, which are classified by an integer, the winding number.*

The winding number is given by the quantized values of the *topological charge* (here written in $SO(3)$ and in four-dimensional relativistic notation),

$$Q(\mathbf{A}_\mu) = \frac{1}{32\pi^2} \int \mathrm{d}^4 x \, \mathbf{F}_{\mu\nu}(x) \cdot \tilde{\mathbf{F}}_{\mu\nu}(x), \tag{13.29}$$

which is directly related to the *axial anomaly* (Section 17.3). One verifies that the integrand is a total derivative and thus the $Q(\mathbf{A}_\mu)$ depends only on the behaviour of the gauge field at large distance.

Thus, the *semi-classical vacuum has a periodic structure* analogous to the structure of the cosine potential in quantum mechanics, with an infinite number of classical degenerate minima.

Instantons. Like in the case of the periodic cosine potential (Section 18.1), the degeneracy between minima is lifted by quantum tunnelling. In the framework of functional integrals, barrier penetration effects are related to *instantons*, finite action solutions of imaginary time field equations. They satisfy the self-duality equations

$$\mathbf{F}_{\mu\nu}(x) = \pm \tilde{\mathbf{F}}_{\mu\nu}(x). \tag{13.30}$$

The simplest instanton solution, which depends on an arbitrary scale parameter λ, is

$$A_m^i = \frac{2}{r^2 + \lambda^2} (x_4 \delta_{im} + \epsilon_{imk} x_k), \quad m = 1, 2, 3, \quad A_4^i = -\frac{2x_i}{r^2 + \lambda^2}. \tag{13.31}$$

Finally, as in the example of the cosine potential, the lowest eigenstates have a band structure, an angle θ characterizing the different states: the so-called θ-vacua. To project onto a θ-*vacuum*, one adds a term to the classical action of gauge theories,

$$S_\theta(\mathbf{A}_\mu) = S(\mathbf{A}_\mu) + \frac{i\theta}{32\pi^2} \int d^4x\, \mathbf{F}_{\mu\nu}(x) \cdot \tilde{\mathbf{F}}_{\mu\nu}(x). \qquad (13.32)$$

One can then integrate over all fields $\mathbf{A}_\mu$ without restriction.

At least in the semi-classical approximation, the gauge theory depends on an additional parameter, the angle θ (for details, see Section 18.3). For non-vanishing values of θ, the additional term violates CP conservation and is at the origin of the *strong CP violation problem*: experimental bounds imply $|\theta| < 10^{-9}$ and, thus, unnaturally small values (another example of *fine tuning*).

The existence of a new very light particle, the pseudoscalar *axion* (which has not been discovered yet), has been proposed to solve the CP violation problem [163]. This conjectured particle is also considered as a possible candidate for dark matter.

Quarks in the instanton background. We consider now the complete QCD gauge action,

$$S(\mathbf{A}_\mu, \bar{\mathbf{Q}}, \mathbf{Q}) = -\int d^4x \left[\frac{1}{4e^2}\, \mathrm{tr}\, \mathbf{F}_{\mu\nu}^2(x) + \sum_{f=1}^{N_f} \bar{\mathbf{Q}}_f(x)\,(\not{D} + m_f)\,\mathbf{Q}_f(x) \right].$$

First, if the θ term in equation (13.32) contributes and one fermion field is massless, the Dirac operator has at least one vanishing eigenvalue and the determinant resulting from the fermion integration vanishes. Then, instantons do not contribute to the field integral and the strong CP violation problem is solved. However, such an hypothesis seems to be inconsistent with the quark masses extracted from experimental data.

Moreover, if the instantons contribute, *they solve the $U(1)$ problem, that is, the absence of a Goldstone boson associated with the almost spontaneous breaking of the axial $U(1)$ current* [12].

13.7.2 Physics application: The solution of the $U(1)$ problem

In a theory in which the quarks are massless and interact through a colour gauge group, the action has a chiral $U(N_F) \times U(N_F)$ symmetry, in which N_F is the number of flavours. The spontaneous breaking of the chiral group into its diagonal subgroup $U(N_F)$ leads one to expect N_F^2 Goldstone bosons associated with the axial currents.

Of course, the axial current corresponding to the $U(1)$ Abelian subgroup has an anomaly but the WT identities that imply the existence of Goldstone bosons correspond to constant group transformations. Therefore, they involve only the space integral of the divergence of the current. Since the anomaly is a total derivative, one might have expected the integral to vanish.

However, non-Abelian gauge theories admit instanton solutions. These instanton solutions correspond to gauge configurations that approach non-trivial pure gauges at infinity and give the set of discrete non-vanishing values one expects from equation (17.56) to the space integral of the anomaly (17.49) (see Sections 17.3, particularly 17.3.2).

This indicates (although no complete satisfactory calculation of the instanton contribution has been performed) that, for small, but non-vanishing, quark masses, the $U(1)$ axial current is far from being conserved and, therefore, no light would-be Goldstone boson is generated. This observation has resolved a long standing puzzle, since, experimentally, no corresponding light pseudoscalar boson is found for $N_F = 2, 3$.

13.8   Lattice gauge theories: Generalities

Lattice gauge theories play a double role: they provide both a regularization for perturbative gauge theories in the continuum and, moreover, the only known non-perturbative definition. They can be used for theoretical purposes and also for the study of properties of gauge theories by numerical methods [44].

Lattice gauge theories provide a lattice regularization of the continuum gauge theories: the low temperature or small coupling expansion of the lattice model is a regularized continuum perturbation theory.

However, other analytic results can be obtained in the high temperature or *strong coupling limit* and in the mean field approximation.

Lattice gauge theories have properties quite different from those of usual lattice models in statistical physics. The *phase structure cannot be discussed in terms of a local order parameter* (a local field).

For example, in pure gauge theories (without matter fields) on the lattice, to distinguish between the confined and deconfined phases, it is necessary to examine the behaviour of a non-local quantity, a functional of loops called the *Wilson loop*.

Notation. In the sections devoted to lattice gauge theories, *we abandon the rule of summation over repeated indices.*

13.8.1   *Gauge invariance and parallel transport on the lattice*

The construction of lattice gauge theories is directly based on the geometric idea of *parallel transport* as described in Section 13.1.

To each site i (i represents the set of lattice coordinates) of a lattice, we associate a set of dynamic variables, $\{\phi_i, \phi_i^*\}$, representing matter fields, on which acts a *unitary representation* $\mathcal{D}(G)$ of a Lie group G (*e.g.*, $SU(N)$):

$$\phi_{\mathbf{g}} = \mathbf{g}\phi\,, \quad \phi_{\mathbf{g}}^* = \mathbf{g}^\dagger \phi^*\,, \quad \mathbf{g} \in \mathcal{D}(G)\,.$$

We assume that the lattice action and the integration measure over the dynamic variables are invariant under global (site-independent) transformations of $\mathcal{D}(G)$.

A model is gauge invariant (local invariance) if it is invariant under independent group transformation $\mathbf{g}_i$ *on each lattice site i.*

For the (ϕ, ϕ^*) integration measure, as well as for all the terms in the lattice action which depend only on one site, global invariance implies local invariance. However, field derivatives are replaced on the lattice by finite differences that involve products of fields at different sites.

To render a product $\phi_i^* \phi_j$, i and j being different sites, invariant, it is necessary to introduce new quantities, parallel transporters in the form of matrices $\mathbf{U}[C_{ij}]$ belonging to the representation $\mathcal{D}(G)$, which depends on curves joining two sites i, j and transforming like

$$\mathbf{U}[C_{ij}] \mapsto \mathbf{g}_i \mathbf{U}[C_{ij}] \mathbf{g}_j^{-1} . \tag{13.33}$$

Then, the quantity

$$\phi_i^* \mathbf{U}[C_{ij}] \phi_j \tag{13.34}$$

is gauge invariant.

The parallel transporters satisfy the rules described in Section 13.1.1. For example, the rules (13.2) imply

$$\mathbf{U}[C_{ij}] = \mathbf{U}^{-1}[C_{ji}] = \mathbf{U}^\dagger[C_{ji}], \tag{13.35}$$

if C_{ij} and C_{ji} are the same curves but with opposite orientation. Also,

$$\mathbf{U}[C_{ij}] \mathbf{U}[C'_{jk}] = \mathbf{U}[(C_{ij} \cup C'_{jk})]. \tag{13.36}$$

In the continuum, a parallel transporter can be obtained by integrating transport along infinitesimal curves and, thus, can be expressed in terms of a gauge field or connection, element of the representation of the Lie algebra.

On the lattice curves follow links, the segments that connect adjacent sites. One can thus generate a parallel transport between two sites by the product of parallel transporters associated to successive links along a curve C drawn on the lattice between the two sites:

$$\mathbf{U}[C_{ij}] \equiv \prod_{\text{all links } \ell \in C_{ij}} \mathbf{U}_\ell ,$$

where the product is ordered along the path C_{ij}.

One can thus take as *dynamic variables elements* $\mathbf{U}_\ell$ *of the group representation, associated with parallel transport along oriented links of the lattice.*

13.9 Pure lattice gauge theory

We first consider pure lattice gauge theories (without matter fields). The introduction of fermion fields requires a discussion of the chiral properties of the lattice approximation of the fermion action (see Section 13.11).

In this approximation, realistic properties of QCD cannot be determined, but one can still investigate one essential question: Is the *confinement property* supported by numerical calculations in lattice gauge theory, that is, does the force between charged particles increase at large distances, so that heavy quarks in the fundamental representation cannot be separated?

Other problems have also been discussed in this framework: for example, the appearance of massive group singlet bound states in the spectrum (gluonium), the *question of a deconfinement transition at finite physical temperature in QCD* by treating fermions in the so-called quenched approximation.

We discuss the pure gauge theory and its formal continuum limit as obtained from a low temperature, strong coupling expansion.

Notation. In what follows, we consider only link variables and, therefore, we simplify the notation by setting $\mathbf{U}[C_{ij}] \equiv \mathbf{U}_{ij}$ for i, j adjacent sites on the lattice.

13.9.1 Gauge invariant action and partition function

First, one must define a gauge invariant interaction for the link variables. It follows from the transformation (13.33) that only the traces of the products of link variables along closed loops are gauge invariant. On a hypercubic lattice, the shortest loop is a square, called hereafter a *plaquette*. In what follows, we thus consider a plaquette action of the form (i, j, k, l form a square on the lattice)

$$\mathcal{S}(\mathbf{U}) = -\beta_p \sum_{\substack{\text{all plaquettes} \\ i,j,k,l}} \operatorname{tr} \mathbf{U}_{ij} \mathbf{U}_{jk} \mathbf{U}_{kl} \mathbf{U}_{li}\,, \qquad (13.37)$$

in which β_p is the plaquette coupling.

The appearance of products of parallel transporters along closed loops is not surprising since the pure gauge action of the continuum theory is associated with infinitesimal transport along a closed loop.

Note that each plaquette appears with both orientations in such a way that the sum is real when the group is unitary.

The quantum partition function. A quantum partition function corresponding to the action (13.37) can then be defined as

$$\mathcal{Z} = \int \prod_{\text{links}\,\ell} \mathrm{d}\mathbf{U}_\ell\, \mathrm{e}^{-\mathcal{S}(\mathbf{U})}, \qquad (13.38)$$

where the integration measure is the group invariant (de Haar) measure associated with the group G (denoted symbolically $\mathrm{d}U$).

In contrast to continuum gauge theories, the expression (13.38) *is defined on the lattice* (at least as long as the volume is finite) *because the group is compact and thus the volume of the group is finite*. Therefore, gauge fixing is not required and a completely gauge invariant formulation of the theory is possible.

13.9.2 Low coupling analysis

To understand the precise connection between the lattice theory (13.38) and the continuum field theory, one must investigate the lattice theory at low coupling, that is, at large positive β_p. In this limit, the partition function is dominated by minimal action configurations.

Let us show that the minimum of the action corresponds to the matrices $\mathbf{U}$ gauge transform of the identity. We start from a first plaquette 1234. Without loss of generality, we can set

$$\mathbf{U}_{12} = \mathbf{g}_1^{-1}\mathbf{g}_2\,, \quad \mathbf{g}_1, \mathbf{g}_2 \in \mathcal{D}(G).$$

The matrix $\mathbf{g}_1$ is arbitrary and $\mathbf{g}_2$ is calculated from $\mathbf{U}_{12}$ and $\mathbf{g}_1$. Then, we can also set

$$\mathbf{U}_{23} = \mathbf{g}_2^{-1}\mathbf{g}_3, \quad \mathbf{U}_{34} = \mathbf{g}_3^{-1}\mathbf{g}_4.$$

These relations define first $\mathbf{g}_3$, and then $\mathbf{g}_4$. The minimum of the action is obtained when the real part of all traces is maximum, that is, when the products of the group elements on a plaquette are equal to 1. In particular,

$$\mathbf{U}_{12}\mathbf{U}_{23}\mathbf{U}_{34}\mathbf{U}_{41} = 1,$$

which yields

$$\mathbf{U}_{41} = \mathbf{g}_4^{-1}\mathbf{g}_1.$$

If we now take an adjacent plaquette, the argument can be repeated for all links but one, which has already been fixed. In this way, *one can show that the minimum of the action is a pure gauge.*

Thus, when the coupling constant β_p becomes large, all group elements are constrained to stay, up to a gauge transformation, close to the identity (in a finite volume with consistent boundary conditions).

From this analysis, one learns that *the minimum of the lattice action is highly degenerate at low coupling*, since it is parametrized by a gauge transformation, which corresponds to a finite number of degrees of freedom per site.

This unusual property of lattice gauge theories corresponds to the property that the gauge action in classical mechanics determines the motion only up to a gauge transformation. *To perform a low coupling expansion, it becomes necessary to 'fix' the gauge in order to sum over all minima.*

Low coupling expansion. We choose a gauge such that the minimum of the action corresponds to all matrices $\mathbf{U} = 1$. At low coupling, the matrices $\mathbf{U}$ are then close to the identity:

$$\mathbf{U}(x, x + an_\mu) = 1 - a\mathbf{A}_\mu(x) + O\left(a^2\right),$$

in which a is the lattice spacing, x is the point on the lattice, and n_μ is the unit vector in the direction μ.

In the matrix $\mathbf{A}_\mu(x)$, we recognize the connection or gauge field. One can then expand the lattice action for small fields.

At leading order, one finds

$$\sum_{\mu,\nu,x} \operatorname{tr} e^{-a^2 \mathbf{F}_{\mu\nu}(x)} - \operatorname{tr} 1 = a^4 \sum_{\mu,\nu,x} \operatorname{tr} \mathbf{F}_{\mu\nu}^2(x) + O\left(a^6\right),$$

where $\mathbf{F}_{\mu\nu}(x)$ is the curvature tensor

$$\mathbf{F}_{\mu\nu}(x) = \partial_\mu \mathbf{A}_\nu(x) - \partial_\nu \mathbf{A}_\mu(x) + [\mathbf{A}_\mu(x), \mathbf{A}_\nu(x)] + O\left(a\right).$$

This result shows that the leading term of the small field expansion of the plaquette action (13.37) is the standard gauge action. The relation between β_p and the bare coupling constant g_0 of continuum gauge theories is then

$$a^4 \beta_p \sim g_0^{-2}. \tag{13.39}$$

Therefore, *the low coupling expansion, in a fixed gauge, of lattice gauge theories indeed provides a regularization of continuum gauge theories.*

We have here discussed only the pure gauge action but the generalization to matter fields is simple.

Higher order terms in the small field expansion yield additional interactions needed to maintain gauge invariance on the lattice. This is not surprising: we have already shown that the gauge invariant extension of Pauli–Villars's regularization also introduces additional interactions.

13.10 Wilson loop and the confinement property

The form of the RG β-function shows that gauge theories are asymptotically free in four dimensions, which means that the origin in the coupling constant space is a UV fixed point, and also implies that the effective interaction increases at large distance. Therefore, *the spectrum of a non-Abelian gauge theory cannot be determined from perturbation theory.* To explain the non-observation of free quarks, it has been conjectured that the spectrum of the symmetric phase consists only in neutral states, that is, states which are singlets for the group transformations (the confinement property).

To discuss the confinement issue, it has been suggested by Wilson [44] to study, in pure gauge theories, *a gauge invariant non-local quantity, the energy of the vacuum in the presence of largely separated static charges.*

We examine this quantity first in pure Abelian gauge theories because, in the continuum, only analytic calculations are involved.

13.10.1 Wilson's loop in continuum: d-Dimensional Abelian gauge theories

In continuum field theory, in order to calculate the average energy, it is necessary to introduce the gauge Hamiltonian, and, therefore, it is convenient to work in the temporal gauge $A_d = 0$.

A wave function for two static point-like charges, in the temporal gauge, is

$$\psi(A) = \exp\left[-ie \oint_{C_0} \mathbf{A}(s) \cdot \mathrm{d}\mathbf{s}\right],$$

in which the charges are located at both ends of the curve $C_0 \subset \mathbb{R}^{d-1}$.

By evaluating the behaviour for large time T of the matrix element

$$W(C_0) = \langle\psi|\,\mathrm{e}^{-HT}\,|\psi\rangle,$$

in which H is the gauge Hamiltonian in the temporal gauge, one obtains the energy $E(C_0)$ of the vacuum in presence of static charges:

$$W(C_0) \underset{T\to\infty}{\sim} \mathrm{e}^{-TE(C_0)}.$$

If the charges are separated by a distance R, one expects E to depend only on R and not on C_0.

The loop functional $W(C_0)$ can be calculated from the field integral,

$$W(C_0) = \left\langle \exp\left[-ie \oint_{C_0'} \mathbf{A}(s) \cdot \mathrm{d}\mathbf{s} \right] \right\rangle .$$

The curve C_0', which is now defined in space and time, is the union of two curves, which coincide with C_0 at time 0, and with $-C_0$ at time T, respectively. The expectation value here means average over gauge field configurations with the gauge action.

Since, in the temporal gauge, the time component of A_μ vanishes, we can add to C_0' two straight lines in the time direction, which join the ends of the curves $C_0(t = 0)$ and $C_0(t = T)$. $W(C_0)$ then becomes a functional of a closed loop C (see Fig. 13.2):

$$W(C_0) \equiv W(C) = \left\langle \exp\left[-ie \oint_{C} \mathbf{A}(s) \cdot \mathrm{d}\mathbf{s} \right] \right\rangle . \qquad (13.40)$$

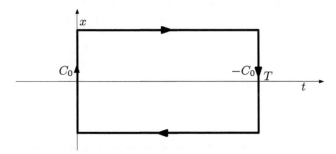

Fig. 13.2   The Wilson loop C.

The advantage of the representation (13.40) is that it is explicitly gauge invariant, since it is the expectation value of the parallel transporter corresponding to a closed loop in space-time. It can be evaluated in any gauge.

The question of confinement is related to the behaviour of the energy E when the separation R between charges becomes large.

In a pure Abelian gauge theory, in the continuum, which is a free field theory, the expression (13.40) can be evaluated explicitly. To simplify calculations, we take for C_0 also a straight line and use Feynman's gauge. The quantity $W(C)$ then is given by

$$W(C) = \int [\mathrm{d}A_\mu] \exp\left[-\mathcal{S}(A_\mu) + \int \mathrm{d}^d x\, \mathbf{J}(x) \cdot \mathbf{A}(x) \right]$$

with

$$\mathcal{S}(A_\mu) = \tfrac{1}{2} \int \mathrm{d}^d x \sum_{\mu,\nu} \partial_\mu A_\nu(x) \partial_\mu A_\nu(x)$$

and

$$J_\mu(x) = -ie \oint_{C} \delta(x - s)\mathrm{d}s_\mu .$$

The result is

$$\ln W(C) = -\frac{\Gamma(d/2 - 1)}{8\pi^{d/2}} e^2 \oint_{C \times C} d\mathbf{s}_1 \cdot d\mathbf{s}_2 \, |\mathbf{s}_1 - \mathbf{s}_2|^{2-d}. \qquad (13.41)$$

The integral in the right-hand side exhibits a short distance singularity, which requires a cut-off. Moreover, to normalize the right-hand side of equation (13.41), we divide it by the value of $W(C)$ taken at a fixed distance $R = a$. We now write the integrals more explicitly as

$$\oint_{C \times C} \frac{d\mathbf{s}_1 \cdot d\mathbf{s}_2}{2\,|\mathbf{s}_1 - \mathbf{s}_2|^{d-2}} = \int_0^T |u - t|^{2-d} \, du \, dt + \int_0^R |x - y|^{2-y} \, dx \, dy$$

$$- \int_0^R \left[(x - y)^2 + T^2\right]^{1-d/2} dx \, dy$$

$$- \int_0^T \left[(t - u)^2 + R^2\right]^{1-d/2} dt \, du.$$

The first term in the right-hand side is cancelled by the normalization. The second term is independent of T and, therefore, negligible for large T. The third term decreases with T for $d > 2$, which we now assume.

Only the last term increases with T:

$$\int_0^T \left\{ \left[(t - u)^2 + R^2\right]^{1-d/2} - \left[(t - u)^2 + a^2\right]^{1-d/2} \right\} dt \, du$$

$$\sim \sqrt{\pi} \frac{\Gamma((d - 3)/2)}{\Gamma(d/2 - 1)} \left(R^{3-d} - a^{3-d}\right) T.$$

Therefore, the vacuum energy $E(R)$ in presence of the static charges has the form

$$E(R) - E(a) = \frac{e^2}{4\pi^{(d-1)/2}} \Gamma((d - 3)/2) \left(a^{3-d} - R^{3-d}\right).$$

One recognizes the Coulomb potential between two charges.

For $d \leq 3$, the energy of the vacuum increases without bound when the charges are separated, and free charges cannot exist.

For $d = 3$, the potential increases logarithmically.

For $d = 2$, the Coulomb potential increases linearly with distance.

In more general situations, the method that we have used above to determine the energy is complicated because the large T limit has to be taken first, followed by an evaluation of the large R behaviour. It is more convenient to take a square loop, $T = R$, and evaluate the large R behaviour of $W(C)$. Here, one obtains

$$\ln W\left[C(R)\right] - \ln W\left[C(a)\right]$$

$$= \frac{1}{2\pi^{d/2}} \Gamma(d/2 - 1)e^2 \left\{ \int_0^R \left[(u - t)^2 + R^2\right]^{1-d/2} du \, dt \right.$$

$$\left. - \int_0^a \left[(u - t)^2 + a^2\right]^{1-d/2} du \, dt - \int_a^R |u - t|^{2-d} du \, dt \right\}.$$

For $d > 3$, dimensions in which the Coulomb potential decreases, the right-hand side is dominated by terms that correspond to the region $|s_1 - s_2| \ll R$ in equation (13.41):

$$\ln W\left[C(R)\right] - \ln W\left[C(a)\right] \sim \text{const.} \times R \,.$$

This behaviour is called the *perimeter law* since $\ln W(C)$ is proportional to the perimeter of C and is, therefore, relevant to the $d = 4$ Coulomb phase.

By contrast, for $d \le 3$, $\ln W(C)$ increases as R^{4-d}. The reason is that two charges separated on C by a distance of order R, feel a potential of order R^{3-d}.

In particular, for $d = 2$, $\ln W(C)$ increases like R^2, that is, like the area of the surface enclosed by C: this is the *area law* expected in confinement situations.

13.10.2 Non-Abelian gauge theories

In the temporal gauge, the wave function corresponding to two opposite point-like static charges is also related to a parallel transporter along a curve joining the charges.

The arguments already given in the Abelian case, show that the expectation value of the operator e^{-TH} in the corresponding state is given by the average, in the sense of a field integral, of the parallel transporter along a closed loop:

$$W(C) = \left\langle \mathrm{P}\exp\left[-i\oint_C \sum_\mu \mathbf{A}_\mu(s)\mathrm{d}s_\mu\right]\right\rangle,$$

in which we recall that the symbol P means path ordering, since the matrices $\mathbf{A}_\mu(s)$ at different points do not commute.

If we calculate $W(C)$ in perturbation theory, at leading order, we find the same results as in the Abelian case.

However, we know from RG arguments, that perturbation theory cannot be trusted at large distances. Therefore, to get a qualitative understanding of the phase structure, one can use the lattice gauge theory to calculate $W(C)$ in the large coupling or high temperature limit $\beta_p \to 0$.

Strong coupling expansion for the Wilson loop. We assume that the group we consider has a *non-trivial centre* and take the explicit example of gauge elements on the lattice belonging to the fundamental representation of $SU(N)$ (whose centre is $\mathbb{Z}_N$, with elements the identity multiplied by roots z of unity, $z^N = 1$).

We calculate $W(C)$ by expanding the integrand in expression (13.38) in powers of β_p. We choose for simplicity for the loop C a rectangle although the generalization to other contours is simple.

Any non-vanishing contribution must be invariant by the change of variables $U_\ell \mapsto z_\ell U_\ell$, where z_ℓ belongs to the centre. Let us consider one link belonging to the loop and multiply the corresponding link variable $\mathbf{U}(x, x + an_\mu)$ by z_0. We now multiply all link variables $\mathbf{U}(x+y, x+y+an_\mu)$, which are obtained by a translation y in the hyperplane perpendicular to n_μ, by z_y. Another link belonging to the loop belongs to the set but with opposite orientation.

Plaquettes involving such variables involve them in pairs. For a result to be invariant and thus non-vanishing, the number of times each link variable appears in the direction n_μ minus the number of times it appears in the direction $-n_\mu$ must vanish (mod N). Thus, we start adding plaquettes to satisfy this condition at point x. However, the addition of one plaquette does not change the total difference between the numbers of links in the $+n_\mu$ and $-n_\mu$ directions.

Therefore, at least one condition remains always unsatisfied until the plaquettes reach the other link of the loop.

We can repeat the arguments for the remaining links of the loop and the new non integrated remaining links of the plaquettes. The number of required plaquette variables to get a non-vanishing result, is at least equal to the area of the rectangle, the minimal area surface having the loop as boundary (see Fig. 13.3).

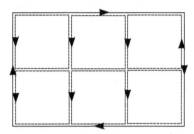

Fig. 13.3 The inside of loop C covered with plaquettes: Area law for strong coupling.

We can then perform the integrations, which are just factorized group integrations. In this way, we find a contribution to $W(C)$ proportional to $(\beta_p)^A$, in which A is the number of plaquettes.

The largest contribution corresponds to plaquettes covering the minimal area surfaces bounded by the loop. Indeed, it is obtained by covering the rectangle with plaquettes in such a way that each link variable appears only twice in either orientation.

For a rectangular loop $R \times T$, one obtains

$$W(C) \sim e^{RT \ln \beta_p} . \qquad (13.42)$$

This result indicates that *the potential between the static charges is linearly rising at large distance*. Static charges creating the loop cannot be screened by the gauge field, otherwise one would have found a perimeter law.

Remarks

(i) If the group centre is trivial, it is possible to form a tube along the loop, and this implies a perimeter law. If, for example, the group is $SO(3)$, in the decomposition of a product of two spin 1 representations, we again find a spin 1 which can be coupled to a third spin 1 to form a scalar. Thus, two plaquettes can be glued to the same link of the loop without constraint on the orientation of the plaquette.

(ii) The asymptotic form (13.42) is also valid for the Abelian $U(1)$ lattice gauge theory. Therefore, in four dimensions, the Wilson loop has a perimeter law in the weak coupling expansion, and an area law at large coupling.

One thus expects a phase transition between a low coupling Coulomb phase, described by a free field theory, and a strong coupling confined phase.

The string tension. The coefficient in front of the area

$$\sigma(\beta_p) \underset{\beta_p \to 0}{\sim} -\ln \beta_p \,,$$

is called the *string tension*. If no phase transition occurs when β_p varies from zero to infinity, the gauge theory leads to confinement. In this case, the behaviour of the string tension for β_p small is predicted by the RG. Since σ has the dimension of a mass squared, one finds

$$\sigma(g_0) \sim (g_0^2)^{-\beta_2/\beta_3^2} \exp\left(-1/\beta_2 g_0^2\right), \qquad (13.43)$$

in which g_0^2 is related to β_p by equation (13.39), and β_2, β_3 are the two first coefficients of the small coupling expansion of the RG β-function. A physical quantity relevant to the continuum limit can then be obtained by dividing $\sqrt{\sigma}$ by its asymptotic behaviour. Let us define Λ_L as

$$\Lambda_L = a^{-1}(\beta_2 g_0^2)^{-\beta_3/2\beta_2^2} \exp\left(-1/2\beta_2 g_0^2\right).$$

Then, $\Lambda_L/\sqrt{\sigma}$ has a continuum limit. When one calculates σ by non-perturbative lattice methods, the verification of the scaling behaviour (13.43) indicates that the result is relevant to the continuum field theory and is not just a lattice artefact.

It is possible to systematically expand σ in powers of β_p. The possibility of verifying that confinement is realized in the continuum limit, depends on the possibility of analytically continuing the strong coupling expansion up to the origin.

Unfortunately, theoretical arguments lead to believe that, independently of the group, the string tension is affected by a singularity associated with the roughening transition, a transition which, however, is not related to bulk properties. At strong coupling, the contributions to the string tension come from smooth surfaces. When g_0^2 decreases (β_p increases), one passes through a critical point g_{0R}^2, after which the relevant surfaces become rough. At the singular coupling g_{0R}^2, the string tension does not vanish but has a weak singularity.

Still, at this point, the strong coupling expansion diverges. Therefore, it is impossible to extrapolate to arbitrarily small coupling. The usefulness of the strong coupling expansion then depends on the position of the roughening transition with respect to the onset of weak coupling behaviour. Notice that, numerically, in the neighbourhood of the roughening transition, rotational symmetry is approximately restored (at least at large enough distance).

One can also calculate other quantities which are associated to bulk properties, and are, therefore, not affected by roughening singularities, such as the free energy (the connected vacuum amplitude) or the plaquette-plaquette correlation function. However, even for these quantities, the extrapolation is not easy because the transition between strong and weak coupling behaviours is, in general, very sharp.

From the numerical point of view, it seems that the plaquette-plaquette correlation function is the most promising case for strong coupling expansion.

Confinement: Intuitive picture. The potential between static charges increases linearly in the same way as the Coulomb potential in one space dimension. This leads to the following physical picture: in QED the gauge field responsible for the potential has no charge and propagates essentially like a free field isotropically in all space directions. Conservation of flux on a sphere then yields the R^{2-d} force between the charges. However, in the non-Abelian case, the attractive force between the gauge particles generates instead a flux tube between static charges in such a way that the force remains the same as in one space dimension.

13.11 Fermions on the lattice. Chiral symmetry

A straightforward lattice action for fermions is

$$S(\mathbf{U}, \bar{\psi}, \psi) = \frac{\kappa}{2a} \sum_{x \in (a\mathbb{Z})^4} \sum_{\mu=1}^{4} [\bar{\psi}(x + an_\mu)\mathbf{U}(x + an_\mu, x)\boldsymbol{\gamma} \cdot n_\mu \psi(x)$$
$$- \bar{\psi}(x - an_\mu)\mathbf{U}(x - an_\mu, x)\boldsymbol{\gamma} \cdot n_\mu \psi(x)],$$

where n_μ is the unit vector in the μ direction, $\boldsymbol{\gamma}$ a vector of generalized Dirac γ matrices and κ a coupling constant.

Indeed, expanding to first order in the lattice spacing a, one finds

$$S(\mathbf{U}, \bar{\psi}, \psi) = \kappa \sum_{x \in (a\mathbb{Z})^4} \sum_{\mu=1}^{4} [n_\mu \cdot \nabla \bar{\psi}(x)\boldsymbol{\gamma} \cdot n_\mu \psi(x) + \bar{\psi}(x)\mathbf{A} \cdot n_\mu \boldsymbol{\gamma} \cdot n_\mu \psi(x)].$$

One important qualitative feature of strong interaction physics is the approximate spontaneous breaking of chiral symmetry. However, non-trivial problems arise when one tries to implement chiral symmetry on the lattice.

One has the choice between writing a simple action which is not explicitly chiral symmetric and in which one tries to restore chiral symmetry by adjusting the fermion mass term (Wilson's fermions), writing a chiral symmetric action with too many fermions (staggered or Kogut–Susskind fermions [164]), or, as it has been more recently discovered, introducing various Dirac operators satisfying the Ginsparg–Wilson relation [165], called *overlap* [166] or *domain wall* [167] fermions.

In the latter method, several implementations can be interpreted as making the fermions depend on an extra space dimension, which makes computer simulations more difficult.

13.11.1 Numerical methods: Computer simulations

We do not describe here the numerical methods that have been used in lattice gauge theories [168]. In pure gauge theories, the existence of phase transitions has been investigated for many lattice actions. For the gauge group $SU(3)$, relevant to the physics of strong interactions, the string tension has been carefully measured and the plaquette-plaquette correlation function has been studied to determine the mass of low-lying gluonium states.

Finally, calculations have been performed at finite physical temperature, that is, on a (3+1)-dimensional lattice in the limit in which the size of the lattice remains finite in one dimension, this size being related to the temperature. In this way, the temperature of a deconfinement transition has been determined.

Fermions: Numerical issues. A serious practical difficulty arises with fermions: because it is impossible to simulate fermions numerically, it is necessary to integrate out fermions explicitly. This generates an effective gauge field action which contains a contribution proportional to the fermion determinant, and is, therefore, no longer local. Unfortunately, the speed of numerical methods is extremely sensitive to the locality of the action.

This explains why most numerical simulations with fermions have been, for a long time, performed only in the so-called quenched approximation, in which the fermion determinant is neglected. This approximation corresponds to the neglect of all fermion loops and bears some similarity to the eikonal approximation. In this approximation, the approximate spontaneous breaking of chiral symmetry has been verified by measuring the decrease of the pion mass for decreasing quark masses.

Owing to the difficulty of the problem, the numerical study of the effect of dynamical fermions for realistic lattice sizes that are close enough to the chiral limit has begun only in last decade. The physical spectrum of hadrons can now be reproduced, confirming again the relevance of QCD [169].

14 From BRST symmetry to the Zinn-Justin equation

Quantum field theories in a naive formulation lead to physics results plagued with infinities due to short distance singularities and require a *regularization*, an operation by which their short distance structure below a *cut-off* scale $1/\Lambda$ is modified in an *non-physical way* to render the theory finite. For a class of quantum theories called, therefore, *renormalizable*, it is possible to construct a theory whose perturbative expansion remains finite when the cut-off is removed by choosing the parameters of the initial Lagrangian suitably cut-off dependent (a fine tuning that may be physically problematic and even sometimes impossible beyond perturbation theory), a mathematical procedure called *renormalization* and whose deeper meaning can only be understood in the framework of the *renormalization group*.

Non-Abelian gauge theories are relativistic quantum theories at the basis of the Standard Model of particle physics (which describes fundamental interactions at the microscopic scale). In non-Abelian gauge theories, as in quantum electrodynamics (QED), gauge invariance implies that not all components of the gauge field are dynamical and that a simple canonical quantization is impossible. The quantization requires a *gauge fixing* that confines the gauge field to stay on or near some section of the gauge field space.

Note that we consider here only relativistically covariant gauges leading to theories renormalizable in the sense of power counting (a form of dimensional analysis).

However, the mathematical consistency and physical relevance of quantized gauge theories require a more specific form of renormalizability, which preserves some complex form of gauge invariance in the renormalization process.

At the beginning of the 1970s, much effort was devoted to the proof of the perturbative renormalizability of non-Abelian gauge theories. Initial arguments by 't Hooft and Veltman [75] were based on Feynman diagrams, while Lee–Zinn-Justin's proof [76], based on Slavnov–Taylor identities [154], used powerful functional techniques. After it was realized that the quantized action could be written in local form with the help of non-physical fermions (the *Faddeev–Popov ghost fields*), Becchi, Rouet and Stora, and Tyutin, [43], discovered that the action had an unexpected fermion-type symmetry, now called the *BRST symmetry*, which, as a technical tool, replaces gauge invariance in the context of quantized gauge theories.

A completely general and simpler proof of renormalizability of non-Abelian gauge theories followed, based on a quadratic master equation [156], which is also called Zinn-Justin (ZJ) equation [170]. The ZJ equation is satisfied both by the generating functional of vertex functions, which are also called one particle irreducible Green's functions, and by the quantized action. It can be shown that gauge theories can be renormalized while preserving the ZJ equation. The general solution of this equation, taking into account *locality*, *power counting* and *ghost number conservation*, gives then the general form of the *renormalized action*.

From Random Walks to Random Matrices. Jean Zinn-Justin, Oxford University Press (2019).
© Jean Zinn-Justin. DOI: 10.1093/oso/9780198787754.001.0001

Moreover, the ZJ equation implies independence of physical observables from the specific *gauge-fixing* procedure required to construct the quantum theory [157].

14.1 Non-Abelian gauge theories: Classical field theory

A classical non-Abelian gauge theory is a generalization of Maxwell's electrodynamics, in which the gauge invariance is based on a non-Abelian gauge group G in place of the Abelian $U(1)$ group underlying Maxwell's theory. We consider here only *local field theories*, that is, theories in which the action is the space-time integral of a function of fields and their derivatives (the Lagrangian density, or simply *Lagrangian*).

A *classical gauge theory* is a classical field theory whose action is invariant under gauge transformations (see Chapter 11). A gauge transformation is a space-time dependent representation of a matrix complex Lie group G, acting on the fields of the theory. If the group G is non-Abelian, one speaks of *non-Abelian gauge transformations* and *non-Abelian gauge theories*. In what follows, in view of physics applications, we restrict the discussion to unitary groups (but orthogonal groups require straightforward modifications) and $(3 + 1)$ space-time dimensions.

14.1.1 *Gauge transformations and gauge fields*

We assume that matter fields $\phi(x)$ form complex vectors that, in a *gauge transformation*, transform like (see also Sections 11.8 and 13.1.2)

$$\phi(x) \mapsto \phi^{\mathbf{g}}(x) = \mathbf{g}(x)\phi(x), \quad \mathbf{g}(x) \in G \quad \forall x, \tag{14.1}$$

where $\mathbf{g}(x)$ smoothly maps the space-time to the group G.

To construct a gauge theory, it is necessary to introduce a *gauge field* (or connection) $\mathbf{A}_\mu(x)$, also known as a *Yang–Mills field*, which, for each space-time index $\mu = 0, 1, 2, 3$, takes values in the Lie algebra of G. Gauge transformations act on the gauge field as

$$\mathbf{A}_\mu(x) \mapsto \mathbf{A}_\mu^{\mathbf{g}}(x) = \mathbf{g}(x)\mathbf{A}_\mu(x)\mathbf{g}^{-1}(x) + \mathbf{g}(x)\partial_\mu \mathbf{g}^{-1}(x). \tag{14.2}$$

When $\mathbf{g}(x)$ is constant for all x, the gauge field $\mathbf{A}_\mu(x)$ transforms under the adjoint representation of the group G. For generic $\mathbf{g}(x)$, the transformation (14.2) is no longer linear in $\mathbf{A}_\mu(x)$ but only affine. The form of the transformation ensures that $\mathbf{A}_\mu^{\mathbf{g}}(x)$ is still valued in the Lie algebra of G.

14.1.2 *Covariant derivatives and curvature*

In a local, gauge invariant, field theory, ordinary derivatives must be replaced by covariant derivatives that, because they transform linearly under gauge transformations, ensure the gauge invariance of the action. Covariant derivatives are constructed using the gauge connection $\mathbf{A}_\mu$. Their explicit form depends on the representation under which fields are transforming.

For example, for the matter fields that transform as in equation (14.1), the co-variant derivative takes the form $\mathbf{D}_\mu = \mathbf{1}\,\partial_\mu + \mathbf{A}_\mu$ and transform like

$$\mathbf{D}_\mu \mapsto \mathbf{D}_\mu^{\mathbf{g}} = \mathbf{1}\,\partial_\mu + \mathbf{A}_\mu^{\mathbf{g}} = \mathbf{g}(x)\mathbf{D}_\mu \mathbf{g}^{-1}(x), \tag{14.3}$$

where the product has to be understood as a product of differential and multiplicative operators. Covariant derivatives of fields then transforms as

$$\mathbf{D}_\mu^{\mathbf{g}}\,\mathbf{g}(x)\phi(x) = \mathbf{g}(x)\mathbf{D}_\mu\phi(x).$$

As a consequence of the property (14.3), the commutator

$$\mathbf{F}_{\mu\nu}(x) = [\mathbf{D}_\mu, \mathbf{D}_\nu] = \partial_\mu\mathbf{A}_\nu(x) - \partial_\nu\mathbf{A}_\mu(x) + [\mathbf{A}_\mu(x), \mathbf{A}_\nu(x)]\,,$$

which is no longer a differential operator, is also a tensor (the curvature of the connection) for gauge transformations:

$$\mathbf{F}_{\mu\nu}(x) \mapsto \mathbf{F}_{\mu\nu}^{\mathbf{g}}(x) = \mathbf{g}(x)\mathbf{F}_{\mu\nu}(x)\mathbf{g}^{-1}(x). \tag{14.4}$$

It then follows from equation (14.4) that the local action for the gauge field,

$$\mathcal{S}_{\text{class.}}(\mathbf{A}) = \frac{1}{4e^2}\int \mathrm{d}^4x\,\mathrm{tr}\sum_{\mu,\nu}\mathbf{F}_{\mu\nu}(x)\mathbf{F}^{\mu\nu}(x)\,, \tag{14.5}$$

is gauge invariant. (e in equation (14.5) characterizes the strength of the interaction in the presence of matter fields.) More generally, it should be emphasized that only gauge invariant polynomials in the fields are related to local physical observables.

When, in the transformation (14.2), $\mathbf{g}(x)$ is close to the identity, that is, when $\mathbf{g}(x) = \mathbf{1} + \boldsymbol{\omega}(x) + O(\|\boldsymbol{\omega}\|)^2$, $\boldsymbol{\omega}(x)$ being a 'small' smooth map valued in the Lie algebra of G, the gauge transformation takes the form

$$\mathbf{A}_\mu^{\mathbf{g}}(x) - \mathbf{A}_\mu(x) = -\mathbf{D}_\mu\boldsymbol{\omega}(x) + O(\|\boldsymbol{\omega}\|)^2,$$
$$\text{with}\quad \mathbf{D}_\mu\boldsymbol{\omega}(x) \equiv \partial_\mu\boldsymbol{\omega}(x) + [\mathbf{A}_\mu(x), \boldsymbol{\omega}(x)]. \tag{14.6}$$

Equation (14.6) gives the form of the *covariant derivative* $\mathbf{D}_\mu$ when it is applied to fields valued in the Lie algebra of G, like $\boldsymbol{\omega}(x)$.

14.2 Non-Abelian gauge theories: The quantized action

In gauge theories, gauge invariance implies that not all components of the gauge field are dynamical and the usual canonical quantization procedure is not applicable. Moreover, in non-Abelian gauge theories, the methods required for the construction of a quantized theory are more involved than in QED.

The construction of *local, relativistic-covariant* quantum non-Abelian gauge theories relies on functional integration techniques. It necessitates a gauge fixing and generates a non-trivial measure in the gauge field space. The measure has the form of the so-called Faddeev–Popov determinant, which can be represented as an integral over non-physical fermion 'ghost fields' [40]. BRST symmetry [43] emerges from this formalism.

14.2.1   Quantized gauge action: The field integral viewpoint

In what follows, as a minor simplification, we discuss only gauge theories without matter, the modifications due to the addition of matter fields being simple (even though, in the case of fermions and chiral gauge invariance, it may lead to obstruction to quantization in the form of *chiral anomalies*).

Following Feynman [4], one would naively expect the quantum evolution operator to be given by an integral over classical gauge fields of the form

$$\mathcal{U} = \int [\mathrm{d}\mathbf{A}_\mu] \exp\left(\frac{i}{\hbar}\mathcal{S}_{\mathrm{class.}}(\mathbf{A})\right),$$

where $\mathcal{S}_{\mathrm{class.}}(\mathbf{A})$ is the classical action (14.5). However, as a consequence of gauge invariance, the integrand is constant along gauge orbits (the trajectories obtained by starting from one gauge field and acting on it with all gauge transformations) and thus the integral is not defined.

The idea is then to introduce a surface (section) that cuts all gauge orbits once and to restrict the integral to this section with a measure that ensures that all choices of sections are equivalent. This surface is defined by a (Lie-algebra-valued) constraint of the form $\mathbf{G}(\mathbf{A}, x) = 0$, known as the gauge-fixing condition, or simply *gauge fixing*. The appropriate integration measure has been identified by Faddeev and Popov and involves the determinant of a linear operator $\mathbf{M}(\mathbf{A})$ defined by

$$[\mathbf{M}(\mathbf{A})\boldsymbol{\omega}](x) = \Delta_\omega \mathbf{G}(\mathbf{A}, x),$$

where $\Delta_\omega \mathbf{G}$ is the variation of $\mathbf{G}$ at first order in $\boldsymbol{\omega}$ corresponding to the variation of $\mathbf{A}_\mu$ in (14.6):

$$\Delta_\omega \mathbf{G}(\mathbf{A}, x) = \mathbf{G}(\mathbf{A} - \mathbf{D}\boldsymbol{\omega}, x) - \mathbf{G}(\mathbf{A}, x) + O(\|\boldsymbol{\omega}\|)^2.$$

In relativistic-covariant gauges $\mathbf{M}(\mathbf{A})$ typically is a differential operator. For example, in Landau's gauge,

$$\mathbf{G}(\mathbf{A}, x) \equiv \sum_\mu \partial_\mu \mathbf{A}^\mu(x) = 0 \;\Rightarrow\; [\mathbf{M}(\mathbf{A})\bar{\mathbf{C}}](x) = -\sum_\mu \partial_\mu \mathbf{D}^\mu \bar{\mathbf{C}}(x). \qquad (14.7)$$

It follows from the rules of integration in Grassmann algebras that determinants can be represented by integrals over generators of Grassmann algebras, which are anticommuting variables (see Section 14.6, and equation (14.32) in particular).

In gauge theories, this corresponds to introducing two *spinless* unphysical fermion fields $\mathbf{C}(x)$ and $\bar{\mathbf{C}}(x)$, often called Faddeev–Popov ghost fields, which are matrices belonging to the Lie algebra of G. Such spinless fermions are *non-physical* because they violate the spin–statistics relation. In addition, one introduces a scalar field $\boldsymbol{\lambda}(x)$, again belonging to the Lie algebra, in order to enforce the gauge condition $\mathbf{G}(\mathbf{A}, x) = 0$ in the field integral by a functional δ-function in a Fourier representation.

The evolution of the quantized gauge theory is then described in terms of an integral over four types of fields, $\mathbf{A}_\mu(x), \boldsymbol{\lambda}(x), \mathbf{C}(x), \bar{\mathbf{C}}(x)$, of the form

$$\mathcal{U} = \int [\mathrm{d}\mathbf{A}_\mu][\mathrm{d}\boldsymbol{\lambda}][\mathrm{d}\mathbf{C}\,\mathrm{d}\bar{\mathbf{C}}] \exp\left(\frac{i}{\hbar}\mathcal{S}_{\text{quant.}}(\mathbf{A}, \bar{\mathbf{C}}, \mathbf{C}, \boldsymbol{\lambda})\right), \qquad (14.8)$$

where $\mathcal{S}_{\text{quant.}}$ is a *local* action that reads

$$\mathcal{S}_{\text{quant.}}(\mathbf{A}, \bar{\mathbf{C}}, \mathbf{C}, \boldsymbol{\lambda}) = \mathcal{S}_{\text{class.}}(\mathbf{A}) + \hbar\mathcal{S}_{\text{gauge}}(\mathbf{A}, \bar{\mathbf{C}}, \mathbf{C}, \boldsymbol{\lambda}) \qquad (14.9)$$

with

$$\mathcal{S}_{\text{gauge}}(\mathbf{A}, \bar{\mathbf{C}}, \mathbf{C}, \boldsymbol{\lambda}) = \int \mathrm{d}^4 x \operatorname{tr}\left[\boldsymbol{\lambda}(x)\mathbf{G}(\mathbf{A}, x) - \mathbf{C}(x)\mathbf{M}(\mathbf{A})\bar{\mathbf{C}}(x)\right]. \qquad (14.10)$$

Finally, instead of using a strict constraint of the form $\mathbf{G}(\mathbf{A}, x) = 0$, it is often convenient to impose $\mathbf{G}(\mathbf{A}, x) = \mathbf{s}(x)$ and to integrate over the field $\mathbf{s}(x)$ with a Gaussian distribution of width ξ. This procedure corresponds to extending the field integral to a whole neighbourhood of the gauge section and, after the integration over $\mathbf{s}(x)$, it amounts to adding a term quadratic in $\boldsymbol{\lambda}$ to the gauge action, which becomes

$$\mathcal{S}_{\text{gauge}}(\mathbf{A}, \bar{\mathbf{C}}, \mathbf{C}, \boldsymbol{\lambda}) = \int \mathrm{d}^4 x \operatorname{tr}\left[\boldsymbol{\lambda}(x)\mathbf{G}(\mathbf{A}, x) - \mathbf{C}(x)\mathbf{M}(\mathbf{A})\bar{\mathbf{C}}(x) + \frac{\xi}{2}\boldsymbol{\lambda}^2(x)\right]. \qquad (14.11)$$

In the limit $\xi \to 0$, Landau's type gauge fixing and the correspondent gauge action (14.10) are recovered.

Perturbative expansion. From the point of view of perturbative calculations, it is convenient to perform the Gaussian integration over the field $\boldsymbol{\lambda}$. However, this renders the BRST transformations more complicated.

14.3 BRST symmetry of the quantized action

As a consequence of the quantization procedure, the gauge symmetry of the classical action is destroyed and the structure of the quantized theory no longer has any obvious geometric property.

However, remarkably enough, the quantized action (equations (14.9) and (14.11)) is invariant, *independently of the choice of the gauge condition* $\mathbf{G}(\mathbf{A}, x)$, under a fermion-like symmetry called BRST symmetry.

14.3.1 BRST symmetry

The quantized action (equations (14.9) and (14.11)) is invariant under the *BRST transformations* (simplified by the introduction of the $\boldsymbol{\lambda}$ field [156])

$$\begin{cases} \mathbf{A}_\mu(x) \mapsto \mathbf{A}_\mu(x) + \varepsilon\Delta\mathbf{A}_\mu(x), & \bar{\mathbf{C}}(x) \mapsto \bar{\mathbf{C}}(x) + \varepsilon\Delta\bar{\mathbf{C}}(x), \\ \mathbf{C}(x) \mapsto \mathbf{C}(x) + \varepsilon\Delta\mathbf{C}(x), & \boldsymbol{\lambda}(x) \mapsto \boldsymbol{\lambda}(x) + \varepsilon\Delta\boldsymbol{\lambda}(x), \end{cases} \qquad (14.12)$$

where ε is a Grassmann constant (it anticommutes with $\bar{\mathbf{C}}(x)$ and $\bar{\mathbf{C}}(x)$, and $\varepsilon^2 = 0$) and

$$\Delta\mathbf{A}_\mu(x) = -\mathbf{D}_\mu\bar{\mathbf{C}}(x), \quad \Delta\bar{\mathbf{C}}(x) = \bar{\mathbf{C}}^2(x), \quad \Delta\mathbf{C}(x) = \boldsymbol{\lambda}(x), \quad \Delta\boldsymbol{\lambda}(x) = 0.$$

In these transformations, bosons are transformed into fermions, and vice-versa. A short calculation then shows that $\Delta\mathbf{A}_\mu(x)$ and $\Delta\bar{\mathbf{C}}(x)$ are BRST invariant.

Another way of expressing the symmetry is based on the introduction of a functional differential operator. We expand all fields on the basis of (anti-Hermitian) matrices $\mathbf{t}^a$ generating the Lie algebra of G,

$$
\begin{cases}
\mathbf{A}_\mu(x) = \sum_a A_\mu^a(x)\mathbf{t}^a, & \bar{\mathbf{C}}(x) = \sum_a \bar{C}^a(x)\mathbf{t}^a, \\[2mm]
\mathbf{C}(x) = \sum_a C^a(x)\mathbf{t}^a, & \boldsymbol{\lambda}(x) = \sum_a \lambda^a(x)\mathbf{t}^a,
\end{cases}
\tag{14.13}
$$

as well as

$$
\Delta\mathbf{A}_\mu(x) = \sum_a \Delta A_\mu^a(x)\mathbf{t}^a, \quad \Delta\bar{\mathbf{C}}(x) = \sum_a \Delta\bar{C}^a(x)\mathbf{t}^a.
$$

We define the functional differential operator (δ denotes *functional differentiation*)

$$
\mathcal{D} = \int \mathrm{d}^4x \sum_a \left[\sum_\mu \Delta A_\mu^a(x)\frac{\delta}{\delta A_\mu^a(x)} + \Delta\bar{C}^a(x)\frac{\delta}{\delta\bar{C}^a(x)} + \lambda^a(x)\frac{\delta}{\delta C^a(x)} \right]. \tag{14.14}
$$

In terms of the differential operator (14.14), the BRST invariance of the quantized action can be expressed in the form

$$
\mathcal{D}\mathcal{S}_{\text{quant.}} = 0. \tag{14.15}
$$

A short calculation, which uses the property that the BRST transforms of $\mathbf{A}, \bar{\mathbf{C}}, \boldsymbol{\lambda}$ are BRST invariant, yields

$$
\mathcal{D}^2 = 0. \tag{14.16}
$$

This very important property of the differential operator $\mathcal{D}$ shows that it can be identified with a cohomology operator. In cohomological terminology, equation (14.15) states that the quantized action $\mathcal{S}_{\text{quant.}}$ is *BRST closed*.

14.3.2 Gauge independence

Quantities of the form $\mathcal{D}\Phi$ are said to be *BRST exact*. The property (14.16) of $\mathcal{D}$ implies that BRST exact quantities are also BRST closed. A short calculation shows that the additional contribution to the quantum action generated by the quantization procedure is BRST exact. Indeed,

$$
\mathcal{S}_{\text{gauge}} = \mathcal{D}\Phi(\mathbf{A}, \mathbf{C}, \boldsymbol{\lambda}), \quad \text{with} \quad \Phi = \int \mathrm{d}^4x\, \mathrm{tr}\, \mathbf{C}(x)\left(\mathbf{G}(\mathbf{A}, x) + \tfrac{\xi}{2}\boldsymbol{\lambda}(x)\right). \tag{14.17}
$$

In the field integral implementation of BRST symmetry, one must verify the invariance of the integration measure in equation (14.8) with respect to change of gauges.

We thus consider an infinitesimal change,

$$
\mathbf{G}(\mathbf{A}, x) \mapsto \mathbf{G}(\mathbf{A}, x) + \delta\mathbf{G}(\mathbf{A}, x) \;\Rightarrow\; \Phi \mapsto \Phi + \delta\Phi,
$$

to first order in $\delta\boldsymbol{G}$.

The variation of the partition function (or vacuum amplitude) is then

$$\delta\mathcal{U} = \int [\mathrm{d}\mathbf{A}_\mu][\mathrm{d}\boldsymbol{\lambda}][\mathrm{d}\mathbf{C}\,\mathrm{d}\bar{\mathbf{C}}]\mathcal{D}\Phi(\mathbf{A},\mathbf{C},\boldsymbol{\lambda})\exp\left(\frac{i}{\hbar}\mathcal{S}_{\text{quant.}}(\mathbf{A},\bar{\mathbf{C}},\mathbf{C},\boldsymbol{\lambda})\right).$$

However, integration by parts implies acting with $\mathcal{D}$ on the left and, thus, commuting derivatives with coefficients in $\mathcal{D}$. This commutation yields again the conditions

$$\sum_{a,\mu}\left[\frac{\delta}{\delta A_\mu^a(x)},\Delta A_\mu^a(x)\right] = \sum_{a,\mu}\frac{\delta\Delta A_\mu^a(x)}{\delta A_\mu^a(x)} = 0\,, \tag{14.18a}$$

$$\sum_{a}\left[\frac{\delta}{\delta \bar{C}^a(x)},\Delta \bar{C}^a(x)\right] = \sum_{a}\frac{\delta\Delta\bar{C}^a(x)}{\delta\bar{C}^a(x)} = 0\,, \tag{14.18b}$$

that is, that the traces of the generators of the Lie algebra of the group G in the adjoint representation (corresponding to the Lie algebra structure constants) must vanish, a property satisfied by compact Lie groups. However, note that in the presence of matter fields, this condition may apply to generators in other representations and is then satisfied only for *semi-simple Lie algebras*.

After an integration by parts, $\mathcal{D}$ acts on $e^{i\mathcal{S}_{\text{quant.}}/\hbar}$, which is BRST invariant and, thus,

$$\delta\mathcal{U} = 0\,.$$

Note that this result extends to the expectation value of all gauge invariant quantities [160, 171].

Gauge independence, in particular, is essential in the proof that physical observables satisfy the *unitarity* requirement (necessary for probability conservation), a property that is not obvious for a quantized gauge theory.

14.3.3 BRST invariant solutions

Let us briefly outline a method that facilitates the determination of general BRST invariant polynomials in the fields. Decomposing $\mathcal{D}$ into

$$\mathcal{D} = \mathcal{D}_+ + \mathcal{D}_-\,,$$

with

$$\mathcal{D}_+ = \int \mathrm{d}^4x \sum_a \left[\Delta A_\mu^a(x)\frac{\delta}{\delta A_\mu^a(x)} + \Delta\bar{C}^a(x)\frac{\delta}{\delta\bar{C}^a(x)}\right],$$

$$\mathcal{D}_- = \int \mathrm{d}^4x \sum_a \lambda^a(x)\frac{\delta}{\delta C^a(x)}\,,$$

one finds

$$\mathcal{D}_+^2 = \mathcal{D}_-^2 = 0\,,\quad \mathcal{D}_+\mathcal{D}_- + \mathcal{D}_-\mathcal{D}_+ = 0\,.$$

One then uses the property that Grassmann algebras are graded algebras and that $\mathcal{D}_+$ increases the degree in $\bar{\mathbf{C}}$ while $\mathcal{D}_-$ leaves it unchanged, to expand an equation of the form $\mathcal{D}\mathbf{F} = 0$ in powers of $\bar{\mathbf{C}}$.

14.4 The ZJ equation and remormalization

The initial field integral (14.8) of the quantized gauge theory is not defined, even in the sense of a perturbative expansion: the perturbative expansion is generated by keeping in the exponential the part of the classical action that is quadratic in the fields (free action) and expanding in a formal power series the exponential of the remaining part. This expansion is usefully described in terms of Feynman diagrams, each one representing a Feynman integral contribution to the perturbative series.

In perturbation theory, the field integral (14.8) is not defined because, at space-time dimension $d = 4$, divergent Feynman integrals (one also speaks of ultraviolet (UV) divergences) arise at all orders, corresponding to short distance or large momentum singularities. These divergences reflect the property that the Gaussian or free measure does not restrict the contributions to the field integral to fields regular enough, here differentiable at least once.

14.4.1 Regularization

A first necessary step is called *regularization*, in which one modifies in some *non-physical way* the classical action to render the perturbative expansion finite. The construction of quantum gauge theories and practical calculations is much simplified if the regularization preserves the BRST symmetry, even if, strictly speaking, this requirement is not mandatory.

Different methods are available: one can modify the theory at short distance, in the continuum, at a scale specified by a momentum cut-off or by introducing a space-time lattice (*lattice gauge theories*). In these schemes, one replaces an unknown short distance structure by an artificial, and presumably simpler, short distance structure. Renormalization and renormalization group are then used to prove some relative *short distance insensitivity* (a form of universality).

In theories without chiral fermion fields, a BRST invariant regularization can also be achieved by using *dimensional regularization*, a regularization method based on a formal analytic continuation of Feynman diagrams to arbitrary complex values of the space-time dimension d. Poles then appear at $d = 4$ in the dimensionally regularized Feynman integrals: these singularities are related to the initial short distance divergences.

This regularization leads to the simplest perturbative calculations: however, it lacks physical interpretation. Also, *it performs a partial renormalization by cancelling power-law divergences*, which may lead to wrong physics conclusions.

14.4.2 Renormalization with counter-terms

Renormalization can be performed by adding, order by order in a loop structured perturbative expansion (formally, an expansion in powers of $\hbar$), local terms (space-time integrated monomials in the fields and their derivatives) called *counter-terms* to the initial action, which cancel the UV divergences.

These local terms have divergent coefficients: for example, in dimensional regularization, they are multiplied by constants that diverge when $d \to 4$.

When the number of different field monomials, required to render finite the perturbative expansion, is bounded at all orders by some fixed number, one calls the quantum field theory *(perturbatively) renormalizable by power counting*. Gauge theories with properly chosen gauge sections, like, for example, Landau's gauge in equation (14.7), satisfy this criterion.

However, in gauge theories, one still has to prove that renormalization can be achieved without destroying the geometric structure that ensures that physics results do not depend on the choice of the gauge section.

Initial proofs of the gauge independence of the renormalized gauge theory were based on Slavnov–Taylor identities [154], suitably extended to the situation of spontaneous symmetry breaking by Lee–Zinn-Justin [76].

A more general and more transparent proof was then given by Zinn-Justin using BRST symmetry and the ZJ equation. Note that all these proofs apply to compact Lie groups with semi-simple Lie algebras; the simpler Abelian case has to be dealt with by different methods.

If renormalization would preserve BRST symmetry in the explicit form (14.12), gauge independence could be proved easily. However, this is not the case because counter-terms have the effect of rescaling fields. Consider, for example, $\mathbf{A}_\mu \mapsto Z_A^{1/2}\mathbf{A}_\mu$, where Z_A is a divergent constant: because the gauge transformation of $\mathbf{A}_\mu$ is affine, the form of the gauge transformation is modified for the rescaled fields. Other fields are similarly renormalized.

Moreover, if the gauge-fixing function $\mathbf{G}(\mathbf{A}, x)$ is not linear in the fields, the gauge-fixing equation is also modified and counter-terms *quartic in the ghost fields* are generated.

14.4.3 ZJ equation

In non-Abelian gauge theories, two BRST variations, $\Delta\mathbf{A}_\mu$ and $\Delta\bar{\mathbf{C}}$, are not linear in the dynamical fields. Local polynomials in the fields of degree larger than one are called composite operators. They generate new divergences and require new types of counter-terms. The renormalization of composite operators can be best discussed by introducing source fields that generate, by functional differentiation, their multiple insertions in correlation (or Green's) functions. We denote by $\mathbf{K}^\mu$ and $\mathbf{L}$ the sources for the $\Delta\mathbf{A}_\mu$ and $\Delta\bar{\mathbf{C}}$, respectively. Then, we consider

$$\mathcal{S} = \mathcal{S}_{\text{quant.}} - \hbar\int \mathrm{d}^4x\,\mathrm{tr}\left(\sum_\mu \mathbf{K}^\mu(x)\Delta\mathbf{A}_\mu(x) + \mathbf{L}(x)\Delta\bar{\mathbf{C}}(x)\right),$$

such that

$$\frac{\delta\mathcal{S}}{\delta\mathbf{K}^\mu(x)} = -\hbar\Delta\mathbf{A}_\mu(x), \qquad \frac{\delta\mathcal{S}}{\delta\mathbf{L}(x)} = -\hbar\Delta\bar{\mathbf{C}}(x). \tag{14.19}$$

The source $\mathbf{K}^\mu(x)$ is a Grassmann (anticommuting) field, while $\mathbf{L}(x)$ is a complex field, and both are matrices that belong to the Lie algebra.

The properties $\Delta\mathbf{A}_\mu = \mathcal{D}\mathbf{A}$, $\Delta\bar{\mathbf{C}} = \mathcal{D}\bar{\mathbf{C}}$, $\mathcal{D}^2 = 0$ and the invariance of the sources under BRST transformation imply that still

$$\mathcal{D}\mathcal{S} = 0\,. \tag{14.20}$$

We can expand $\mathbf{K}^\mu(x)$ and $\mathbf{L}(x)$ on the basis of generators of the Lie algebra, as in expression (14.13). The ZJ equation then follows directly from the BRST symmetry of $\mathcal{S}$ and from relations (14.19), and can be written in the form

$$\int \mathrm{d}^4 x \sum_a \left(\sum_\mu \frac{\delta \mathcal{S}}{\delta K^{\mu a}(x)} \frac{\delta \mathcal{S}}{\delta A_\mu^a(x)} + \frac{\delta \mathcal{S}}{\delta L^a(x)} \frac{\delta \mathcal{S}}{\delta \bar{C}^a(x)} - \lambda^a(x) \frac{\delta \mathcal{S}}{\delta C^a(x)} \right)$$
$$= 0 . \tag{14.21}$$

To discuss equation (14.21) in simpler terms, it is sometimes convenient to add to $\mathcal{S}$ the source for the $\boldsymbol{\lambda}$-field: $\mathcal{S} \mapsto \mathcal{S} - \hbar \int \mathrm{d}^4 x \, \mathrm{tr} \boldsymbol{\lambda}(x) \boldsymbol{\mu}(x)$. Equation (14.21) then takes the homogeneous quadratic form

$$\int \mathrm{d}^4 x \sum_a \left(\sum_\mu \frac{\delta \mathcal{S}}{\delta K^{\mu a}(x)} \frac{\delta \mathcal{S}}{\delta A_\mu^a(x)} + \frac{\delta \mathcal{S}}{\delta L^a(x)} \frac{\delta \mathcal{S}}{\delta \bar{C}^a(x)} + \frac{\delta \mathcal{S}}{\delta \mu^a(x)} \frac{\delta \mathcal{S}}{\delta C^a(x)} \right)$$
$$= 0 . \tag{14.22}$$

In contrast to equation (14.15), where BRST transformations (14.12) are explicit, equations (14.21) and (14.22) can be proved to be *stable under renormalization*, that is, the *renormalized action* $\mathcal{S}_{\mathrm{ren.}} = \mathcal{S} + \mathcal{S}_{\mathrm{CT}}$, the sum of the initial quantized action and properly chosen counter-terms, still satisfies equation (14.21) (counter-terms are determined up to an addition of finite contributions).

The result relies on the property that, when the action satisfies equation (14.21), then the generating functional of vertex (or one-line irreducible correlation) functions satisfies the same equation. This implies constraints on the divergences in the perturbative expansions and, therefore, on the required counter-terms.

The explicit solution of the ZJ equation, satisfying *locality* (the action is a space-time integral over functions of fields and derivatives), *power counting*, which is a form of dimensional analysis, and *ghost number conservation* (if one assigns a ghost charge of $+1$ to C and -1 to $\bar{C}$, so that the action has total charge 0), yields the general form of the renormalized action. In particular, the renormalized action has still a BRST symmetry but with renormalized fields and parameters.

In the example of gauge-fixing-functions $\mathbf{G}$ linear in the fields, the $\boldsymbol{\lambda}$-quantum equation of motion yields additional relations between counter-terms. In the simple example of the gauge (14.7) and in the absence of matter fields, the renormalized action $\mathcal{S}_{\mathrm{ren.}}$ is then obtained from $\mathcal{S}$ by the simple substitutions

$$e \mapsto Z_e^{1/2} e , \quad \mathbf{A}_\mu \mapsto Z_A^{1/2} \mathbf{A}_\mu , \quad \mathbf{C}\bar{\mathbf{C}} \mapsto Z_C \mathbf{C}\bar{\mathbf{C}} , \quad \xi \mapsto Z_A \xi .$$

(Since only the product $\mathbf{C}\bar{\mathbf{C}}$ always appears, only the renormalization of $\mathbf{C}\bar{\mathbf{C}}$ is defined.)

By contrast, when the gauge-fixing function $\mathbf{G}$ contains quadratic terms in the fields, renormalization generates terms quartic in the ghost fields and, thus, the integration over $\mathbf{C}$ and $\bar{\mathbf{C}}$ no longer yields a simple determinant. Nevertheless, the ZJ equation still implies *gauge independence* (*i.e.*, independence of the gauge section) of physical quantities and, thus, the field theory has the same physical properties. The ZJ equation can also be used to discuss the renormalization properties of gauge invariant operators, which are related to physical observables [160].

Finally, note that a (simpler) quadratic equation, somewhat analogous to equations (14.21) or (14.22), appears in the renormalization of the non-linear σ-model, a quantum field theory renormalizable in $(1+1)$ space-time dimensions with an $O(N)$ orthogonal symmetry, in which the field belongs to a sphere S_{N-1} (see Section 10.6).

14.5 The ZJ equation: A few general properties

The main formal difference between the ZJ equation (14.22) (or equation (14.21)) and equation (14.20) is its quadratic structure, because the transformations of fields depend on the action itself. Thus, several properties of the equation can be understood by studying the more general equation

$$\int \mathrm{d}^4 x \operatorname{tr} \left(\frac{\delta S}{\delta \mathbf{Q}(x)} \frac{\delta S}{\delta \mathbf{\Pi}(x)} \right) = 0 \,, \qquad (14.23)$$

where, with respect to the Section 14.4.3, the field $\mathbf{Q}(x)$ plays the role of the set of commuting fields $\{\mathbf{A}_\mu, \mathbf{L}, \boldsymbol{\mu}\}$, and the field $\mathbf{\Pi}(x)$ the role of the set of anticommuting fields $\{\mathbf{K}_\mu, \bar{\mathbf{C}}, \mathbf{C}\}$.

For notational simplicity, we replace equation (14.23) by a formally identical but simpler equation; the generalization to field theory is then straightforward. We assume that S is a smooth function of N real variables q_i and N generators π_i of a Grassmann algebra: $\pi_i \pi_j + \pi_j \pi_i = 0$, which belongs to the commuting subalgebra and satisfies the equation

$$\sum_i \frac{\partial S}{\partial q_i} \frac{\partial S}{\partial \pi_i} = 0 \,. \qquad (14.24)$$

The index i plays the role of space-time coordinate together with Lorentz and group indices in equation (14.23). Summation over i replaces integration and summation.

We also consider the differential operator

$$\mathcal{D} = \sum_i \left(\frac{\partial S}{\partial \pi_i} \frac{\partial}{\partial q_i} + \frac{\partial S}{\partial q_i} \frac{\partial}{\partial \pi_i} \right) \,. \qquad (14.25)$$

It is simple to verify that equation (14.24) is the necessary and sufficient condition for $\mathcal{D}$ to be a BRST cohomology operator, that is, for $\mathcal{D}$ to satisfy $\mathcal{D}^2 = 0$.

14.5.1 Special solutions

It is possible to characterize all solutions of equation (14.24) of the special form

$$S = \Sigma^{(0)}(q) + \sum_{i,j} \Sigma_{ij}^{(2)}(q) \pi_i \pi_j \,,$$

where $\Sigma^{(0)}$ and $\Sigma_{ij}^{(2)} = -\Sigma_{ji}^{(2)}$ are analytic functions of the q_i's. Equation (14.24) is equivalent to a system of two equations, corresponding to the vanishing of the terms of degrees 1 and 3 in the generators π_i in equation (14.24).

The first equation, coming from the linear term, can be more easily expressed by introducing the differential operator

$$\mathrm{d}_i := \sum_k \Sigma_{ki}^{(2)}(q) \frac{\partial}{\partial q_k}.$$

It then takes the form

$$\mathrm{d}_i \Sigma^{(0)}(q) = 0 \quad \forall\, i. \tag{14.26}$$

The second equation takes the form

$$\sum_l \left\{ \frac{\partial \Sigma_{jk}^{(2)}}{\partial q_l} \Sigma_{li}^{(2)} \right\}_{ijk} = 0 \quad \forall\, i, j, k, \tag{14.27}$$

where the notation means antisymmetrized over i, j, k. From the latter equation, one derives the commutation relation

$$[\mathrm{d}_i, \mathrm{d}_j] = \sum_k \frac{\partial \Sigma_{ij}^{(2)}}{\partial q_k} \mathrm{d}_k. \tag{14.28}$$

Equation (14.28) is the *compatibility condition* for the linear differential system (14.26). It also implies that the operators d_i are the generators of a Lie algebra in some non-linear representation. Finally, if $\Sigma_{ij}^{(2)}$ is a first degree polynomial, $\partial \Sigma_{ij}^{(2)} / \partial q_k$ are the structure constants of the Lie algebra, and equation (14.27) contains the corresponding Jacobi identity.

14.5.2 Perturbative solutions

We assume that we have found a solution $\mathcal{S}^{(0)}$ of equation (14.24), to which is associated a BRST operator $\mathcal{D}_0$ as in equation (14.25), and we look for solutions that can be expanded in terms of a real parameter κ in the form

$$\mathcal{S}(\kappa) = \sum_{n \geq 0} \kappa^n \mathcal{S}^{(n)}.$$

Expanding equation (14.24) at order κ, one obtains the condition $\mathcal{D}_0 \mathcal{S}_1 = 0$. Thus, one has to find $\mathcal{D}_0$ closed solutions. More generally, at order κ^n the equation can be written as

$$\mathcal{D}_0 \mathcal{S}^{(n)} = - \sum_{1 \leq m \leq n-1} \sum_i \left(\frac{\partial \mathcal{S}^{(m)}}{\partial q_i} \frac{\partial \mathcal{S}^{(n-m)}}{\partial \pi_i} \right).$$

This reduces the problem of the recursive determination of the coefficients $\mathcal{S}^{(n)}$ to an investigation of the properties of the $\mathcal{D}_0$ operator.

14.5.3 Canonical invariance

Equation (14.24) has properties reminiscent of those of the symplectic form $\mathrm{d}p \wedge \mathrm{d}q$ of classical mechanics; in particular, it is invariant under some generalized canonical transformations. Indeed, after the change of variables $(\pi, q) \mapsto (\pi', q')$,

$$q_i = \frac{\partial \varphi}{\partial \pi_i}(\pi, q'), \quad \pi_i' = \frac{\partial \varphi}{\partial q_i'}(\pi, q'), \tag{14.29}$$

in which $\varphi(\pi, q')$ is a function belonging to the anticommuting part of the Grassmann algebra, one recovers equation (14.24) in the new variables:

$$\sum_i \frac{\partial S}{\partial \pi_i'} \frac{\partial S}{\partial q_i'} = 0 .$$

The proof goes in two steps, which both involve the anticommutation of π_i and π_j or of the corresponding derivatives. One first goes from q_i to q_i' at π_i fixed. One finds

$$\left.\frac{\partial S}{\partial q_i'}\right|_\pi = \sum_j \frac{\partial q_j}{\partial q_i'} \left.\frac{\partial S}{\partial q_j}\right|_\pi = \sum_j \frac{\partial^2 \varphi}{\partial q_i' \partial \pi_j} \left.\frac{\partial S}{\partial q_j}\right|_\pi , \quad \left.\frac{\partial S}{\partial \pi_i}\right|_{q'} = \left.\frac{\partial S}{\partial \pi_i}\right|_q .$$

Then one changes from π_i to π_i':

$$\left.\frac{\partial S}{\partial \pi_i}\right|_{q'} = \sum_j \frac{\partial^2 \varphi}{\partial \pi_i \partial q_j'} \left.\frac{\partial S}{\partial \pi_j'}\right|_{q'} , \quad \left.\frac{\partial S}{\partial q_i'}\right|_\pi = \left.\frac{\partial S}{\partial q_i'}\right|_{\pi'} + \sum_j \frac{\partial^2 \varphi}{\partial \pi_i \partial q_j'} \left.\frac{\partial S}{\partial \pi_j'}\right|_{q'} .$$

By collecting all terms, it is possible to the property.

14.5.4 Infinitesimal canonical transformations

We now consider infinitesimal canonical transformations of type (14.29). The function $\varphi(\pi, q') = \sum_i \pi_i q_i'$ corresponds to the identity. We then write the function φ in terms of a real parameter κ as

$$\varphi(\pi, q') = \sum_i \pi_i q_i' + \kappa \psi(\pi, q') .$$

The variation of S at first order in κ is

$$\mathcal{S}(\pi', q') - \mathcal{S}(\pi, q) = \kappa \sum_i \left(\frac{\partial \psi}{\partial q_i} \frac{\partial S}{\partial \pi_i} - \frac{\partial \psi}{\partial \pi_i} \frac{\partial S}{\partial q_i} \right) + O\left(\kappa^2\right) = -\kappa \, \mathcal{D}\psi + O\left(\kappa^2\right) .$$

One thus finds that an infinitesimal canonical transformation generates a BRST exact contribution and, conversely, any infinitesimal addition to $\mathcal{S}$ of a BRST exact term can be generated by a canonical transformation acting on $\mathcal{S}$. One then verifies that, indeed, the additional contributions to the action due to the gauge-fixing procedure can also be generated by such a canonical transformation with

$$\kappa \, \psi(\pi, q') \mapsto \int \mathrm{d}^4x \, \mathrm{tr} \, \mathbf{C}(x) \left(\mathbf{G}(\mathbf{A}, x) + \tfrac{\xi}{2}\lambda(x) \right) ,$$

acting on

$$\mathcal{S}_{\mathrm{class.}}(\mathbf{A}_\mu) - \hbar \int \mathrm{d}^4x \, \mathrm{tr} \left(\sum_\mu \mathbf{K}^\mu(x)\Delta\mathbf{A}_\mu(x) + \mathbf{L}(x)\Delta\bar{\mathbf{C}}(x) + \boldsymbol{\mu}(x)\boldsymbol{\lambda}(x) \right) .$$

14.6 BRST symmetry: The algebraic origin

One might be surprised that quantized gauge theories have this peculiar BRST symmetry. In fact, BRST symmetry is an automatic property of constraint systems handled in a specific way, as we explain now. In particular, in gauge theories, it is induced by the constraint of the gauge section (14.7) but its form is complicated by the choice of coordinates because the equation of the section applies to group elements $\mathbf{g}(x)$, the gauge transformations.

Let φ^a be a set of real quantities satisfying a system of real equations,

$$E_a(\varphi) = 0 \,, \tag{14.30}$$

where the functions $E_a(\varphi)$ are smooth, and $E_a = E_a(\varphi)$ is a one-to-one map in some neighbourhood of $E_a = 0$, which can be inverted in $\varphi^a = \varphi^a(E)$. In particular, this implies that equation (14.30) has a unique solution $\varphi_s^a \equiv \varphi^a(0)$. In the neighbourhood of φ_s, the determinant $\det \mathbf{E}$ of the matrix $\mathbf{E}$ with elements

$$E_{ab} \equiv \frac{\partial E_a}{\partial \varphi^b} \,,$$

does not vanish and thus we choose $E_a(\varphi)$ such that it is positive.

For any continuous function $F(\varphi)$, we now derive a formal expression for $F(\varphi_s)$ that does not involve solving equation (14.30) explicitly. We start from the simple identity

$$F(\varphi_s) = \int \left\{ \prod_a \mathrm{d}E_a \, \delta(E_a) \right\} F\big(\varphi(E)\big),$$

where $\delta(E)$ is Dirac's δ-function. We then change variables $E \mapsto \varphi$. This generates the Jacobian $\mathcal{J}(\varphi) = \det \mathbf{E} > 0$. Thus,

$$F(\varphi_s) = \int \left\{ \prod_a \mathrm{d}\varphi^a \, \delta\left[E_a(\varphi)\right] \right\} \mathcal{J}(\varphi) \, F(\varphi). \tag{14.31}$$

We replace the δ-function by its Fourier representation:

$$\prod_a \delta\left[E_a(\varphi)\right] = \int \prod_a \frac{\mathrm{d}\bar{\varphi}^a}{2i\pi} \exp\left(-\sum_a \bar{\varphi}^a E_a(\varphi)\right),$$

where the $\bar{\varphi}$ integration runs along the imaginary axis. Moreover, a determinant can be written as an integral over Grassmann variables (*i.e.*, generators of a Grassmann or exterior algebra) $\bar{c}^a$ and c^a in the form

$$\det \mathbf{E} = \int \prod_a (\mathrm{d}\bar{c}^a \mathrm{d}c^a) \exp\left(\sum_{a,b} c^a E_{ab} \bar{c}^b\right). \tag{14.32}$$

Expression (14.31) then becomes

$$F(\varphi_s) = \mathcal{N} \int \prod_a (d\varphi^a d\bar{\varphi}^a d\bar{c}^a dc^a) \, F(\varphi) \exp\left[-S(\varphi, \bar{\varphi}, c, \bar{c})\right],\qquad (14.33)$$

in which $\mathcal{N}$ is a constant normalization factor and

$$S(\varphi, \bar{\varphi}, c, \bar{c}) = \sum_a \bar{\varphi}^a E_a(\varphi) - \sum_{a,b} c^a E_{ab}(\varphi) \bar{c}^b,\qquad (14.34)$$

is a commuting element of the Grassmann algebra.

Somewhat surprisingly, the function S has a symmetry identical to the BRST symmetry but in a different coordinate system, which we now describe.

14.6.1 BRST symmetry

The function S defined by equation (14.34) is invariant under the BRST transformations,

$$\begin{cases} \varphi^a \mapsto \varphi^a + \delta_{\text{BRST}}\varphi^a, \text{ with } \delta_{\text{BRST}}\varphi^a = \varepsilon \bar{c}^a, & \bar{c}^a \mapsto \bar{c}^a, \\ c^a \mapsto c^a + \delta_{\text{BRST}}c^a, \text{ with } \delta_{\text{BRST}}c^a = \varepsilon \bar{\varphi}^a, & \bar{c}^a \mapsto \bar{c}^a, \end{cases}\qquad (14.35)$$

where ε is an anticommuting constant, an additional generator of the Grassmann algebra such that

$$\varepsilon^2 = 0, \quad \varepsilon \bar{c}^a + \bar{c}^a \varepsilon = 0, \quad \varepsilon c^a + \varepsilon c^a = 0.$$

Moreover, the integration measure in equation (14.33) is also invariant.

The apparent simplicity of these transformations compared to the expressions (14.12) is largely due to a different parametrization (see Section 14.6.2).

The BRST transformation is clearly *nilpotent* (with a vanishing square) since the variation of the variation always vanishes. Note that the transformations of c^a and $\bar{\varphi}^a$ are identical to the transformations of the fields $\mathbf{C}(x)$ and $\boldsymbol{\lambda}(x)$ in expressions (14.12).

The BRST transformation can also be represented by a Grassmann differential operator $\mathcal{D}$ acting on functions of $\{\varphi, \bar{\varphi}, c, \bar{c}\}$,

$$\mathcal{D} = \mathcal{D}_+ + \mathcal{D}_-, \quad \mathcal{D}_+ = \sum_a \bar{c}^a \frac{\partial}{\partial \varphi^a}, \quad \mathcal{D}_- = \sum_a \bar{\varphi}^a \frac{\partial}{\partial c^a}.$$

Then,

$$\mathcal{D}S = 0.$$

$\mathcal{D}_+$ and $\mathcal{D}_-$ satisfy the relations

$$\mathcal{D}_+^2 = \mathcal{D}_-^2 = \mathcal{D}_+\mathcal{D}_- + \mathcal{D}_-\mathcal{D}_+ = 0.$$

The nilpotency of the BRST transformation follows:

$$\mathcal{D}^2 = 0.\qquad (14.36)$$

The differential operator $\mathcal{D}$ is a *cohomology: operator*, a generalization of the exterior differentiation of differential forms. In particular, the first term $\mathcal{D}_+$ in the BRST operator is identical to the exterior derivative of differential forms in a formalism in which the Grassmann variables $\bar{c}^a$ generate the corresponding exterior algebra.

All quantities satisfying $Q(\varphi, \bar{\varphi}, c, \bar{c}) = 0$ are called *BRST closed*. All quantities of the form $\mathcal{D}Q(\varphi, \bar{\varphi}, c, \bar{c})$ are called *BRST exact*. Equation (14.36) implies that all quantities BRST exact are BRST closed, since $\mathcal{D}(\mathcal{D}Q(\varphi, \bar{\varphi}, c, \bar{c})) = 0$. The function S (defined in equation (14.34)) is not only BRST closed but also BRST exact:

$$S = \mathcal{D}\left(\sum_a c^a E_a(\varphi)\right).$$

The reciprocal property and the meaning and implications of the BRST symmetry follow from some simple arguments based on BRST cohomology.

14.6.2 BRST symmetry and group elements

We now assume that the variables φ^a parametrize locally elements $\mathbf{g}(\varphi)$ of a group G in some matrix representation. It is then natural to parametrize the BRST variation of $\mathbf{g}$ in terms of an element $\bar{\mathbf{C}}$ of the Lie algebra (which is also a generator of a Grassmann algebra) in the form

$$\delta_{\mathrm{BRST}}\,\mathbf{g} = \varepsilon\,\bar{\mathbf{C}}\,\mathbf{g}\,. \tag{14.37}$$

Calculating directly the variation of $\mathbf{g}$ from the variation (14.35) of φ^a, one obtains the relation

$$\bar{\mathbf{C}}\mathbf{g} = \sum_a \frac{\partial \mathbf{g}}{\partial \varphi^a}\bar{c}^a \;\Rightarrow\; \bar{\mathbf{C}} = \sum_a \frac{\partial \mathbf{g}}{\partial \varphi^a}\mathbf{g}^{-1}\bar{c}^a\,.$$

The BRST variation of the matrix $\bar{\mathbf{C}}$ is

$$\delta_{\mathrm{BRST}}\bar{\mathbf{C}} = \varepsilon \sum_{a,b}\left(\frac{\partial^2 \mathbf{g}}{\partial \varphi^a \partial \varphi^b}\mathbf{g}^{-1} - \frac{\partial \mathbf{g}}{\partial \varphi^a}\mathbf{g}^{-1}\frac{\partial \mathbf{g}}{\partial \varphi^b}\mathbf{g}^{-1}\right)\bar{c}^b\bar{c}^a = \varepsilon\,\bar{\mathbf{C}}^2,$$

where the anticommutation of $\bar{c}^a$ and $\bar{c}^b$ has been used. One recognizes the transformation of the ghost fields in non-Abelian gauge theories (the second equation in expressions (14.12)). Applying then the transformation (14.37) to $\mathbf{g}(x)$ (at $\mathbf{A}_\mu(x)$ fixed) in $\mathbf{A}_\mu^{\mathbf{g}}(x)$ in expression (14.2), one derives the variation of $\mathbf{A}_\mu(x)$ in the first equation of expressions (14.12).

15 Quantum field theory: Asymptotic safety

In quantum field theories (QFTs), *the existence of ultraviolet (UV) fixed points of the renormalization group (RG) seems to be a necessary condition for the existence of a renormalized QFT consistent on all scales* [97]. The condition of existence of UV fixed points has been called condition of *asymptotic safety.*

However, the meaning and importance of UV fixed points have themselves evolved with a deeper understanding of QFT, in particular, as an effective field theory. Moreover, we argue here that the notion of asymptotic safety also applies to the existence of *infrared (IR, i.e., small momentum) fixed points* in massless theories.

The existence of renormalized theories consistent on all scales implies the existence of several scales: the shortest distance at which local field theory starts to become relevant but where the specific properties of the microscopic model are still apparent, a short distance scale (which is much larger than the microscopic scale) where universal properties emerge, a *crossover scale*, and the universal large distance scale.

By contrast, in IR free field theories like the ϕ^4 field theory in four dimensions, the universal short distance scale is absent and the renormalized field theory becomes free in the infinite cut-off limit (the triviality issue).

To illustrate the problem, we review a few classical examples and then try to draw some general conclusions.

Although eventually we focus on four-dimensional theories, we find it instructive to first also discuss other dimensions.

15.1 RG and consistency

Since the naive perturbative expansion of QFT does not exist, due to unavoidable large momentum (or UV) divergences, in order to give a meaning to the expansion an artificial, non-physical structure at a short distance scale $1/\Lambda$ (Λ is the corresponding momentum cut-off) has, at least in an intermediate step, to be introduced. Then, in the case of renormalizable QFTs, it is possible to tune the initial (*bare*) parameters of the theory as functions of the cut-off in such a way that the *renormalized perturbative expansion* has a finite limit for infinite cut-off.

The question then becomes, what is the physical meaning of a (bare) Lagrangian with its cut-off dependent tuned parameters

Quantum electrodynamics. After several authors noticed that in 'renormalized' massless quantum electrodynamics (QED) an RG could be formulated, Gell-Mann and Low [66], in the context of QED, pointed out that, in the bare Lagrangian, the bare coupling constant could have a finite limit at infinite cut-off *if QED would exhibit a stable UV fixed point*. But, as it was found, in QED the origin $\alpha = 0$ is an *IR stable fixed point.*

From Random Walks to Random Matrices. Jean Zinn-Justin, Oxford University Press (2019).
© Jean Zinn-Justin. DOI: 10.1093/oso/9780198787754.001.0001

A possible UV fixed point could only be non-perturbative. Such a fixed point has never been found in QED and does not seem to exist, leading to the *triviality issue*.

At the time, this negative result supported another scheme, whihc was based on the BPHZ technical developments, and claimed that *only renormalized perturbation theory is meaningful*, and *the bare theory, with its bare cut-off dependent Lagrangian, is only a bookkeeping device*.

One important apparent consequence was that QFT could only deal with *weakly coupled* field theories. For example, the *strong nuclear force* could not be described by QFT and this led in the 1960s to the alternative development of the *S-matrix theory* to describe the physics of strongly interacting particles (although, alternatively, application of *series summation methods*, like *Padé approximants* [172], to the perturbative expansion were eventually proposed).

15.1.1   *The non-Abelian gauge theory revolution: Asymptotic freedom*

The end of the 1960s witnessed a series of spectacular new results. Deep inelastic experiments at the SLAC National Accelerator Laboratory confirmed the quark idea and indicated that strong interactions were, in fact, weak at short distance. The possible relevance of *RG ideas* was then emphasized [87].

Non-Abelian gauge theories were quantized and made it possible to complete the construction of a consistent model describing weak and electromagnetic interactions. Moreover, the discovery that a *class of non-Abelian gauge theories is asymptotically free*, that is, that zero coupling is a *UV stable fixed point*, yielded a natural explanation for the weak interaction of quarks at short distance. Quantum chromodynamics (QCD) became part of the Standard Model of all interactions, but gravity. This achievement was a triumph of the concept of *renormalizable QFTs*.

However, in QCD the problem of *quark confinement*, that is, that conversely interactions become large at large distance, made it necessary to confront the problem of *the global existence of a QFT, which no longer could be entirely reduced to the renormalized perturbative expansion*.

Other theories, like the Gross–Neveu model in the $1/N$ expansion or the non-linear σ-model in both the $1/N$ and $(d-2)$ expansions, were discovered that are *non-renormalizable by power counting* but, nevertheless, *behave like renormalizable theories thanks to the presence of non-trivial UV fixed points*.

The concept of *asymptotic safety* [97] eventually emerged from such examples: the existence of UV fixed points seems to be a sufficient condition for the *existence of renormalized field theories consistent on all scales*. Later, one tried to apply this idea to *quantum gravity*, and the question still remains an active field of research.

15.1.2   *Wilson's theory of critical phenomena*

An important development strongly influenced our view of QFT: *Wilson's theory of critical phenomena* [78, 100]. It showed how renormalizable QFTs emerge in a completely different context: the large distance properties of a large class of statistical systems with short range interactions near the critical temperature of a continuous phase transitions can be described by statistical field theories (QFT in Euclidean or imaginary time).

There, physics is initially described by some microscopic model, like a lattice model, which is not of field theory type and has no infinities.

However, to describe its large distance properties near the phase transition, the initial model can be replaced by an *effective local QFT*. The *effective field theory may contain all possible local interactions* (products of fields and their derivatives at the same point) consistent with symmetries, but the *renormalizable or super-renormalizable terms are distinguished by the property that they dominate long distance physics* (at least in some *neighbourhood of the Gaussian fixed point*).

Such a QFT is equipped with a *cut-off*, simplified, *ad hoc* substitute for the initial inverse microscopic scale (like the range of interactions or the lattice spacing).

In this framework, bare field theory and cut-off have some physical reality: the usual *bare theory* is the *effective field theory at cut-off scale, restricted to the renormalizable and super-renormalizable interactions*.

Then, the large distance properties of the bare theory can be derived directly from the RG of the bare QFT equations [103] or, at leading order, from the renormalized theory, provided a non-Gaussian IR fixed point exists.

The existence of the *Wilson–Fisher IR fixed point* [79] implies a universal large distance behaviour and, thus, in particular, insensitivity to the short distance regularization: the specific cut-off implementation does not affect the large distance properties.

Finally, one can also introduce a QFT renormalized at a suitable large distance scale (like the correlation length scale) and employ, for example, Callan–Symanzik RG equations (*e.g.*, as in Ref. [28]). The *renormalized QFT is an extremely efficient tool* for determining the leading *asymptotic*, universal, large distance properties.

15.2   Super-renormalizable effective field theories: The $(\phi^2)^2$ example

In dimensions $d < 4$, the large distance properties of a large class of continuous, macroscopic phase transitions can be described by the $O(N)$ symmetric $(\phi^2)^2$ scalar field theory. The Euclidean action can be written as

$$\mathcal{S}(\phi) = \int \left\{ \frac{1}{2} \left[\nabla_\Lambda \phi(x) \right]^2 + \frac{1}{2} \Lambda^2 r \phi^2(x) + \frac{1}{4!} g \Lambda^{4-d} \left[\phi^2(x) \right]^2 \right\} \mathrm{d}^d x \,, \qquad (15.1)$$

where, for example,

$$\nabla_\Lambda = \nabla \left(1 - \nabla^2/\Lambda^2 + O(\nabla^4/\Lambda^4) \right),$$

Λ being a large momentum cut-off, and the parameters g, r being dimensionless.

15.2.1   The renormalized field theory and Callan–Symanzik equations

Technically, for $d < 4$ fixed, the theory is *super-renormalizable* but only the massive theory has a perturbative expansion. *At $g\Lambda^{4-d}$ fixed*, it requires only the *fine tuning* of $\Lambda^2 r$, the coefficient of ϕ^2 (a mass renormalization), in such a way that the physical mass m remains finite but non-vanishing when the cut-off Λ goes to infinity.

However, from the viewpoint of an effective field theory, a limit $\Lambda \to \infty$ at $g\Lambda^{4-d}$ fixed corresponds to an *additional fine tuning* by choosing the coupling constant g close to the UV fixed point $g = 0$ and assuming the *absence of other, less IR relevant interactions* like ϕ^6, a situation that, remarkably enough, is physically realized in the *weakly interacting dilute Bose gas* (see Section 20.4).

In the super-renormalizable limit, correlation functions have a $\Lambda \to \infty$ limit, and the existence of a renormalized field theory can be proved beyond perturbation theory.

In this limit, the coupling constant $g\Lambda^{4-d}$ has to be considered of order m^{4-d}, and a standard parametrization of the renormalized vertex functions (or correlation functions), after a finite field renormalization $Z(g_{\mathrm{r}})$ (written here for $N = 1$), is

$$\Gamma_{\mathrm{r}}^{(2)}(p; m, g_{\mathrm{r}}) = m^2 + p^2 + O(p^4),$$
$$\Gamma_{\mathrm{r}}^{(4)}(0, 0, 0, 0; mg_{\mathrm{r}}) = m^{4-d}g_{\mathrm{r}},$$

where $g\Lambda^{4-d}$ is related to g_{r} by an equation of the form,

$$g\Lambda^{4-d} = m^{4-d}G(g_{\mathrm{r}}). \tag{15.2}$$

These renormalized vertex functions satisfy inhomogeneous RG equations (Callan–Symanzik equations [87, 83]) of the form

$$\left[m\frac{\partial}{\partial m} + \beta_r(g_{\mathrm{r}})\frac{\partial}{\partial g_{\mathrm{r}}} - \tfrac{1}{2}n\eta_{\mathrm{r}}(g_{\mathrm{r}}) \right] \Gamma^{(n)}(p_i; m, g_{\mathrm{r}}) = m^2\big(2 - \eta_{\mathrm{r}}(g_{\mathrm{r}})\big)\Gamma_{\mathrm{r}}^{(1,n)}(0, p_i; m, g_{\mathrm{r}}),$$

where $\Gamma_{\mathrm{r}}^{(1,n)}$ is vertex function with $\int \mathrm{d}^d x\, \phi^2(x)$ inserted and where

$$\beta_{\mathrm{r}}(g_{\mathrm{r}}) = -(4 - d)\,/\,(\mathrm{d}\ln G/\mathrm{d}g_{\mathrm{r}}), \quad \eta_{\mathrm{r}}(g_{\mathrm{r}}) = \beta_{\mathrm{r}}(g_{\mathrm{r}})\frac{\mathrm{d}\ln Z(g_{\mathrm{r}})}{\mathrm{d}g_{\mathrm{r}}}.$$

For $d < 4$, $\beta_{\mathrm{r}}(g_{\mathrm{r}}) = -(4 - d)g_{\mathrm{r}} + O(g_{\mathrm{r}}^2)$.

For $|p| \gg m$, one can argue that the right=hand side of Callan–Symanzik equations is negligible and the equations become homogeneous. Moreover, for small enough g_{r}, the effective coupling flows into the Gaussian fixed point $g_{\mathrm{r}} = 0$: the large momentum behaviour is governed by the UV fixed point $g_{\mathrm{r}} = 0$. The renormalized theory satisfies the criterion of asymptotic safety and is, indeed, consistent on all scales.

For $d < 4$, to second order one finds

$$\beta_{\mathrm{r}}(g_{\mathrm{r}}) = -(4 - d)g_{\mathrm{r}} + \beta_2(d)g_{\mathrm{r}}^2 + O(g_{\mathrm{r}}^3), \quad \beta_2 > 0.$$

For $d < 4$, the RG β-function has another zero $g_{\mathrm{r}}^* = O(d - 4)$ with a positive slope ($\beta_{\mathrm{r}}'(g_{\mathrm{r}}^*) > 0$), which thus has the property of an IR fixed point (and is UV repulsive). It still exists for $d = 3$.

For $g_r = g_{\mathrm{r}}^*$, the function $G(g_{\mathrm{r}})$ in equation (15.2) diverges, as expected for a limit where Λ goes to infinity at generic g fixed. Moreover, correlation functions have then a non-trivial large momentum behaviour, which corresponds to a massless or critical theory and is directly related to Wilson–Fisher's fixed point [79].

15.2.2 The effective critical field theory

We now examine the same problem from the viewpoint of the initial (bare) theory. Within the framework of the $\varepsilon = 4 - d$ expansion, one can define a perturbative expansion directly for a massless or critical theory, which requires choosing a critical value $r = r_c$ in action (15.1), but there is convincing evidence that the massless theory also exists at fixed dimension for any N in $d = 3$ and for $N \leq 2$ in $d = 2$.

RG analysis. By contrast with the framework of renormalization theory, from the viewpoint of effective field theories, the $\Lambda \to \infty$ limit has to be taken at g fixed. Within the ε-expansion, one proves that the inverse two-point function (or vertex function) $\tilde{\Gamma}^{(2)}(p)$ at the transition point $r = r_c$ (thus, $\tilde{\Gamma}^{(2)}(0) = 0$), satisfies the RG equation [103],

$$\left(\Lambda\frac{\partial}{\partial\Lambda} + \beta(g)\frac{\partial}{\partial g} - \eta(g)\right)\tilde{\Gamma}^{(2)}(p, \Lambda, g) = 0,$$

where $\eta(g) = O(g^2)$ and

$$\beta(g) = -(4 - d)g + \beta_2(d)g^2 + O(g^3), \quad \beta_2(d) > 0.$$

The β-function has a trivial zero $g = 0$, which is the UV fixed point, and we know that it also has a non-trivial IR stable zero $g = g_{\text{IR}}^*$, which for $\varepsilon = 4 - d > 0$ small behaves like,

$$g_{\text{IR}}^* = \frac{48\pi^2}{(N + 8)}\varepsilon + O(\varepsilon^2).$$

Together with dimensional analysis, the RG equation implies that $\tilde{\Gamma}^{(2)}$ has the general form

$$\tilde{\Gamma}^{(2)}(p, \Lambda, g) = p^2 Z(g) F(p/\Lambda(g))$$

with

$$\beta(g)\frac{\partial \ln Z(g)}{\partial g} = \eta(g), \quad \beta(g)\frac{\partial \ln \Lambda(g)}{\partial g} = -1.$$

On dimensional grounds, $\Lambda(g)$ is proportional to Λ. The function $\Lambda(g)$ is then obtained by integration:

$$\Lambda(g) = \Lambda g^{1/(4-d)} \exp\left[-\int_0^g dg'\left(\frac{1}{\beta(g')} + \frac{1}{(4-d)g'}\right)\right].$$

In a generic situation, g is of order unity and, thus, $\Lambda(g)$ is of order Λ: *universality can only be observed in the IR or large distance behaviour of correlation functions.*

By contrast, $g \ll 1$ implies $\Lambda(g) \sim g^{1/(4-d)}\Lambda \ll \Lambda$. Then, the intermediate scale $\Lambda(g)$ *becomes a mass crossover scale* separating a *universal long distance regime* governed by the non-trivial zero $g_{\text{IR}}^* > 0$ of the β-function, from a *universal short distance regime* governed by the Gaussian fixed point $g = 0$.

One finds

$$\tilde{\Gamma}^{(2)}(p) \propto p^{2-\eta}, \text{ for } p \ll \Lambda(g),$$

where $\eta > 0$ is a critical exponent, and

$$\tilde{\Gamma}^{(2)}(p) \propto p^2, \text{ for } \Lambda(g) \ll p \ll \Lambda.$$

Due to existence of both an IR and a UV fixed point, by tuning the coupling constant g, it is possible to define a renormalized massless field theory consistent on all scales.

15.3 A renormalizable field theory: The $(\phi^2)^2$ theory in dimension 4

We consider again the N-component $(\phi^2)^2$ QFT, whose action in dimension 4 can be written as (equation (15.1))

$$\mathcal{S}(\phi) = \int \left\{ \frac{1}{2} \left[\nabla_\Lambda \phi(x) \right]^2 + \frac{1}{2} \Lambda^2 r \phi^2(x) + \frac{1}{4!} g \left[\phi^2(x) \right]^2 \right\} \mathrm{d}^4 x \,.$$

The theory is renormalizable, which means that one can formally define a cut-off independent renormalized theory, order by order in the perturbative expansion. Restricting again the analysis to the massless theory, one constructs *renormalized connected correlation functions* $W_{\mathrm{r}}^{(n)}$ functions of a renormalization scale $\mu \ll \Lambda$ (Λ is the cut-off) and a renormalized coupling constant g_{r} by

$$W_{\mathrm{r}}^{(n)}(x_i, g_{\mathrm{r}}, \mu) = \lim_{\substack{\Lambda \to \infty \\ \mu, g_{\mathrm{r} \text{ fixed}}}} Z^{-n/2}(g_{\mathrm{r}}, \Lambda/\mu) W^{(n)}(x_i, g, \Lambda), \tag{15.3}$$

where Z is a properly adjusted field renormalization.

While the existence of such a limit suggests that a field theory consistent on all scales can be defined, the question is whether this statement is valid beyond perturbation theory.

First, the perturbative expansion is divergent for all values of the coupling constant and, moreover, unlike what happens in dimensions smaller than 4, it is not Borel summable due to *renormalon* effects. Therefore, the perturbative expansion does not define a unique function.

From equation (15.3), one derives two RG equations, one for the renormalized functions and another one for the initial (bare) functions. The renormalized correlation functions satisfy

$$\left(\mu \frac{\partial}{\partial \mu} + \beta_{\mathrm{r}}(g_{\mathrm{r}}) \frac{\partial}{\partial g_r} + \tfrac{1}{2} n \eta_{\mathrm{r}}(g_{\mathrm{r}}) \right) W^{(n)}(x_i, g_{\mathrm{r}}, \mu) = 0$$

but the equation, valid to all orders in perturbation theory, remains valid beyond perturbation theory only if the infinite cut-off limit really exist.

To investigate the very existence of the renormalized theory, we can use the bare RG equations. The initial (or bare) correlation functions satisfy

$$\left(\Lambda \frac{\partial}{\partial \Lambda} + \beta(g) \frac{\partial}{\partial g} + \tfrac{1}{2} n \eta(g) \right) W^{(n)}(x_i, g, \Lambda) = 0 \,,$$

where $|x_i| \gg 1/\Lambda$, and where perturbative contributions decreasing like $(\ln \Lambda)^\ell / \Lambda^2$ have been neglected, and with different β and η functions beyond leading order.

The RG β-function for g small is

$$\beta(g) = \frac{N+8}{48\pi^2} g^2 + O(g^3),$$

and the theory is IR free, like QED, because the RG β-function is positive for small coupling g.

From the β-function one defines the running or effective coupling constant $g(\lambda)$ at scale Λ/λ,

$$\ln \lambda = \int_g^{g(\lambda)} \frac{\mathrm{d}g'}{\beta(g')}.$$

The effective coupling constant $g(\mu/\Lambda)$ is one possible definition of the renormalized coupling constant g_{r}. If $\beta(g)$ remains positive for $g > 0$, $g(\lambda)$ is an increasing function of λ. If one insists on taking the infinite cut-off limit, then $\lambda \to 0$. For $\lambda \to 0$, $g(\lambda)$ then goes to zero like

$$\ln \lambda \sim \frac{48\pi^2}{N+8} \int^{g(\lambda)} \frac{\mathrm{d}g'}{g'^2} \;\Rightarrow\; g(\lambda) \sim \frac{N+8}{48\pi^2} \frac{1}{|\ln \lambda|}.$$

This leads to the well-known *triviality issue*: beyond perturbation theory, the renormalized theory does not exist for $g_{\mathrm{r}} \neq 0$.

A way out would be that the β-function has another zero $g^* > 0$ which would then be an IR repulsive or UV attractive fixed point. For $g = g^*$, the theory would be non-trivial and, by fine tuning g close to g^*, one could define a renormalized theory consistent on all scales (an example of asymptotic safety).

However, the β-function does not seem to have another zero, and this way out does not seem to exist.

Then, in the absence of a UV fixed point, correlation functions have no universal large momentum behaviour.

By contrast, in the framework of effective field theory, one does not insist on taking the infinite cut-off limit, and one simply concludes that the $(\phi^2)^2$ QFT has a *limited distance or momentum range of validity* and that this *range decreases when the renormalized coupling increases*:

$$\ln(\Lambda/\mu) \leq \frac{48\pi^2}{N+8} \frac{1}{g(\mu/\Lambda)}.$$

In the example of 'trivial' QED, the small value of α implies a validity even beyond the Planck scale and, thus, sufficient for any physical purpose. Indeed, taking into account all leptons and quarks, *neglecting weak interactions* (which, however, strongly modify the RG flow), one finds

$$\beta(\alpha) = \frac{16}{3\pi}\alpha^2 + O(\alpha^3) \;\Rightarrow\; \ln(\Lambda/\mu) \leq \frac{3\pi}{16\alpha(\mu)}.$$

If Λ is the *Planck mass* and μ the *top mass*, the inequality implies

$$\alpha(\mu) \leq \frac{3\pi}{16\ln(\Lambda/\mu)} \approx 0.016,$$

which is satisfied by the physical value and, in addition, yields the right order of magnitude.

15.4 The non-linear σ-model

With the non-linear σ-model, one approaches the QCD situation. The model is *renormalizable, asymptotically free in $d = 2$ dimensions and non-renormalizable in higher dimensions.* We consider here only the $O(N)$ symmetric vector model, which has been more extensively studied.

The action can then be written, for example, as

$$\mathcal{S}(\phi) = \frac{\Lambda^{d-2}}{2g} \int^{\Lambda} \mathrm{d}^d x \, [\nabla_x \phi(x)]^2$$

with the constraint

$$\phi^2(x) = 1 \,.$$

Here, Λ is the cut-off and g is dimensionless (and has the interpretation of a temperature in classical statistical physics).

The perturbative phase corresponds to a spontaneous breaking of the $O(N)$ symmetry with $(N-1)$ massless Goldstone modes.

15.4.1 Dimension 2

In dimension 2, the massless modes are responsible for IR divergent contributions, and an IR cut-off, which necessarily breaks the $O(N)$ symmetry explicitly, is required. For example, one can add to the action a term of the form

$$\mathbf{c} \cdot \int \mathrm{d}^2 x \, \phi(x)$$

and renders all fields massive.

A few terms of the RG β-function have been calculated. At leading order, one finds

$$\beta(g) = -(N-2)\frac{g^2}{2\pi} + O(g^3). \tag{15.4}$$

For $N > 2$ (the non-Abelian situation), $g = 0$ is a UV fixed point. The model shares with QCD, the property of *asymptotic freedom* and thus of *asymptotic safety. With a suitable regularization, one can construct a renormalized field theory consistent on all scales.* Unlike what perturbation theory suggests, in the symmetric limit $\mathbf{c} = 0$, the $O(N)$ symmetry is unbroken and the spectrum consists in N massive states.

In the symmetric $\mathbf{c} = 0$ limit, all masses are proportional to the RG invariant mass scale

$$m(g) = \Lambda \exp\left[-\int^g \frac{\mathrm{d}g'}{\beta(g')}\right] \underset{g\to 0}{\propto} \Lambda \, \mathrm{e}^{-2\pi/(N-2)g} \,.$$

Note that the *physical condition $m(g) \ll \Lambda$* here again requires some (modest) *fine tuning* of the coupling constant, since

$$g(m) \sim \frac{2\pi}{(N-2)\ln(\Lambda/m)} \ll 1 \,. \tag{15.5}$$

The fine tuning of dimensionless coupling constants in renormalizable field theories is mild because their running is only logarithmic.

15.4.2 Higher dimensions

In higher dimensions, the non-linear σ-model is not renormalizable by power counting. However, one can find $O(N)$ symmetric regularizations such that the perturbative expansion is defined and, at least for small coupling, describes a spontaneously broken phase with Goldstone particles. As the action

$$\mathcal{S}(\boldsymbol{\phi}) = \frac{\Lambda^{d-2}}{2g} \int \mathrm{d}^d x \left[\nabla_x \phi(x) \right]^2,$$

shows, for $d > 2$, the coupling constant $g\Lambda^{2-d}$ is *generically small* in the physical domain. Therefore, generically, one finds a model with $(N-1)$ *weakly interacting* Goldstone particles, and the physics is *completely perturbative*.

However, in the case of the $O(N)$ non-linear σ-model, various consistent arguments (large N expansion, $(d-2)$ expansion, lattice regularization) indicate that, with a suitable symmetric regularization, from the non-linear σ-model can be inferred a renormalized field theory, at least in dimension 3 [86].

An example: The $\varepsilon = d - 2$ expansion. To all orders in a double g and $\varepsilon = d - 2$ expansions, it is possible to prove that the vertex (or one-particle-irreducible (1PI)) functions in Fourier representation satisfy the RG equations

$$\left[\Lambda \frac{\partial}{\partial \Lambda} + \beta(g) \frac{\partial}{\partial g} - \frac{n}{2} \zeta(g) \right] \tilde{\Gamma}^{(n)}(p_i; g, \Lambda) = 0,$$

where

$$\beta(g) = \varepsilon g - \frac{(N-2)}{2\pi} g^2 + O\left(g^3, g^2\varepsilon\right),$$

$$\zeta(g) = \frac{(N-1)}{2\pi} g + O\left(g^2, g\varepsilon\right).$$

The RG β-function has two zeros, $g = 0$, which is an IR fixed point, and a non-trivial value

$$g = g^*_{\mathrm{UV}} = \frac{2\pi\varepsilon}{(N-2)} + O(\varepsilon^2),$$

which is a *UV fixed point* and, from the point of view of classical statistical physics, a *critical temperature* separating a broken phase for $g < g^*_{\mathrm{UV}}$ from a symmetric phase for $g > g^*_{\mathrm{UV}}$.

The two-point vertex function, the solution of the RG equation, can then be written as

$$\tilde{\Gamma}^{(2)}(p; g, \Lambda) = [\Lambda(g)]^d Z(g) F(p/\Lambda(g))$$

with

$$\Lambda(g) = \Lambda \exp\left[-\int^g \frac{\mathrm{d}g'}{\beta(g')} \right], \quad Z(g) = \exp\left[\int^g \frac{\mathrm{d}g' \zeta(g')}{\beta(g')} \right].$$

$\Lambda(g)$ is thus the physical mass scale. For generic values of g, the integral is finite and $\Lambda(g)/\Lambda$ is thus finite. For $g < g^*_{\mathrm{UV}}$, this means that one finds weakly interacting Goldstone bosons and, for $g > g^*_{\mathrm{UV}}$, no particle propagates.

Only for $|g - g^*_{\text{UV}}| \ll 1$, and this again implies some *fine tuning*, does the integral diverge. From the statistical viewpoint, this is the *critical domain*. Then, $\Lambda(g) \ll \Lambda$ and a new physical mass scale is generated:

$$\Lambda(g) \propto \Lambda |g - g^*_{\text{UV}}|^\nu \ll \Lambda \quad \text{with} \quad \nu = -\frac{1}{\beta'(g^*_{\text{UV}})}.$$

The physics is then the same as for the generic $(\phi^2)^2$ field theory.

For $g > g^*_{\text{UV}}$, $\Lambda(g)$ provides a physical mass scale, which is now much smaller than the cut-off.

For $g < g^*_{\text{UV}}$, $\Lambda(g)$ is a *crossover scale* between a universal small momentum free behaviour and a universal large momentum, critical, behaviour.

For $4 > d \geq 2$, thanks to the existence of an UV fixed point, with some fine tuning, a QFT consistent on all scales can be defined.

15.5 The Gross–Neveu model

The Gross–Neveu model describes a set of $N > 1$ massless fermions $\{\psi^i, \bar{\psi}^i\}$, interacting through a four-fermion self-interaction corresponding to the $U(N)$ symmetric action,

$$\mathcal{S}(\bar{\psi}, \psi) = -\int \mathrm{d}^d x \left[\bar{\psi}(x) \cdot \partial\!\!\!/\, \psi(x) + \tfrac{1}{2}\Lambda^{2-d} G \left(\bar{\psi}(x) \cdot \psi(x) \right)^2 \right], \quad G > 0. \quad (15.6)$$

In addition to the $U(N)$ symmetry, the GRoss–Neveu model has a discrete reflection symmetry (a discrete chiral symmetry in even dimensions: see equation (8.19)), which prevents the addition of a fermion mass term. The sign of the *dimensionless* coupling constant G is such that the interaction is *attractive*.

The model can be analysed in $d \geq 2$ dimensions with arguments analogous to those used for the non-linear σ-model.

The model requires a symmetric UV cut-off, for example,

$$\partial\!\!\!/ \mapsto \partial\!\!\!/(1 - \nabla^2_x/\Lambda^2 + \cdots).$$

Two dimensions. The model is renormalizable in two dimensions, and the RG β-function has the form

$$\beta(G) = -(N-1)\frac{G^2}{\pi} + O(G^3).$$

For $N > 1$, the origin is a UV fixed point, and the model has the property of *asymptotic freedom* like QCD. It satisfies the condition of asymptotic safety and is a candidate for a renormalized field theory consistent on all scales. The spectrum is non-perturbative with spontaneous chiral symmetry breaking and mass generation.

$d = 2 + \varepsilon$ *dimension.* The RG β-function reads

$$\beta(G) = \varepsilon G - (N-1)\frac{G^2}{\pi} + (N-1)\frac{G^3}{2\pi^2} + O\left(G^4\right).$$

The origin $G = 0$ now is an IR fixed point and, for G small, the physics is per-turbative with chiral symmetry and massless fermions. However, for ε small, the β-function has another IR repulsive zero,

$$G_c = \frac{\pi \varepsilon}{N-1} + O(\varepsilon^2) \;\Rightarrow\; \beta'(G_c) = -\varepsilon + O(\varepsilon^2).$$

From an RG viewpoint, the value G_c corresponds to a *UV fixed point*, which is repulsive for large distance physics. Above G_c, the condition of small masses implies a fine tuning of G such that $G - G_c \ll 1$.

The existence of a UV fixed point realizes the condition of *asymptotic safety*. In the limit $G - G_c \ll 1$, below dimension 4, it yields a field theory that has both universal large distance and short distance physics. It makes it possible to define a renormalized theory consistent on all scales.

15.6   QCD

QCD is the part of the Standard Model of fundamental interactions at the micro-scopic scale that generates the strong nuclear force. It provides an outstanding physical example of a field theory where the asymptotic safety problem has been discussed.

For a non-Abelian gauge theory corresponding, for example, to the $SU(N)$ group with N_F fermions, the RG β-function has, in terms of the gauge coupling constant g, the expansion (see Section 13.5),

$$\beta(g^2) = -\left(\frac{11N}{3} - \frac{2N_F}{3}\right)\frac{g^4}{8\pi^2} + O\left(g^6\right). \tag{15.7}$$

For the $SU(3)$ group and six quark flavours, the β-function is negative for g^2 small and, thus, QCD has the property of asymptotic freedom in four dimensions, provid-ing a natural explanation to the weak interaction between quarks at short distance.

As an additional consequence, QCD has the property of *asymptotic safety* in the same way as the two-dimensional non-linear σ-model.

Thus, after some mild fine tuning of the coupling constant (see equations (15.4) and (15.5)), one can define a renormalized theory, *which could be consistent on all scales*.

Whether QCD is really consistent on all scales remains a very difficult open mathematical question but, to a large extent, it is only a mathematical question.

Indeed, first, *due its initial infinities, QCD has to be embedded in another theory* to provide the cut-off that renders the theory finite.

Then, *consistency on all scales is not a physical requirement*. One only needs consistency up to some energy scale where physics will be modified anyway, at latest, the Planck scale.

As for any QFT, the real physical question is: up to which scale is the theory physically consistent and relevant?

15.7 General interactions and summary

From the viewpoint of *critical phenomena*, which we now consider also to be the
relevant viewpoint for any QFT, all QFTs are *effective large distance theories*. This
applies, in particular, to the theory of fundamental interactions at the microscopic
scale (called the Standard Model). In a generic effective field theory, all local
interactions consistent with field content, symmetries, and so on, are present. All
monomials are affected by *powers of the cut-off* (which represents the energy scale
of some new physics) dictated by *power counting*.

Leading effects of terms with positive powers of the cut-off either are removed
by *fine tuning* (this necessarily includes scalar mass terms) or generate infinite couplings, as in the example of the three-dimensional $(\phi^2)^2$ field theory. *IR asymptotic
safety then requires the existence of IR fixed points.*

The coefficients in front of renormalizable interactions have, by definition, no
cut-off dependence and, under RG, have a logarithmic flow.

In IR free theories, the effective or renormalized interactions vanish for large cut-
off (triviality), and these theories have a limited energy range of validity. Triviality
can be avoided only if non-trivial UV fixed points *(UV asymptotic safety)* can be
found. However, some *fine tuning* is then required.

UV asymptotically free theories, like four-dimensional QCD or the two-dimensional non-linear σ-model, require some mild *fine tuning* to generate a mass scale
much smaller than the cut-off. They are then *asymptotically safe* and, with ade-
quate regularization, after renormalization, mathematical candidates for being fully
consistent renormalized theories.

In the case of several independent coupling constants, this does not exhaust all
possible situations. In a general situation, a more detailed analysis is required. For
example, situations with *weak first order phase transitions* can appear, where the
continuum theory does not even exist (IR asymptotically unsafe theories).

Non-renormalizable interactions are multiplied by negative powers of the cut-off.
The theory of renormalization of composite operators (*i.e.,* local monomials in the
fields) tells us that adding a given monomial to the effective action has the same
effect as renormalizing terms of lower dimension (in the sense of power counting)
and, in addition, generates corrections that vanish at large cut-off [104, 18].

At least, this is the generic situation. However, the examples of the non-linear σ-
model or the Gross–Neveu model in three dimensions, indicate that, after some *fine
tuning*, the effect of these operators can be enhanced *if a UV fixed point can be found.
The theory is then UV asymptotically safe.* In this situation, *non-renormalizable
interactions can generate non-trivial interacting field theories.*

Finally, the concept of effective theories applied to quantum gravity, obviously
suggests that Einstein–Hilbert's action is just the first, most important at large
distance, term of a general local expansion. Whether such a theory possesses a
non-trivial UV fixed point remains an open and quite challenging question.

*A UV fixed point would imply a crossover scale, much larger than Planck's length,
between long distance weak gravity and shorter distance stronger gravity. It would
also raise the question of fine tuning.*

16 Symmetries: From classical to quantum field theories

When in the tree or classical approximation, a local field theory has a symmetry, whether global or local as in gauge theories, formal manipulations suggest that the symmetry can also be implemented in the full quantum theory, provided one uses the proper quantization rules. While this is often true, one can find counterexamples and, therefore, the question requires a thorough investigation.

Indeed, when the classical Hamiltonian contains products of position and conjugate momentum variables, it does not fully specify the quantum Hamiltonian, due to the problem of ordering quantum position and momentum operators in products. It may happen that no ordering is consistent with the implementation of some classical symmetry.

In the perturbative expansion, the difficulty translates into the problem that simple formal manipulations ignore the unavoidable ultraviolet (UV) divergences of perturbation theory.

Similarly, the field integral formulation of quantum field theory, which involves only the classical action, seems to indicate that symmetries of the classical action translate automatically into symmetries of the full quantum theory.

However, the fields contributing to the field integral, as specified by the classical action, are too singular and the naive field integral is not defined.

The existence of symmetric *regularizations*, that is, large momentum modifications of the theory to render perturbation theory finite, makes it possible to solve the problem in most cases but a physically very relevant exception is provided by the combination of gauge symmetry and chiral fermions. Depending on the specific group and field content, obstructions are found to the quantization of chiral gauge symmetries [173–176], called *anomalies*.

In this chapter, we review, and discuss from the viewpoint of symmetries, the advantages and shortcomings of three regularization techniques: momentum cut-off regularizations, dimensional regularization and lattice regularization. We show that all three leave room for possible anomalies when both gauge fields and chiral fermions are present.

Conventions. Throughout the chapter, we work in *Euclidean space* (with imaginary or Euclidean time), and this also implies a formalism of *Euclidean fermions*. For details see, Ref. [18].

16.1 Symmetries and regularization

Local quantum field theories (the action is a space-time integral of a function of the field and its derivatives) are affected by large momentum (also called UV) divergences (equivalently, short distance singularities) that invalidate formal algebraic proofs of symmetry properties.

From Random Walks to Random Matrices. Jean Zinn-Justin, Oxford University Press (2019).
© Jean Zinn-Justin. DOI: 10.1093/oso/9780198787754.001.0001

16.1.1 UV divergences

In quantum field theory, UV divergences may have two different origins.

Quite generally, a field corresponds to an infinite number of degrees of freedom, the values of the field at every point in space. In local theories, this leads to divergences that have renormalization group (RG) interpretations. These divergences can, in general, be regularized (*i.e.*, suppressed) by adding terms quadratic in the fields with higher derivatives.

From a deeper perspective, field degrees of freedom remain coupled on all scales and an artificial short distance structure, initially absent, has to be introduced. RG methods are then required to demonstrate the insensitivity of physics results to these short distance modifications.

However, divergences of another origin may appear, which are related to the transition between classical and quantum Hamiltonians and the problem of ordering non-commuting quantum operators in products. These divergences reflect the property that the knowledge of the classical theory is not sufficient, in general, to fully determine the quantized theory.

Such divergences are already present in ordinary quantum mechanics in perturbation theory, for instance, in the quantization of the geodesic motion of a particle on a manifold (like a sphere). From the viewpoint of path integrals, the naive path integral formulation that involves only the classical Lagrangian is not defined.

Even in the case of forces linear in the velocities (like a coupling to a magnetic field) or, equivalently, Hamiltonians with interactions linear in the conjugate momentum, finite ambiguities are found. In the same situation, in local quantum field theories the problem is even more severe.

The problem can be understood by considering the commutator of a scalar field operator $\hat{\phi}$ and its conjugate momentum $\hat{\pi}$, in the Schrödinger picture (in d space-time dimensions). It takes the form (canonical commutation relations)

$$[\hat{\phi}(x), \hat{\pi}(y)] = i\hbar\, \delta^{d-1}(x - y). \tag{16.1}$$

(For fermions, the commutator is replaced by the anticommutator.) As soon as derivative couplings are involved (in covariant theories), or when fermions are present, local Hamiltonians contain products of fields and conjugate momenta at the same point. These products lead to divergences, except in quantum mechanics ($d = 1$ with our conventions), where they reduce to finite ambiguities.

Field integrals. From the viewpoint of field integrals, UV divergences have a technical interpretation: the fields that contribute to the field integral, as specified by the classical action, are not regular enough, and the expectation values of the local polynomials in the fields that appear in the interactions thus are not defined. Adding quadratic terms with higher derivatives has the effect of selecting more regular fields.

16.1.2 Symmetries and regularization

Regularization, which renders the perturbative expansion finite, is a very useful intermediate step in the renormalization programme. It consists in modifying the initial theory at short distance, at large momentum or by continuation in the dimension of space, to render the perturbative expansion finite. Since these modifications, from the point of view of particle physics, give non-physical properties to the quantum field theory, they are often considered to be only intermediate steps in the removal of divergences. The proof of the independence from the regularization of physics observables then relies on renormalization theory.

When a regularization can be found that preserves the symmetry of the initial classical action, a symmetric quantum field theory can be constructed.

Momentum cut-off regularization, based on modifying propagators at large momenta, is specifically designed to cut the infinite number of degrees of freedom. With some care, it can be chosen as to preserve formal symmetries of the unrenormalized theory that correspond to global (space-independent) linear group transformations.

However, problems arise when the symmetries are realized in the form of non-linear or space-dependent transformations, as in the example of non-linear σ models or in gauge theories, due to the unavoidable presence of derivative couplings and the related problem of quantum operator ordering in products. It is easy to verify that, in this situation, cutting only large momenta fails, in general, to provide a full regularization because *one-loop divergences are not suppressed.*

The addition of *auxiliary regulator fields* has, often, the same effect as modifying propagators but offers a few additional possibilities, in particular, because regulator fields may have the wrong spin–statistics connection. Fermion loops in a gauge background can be regularized by such a method.

Other methods require a specific analysis. In many examples, dimensional regularization (based on the continuation of Feynman diagrams to complex values of the dimension) can be used because then the commutator between field and conjugated momentum taken at the same point vanishes automatically. However, in the case of chiral fermions, dimensional regularization is not available because chiral symmetries are specific to even space-time dimensions.

Of particular interest is the method of lattice regularization, because it also provides a non-perturbative definition of quantum field theories: it makes it possible to discuss the existence of a quantum field theory and to determine its physical properties by non-perturbative numerical techniques. Moreover, lattice regularization, indeed, specifies an order between quantum operators.

However, it leads to a problem in presence of chiral fermions: the manifestation of this difficulty takes the form of a doubling of the fermion degrees of freedom on the lattice in the simplest discretization schemes.

The more recent methods of overlap [166] and domain wall fermions [167], in the framework of Ginsparg–Wilson's relation [165], provide an unconventional solution to the problem of lattice regularization in gauge theories with chiral fermions. They evade the fermion doubling problem because chiral transformations are no longer strictly local on the lattice [179], although they become local in the continuum limit.

They relate the problem of anomalies with the invariance of the fermion integration measure in chiral transformations, as in some continuum formulations [180].

Anomalies. That no conventional regularization scheme can be found in the case of gauge theories with chiral fermions is not surprising, since examples of theories with anomalies can be found. In these theories, some local symmetry of the tree or classical approximation cannot be implemented in the full quantum theory [173–176]. This may create obstructions to the construction of chiral gauge theories because exact gauge symmetry and, thus, the absence of anomalies, is essential for the physical consistency of a gauge theory (see Chapter 17).

16.2 Higher derivatives and momentum cut-off regularization

Regularization by adding higher derivatives, which generates in Fourier representation a momentum cut-off, is a method that works in the continuum (compared to lattice methods) and at fixed dimension (unlike dimensional regularization). The basic idea is to render the fields that contribute to the field integral more regular in such a way that all local monomials that contribute to the interaction have a finite expectation value. Correspondingly, the field propagators are modified beyond a large momentum cut-off and render all Feynman diagrams convergent.

From the viewpoint of particle physics, such modifications result in non-physical properties of the quantum field theory at the cut-off scale. In the formal renormalization theory, they are intermediate steps in the renormalization programme (physical properties are recovered in the infinite cut-off limit). In the more physical modern thinking, the necessity of a regularization indicates that all quantum field theories have eventually to be embedded in a more complete, non-local theory where divergences are absent.

The regularization replaces the physical short distance structure, which is sometimes too complicated or is unknown (as in particle physics), by a simpler artificial structure. The regularized theory is thus consistent only much below the cut-off scale. Moreover, renormalization and RG arguments are needed to prove that the large distance, small momentum properties are asymptotically independent of the regularization scheme.

16.2.1 *Scalar fields: Higher derivative regularization*
Simple modifications of the propagator improve the convergence of Feynman diagrams at large momentum. For example, in the case of the action of a self-coupled scalar field $\phi(x)$ of the form,

$$\mathcal{S}(\phi) = \int \mathrm{d}^d x \left[\tfrac{1}{2}\phi(x)(-\nabla_x^2 + m^2)\phi(x) + V_{\mathrm{I}}\big(\phi(x)\big) \right], \qquad (16.2)$$

one substitutes

$$\nabla_x^2 \mapsto \nabla_x^2 \prod_{i=1}^{n} (1 - \nabla_x^2/M_i^2), \qquad (16.3)$$

where the masses M_i are proportional to the momentum cut-off Λ:

$$M_i = \alpha_i \Lambda, \quad \alpha_i > 0. \qquad (16.4)$$

In the formal $\Lambda \to \infty$ limit, at all parameters α_i fixed, the initial action is recovered.

The initial inverse propagator in the Fourier representation

$$\tilde{\Delta}_{\mathrm{B}}^{-1}(p) = m^2 + p^2 \,,$$

then becomes

$$[\tilde{\Delta}_{\mathrm{B}}]_{\mathrm{reg.}}^{-1}(p) = m^2 + p^2 \prod_{i=1}^{n}(1 + p^2/M_i^2). \tag{16.5}$$

In the absence of derivative couplings, the perturbative expansion is finite if the fields contributing to the field integral are continuous, a condition realized if the degree n satisfies $2n + 2 > d$. In the case of a vertex with $2k$ derivatives, the contributing fields must be sufficiently differentiable, and the condition becomes $2n + 2 > d + 2k$.

The inverse propagator (16.5) cannot be derived from a Hermitian Hamiltonian. Indeed, Hermiticity of the Hamiltonian implies that, if the propagator is, as above, a rational function, it must be a sum of poles in p^2 with positive residues (as a sum over intermediate states of the two-point function shows) and thus cannot decrease faster than $1/p^2$.

16.2.2 Schwinger's proper time regularization

While the modification (16.5) can be implemented also in Minkowski space because the regularized propagators decrease in all complex p^2 directions (except real negative), in Euclidean (or imaginary) time more general modifications are possible. Schwinger's proper time representation suggests

$$\tilde{\Delta}_{\mathrm{B}}(p) = \int_0^\infty \mathrm{d}t \, \rho(t\Lambda^2) \, \mathrm{e}^{-t(p^2 + m^2)}, \tag{16.6}$$

in which the function $\rho(t)$ is positive (to ensure that $\tilde{\Delta}_{\mathrm{B}}(p)$ does not vanish and thus is invertible) and satisfies the condition

$$|1 - \rho(t)| < C \, \mathrm{e}^{-\sigma t} \ (\sigma > 0) \text{ for } t \to +\infty \,.$$

By choosing a function $\rho(t)$ that decreases fast enough for $t \to 0$, the behaviour of the propagator can be arbitrarily improved. If $\rho(t) = O(t^n)$, the behaviour (16.5) is recovered. Another example is

$$\rho(t) = \theta(t - 1), \tag{16.7}$$

$\theta(t)$ being the step function, which leads to an exponential decrease:

$$\tilde{\Delta}_{\mathrm{B}}(p) = \frac{\mathrm{e}^{-(p^2 + m^2)/\Lambda^2}}{p^2 + m^2}. \tag{16.8}$$

As the example shows, it is thus possible to find, in this more general class, propagators without non-physical singularities, but they do not follow from a Hamiltonian formalism because continuation to real time becomes impossible.

16.2.3 Regularization: Spin 1/2 fermions

We consider here Euclidean fermions $\psi, \bar{\psi}$, the analytic continuation to imaginary or euclidean time of the usual Dirac spin 1/2 fermions. We also use the standard notation $\not{p} \equiv \sum_\mu p_\mu \gamma_\mu$ or $\not{\partial} = \sum_\mu \gamma_\mu \partial_\mu$ with Hermitian Dirac matrices γ_μ.

For spin 1/2 fermions analogous regularization methods are applicable. To the free Dirac fermion action,

$$\mathcal{S}_{\mathrm{F0}}(\bar{\psi}, \psi) = -\int \mathrm{d}^d x\, \bar{\psi}(x)(\not{\partial} + m)\psi(x), \tag{16.9}$$

where m is the fermion mass, corresponds in Fourier representation the propagator

$$\tilde{\Delta}_{\mathrm{F}}(p) = \frac{1}{m + i\not{p}}.$$

To regularize the action, one can substitute,

$$\not{\partial} \mapsto \not{\partial} \prod_{i=1}^{n}(1 - \nabla_x^2/M_i^2).$$

The propagator is replaced by the regularized propagator $\tilde{\Delta}_{\mathrm{F}}(p)$, where

$$\tilde{\Delta}_{\mathrm{F}}^{-1}(p) = m + i\not{p} \prod_{i=1}^{n}(1 + p^2/M_i^2). \tag{16.10}$$

Fermion correlation functions at coinciding points are defined if $2n + 1 > d$.

16.2.4 Regularization and determinants

Regularizations based on a momentum cut-off have several advantages: one can work at fixed dimension and in the continuum. In statistical physics, in some sense, these regularizations are the closest to a true physical cut-off.

However, the following potential weakness has to be stressed.

The generating functional of correlation functions (or partition function in an external field) is obtained by adding to the action (16.2) a source term for the fields,

$$\mathcal{S}(\phi) \mapsto \mathcal{S}(\phi) - \int \mathrm{d}^d x\, J(x)\phi(x).$$

After some algebraic transformations, the generating functional can be written as

$$\mathcal{Z}(J) = \det^{1/2}(\Delta_{\mathrm{B}})$$
$$\times \exp\left[-\mathcal{V}_{\mathrm{I}}(\delta/\delta J)\right] \exp\left(\tfrac{1}{2} \int \mathrm{d}^d x\, \mathrm{d}^d y\, J(x)\Delta_{\mathrm{B}}(x-y)J(y)\right), \tag{16.11}$$

where $\delta/\delta J(x)$ denotes a functional derivative (see Section 6.3), the determinant is generated by a Gaussian integration, and

$$\mathcal{V}_{\mathrm{I}}(\phi) \equiv \int \mathrm{d}^d x\, V_{\mathrm{I}}(\phi(x)).$$

It is simple to verify that the momentum cut-off regularization described so far cannot deal with the determinant (which is related to *one-loop Feynman diagrams*).

As long as the determinant is a divergent factor that cancels in normalized correlation functions, the problem can be ignored but, in the case of a determinant in the background of an external field (which generates a set of field-dependent one-loop diagrams), this signals a difficulty.

Quantum field theories where this problem occurs include models with non-linearly realized (as in the non-linear σ-model) or gauge symmetries due to the presence of derivative couplings.

Supersymmetry. In the theory containing both bosons and fermions, boson integration generates inverse determinants while fermion integration generates determinants. With proper adjustments, both factors can cancel. An example of such a situation is provided by supersymmetric field theories.

16.2.5 *Application to global linear symmetries*

To implement symmetries of the classical action in the quantum theory, one needs a regularization scheme that preserves the symmetry. This requires some care but can always be achieved for linear global symmetries, that is, symmetries that correspond to transformations of the fields of the form

$$\phi_R(x) = \mathbf{R}\,\phi(x)\,,$$

where $\mathbf{R}$ is a constant matrix. The main reason is that in the quantum Hamiltonian field operators and conjugate momenta are not mixed by the transformation and, therefore, the order of operators is not important. To take an example directly relevant to this chapter, in four dimensions a theory with massless fermions may have a chiral symmetry corresponding to the transformations

$$\psi_\theta(x) = e^{i\theta\gamma_5}\psi(x), \quad \bar{\psi}_\theta(x) = \bar{\psi}(x)\,e^{i\theta\gamma_5}\,.$$

The substitution (16.10) (for $m = 0$) preserves chiral symmetry. Note the importance here of being able to work at fixed dimension 4 because chiral symmetry is defined only in even dimensions. In particular, the invariance of the integration measure $[\mathrm{d}\bar{\psi}(x)\mathrm{d}\psi(x)]$ relies on the property that $\operatorname{tr}\gamma_5 = 0$.

16.3 Regulator fields

With the introduction of regulator fields, one can reproduce the regularizations in the form (16.5) or (16.10), as we indicate below, but regulator fields offer new possibilities in the case of gauge theories [177].

Necessarily, some of the regulator fields have non-physical properties. The regularized quantum field theory is thus physically acceptable only for momenta much smaller than the masses of the regulator fields.

16.3.1 Scalar fields

In the case of scalar fields, to regularize the action (16.2) for the scalar field ϕ, one introduces additional dynamical fields ϕ_r, $r = 1, \ldots, r_{\max}$, and considers the modified action

$$
\mathcal{S}_{\text{reg.}}(\phi, \phi_r) = \tfrac{1}{2} \int \mathrm{d}^d x \Big[\phi(x) \left(-\nabla^2 + m^2 \right) \phi(x)
$$

$$
+ \sum_r \frac{1}{z_r} \phi_r(x) \left(-\nabla^2 + M_r^2 \right) \phi_r(x) \Big] + V_{\mathrm{I}} \left(\phi + \textstyle\sum_r \phi_r \right), \quad (16.12)
$$

where the masses of the regulator fields are again proportional to the cut-off (equation (16.4)). With the action (16.12), any internal ϕ propagator is replaced by the sum of the ϕ propagator and all the ϕ_r propagators,

$$
\frac{1}{p^2 + m^2} \mapsto \frac{1}{p^2 + m^2} + \sum_r \frac{z_r}{p^2 + M_r^2} \,.
$$

Expanding in powers of $1/p^2$ for p large, one can choose the constants z_r to cancel as many terms as possible. For such a choice, after integration over the regulator fields, the form (16.5) is recovered. Note that the condition of cancellation of the $1/p^2$ contribution implies

$$
1 + \sum_r z_r = 0 \,.
$$

Therefore, not all z_r can be positive. To ensure the existence of the field integral, one has to integrate over imaginary values of the fields ϕ_r that correspond to the negative values, showing that these fields are non-physical.

16.3.2 Fermions

The fermion inverse propagator (16.10) can be written as

$$
\Delta_{\mathrm{F}}^{-1}(p) = (m + i\not{p}) \prod_{r=1}^{r_{\max}} (1 + i\not{p}/M_r)(1 - i\not{p}/M_r).
$$

This indicates that, again, the same form can be obtained by a set of regulator fields $\{\bar{\psi}_{r\pm}, \psi_{r\pm}\}$. One replaces the kinetic part of the action by

$$
\int \mathrm{d}^d x \, \bar{\psi}(x)(\not{\partial} + m)\psi(x) \mapsto \int \mathrm{d}^d x \, \bar{\psi}(x)(\not{\partial} + m)\psi(x)
$$

$$
+ \sum_{\epsilon = \pm 1, r} \frac{1}{z_{r\epsilon}} \int \mathrm{d}^d x \, \bar{\psi}_{r\epsilon}(x)(\not{\partial} + \epsilon M_r)\psi_{r\epsilon}(x). \quad (16.13)
$$

Moreover, in the interaction terms, the fields ψ and $\bar{\psi}$ are replaced by the sums

$$
\psi \mapsto \psi + \sum_{r,\epsilon} \psi_{r\epsilon} \,, \qquad \bar{\psi} \mapsto \bar{\psi} + \sum_{r,\epsilon} \bar{\psi}_{r\epsilon} \,.
$$

For a proper choice of the constants z_r, after integration over the regulator fields, the form (16.10) is recovered.

Chiral symmetry. For $m = 0$, the propagator (16.10) is chiral invariant. Chiral transformations change the sign of mass terms. Here, chiral symmetry can be maintained only if, in addition to normal chiral transformations, $\psi_{r,+}$ and ψ_{-r} are exchanged (which implies $z_{r+} = z_{r-}$). Thus, chiral symmetry is preserved by the regularization, even though the regulators are massive, by *fermion doubling*. The fermions ψ_+ and ψ_- are chiral partners. For a pair $\psi \equiv (\psi_+, \psi_-)$, $\bar{\psi} \equiv (\bar{\psi}_+, \bar{\psi}_-)$, the action can be written as

$$\int \mathrm{d}^d x\, \bar{\psi}(x)\, (\slashed{\partial} \otimes \mathbf{1} + M\mathbf{1} \otimes \sigma_3)\, \psi(x),$$

where the matrix $\mathbf{1}$ in the first term and the Pauli matrix σ_3 in the second term act in $\pm$ space. The spinors then transform like

$$\psi_\theta(x) = \mathrm{e}^{i\theta\gamma_5 \otimes \sigma_1}\psi(x), \qquad \bar{\psi}_\theta(x) = \bar{\psi}(x)\, \mathrm{e}^{i\theta\gamma_5 \otimes \sigma_1},$$

because σ_1 anticommutes with σ_3.

16.4 Abelian gauge theory, the theoretical framework of QED

The problem of divergences of charged matter coupled to a gauge field can be decomposed into two steps: first, matter in an external gauge field and, then, integration over the gauge field. For gauge fields, we assume a relativistic-covariant gauge fixing, like Landau's gauge, in such a way that power counting is the same as for scalar fields.

16.4.1 Charged fermions in a gauge field background

A new problem arises in the presence of a gauge field: only covariant derivatives are allowed because gauge invariance is essential for the physical consistency of the theory. The action for fermion matter in a gauge background with a momentum regularization thus reads

$$\mathcal{S}(\bar{\psi}, \psi, A) = \int \mathrm{d}^d x\, \bar{\psi}(x)\, (m + \slashed{D}) \prod_r \left(1 - \slashed{D}^2/M_r^2\right) \psi(x), \qquad (16.14)$$

where the covariant derivative for a charge e is

$$\mathrm{D}_\mu = \partial_\mu + ieA_\mu, \quad \text{and} \quad \slashed{D} = \sum_\mu \gamma_\mu \partial_\mu.$$

In even dimensions, unlike dimensional or lattice regularizations, this regularization preserves a possible chiral symmetry for $m = 0$.

However, the higher order covariant derivatives generate new, more singular, gauge–matter interactions and it is no longer clear whether the theory can in this way be rendered finite.

Fermion correlation functions in the gauge background are generated by

$$\mathcal{Z}(\bar{\eta}, \eta; A) = \int \left[\mathrm{d}\psi(x) \mathrm{d}\bar{\psi}(x) \right]$$

$$\times \exp \left[-\mathcal{S}(\bar{\psi}, \psi, A) + \int \mathrm{d}^d x \left(\bar{\eta}(x)\psi(x) + \bar{\psi}(x)\eta(x) \right) \right], \quad (16.15)$$

where $\bar{\eta}, \eta$ are external Grassmann fields (*i.e.*, belonging to a Grassmann algebra). Integrating over fermions explicitly, one obtains

$$\mathcal{Z}(\bar{\eta}, \eta; A) = \mathcal{Z}_0(A) \exp \left[-\int \mathrm{d}^d x \, \mathrm{d}^d y \, \bar{\eta}(y) \Delta_{\mathrm{F}}(A; y, x)\eta(x) \right], \quad (16.16)$$

$$\mathcal{Z}_0(A) = \mathcal{N} \det \left[(m + \slashed{D}) \prod_r \left(1 - \slashed{D}^2/M_r^2 \right) \right], \quad (16.17)$$

where $\mathcal{N}$ is a normalization factor such that $\mathcal{Z}_0(0) = 1$, and $\Delta_{\mathrm{F}}(A; y, x)$ is the fermion propagator in an external gauge field A (see the Feynman diagrams of Fig. 16.1).

Fig. 16.1 Contributions to $\Delta_{\mathrm{F}}(A; y, x)$ (the fermions and gauge fields correspond to continuous and dotted lines, respectively).

Fermion propagator in a gauge field. Diagrams constructed from $\Delta_{\mathrm{F}}(A; y, x)$ belong to loops with gauge field propagators and, therefore, can be rendered finite provided the gauge field propagator can be improved, a condition that we check below.

The determinant. The second problem involves the determinant, which generates one-loop diagrams of the form of closed fermion loops with external gauge fields (see Fig. 16.2). Using $\ln \det = \mathrm{tr} \ln$, one finds

$$\ln \mathcal{Z}_0(A) = \mathrm{tr} \ln (m + \slashed{D}) + \sum_r \mathrm{tr} \ln \left(1 - \slashed{D}^2/M_r^2 \right) - (A = 0).$$

In even dimensions, the matrix γ_5 and its equivalent (σ_3 for $d = 2$) anticommute with $\slashed{D}$. This leads to the identity, written here for $d = 4$,

$$\det(\slashed{D} + m) = \det \gamma_5 (\slashed{D} + m)\gamma_5 = \det(m - \slashed{D})$$

and, thus,

$$\ln \mathcal{Z}_0(A) = \tfrac{1}{2} \mathrm{tr} \ln \left(m^2 - \slashed{D}^2 \right) + \sum_r \mathrm{tr} \ln \left(1 - \slashed{D}^2/M_r^2 \right) - (A = 0).$$

Fig. 16.2 Contributions to the determinant (the fermions and gauge fields correspond to continuous and dotted lines, respectively).

The regularization with higher derivatives has no effect, from the point of view of power counting, on the determinant because all contributions add. Therefore, the determinant requires an additional regularization.

16.4.2 *The fermion determinant*

The fermion determinant (16.17) can be regularized by adding to the action a boson regulator field with fermion spin (which is non-physical since it violates the spin–statistics connection) and, therefore, a propagator similar to Δ_{F} but with different masses:

$$\mathcal{S}_{\mathrm{B}}(\bar{\phi}, \phi; A) = \int \mathrm{d}^d x\, \bar{\phi}(x) \left(M_0^{\mathrm{B}} + \slashed{D}\right) \prod_{r=1} \left(1 - \slashed{D}^2/(M_r^{\mathrm{B}})^2\right) \phi(x). \tag{16.18}$$

The integration over the non-physical boson fields $\bar{\phi}, \phi$ adds to $\ln \mathcal{Z}_0$ the quantity

$$\delta \ln \mathcal{Z}_0(A) = -\tfrac{1}{2}\,\mathrm{tr}\ln\left(({M_0^{\mathrm{B}}})^2 - \slashed{D}^2\right) - \sum_{r=1} \mathrm{tr}\ln\left(1 - \slashed{D}^2/(M_r^{\mathrm{B}})^2\right) - (A = 0).$$

Expanding the sum $\ln \mathcal{Z}_0 + \delta \ln \mathcal{Z}_0$ in inverse powers of $\slashed{D}$, one adjusts the masses to cancel as many powers of $\slashed{D}$ as possible.

Chiral limit. The unpaired initial fermion mass m is the source of a problem. The corresponding determinant can only be regularized with an unpaired boson M_0^{B}. In the chiral limit $m = 0$, two options are available: either one gives a chiral charge to the boson field and the mass M_0^{B} breaks chiral symmetry, or one leaves it invariant in a chiral transformation. In the latter case, one finds the determinant of the transformed operator

$$\mathrm{e}^{i\theta(x)\gamma_5}\, \slashed{D}\, \mathrm{e}^{i\theta(x)\gamma_5} \left(\slashed{D} + M_0^{\mathrm{B}}\right)^{-1}.$$

For $\theta(x)$ constant $\mathrm{e}^{i\theta\gamma_5}\slashed{D} = \slashed{D}\,\mathrm{e}^{-i\theta\gamma_5}$ and the θ-dependence cancels. Otherwise, a non-trivial contribution remains. This analysis thus indicates possible difficulties with *space-dependent chiral transformations*.

Actually, since the problem reduces to the study of a determinant in an external background, one can study it directly (see Section 17.1). One examines whether it is possible to define some regularized form in a way consistent with chiral symmetry. When this is possible, one then inserts the one-loop renormalized diagrams in the general diagrams regularized by the preceding cut-off methods.

16.4.3 Boson determinant in a gauge background

The boson determinant can be regularized by introducing a massive spinless charged fermion (again non-physical, since it violates the spin–statistics connection). Alternatively, it can be expressed in terms of the statistical operator, using Schwinger's representation ($\operatorname{tr}\ln = \ln\det$)

$$\ln\det H - \ln\det H_0 = \operatorname{tr}\int_0^\infty \frac{dt}{t}\left[e^{-tH_0} - e^{-tH}\right],$$

where the operator H is analogous to a non-relativistic Hamiltonian in a magnetic field,

$$H = -\sum_\mu D_\mu D_\mu + m^2, \quad H_0 = -\nabla^2 + m^2.$$

UV divergences then arise from the small t integration. The integral over time can thus be regularized by cutting it for t small, integrating, for example, over $t \geq 1/\Lambda^2$.

16.4.4 The gauge field propagator

For the free gauge action in a covariant gauge, ordinary derivatives can be used because, in an Abelian theory, the gauge field is neutral. The tensor $F_{\mu\nu}$ is gauge invariant and the action for the scalar combination $\sum_\mu \partial_\mu A_\mu$ is arbitrary. Therefore, the large momentum behaviour of the gauge field propagator can be arbitrarily improved by the substitution

$$\sum_{\mu,\nu} F_{\mu\nu}(x)F_{\mu\nu}(x) \mapsto \sum_{\mu,\nu} F_{\mu\nu}(x)P(-\nabla^2/\Lambda^2)F_{\mu\nu}(x),$$

$$\left(\sum_\mu \partial_\mu A_\mu(x)\right)^2 \mapsto \sum_\mu \partial_\mu A_\mu(x)P(-\nabla^2/\Lambda^2)\sum_\mu \partial_\mu A_\mu(x),$$

where $P(z)$ is a polynomial of degree high enough, positive for $z \geq 0$, and $P(0) = 1$.

16.5 Non-Abelian gauge theories

We consider here the $SU(N)$ group and the gauge fields $\mathbf{A}_\mu$ are anti-Hermitian traceless matrices (see Section 11.8). For convenience, we recall here our notation. We introduce the covariant derivative, as acting on a matter field,

$$\mathbf{D}_\mu = \mathbf{1}\partial_\mu + \mathbf{A}_\mu(x), \tag{16.19}$$

and the curvature tensor

$$\mathbf{F}_{\mu\nu}(x) = [\mathbf{D}_\mu, \mathbf{D}_\nu] = \partial_\mu \mathbf{A}_\nu(x) - \partial_\nu \mathbf{A}_\mu(x) + [\mathbf{A}_\mu(x), \mathbf{A}_\nu(x)]. \tag{16.20}$$

The pure gauge action then is

$$\mathcal{S}(\mathbf{A}_\mu) = -\frac{1}{4g^2}\int d^d x \operatorname{tr}\sum_{\mu,\nu} \mathbf{F}_{\mu\nu}(x)\mathbf{F}_{\mu\nu}(x). \tag{16.21}$$

In the covariant gauge

$$\mathcal{S}_{\text{gauge}}(\mathbf{A}) = -\frac{1}{2g^2\xi} \int d^d x \; \text{tr} \left(\sum_\mu \partial_\mu \mathbf{A}_\mu(x) \right)^2,$$

the ghost field action takes the form

$$\mathcal{S}_{\text{ghost}}(\mathbf{A}_\mu, \bar{\mathbf{C}}, \mathbf{C}) = - \int d^d x \; \text{tr} \, \bar{\mathbf{C}}(x) \sum_\mu \partial_\mu \left(\partial_\mu \mathbf{C}(x) + [\mathbf{A}_\mu(x), \mathbf{C}(x)] \right).$$

The ghost fields thus have a simple δ_{ab}/p^2 propagator and canonical dimension 1 in four dimensions.

Compared with the Abelian case, the new features of the non-Abelian gauge action are the presence of gauge field self-interactions and ghost terms.

The problem of regularization in non-Abelian gauge theories shares several features both with the Abelian case and with the non-linear σ-model. After momentum regularization with higher derivatives, the gauge action takes the form

$$\mathcal{S}_{\text{reg.}}(\mathbf{A}_\mu) = - \int d^d x \; \text{tr} \sum_{\mu,\nu} \mathbf{F}_{\mu\nu}(x) P\left(- \mathbf{D}^2 / \Lambda^2 \right) \mathbf{F}_{\mu\nu}(x), \tag{16.22}$$

where $P(z)$ is a polynomial of degree high enough, positive for $z \geq 0$, and $P(0) = 1$. In the same way, the gauge function $(\sum_\mu \partial_\mu \mathbf{A}_\mu)^2$ is changed into

$$\left(\sum_\mu \partial_\mu \mathbf{A}_\mu(x) \right)^2 \longmapsto \sum_\mu \partial_\mu \mathbf{A}_\mu(x) P\left(- \partial^2 / \Lambda^2 \right) \sum_\mu \partial_\mu \mathbf{A}_\mu(x). \tag{16.23}$$

As a consequence, both the gauge field propagator and the ghost propagator can be arbitrarily improved. However, as in the Abelian case, the covariant derivatives generate new interactions that are more singular. It is easy to verify that the power counting of one-loop diagrams is unchanged while higher order diagrams can be rendered convergent by taking the degree of P large enough.

Regularization by higher derivatives takes care of all diagrams except, as in non-linear σ-models, some one-loop diagrams (and thus subdiagrams).

As with charged matter, the one-loop diagrams have to be examined separately. For fermion matter, it is again possible, as in the Abelian case, to add a set of regulator fields, massive fermions and bosons with fermion spin.

In the chiral situation, the problem of the compatibility between the gauge symmetry and the quantization is reduced to an explicit verification of the Ward–Takahashi identities (expressing the consequences of symmetries on correlation functions) for the one-loop diagrams.

Again, we emphasize that the preservation of gauge symmetry is necessary for the cancellation of non-physical states in physical amplitudes and, thus, essential to ensure the physical consistency of the quantum field theory.

16.6 Dimensional regularization and chiral symmetry

The other regularization schemes we now discuss, have the common property that they modify in some essential way the structure of space-time: dimensional regularization, because it relies on defining Feynman diagrams for non-integer dimensions, and lattice regularization, because continuum space is replaced by a discrete lattice.

16.6.1 Dimensional regularization

Dimensional regularization involves the continuation of Feynman diagrams in the parameter d (d is the space or space-time dimension) to arbitrary complex values [102] and, therefore, seems to have no meaning outside perturbation theory. However, this regularization often leads to the simplest perturbative calculations.

It solves the problem of the commutation of quantum operators in local field theories. Indeed, the canonical commutation relations, in the example of a scalar boson, take (in the Schrödinger picture) the form (equation (16.1))

$$[\hat{\phi}(x), \hat{\pi}(y)] = i\hbar\, \delta^{d-1}(x-y) = i\hbar(2\pi)^{1-d} \int \mathrm{d}^{d-1}p\; \mathrm{e}^{ip(x-y)}\,,$$

where $\hat{\pi}(x)$ is the momentum conjugate to the field $\hat{\phi}(x)$. As we have already stressed, in a local theory all operators are taken at the same point and, therefore, a commutation in the product $\hat{\phi}(x)\hat{\pi}(x)$ generates a divergent contribution (for $d > 1$) proportional to

$$\delta^{d-1}(0) = (2\pi)^{1-d} \int \mathrm{d}^{d-1}p\,.$$

(For $d = 1$, it generates a finite ambiguity.) The rules of dimensional regularization imply the consistency of the change of variables $p \mapsto \lambda p$ and thus

$$\int \mathrm{d}^d p = \lambda^d \int \mathrm{d}^d p \;\Rightarrow\; \int \mathrm{d}^d p = 0\,,$$

in contrast to momentum regularization, where it is proportional to a power of the cut-off. Therefore, the order between operators becomes irrelevant because the commutator vanishes. Dimensional regularization thus is applicable to geometric models where these problems of quantization occur, like non-linear σ-models or gauge theories. On the other hand, this property suggests that *the field theory is not completely defined by the perturbative expansion*.

Moreover, the scaling argument also applies to the integral $\int \mathrm{d}^d p/p^2$, which also vanishes in dimensional regularization. The use of this property requires some care. In a massless theory in two dimensions, it leads to an unwanted cancellation between UV and infrared logarithmic divergences.

More generally, from the scaling argument, one infers that all divergences that take the form of powers of the cut-off in momentum cut-off regularization, are cancelled. This automatic *partial renormalization may lead to erroneous physics conclusions*, like the absence of fine tuning.

Finally, it is not applicable when some essential property of the field theory is specific to the initial dimension. An example is provided by theories containing fermions in which parity symmetry is violated.

16.6.2   *The problem with fermions: The example of dimension 4*

The evaluation of diagrams with fermions can be reduced to the calculation of traces of γ matrices and to scalar integrals. Therefore, only one additional prescription for the trace of the unit matrix is needed. There is no natural continuation, since odd and even dimensions behave differently. Since no algebraic manipulation depends on the explicit value of the trace, any smooth continuation in the neighbourhood of the relevant dimension is satisfactory. A convenient choice is to take the trace constant. In even dimensions, as long as only γ_μ matrices are involved, no other problem arises.

Four dimensions: The problem of the matrix γ_5. However, one cannot find a dimensional continuation that preserves all properties of the matrix γ_5, which is the product of all other γ matrices:

$$\gamma_5 = -\frac{1}{4!} \sum_{\mu_1,\mu_2,\mu_3,\mu_4} \epsilon_{\mu_1\ldots\mu_4} \gamma_{\mu_1} \cdots \gamma_{\mu_4} \,. \tag{16.24}$$

($\epsilon_{\mu_1\ldots\mu_4}$ is the complete antisymmetric tensor and $\epsilon_{1234} = 1$.) This leads to serious difficulties if γ_5 in the calculation of Feynman diagrams has to be replaced by its explicit expression in terms of the other γ matrices [178].

Therefore, problems arise in the case of gauge theories with chiral fermions, because the special properties of γ_5 are involved, as we recall below. This difficulty is the source of chiral anomalies.

Since perturbation theory involves the calculation of traces, one possibility is to define γ_5 near four dimensions by

$$\gamma_5 = \sum_{\mu_1,\mu_2,\mu_3,\mu_4} E_{\mu_1\ldots\mu_4} \gamma_{\mu_1} \cdots \gamma_{\mu_4} \,, \tag{16.25}$$

where $E_{\mu\nu\rho\sigma}$ is a completely antisymmetric tensor, which reduces to $-\epsilon_{\mu\nu\rho\sigma}/4!$ in four dimensions. It is simple to verify that, with this definition, γ_5 then anticommutes with the other γ_μ matrices only in four dimensions. For example, if one evaluates the product $\gamma_\nu \gamma_5 \gamma_\nu$ in d dimensions, replacing γ_5 by expression (16.25) and using systematically the anticommutation relations $\gamma_\mu \gamma_\nu + \gamma_\nu \gamma_\mu = 2\delta_{\mu\nu}$, one finds

$$\gamma_\nu \gamma_5 \gamma_\nu = (d - 8)\gamma_5 \,.$$

By contrast, anticommuting properties of the γ_5 would have led to a factor $-d$. This additional contribution, proportional to $d - 4$, if it is multiplied by a factor $1/(d - 4)$ as the consequence of UV divergences in one-loop diagrams, will lead to a finite difference with the formal result.

The alternative option of keeping the anticommuting property of γ_5 contradicts the form (16.25). Actually, it is possible to verify that the only consistent prescription for generic dimensions then is that the traces of γ_5 with any product of γ_μ matrices vanish and, thus, this prescription is useless.

Finally, an alternative possibility consists of breaking $O(d)$ symmetry and keeping the four γ matrices of $d = 4$.

16.7 Lattice regularization

Higher derivative cut-off regularization does not regularize completely field theories in which geometric properties generate also interactions. Examples are provided by the general non-linear σ-models or non-Abelian gauge theories. In such theories some divergences are related to the problem of quantization when in products classical variables are replaced by non-commuting operators (the problem already appears in simple quantum mechanics). Other regularization methods are then needed, like lattice regularization.

Lattice regularization The Euclidean real space $\mathbb{R}^d$ (time is necessarily *imaginary*) is replaced by a regular lattice. Although several lattices are possible, we consider below only the hypercubic lattice $\mathbb{Z}^d$ of points with integer coordinate.

To each site x of the lattice are attached variables corresponding to fields in the continuum. To the action S in the continuum corresponds a lattice action, the energy of lattice field configurations in the language of classical statistical physics. Locality can be implemented by considering lattice Lagrangian densities that depend only on a site and its neighbours. The regularized partition function becomes a sum over lattice configurations. The advantages of a lattice regularization are:

(i) provided a continuum limit exists, lattice regularization provides an unambiguous and non-perturbative definition of a quantum theory;

(ii) it preserves most global and local symmetries with the exception of the space $O(d)$ symmetry, which is replaced by a hypercubic symmetry (but this turns out not to be a major difficulty), and fermion chirality, which turns out to be a more serious problem, as we will show;

(iii) therefore, it is the only established regularization that provides a non-perturbative definition for gauge theories and other geometric models (see Sections 10.1, 13.8 and 13.9); as a consequence, the regularized partition function can be evaluated by numerical methods, like stochastic integration methods.

It has one serious disadvantage: perturbative calculations are very complicated.

16.7.1 *Scalar bosons on the lattice*

To the action (16.2) for a scalar field ϕ in the continuum corresponds a lattice action. The Euclidean Lagrangian density becomes a function of lattice variables $\phi(x)$, where x now is a lattice site. Derivatives $\partial_\mu \phi$ of the continuum are replaced by finite differences, for example:

$$\partial_\mu \phi(x) \mapsto \nabla_\mu^{\mathrm{lat.}} \phi(x) = \left[\phi(x + a\epsilon_\mu) - \phi(x) \right]/a, \qquad (16.26)$$

where a is the lattice spacing and ϵ_μ is the unit vector in the μ direction. Then,

$$S_{\mathrm{lat.}}(\phi) = a^d \sum_x \left[\tfrac{1}{2} \sum_\mu \left(\nabla_\mu^{\mathrm{lat.}} \phi(x) \right)^2 + \tfrac{1}{2} m^2 \phi^2(x) + V\left(\phi(x)\right) \right].$$

The propagator $\Delta_a(p)$ for the Fourier components of a massive scalar field is given by

$$\Delta_a^{-1}(\mathbf{p}) = m^2 + \frac{2}{a^2} \sum_\mu \left(1 - \cos(ap_\mu) \right). \qquad (16.27)$$

It is a periodic function of the components p_μ of the momentum $\mathbf{p}$ with period $2\pi/a$. In the small lattice spacing limit,

$$\Delta_a^{-1}(\mathbf{p}) = m^2 + \mathbf{p}^2 - \tfrac{1}{12} \sum_\mu a^2 p_\mu^4 + O\left(p_\mu^6\right), \tag{16.28}$$

and the small momentum behaviour is given by p^2. Thus, hypercubic symmetry implies an emergent $O(d)$ symmetry and the continuum propagator is recovered.

16.7.2   Fermions, chiral symmetry and the doubling problem

The definition of relativistic fermions on the lattice leads to the problem of *fermion doubling* and *chiral symmetry implementation*.

A lattice regularization of the free action for a Dirac fermion

$$S(\bar\psi, \psi) = \int \mathrm{d}^d x \, \bar\psi(x) \left(\slashed\partial + m\right) \psi(x),$$

which preserves chiral properties in the massless limit, can be obtained by replacing the derivative $\partial_\mu \psi(x)$, for example, by the symmetric combination

$$\nabla_\mu^{\text{lat.}} \psi(x) = \left[\psi(x + a\epsilon_\mu) - \psi(x - a\epsilon_\mu)\right]/2a\,.$$

The Dirac operator is then replaced by the lattice operator (in Fourier representation):

$$m + i\slashed p \mapsto D_{\text{lat.}}(p) = m + i \sum_\mu \gamma_\mu \frac{\sin a p_\mu}{a}\,, \tag{16.29}$$

a periodic function of the components p_μ of the momentum vector.

A problem then arises: the equations relevant to the small lattice spacing limit,

$$\sin(a\,p_\mu) = 0 \;\Rightarrow\; p_\mu = 0 \quad (\text{mod } \pi/a)\,,$$

have each the two solutions $p_\mu = 0$ and $p_\mu = \pi/a$ within one period, that is, 2^d solutions within what is called the Brillouin zone, a problem that cannot be solved by adding terms connecting fermions separated by more than one lattice spacing.

Therefore, the propagator (16.29) propagates 2^d fermions. To remove this degeneracy, it is possible to add to the regularized action an additional scalar term δS involving second derivatives (the recipe of *Wilson's fermions*):

$$\delta S(\bar\psi, \psi) = \tfrac{1}{2} M \sum_{x,\mu} \left[2\bar\psi(x)\psi(x) - \bar\psi\left(x + a\epsilon_\mu\right)\psi(x) - \bar\psi(x)\psi\left(x + a\epsilon_\mu\right)\right]. \tag{16.30}$$

The modified Dirac operator for the Fourier components of the field reads

$$D_W(p) = m + M \sum_\mu (1 - \cos a p_\mu) + \frac{i}{a} \sum_\mu \gamma_\mu \sin a p_\mu\,. \tag{16.31}$$

The fermion propagator becomes

$$\Delta(p) = D_W^\dagger(p) \left(D_W(p) D_W^\dagger(p)\right)^{-1}.$$

Explicitly, the denominator reads

$$D_W(p)D_W^\dagger(p) = \left[m + M\sum_\mu (1 - \cos ap_\mu)\right]^2 + \frac{1}{a^2}\sum_\mu \sin^2 ap_\mu\,.$$

Therefore, the degeneracy between the different states is lifted. For each component p_μ that takes the value π/a, the mass is increased by $2M$. If M is of order $1/a$, the spurious states are eliminated in the continuum limit.

However, a problem arises if one wants to construct a theory with chiral symmetry and massless fermions. Chiral symmetry implies that the Dirac operator $D(p)$ anti-commutes with γ_5,

$$D(p)\gamma_5 + \gamma_5 D(p) = 0\,, \tag{16.32}$$

and, therefore, both the mass term and the term (16.30) are excluded. It remains possible to add various counter-terms and try to adjust them to recover chiral symmetry in the continuum limit. But there is no *a priori* guarantee that this is indeed possible and, moreover, calculations are plagued by additional fine tuning problems and cancellations of unnecessary UV divergences.

It can be shown that modifying the fermion propagator by adding terms connecting fermions separated by more than one lattice spacing does not solve the doubling problem. In fact, this doubling of the number of fermion degrees of freedom is directly related to the problem of anomalies.

Since the simplest form of the propagator yields 2^d fermion states, one tries in numerical simulations to reduce this number to a smaller multiple of 2, using, for instance, the idea of staggered fermions introduced by Kogut and Susskind [164].

Another class of theoretical solutions with two types of implementations, overlap fermions [166] and domain wall fermions [167], has been found more recently based on chiral transformations spread over several lattice sites [179] and solutions of the Ginsparg–Wilson relation [165]. These topics are further discussed in Section 17.7.

16.7.3 *Gauge theories: Gauge fields and scalar bosons*

The basic construction of lattice gauge theory is described in Sections 13.8 and13.9 and we recall here only a few elements for convenience.

Gauge transformations are independent group transformations on each lattice site. Gauge fields are replaced by link variables $\mathbf{U}_{xy}$, group elements associated with the links joining the sites x and y on the lattice [44]. A typical gauge invariant lattice action corresponding to the continuum action of a gauge field coupled to complex scalar bosons ϕ^*, ϕ then has the form

$$\mathcal{S}(\mathbf{U}, \phi^*, \phi) = \beta \sum_{\substack{\text{all} \\ \text{plaquettes}}} \text{tr}\,\mathbf{U}_{xy}\mathbf{U}_{yz}\mathbf{U}_{zt}\mathbf{U}_{tx} + \kappa \sum_{\substack{\text{all} \\ \text{links}}} \phi_x^* \mathbf{U}_{xy}\phi_y$$

$$+ \sum_{\substack{\text{all} \\ \text{sites}}} V(\phi_x^* \phi_x), \tag{16.33}$$

where x, y,... denote lattice sites, a plaquette is a closed loop in the form of a square, and β and κ are coupling constants.

17 Quantum anomalies: A few physics applications

In Chapter 16, we have discussed the problem of regularization of quantum field theories, methods to render theories finite without changing their large scale properties, with a few standard examples. We have shown that none of the standard regularization methods can deal in a straightforward way with one-loop diagrams in the case of gauge fields coupled to chiral fermions and verified that, in the case of gauge theories with chiral fermion matter, obstructions to regularization appear in all schemes, leaving room for *quantum anomalies*, that is, the impossibility of implementing symmetries of the classical field theory in the full quantum field theory.

The existence of anomalies has important physics consequences. For example, they may lead to obstructions to the construction of gauge theories when the gauge field couples differently to the two fermion chiral components. They are responsible for the non-vanishing of the electromagnetic decay of the π_0 meson and solve the $U(1)$ problem (see also Sections 18.4 and 18.5 for additional considerations).

We study only anomalies that are local quantities because they result from short distance singularities. They can be determined by perturbative calculations. (Peculiar global non-perturbative anomalies have also been exhibited.) They take the form of local polynomials in the fields. Since anomalies are responses to local (space-dependent) group transformations but vanish for a class of space-independent transformations, they have a *topological* character.

In this chapter, inspired by the lectures in [181], first we verify the existence of anomalies [173–175] and derive their form explicitly, beginning with the simplest example of the so-called Abelian anomaly, that is, the anomaly in the conservation of the Abelian axial current in gauge theories.

We relate anomalies to the index of a covariant Dirac operator in the background of a gauge field [180], confirming their *topological* character.

From the anomaly for a general axial current, we infer conditions for gauge theories that couple differently to fermion chiral components to be anomaly-free [176].

For a number of years, the problem of chiral anomalies has been discussed mainly within the framework of perturbation theory. New non-perturbative formulations based on lattice regularization and the Ginsparg–Wilson relation [165] have been more recently proposed. We describe here the so-called overlap [166] and domain wall fermion [167] regularizations. The method of domain wall fermions is related to the mechanism of zero modes in *supersymmetric quantum mechanics*.

The absence of anomalies can then be verified directly on the lattice. The analysis confirms that field theories that had been discovered to be anomaly-free in perturbation theory, are also anomaly-free in a non-perturbative lattice construction. Hence, the specific problem of lattice chiral fermions was technical in essence rather than reflecting an inconsistency of chiral gauge theories beyond perturbation theory, as one might have feared.

From Random Walks to Random Matrices. Jean Zinn-Justin, Oxford University Press (2019).
© Jean Zinn-Justin. DOI: 10.1093/oso/9780198787754.001.0001

Finally, since these regularization schemes have a natural implementation in five dimensions in the form of domain wall fermions, this again opens the door to speculations about additional space dimensions.

17.1 Electromagnetic decay of the neutral pion and Abelian anomaly

We now show that, indeed, chiral symmetric gauge theories, involving gauge fields coupled to massless fermions, can be found where the axial current is not conserved. The divergence of the axial current in a chiral quantum field theory, when it does not vanish, is called an *anomaly*.

First, we discuss the Abelian axial current, in four dimensions (the generalization to all even dimensions then is simple) and its implication for the electromagnetic decay of the neutral pion and, we then discuss the general non-Abelian situation.

17.1.1 *Abelian axial current and Abelian vector gauge field*

The only possible source of anomalies is one-loop fermion diagrams in gauge theories when chiral symmetry is involved. This reduces the problem to a discussion of fermions in the background of gauge fields or, equivalently, to the properties of the determinant of the gauge covariant Dirac operator.

Thus, we consider a quantum electrodynamics (QED)-like fermion action for massless Dirac fermions $\psi, \bar\psi$, coupled to an Abelian gauge field A_μ, of the form,

$$\mathcal{S}(\bar\psi, \psi; A) = -\int \mathrm{d}^4 x\, \bar\psi(x) \slashed{D}\, \psi(x) \tag{17.1}$$

with

$$\mathrm{D}_\mu = \partial_\mu + ieA_\mu\,, \quad \slashed{D} \equiv \sum_\mu \gamma_\mu \mathrm{D}_\mu = \slashed\partial + ie\slashed{A}(x).$$

To the covariant derivative D_μ corresponds the electromagnetic tensor (with an *unusual normalization*)

$$F_{\mu\nu}(x) = [\mathrm{D}_\mu, \mathrm{D}_\nu] = ie[\partial_\mu A_\nu(x) - \partial_\nu A_\mu(x)]. \tag{17.2}$$

We want to study the properties of the fermion field integral

$$\mathcal{Z}(A_\mu) = \int [\mathrm{d}\psi \mathrm{d}\bar\psi] \exp\left[-\mathcal{S}(\psi, \bar\psi; A)\right] = \det \slashed{D}. \tag{17.3}$$

One can find regularizations that preserve invariance with respect to the gauge transformations,

$$\psi(x) = \mathrm{e}^{ie\Lambda(x)}\, \psi'(x), \quad \bar\psi(x) = \mathrm{e}^{-ie\Lambda(x)}\, \bar\psi'(x), \quad A_\mu(x) = -\partial_\nu\Lambda(x) + A'_\mu(x), \tag{17.4}$$

and, since the fermions are massless, chiral symmetry. Therefore, one would naively expect the corresponding axial current to be conserved (continuous symmetries are classically associated with current conservation).

However, the proof of current conservation involves *space-dependent chiral transformations* and, therefore, steps that cannot be regularized without breaking local chiral symmetry.

Chiral gauge transformations and axial current. After the space-dependent chiral, or chiral gauge transformations

$$\psi_\theta(x) = e^{i\theta(x)\gamma_5}\,\psi(x), \quad \bar\psi_\theta(x) = \bar\psi(x)\,e^{i\theta(x)\gamma_5}, \tag{17.5}$$

the action becomes

$$\mathcal{S}_\theta(\bar\psi,\psi;A) = -\int \mathrm{d}^4x\,\bar\psi_\theta(x)\,\slashed{D}\,\psi_\theta(x) = -\int \mathrm{d}^4x\,\bar\psi(x)\,e^{i\theta(x)\gamma_5}\,\slashed{D}\,e^{i\theta(x)\gamma_5}\,\psi(x)$$

$$= \mathcal{S}(\bar\psi,\psi;A) + \int \mathrm{d}^4x\,\sum_\mu \partial_\mu\theta(x)J^5_\mu(x), \tag{17.6}$$

where the coefficient of $\partial_\mu\theta(x)$, (with the convention $\gamma_5 = -\gamma_1\gamma_2\gamma_3\gamma_4$)

$$J^5_\mu(x) \equiv i\bar\psi(x)\gamma_5\gamma_\mu\psi(x), \tag{17.7}$$

is the *axial current*.

Since

$$\int \mathrm{d}^4x\,\sum_\mu \partial_\mu\theta(x)J^5_\mu(x) = -\int \mathrm{d}^4x\,\theta(x)\sum_\mu \partial_\mu J^5_\mu(x),$$

the field integral (17.3) where $\mathcal{S}$ is replaced by $\mathcal{S}_\theta$,

$$\mathcal{Z}(A_\mu,\theta) = \det\left(e^{i\gamma_5\theta(x)}\,\slashed{D}\,e^{i\gamma_5\theta(x)}\right), \tag{17.8}$$

is the generating functional of the divergence of the axial current $\sum_\mu \partial_\mu J^5_\mu(x)$ correlation functions in an external field A_μ.

Since the determinant of $e^{i\gamma_5\theta}$ is 1 (γ_5 has eigenvalues $(1,1,-1,-1)$), one would naively conclude that $\mathcal{Z}(A_\mu,\theta) = \mathcal{Z}(A_\mu)$ and, therefore, that the current $J^5_\mu(x)$ is conserved. This is a conclusion we now check by an explicit calculation of the expectation value of $\sum_\mu \partial_\mu J^5_\mu(x)$ in the case of the action (17.1).

Chiral properties of the fermion determinant
(i) For any regularization that is consistent with the Hermiticity of γ_5,

$$|\mathcal{Z}(A_\mu,\theta)|^2 = \det\left[e^{i\gamma_5\theta(x)}\,\slashed{D}\,e^{i\gamma_5\theta(x)}\right]\det\left[e^{-i\gamma_5\theta(x)}\,\slashed{D}^\dagger\,e^{-i\gamma_5\theta(x)}\right] = \det\left(\slashed{D}\slashed{D}^\dagger\right),$$

and thus $|\mathcal{Z}(A_\mu,\theta)|$ is independent of θ. Therefore, an anomaly can affect only the imaginary part of $\ln\mathcal{Z}$.

(ii) We have shown that regularizations with regulator fields such that gauge invariance is maintained and the determinant is independent of θ for $\theta(x)$ constant, can be found (Section 16.4).

(iii) If the regularization is gauge invariant, $\mathcal{Z}(A_\mu, \theta)$ is also gauge invariant. Therefore, a possible anomaly is also gauge invariant.

(iv) A possible anomaly is a short distance effect (equivalently a large momentum effect). As $\ln \mathcal{Z}(A_\mu, \theta)$ receives only contributions from one-particle-irreducible (1PI) diagrams, short distance contributions generated by the singularities of one-loop diagrams and, thus, the anomaly take the form of local polynomials in the field A_μ and the source $\partial_\mu \theta$, constrained by parity and power counting. We set,

$$\ln \mathcal{Z}(A_\mu, \theta) - \ln \mathcal{Z}(A_\mu, 0) = i \int \mathrm{d}^4 x \, \mathcal{L}(A, \partial\theta; x),$$

where $\mathcal{L}$ is the sum of monomials of dimension 4.

The field A_μ and $\partial_\mu \theta$ have dimension 1, and no mass parameter is available. Thus, at order θ, only one monomial is available,

$$\mathcal{L}(A, \partial\theta; x) \propto e^2 \sum_{\mu,\nu,\rho,\sigma} \epsilon_{\mu\nu\rho\sigma} \partial_\mu \theta(x) A_\nu(x) \partial_\rho A_\sigma(x),$$

where $\epsilon_{\mu\nu\rho\sigma}$ is the complete antisymmetric tensor with $\epsilon_{1234} = 1$. An integration by parts and antisymmetrization leads to the gauge invariant expression

$$\int \mathrm{d}^4 x \, \mathcal{L}(A, \partial\theta; x) \propto \sum_{\mu,\nu} \int \mathrm{d}^4 x \, F_{\mu\nu}(x) \tilde{F}_{\mu\nu}(x) \theta(x),$$

where $F_{\mu\nu}$ is the electromagnetic tensor (17.2) and we have introduced its dual,

$$\tilde{F}_{\mu\nu}(x) = \sum_{\rho,\sigma} \epsilon_{\mu\nu\rho\sigma} F_{\rho\sigma}(x). \tag{17.9}$$

The coefficient of $\theta(x)$ is the expectation value in an external gauge field of $\nabla \cdot \mathbf{J}^5(x)$, the divergence of the axial current. It is determined up to a multiplicative constant:

$$\left\langle \nabla \cdot \mathbf{J}^5(x) \right\rangle \propto \sum_{\mu,\nu} F_{\mu\nu}(x) \tilde{F}_{\mu\nu}(x) = -4e^2 \sum_{\mu,\nu,\rho,\sigma} \epsilon_{\mu\nu\rho\sigma} \partial_\mu A_\nu(x) \partial_\rho A_\sigma(x),$$

where we denote by $\langle \bullet \rangle$ expectation values with respect to the measure $\mathrm{e}^{-\mathcal{S}(\bar{\psi},\psi;A)}$.

Since the possible anomaly is independent up to a multiplicative factor of the regularization, it must indeed be a gauge invariant local function of A_μ.

To find the multiplicative factor, it suffices to calculate the coefficient of the term quadratic in A in the expansion of $\langle \nabla \cdot \mathbf{J}^5(x) \rangle$ in powers of A. In terms of Feynman diagrams, this corresponds to evaluating the one-loop axial-vector-vector current diagrams shown in Fig. 17.1.

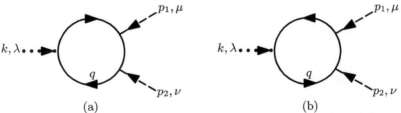

(a)   (b)

Fig. 17.1 Potentially anomalous diagrams: k, λ corresponds to the axial current.

17.1.2 Regulator fields and explicit anomaly calculation

Rather than evaluating the Feynman diagrams in the Fourier representation, we evaluate the anomaly more directly. We regularize the action (17.1) by adding the contribution of a charged boson field ϕ with spin $1/2$ and a large mass M (see Section 16.4). We obtain the *gauge invariant action*,

$$\mathcal{S}(\bar{\psi}, \psi, \bar{\phi}, \phi) = \int \mathrm{d}^4 x \left[-\bar{\psi}(x)\slashed{D}\,\psi(x) + \bar{\phi}(x)(\slashed{D} + M)\phi(x) \right].$$

We perform in the corresponding regularized field integral (17.3) (now an integral over $\psi, \bar{\psi}, \phi, \bar{\phi}$) a change of variables of the form of a space-dependent chiral transformation (17.5) acting identically on the fermion and boson fields,

$$\psi_\theta(x) = e^{i\theta(x)\gamma_5}\,\psi(x), \quad \bar{\psi}_\theta(x) = \bar{\psi}(x)\,e^{i\theta(x)\gamma_5},$$
$$\phi_\theta(x) = e^{i\theta(x)\gamma_5}\,\phi(x), \quad \bar{\phi}_\theta(x) = \bar{\phi}(x)\,e^{i\theta(x)\gamma_5}.$$

In this transformation, the integration measure $[\mathrm{d}\phi\mathrm{d}\bar{\phi}\mathrm{d}\psi\mathrm{d}\bar{\psi}]$ is invariant but the boson mass term is affected.

Then,

$$e^{i\theta(x)\gamma_5}\slashed{D}\,e^{i\theta(x)\gamma_5} = -i\sum_\mu \gamma_5\gamma_\mu\partial_\mu\theta(x). \tag{17.10}$$

Therefore, the variation $\delta\mathcal{S}$ of the action at first order in θ is

$$\delta\mathcal{S} = \int \mathrm{d}^4 x \left[\sum_\mu \partial_\mu\theta(x)J_\mu^5(x) + 2iM\theta(x)\bar{\phi}(x)\gamma_5\phi(x) \right] \tag{17.11}$$

with

$$J_\mu^5(x) = i\bar{\psi}(x)\gamma_5\gamma_\mu\psi(x) - i\bar{\phi}(x)\gamma_5\gamma_\mu\phi(x).$$

Integrating by parts in expression (17.11) to factorize $\theta(x)$, expanding then the integrand in the field integral in powers of θ and integrating, we express that the field integral is not modified by a change of variables. Identifying the coefficient of $\theta(x)$, one thus obtains the equation

$$\langle \nabla \cdot \mathbf{J}^5(x) \rangle = 2iM \langle \bar{\phi}(x)\gamma_5\phi(x) \rangle$$
$$= iM \operatorname{tr} \langle x | (\gamma_5(\slashed{D} + M)^{-1} + (\slashed{D} + M)^{-1}\gamma_5) | x \rangle, \tag{17.12}$$

where the bra-ket notation of quantum mechanics has been used in the second line to specify matrix elements of operators, and the trace refers to γ matrices.

The divergence of the axial current comes here from the *boson contribution*.

The expression (17.12) can be transformed using the following identities

$$(\slashed{D} + M)^{-1}\gamma_5 = -\gamma_5(\slashed{D} - M)^{-1},$$
$$(\slashed{D} + M)^{-1} - (\slashed{D} - M)^{-1} = 2M(M^2 - \slashed{D}^2)^{-1}$$

and setting $\gamma_\mu\gamma_\nu = \delta_{\mu\nu} + \sigma_{\mu\nu}$ with $\sigma_{\mu\nu} = \frac{1}{2}[\gamma_\mu, \gamma_\nu]$,

$$\slashed{D}^2 = \mathrm{D}^2 + \frac{1}{2}\sum_{\mu,\nu} \sigma_{\mu\nu} F_{\mu\nu}.$$

The contribution of order A^2 is obtained by expanding up to second order in $F_{\mu\nu}$ the expression

$$\left(M^2 - \slashed{D}^2\right)^{-1} = (M^2 - D^2)^{-1} + \tfrac{1}{2}(M^2 - D^2)^{-1}\sum_{\mu,\nu} F_{\mu\nu}\sigma_{\mu\nu}(M^2 - D^2)^{-1}$$

$$+ \tfrac{1}{4}(M^2 - D^2)^{-1}\sum_{\mu,\nu} F_{\mu\nu}\sigma_{\mu\nu}(M^2 - D^2)^{-1}\sum_{\rho,\sigma} F_{\rho\sigma}\sigma_{\rho\sigma}(M^2 - D^2)^{-1}$$

$$+ O(A^3). \tag{17.13}$$

The trace with γ_5 of the two first terms vanishes. In the third term, one uses (definition (16.24))

$$\operatorname{tr}\gamma_5\gamma_\mu\gamma_\nu\gamma_\rho\gamma_\sigma = -4\,\epsilon_{\mu\nu\rho\sigma}$$

and, therefore,

$$\operatorname{tr}\gamma_5\sigma_{\mu\nu}\sigma_{\rho\sigma} = -4\,\epsilon_{\mu\nu\rho\sigma},$$

where $\epsilon_{\mu\nu\rho\sigma}$ is the completely antisymmetric tensor with $\epsilon_{1234} = 1$. In the denominators, one can now substitute $D^2 \mapsto \nabla^2$. One obtains,

$$\langle \nabla \cdot \mathbf{J}^5(x)\rangle = -2iM^2 \sum_{\mu,\nu,\rho,\sigma}\epsilon_{\mu\nu\rho\sigma}$$

$$\times \langle x|\left[M^2 - \nabla^2\right]^{-1}F_{\mu\nu}\left[M^2 - \nabla^2\right]^{-1}F_{\rho\sigma}\left[M^2 - \nabla^2\right]^{-1}|x\rangle.$$

In the large M limit, the support of $\langle x|[M^2 - \nabla^2]^{-1}|y\rangle$ converges to $x = y$. The argument of $F_{\mu\nu}$ can be replaced everywhere by x. The expression reduces to

$$\langle \nabla \cdot \mathbf{J}^5(x)\rangle = -2iM^2 \sum_{\mu,\nu,\rho,\sigma}\epsilon_{\mu\nu\rho\sigma}F_{\mu\nu}(x)F_{\rho\sigma}(x)\langle x|\left[M^2 - \nabla^2\right]^{-3}|x\rangle.$$

Finally,

$$\langle x|\left[M^2 - \nabla^2\right]^{-3}|x\rangle = \frac{1}{16\pi^4}\int \frac{\mathrm{d}^4 k}{(k^2 + M^2)^3} = \frac{1}{32\pi^2 M^2}.$$

The divergence of the axial current is thus (using the definition (17.9)),

$$\langle \nabla \cdot \mathbf{J}^5(x)\rangle = -\frac{i}{16\pi^2}\sum_{\mu,\nu}F_{\mu\nu}(x)\tilde{F}_{\mu\nu}(x). \tag{17.14}$$

With the more standard normalization of the electromagnetic tensor, the coefficient becomes $ie^2/16\pi^2 = i\alpha/4\pi$.

Therefore, in a QED-like gauge invariant field theory with massless fermions, the axial current is not conserved: this is called the *chiral anomaly*.

Since global chiral symmetry is not broken, the integral over the whole space of the anomalous term must vanish at least for a class of 'small' gauge fields. This condition is verified since the anomaly can be written as a total derivative,

$$\sum_{\mu,\nu} F_{\mu\nu}(x)\tilde{F}_{\mu\nu}(x) = -4e^2 \sum_{\mu,\nu,\rho,\sigma} \partial_\mu\big(\epsilon_{\mu\nu\rho\sigma}A_\nu(x)\partial_\rho A_\sigma(x)\big). \tag{17.15}$$

The space integral of the anomalous term depends only on the behaviour of the gauge field at the boundaries, and this property indicates a connection between *topology and anomalies*.

Equation (17.14) also implies

$$\ln\det\Big[e^{i\gamma_5\theta(x)}\,\slashed{D}\,e^{i\gamma_5\theta(x)}\Big] = \ln\det\slashed{D} + \frac{i}{16\pi^2}\int \mathrm{d}^4x\,\theta(x)\sum_{\mu,\nu} F_{\mu\nu}(x)\tilde{F}_{\mu\nu}(x). \tag{17.16}$$

General even dimensions. The generalization to other even dimensions $2n$ is rather straightforward. The result can be derived by expanding an expression analogous to expression (17.12) up to degree n in A. If again gauge invariance is imposed, the anomaly in the divergence of the axial current $J_\lambda^{2n+1}(x)$ becomes

$$\langle \nabla \cdot \mathbf{J}^{2n+1}(x)\rangle = \frac{2i}{(4i\pi)^n n!} \sum_{\mu_1,\nu_1,\ldots,\mu_n,\nu_n} \epsilon_{\mu_1\nu_1\ldots\mu_n\nu_n} F_{\mu_1\nu_1}(x)\cdots F_{\mu_n\nu_n}(x)\,, \tag{17.17}$$

where $\epsilon_{\mu_1\nu_1\ldots\mu_n\nu_n}$ is the completely antisymmetric tensor with $\epsilon_{1,2,\ldots,2n} = 1$.

Calculation by point splitting. Another calculation, based on regularization by point splitting, gives further insight into the mechanism that generates the anomaly. One considers the non-local operator

$$J_\mu^5(x,a) = i\bar{\psi}(x - a/2)\gamma_5\gamma_\mu\psi(x + a/2)\exp\left[ie\int_{x-a/2}^{x+a/2}\mathbf{A}(s)\cdot \mathrm{ds}\right], \tag{17.18}$$

in the limit $|a| \to 0$. To avoid breaking of rotation symmetry by the regularization, before taking the limit $|a| \to 0$, one has to average over all orientations of the vector a. The multiplicative gauge factor (parallel transporter) ensures gauge invariance of the regularized operator (transformations (17.4)).

Calculating the divergence of the axial current and using the quantum field equations, one recovers the result (17.14).

17.1.3 The electromagnetic decay of the neutral pion

In a model for low energy hadron physics where hadrons are considered elementary particles, the low mass of the pion particles is explained by an approximate spontaneously broken $SU(2) \times SU(2)$ chiral symmetry, the pion being an approximate Goldstone scalar boson. This low energy physics can be described by an effective field theory, the σ-model. The action is the sum of a $SU(2) \times SU(2)$ symmetric contribution and a symmetry breaking term linear in the σ field, which, together with the π field, transforms under a representation of the symmetry group.

The non-conservation of the axial current $\mathbf{J}_\mu^5$ is at leading order expressed by the equation

$$\nabla \cdot \mathbf{J}^5(x) = m_\pi^2 f_\pi \boldsymbol{\pi}(x), \tag{17.19}$$

where $\boldsymbol{\pi}(x)$ is the pion field, f_π is the pion decay rate constant, and m_π is the pion mass. Moreover, we use here the conventional normalization of the axial current, which differs by a factor 2 from our convention elsewhere.

We consider now the third component of the current, $[J_\mu^5]_3$, which corresponds in the right-hand side to the neutral pion π_0 component. After the introduction of electromagnetic interactions in the model, the relation between the divergence of the axial current and the π_0-field makes it possible to calculate the electromagnetic decay rate of the neutral pion when the four-momentum $\mathbf{k}$ of the pion vanishes. In the absence of anomalies, the expectation value of relation (17.19) multiplied by two photon fields implies that the decay rate vanishes for $\mathbf{k} = 0$, in contradiction to reasonable smoothness assumptions and experimental results. By contrast, taking into account the anomaly equation (17.14), one finds

$$\nabla \cdot [\mathbf{J}^5]_3(x) = m_\pi^2 f_\pi \pi_0(x) + \frac{i}{32\pi^2} \sum_{\mu,\nu,\rho,\sigma} \epsilon_{\mu\nu\rho\sigma} F_{\mu\nu}(x) F_{\rho\sigma}(x). \tag{17.20}$$

Multiplying the equation by two photon fields, calculating the expectation value and taking the limit $\mathbf{k} = 0$ to eliminate the left-hand side, one now obtains a non-vanishing decay amplitude for a non-physical π_0 at zero total momentum. In the σ-model, at leading order, one can then extrapolate to $k^2 = -m_\pi^2$.

The theoretical rate Γ is then given by [182]

$$\Gamma_{\text{theor.}} = \frac{\alpha^2 m_\pi^3}{64\pi^3 f_\pi^2} = 7.6 \text{ eV},$$

in very good agreement with the experimental value $\Gamma_{\text{exp}} = (7.63 \pm 0.16)$ eV. An analogous estimate was first derived by Steinberger [183] from direct Feynman graph calculation, before its relation to anomalies was discovered.

Note that a similar theoretical estimate is obtained in the quark model with massless quarks, for *three colours*.

17.1.4 Chiral gauge theories

A gauge theory is consistent only if the gauge field is coupled to a conserved current. An anomaly that affects the current destroys gauge invariance in the full quantum theory. Therefore, the theory with axial gauge symmetry, where the action in the fermion sector reads

$$\mathcal{S}(\bar{\psi}, \psi; B) = -\int \mathrm{d}^4 x \, \bar{\psi}(x)[\slashed{\partial} + ig\gamma_5 \slashed{B}(x)]\psi(x), \tag{17.21}$$

is inconsistent because the axial current is not conserved.

Indeed, current conservation applies to the BBB vertex at one-loop order. Because now the three point vertex is symmetric, the divergence does not vanish.

More generally, the anomaly prevents the construction of a theory that would have both an Abelian gauge vector and axial symmetry, where the action in the fermion sector would read

$$\mathcal{S}(\bar{\psi}, \psi; A, B) = - \int \mathrm{d}^4 x \, \bar{\psi}(x)[\slashed{\partial} + ie\slashed{A}(x) + i\gamma_5 g\slashed{B}(x)]\psi(x). \tag{17.22}$$

A way to solve both problems is to cancel the anomaly by introducing another fermion of opposite chirality. With more fermions, other combinations of couplings are possible. However, a purely axial gauge theory with two fermions of opposite chiral charges can be rewritten as a vector theory by combining differently the chiral components of both fermions.

17.2 A two-dimensional illustration: The Schwinger model

We consider *two-dimensional QED*, with one *massless Dirac fermion* coupled to an Abelian gauge field. The model, first discussed by Schwinger [145], exhibits the simplest example of a *chiral anomaly*.

It also exhibits, in two dimensions, both a *confinement property* (see Section 13.10) and a *spontaneous breaking of chiral symmetry*.

The model can be solved exactly by an elegant bosonization technique.

Notation. In two dimensions, the three γ matrices can be identified with the three Pauli σ matrices. One representation is

$$\sigma_1 = \begin{pmatrix} 0 & 1 \\ 1 & 0 \end{pmatrix}, \quad \sigma_2 = \begin{pmatrix} 0 & -i \\ i & 0 \end{pmatrix}, \quad \sigma_3 = \begin{pmatrix} 1 & 0 \\ 0 & -1 \end{pmatrix}. \tag{17.23}$$

17.2.1 The classical theory

The action reads

$$\mathcal{S}(\bar{\psi}, \psi, A_\mu) = - \int \mathrm{d}^2 x \left[\frac{1}{4e^2} \sum_{\mu,\nu} F_{\mu\nu}^2(x) + \bar{\psi}(x)\slashed{D}\,\psi(x) \right] \tag{17.24}$$

with

$$\slashed{D} \equiv \slashed{\partial} + ie\slashed{A} = \sum_\mu \gamma_\mu \mathrm{D}_\mu\,, \quad F_{\mu\nu}(x) = [\mathrm{D}_\mu, \mathrm{D}_\nu] = ie\big(\partial_\mu A_\nu(x) - \partial_\nu A_\mu(x)\big). \tag{17.25}$$

A peculiarity of two dimensions is that the classical fermion action has two $U(1)$ gauge symmetries corresponding to both phase and chiral phase transformations,

$$
\begin{aligned}
\psi(x) &= \mathrm{e}^{ie(\theta_3(x)\gamma_3 + \theta(x))}\,\psi'(x), \\
\bar{\psi}(x) &= \bar{\psi}'(x)\,\mathrm{e}^{ie(\gamma_3\theta_3(x) - \theta(x))}\,.
\end{aligned}
\tag{17.26}
$$

Indeed,

$$e^{i(e\gamma_3\theta_3(x)-\theta(x))}\,\slashed{\partial}\,e^{ie(\theta_3(x)\gamma_3+\theta(x))} = \slashed{\partial} + ie\left(-\gamma_3\slashed{\partial}\theta_3(x) + \slashed{\partial}\theta(x)\right).$$

Then, since

$$\gamma_3\gamma_\mu = i\sum_\nu \epsilon_{\mu\nu}\gamma_\nu\,, \tag{17.27}$$

$(\epsilon_{\mu\nu} = -\epsilon_{\nu\mu}, \epsilon_{12} = 1)$, one can rewrite the variation as

$$ie\left(-\gamma_3\slashed{\partial}\theta_3(x) + \slashed{\partial}\theta(x)\right) = e\sum_\mu \gamma_\mu\left[-\sum_\nu \epsilon_{\mu\nu}\partial_\nu\theta_3(x) + i\partial_\mu\theta(x)\right]. \tag{17.28}$$

Both terms can be cancelled by the shift of the gauge field,

$$A_\mu(x) \mapsto A_\mu(x) - i\sum_\nu \epsilon_{\mu\nu}\partial_\nu\theta_3(x) - \partial_\mu\theta(x).$$

This shows that, if one parametrizes the gauge field in terms of the two scalar fields $\{\theta(x), \theta_3(x)\}$ as

$$A_\mu(x) = i\sum_\nu \epsilon_{\mu\nu}\partial_\nu\theta_3(x) + \partial_\mu\theta(x), \tag{17.29}$$

after the double gauge transformation (17.26), one can decouple the fermions and the gauge field.

The gauge action is not chiral invariant but with the parametrization (17.29) depends only on θ_3 (this realizes a gauge fixing) and becomes

$$-\frac{1}{4e^2}\int \mathrm{d}^2x \sum_{\mu,\nu} F_{\mu\nu}^2(x) = -\frac{1}{2}\int \mathrm{d}^2x\,\left(\nabla^2\theta_3(x)\right)^2.$$

17.2.2 Quantum theory: Spectrum and anomaly

We have shown that the quantum theory is affected by the chiral anomaly (equation (17.17)) and this necessarily affects the analysis.

The field theory is super-renormalizable by power counting in a covariant gauge. The only divergent diagram corresponds to the one-loop contribution to the gauge field two-point function (Fig. 17.2) and comes from the expansion of the fermion determinant to order A^2.

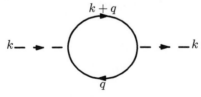

Fig. 17.2 The one-loop contribution to the current two-point function.

To avoid temporary infrared (IR) divergences, we give a mass m to the fermions. We use dimensional regularization, which maintains the vector gauge invariance necessary for the consistency of the gauge theory.

The action becomes (before gauge fixing)

$$S(\bar{\psi}, \psi, A_\mu) = -\int \mathrm{d}^d x \left[\frac{1}{4e^2} \sum_{\mu,\nu} F_{\mu\nu}^2(x) + \bar{\psi}(x) \left(\slashed{D} + m \right) \psi(x) \right]. \qquad (17.30)$$

At one-loop order, the vertex (or inverse) two-point function of the gauge field then reads

$$\tilde{\Gamma}_{\mu\nu}^{(2)}(k) = \left(\delta_{\mu\nu} p^2 - k_\mu k_\nu \right) + \Sigma_{\mu\nu}(k) \qquad (17.31)$$

with

$$\Sigma_{\mu\nu}(k) = e^2 \int \frac{\mathrm{d}^d q}{(2\pi)^d} \frac{\mathrm{tr} \left[\gamma_\mu \left(\slashed{q} + im \right) \gamma_\nu \left(\slashed{q} - \slashed{k} + im \right) \right]}{(q^2 + m^2) \left[(q - k)^2 + m^2 \right]}.$$

A short calculation shows that, as expected, the regularized one-loop contribution is transverse. Setting

$$\int \frac{\mathrm{d}^d q}{(2\pi)^d} \mathrm{tr} \left[\gamma_\mu \frac{\slashed{q} + im}{q^2 + m^2} \gamma_\nu \frac{\left(\slashed{q} - \slashed{k} + im \right)}{(q - k)^2 + m^2} \right] = S(k) \left(\delta_{\mu\nu} - \frac{k_\mu k_\mu}{k^2} \right), \qquad (17.32)$$

summing over $\mu = \nu$ on both sides, using

$$\sum_\mu \gamma_\mu \slashed{q} \gamma_\mu = (2 - d)\slashed{q},$$

and taking the trace over γ matrices, one obtains

$$S(k)(d - 1) = \mathrm{tr}\,\mathbf{1} \int \frac{\mathrm{d}^d q}{(2\pi)^d} \frac{-dm^2 + (d - 2)\left(q \cdot k - q^2 \right)}{(q^2 + m^2)\left[(q - k)^2 + m^2 \right]}. \qquad (17.33)$$

The right-hand side contains a term that is proportional to m^2, ultraviolet (UV) convergent and which vanishes with m for $\mathbf{k} \neq 0$. The second term is proportional to $(d - 2)$. In the $d = 2$ limit, only the divergent part of the integral survives:

$$\int \frac{\mathrm{d}^d q}{(2\pi)^d} \frac{q^2}{(q^2 + m^2)^2} \underset{d \to 2}{\sim} \frac{1}{2\pi(2 - d)} \quad \Rightarrow \quad S(k) = \frac{1}{\pi}. \qquad (17.34)$$

Therefore,

$$\Sigma_{\mu\nu}(p) = \frac{e^2}{\pi} \left(\delta_{\mu\nu} - k_\mu k_\nu / k^2 \right),$$

a non-vanishing result directly related to the *chiral anomaly*. The gauge field two-point vertex function at one-loop order follows:

$$\tilde{\Gamma}_{\mu\nu}^{(2)}(k) = \left(k^2 \delta_{\mu\nu} - k_\mu k_\nu \right) \left(1 + \frac{e^2}{\pi k^2} \right). \qquad (17.35)$$

Higher order contributions vanish because they are UV finite and, thus, not affected by the anomaly. The effect of the anomaly is to substitute

$$\tfrac{1}{4} \sum_{\mu,\nu} F_{\mu\nu}^2(x) \mapsto \tfrac{1}{4} \sum_{\mu,\nu} F_{\mu\nu}(x)(1 - e^2 \nabla^{-2}/\pi) F_{\mu\nu}(x)$$

and, therefore, in the parametrization (17.29), leads to the contribution to the action,

$$\frac{1}{2e^2} \int \mathrm{d}^2 x\, \theta_3(x) \nabla^2 (-\nabla^2 + e^2/\pi) \theta_3(x).$$

In the gauge $\theta(x) = 0$, in terms of the fields $\bar\psi, \psi, \theta_3$, the Schwinger model action then takes the free field form

$$\mathcal{S}(\bar\psi, \psi, \theta_3) = \int \mathrm{d}^2 x \left[-\bar\psi(x)\,\slashed{\partial}\psi(x) - \frac{1}{2\pi}\left(\nabla\theta_3(x)\right)^2 - \frac{1}{2e^2}\left(\nabla^2\theta_3(x)\right)^2 \right]. \quad (17.36)$$

One infers the θ_3-field propagator in the Fourier representation:

$$\Delta_{\theta_3}(p) = -\frac{e^2}{p^2(p^2 + e^2/\pi)} = \pi \left(\frac{1}{p^2 + e^2/\pi} - \frac{1}{p^2} \right). \quad (17.37)$$

The expression shows that the boson θ_3-field propagates two modes corresponding to a positive metric neutral massive field with mass $e/\sqrt{\pi}$ and a massless non-physical field with negative metric. The generation of a non-vanishing mass is a direct consequence of the chiral anomaly.

We now introduce two auxiliary fields φ and χ with $\theta_3(x) = \varphi(x) - \chi(x)$ and rewrite the action equivalently in terms of the two boson fields as

$$\mathcal{S}(\bar\psi, \psi, \varphi, \chi) = -\int \mathrm{d}^2 x\, \bar\psi(x)\,\slashed{\partial}\psi(x)$$
$$+ \frac{1}{2\pi} \int \mathrm{d}^2 x \left[\left(\nabla\varphi(x)\right)^2 + m_\varphi^2 \varphi^2(x) - \left(\nabla\chi(x)\right)^2 \right], \quad (17.38)$$

where now φ corresponds to the physical massive field with mass

$$m_\varphi = e/\sqrt{\pi} \quad (17.39)$$

and χ to the non-physical massless field.

Note that a massless field in two dimensions has a divergent two-point function and only the derivative and the exponential have defined correlation functions.

Finally, if one integrates over both the fermions and the massless field χ, the determinants cancel and the action reduces to

$$\mathcal{S}(\varphi) = \frac{1}{2\pi} \int \mathrm{d}^2 x \left[\left(\nabla\varphi(x)\right)^2 + m_\varphi^2 \varphi^2(x) \right]. \quad (17.40)$$

17.2.3 Two dimensions: The chiral anomaly

As an exercise, we now verify by explicit calculation the general expression (17.17) in the special example of dimension 2, where it reads

$$\langle \nabla \cdot \mathbf{J}^3(x) \rangle = -\frac{1}{2\pi} \sum_{\mu,\nu} \epsilon_{\mu\nu} F_{\mu\nu}(x) \,. \tag{17.41}$$

We then use it in the next section, to relate more directly the generated mass and the chiral anomaly.

The expression of the anomaly is automatically a total derivative since

$$\sum_{\mu,\nu} \epsilon_{\mu\nu} F_{\mu\nu}(x) = ie \sum_{\mu} \partial_\mu \sum_{\nu} 2\epsilon_{\mu\nu} A_\nu(x) \,.$$

The general form of the right-hand side is again dictated by locality, parity, gauge invariance and power counting: the anomaly must have mass dimension 2.

The explicit verification requires some care because, in two dimensions, massless fields may lead to IR divergences. One thus gives a mass m to fermions, which also breaks chiral symmetry explicitly, and takes the massless limit at the end of the calculation. The calculation follows closely the lines of Section 17.1.2. In particular, the expansion (17.13) is directly useful.

With a mass m for the fermion and a mass M for the regulator boson, one finds the non-conservation equation,

$$\langle \nabla \cdot \mathbf{J}^3(x) \rangle = -2im \langle \bar{\psi}(x)\gamma_3\psi(x) \rangle = -im \operatorname{tr} \langle x| \left(\gamma_3(\slashed{D} + m)^{-1} + (\slashed{D} + m)^{-1}\gamma_3 \right) |x\rangle$$
$$- (m \to M),$$

where, in two dimensions, the Dirac γ matrices are the *three Pauli matrices* (expressions (17.23)): $\gamma_\mu \equiv \sigma_\mu$.

Then,

$$\gamma_3(\slashed{D} + m)^{-1} + (\slashed{D} + m)^{-1}\gamma_3 = 2m\gamma_3 \left(m^2 - \slashed{D}^2 \right)^{-1} \,.$$

With the normalization (17.25) of $F_{\mu\nu}$,

$$\slashed{D}^2 = D^2 + \tfrac{1}{2} \sum_{\mu,\nu} \sigma_{\mu\nu} F_{\mu\nu} \,,$$

where

$$\gamma_\mu\gamma_\nu = \delta_{\mu\nu} + \sigma_{\mu\nu} \,, \qquad \sigma_{\mu\nu} = \tfrac{1}{2}[\gamma_\mu, \gamma_\nu] = i\epsilon_{\mu\nu}\gamma_3 \,,$$

($\epsilon_{\mu\nu}$ is the antisymmetric tensor, with $\epsilon_{12} = 1$).

One needs the two first terms of the expansion (17.13),

$$\left(m^2 - \not{D}^2\right)^{-1} = (m^2 - D^2)^{-1} + \tfrac{1}{2}(m^2 - D^2)^{-1}\sum_{\mu,\nu} F_{\mu\nu}\sigma_{\mu\nu}(m^2 - D^2)^{-1} + O(A^2).$$

One then evaluates the trace, using

$$\operatorname{tr}\gamma_3\sigma_{\mu\nu} = 2i\epsilon_{\mu\nu}.$$

Only the second term contributes and, since it is linear in A, one can set $A = 0$ in the denominators, replacing D^2 by ∇^2. One finds

$$\langle\nabla\cdot\mathbf{J}^3(x)\rangle = 2m^2\sum_{\mu,\nu}\epsilon_{\mu\nu}\langle x|(m^2 - \nabla^2)^{-1}F_{\mu\nu}(m^2 - \nabla^2)^{-1}|x\rangle - (m \to M).$$

The two contributions are separately convergent. When $m \to 0$, in the first term, the m^2 factor dominates the logarithmic IR divergence and the contribution vanishes. For $M \to \infty$, the support of the propagator in the second term reduces to a point. Thus, one can replace $F_{\mu\nu}$ by $F_{\mu\nu}(x)$. The expression becomes

$$\langle\nabla\cdot\mathbf{J}^3(x)\rangle = -2M^2\sum_{\mu,\nu}\epsilon_{\mu\nu}F_{\mu\nu}(x)\langle x|(m^2 - \nabla^2)^{-2}|x\rangle.$$

Finally,

$$\langle x|(M^2 - \nabla^2)^{-2}|x\rangle = \frac{1}{4\pi^2}\int\frac{\mathrm{d}^2 k}{(k^2 + M^2)^2} = \frac{1}{4\pi M^2},$$

confirming the result (17.41).

17.2.4 Currents: Field equations, anomaly and spectrum

We now establish a more direct relation between the chiral anomaly and the mass $m_\varphi = e/\sqrt{\pi}$ of the remaining boson field.

The vector and axial currents are given, respectively, by (equation (17.28)),

$$J_\mu(x) = -i\bar{\psi}(x)\gamma_\mu\psi(x), \quad J_\mu^3(x) = i\bar{\psi}(x)\gamma_3\gamma_\mu\psi(x),$$

and, thus, in dimension 2, the currents are related (equation (17.27)):

$$J_\mu^3(x) = -i\sum_\nu\epsilon_{\mu\nu}J_\nu(x). \tag{17.42}$$

The anomaly of the axial current is (equation (17.17))

$$\nabla\cdot\mathbf{J}^3(x) = -\frac{ie}{2\pi}\sum_{\mu,\nu}\epsilon_{\mu\nu}F_{\mu\nu}(x).$$

On the other hand, the vector current is exactly conserved:

$$\nabla \cdot \mathbf{J}(x) = 0 \,.$$

Using the relation (17.42), one infers

$$\partial_\mu J_\nu(x) - \partial_\nu J_\mu(x) = \frac{e}{\pi} F_{\mu\nu}(x) \;\Rightarrow\; \nabla^2 J_\nu(x) = \frac{e}{\pi} \sum_\mu \partial_\mu F_{\mu\nu}(x) \,.$$

The gauge field equation yields

$$e J_\mu(x) - \sum_\nu \partial_\nu F_{\nu\mu}(x) = 0 \,.$$

Combining these equations, one finds

$$\left(-\nabla^2 + (e^2/\pi)\right) J_\mu(x) = 0 \,.$$

This equation shows that the current J_μ and thus the curvature F_{12} are free fields of mass $m_\varphi = e/\sqrt{\pi}$, in agreement with the result (17.39).

17.2.5 Confinement and chiral symmetry breaking

Using the action (17.38) with the parametrization in equations (17.26) and (17.29), one can calculate the $\bar{\psi}\psi$ correlation functions. For example, one finds

$$\langle \bar{\psi}(x)\psi(x) \rangle = \frac{\mathrm{e}^\gamma}{2\pi} m_\varphi \,, \tag{17.43}$$

where $m_\varphi = e/\sqrt{\pi}$ and γ is Euler's constant. Since the composite field $\bar{\psi}\psi$ is not chiral invariant, the non-vanishing result shows that the global chiral symmetry is spontaneously broken. Note that spontaneous symmetry breaking with order is here possible because the electromagnetic interaction generates a linear rising confining potential between charges.

Similarly, the two-point functions are given by

$$\langle \bar{\psi}(x)\psi(x)\bar{\psi}(0)\psi(0) \rangle = \langle \bar{\psi}\psi \rangle^2 \cosh\left(4\pi\Delta(x, m_\varphi)\right) , \tag{17.44a}$$

$$\langle \bar{\psi}(x)\gamma_3\psi(x)\bar{\psi}(0)\gamma_3\psi(0) \rangle = \langle \bar{\psi}\psi \rangle^2 \sinh\left(4\pi\Delta(x, m_\varphi)\right) , \tag{17.44b}$$

where $\Delta(x, m_\varphi)$ is the propagator of the massive boson field. These expressions have several remarkable properties: the two-point functions have only singularities associated with the massive boson field. If one expands the exponentials in powers of the propagator and the Fourier transform, one finds that the expression (17.44b) has a pole at $k^2 = -m_\varphi^2$ and cuts at $k^2 = -(2n-1)^2 m_\varphi^2$, $n > 1$, in momentum space, while the expression (17.44a) has cuts at $k^2 = -(2n)^2 m_\varphi^2$.

Only the massive neutral boson appears in the intermediate states but: there are no charged fermions (the confinement property) and, moreover, the boson is a pseudoscalar particle, since it appears as a simple pole only in the $\bar{\psi}\gamma_3\psi$ two-point function.

Finally, the field theory, being super-renormalizable, is also asymptotically (UV) free.

These are properties one also expects in the true physical world with quarks and gluons (Chapter 13).

17.3 Abelian axial current and non-Abelian gauge fields

We still consider an Abelian axial current but now in the framework of a non-Abelian gauge theory. First, we derive the anomaly and then discuss the relation between anomaly, and the spectrum of the Dirac operator in a gauge field background.

We consider a field theory with fermion matter invariant under a unitary gauge group G, with $\mathbf{A}_\mu$ being the corresponding gauge field. We assume that the fermion fields transform non-trivially under a gauge group. The fermion matter action then takes the form,

$$S(\bar{\psi}, \psi; A) = -\int \mathrm{d}^4 x\, \bar{\psi}(x)\, \slashed{D}\, \psi(x) \tag{17.45}$$

with

$$\mathbf{D}_\mu = \mathbf{1}\partial_\mu + \mathbf{A}_\mu, \quad \slashed{D} = \sum_\mu \gamma_\mu \mathbf{D}_\mu \equiv \slashed{\partial} + \slashed{A}\,. \tag{17.46}$$

We define the curvature tensor by

$$\mathbf{F}_{\mu\nu} = [\mathbf{D}_\mu, \mathbf{D}_\nu] = \partial_\mu \mathbf{A}_\nu - \partial_\nu \mathbf{A}_\mu + [\mathbf{A}_\mu, \mathbf{A}_\nu]. \tag{17.47}$$

In a gauge transformation represented by a unitary matrix $\mathbf{g}(x)$, the gauge field $\mathbf{A}_\mu$ and the Dirac operator become

$$\mathbf{A}_\mu(x) \mapsto \mathbf{g}(x)\partial_\mu \mathbf{g}^{-1}(x) + \mathbf{g}(x)\mathbf{A}_\mu(x)\mathbf{g}^{-1}(x) \;\Rightarrow\; \slashed{D} \mapsto \mathbf{g}^{-1}(x)\slashed{D}\mathbf{g}(x)\,. \tag{17.48}$$

17.3.1 The axial anomaly

The axial current

$$J_\mu^5(x) = i\bar{\psi}(x) \cdot \gamma_5 \gamma_\mu \psi(x)$$

is still gauge invariant.

The derivation of the anomaly is then formally identical to the derivation of Section 17.1.2. One infers

$$\langle \nabla \cdot \mathbf{J}^5(x) \rangle = -\frac{i}{16\pi^2} \sum_{\mu,\nu,\rho,\sigma} \epsilon_{\mu\nu\rho\sigma}\, \mathrm{tr}\, \mathbf{F}_{\mu\nu}(x)\mathbf{F}_{\rho\sigma}(x)\,, \tag{17.49}$$

in which $\mathbf{F}_{\mu\nu}$ now is the curvature tensor (17.47). The result is completely determined by dimensional analysis, gauge invariance and the preceding Abelian calculation that yields the term of order $\mathbf{A}^2$.

Moreover, it is simple to verify that

$$\sum_{\mu,\nu,\rho,\sigma} \epsilon_{\mu\nu\rho\sigma}\, \mathrm{tr}\, \mathbf{F}_{\mu\nu}(x)\mathbf{F}_{\rho\sigma}(x)$$

$$= 4 \sum_{\mu,\nu,\rho,\sigma} \epsilon_{\mu\nu\rho\sigma} \partial_\mu \, \mathrm{tr}[\mathbf{A}_\nu(x)\partial_\rho \mathbf{A}_\sigma(x) + \tfrac{2}{3}\mathbf{A}_\nu(x)\mathbf{A}_\rho(x)\mathbf{A}_\sigma(x)], \tag{17.50}$$

a confirmation that, as expected, the anomaly is again a total derivative.

17.3.2 Anomaly and eigenvalues of the Dirac operator

We assume that the spectrum of $\rlap{\,/}D$, the Dirac operator in a non-Abelian gauge field (equation (17.46)), is discrete (enclosing, if necessary, the fermions temporarily in a box) and denote by d_n and $\varphi_n(x)$ the corresponding eigenvalues and eigenvectors,

$$\rlap{\,/}D\varphi_n(x) = d_n\varphi_n(x). \tag{17.51}$$

For a unitary or orthogonal group, the massless Dirac operator is anti-Hermitian; therefore, the eigenvalues are imaginary and the eigenvectors orthogonal. In addition, we choose them with unit norm.

The eigenvalues are gauge invariant because, in a gauge transformation characterized by a unitary matrix $\mathbf{g}(x)$, the Dirac operator transforms as in equation (17.48), and thus simply

$$\varphi_n(x) \mapsto \mathbf{g}(x)\varphi_n(x).$$

The anticommutation $\rlap{\,/}D\gamma_5 + \gamma_5\rlap{\,/}D = 0$ implies

$$\rlap{\,/}D\gamma_5\varphi_n(x) = -d_n\gamma_5\varphi_n(x). \tag{17.52}$$

Therefore, either d_n is different from 0 and $\gamma_5\varphi_n$ is an eigenvector of $\rlap{\,/}D$ with eigenvalue $-d_n$, or d_n vanishes. The eigenspace corresponding to the eigenvalue 0 then is invariant under γ_5, which can be diagonalized: the eigenvectors of $\rlap{\,/}D$ can be chosen as eigenvectors of definite chirality, that is, eigenvectors of γ_5 with eigenvalue ± 1:

$$\rlap{\,/}D\varphi_n(x) = 0\,, \quad \gamma_5\varphi_n(x) = \pm\varphi_n(x).$$

We call n_+ and n_- the dimensions of the eigenspaces of positive and negative chirality, respectively.

We now consider the determinant of the operator $\rlap{\,/}D + m$ regularized by mode truncation (a mode regularization):

$$\det{}_N(\rlap{\,/}D + m) = \prod_{n \leq N}(d_n + m), \tag{17.53}$$

keeping the N lowest eigenvalues of $\rlap{\,/}D$ (in modulus), with $N - n_+ - n_-$ even, in such a way that the corresponding subspace remains γ_5 invariant.

The regularization is gauge invariant because the eigenvalues of $\rlap{\,/}D$ are gauge invariant.

Note that, in the truncated space,

$$\operatorname{tr}\gamma_5 = n_+ - n_- . \tag{17.54}$$

The trace of γ_5 equals $n_+ - n_-$, the *index* of the Dirac operator $\rlap{\,/}D$ [184]. A non-vanishing index thus signals axial current non-conservation.

In a chiral transformation (17.5) with constant θ, the regularized determinant of $(\rlap{\,/}D + m)$ becomes

$$\det{}_N(\rlap{\,/}D + m) \mapsto \det{}_N\left(\mathrm{e}^{i\theta\gamma_5}(\rlap{\,/}D + m)\,\mathrm{e}^{i\theta\gamma_5}\right).$$

We now consider the various eigenspaces.

If $d_n \neq 0$, the matrix γ_5 is represented by the Pauli matrix σ_1 in the sum of eigenspaces corresponding to the two eigenvalues $\pm d_n$ and $\slashed{D} + m$ by $d_n \sigma_3 + m$. The determinant in the subspace then is

$$\det\left(e^{i\theta\sigma_1}(d_n\sigma_3 + m)\,e^{i\theta\sigma_1}\right) = \det e^{2i\theta\sigma_1}\det(d_n\sigma_3 + m) = m^2 - d_n^2,$$

because σ_1 is traceless.

In the eigenspace of dimension n_+ of vanishing eigenvalues d_n with eigenvectors with positive chirality, γ_5 is diagonal with eigenvalue 1 and, thus,

$$m^{n_+} \mapsto m^{n_+}\,e^{2i\theta n_+}\,.$$

Similarly, in the eigenspace of chirality -1 and dimension n_-,

$$m^{n_-} \mapsto m^{n_-}\,e^{-2i\theta n_-}\,.$$

One infers

$$\det{}_N\left(e^{i\theta\gamma_5}(\slashed{D} + m)\,e^{i\theta\gamma_5}\right) = e^{2i\theta(n_+ - n_-)}\det{}_N(\slashed{D} + m).$$

The ratio of the two determinants is independent of N. Taking the limit $N \to \infty$, one concludes

$$\det\left[\left(e^{i\gamma_5\theta}(\slashed{D} + m)\,e^{i\gamma_5\theta}\right)(\slashed{D} + m)^{-1}\right] = e^{2i\theta(n_+ - n_-)}\,. \tag{17.55}$$

Note that the left-hand side of equation (17.55) is obviously 1 when $\theta = n\pi$, which implies that the coefficient of 2θ in the right-hand side must indeed be an integer.

The variation of $\ln\det(\slashed{D} + m)$ at first order in θ,

$$\ln\det\left[\left(e^{i\gamma_5\theta}(\slashed{D} + m)\,e^{i\gamma_5\theta}\right)(\slashed{D} + m)^{-1}\right] = 2i\theta\,(n_+ - n_-)\,,$$

is related to the variation of the action (17.1) (see equation (17.16)) and, thus, to the expectation value of the integral of the divergence of the axial current, $\int d^4x \left\langle \nabla \cdot \mathbf{J}_\mu^5(x) \right\rangle$ in four dimensions. In the limit $m = 0$, it is then related to the space integral of the chiral anomaly (17.49).

The index of the Dirac operator is thus given by the local expression

$$-\frac{1}{32\pi^2}\sum_{\mu,\nu}\int d^4x\,\operatorname{tr}\mathbf{F}_{\mu\nu}(x)\tilde{\mathbf{F}}_{\mu\nu}(x) = n_+ - n_- \tag{17.56}$$

with

$$\tilde{\mathbf{F}}_{\mu\nu}(x) = \sum_{\rho,\sigma}\epsilon_{\mu\nu\rho\sigma}\mathbf{F}_{\rho\sigma}(x)\,.$$

The result calls for several comments:

(i) At first order in θ, in the absence of regularization, we have calculated ($\ln \det = \operatorname{tr} \ln$)

$$\ln \det \left[1 + i\theta \left(\gamma_5 + (\slashed{D} + m)\gamma_5(\slashed{D} + m)^{-1} \right) \right] \sim 2i\theta \operatorname{tr} \gamma_5 \,,$$

where the cyclic property of the trace has been used. Since the trace of the matrix γ_5 in the full space vanishes, one could expect, naively, a vanishing result. But 'trace' here means trace in matrix space and in coordinate space, and γ_5 really stands for $\gamma_5 \delta(x - y)$. The mode regularization gives a well-defined finite result for the undefined formal product $0 \times \delta^d(0)$.

(ii) The property that the integral (17.56) is quantized shows that the form of the anomaly is related to the topological properties of the gauge field, since the integral does not change when the gauge field is deformed continuously. The integral of the anomaly over the whole space, thus, depends only on the behaviour at large distances of the curvature tensor $\mathbf{F}_{\mu\nu}$, and the anomaly must be a total derivative, as equation (17.50) confirms.

(iii) One might be surprised that $\det \slashed{D}$ is not invariant under global chiral transformations. However, we have just established that, when the integral of the anomaly does not vanish, $\det \slashed{D}$ vanishes. Therefore, to give a meaning to the left-hand side of equation (17.55), we have been forced to add a mass to the Dirac operator.

The determinant of $\slashed{D}$ in the subspace orthogonal to eigenvectors with vanishing eigenvalue, even in presence of a mass, is chiral invariant by parity doubling. But, for $n_+ \neq n_-$, this is not the case for the determinant in the eigenspace of eigenvalue 0, because the trace of γ_5 does not vanish in this eigenspace (equation (17.54)). In the limit $m \to 0$, the complete determinant vanishes but not the ratio of determinants for different values of θ, because the powers of m cancel.

(iv) The discussion of the index of the Dirac operator is valid in any even dimension. Therefore, the topological character and the quantization of the space integral of the anomaly are general.

17.4   Non-Abelian anomaly and chiral gauge theories

We now discuss the problem of the conservation of a general axial current in a non-Abelian vector gauge theory.

17.4.1   General axial current

We consider the action for N massless Dirac fermions $\psi, \bar{\psi}$ in the background of a non-Abelian vector gauge field. The corresponding action can be written as

$$\mathcal{S}(\psi, \bar{\psi}; A) = - \int \mathrm{d}^4 x \, \bar{\psi}(x) \cdot \slashed{D} \, \psi(x). \tag{17.57}$$

When the gauge field vanishes, the action $\mathcal{S}(\psi, \bar{\psi}; 0)$ has a $U(N) \times U(N)$ chiral symmetry corresponding to the transformations

$$\psi'(x) = \left[\tfrac{1}{2}(1 + \gamma_5)\mathbf{U}_+ + \tfrac{1}{2}(1 - \gamma_5)\mathbf{U}_- \right] \psi(x) \,, \tag{17.58}$$

$$\bar{\psi}'(x) = \bar{\psi}(x) \left[\tfrac{1}{2}(1 + \gamma_5)\mathbf{U}_-^{\dagger} + \tfrac{1}{2}(1 - \gamma_5)\mathbf{U}_+^{\dagger} \right] \,, \tag{17.59}$$

where $\mathbf{U}_\pm$ are $N \times N$ unitary matrices.

We denote by $\mathbf{t}^\alpha$ the anti-Hermitian generators of $U(N)$:

$$\mathbf{U} = \mathbf{1} + \sum_\alpha \theta_\alpha \mathbf{t}^\alpha + O(|\theta|^2).$$

Vector currents correspond to the diagonal $U(N)$ subgroup of $U(N) \times U(N)$ of elements such that $\mathbf{U}_+ = \mathbf{U}_-$, as the equations (17.58) and (17.59) imply.

We couple a gauge field A_μ^α to all vector currents and define

$$\mathbf{A}_\mu(x) = \sum_\alpha \mathbf{t}^\alpha A_\mu^\alpha(x).$$

We define axial currents in terms of the infinitesimal space-dependent chiral transformations

$$\mathbf{U}_\pm(x) = \mathbf{1} \pm \sum_\alpha \theta_\alpha(x)\mathbf{t}^\alpha + O(\theta^2)$$

and, thus,

$$\delta\psi(x) = \sum_\alpha \theta_\alpha(x)\gamma_5 \mathbf{t}^\alpha \psi(x), \quad \delta\bar{\psi}(x) = \sum_\alpha \theta_\alpha(x)\bar{\psi}(x)\gamma_5 \mathbf{t}^\alpha.$$

The variation of the action then is

$$\delta\mathcal{S} = \int \mathrm{d}^4x \sum_{\alpha,\mu} \left\{ J_\mu^{5\alpha}(x)\partial_\mu\theta_\alpha(x) + \theta_\alpha(x)\bar{\psi}(x)\gamma_5\gamma_\mu[\mathbf{A}_\mu, \mathbf{t}^\alpha]\psi(x) \right\}, \qquad (17.60)$$

where $J_\mu^{5\alpha}(x)$ is the axial current:

$$J_\mu^{5\alpha}(x) = \bar{\psi}(x)\gamma_5\gamma_\mu \mathbf{t}^\alpha \psi(x). \qquad (17.61)$$

Since the gauge group has a non-trivial intersection with the chiral group, the commutator $[\mathbf{A}_\mu, \mathbf{t}^\alpha]$ does not vanish:

$$[\mathbf{A}_\mu(x), \mathbf{t}^\alpha] = \sum_{\beta,\gamma} A_\mu^\beta(x) f_{\beta\alpha\gamma} \mathbf{t}^\gamma,$$

where the $f_{\beta\alpha\gamma}$ are the totally antisymmetric structure constants of the Lie algebra of $U(N)$. Thus, after an integration by parts,

$$\delta\mathcal{S} = \int \mathrm{d}^4x \sum_{\alpha,\mu} \theta_\alpha(x) \left[-\partial_\mu J_\mu^{5\alpha}(x) + \sum_{\beta,\gamma} f_{\beta\alpha\gamma} A_\mu^\beta(x) J_\mu^{5\gamma}(x) \right]$$

$$\equiv -\int \mathrm{d}^4x \sum_\alpha \theta_\alpha(x) \sum_\mu [\mathbf{D}_\mu J_\mu^5]^\alpha(x). \qquad (17.62)$$

The classical current conservation equation is replaced by the gauge covariant conservation equation

$$\sum_\mu [\mathbf{D}_\mu J_\mu^5]^\alpha(x) = 0 \,, \tag{17.63}$$

where $[\mathbf{D}_\mu]^\beta_\alpha$ is the covariant derivative.

In the contribution to the anomaly, the terms quadratic in the gauge fields are modified, compared to the expression (17.49), only by the appearance of a new geometric factor. Then, the complete form of the anomaly is dictated by gauge covariance. One infers

$$\sum_\lambda [\mathbf{D}_\lambda J_\lambda^5]^\alpha(x) = -\frac{1}{16\pi^2} \sum_{\mu,\nu} \text{tr}[\mathbf{t}^\alpha \mathbf{F}_{\mu\nu}(x)\tilde{\mathbf{F}}_{\mu\nu}(x)]. \tag{17.64}$$

The expression can again be derived directly by the method of Section 17.1.2.

This is the result for the most general chiral and gauge transformations. If we restrict both groups in such a way that the gauge group has an empty intersection with the chiral group, the anomaly becomes proportional to $\text{tr}\,\mathbf{t}^\alpha$, where $\mathbf{t}^\alpha$ are the generators of the chiral group $G \times G$ and is, therefore, different from zero only for the Abelian factors of the group G.

17.5 Weak and electromagnetic interactions: Anomaly cancellation

Quantum anomalies in the context of chiral gauge theories lead to obstruction to gauge invariance and may lead to inconsistency of gauge theories [176].

17.5.1 Obstruction to gauge invariance

We now consider left-handed (or right-handed) fermions coupled to a non-Abelian gauge field. The fermion action takes the form

$$\mathcal{S}_{\text{F}}(\bar{\psi}, \psi; \mathbf{A}) = -\int \mathrm{d}^4x \, \bar{\psi}(x)\tfrac{1}{2}\,(1 + \gamma_5)\,\slashed{\mathbf{D}}\,\psi(x) \tag{17.65}$$

(the discussion with $\frac{1}{2}(1 - \gamma_5)$ is similar).

The gauge theory is consistent only if the partition function

$$\mathcal{Z}(\mathbf{A}) = \int \left[\mathrm{d}\psi\mathrm{d}\bar{\psi}\right] \exp\left[-\mathcal{S}_{\text{F}}(\bar{\psi}, \psi; \mathbf{A})\right], \tag{17.66}$$

is gauge invariant.

We introduce the generators $\mathbf{t}^\alpha$ of the gauge group in the fermion representation and define the corresponding current by

$$J_\mu^\alpha(x) = \bar{\psi}(x)\tfrac{1}{2}\,(1 + \gamma_5)\,\gamma_\mu \mathbf{t}^\alpha \psi(x). \tag{17.67}$$

Again, the invariance of $\mathcal{Z}(\mathbf{A})$ under an infinitesimal gauge transformation implies, for the current $\mathbf{J}_\mu = \sum_\alpha J_\mu^\alpha t^\alpha$, the covariant conservation equation

$$\sum_\mu \langle \mathbf{D}_\mu \mathbf{J}_\mu(x) \rangle = 0 \,,$$

with

$$\mathbf{D}_\mu = \partial_\mu + [\mathbf{A}_\mu, \bullet] \,.$$

The calculation of the quadratic contribution to the anomaly is a straightforward generalization of preceding calculations. The group structure is reflected by a simple geometric factor. The global factor can be taken from the Abelian calculation. It differs from result (17.14) by a factor of $1/2$ that comes from the projector $\frac{1}{2}(1+\gamma_5)$.

The general form of the term of degree 3 in the gauge field can also easily be found while the calculation of the global factor is somewhat tedious. We show in Section 17.6 that it can be obtained from consistency conditions. The complete expression then reads

$$\sum_\mu (\mathbf{D}_\mu \mathbf{J}_\mu(x))^\alpha = -\frac{1}{24\pi^2} \sum_{\mu,\nu,\rho,\sigma} \partial_\mu \epsilon_{\mu\nu\rho\sigma} \, \mathrm{tr} \left[t^\alpha \left(\mathbf{A}_\nu \partial_\rho \mathbf{A}_\sigma + \tfrac{1}{2} \mathbf{A}_\nu \mathbf{A}_\rho \mathbf{A}_\sigma \right) \right] \,. \quad (17.68)$$

If the projector $\frac{1}{2}(1+\gamma_5)$ is replaced by $\frac{1}{2}(1-\gamma_5)$, the sign of the anomaly changes.

Unless the anomaly vanishes identically, there is an obstruction to the construction of the gauge theory. The first term is proportional to

$$d_{\alpha\beta\gamma} = \tfrac{1}{2} \, \mathrm{tr} \left[t^\alpha \left(t^\beta t^\gamma + t^\gamma t^\beta \right) \right] \,. \quad (17.69)$$

The second term involves the product of four generators, but, taking into account the antisymmetry of the ϵ tensor, one product of two consecutive generators can be replaced by a commutator. Therefore, the term is again proportional to $d_{\alpha\beta\gamma}$.

For a unitary representation, the generators t^α are, with our conventions, anti-Hermitian. Therefore, the coefficients $d_{\alpha\beta\gamma}$ are purely imaginary:

$$d_{\alpha\beta\gamma}^* = \tfrac{1}{2} \, \mathrm{tr} \left[t^\alpha \left(t^\beta t^\gamma + t^\gamma t^\beta \right) \right]^\dagger = -d_{\alpha\beta\gamma} \,. \quad (17.70)$$

These coefficients vanish for all representations that are real,: that is, the t^α are antisymmetric, or pseudo-real, that is $t^\alpha = -S\,^T t^\alpha S^{-1}$. It follows that the only non-Abelian groups that can lead to anomalies in four dimensions are $SU(N)$ for $N \geq 3$: $SO(6)$: and E_6.

17.5.2 An application: Weak–electromagnetic interactions

The condition of anomaly cancellation discussed in Section 17.5.1 constrains the sector of weak and electromagnetic interactions of the Standard Model of particle physics. In the Standard Model, the anomalous contributions cancel between leptons and quarks. This cancellation occurs within each generation, as we now show, provided that, for each flavour, quarks exist in three states.

In the weak and electromagnetic group $SU(2) \times U(1)$, $SU(2)$ alone is a safe group. Therefore, the potential problems can come only from the $U(1)$ factor.

One expects *a priori* two conditions coming from vertices with one $U(1)$ and two $SU(2)$ gauge fields and those with three $U(1)$ gauge fields. Actually, one discovers that both conditions are equivalent. If one considers two $SU(2)$ and one $U(1)$ gauge fields, only $SU(2)$ doublets contribute and equation (17.69) leads to the condition

$$\sum_{\text{all doublets}} Y_{\text{L}} \operatorname{tr} \mathbf{t}^{\alpha} \mathbf{t}^{\beta} = 0 \,,$$

in which Y_{L} is the $U(1)$ charge. This condition reduces to

$$\sum_{\text{all doublets}} Y_{\text{L}} = 0 \,. \tag{17.71}$$

The vertex with three $U(1)$ gauge fields yields the condition:

$$\sum_{\text{left-handed parts}} Y_{\text{L}}^3 - \sum_{\text{right-handed parts}} Y_{\text{R}}^3 = 0 \,,$$

because the contributions to the anomaly of right-handed and left-handed couplings have opposite signs.

In the Standard Model, the left and right charges are related. Summing the charges of one doublet and the corresponding singlets, one obtains

$$\sum_{\text{all doublets}} (Y_{\text{L}} + 1)^3 + (Y_{\text{L}} - 1)^3 - 2Y_{\text{L}}^3 = 0 \,,$$

a condition that reduces to equation (17.71).

In one generation, the lepton doublet has $Y_{\text{L}} = -1$, and the quark has $Y_{\text{L}} = 1/3$. Therefore, a cancellation requires that the quarks exist in three states. These states are provided by the *colour quantum number*.

17.6   Wess–Zumino consistency conditions

In Section 17.5.1, we determined the part of the anomaly that is quadratic in the gauge field and asserted that the remaining non-quadratic contributions could be derived from algebraic considerations.

The anomaly is the variation of a functional under an infinitesimal gauge transformation. This implies compatibility conditions, which here are constraints on the general form of the anomaly, the Wess–Zumino consistency conditions [186]. One convenient method for deriving these constraints is based on BRS transformations: one expresses that the BRS transformations are nilpotent with vanishing square (see Sections 13.3.5 and 14.3).

In a BRS transformation, the variation of the gauge field $\mathbf{A}_{\mu}$ takes the form (equation (13.22))

$$\delta_{\text{BRS}} \mathbf{A}_{\mu}(x) = -\varepsilon \mathbf{D}_{\mu} \bar{\mathbf{C}}(x) \,, \tag{17.72}$$

where $\bar{\mathbf{C}}$ is a spinless fermion 'ghost' field, and ε is an anticommuting constant.

The corresponding variation of $\ln \mathcal{Z}(\mathbf{A})$ is

$$\delta_{\mathrm{BRS}} \ln \mathcal{Z}(\mathbf{A}) = \varepsilon \int \mathrm{d}^4 x \sum_\mu \langle \mathbf{J}_\mu(x) \rangle \, \mathbf{D}_\mu \bar{\mathbf{C}}(x) \,. \tag{17.73}$$

The anomaly equation has the general form

$$\sum_\mu \langle \mathbf{D}_\mu \mathbf{J}_\mu(x) \rangle = \mathcal{A}(\mathbf{A}; x) \,. \tag{17.74}$$

In terms of $\mathcal{A}$, equation (17.73), after an integration by parts, can be rewritten as

$$\delta_{\mathrm{BRS}} \ln \mathcal{Z}(\mathbf{A}) = -\varepsilon \int \mathrm{d}^4 x \, \mathcal{A}(\mathbf{A}; x) \, \bar{\mathbf{C}}(x). \tag{17.75}$$

Since the right-hand side is a BRS variation, it satisfies non-trivial constraints, called the Wess–Zumino consistency conditions, obtained by expressing that the square of the BRS operator δ_{BRS} vanishes:

$$\delta_{\mathrm{BRS}}^2 = 0 \,.$$

(The BRS operator has the property of a cohomology operator), The calculation of the BRS variation of $\mathcal{A}\bar{\mathbf{C}}$, also involves the BRS transformation of the fermion ghost field $\bar{\mathbf{C}}(x)$:

$$\delta_{\mathrm{BRS}} \bar{\mathbf{C}}(x) = -\varepsilon \, \bar{\mathbf{C}}^2(x). \tag{17.76}$$

The condition that $\mathcal{A}\bar{\mathbf{C}}$ is BRS invariant,

$$\delta_{\mathrm{BRS}} \int \mathrm{d}^4 x \, \mathcal{A}(\mathbf{A}; x) \, \bar{\mathbf{C}}(x) = 0 \,,$$

yields a constraint on the possible form of the anomalies that completely determines the cubic term in $\mathbf{A}$ in the right-hand side of equation (17.68). One can verify that

$$\delta_{\mathrm{BRS}} \sum_{\mu, \nu, \rho, \sigma} \epsilon_{\mu\nu\rho\sigma} \int \mathrm{d}^4 x \, \mathrm{tr} \left[\bar{\mathbf{C}}(x) \partial_\mu \left(\mathbf{A}_\nu \partial_\rho \mathbf{A}_\sigma + \tfrac{1}{2} \mathbf{A}_\nu \mathbf{A}_\rho \mathbf{A}_\sigma \right) \right] = 0 \,.$$

Explicitly, after an integration by parts, the equation takes the form

$$\sum_{\mu, \nu, \rho, \sigma} \epsilon_{\mu\nu\rho\sigma} \, \mathrm{tr} \int \mathrm{d}^4 x \, \{ \partial_\mu \bar{\mathbf{C}}^2(x) \mathbf{A}_\nu \partial_\rho \mathbf{A}_\sigma + \partial_\mu \bar{\mathbf{C}} \mathbf{D}_\nu \bar{\mathbf{C}} \partial_\rho \mathbf{A}_\sigma + \partial_\mu \bar{\mathbf{C}} \mathbf{A}_\nu \partial_\rho \mathbf{D}_\sigma \bar{\mathbf{C}}$$
$$+ \tfrac{1}{2} \partial_\mu \bar{\mathbf{C}}^2(x) \mathbf{A}_\nu \mathbf{A}_\rho \mathbf{A}_\sigma + \tfrac{1}{2} \partial_\mu \bar{\mathbf{C}} \left(\mathbf{D}_\nu \bar{\mathbf{C}} \mathbf{A}_\rho \mathbf{A}_\sigma + \mathbf{A}_\nu \mathbf{D}_\rho \bar{\mathbf{C}} \mathbf{A}_\sigma + \mathbf{A}_\nu \mathbf{A}_\rho \mathbf{D}_\sigma \bar{\mathbf{C}} \right) \}$$
$$= 0 \,.$$

The linear terms in $\mathbf{A}$, after integrating by parts the first term, and using the antisymmetry of the ϵ symbol, cancel automatically:

$$\sum_{\mu, \nu, \rho, \sigma} \epsilon_{\mu\nu\rho\sigma} \, \mathrm{tr} \int \mathrm{d}^4 x \, \left(\partial_\mu \bar{\mathbf{C}} \partial_\nu \bar{\mathbf{C}} \partial_\rho \mathbf{A}_\sigma + \partial_\mu \bar{\mathbf{C}} \mathbf{A}_\nu \partial_\rho \partial_\sigma \bar{\mathbf{C}} \right) = 0 \,.$$

Similarly, the cubic terms cancel (the anticommuting properties of $\bar{\mathbf{C}}$ have to be used):

$$\sum_{\mu,\nu,\rho,\sigma} \epsilon_{\mu\nu\rho\sigma} \operatorname{tr} \int \mathrm{d}^4 x \left\{ \left(\partial_\mu \bar{\mathbf{C}} \bar{\mathbf{C}} + \bar{\mathbf{C}} \partial_\mu \bar{\mathbf{C}} \right) \mathbf{A}_\nu \mathbf{A}_\rho \mathbf{A}_\sigma + \partial_\mu \bar{\mathbf{C}} \left([\mathbf{A}_\nu, \bar{\mathbf{C}}] \bar{\mathbf{C}} \mathbf{A}_\rho \mathbf{A}_\sigma \right. \right.$$
$$\left. \left. + \mathbf{A}_\nu [\mathbf{A}_\rho, \bar{\mathbf{C}}] \mathbf{A}_\sigma + \mathbf{A}_\nu \mathbf{A}_\rho [\mathbf{A}_\sigma, \bar{\mathbf{C}}] \right) \right\} = 0 \,.$$

Only the quadratic terms give a non-trivial relation between the quadratic and cubic terms in the anomaly, both contributions being proportional to

$$\sum_{\mu,\nu,\rho,\sigma} \epsilon_{\mu\nu\rho\sigma} \operatorname{tr} \int \mathrm{d}^4 x \, \partial_\mu \bar{\mathbf{C}}(x) \partial_\nu \bar{\mathbf{C}}(x) \mathbf{A}_\rho(x) \mathbf{A}_\sigma(x).$$

17.7   Lattice fermions: Ginsparg–Wilson relation

Notation. For convenience, we set the lattice spacing $a = 1$ and use for the fields the notation $\psi(x) \equiv \psi_x$.

In Section 16.7.2, we showed that the implementation on the lattice of fermions with chiral symmetry leads to the famous fermion doubling problem. We explain now how this problem can be solved using some specific class of Dirac operators satisfying the Ginsparg–Wilson relation.

Ginsparg–Wilson relation. Quite early, it had been noted that a potential solution to the fermion doubling problem while still retaining chiral properties in the continuum limit, was to find lattice Dirac operators $\mathbf{D}$ that, instead of anticommuting with γ_5, satisfy the relation [165],

$$\mathbf{D}^{-1}\gamma_5 + \gamma_5 \mathbf{D}^{-1} = \gamma_5 \mathbf{1} \,, \tag{17.77}$$

where $\mathbf{1}$ stands for the identity, both for lattice sites and in the algebra of γ-matrices. More explicitly,

$$(\mathbf{D}^{-1})_{xy}\gamma_5 + \gamma_5 (\mathbf{D}^{-1})_{xy} = \gamma_5 \delta_{xy} \,.$$

More generally, the right-hand side can be replaced by any local positive operator on the lattice: locality of a lattice operator is defined by an at least exponential decrease of its matrix elements when the points x, y are separated. The anticommutator being local, it is expected that it does not affect correlation functions at large distance and that chiral properties are recovered in the continuum limit. Note that, when $\mathbf{D}$ is the Dirac operator in a gauge background, the condition (17.77) is gauge invariant.

However, lattice Dirac operators satisfying the Ginsparg–Wilson relation (17.77) were only discovered much later, because the condition that both $\mathbf{D}$ and the anticommutator $\{\mathbf{D}^{-1}, \gamma_5\}$ should be local seemed difficult to reconcile, especially in the most interesting cases of gauge theories.

Note that, while relation (17.77) implies some generalized form of chirality on the lattice, it does not guarantee the absence of doublers. However, within this class, solutions can be found without doublers.

17.7.1 Lattice: New implementation of chiral symmetry

First, we discuss the main properties of a Dirac operator satisfying relation (17.77): we then exhibit a generalized form of chiral transformations on the lattice.

Using the relation, which is quite generally true for any Euclidean Dirac operator satisfying Hermiticity and reflection symmetry (for Euclidean fermions see, for example, Chapter 8 of Ref. [18]),

$$\mathbf{D}^\dagger = \gamma_5 \mathbf{D} \gamma_5 \,, \tag{17.78}$$

one can rewrite relation (17.77), after multiplication by γ_5, as

$$\mathbf{D}^{-1} + \left(\mathbf{D}^{-1}\right)^\dagger = \mathbf{1}$$

and, therefore,

$$\mathbf{D} + \mathbf{D}^\dagger = \mathbf{D}\mathbf{D}^\dagger = \mathbf{D}^\dagger \mathbf{D} \,. \tag{17.79}$$

This implies that the lattice operator $\mathbf{D}$ has an *index* and, in addition, that

$$\mathbf{S} = \mathbf{1} - \mathbf{D} \tag{17.80}$$

is unitary,

$$\mathbf{S}\mathbf{S}^\dagger = \mathbf{1} \,. \tag{17.81}$$

The eigenvalues of $\mathbf{S}$ lie on the unit circle. The eigenvalue 1 corresponds to the pole of the Dirac propagator.

Note also the relations

$$\gamma_5 \mathbf{S} = \mathbf{S}^\dagger \gamma_5 \,, \quad (\gamma_5 \mathbf{S})^2 = \mathbf{1} \,. \tag{17.82}$$

The matrix $\gamma_5 \mathbf{S}$ is Hermitian and $\frac{1}{2}(1 \pm \gamma_5 \mathbf{S})$ are two orthogonal projectors. If $\mathbf{D}$ is a Dirac operator in a gauge background, these projectors depend on the gauge field.

Chiral transformations on the lattice. It is then possible to construct lattice actions that have a chiral symmetry that corresponds to local but not point-like transformations [179]. In the Abelian example,

$$\psi'_x = \sum_y \left(e^{i\theta\gamma_5 \mathbf{S}}\right)_{xy} \psi_y \,, \quad \bar{\psi}'_x = \bar{\psi}_x \, e^{i\theta\gamma_5} \,. \tag{17.83}$$

(In the formalism of field integrals, ψ and $\bar{\psi}$ are independent integration variables and, thus, can transform independently.) Indeed, the invariance of the lattice action

$$\mathcal{S}(\bar{\psi}, \psi) = \sum_{x,y} \bar{\psi}_x \mathbf{D}_{xy} \psi_y = \mathcal{S}(\bar{\psi}', \psi') \,,$$

is implied by

$$e^{i\theta\gamma_5} \, \mathbf{D} \, e^{i\theta\gamma_5 \mathbf{S}} = \mathbf{D} \quad \Leftrightarrow \quad \mathbf{D} \, e^{i\theta\gamma_5 \mathbf{S}} = e^{-i\theta\gamma_5} \, \mathbf{D} \,.$$

Using the second relation in equation (17.82), we expand the exponentials and reduce the equation to

$$\mathbf{D}\gamma_5 \mathbf{S} = -\gamma_5 \mathbf{D} \,, \tag{17.84}$$

which is another form of relation (17.77).

The problem of the fermion measure. However, the transformations (17.83), no longer leave the integration measure of the fermion fields,

$$\prod_x d\psi_x d\bar\psi_x \,,$$

automatically invariant. The Jacobian of the change of variables $\psi \mapsto \psi'$ is

$$J = \det e^{i\theta\gamma_5}\, e^{i\theta\gamma_5 \mathbf{S}} = \det e^{i\theta\gamma_5(\mathbf{2}-\mathbf{D})} = 1 + i\theta \operatorname{tr} \gamma_5(\mathbf{2} - \mathbf{D}) + O(\theta^2), \qquad (17.85)$$

where 'trace' means trace in the space of γ matrices and in the lattice indices. This leaves open the possibility of generating the expected anomalies, when the Dirac operator of the free theory is replaced by the covariant operator in the background of a gauge field, as we now show.

17.7.2   *Eigenvalues and index of the Dirac operator in a gauge background*

We briefly discuss the index of a lattice Dirac operator $\mathbf{D}$ satisfying relation (17.77), in a gauge background. We assume that its spectrum is discrete (this is certainly true on a finite lattice where $\mathbf{D}$ is a matrix). The operator $\mathbf{D}$ is related by equation (17.80) to a unitary operator $\mathbf{S}$ whose eigenvalues have modulus 1. Therefore, if we denote by $|n\rangle$ its n-th eigenvector,

$$\mathbf{D}\,|n\rangle = (\mathbf{1} - \mathbf{S})\,|n\rangle = (1 - e^{i\theta_n})\,|n\rangle \ \Rightarrow \ \mathbf{D}^\dagger\,|n\rangle = (1 - e^{-i\theta_n})\,|n\rangle \,.$$

Then, using equation (17.78), one infers

$$\mathbf{D}\gamma_5\,|n\rangle = (1 - e^{-i\theta_n})\gamma_5\,|n\rangle \,.$$

The discussion that follows is then analogous to the discussion in Section 17.3.2, to which we refer for details. We note that when the eigenvalues are not real, $\theta_n \neq 0$ (mod π), $\gamma_5\,|n\rangle$ is an eigenvector different from $|n\rangle$, because the eigenvalues are different. By contrast, in the two subspaces corresponding to the eigenvalues 0 and 2, one can choose eigenvectors with definite chirality

$$\gamma_5\,|n\rangle = \pm\,|n\rangle \,.$$

We denote below by $n_\pm$ the number of eigenvalues 0, and $\nu_\pm$ the number of eigenvalues 2 with chirality ± 1.

Note that, on a finite lattice, δ_{xy} is a finite matrix and, thus,

$$\operatorname{tr} \gamma_5 \delta_{xy} = 0 \,.$$

Therefore,

$$\operatorname{tr} \gamma_5(\mathbf{2} - \mathbf{D}) = -\operatorname{tr} \gamma_5 \mathbf{D} \,,$$

which implies

$$\sum_n \langle n|\,\gamma_5(\mathbf{2} - \mathbf{D})\,|n\rangle = -\sum_n \langle n|\,\gamma_5 \mathbf{D}\,|n\rangle \,. \qquad (17.86)$$

In the equation, all complex eigenvalues cancel because the vectors $|n\rangle$ and $\gamma_5 |n\rangle$ are orthogonal. The sum reduces to the subspace of real eigenvalues, where the eigenvectors have definite chirality. In the left-hand side, only the eigenvalue 0 contributes and, in the right-hand side, only the eigenvalue 2. One finds

$$n_+ - n_- = -(\nu_+ - \nu_-).$$

The equation tells us that the difference between the number of states of different chirality in the 0 eigenvalue sector is cancelled by the difference in the sector of eigenvalue 2 (which corresponds to very massive states).

Remark. It is interesting to note the relation between the spectrum of $\mathbf{D}$ and the spectrum of $\gamma_5 \mathbf{D}$, which from relation (17.78) is a Hermitian matrix,

$$\gamma_5 \mathbf{D} = \mathbf{D}^\dagger \gamma_5 = (\gamma_5 \mathbf{D})^\dagger,$$

and, thus, diagonalizable with real eigenvalues. It is simple to verify the following two equations, of which the second one is obtained by changing θ into $\theta + 2\pi$:

$$\gamma_5 \mathbf{D}(1 - i\, e^{i\theta_n/2}\, \gamma_5) \, |n\rangle = 2 \sin(\theta_n/2)(1 - i\, e^{i\theta_n/2}\, \gamma_5) \, |n\rangle \,,$$
$$\gamma_5 \mathbf{D}(1 + i\, e^{i\theta_n/2}\, \gamma_5) \, |n\rangle = -2 \sin(\theta_n/2)(1 + i\, e^{i\theta_n/2}\, \gamma_5) \, |n\rangle \,.$$

These equations imply that the eigenvalues $\pm 2 \sin(\theta_n/2)$ of $\gamma_5 \mathbf{D}$ are paired, except for $\theta_n = 0 \pmod{\pi}$, where $|n\rangle$ and $\gamma_5 |n\rangle$ are proportional. For $\theta_n = 0$, $\gamma_5 \mathbf{D}$ has also eigenvalue 0. For $\theta_n = \pi$, $\gamma_5 \mathbf{D}$ has eigenvalue ± 2, depending on the chirality of $|n\rangle$.

In the same way,

$$\gamma_5(2 - \mathbf{D})(1 + e^{i\theta_n/2}\, \gamma_5) \, |n\rangle = 2 \cos(\theta_n/2)(1 + e^{i\theta_n/2}\, \gamma_5) \, |n\rangle \,,$$
$$\gamma_5(2 - \mathbf{D})(1 - e^{i\theta_n/2}\, \gamma_5) \, |n\rangle = -2 \cos(\theta_n/2)(1 - e^{i\theta_n/2}\, \gamma_5) \, |n\rangle \,.$$

Jacobian and anomaly on the lattice. The variation of the Jacobian (17.85) can now be evaluated. Opposite eigenvalues of $\gamma_5(2 - \mathbf{D})$ cancel. The eigenvalues for $\theta_n = \pi$ give factors 1. Only $\theta_n = 0$ gives a non-trivial contribution:

$$J = \det e^{i\theta \gamma_5 (2 - \mathbf{D})} = e^{2i\theta(n_+ - n_-)}.$$

The quantity $\operatorname{tr} \gamma_5(2 - \mathbf{D})$, which is the coefficient of the term of order θ, is a sum of terms that are local, gauge invariant, pseudoscalar, and topological, as the continuum anomaly (17.49), since

$$\operatorname{tr} \gamma_5(2 - \mathbf{D}) = \sum_n \langle n| \gamma_5(2 - \mathbf{D}) |n\rangle = 2(n_+ - n_-).$$

17.7.3 Chiral transformations on the lattice: Non-Abelian generalization

We now consider the non-Abelian chiral transformations

$$\psi_U = \left[\tfrac{1}{2}(1+\gamma_5 \mathbf{S})\mathbf{U}_+ + \tfrac{1}{2}(1-\gamma_5 \mathbf{S})\mathbf{U}_-\right]\psi\,, \qquad (17.87)$$

$$\bar{\psi}_U = \bar{\psi}\left[\tfrac{1}{2}(1+\gamma_5)\mathbf{U}_-^{\dagger} + \tfrac{1}{2}(1-\gamma_5)\mathbf{U}_+^{\dagger}\right]\,, \qquad (17.88)$$

where $\mathbf{U}_\pm$ are matrices belonging to some unitary group G. Near the identity

$$\mathbf{U} = \mathbf{1} + \boldsymbol{\Theta} + O(\boldsymbol{\Theta}^2),$$

where $\boldsymbol{\Theta}$ is an element of the Lie algebra.

We note that this amounts using different definitions for the chiral components of $\bar\psi$ and ψ, for ψ, the definition depending even on the gauge field.

We assume that G is a vector symmetry of the fermion action and, thus, the Dirac operator commutes with all elements of the Lie algebra:

$$[\mathbf{D}, \boldsymbol{\Theta}] = 0\,.$$

Then, again, the relation (17.77) in the form (17.84) implies the invariance of the fermion action:

$$\bar{\psi}_U\, \mathbf{D}\, \psi_U = \bar{\psi}\, \mathbf{D}\, \psi\,.$$

The Jacobian of an infinitesimal chiral transformation $\boldsymbol{\Theta} = \boldsymbol{\Theta}_+ = -\boldsymbol{\Theta}_-$ is

$$J = 1 + \operatorname{tr}\gamma_5\boldsymbol{\Theta}(2 - \mathbf{D}) + O(\boldsymbol{\Theta}^2)\,.$$

Wess–Zumino consistency conditions. To determine anomalies in the case of gauge fields coupling differently to fermion chiral components, one can, on the lattice, also use the property that BRS transformations are nilpotent. They take the form

$$\delta \mathbf{U}_{xy} = \bar{\varepsilon}\left(\bar{\mathbf{C}}_x \mathbf{U}_{xy} - \mathbf{U}_{xy}\bar{\mathbf{C}}_y\right),$$

$$\delta \mathbf{C}_x = \bar{\varepsilon}\bar{\mathbf{C}}_x^2\,,$$

instead of equations (17.72) and (17.76). Moreover, the matrix elements $\mathbf{D}_{xy}$ of the gauge covariant Dirac operator transform like $\mathbf{U}_{xy}$.

17.7.4 Explicit realization: Overlap fermions

An explicit solution of the Ginsparg–Wilson relation without doublers can be derived from operators $\mathbf{D}_W$ that share the properties of the Wilson–Dirac operator defined by equation (16.31), which avoids doublers at the price of breaking chiral symmetry explicitly.

Setting

$$\mathbf{A} = 1 - \mathbf{D}_\mathrm{W}/M \,, \tag{17.89}$$

where $M > 0$ is a mass parameter that, in particular, must chosen such that $\mathbf{A}$ has no 0 eigenvalue, one considers

$$\mathbf{S} = \mathbf{A} \left(\mathbf{A}^\dagger \mathbf{A}\right)^{-1/2} \;\Rightarrow\; \mathbf{D} = 1 - \mathbf{A} \left(\mathbf{A}^\dagger \mathbf{A}\right)^{-1/2}. \tag{17.90}$$

The matrix $\mathbf{A}$ is such that

$$\mathbf{A}^\dagger = \gamma_5 \mathbf{A} \gamma_5 \;\Rightarrow\; \mathbf{B} = \gamma_5 \mathbf{A} = \mathbf{B}^\dagger.$$

The Hermitian matrix $\mathbf{B}$ has real eigenvalues. Moreover, c

$$\mathbf{B}^\dagger \mathbf{B} = \mathbf{B}^2 = \mathbf{A}^\dagger \mathbf{A} \;\Rightarrow\; \left(\mathbf{A}^\dagger \mathbf{A}\right)^{1/2} = |\mathbf{B}|.$$

One concludes

$$\gamma_5 \mathbf{S} = \operatorname{sgn} \mathbf{B} \,,$$

where $\operatorname{sgn} \mathbf{B}$ is the matrix with the same eigenvectors as $\mathbf{B}$, but where all the eigenvalues have been replaced by their respective signs. In particular, this shows that $(\gamma_5 \mathbf{S})^2 = \mathbf{1}$.

With this ansatz, $\mathbf{D}$ has a 0 eigenmode when $\mathbf{A} \left(\mathbf{A}^\dagger \mathbf{A}\right)^{-1/2}$ has the eigenvalue 1. This can happen when $\mathbf{A}$ and $\mathbf{A}^\dagger$ have the same eigenvector with a *positive* eigenvalue.

This is the idea of overlap fermions [166], the name 'overlap' referring to the way this Dirac operator was initially introduced.

An example: Free lattice fermions. We verify the absence of doublers for vanishing gauge fields. The Fourier representation of a Wilson–Dirac operator has the general form

$$D_W(p) = \alpha(p) + i\gamma_\mu \beta_\mu(p), \tag{17.91}$$

where $\alpha(p)$ and $\beta_\mu(p)$ are real, periodic, smooth functions. In the continuum limit, one must recover the usual massless Dirac operator, which implies

$$\beta_\mu(p) \underset{|p|\to 0}{\sim} p_\mu \,, \quad \alpha(p) \geq 0 \,, \; \alpha(p) \underset{|p|\to 0}{=} O(p^2),$$

and $\alpha(p) > 0$ for all values of p_μ such that $\beta_\mu(p) = 0$ for $|p| \neq 0$ (*i.e.*, all values that correspond to doublers). Equation (16.31) in the limit $m = 0$ provides an explicit example.

Doublers appear if the determinant of the overlap operator $\mathbf{D}$ (equations (17.89) and (17.90)) vanishes for $|\mathbf{p}| \neq 0$. In the example of the operator (17.91), a short calculation shows that this happens when

$$\left[\sqrt{\left(M - \alpha(p)\right)^2 + \beta_\mu^2(p)} - M + \alpha(p) \right]^2 + \beta_\mu^2(p) = 0 \,.$$

This implies $\beta_\mu(p) = 0$, an equation that necessarily admits doubler solutions, and

$$|M - \alpha(p)| = M - \alpha(p).$$

The solutions to the equation depend on the value of $\alpha(p)$ with respect to M for the doubler modes, that is, for the values of p such that $\beta_\mu(p) = 0$. If $\alpha(p) \le M$, the equation is automatically satisfied and the corresponding doubler survives. As mentioned in this introduction to the section, the relation (17.77) by itself does not guarantee the absence of doublers. By contrast, if $\alpha(p) > M$, the equation implies $\alpha(p) = M$, which is impossible. Therefore, by rescaling $\alpha(p)$, if necessary, one can keep the wanted $p_\mu = 0$ mode while eliminating all doublers. The modes associated to doublers for $\alpha(p) \le M$ then correspond to the eigenvalue 2 for $\mathbf{D}$, and the doubling problem is solved, at least in a free theory.

In the presence of a gauge field, the argument can be generalized, provided the plaquette terms (the pure gauge actions) in the lattice action are constrained to remain sufficiently close to 1.

Overlap fermions and numerical simulations. Let us emphasize that, if the doubling problem has been solved from the theoretical point of view, from the numerical point of view, the calculation of the operator $(\mathbf{A}^\dagger\mathbf{A})^{-1/2}$ in a gauge background represents a major numerical challenge and has severely restricted the practical usage of such fermions.

17.8 Supersymmetric quantum mechanics and domain wall fermions

Because the construction of lattice fermions without doublers we have described in Section 17.7.4 is somewhat artificial, one may wonder whether there is a context in which they would appear more naturally. Therefore, we now outline how a similar lattice Dirac operator can be generated by embedding four-dimensional space in a larger five-dimensional space. This is the method of using *domain wall* fermions [167].

Because the general idea behind domain wall fermions has first emerged in another context, as a preparation, we recall a few properties of the spectrum of the Hamiltonian in supersymmetric quantum mechanics, a topic also related to the index of the Dirac operator (Section 17.3.2), and very directly to stochastic dynamics in the form of Langevin or Fokker–Planck equations (see, *e.g.*, Ref. [18] and Section 22.5).

17.8.1 Supersymmetric quantum mechanics

One can construct quantum theories that exhibit the simplest form of supersymmetry where space-time reduces to time only. Then, quantum field theory reduces simple quantum mechanics, and physical quantities are given by path integrals [185].

We introduce a first order differential operator $\mathfrak{D}$ acting on functions of one real variable, which is a 2×2 matrix (σ_i are the three Pauli matrices (17.23)):

$$\mathfrak{D} \equiv \sigma_1 \mathrm{d}/\mathrm{d}x - i\sigma_2 A(x). \tag{17.92}$$

The function $A(x)$ is smooth, except when stated otherwise, and real. Thus, the operator $\mathfrak{D}$ is anti-Hermitian.

The operator $\mathfrak{D}$ shares several properties with the Dirac operator described in Section 17.3.2. In particular, it satisfies

$$\sigma_3 \mathfrak{D} + \mathfrak{D}\sigma_3 = 0\,,$$

and, thus, has an index (σ_3 playing the role of γ_5). If $(\psi_+(x), \psi_-(x))$ is an eigenvector of $\mathfrak{D}$ with eigenvalue d, either $d = 0$, and $(\psi_+(x), \psi_-(x))$ can be chosen as

$$\sigma_3(\psi_+(x), \psi_-(x)) = \pm(\psi_+(x), \psi_-(x)),$$

or $\sigma_3(\psi_+(x), \psi_-(x))$ is an eigenvector with the eigenvalue $-d = d^*$. This observation is consistent with the analysis of Section 17.3.2 applied to the operator $\mathfrak{D}$.

We introduce the operators $(\mathrm{d}_x \equiv \mathrm{d}/\mathrm{d}x)$

$$\mathrm{D} = \mathrm{d}_x + A(x) \;\Rightarrow\; \mathrm{D}^\dagger = -\mathrm{d}_x + A(x) \tag{17.93}$$

and

$$Q = \tfrac{1}{2}\mathrm{D}\,(\sigma_1 - i\sigma_2) \;\Rightarrow\; Q^\dagger = \tfrac{1}{2}\mathrm{D}^\dagger\,(\sigma_1 + i\sigma_2)\,.$$

Then,

$$\mathfrak{D} = Q - Q^\dagger\,,$$
$$Q^2 = (Q^\dagger)^2 = 0\,. \tag{17.94}$$

We consider the Hamiltonian H, the anticommutator of Q and $Q^\dagger$,

$$H = QQ^\dagger + Q^\dagger Q = -\mathfrak{D}^2 = \begin{pmatrix} \mathrm{D}^\dagger\mathrm{D} & 0 \\ 0 & \mathrm{D}\mathrm{D}^\dagger \end{pmatrix}, \tag{17.95}$$

where

$$\mathrm{D}^\dagger\mathrm{D} = -\mathrm{d}_x^2 + A^2(x) - A'(x), \quad \mathrm{D}\mathrm{D}^\dagger = -\mathrm{d}_x^2 + A^2(x) + A'(x).$$

The Hamiltonian is positive semi-definite. The relations (17.94) imply that

$$[H, Q] = [H, Q^\dagger] = 0\,.$$

The operators $Q, Q^\dagger$ are the generators of the simplest form of a *supersymmetric algebra*, and the Hamiltonian H is supersymmetric.

The eigenvectors of H have the form $(\psi_+(x), 0)$ and $(0, \psi_-(x))$ and satisfy, respectively,

$$\mathrm{D}^\dagger\mathrm{D}\,|\psi_+\rangle = \varepsilon_+\,|\psi_+\rangle, \quad \text{and} \quad \mathrm{D}\mathrm{D}^\dagger\,|\psi_-\rangle = \varepsilon_-\,|\psi_-\rangle, \quad \varepsilon_\pm \geq 0\,, \tag{17.96}$$

Moreover, if x belongs to a bounded interval or $A(x) \to \infty$ for $|x| \to \infty$, then the spectrum of H is discrete.

Multiplying the first equation in equations (17.96) by D, one concludes that, if $D|\psi_+\rangle \neq 0$ and, thus, ϵ_+ does not vanish, it is an eigenvector of $DD^\dagger$ with eigenvalue ε_+, and conversely. Therefore, except for a possible ground state with one vanishing eigenvalue, the spectrum of H is doubly degenerate, with the degenerate eigenvalues corresponding to the two eigenvalues $\pm i\sqrt{\varepsilon}$ of $\mathfrak{D}$.

It is convenient to introduce a function $S(x)$ defined by

$$S'(x) = A(x), \tag{17.97}$$

and, for simplicity, to discuss only the situation of operators on the entire real line.

We assume that

$$S(x)/|x| \underset{x \to \pm\infty}{\geq} \ell > 0 \,.$$

Then, the function $S(x)$ is such that $e^{-S(x)}$ is a normalizable wave function:

$$\int \mathrm{d}x \, e^{-2S(x)} < \infty.$$

In the stochastic interpretation where $D^\dagger D$ has the interpretation of a Fokker–Planck Hamiltonian generating the time evolution of some probability distribution, $e^{-2S(x)}$ is the equilibrium distribution.

When $e^{-S(x)}$ is normalizable, one knows one eigenvector with vanishing eigenvalue and chirality $+1$, which corresponds to the isolated ground state of $D^\dagger D$,

$$\mathfrak{D}|\psi_+, 0\rangle = 0 \;\Leftrightarrow\; D|\psi_+\rangle = 0 \,, \quad \sigma_3|\psi_+, 0\rangle = |\psi_+, 0\rangle$$

with

$$\psi_+(x) = e^{-S(x)} \,.$$

On the other hand, the formal solution of $D^\dagger|\psi_-\rangle = 0$,

$$\psi_-(x) = e^{S(x)},$$

is not normalizable and, therefore, no eigenvector with negative chirality is found.

One concludes that the operator $\mathfrak{D}$ has only one eigenvector with eigenvalue 0 corresponding to positive chirality: the index of $\mathfrak{D}$ is one. Note that expressions for the index of the Dirac operator in a general background have been derived. In the present example, they yield

$$\text{Index} = \tfrac{1}{2}\left[\operatorname{sgn} A(+\infty) - \operatorname{sgn} A(-\infty)\right],$$

in agreement with the explicit calculation.

A few examples

(i) In the example of the function $S(x) = \tfrac{1}{2}x^2$, the two components of the Hamiltonian H become

$$D^\dagger D = -\mathrm{d}_x^2 + x^2 - 1 \,, \quad DD^\dagger = -\mathrm{d}_x^2 + x^2 + 1 \,.$$

The Hamiltonian decomposes into two shifted harmonic oscillators with eigenvalues $\varepsilon_{+n} = 2n$, $\varepsilon_{-n} = 2n + 2$, $n \geq 0$, and the spectrum of $\mathfrak{D}$ contains one eigenvalue 0, and a spectrum of opposite eigenvalues, $\pm i\sqrt{2n}$, $n \geq 1$.

(ii) Another example useful for later purpose is $S(x) = |x|$. Then, $A(x) = \mathrm{sgn}(x)$ is singular, and $A'(x) = 2\delta(x)$. The two components of the Hamiltonian H become

$$\mathrm{D}^\dagger \mathrm{D} = -\mathrm{d}_x^2 + 1 - 2\delta(x), \quad \mathrm{D}\mathrm{D}^\dagger = -\mathrm{d}_x^2 + 1 + 2\delta(x). \tag{17.98}$$

Here, one finds one isolated eigenvalue 0 and a degenerate continuous spectrum $\varepsilon \geq 1$.

(iii) A less singular but similar example that can be solved analytically corresponds to $A(x) = \mu \tanh(x)$, with μ a positive constant. It leads to the potentials

$$V(x) = A^2(x) \pm A'(x) = \mu^2 - \frac{\mu(\mu \mp 1)}{\cosh^2(x)}.$$

The two operators have a continuous spectrum starting at μ^2 and a discrete spectrum

$$\mu^2 - (\mu - n)^2, \quad n \in \mathbb{N} \leq \mu, \qquad \mu^2 - (\mu - n - 1)^2, \quad n \in \mathbb{N} \leq \mu - 1.$$

(iv) Finally, for the function $A(x) = x^2$, one finds no eigenvalue 0 because $S(x) = \pm\frac{1}{3}x^3$ does not correspond to a normalizable distribution. Consistently, the index vanishes.

17.8.2 Domain wall fermions: Continuum formulation

We now consider four-dimensional space (but the strategy applies to all even dimensional spaces) as a surface embedded in five-dimensional space. We denote by x_μ the usual four coordinates, and by t the coordinate in the fifth dimension. Physical space corresponds to $t = 0$. We then study the five-dimensional Dirac operator $\mathcal{D}$ in the background of a classical scalar field $\varphi(t)$ that depends only on t. The fermion action reads

$$\mathcal{S}(\bar{\psi}, \psi) = -\int \mathrm{d}t\, \mathrm{d}^4x\, \bar{\psi}(t, x)\, \mathcal{D}\, \psi(t, x),$$

with

$$\mathcal{D} = \slashed{\partial} + \gamma_5 \partial_t + M\varphi(Mt),$$

where the parameter M is a mass large with respect to the masses of all physical particles.

Since translation symmetry in 4-space is not broken, one can expand the fields $\psi, \bar{\psi}$ in Fourier components:

$$\psi(t, x) = \int \mathrm{e}^{ipx}\, \tilde{\psi}(t, p)\mathrm{d}^4p, \quad \bar{\psi}(t, x) = \int \mathrm{e}^{-ipx}\widetilde{\bar{\psi}}(t, p)\mathrm{d}^4p.$$

The operator $\mathcal{D}$ then becomes

$$\mathcal{D} = i\sum_\mu p_\mu\gamma_\mu + \gamma_5\partial_t + M\varphi(Mt).$$

To find the mass spectrum corresponding to $\mathcal{D}$, it is convenient to write it as

$$\mathcal{D} = \gamma_\mathbf{p}\left[i|\mathbf{p}| + \gamma_\mathbf{p}\gamma_5\partial_t + \gamma_\mathbf{p}M\varphi(Mt)\right],$$

where $\gamma_\mathbf{p} = \sum_\mu p_\mu\gamma_\mu/|\mathbf{p}|$, and thus $\gamma_\mathbf{p}^2 = 1$. The eigenvectors with vanishing eigenvalue of $\mathcal{D}$ are also those of the operator

$$\mathfrak{D} = i\gamma_\mathbf{p}\mathcal{D} + |\mathbf{p}| = i\gamma_\mathbf{p}\gamma_5\partial_t + i\gamma_\mathbf{p}M\varphi(Mt),$$

with eigenvalue $|\mathbf{p}|$.

We then note that $i\gamma_\mathbf{p}\gamma_5$, $\gamma_\mathbf{p}$, and $-\gamma_5$ are Hermitian matrices that form a representation of the algebra of Pauli matrices. The operator $\mathfrak{D}$ can then be compared to the operator (17.92), with $M\varphi(Mt)$ corresponding to $A(x)$. Under the same conditions, $\mathfrak{D}$ has an eigenvector with an isolated vanishing eigenvalue corresponding to an eigenvector with positive chirality. All other eigenvalues, for dimensional reasons, are proportional to M and thus correspond to fermions of large masses. Moreover, the eigenfunction with eigenvalue 0 decays on a scale $t = O(1/M)$. Therefore, for M large, one is left with a fermion that has a single chiral component and is confined onto the $t = 0$ surface.

One can imagine for the function $\varphi(t)$ some physical interpretation: φ may be an additional scalar field, and $\varphi(t)$ may be a solution of the corresponding field equations that connects the two minima $\varphi = \pm 1$ of the φ potential. In the limit of a very sharp transition, one is led to the Hamiltonian (17.98).

Note that such an interpretation is possible only for even dimensions $d \geq 4$; in dimension 2, zero modes related to breaking of the translation symmetry due to the presence of the wall, would lead to IR divergences. These potential divergences thus forbid a static wall, a property analogous to the one encountered in the quantization of *solitons*.

More precise results follow from the study of Section 17.8.1. There, we have noticed that $\mathfrak{G}(t_1, t_2; p)$, the inverse of the Dirac operator in Fourier representation, has a short distance singularity for $t_2 \to t_1$ in the form of a discontinuity. Here, this is an artefact of treating the fifth dimension differently from the four others. In real space for the function $\mathfrak{G}(t_1, t_2; x_1 - x_2)$ with separate points on the surface, $x_1 \neq x_2$, the limit $t_1 = t_2$ corresponds to points in five dimensions that do not coincide, and this singularity is absent.

A short analysis shows that this amounts in Fourier representation taking the average of the limiting values (a property that can easily be verified for the free propagator). If $\varphi(t)$ is a generic function, one finds

$$\mathcal{D}^{-1} = \frac{i}{2\not{p}}\left[d_1(p^2)(1+\gamma_5) + (1-\gamma_5)p^2 d_2(p^2)\right] + d_3(p^2),$$

where d_3 is regular. As a consequence,

$$\gamma_5\mathcal{D}^{-1} + \mathcal{D}^{-1}\gamma_5 = 2d_3(p^2)\gamma_5,$$

which is a form of Ginsparg–Wilson's relation because the right-hand side is local.

As an exception, if $\varphi(t)$ is an odd function, for $t_1 = t_2 = 0$ one finds

$$\mathcal{D}^{-1}(p) = \frac{i}{2\slashed{p}} \left[d_1(p^2)(1 + \gamma_5) + (1 - \gamma_5)p^2 d_2(p^2) \right],$$

where d_1, d_2 are regular functions of p^2. Therefore, $\mathcal{D}^{-1}$ anticommutes with γ_5, and chiral symmetry is realized in the usual way.

17.8.3 Domain wall fermions: Lattice

We now replace four-dimensional continuum space by a lattice but keep the fifth dimension continuous [167]. We replace the Dirac operator by the Wilson–Dirac operator (17.91) to avoid doublers. In Fourier representation, one finds

$$\mathcal{D} = \alpha(\mathbf{p}) + i\beta_\mu(\mathbf{p})\gamma_\mu + \gamma_5\partial_t + M\varphi(Mt).$$

This has the effect of replacing p_μ by $\beta_\mu(\mathbf{p})$ and shifting $M\varphi(Mt) \mapsto M\varphi(Mt) + \alpha(\mathbf{p})$. To ensure the absence of doublers, one requires that, for the values for which $\beta_\mu(\mathbf{p}) = 0$ and $\mathbf{p} \neq 0$, none of the solutions to the eigenvalue 0 equation are normalizable. This is realized if $M < |\alpha(\mathbf{p})|$, and $\varphi(t)$ is bounded for $|t| \to \infty$, for instance,

$$|\varphi(t)| \leq 1.$$

The inverse Dirac operator on the surface $t = 0$ takes the general form

$$\mathcal{D}^{-1} = i\slashed{\beta} \left[\delta_1(p^2)(1 + \gamma_5) + (1 - \gamma_5)\delta_2(p^2) \right] + \delta_3(p^2),$$

where δ_1 is the only function that has a pole for $p = 0$, and where δ_2, δ_3 are regular. The function δ_3 does not vanish even if $\varphi(t)$ is odd, because the addition of $\alpha(p^2)$ breaks the symmetry. Thus, one always finds Ginsparg–Wilson's relation,

$$\gamma_5\mathcal{D}^{-1} + \mathcal{D}^{-1}\gamma_5 = 2\delta_3(p^2)\gamma_5.$$

More explicit expressions can be obtained in the limit $\varphi(t) = \text{sgn}(t)$ (a situation analogous to equation (17.98)).

Of course, computer simulations of domain walls require also discretizing the fifth dimension.

18 Periodic semi-classical vacuum, instantons and anomalies

In this chapter, we discuss three types of quantum theories in which the classical Euclidean action has a periodic structure with an infinite set of degenerate classical minima. In all these examples, due to barrier penetration effects, the minima do not correspond to ground states of the quantum theory. In the semi-classical approximation, tunnelling between classical minima is related to the existence of *instantons, finite action solutions of the Euclidean equation of motion or field equations.*

Instantons imply that the eigenstates of the Hamiltonian depend on an angle θ. In quantum field theory, this leads to the concept of θ-*vacua.* We exhibit the topological character of these instantons and, in the example of gauge theories, relate them to anomalies and the index of the Dirac operator as described in Chapter 17.

We first discuss the ground state structure of the classical example of the cosine potential in one-dimensional quantum mechanics. We then describe the vacuum structure and exhibit instantons in the example of the two-dimensional CP^{N-1} models [187], which have been extensively studied within the $1/N$ expansion, and four-dimensional $SU(2)$ gauge theories.

We show that instantons can be classified in terms of a topological charge, the space integral of the chiral anomaly. The existence of gauge field configurations that contribute to the anomaly has direct physical implications, like a possible unwanted strong CP violation but, conversely, the solution to the $U(1)$ problem.

18.1 The periodic cosine potential

The simplest systems that involve instantons and topology are quantum Hamiltonians with one-dimensional periodic potentials. An example is

$$H = -\frac{g}{2}\left(\frac{\mathrm{d}}{\mathrm{d}x}\right)^2 + \frac{1}{4g}\left(1 - \cos(2x)\right), \quad g > 0, \qquad (18.1)$$

where g plays the role of $\hbar$ and, thus, we set $\hbar = 1$.

The potential has an infinite number of degenerate minima for $x = n\pi$, $n \in \mathbb{Z}$. Each minimum is an equivalent starting point for a perturbative calculation of the eigenvalues of H. Periodicity implies that the perturbative expansions, expansions in powers of the parameter g, are identical to all orders, a property that seems to imply that the low lying (discrete) spectrum of the quantum Hamiltonian is infinitely degenerate. In fact, due to *quantum tunnelling*, the spectrum is not degenerate and is continuous: it has a *band structure.*

From Random Walks to Random Matrices. Jean Zinn-Justin, Oxford University Press (2019).
© Jean Zinn-Justin. DOI: 10.1093/oso/9780198787754.001.0001

18.1.1   The structure of the ground state

To characterize the structure of the spectrum of the Hamiltonian (18.1), we introduce the unitary operator T that, acting on wave functions, generates an elementary translation of one period π:

$$T\psi(x) = \psi(x+\pi).$$

Since T commutes with the Hamiltonian,

$$[T, H] = 0\,, \tag{18.2}$$

the operators T and H can be diagonalized simultaneously. Each eigenfunction ψ_n of H thus is characterized by an angle θ (pseudo-momentum) associated with an eigenvalue of T:

$$H\psi_n(\theta) = E_n(\theta)\psi_n(\theta), \quad T\psi_n(\theta) = \mathrm{e}^{i\theta}\,\psi_n(\theta). \tag{18.3}$$

The eigenvalues $E_n(\theta)$ are periodic functions of θ and, for $g \to 0$, are close to the eigenvalues of the harmonic oscillator:

$$E_n(\theta) = n + 1/2 + O(g).$$

For g small, all eigenvalues can be calculated as a power series in g by expanding the potential around a minimum. Every minimum generates the same series. Therefore, to all orders in an expansion in powers of g, $E_n(\theta)$ is independent of θ and the spectrum of H is infinitely degenerate.

However, additional small contributions due to barrier penetration, exponential in $1/g$, lift the degeneracy and introduce a θ-dependence. In the path integral framework, they are due to instantons. For g small, to each value of n then corresponds a band (a continuous spectrum) when θ varies in $[0, 2\pi]$.

18.1.2   Path integral representation and topology

The spectrum of H can be extracted from the calculation of the quantity

$$\mathcal{Z}_\ell(\beta) = \operatorname{tr} T^\ell\, \mathrm{e}^{-\beta H} = \frac{1}{2\pi}\sum_{n=0}^{\infty}\int_0^{2\pi} \mathrm{d}\theta\, \mathrm{e}^{i\ell\theta}\, \mathrm{e}^{-\beta E_n(\theta)}\,.$$

Indeed,

$$\mathcal{Z}(\theta, \beta) \equiv \sum_\ell \mathrm{e}^{-i\ell\theta}\, \mathcal{Z}_\ell(\beta) = \sum_{n=0}^{\infty} \mathrm{e}^{-\beta E_n(\theta)}, \tag{18.4}$$

where $\mathcal{Z}(\theta, \beta)$ is the partition function restricted to eigenstates of the translation operator T corresponding to a fixed angle θ.

The path integral representation of $\mathcal{Z}_\ell(\beta)$ differs from the representation of the partition function $\mathcal{Z}_0(\beta)$ only by boundary conditions. The operator T has the effect of translating the argument x in the matrix element $\langle x' | \operatorname{tr} e^{-\beta H} | x \rangle$ before taking the trace. It follows that

$$\mathcal{Z}_\ell(\beta) = \int [\mathrm{d}x(t)] \exp\left[-\mathcal{S}(x)\right], \quad \text{with} \tag{18.5}$$

$$\mathcal{S}(x) = \frac{1}{2g} \int_{-\beta/2}^{\beta/2} \left[\dot{x}^2(t) + \sin^2\left(x(t)\right)\right] \mathrm{d}t, \tag{18.6}$$

where one integrates over all paths satisfying the boundary condition $x(\beta/2) = x(-\beta/2) + \ell\pi$. A careful study of the trace operation in the case of periodic potentials shows that $x(-\beta/2)$ varies over only one period.

Therefore, from equation (18.4), one derives the path integral representation

$$\mathcal{Z}(\theta, \beta) = \sum_\ell \int_{x(\beta/2)=x(-\beta/2)+\ell\pi} [\mathrm{d}x(t)] \exp\left[-\mathcal{S}(x) - i\ell\theta\right]$$

$$= \int_{x(\beta/2)=x(-\beta/2) \ (\mathrm{mod} \ \pi)} [\mathrm{d}x(t)] \exp\left[-\mathcal{S}(x) - i\frac{\theta}{\pi} \int_{-\beta/2}^{\beta/2} \mathrm{d}t \, \dot{x}(t)\right]. \tag{18.7}$$

Note that

$$Q = \frac{1}{\pi} \int_{-\beta/2}^{\beta/2} \mathrm{d}t \, \dot{x}(t) \in \mathbb{Z}, \tag{18.8}$$

is a *topological charge*: it depends on the trajectory only through the boundary conditions and characterizes the continuous mapping of the circle $[-\beta/2, \beta/2]$ onto the multiply covered circle $[x(-\beta/2), x(\beta/2)]$. Because it takes only integer values, two trajectories corresponding to different integer values cannot be related by continuous transformations.

18.1.3 Instantons

For $\beta \to \infty$ large and $g \to 0$, the path integral (18.7) is dominated by the constant solutions $x_c(t) = 0 \mod \pi$ contributing to the $\ell = 0$ sector. A non-trivial θ-dependence comes from instanton (non-constant finite action solutions of the imaginary time equations of motion) contributions corresponding to quantum tunnelling. The action (18.6) is finite for $\beta \to \infty$ only if $x(\pm\infty) = 0 \mod \pi$.

A basic inequality. Quite generally, for finite action solutions,

$$\int_{-\infty}^{+\infty} \mathrm{d}t \left[\dot{x}(t) \pm \sin\left(x(t)\right)\right]^2 \geq 0 \ \Rightarrow \ \mathcal{S} \geq \left|\cos\left(x(+\infty)\right) - \cos\left(x(-\infty)\right)\right|/g. \tag{18.9}$$

The expression on the right-hand side of equation (18.9) is different from 0 only for trajectories that relate 0 to π. Its value is then $2/g$. This minimum is reached for trajectories x_c that are solutions of

$$\dot{x}_c(t) \pm \sin\left(x_c(t)\right) = 0 \ \Rightarrow \ x_c(t) = 2\arctan e^{\pm(t-t_0)}, \tag{18.10}$$

where t_0 is an integration constant, and the corresponding classical action then is indeed

$$\mathcal{S}(x_c) = 2/g. \tag{18.11}$$

The instanton solutions belong to the $\ell = \pm 1$ sector and connect two consecutive minima of the potential. They yield the leading contribution to barrier penetration for $g \to 0$. An explicit calculation yields

$$E_0(\theta, g) = E_{\text{pert.}}(g) - \cos\theta \frac{4}{\sqrt{\pi g}} \, \mathrm{e}^{-2/g}[1 + O(g)],$$

where $E_{\text{pert.}}(g)$ is the sum of the perturbative expansion in powers of g.

18.2 Instantons, anomalies and θ-vacua: CP^{N-1} models

We now consider CP^{N-1} models, a set of two-dimensional field theories where, again, instantons and topology play a role and the semi-classical vacuum has a similar periodic structure. The new feature is the relation between topological charge and the two-dimensional chiral anomaly.

Here, we describe mainly the semi-classical structure and refer the reader to the literature for a detailed analysis.

The complete calculation of instanton contributions in the small coupling limit in the CP^{N-1} models, as well as in the non-Abelian gauge theories discussed in Section 18.3, is a non-trivial issue. Due to the scale invariance of the classical theory, instantons depend on a scale (or size) parameter. Instanton contributions then involve the running coupling constant at the instanton size.

Both families of theories are ultraviolet asymptotically free. Therefore, the running coupling is small for small instantons and the semi-classical approximation is justified. However, in the absence of an infrared cut-off, the running coupling becomes large for large instantons, and it is unclear whether a semi-classical approximation remains valid. However, recently some solutions have been proposed [224].

18.2.1 The CP^{N-1} models

The CP^{N-1} manifolds. We consider an N-component complex vector φ of unit length,

$$\bar{\varphi} \cdot \varphi = 1 \, . \tag{18.12}$$

This condition defines a space also isomorphic to the quotient space $U(N)/U(N-1)$. In addition, two vectors φ and φ' are considered equivalent (notation $\varphi' \sim \varphi$) if

$$\varphi'_\alpha = \mathrm{e}^{i\Lambda} \, \varphi_\alpha \, , \quad \Lambda \text{ real} \, . \tag{18.13}$$

This equivalence characterizes the symmetric space and complex Grassmannian manifold $U(N)/U(1)/U(N-1)$. It is isomorphic to the manifold CP^{N-1} (for $(N-1)$-dimensional complex projective), which is obtained from $\mathbb{C}^N$ by the equivalence relation

$$z_\alpha \sim z'_\alpha \quad \text{if} \quad z'_\alpha = \lambda z_\alpha$$

where λ belongs to the Riemann sphere (a compactified complex plane).

18.2.2 The CP^{N-1} action

A symmetric space admits a unique invariant metric and this determines a unique action with two derivatives, up to a multiplicative factor. It is convenient to introduce an auxiliary vector field $A_\mu(x)$ to implement the condition (18.13). The unique CP^{N-1} classical action can then be written as

$$\mathcal{S}(\varphi, A_\mu) = \frac{1}{g} \int \mathrm{d}^2 x \sum_\mu \overline{\mathrm{D}_\mu \varphi}(x) \cdot \mathrm{D}_\mu \varphi(x), \qquad (18.14)$$

in which $g > 0$ is a coupling constant and D_μ is the covariant derivative,

$$\mathrm{D}_\mu = \partial_\mu + i A_\mu. \qquad (18.15)$$

The field A_μ is a gauge field for the $U(1)$ transformations:

$$\varphi'(x) = \mathrm{e}^{i\Lambda(x)} \varphi(x), \quad A'_\mu(x) = A_\mu(x) - \partial_\mu \Lambda(x). \qquad (18.16)$$

The action (18.14) is $U(N)$ invariant and the gauge symmetry ensures the equivalence (18.13).

We define also the associated curvature tensor,

$$F_{\mu\nu}(x) = [\mathrm{D}_\mu, \mathrm{D}_\nu] = \partial_\mu A_\nu(x) - \partial_\nu A_\mu(x). \qquad (18.17)$$

Gauge field elimination. Since the action contains no kinetic term for A_μ, the gauge field is not a dynamic field and can be integrated out. The action being quadratic in A, the Gaussian integration results in replacing in the action A_μ by the solution of the A-field equation,

$$A_\mu(x) = i\bar\varphi(x) \cdot \partial_\mu \varphi(x), \qquad (18.18)$$

where equation (18.12) has been used. After the substitution, the composite field $\bar\varphi(x) \cdot \partial_\mu \varphi(x)$ acts as a composite gauge field.

However, for what follows, we find it convenient to keep A_μ as an independent field.

18.2.3 The structure of the semi-classical vacuum

The action (18.14) is minimum when $\mathrm{D}_\mu \varphi$ vanishes and, therefore,

$$\mathrm{D}_\mu \varphi(x) = 0 \;\Rightarrow\; [\mathrm{D}_\mu, \mathrm{D}_\nu]\varphi(x) = F_{\mu\nu}(x)\varphi(x) = 0.$$

Since $\varphi \neq 0$, the equation implies that $F_{\mu\nu}$ vanishes, that A_μ is a pure gauge and that φ is a gauge transform of a constant vector.

To emphasize the analogy with non-Abelian gauge theories (Section 18.3), we now analyse more precisely the structure of the semi-classical vacuum in the temporal gauge $A_2 = 0$. In this gauge, the action is still invariant under space-dependent gauge transformations.

The minima of the classical action then correspond to fields $\varphi(x_1)$, where x_1 is the space variable, of the form

$$\varphi(x_1) = e^{i\Lambda(x_1)}\,\mathbf{v}\,, \quad \bar{\mathbf{v}}\cdot\mathbf{v} = 1\,.$$

Moreover, if the theory is quantized in a way that compactifies topologically the space dimension (*e.g.*, the thermodynamic limit is taken with periodic boundary conditions), then $\Lambda(x_1)$ defines a mapping of the circle S_1 onto a circle S_1:

$$x_1 \in S_1 \;\mapsto\; \Lambda(x_1) \in S_1\,.$$

One is thus led to the consideration of the homotopy classes of mappings from S_1 to S_1, which are characterized by an integer ν, the *winding number*.

This is equivalent to the statement that the homotopy group $\pi_1(S_1)$ is isomorphic to the additive group of integers $\mathbb{Z}$.

Again, the semi-classical vacuum has a periodic structure. This analysis is consistent with Gauss's law, which implies that states are invariant only under infinitesimal gauge transformations and, thus, under gauge transformations of the class $\nu = 0$ that are continuously connected to the identity.

Topological charge. We consider the quantity

$$Q(\varphi) = \frac{1}{2\pi} \lim_{R\to\infty} \oint_{|x|=R} \sum_{\mu} \mathrm{d}x_{\mu}\, A_{\mu}(x)\,. \tag{18.19}$$

$Q(\varphi)$ depends only on the behaviour of the classical field for $|x|$ large. Finiteness of the action demands that, at large distances, $\mathrm{D}_{\mu}\varphi$ vanishes and, therefore, A_{μ} is a pure gauge:

$$A_{\mu}(x) = \partial_{\mu}\Lambda(x) \;\Rightarrow\; Q(\varphi) = \frac{1}{2\pi} \lim_{R\to\infty} \oint_{|x|=R} \sum_{\mu} \mathrm{d}x_{\mu}\partial_{\mu}\Lambda(x)\,. \tag{18.20}$$

In the $A_2 = 0$ gauge and compactified space, we conclude

$$Q(\varphi) = \frac{1}{2\pi} \int \mathrm{d}x_1 \frac{\mathrm{d}\Lambda}{\mathrm{d}x_1} = \nu \in \mathbb{Z}\,.$$

Thus, $Q(\varphi)$ is the *topological charge* associated with vacuum degeneracy.

18.2.4 Instantons and topology

We prove now that, as in the example of the cosine potential, the degeneracy of classical vacua is lifted by instantons.

To prove the existence of locally stable non-trivial minima of the action, the following Bogomolnyi's inequality [189] can be used (note the analogy with equation (18.9)):

$$\int \mathrm{d}^2x \sum_{\mu} \left| \mathrm{D}_{\mu}\varphi(x) \mp i \sum_{\nu} \epsilon_{\mu\nu}\mathrm{D}_{\nu}\varphi(x) \right|^2 \geq 0\,, \tag{18.21}$$

($\epsilon_{\mu\nu}$ being the antisymmetric tensor, with $\epsilon_{12} = 1$).

After expansion, the inequality can be cast into the form

$$\mathcal{S}(\varphi) \geq 2\pi |\Sigma(\varphi)|/g, \tag{18.22}$$

with

$$\Sigma(\varphi) = -\frac{i}{2\pi} \sum_{\mu,\nu} \epsilon_{\mu\nu} \int \mathrm{d}^2x \, \mathrm{D}_\mu\varphi(x) \cdot \overline{\mathrm{D}_\nu\varphi}(x)$$

$$= \frac{i}{2\pi} \int \mathrm{d}^2x \sum_{\mu,\nu} \epsilon_{\mu\nu} \bar\varphi(x) \cdot \mathrm{D}_\nu\mathrm{D}_\mu\varphi(x). \tag{18.23}$$

Then,

$$i\sum_{\mu,\nu} \epsilon_{\mu\nu}\mathrm{D}_\nu\mathrm{D}_\mu = \tfrac{1}{2}i\sum_{\mu,\nu} \epsilon_{\mu\nu}[\mathrm{D}_\nu,\mathrm{D}_\mu] = -\tfrac{1}{2}\sum_{\mu,\nu} \epsilon_{\mu\nu}F_{\mu\nu}. \tag{18.24}$$

Therefore, using equation (18.12), one obtains

$$\Sigma(\varphi) = \frac{1}{4\pi} \int \mathrm{d}^2x \sum_{\mu,\nu} \epsilon_{\mu\nu}F_{\mu\nu}(x). \tag{18.25}$$

The integrand is proportional to the two-dimensional Abelian chiral anomaly (equation (17.41)). It is a total divergence,

$$\tfrac{1}{2}\sum_{\mu,\nu} \epsilon_{\mu\nu}F_{\mu\nu}(x) = \sum_{\mu,\nu} \partial_\mu\epsilon_{\mu\nu}A_\nu(x).$$

Substituting this form into equation (18.25) and integrating over a large disc of radius R, one obtains

$$\Sigma(\varphi) = \frac{1}{2\pi} \lim_{R\to\infty} \oint_{|x|=R} \sum_\mu \mathrm{d}x_\mu \, A_\mu(x). \tag{18.26}$$

We recognize the topological charge $Q(\varphi)$ defined by equation (18.19). We infer

$$\mathcal{S}(\varphi) \geq 2\pi|\nu|/g, \quad \nu \in \mathbb{Z}. \tag{18.27}$$

The equality $\mathcal{S}(\varphi) = 2\pi|\nu|/g$ corresponds to a local minimum and implies that the classical solutions satisfy the first order partial differential (self-duality) equations,

$$\mathrm{D}_\mu\varphi(x) = \pm i\sum_\nu \epsilon_{\mu\nu}\mathrm{D}_\nu\varphi(x). \tag{18.28}$$

For each sign, there is really only one equation, for instance $\mu = 1, \nu = 2$. It is easy to verify that both equations imply the φ-field equations and, combined with the constraint (18.12), the A-field equation (18.18).

In complex coordinates $z = x_1 + ix_2$, $\bar z = x_1 - ix_2$, and $A_z = (A_1 - iA_2)/2$, $A_{\bar z} = (A_1 + iA_2)/2$), they can be written as

$$\partial_z\varphi_\alpha(z,\bar z) = -iA_z(z,\bar z)\varphi_\alpha(z,\bar z),$$
$$\partial_{\bar z}\varphi_\alpha(z,\bar z) = -iA_{\bar z}(z,\bar z)\varphi_\alpha(z,\bar z).$$

Exchanging the two equations just amounts to exchanging φ and $\bar{\varphi}$. Therefore, we solve only the second equation, which yields

$$\varphi_\alpha(z, \bar{z}) = \kappa(z, \bar{z}) P_\alpha(z),$$

where $\kappa(z, \bar{z})$ is one solution of

$$\partial_{\bar{z}} \kappa(z, \bar{z}) = -i A_{\bar{z}}(z, \bar{z}) \kappa(z, \bar{z}).$$

Vector solutions of equation (18.28) are proportional to holomorphic or anti-holomorphic (depending on the sign) vectors (this reflects the *conformal invariance* of the classical field theory).

In a special gauge (which corresponds to the $\sum_\mu \partial_\mu A_\mu = 0$ gauge), the function $\kappa(z, \bar{z})$ can be chosen real and is then constrained by the condition (18.12), which becomes

$$\kappa^2(z, \bar{z}) \, P \cdot \bar{P} = 1.$$

The asymptotic conditions constrain the functions $P_\alpha(z)$ to be polynomials. Common roots to all P_α would correspond to non-integrable singularities for φ_α and, therefore, are excluded by the condition of finiteness of the action. Finally, if the polynomials have maximal degree n, asymptotically

$$P_\alpha(z) \sim c_\alpha z^n \;\Rightarrow\; \varphi_\alpha \sim \frac{c_\alpha}{\sqrt{\mathbf{c} \cdot \bar{\mathbf{c}}}} (z/\bar{z})^{n/2}.$$

When the phase of z varies by 2π, the phase of φ_α varies by $2n\pi$, showing that n is the corresponding *winding number*.

Instantons and semi-classical vacua. We now consider a large rectangle with extension R in the space direction and T in the Euclidean time direction and, by a smooth gauge transformation, continue the instanton solution to the temporal gauge. Then, the variation of the pure gauge comes entirely from the sides at fixed time. For $R \to \infty$, one finds

$$\Lambda(+\infty, 0) - \Lambda(-\infty, 0) - [\Lambda(+\infty, T) - \Lambda(-\infty, T)] = 2n\pi.$$

Therefore, instantons interpolate between different classical minima. As in the case of the cosine potential and in analogy with expression (18.7), to project onto a proper quantum eigenstate, the 'θ-vacuum' corresponding to an angle θ, one adds a topological term to the classical action. Here,

$$S(\varphi) \mapsto S(\varphi) + i\frac{\theta}{4\pi} \int \mathrm{d}^2 x \sum_{\mu, \nu} \epsilon_{\mu\nu} F_{\mu\nu}(x). \tag{18.29}$$

Exterior differential calculus formalism. Replacing the gauge field in the topological charge Q by the explicit expression (18.18), one finds

$$Q(\varphi) = \frac{i}{2\pi} \int \mathrm{d}^2 x \sum_{\mu, \nu} \epsilon_{\mu\nu} \partial_\mu \bar{\varphi}(x) \cdot \partial_\nu \varphi(x) = \frac{i}{2\pi} \int \mathrm{d}\bar{\varphi}_\alpha \wedge \mathrm{d}\varphi_\alpha,$$

where the notation of exterior differential calculus is used.

One recognizes the integral of a 2-form, a symplectic form, and $4\pi Q$ is the area of a 2-surface embedded in CP^{N-1}. A symplectic form is always closed. Here, it is also exact, so that Q is the integral of a 1-form (*cf.*, equation (18.19)):

$$Q(\varphi) = \frac{i}{2\pi} \sum_\alpha \int \bar{\varphi}_\alpha \mathrm{d}\varphi_\alpha = \frac{i}{4\pi} \int \sum_\alpha (\bar{\varphi}_\alpha \mathrm{d}\varphi_\alpha - \varphi_\alpha \mathrm{d}\bar{\varphi}_\alpha) \,.$$

18.2.5 CP^1 and $O(3)$ non-linear σ-models

The CP^1 model is locally isomorphic to the $O(3)$ non-linear σ-model, with the identification

$$\phi^i(x) = \sum_{\alpha,\beta=1,2} \bar{\varphi}_\alpha(x)\sigma^i_{\alpha\beta}\varphi_\beta(x) \,, \tag{18.30}$$

where σ^i are the three Pauli matrices (expressions (17.23)).

Using then the explicit representation,

$$\sigma_1 = \begin{pmatrix} 0 & 1 \\ 1 & 0 \end{pmatrix}, \quad \sigma_2 = \begin{pmatrix} 0 & -i \\ i & 0 \end{pmatrix}, \quad \sigma_3 = \begin{pmatrix} 1 & 0 \\ 0 & -1 \end{pmatrix},$$

it is simple to verify that

$$\sum_i \phi^i(x)\phi^i(x) = 1 \,, \quad \sum_i \nabla\phi^i(x) \cdot \nabla\phi^i(x) = 4 \sum_\mu \overline{D_\mu\varphi}(x) \cdot D_\mu\varphi(x) \,.$$

Therefore, the field theory can be expressed in terms of the fields ϕ^i and takes the form of the non-linear σ-model.

The fields ϕ^i are gauge invariant and the whole physical picture is a picture of confinement of the charged scalar 'quarks' $\varphi_\alpha(x)$ and the propagation of neutral bound states corresponding to the fields ϕ^i.

Instantons in the ϕ description take the form of ϕ configurations with uniform limit for $|x| \to \infty$. Thus, they define a mapping from the compactified plane topologically equivalent to S_2 to the sphere S_2 (the ϕ^i configurations). Since $\pi_2(S_2) = \mathbb{Z}$, the φ and ϕ pictures are consistent.

In the example of CP^1, a solution of winding number 1 is

$$\varphi_1 = \frac{1}{\sqrt{1+z\bar{z}}} \,, \quad \varphi_2 = \frac{z}{\sqrt{1+z\bar{z}}} \,.$$

Translating the CP^1 minimal solution into the $O(3)$ σ-model language, one finds

$$\phi_1 = \frac{z+\bar{z}}{1+\bar{z}z} \,, \quad \phi_2 = \frac{1}{i}\frac{z-\bar{z}}{1+\bar{z}z} \,, \quad \phi_3 = \frac{1-\bar{z}z}{1+zz} \,.$$

This defines a stereographic mapping of the plane onto the sphere S_2, as one can verify by setting $z = \tan(\eta/2)\,\mathrm{e}^{i\theta}$, $\eta \in [0,\pi]$.

In the $O(3)$ representation,

$$Q = \frac{i}{2\pi} \int \sum_\alpha \mathrm{d}\bar{\varphi}_\alpha \wedge \mathrm{d}\varphi_\alpha$$

$$= \frac{1}{8\pi} \sum_{i,j,k} \epsilon_{ijk} \int \phi_i \mathrm{d}\phi_j \wedge \phi_k \equiv \frac{1}{8\pi} \sum_{\mu,\nu,i,j,k} \epsilon_{\mu\nu}\epsilon_{ijk} \int \mathrm{d}^2x \, \phi_i(x)\partial_\mu \phi_j(x)\partial_\nu \phi_k(x).$$

The topological charge $4\pi Q$ can be interpreted as the area of the sphere S_2, multiply covered, and embedded in $\mathbb{R}^3$. Its value is a multiple of the area of S_2, which, in this interpretation, explains the quantization.

18.3 Non-Abelian gauge theories: Instantons and anomalies

We now consider non-Abelian gauge theories in four dimensions. Gauge field configurations can be found that contribute to the chiral anomaly and for which, therefore, the right-hand side of equation (17.56) does not vanish. An important example is provided by instantons, that is, finite action solutions of Euclidean field equations. They are related to quantum tunnelling between different classical minima.

To discuss this problem, it is sufficient to consider matterless gauge theories and the gauge group $SU(2)$, since a general theorem states that, for a Lie group containing $SU(2)$ as a subgroup, the instantons are those of the $SU(2)$ subgroup.

In the absence of matter fields, it is convenient to use an $SO(3)$ notation. The gauge field $\mathfrak{A}_\mu$ element of the Lie algebra of $SU(2)$ (anti-Hermitian traceless 2×2 matrix, : see Section 13.1.2) can be parametrized in terms of 3-vectors as

$$\mathfrak{A}_\mu = -\tfrac{1}{2}i\mathbf{A}_\mu \cdot \boldsymbol{\sigma}, \tag{18.31}$$

where σ_i are the three Pauli matrices.

The gauge action can then be written as

$$\mathcal{S}(\mathbf{A}_\mu) = \frac{1}{4g^2} \int \sum_{\mu,\nu} [\mathbf{F}_{\mu\nu}(x)]^2 \, \mathrm{d}^4x, \tag{18.32}$$

(g is the gauge coupling constant), where the curvature,

$$\mathbf{F}_{\mu\nu}(x) = \partial_\mu \mathbf{A}_\nu(x) - \partial_\nu \mathbf{A}_\mu(x) + \mathbf{A}_\mu(x) \times \mathbf{A}_\nu(x), \tag{18.33}$$

is also a $SO(3)$ vector.

18.3.1 Instantons, chiral anomaly and topology

The existence and some properties of instantons in non-Abelian gauge theories follow from considerations analogous to those presented for the CP^{N-1} models.

We define the dual of the tensor $\mathbf{F}_{\mu\nu}$ by

$$\tilde{\mathbf{F}}_{\mu\nu}(x) = \tfrac{1}{2} \sum_{\rho,\sigma} \epsilon_{\mu\nu\rho\sigma} \mathbf{F}_{\rho\sigma}(x), \tag{18.34}$$

where $\epsilon_{\mu\nu\rho\sigma}$ is the completely antisymmetric tensor, with $\epsilon_{1234} = 1$.

Then, we use Bogomolnyi's inequality [189] in the form,

$$\int d^4x \sum_{\mu,\nu} \left[\mathbf{F}_{\mu\nu}(x) \pm \tilde{\mathbf{F}}_{\mu\nu}(x) \right]^2 \geq 0 \,. \tag{18.35}$$

The inequality implies that any finite action configuration satisfies

$$\mathcal{S}(\mathbf{A}_\mu) \geq 8\pi^2 |Q(\mathbf{A}_\mu)|/g^2, \tag{18.36}$$

with

$$Q(\mathbf{A}_\mu) = \frac{1}{32\pi^2} \int d^4x \sum_{\mu,\nu} \mathbf{F}_{\mu\nu}(x) \cdot \tilde{\mathbf{F}}_{\mu\nu}(x) \,. \tag{18.37}$$

The expression $Q(\mathbf{A}_\mu)$ is a *topological charge*, proportional to the integral of the chiral anomaly (17.49), here written in $SO(3)$ notation.

Without explicit calculation we know, from the analysis of the index of the Dirac operator in a gauge background (equation (17.56)), that the topological charge is an integer:

$$Q(\mathbf{A}_\mu) = \frac{1}{32\pi^2} \int d^4x \, \mathbf{F}_{\mu\nu}(x) \cdot \tilde{\mathbf{F}}_{\mu\nu}(x) = n \in \mathbb{Z} \,. \tag{18.38}$$

One notices the analogy with the CP^{N-1} model.

The inequality (18.37) then implies

$$\mathcal{S}(\mathbf{A}_\mu) \geq 8\pi^2 |n|/g^2 \,. \tag{18.39}$$

The equality corresponds to a local minimum of the action and thus to instantons. It is obtained for fields satisfying the first order *self-duality equations*,

$$\mathbf{F}_{\mu\nu}(x) = \pm \tilde{\mathbf{F}}_{\mu\nu}(x). \tag{18.40}$$

These equations, unlike the general classical field equations (which they imply), are first order partial differential equations and, thus, easier to solve. The one-instanton solution, which depends on an arbitrary scale parameter λ, is (ϵ_{ijk} is the totally antisymmetric tensor with $\epsilon_{123} = 1$)

$$A^i_m = \frac{2}{r^2 + \lambda^2} \left(x_4 \delta_{im} + \epsilon_{imk} x_k \right), \text{ for } m = 1, 2, 3 \,, \text{ and } A^i_4 = -\frac{2x_i}{r^2 + \lambda^2}. \tag{18.41}$$

18.3.2 The topological charge: Quantization.

We now prove more directly (rather than referring to the index of the Dirac operator) the topological character and the quantization of $Q(\mathbf{A}_\mu)$, as defined by equation (18.37), for finite action configurations.

The quantity $\sum_{\mu,\nu} \mathbf{F}_{\mu\nu} \cdot \tilde{\mathbf{F}}_{\mu\nu}$ is a divergence (equation (17.50)). It can be written as

$$\sum_{\mu,\nu} \mathbf{F}_{\mu\nu}(x) \cdot \tilde{\mathbf{F}}_{\mu\nu}(x) = \sum_{\mu} \partial_\mu V_\mu(x).$$

The vector V_μ is given by

$$V_\mu = -4 \sum_{\nu,\rho,\sigma} \epsilon_{\mu\nu\rho\sigma} \operatorname{tr} \left(\mathfrak{A}_\nu \partial_\rho \mathfrak{A}_\sigma + \tfrac{2}{3} \mathfrak{A}_\nu \mathfrak{A}_\rho \mathfrak{A}_\sigma\right) \tag{18.42a}$$

$$= 2 \sum_{\nu,\rho,\sigma} \epsilon_{\mu\nu\rho\sigma} \left[\mathbf{A}_\nu \cdot \partial_\rho \mathbf{A}_\sigma + \tfrac{1}{3} \mathbf{A}_\nu \cdot (\mathbf{A}_\rho \times \mathbf{A}_\sigma)\right]. \tag{18.42b}$$

The integral (18.37) thus depends only on the behaviour of the gauge field at large distances. Stokes' theorem implies

$$\int_{\mathcal{D}} \mathrm{d}^4 x\, \nabla \cdot \mathbf{V}(x) = \int_{\partial\mathcal{D}} \mathrm{d}\Omega\, \hat{\mathbf{n}} \cdot \mathbf{V}(x),$$

where $\mathrm{d}\Omega$ is the measure on the boundary $\partial\mathcal{D}$ of the four-volume $\mathcal{D}$, and $\hat{\mathbf{n}}$ is the unit vector normal to $\partial\mathcal{D}$. We take for $\mathcal{D}$ a sphere of large radius R and find for the topological charge

$$Q(\mathbf{A}_\mu) = \frac{1}{32\pi^2} \int \mathrm{d}^4 x\, \operatorname{tr} \mathbf{F}_{\mu\nu}(x) \cdot \tilde{\mathbf{F}}_{\mu\nu}(x) = \frac{1}{32\pi^2} R^3 \int_{|x|=R} \mathrm{d}\Omega\, \hat{\mathbf{n}} \cdot \mathbf{V}(x). \tag{18.43}$$

The finiteness of the action implies that the classical solutions converge asymptotically towards a pure gauge, that is, with our conventions,

$$\mathfrak{A}_\mu = -\tfrac{1}{2} i \mathbf{A}_\mu \cdot \boldsymbol{\sigma} = \mathbf{g}(x) \partial_\mu \mathbf{g}^{-1}(x) + O\left(|x|^{-2}\right), \quad \text{for } |x| \to \infty. \tag{18.44}$$

The element $\mathbf{g}$ of the $SU(2)$ group can be parametrized in terms of Pauli matrices:

$$\mathbf{g} = u_4 \mathbf{1} + i \mathbf{u} \cdot \boldsymbol{\sigma}, \tag{18.45}$$

where the four-component real vector $(\mathbf{u} \equiv (u_1, u_2, u_3), u_4)$ satisfies

$$u_4^2 + \mathbf{u}^2 = 1$$

and, thus, belongs to the unit sphere S_3.

Since $SU(2)$ is topologically equivalent to the sphere S_3, the pure gauge configurations on a sphere of large radius $|x| = R$ define a mapping from S_3 to S_3. Such mappings belong to different homotopy classes that are characterized by an integer called the *winding number*. Here, we identify the homotopy group $\pi_3(S_3)$, which, again, is isomorphic to the additive group of integers $\mathbb{Z}$.

One-to-one mapping. The simplest mapping is a one-to-one mapping and corresponds to an element of the form $(\mathbf{x} \equiv (x_1, x_2, x_3))$

$$\mathbf{g}(x) = \frac{x_4 \mathbf{1} + i \mathbf{x} \cdot \boldsymbol{\sigma}}{r}, \quad r = (x_4^2 + \mathbf{x}^2)^{1/2}, \tag{18.46}$$

and thus (see also equation (18.41))

$$A_m^i \underset{r\to\infty}{\sim} 2 \frac{x_4}{r^2} \delta_{im} + 2 \sum_k \epsilon_{imk} \frac{x_k}{r^2}, \quad A_4^i = -2 x_i r^{-2}. \tag{18.47}$$

The transformation

$$\mathbf{g}(x) \mapsto \mathbf{U}_1 \mathbf{g}(x) \mathbf{U}_2^\dagger = \mathbf{g}(\mathbf{R}x),$$

where $\mathbf{U}_1$ and $\mathbf{U}_2$ are two constant $SU(2)$ matrices, induces an $SO(4)$ rotation of matrix $\mathbf{R}$ of the vector x_μ. Thus,

$$\mathbf{U}_2 \partial_\mu \mathbf{g}^\dagger(x) \mathbf{U}_1^\dagger = \sum_\nu R_{\mu\nu} \partial_\nu \mathbf{g}^\dagger(\mathbf{R}x), \quad \mathbf{U}_1 \mathbf{g}(x) \partial_\mu \mathbf{g}^\dagger(x) \mathbf{U}_1^\dagger = \mathbf{g}(\mathbf{R}x) \sum_\nu R_{\mu\nu} \partial_\nu \mathbf{g}^\dagger(\mathbf{R}x),$$

and, therefore,

$$\mathbf{U}_1 \mathfrak{A}_\mu(x) \mathbf{U}_1^\dagger = \sum_\nu R_{\mu\nu} \mathfrak{A}_\nu(\mathbf{R}x).$$

By introducing this relation into the definition (18.42a) of V_μ, it is simple to verify that the dependence on the matrix $\mathbf{U}_1$ cancels in the trace and, thus, V_μ transforms like a 4-vector. Since only one vector is available, and taking into account dimensional analysis, one concludes that

$$V_\mu(x) \propto x_\mu/r^4 .$$

For $r \to \infty$, $\mathbf{A}_\mu$ approaches a pure gauge (equation (18.44)) and, therefore, V_μ can be transformed into

$$V_\mu \underset{r\to\infty}{\sim} -\tfrac{1}{3} \sum_{\nu,\rho,\sigma} \epsilon_{\mu\nu\rho\sigma} \mathbf{A}_\nu \cdot (\mathbf{A}_\rho \times \mathbf{A}_\sigma).$$

It is sufficient to calculate V_1. We choose $\rho = 3, \sigma = 4$, and multiply by a factor of 6 to take into account all other choices. Then,

$$V_1 \underset{r\to\infty}{\sim} 16\epsilon_{ijk}(x_4\delta_{2i} + \epsilon_{i2l}x_l)(x_4\delta_{3j} + \epsilon_{j3m}x_m)x_k/r^6 = 16x_1/r^4$$

and, thus,

$$V_\mu \sim 16x_\mu/r^4 = 16\hat{n}_\mu/R^3 .$$

The powers of R in equation (18.43) cancel and, since $\int \mathrm{d}\Omega = 2\pi^2$, the value of the topological charge is simply

$$Q(\mathbf{A}_\mu) = 1 . \tag{18.48}$$

The comparison between this result and equation (17.56) shows that the minimal action solution has indeed been found.

General topological charge: Geometric interpretation. As in the case of the CP^{N-1} model, this result has a geometric interpretation. In general, in the parametrization (18.45),

$$V_\mu \underset{r\to\infty}{\sim} \tfrac{8}{3} \sum_{\nu,\rho,\sigma} \epsilon_{\mu\nu\rho\sigma} \sum_{\alpha,\beta,\gamma} \epsilon_{\alpha\beta\gamma\delta} u_\alpha \partial_\nu u_\beta \partial_\rho u_\gamma \partial_\sigma u_\delta .$$

A few algebraic manipulations starting from

$$\int_{S_3} R^3 \mathrm{d}\Omega \, \hat{\mathbf{n}} \cdot \mathbf{V} = \sum_{\mu,\nu,\rho,\sigma} \tfrac{1}{6}\epsilon_{\mu\nu\rho\sigma} \int V_\mu \mathrm{d}u_\nu \wedge \mathrm{d}u_\rho \wedge \mathrm{d}u_\sigma ,$$

then yield

$$Q = \frac{1}{12\pi^2} \sum_{\mu,\nu,\rho,\sigma} \epsilon_{\mu\nu\rho\sigma} \int u_\mu \mathrm{d}u_\nu \wedge \mathrm{d}u_\rho \wedge \mathrm{d}u_\sigma , \tag{18.49}$$

where the notation of exterior differential calculus again is used.

The area Σ_p of the sphere S_{p-1}, in the same notation, can be written as

$$\Sigma_p = \frac{2\pi^{p/2}}{\Gamma(p/2)} = \frac{1}{(p-1)!} \sum_{\mu_1,\dots,\mu_p} \epsilon_{\mu_1\dots\mu_p} \int u_{\mu_1} \mathrm{d}u_{\mu_2} \wedge \cdots \wedge \mathrm{d}u_{\mu_p},$$

when the vector u_μ describes the sphere S_{p-1} only once. In the right-hand side of equation (18.49), one recognizes an expression proportional to the area of the sphere S_3. Because, in general, u_μ describes S_3 n times when x_μ describes S_3 only once, a factor n is generated.

18.4 The semi-classical vacuum and the strong CP violation

We now proceed in analogy with the analysis of the CP^{N-1} model. In the temporal gauge $\mathbf{A}_4 = 0$, the classical minima of the potential correspond to gauge field components $\mathbf{A}_i$, $i = 1, 2, 3$, which are pure gauge functions of the three space variables x_i:

$$\mathfrak{A}_m = -\tfrac{1}{2} i \mathbf{A}_m \cdot \boldsymbol{\sigma} = \mathbf{g}(x_i) \partial_m \mathbf{g}^{-1}(x_i). \tag{18.50}$$

The structure of the classical minima is related to the homotopy classes of mappings of the group elements $\mathbf{g}$ into compactified $\mathbb{R}^3$ (because $\mathbf{g}(x)$ goes to a constant for $|x| \to \infty$), that is, again, of S_3 into S_3, and thus the semi-classical vacuum, as in the CP^{N-1} model, has a periodic structure. It is possible to verify that the instanton solution (18.41), transported into the temporal gauge by a gauge transformation, connects minima with different winding numbers. Therefore, as in the case of the CP^{N-1} model, to project onto a θ-vacuum, one adds a term to the classical action of gauge theories,

$$\mathcal{S}_\theta(\mathbf{A}_\mu) = \mathcal{S}(\mathbf{A}_\mu) + \frac{i\theta}{32\pi^2} \int d^4x \sum_{\mu,\nu} \mathbf{F}_{\mu\nu}(x) \cdot \tilde{\mathbf{F}}_{\mu\nu}(x), \tag{18.51}$$

and then integrates over all fields $\mathbf{A}_\mu$ without restriction. At least in the semi-classical approximation, the gauge theory thus depends on one additional parameter, the angle θ. For non-vanishing values of θ, the additional term violates CP (the product of charge conjugation and parity) conservation and is at the origin of the *strong CP violation* problem. From experimental data, the angle is at most of order 10^{-9}. Unless θ vanishes for some as yet unknown reason, it is unnaturally small. This has led to the *axion* hypothesis [163].

18.5 Fermions in an instanton background: The $U(1)$ problem

We now apply the analysis to quantum chromodynamics (see Chapter 13), the theory of strong interactions, where N_{F} Dirac fermions $\mathbf{Q}$, $\bar{\mathbf{Q}}$, the quark fields which transform under the fundamental representation of the $SU(3)$ colour group, are coupled to non-Abelian gauge fields $\mathbf{A}_\mu$ corresponding to the $SU(3)$ group.

We return here to standard $SU(3)$ notation with generators of the Lie algebra and gauge fields being represented by anti-Hermitian matrices.

The action can then be written as

$$\mathcal{S}(\mathbf{A}_\mu, \bar{\mathbf{Q}}, \mathbf{Q}) = - \int \mathrm{d}^4 x \left[\frac{1}{4g^2} \sum_{\mu,\nu} \operatorname{tr} \mathbf{F}^2_{\mu\nu}(x) + \sum_{F=1}^{N_F} \bar{\mathbf{Q}}_F(x) \left(\slashed{D} + m_f \right) \mathbf{Q}_F(x) \right],$$

where F is the flavour quantum number.

The existence of Abelian anomalies and instantons has several physical consequences. We mention here two of them.

18.5.1 *Solutions to the strong CP problem*

According to the analysis of Section 17.3.2, only configurations with a non-vanishing index of the Dirac operator contribute to the θ-term. Then, the Dirac operator Ds has at least one vanishing eigenvalue.

If one fermion field is massless, the determinant resulting from the fermion integration vanishes, the instantons do not contribute to the field integral and the strong CP violation problem is solved. However, such an hypothesis seems to be inconsistent with experimental data on quark masses (see Chapter 13). Another scheme is based on the introduction of a pseudoscalar field, the *axion* (which could also contribute to dark matter) [163] but no candidate has, up to now, been experimentally discovered.

18.5.2 *The solution of the $U(1)$ problem*

Experimentally, it is observed that the masses of a number of pseudo-scalar mesons are smaller or even much smaller (in the case of pions) than the masses of the corresponding scalar mesons. This strongly suggests that pseudoscalar mesons are almost Goldstone bosons associated with an approximate chiral symmetry realized in a phase of spontaneous symmetry breaking. (When a continuous (non gauge) symmetry is spontaneously broken, the spectrum of the theory exhibits massless scalar particles called Goldstone bosons.) This picture is confirmed by a number of other phenomenological predictions.

In the Standard Model of particle physics, this approximate symmetry is viewed as the consequence of the very small masses of the **u** and **d** quarks and the moderate value of the strange **s** quark mass.

Indeed, in a theory in which the quarks are massless, the action has a chiral $U(N_F) \times U(N_F)$ symmetry, in which N_F is the number of flavours. The spontaneous breaking of the chiral symmetry group into its diagonal subgroup $U(N_F)$ leads one to expect N_F^2 Goldstone bosons associated with all axial currents (corresponding to the generators of $U(N_F) \times U(N_F)$ that do not belong to the remaining $U(N_F)$ symmetry group).

In the physically relevant theory, several quark masses are non-vanishing but small and one expects this picture to survive approximately with, instead of Goldstone bosons, light pseudoscalar mesons.

The experimental mass pattern is consistent with a weakly broken $SU(2) \times SU(2)$ and more badly violated $SU(3) \times SU(3)$ symmetries but not $U(2) \times U(2)$ or $U(3) \times U(3)$.

The Ward–Takahashi (WT) identities (relations satisfied by vertex or correlation functions expressing the symmetry), which imply the existence of Goldstone bosons, correspond to constant group transformations and, thus, involve only the space integral of the divergence of the current. From the preceding analysis, we know that the axial current corresponding to the $U(1)$ Abelian subgroup has an anomaly, causing the integral to vanish.

However, non-Abelian gauge theories have configurations that give non-vanishing values of the form (17.56) to the space integral of the anomaly (17.49). For small couplings, these configurations are in the neighbourhood of instanton solutions (as discussed in Section 18.3).

This indicates (although no satisfactory calculation of the instanton contribution has been performed yet) that, for small, but non-vanishing, quark masses, the $U(1)$ axial current is far from being conserved and, therefore, no corresponding almost-Goldstone light boson should be expected.

Instanton contributions to the anomaly thus resolve a long-standing experimental puzzle, called the $U(1)$ problem.

Note that the usual derivation of WT identities involves only global chiral transformations and, therefore, there is no need to introduce axial currents. In the case of massive quarks, chiral symmetry is explicitly broken by soft mass terms, and WT identities involve insertions of the operators

$$\mathcal{M}_f = m_f \int \mathrm{d}^4 x \, \bar{\mathbf{Q}}_f(x) \gamma_5 \mathbf{Q}_f(x),$$

which are the variations of the mass terms in an infinitesimal chiral transformation. If the contributions of $\mathcal{M}_f$ vanish when $m_f \to 0$, as one would normally expect, then a situation of approximate chiral symmetry is realized (in a symmetric or spontaneously broken phase). However, if one integrates over fermions first, at fixed gauge fields, one finds (disconnected) contributions proportional to

$$\langle \mathcal{M}_f \rangle = m_f \operatorname{tr} \gamma_5 \left(\slashed{D} + m_f \right)^{-1}.$$

We have shown in Section 17.3.2 that, for topologically non-trivial gauge field configurations, $\slashed{D}$ has zero eigenmodes, which, for $m_f \to 0$, give the leading contributions

$$\langle \mathcal{M}_f \rangle = m_f \sum_n \int \mathrm{d}^4 x \, \varphi_n^*(x) \gamma_5 \varphi_n(x) \frac{1}{m_f} + O(m_f)$$

$$= (n_+ - n_-) + O(m_f).$$

These contributions do not vanish for $m_f \to 0$ and are responsible, after integration over gauge fields, for a violation of chiral symmetry.

19 Field theory in a finite geometry: Finite size scaling

In physics, numerical simulations are performed with finite volume systems. For statistical systems, transfer matrix calculations deal with systems of finite size in all dimensions but one. To extrapolate the results to the infinite system, it is thus necessary to understand how the structure of finite size effects.

In particular, in a system in which the forces are short range, no phase transition can occur in a finite volume, or in a geometry in which the size is infinite only in one dimension. This indicates that, in the case of systems close to a second order phase transition, that is, when the correlation length is large, the infinite size extrapolation is somewhat non-trivial [190–194].

For simplicity, in this chapter, we consider only finite sizes with periodic boundary conditions. In the case of boson fields, this implies the existence of zero modes, a problem that we study first in the case of one-dimension systems and then in the example of the ϕ^4 field theory in two-dimensions, with one finite dimension.

We then analyse finite effects in the case of second order phase transitions, in the framework of the effective $O(N)$ symmetric $(\phi^2)^2$ field theory [193] in $d < 4$ dimensions. We establish the existence of a finite size scaling, extending RG arguments to this situation.

We distinguish between the finite volume geometry (in explicit calculations, we take the example of the hypercube) and the cylindrical geometry in which the size is finite in all dimensions except one.

We explain how to generalize the methods used in the case of infinite systems to calculate new universal quantities related to finite size effects, for example, in $d > 4$, $d = 4 - \varepsilon$ and, briefly, $d = 2 + \varepsilon$ dimensions, using the non-linear σ-model.

19.1 Periodic boundary conditions and the problem of the zero mode

In the field theories that we review in this chapter, for finite sizes, we impose periodic boundary conditions (as physicists often do in numerical simulations). This leads to the problem of the zero modes that we illustrate in this section in the example of *one-dimensional* classical systems, in particular, when the size L of the system is small. We consider a partition function given by the path integral

$$\mathcal{Z}(L) = \int [\mathrm{d}q(t)] \exp\left[-\mathcal{S}(q)\right], \tag{19.1}$$

where the paths satisfy periodic boundary conditions, $q(-L/2) = q(L/2)$. We choose the configuration energy $\mathcal{S}(q)$ of the form

$$\mathcal{S}(q) = \int_{-L/2}^{L/2} \mathrm{d}t \left[\tfrac{1}{2}\dot{q}^2(t) + V\left(q(t)\right)\right], \tag{19.2}$$

where $V(q)$ is a polynomial satisfying the conditions $V(q) \geq 0$, and $V''(q) \geq 0$.

From Random Walks to Random Matrices. Jean Zinn-Justin, Oxford University Press (2019).
© Jean Zinn-Justin. DOI: 10.1093/oso/9780198787754.001.0001

19.1.1 *Finite size and finite temperature quantum mechanics*

The same path integral represents also the partition function of a non-relativistic quantum particle at thermal equilibrium at temperature $1/L$,

$$\mathcal{Z}(L) = \operatorname{tr} \mathrm{e}^{-L\hat{H}}, \tag{19.3}$$

where the configuration energy (19.2) is the imaginary time or Euclidean action, and the quantum Hamiltonian is given by (for $\hbar = 1$)

$$\hat{H} = \tfrac{1}{2}\hat{p}^2 + V(\hat{q}),$$

$\hat{q}, \hat{p}$ being the quantum position and conjugate momentum operators with the commutator $[\hat{q}, \hat{p}] = i$. Therefore, the one-dimensional example is a useful toy model since some properties can be inferred by methods of quantum mechanics.

With our assumptions, the spectrum of $\hat{H}$ is discrete with eigenvalues $E_0 < E_1 < \cdots < E_n \cdots$ and

$$\mathcal{Z}(L) = \sum_{n=0}^{\infty} \mathrm{e}^{-LE_n}.$$

The exponential decay of the two-point correlation function for $|t| \to \infty$ is governed by the difference $E_1 - E_0 = \xi^{-1}$, where ξ is the correlation length.

Regime $L \gg \xi$. In this regime,

$$\mathcal{Z}(L)\,\mathrm{e}^{E_0 L} - 1 \sim \mathrm{e}^{(E_0 - E_1)L} = \mathrm{e}^{-L/\xi}.$$

Therefore, for $L \to \infty$, finite size effects vanish exponentially. Partition function and correlation functions can be calculated perturbatively. If $V(q) = \frac{1}{2}\omega^2 q^2 + O(q^3)$, the leading order approximation is given by a Gaussian integral corresponding to the action

$$\mathcal{S}_0(q) = \tfrac{1}{2} \int_{-\infty}^{\infty} \mathrm{d}t \left[\dot{q}^2(t) + \omega^2 q^2(t) \right].$$

Note that the correlation length at leading order is $\xi = 1/\omega$.

19.1.2 *The role of the zero mode: Effective integral*

For L finite, since the paths satisfy periodic boundary conditions, they can be expanded as a Fourier series on an orthonormal basis of periodic functions on the interval $[-L/2, L/2]$ in the form,

$$q(t) = q_0 + \delta q(t), \quad \delta q(t) = \frac{1}{\sqrt{L}} \sum_{n \neq 0} q_n\, \mathrm{e}^{2i\pi nt/L}, \text{ with } q_{-n} = q_n^*.$$

The integral over paths can then be replaced by an integration over the coefficients q_n of the expansion of the path. The *zero mode* q_0 has not been normalized but the corresponding Jacobian contributes only to the normalization of the path integral.

We have separated the zero mode q_0 because it is not damped by the derivative term $\dot{q}^2(t)$, unlike the other modes. Indeed, using the orthogonality of the basis, one infers

$$\int_{-L/2}^{L/2} dt\, \dot{q}^2(t) = \sum_{n\neq 0} \frac{4\pi^2 n^2}{L^2} |q_n|^2.$$

For L small, which at leading order means $L \ll 1/\omega = \xi$, the zero mode has to be treated exactly while the contributions of the other modes, which are of order L, can be calculated perturbatively. We thus expand $V(q)$ in powers of δq up to quadratic order as,

$$V(q_0 + \delta q(t)) = V(q_0) + V'(q_0)\delta q(t) + \tfrac{1}{2} V''(q_0)(\delta q(t))^2 + O((\delta q)^3).$$

Integrating over t, using again the orthogonality of the basis, one finds

$$\int_{-L/2}^{L/2} dt\, V(q_0 + \delta q(t)) = LV(q_0) + \tfrac{1}{2} V''(q_0) \sum_{n\neq 0} |q_n|^2 + O(q_n^3)$$

and, thus,

$$\mathcal{S}(q) = LV(q_0) + \tfrac{1}{2} \sum_{n\neq 0} \left(\frac{4n^2\pi^2}{L^2} + V''(q_0) \right) |q_n|^2 + O(q_n^3).$$

If one is interested only in the expectation values of functions of q_0, one can integrate over the coefficients q_n. Performing the Gaussian integrals and normalizing all integrals by dividing them by the integral for $V''(q) \equiv 0$, one obtains,

$$\prod_{n>0} \int dq_n dq_n^* \exp\left[-\left(\frac{4n^2\pi^2}{L^2} + V''(q_0) \right) |q_n|^2 \right] \propto \prod_{n>0} \left[1 + \frac{L^2}{4\pi^2 n^2} V''(q_0) \right]^{-1}$$

$$= \frac{L\sqrt{V''(q_0)}}{2\sinh\left(L\sqrt{V''(q_0)}/2 \right)}.$$

Writing the q_0 integrand as $e^{-\Sigma}$, one finds

$$\Sigma(q_0) = LV(q_0) + \ln\left[\frac{2\sinh\left(L\sqrt{V''(q_0)}/2 \right)}{L\sqrt{V''(q_0)}} \right].$$

For $L^2 V''(q_0) \ll 1$, one can expand

$$\Sigma(q_0) = LV(q_0) + \tfrac{1}{24} L^2 V''(q_0) + O(L^4).$$

For L small, the path integral is reduced to a simple *effective integral* over the zero mode, an example of *dimensional reduction*.

A comparison of the result with the expression (19.3) makes it also possible to determine the normalization. The partition function is given by

$$\mathcal{Z}(L) = \frac{1}{\sqrt{2\pi L}} \int dq \, e^{-\Sigma(q)} .$$ (19.4)

The higher order contributions of the non-zero modes to the zero-mode integrand can be calculated perturbatively, the leading contribution to Σ being of order L^3.

The harmonic well. In the example,

$$V(q) = \tfrac{1}{2}\omega^2 q^2, \quad \omega > 0,$$ (19.5)

the operator $\hat{H}$ is the quantum Hamiltonian of the harmonic oscillator. Then,

$$\Sigma(q) = \tfrac{1}{2}L\omega^2 q^2 + \ln\left(\frac{2\sinh\left(L\omega/2\right)}{L\omega} \right)$$

and, after integration over the zero mode, the partition function is given by

$$\mathcal{Z}(L) = \frac{1}{2\sinh(L\omega/2)},$$

which is the exact result. Moreover, in the case of the potential (19.5), the two-point function is proportional to $e^{-\omega|t|}$ and, therefore, $\xi = 1/\omega$ is the correlation length. In the limit $L \ll \xi$, $\mathcal{Z}$ has the small size expansion

$$\mathcal{Z}(L) = \frac{\xi}{L} - \frac{1}{24}\frac{L}{\xi} + O((L/\xi)^3).$$

19.2 Cylindrical geometry: Two-dimensional field theory

We now consider a two-dimensional statistical field theory with the Euclidean (or imaginary time) action

$$\mathcal{S}(\phi) = \int dt \, dx \left[\tfrac{1}{2}\big(\partial_t \phi(t,x)\big)^2 + \tfrac{1}{2}\big(\partial_x \phi(t,x)\big)^2 + V\big(\phi(t,x)\big) \right],$$ (19.6)

where $V(\phi)$ is the $\mathbb{Z}_2$ symmetric ($\phi \mapsto -\phi$) quartic polynomial,

$$V(\phi) = \tfrac{1}{2}r\phi^2 + \frac{g}{4!}\phi^4, \ g \geq 0.$$

A cut-off Λ (the inverse of the initial microscopic scale) is required to regularize the perturbative expansion. Generically, the parameters r and g, which have a mass dimension 2, are of order Λ^2. As usual, the parameter r has to be fine tuned to generate a physical ϕ mass, or inverse correlation length, much smaller than Λ.

The model, which belongs to the Ising model universality class, is expected to have a continuous phase transition, with $\mathbb{Z}_2$ spontaneous symmetry breaking when the parameter r, which plays the role of the temperature, varies.

19.2.1 *Super-renormalizable perturbative field theory*

The perturbative expansion in powers of g has only one primitively divergent diagram, the one-loop contribution to the two-point function. At one-loop order, the inverse two-point function in the Fourier representation reads

$$\tilde{\Gamma}^{(2)}(p) = p^2 + r + \tfrac{1}{2}g\Omega_2(\sqrt{r}) + O(g^2)$$

with (a large momentum cut-off Λ on the p integration is implied)

$$\Omega_2(\sqrt{r}) = \frac{1}{4\pi^2}\int \frac{\mathrm{d}^2 p}{p^2 + r}. \tag{19.7}$$

We introduce the parameter m, which characterizes the deviation from the massless or critical theory in the symmetric phase,

$$\tilde{\Gamma}^{(2)}(0) = m^2$$

and express the parameter r as a function of m, g, Λ (fine tuning). At this order,

$$r = m^2 - \tfrac{1}{2}g\Omega_2(m) + O(g^2) = m^2 - \frac{1}{4\pi}g[\ln(\Lambda/m) + O(1)] + O(g^2). \tag{19.8}$$

As a function of m, the perturbative expansion has a finite limit for $\Lambda \to \infty$. For dimensional reason, the expansion makes sense for g at most of order m^2 and this corresponds to an additional fine tuning. By contrast, the physics of critical phenomena is recovered in the limit $g \to \infty$ (see also Section 15.2). However, we adopt the former framework here.

Finally, the critical or massless field theory has no perturbative expansion due to infrared (IR) divergences for $m \to 0$. In particular, the critical value r_c at which m vanishes is non-perturbative.

19.2.2 *Finite size with periodic boundary conditions*

We assume a finite size L with periodic boundary conditions in the Euclidean time t and an infinite size in the space dimension. With only one infinite size direction, the field theory can no longer have a phase transition.

Again, the partition function has also the interpretation of the partition function of a one-dimension quantum theory at a finite temperature $1/L$ (see Section 21.3).

With a finite size, one expects to be able to reduce the non- perturbative calculations in the critical limit $m = 0$ to the study of an effective action for the zero mode. Indeed, the momenta in the finite size are quantized,

$$p = (2\pi n)/L, \; n \in \mathbb{Z},$$

and, for example, the function $\Omega_2(m)$ (equation (19.7)) becomes

$$\Omega_2(m) = \frac{1}{2\pi L}\sum_n \int \frac{\mathrm{d}p}{m^2 + p^2 + 4\pi^2 n^2/L^2}.$$

All contributions in the sum over n have a finite $m = 0$ limit, except the term $n = 0$ (the zero mode).

Therefore, as we have already pointed out in Section 19.1.2, in such a situation, we have to deal with the zero mode separately.

19.2.3 The effective theory of the zero mode

To isolate the zero mode, we expand the field in Fourier components in the time dimension, setting

$$\phi(t,x) = \varphi(x) + \chi(t,x), \quad \chi(t,x) \equiv (2\pi/L) \sum_{n\in\mathbb{Z}\neq 0} e^{2i\pi nt/L} \phi_n(x), \qquad (19.9)$$

with $\phi_{-n}(x) = \phi_n^*(x)$.

We focus below on some of the properties of the partition function and the correlation functions of the zero mode,

$$\varphi(x) = \frac{1}{L}\int_{-L/2}^{L/2} dt\, \phi(t,x).$$

They can be calculated from a path integral involving an effective action $\mathcal{S}_L(\varphi)$ obtained by integrating over the field χ,

$$\exp[-\mathcal{S}_L(\varphi)] = \mathcal{N}^{-1}\int [d\chi]\exp[-\mathcal{S}(\varphi+\chi)],$$

where the normalization $\mathcal{N}$ is now chosen in such a way that $\mathcal{S}_L(\varphi=0)=0$.

Inserting the decomposition (19.9) into the action, one obtains

$$\mathcal{S}(\phi) = \tfrac{1}{2}L\int dx\,(\partial_x\varphi(x))^2 + \frac{2\pi^2}{L}\sum_{n\neq 0}\int dx\left[(4\pi^2 n^2/L^2)|\phi_n(x)|^2 + |\partial_x\phi_n(x)|^2\right]$$
$$+ \int dt\,dx\, V(\varphi(x)+\chi(t,x)).$$

As we have explained in Section 19.1, since the non-zero modes are, for L small (Lm or $L\sqrt{g}$ small), very massive, one can integrate perturbatively over χ after expanding the action in powers of χ. Thus, up to order χ^2,

$$\int dt\,dx\, V(\varphi(x)+\chi(t,x)) = L\int dx\, V(\varphi(x)) + \frac{2\pi^2}{L}\int dx\, V''(\varphi(x))\sum_{n\neq 0}|\phi_n(x)|^2.$$

19.2.4 Leading order calculation

At leading order, one neglects the non-zero modes, and $r = m^2$. For $m=0$, the effective φ action (the critical limit) reduces to

$$\mathcal{S}_L(\varphi) = L\int dx\left[\tfrac{1}{2}(\partial_x\varphi(x))^2 + \frac{g}{4!}\varphi^4(x)\right]. \qquad (19.10)$$

The action corresponds to the quartic anharmonic quantum Hamiltonian (for $\hbar=1$)

$$\hat{H} = -\frac{1}{2L}\left(\frac{d}{d\varphi}\right)^2 + L\frac{g}{4!}\varphi^4 \qquad (19.11)$$

and the decay of the connected φ correlation functions is governed by the difference of the two lowest eigenvalues of the Hamiltonian.

We rescale, $\varphi \mapsto g^{-1/6}L^{-1/3}\varphi$. The Hamiltonian becomes,

$$\hat{H} = L^{-1/3}g^{1/3}\hat{H}_{\text{sc.}}, \text{ with } \hat{H}_{\text{sc.}} = -\frac{1}{2}\left(\frac{\mathrm{d}}{\mathrm{d}\varphi}\right)^2 + \frac{1}{4!}\varphi^4. \tag{19.12}$$

The Hamiltonian $\hat{H}_{\text{sc.}}$ has a discrete spectrum. Denoting by $\mathcal{E}_n$ its eigenvalues, one obtains for the eigenvalues E_n of $\hat{H}$ and the correlation length ξ_L in the infinite direction,

$$E_n = L^{-1/3}g^{1/3}\mathcal{E}_n \Rightarrow \xi_L = \frac{1}{E_1 - E_0} = \left(\frac{L}{g}\right)^{1/3}\frac{1}{\mathcal{E}_1 - \mathcal{E}_0}.$$

For $m = 0$, a limit where the perturbative expansion does not exist for $L = \infty$, and in the expected domain of applicability of the approximation, $L\sqrt{g} \ll 1$, conditions that exclude the regime relevant to critical phenomena, one finds $L \ll \xi_L \ll g^{-1/2}$.

19.2.5 One-loop corrections

The Gaussian integration over the non-zero modes at leading order generates corrections to the action obtained as a product of determinants. We define

$$\Sigma_n(\phi_n) = \frac{2\pi^2}{L}\int \mathrm{d}x \left[(4\pi^2 n^2/L^2)|\phi_n(x)|^2 + |\partial_x\phi_n(x)|^2 + V''(\varphi(x))\sum_{n\neq 0}|\phi_n(x)|^2\right].$$

The integration over (ϕ_n, ϕ_{-n}) with $n > 0$ yields (tr ln = ln det)

$$\int [\mathrm{d}\phi_n \mathrm{d}\phi_{-n}]\,\mathrm{e}^{-2\Sigma_n(\phi_n)} \propto \det^{-1}\left[-(\mathrm{d}/\mathrm{d}x)^2 + (4\pi^2 n^2/L^2) + V''(\varphi(x))\right]$$

$$\propto \exp\left[-\operatorname{tr}\ln(\mathbf{H} - E_n)\right], \tag{19.13}$$

where $\mathbf{H}$ is the Schrödinger operator

$$\mathbf{H} = -(\mathrm{d}/\mathrm{d}x)^2 + \tfrac{1}{2}g\varphi^2(x)$$

and

$$E_n = -4\pi^2 n^2/L^2.$$

The one-loop contribution generates non-local interactions and it would seem that the action no longer corresponds to a quantum Hamiltonian. However, the evaluation of the expression (19.13) for $L \to 0$ is directly related to the classical problem of expanding the resolvent of a Schrödinger operator,

$$G(E) = \operatorname{tr}(\mathbf{H} - E)^{-1},$$

for large negative energies E and it is known that the resolvent has a local expansion involving the potential and its derivatives.

We normalize the expression (19.13) by dividing it by its value for $\varphi = 0$. The one-loop contribution to the action is then the sum over n of expressions of the form

$$\mathcal{S}_n^{(1)} = \operatorname{tr} \ln \left[\mathbf{1} + \mathbf{G}_0(E_n)\mathbf{U}\right],$$

where the matrix elements of $\mathbf{G}_0(E)$, are

$$\mathbf{G}_0(x, x'; E) = \frac{1}{2\pi} \int \frac{\mathrm{d}p \; e^{ip(x-x')}}{p^2 - E}$$

and $\mathbf{U}$ is diagonal with elements $U(x) = \frac{1}{2}g\varphi^2(x)$.

For $E \to -\infty$, $\mathbf{G}_0(x, x'; E)$ vanishes exponentially as $e^{-\sqrt{-E}|x-x'|}$. This explains why $\mathcal{S}_n^{(1)}$ has a local expansion involving U and its derivatives. Moreover, although eventually terms with more than two derivatives will be generated, they are subleading for $L \to 0$ and can be treated as perturbations.

Local expansion: leading term Here, we keep only the leading contribution for $L \to 0$, which is the linear term in U,

$$\operatorname{tr} \ln \left[\mathbf{1} + \mathbf{G}_0(E_n)\mathbf{U}\right] \sim \operatorname{tr} \mathbf{G}_0(E_n)\mathbf{U} = \int \mathrm{d}x \, G_0(x, x; E_n)U(x)$$

$$= \frac{g}{4\pi} \int \frac{\mathrm{d}p}{p^2 + 4\pi^2 n^2/L^2} \int \mathrm{d}x \, \varphi^2(x).$$

The total contribution is divergent and the divergence has to be cancelled by a shift of r (equation (19.8)). In the massless limit, one has to introduce a physical mass scale μ.

To sum over n, we use the identity

$$\sum_{n>0} \frac{1}{n^2 + a^2} = \frac{\pi \cosh(\pi a)}{2a \sinh(\pi a)} - \frac{1}{2a^2}.$$

Identifying $a = L|p|/2\pi$, one finds the contribution to $\mathcal{S}_L$,

$$\delta \mathcal{S}_L(\varphi) = \Sigma^{(1)} \int \mathrm{d}x \, \varphi^2(x)$$

with

$$\Sigma^{(1)} = \frac{gL}{16\pi} \int \frac{\mathrm{d}p}{|p|} \left[\frac{\cosh(L|p|/2)}{\sinh(L|p|/2)} - \frac{2}{L|p|}\right]$$

$$= \frac{gL}{8\pi} \int \frac{\mathrm{d}p}{|p|} \left(\frac{1}{2} + \frac{1}{\exp(L|p|) - 1} - \frac{1}{L|p|}\right).$$

The expression has, as expected a large momentum divergence and requires a cutoff. This regularization can be achieved by subtracting to the propagator an nonphysical propagator $1/(p^2 + \Lambda^2)$ (Pauli–Villars regularization) with $\Lambda \gg 1/L$.

In this limit, the finite size effects can be neglected and the regularized expression can be written as

$$\Sigma_{\text{reg.}}^{(1)} = \frac{gL}{8\pi} \int_0^\infty dp \left[\frac{1}{p}\left(1 + \frac{2}{\exp(Lp)-1} - \frac{2}{Lp}\right) - \frac{1}{\sqrt{p^2 + \Lambda^2}} \right].$$

We can rescale $Lp = q$ and obtain

$$\Sigma_{\text{reg.}}^{(1)} = \frac{gL}{8\pi} \int_0^\infty dq \left[\frac{1}{q}\left(1 + \frac{2}{e^q-1} - \frac{2}{q}\right) - \frac{1}{\sqrt{q^2 + L^2\Lambda^2}} \right].$$

This quantity can be evaluated explicitly for $L\Lambda \gg 1$. One obtains,

$$\Sigma_{\text{reg.}}^{(1)} = \frac{gL}{8\pi}\left(\ln(L\Lambda/4\pi) + \gamma\right),$$

where γ is Euler's constant.

Since the critical value r_c is unknown, we parametrize r in terms of a mass scale μ and set $r = -g\ln(\Lambda/\mu)/4\pi$ (a mass renormalization consistent with expression (19.8)). This generates an additional contribution to the Hamiltonian (19.11),

$$\hat{H} \mapsto \hat{H} + \frac{gL}{8\pi}\left(\ln(L\mu/4\pi) + \gamma\right)\varphi^2.$$

The correction to the Hamiltonian $\hat{H}_{\text{sc.}}$ is of order $L^{4/3}g^{2/3}\ln(\mu L)$, which is small for L small.

Higher order corrections. The action (19.6) satisfies

$$\mathcal{S}(\phi) = \mathcal{S}(\phi\sqrt{g})/g.$$

This shows that g is a loop expansion parameter and, for ℓ-loop contributions, the coefficient of ϕ^{2k} is, for dimensional reasons, proportional to

$$g^{\ell-1}(g^k\phi^{2k})L^{2\ell-2+2k-1},$$

up to logarithmic factors, since g has mass dimension 2 and φ is dimensionless. For L small, increasing ℓ or k leads to increasingly small corrections.

19.3   Effective $(\phi^2)^2$ field theory at criticality in finite geometries

In the Section 19.2, we started investigating finite size effects, using the example of a two-dimensional quantum field theory.

We now explore the *universal properties* of finite size effects, in the critical domain near the critical temperature, in the framework of the $O(N)$-symmetric N-vector model in higher dimensions.

The universal properties of the $O(N)$-symmetric N-vector model can be described by an effective quantum field theory with a $(\phi^2)^2$ interaction. This quantum field theory requires regularizations and, in dimensions $d \leq 4$, is renormalizable. Renormalization group (RG) equations and scaling properties follow.

In this section, we assume *dimensional continuation*, and the dimension d is considered as being a continuous (real or complex) parameter.

19.3.1 The $(\phi^2)^2$ field theory for $2 < d \leq 4$

The universal properties of the $O(N)$-symmetric N-vector model can be described by an effective field theory involving an N-component field ϕ and an action of the form

$$\mathcal{S}(\phi) = \int \left\{ \frac{1}{2} \left[\nabla \phi(x) \right]^2 + \frac{1}{2} r \phi^2(x) + \frac{1}{4!} g \Lambda^{4-d} \left(\phi^2(x) \right)^2 \right\} \mathrm{d}^d x, \qquad (19.14)$$

where a large momentum cut-off Λ, the inverse of the microscopic scale (like the lattice spacing in lattice models), is implied. In the continuum, it can be implemented by the substitution of, for example,

$$\nabla \mapsto \nabla \left[1 - \alpha_1 \nabla^2 + \alpha_2 \nabla^4 + \cdots \right], \text{ with } \alpha_i > 0.$$

The parameter g is dimensionless, and $r = O(\Lambda^2)$, which plays the role of temperature T of some initial statistical model, has mass dimension 2.

At a value $r = r_c$, a phase transition occurs between a disordered phase for $r > r_c$ and an ordered phase for $r < r_c$ in which the $O(N)$ symmetry is spontaneously broken.

We thus set

$$r - r_c = \rho \propto T - T_c \text{ for } |T - T_c| \ll 1.$$

The critical domain, where the field theory describes universal properties is defined by

$$|r - r_c = \rho| \ll \Lambda^2.$$

Moreover, for $N \geq 2$ and $d > 2$, one finds some additional universal properties for all $r < r_c$ due to massless Goldstone modes (see Chapter 10).

19.3.2 RG equations

We recall that, for $d \leq 4$, the corresponding (one-particle-irreducible(1PI)) vertex functions in the Fourier representation satisfy the RG equations [103]

$$\left[\Lambda \frac{\partial}{\partial \Lambda} + \beta(g) \frac{\partial}{\partial g} - \frac{n}{2} \eta(g) - \eta_2(g) \rho \frac{\partial}{\partial \rho} \right] \tilde{\Gamma}^{(n)}(p_i; \rho, g, \Lambda) = 0. \qquad (19.15)$$

They can be solved in the usual way by setting

$$\tilde{\Gamma}^{(n)}(p_i; \rho, g, \Lambda) = Z^{-n/2}(\lambda) \tilde{\Gamma}^{(n)} \left(p_i; \rho(\lambda), g(\lambda), \lambda \Lambda \right), \qquad (19.16)$$

the various functions of λ being defined by

$$\lambda \frac{\mathrm{d}}{\mathrm{d}\lambda} g(\lambda) = \beta \big(g(\lambda) \big), \qquad g(1) = g,$$

$$\lambda \frac{\mathrm{d}}{\mathrm{d}\lambda} \ln \rho(\lambda) = -\eta_2 \big(g(\lambda) \big), \qquad \rho(1) = \rho,$$

$$\lambda \frac{\mathrm{d}}{\mathrm{d}\lambda} Z(\lambda) = \eta \big(g(\lambda) \big), \qquad Z(1) = 1.$$

For $d < 4$, in particular for $d = 4 - \varepsilon$, for $\lambda \to 0$, the effective coupling at scale λ converges towards the non-trivial zero g^* of the β-function (corresponding to Wilson–Fisher's famous IR fixed point), which governs the large distance of correlation functions at and near T_c. Setting $g = g^*$, one obtains the scaling behaviour and critical exponents of thermodynamic quantities.

19.3.3 RG in finite geometries: Finite size scaling

We consider here only systems where the finite sizes are characterized by a unique length L, which is large in the microscopic scale, for example, much larger than the lattice spacing in lattice models.

We consider only boundary conditions that *do not break translation symmetry* to avoid surface effects, which are of a different nature. Periodic boundary conditions certainly satisfy such a criterion. Depending on the specific symmetries of a model, other boundary conditions are also available (like anti-periodic boundary conditions for Ising-like systems).

The crucial observation which explains finite size scaling is that the renormalization theory, which leads to RG equations, is *insensitive to finite size effects*, since renormalizations are entirely due to *short distance singularities*. As a consequence, RG equations are not modified. However, their solutions are different because correlation functions now depend on one additional dimensional parameter L.

We discuss below the solution of RG equations both in the example of the ϕ^4 field theory and, briefly, in Section 19.7 in the example of the non-linear σ-model.

The presence of the new length scale L affects the dimensional analysis. Indeed,

$$\tilde{\Gamma}^{(n)}(p_i; \rho, g, L, \Lambda) = \Lambda^{d-n(d-2)/2}\tilde{\Gamma}^{(n)}(p_i/\Lambda; \rho/\Lambda^2, g, L\Lambda, 1). \tag{19.17}$$

We can use the relation in the right-hand side of equation (19.16) and then choose λ such that

$$\lambda L\Lambda = 1 \quad \text{or} \quad \lambda = 1/L\Lambda. \tag{19.18}$$

The dilatation parameter λ goes to zero for ΛL large and, therefore, $g(\lambda)$ approaches the IR fixed point g^*. This implies for $Z(\lambda)$ and $\rho(\lambda)$, the behaviour ($\eta_2(g^*) = 1/\nu - 2$),

$$Z(\lambda) \propto (L\Lambda)^{-\eta}, \quad \frac{\rho(\lambda)}{\lambda^2\Lambda^2} \propto \frac{\rho}{\Lambda^2}(L\Lambda)^{1/\nu}. \tag{19.19}$$

The scaling form of finite size correlation functions then follows:

$$\tilde{\Gamma}^{(n)}(p_i; \rho, g, L, 1) \propto L^{-d+n(d-2+\eta)/2}\tilde{\Gamma}^{(n)}(Lp_i; \rho L^{1/\nu}, g^*, 1, 1). \tag{19.20}$$

It is characterized by the appearance of a new scaling variable $\rho L^{1/\nu} \propto (\xi/L)^{1/\nu}$, where, for $\rho > 0$, $\xi(\rho)$ is the correlation length and, in general, an RG invariant physical scale.

From equation (19.20), the usual infinite size scaling form is recovered by expressing that the correlation functions have a limit for $L \gg \xi(\rho)$. In the opposite limit, $\xi(\rho) \gg L$, correlation functions have a regular expansion in powers of ρ, even for zero magnetization, in a finite volume or for a cylindrical geometry, because no phase transitions can occur in both situations (for short range interactions).

Note that all combinations which are independent of the normalization of the field ϕ, of the temperature ρ, and of the magnetic field are universal, for the reasons explained in the infinite volume case, once the geometry and boundary conditions are fixed.

Finally, adapting the usual analysis of corrections to the scaling form to the finite size situation, one infers that the leading corrections to relations (19.20) have, near $d = 4$, the form of scaling functions multiplied by a factor $L^{-\omega}$ ($\omega = \beta'(g^*)$).

In Sections 19.6 and 19.5, we illustrate the general use for the finite size analysis to the infinite volume extrapolation of finite size numerical results, by considering a few quantities that have been considered in practical calculations.

19.4 Momentum quantization in finite geometries

Scaling properties (19.20) do not depend on the specific form of boundary conditions, but the explicit universal finite size expressions do.

Even the technical details of the calculation for T close to T_c vary. Indeed, the characteristic feature of all finite geometries is that, in the Fourier representation, momenta that correspond to directions in which the size of the system is finite, are quantized. However, the precise momentum spectrum varies with the boundary conditions even when boundary conditions do not break translation invariance.

Periodic boundary conditions, which we use throughout the chapter, and other non-periodic boundary conditions which also do not break translation symmetry (twisted boundary conditions), lead to different momentum quantizations.

19.4.1 Periodic boundary conditions and the problem of the zero mode

We have already examined the problem of the zero modes in Sections 19.1 and 19.2.

In the example of a d-dimensional hypercube of linear size L with periodic boundary conditions, the quantized momenta $\mathbf{p}$ have the form

$$\mathbf{p} = 2\pi\mathbf{k}/L\,, \quad \mathbf{k} \in \mathbb{Z}^d\,. \tag{19.21}$$

When the product $\rho L^{1/\nu} = (L/\xi)^{1/\nu}$ is positive and not small, the usual methods of calculation of the infinite volume are applicable and finite size effects due to momentum quantization are only quantitative, decreasing like $\exp[-\text{const.}\, L/\xi]$.

When the product $\rho L^{1/\nu}$ is negative and not small (the ordered phase), the physics of the infinite and finite systems are quite different, and this problem will be examined later.

Finally, at the critical temperature $\rho = 0$, in a finite volume, the propagator has an isolated pole at $\mathbf{p} = \mathbf{0}$ which generates IR divergences in perturbation theory, although one expects physical quantities to be regular functions of the temperature at the transition. These divergences simply reflect a disease of the Gaussian model. More generally, the *perturbative expansion is badly behaved for* $\xi \gg L$.

In the cylindrical geometry, one component of the momentum, denoted here by ω, varies continuously and the other components are quantized. At T_c, Feynman diagrams receive divergent contributions of the form $\int d\omega/\omega^2$. Finally, a geometry in which the sizes in two or three dimensions among d are infinite still leads to IR divergences.

As a consequence, even in high dimensions for which in the infinite geometry mean field theory is exact, IR divergences appear at T_c.

To overcome this difficulty, it is necessary to separate the *zero momentum Fourier component of the field*. The components $\mathbf{k} \neq \mathbf{0}$ can be treated by the methods developed in the infinite geometry (perturbation theory and RG). By contrast, it is necessary to integrate exactly over the component $\mathbf{k} = \mathbf{0}$ because its fluctuations are damped at T_c only by interaction terms (see Sections 19.1 and 19.2 for details).

Therefore, in the case of a finite volume, one constructs an effective integral over the component $\tilde{\phi}(\mathbf{p} = \mathbf{0})$ by integrating over all other components. In a cylindrical geometry, the integration over all components except $\tilde{\phi}(\mathbf{p_T} = \mathbf{0}, \tau)$, denoting by τ the coordinate in the infinite direction, leads to an effective quantum Hamiltonian.

We examine in Section 19.5, the two first geometries separately, beginning with the simplest example of the periodic hypercube.

Twisted boundary conditions. For systems with symmetries, additional boundary conditions do not break translation invariance: conditions such that the values of the order parameter at opposite boundaries differ by a constant group transformation (often called twisted boundary conditions).

For instance, for Ising-like systems, one can use anti-periodic boundary conditions; for the N-vector model with $O(N)$ symmetry, one can impose a rotation of a given angle around some axis. In such situations, the quantized momenta p_μ are shifted by some additional constants.

19.5   The $(\phi^2)^2$ field theory in a periodic hypercube

We first study the effective $(\phi^2)^2$ field theory in the dimensions $d > 4$ and $d = 4 - \varepsilon$. As explained in Section 19.4.1, we expand $\phi(x)$ in Fourier components, separating the zero mode:

$$\phi(x) = \varphi + \chi(x),$$
$$\chi(x) = (2\pi/L)^d \sum_{\mathbf{k} \neq 0} e^{i\mathbf{p} \cdot \mathbf{x}} \, \tilde{\phi}(\mathbf{p}), \quad \mathbf{p} = 2\pi \mathbf{k}/L, \quad \mathbf{k} \in \mathbb{Z}^d. \qquad (19.22)$$

The integration over the field $\chi(x)$ is performed as in the infinite geometry limit: this generates a perturbative expansion which has RG properties. An integral over the last $\mathbf{k} = 0$ modes remains, which must be calculated exactly. Note that the first part of the procedure is formally equivalent to the shift of the expectation value of the field $\phi(x)$ in the infinite geometry. The main difference, apart from the replacement of integrals by discrete sums in Feynman diagrams, is that the average

$$\varphi = L^{-d} \int \phi(x) \mathrm{d}^d x = (L/2\pi)^d \tilde{\phi}(0),$$

here remains a fluctuating quantity.

Moments of the averaged field distribution. As an illustration, we calculate expectation values, moments of the distribution of the average field per unit volume. We set

$$\exp[-\Sigma(\varphi)] = \mathcal{N}^{-1} \int [\mathrm{d}\chi] \exp[-\mathcal{S}(\varphi + \chi)], \qquad (19.23)$$

where $\mathcal{S}(\phi)$ is the action (19.14), and the normalization $\mathcal{N}$ is chosen such that $\Sigma(0) = 0$ for $\rho = 0$.

Moments then are given by

$$m_\sigma = \mathcal{Z}^{-1} \int \mathrm{d}\varphi |\varphi|^\sigma \exp[-\Sigma(\varphi)] \text{ with } \mathcal{Z} = \int \mathrm{d}\varphi \exp[-\Sigma(\varphi)]. \tag{19.24}$$

In the infinite volume limit, $\Sigma(\varphi) = \Gamma(\varphi) - \Gamma(\mathbf{0})$, where $\Gamma(\varphi)$ is the thermodynamic potential, as obtained in perturbation theory.

The moments m_{2s}, s integer, involve only powers of φ^2 and thus are related to zero momentum correlation functions:

$$\varphi^2 = L^{-2d} \int \mathrm{d}^d x \, \mathrm{d}^d y \, \phi(x) \cdot \phi(y) = (2\pi/L)^{2d} \tilde{\phi}^2(0). \tag{19.25}$$

19.5.1 Moments: Leading order calculation

At leading order, one neglects the fluctuations of all non-zero modes ($\chi = 0$). The function (19.23) then reduces to

$$\Sigma_0(\varphi) = \mathcal{S}(\varphi) = L^d \left(\frac{1}{2}\rho\varphi^2 + \frac{1}{4!} u \left(\varphi^2 \right)^2 \right), \tag{19.26}$$

where $\mathcal{S}$ is the action (19.14) with $u = g\Lambda^{4-d}$. After the change of variables

$$\varphi \mapsto \left(uL^d \right)^{-1/4} \varphi, \tag{19.27}$$

one finds that the moments m_σ take the form

$$m_\sigma(L, \rho) = \left(uL^d \right)^{-\sigma/4} \mu_\sigma(\rho L^{d/2} u^{-1/2}), \tag{19.28}$$

in which $\mu_\sigma(z)$ is given by

$$\mu_\sigma(z) = \frac{g_{\sigma+N}(z)}{g_N(z)}, \tag{19.29}$$

with

$$g_\sigma(z) = \int_0^\infty \mathrm{d}\varphi \, \varphi^{\sigma-1} \exp\left[-\left(\frac{1}{2}z\varphi^2 + \frac{1}{4!}\varphi^4 \right) \right]. \tag{19.30}$$

Usual perturbation theory amounts to calculating the integral by the steepest descent method, a calculation valid for $|z| \gg 1$. By contrast, the function $g_\sigma(z)$ has a convergent expansion in powers of z.

19.5.2 *Moments: One-loop corrections at T_c*

The one-loop correction Σ_1 to Σ for $\rho = 0$, generated in equation (19.23) by an integration over χ in the Gaussian approximation, is given by (a cut-off Λ is implied)

$$\Sigma_1(\varphi, L, \rho = 0, u)$$

$$= \frac{1}{2} \sum_{\mathbf{k} \neq 0} \mathrm{tr} \ln \left[\delta_{ij} + \frac{1}{(2\pi\mathbf{k}/L)^2} \left(\frac{u}{6} \varphi^2 \delta_{ij} + \frac{u}{3} \varphi_i \varphi_j \right) \right]$$

$$= \frac{1}{2} \sum_{\mathbf{k} \neq 0} \left[\ln \left(1 + \frac{\frac{1}{2} u \varphi^2}{(2\pi\mathbf{k}/L)^2} \right) + (N - 1) \ln \left(1 + \frac{\frac{1}{6} u \varphi^2}{(2\pi\mathbf{k}/L)^2} \right) \right]. \qquad (19.31)$$

Expanding in powers of φ, one finds the leading contribution,

$$\Sigma_1(\varphi, L, \rho = 0, u) = \tfrac{1}{2} a_2 u L^2 \varphi^2 + O\big((\varphi^2)^2\big) \qquad (19.32)$$

with

$$a_2 = \frac{N+2}{6} \sum_{\mathbf{k} \neq 0} \frac{1}{(2\pi\mathbf{k})^2}, \qquad (19.33)$$

a contribution equivalent to a shift $\rho \mapsto \rho + a_2 u L^{2-d}$.

To regularize and calculate a_2, it is convenient to use here Schwinger's regularization, which amounts to the substitution

$$\frac{1}{p^2 + m^2} \mapsto \int_{1/\Lambda^2}^{\infty} \mathrm{d}s \; \mathrm{e}^{-s(p^2 + m^2)}$$

for the propagator in the Fourier representation.

One now introduces the function, related to Jacobi's theta function $\vartheta(z; \tau) \equiv \theta_3$,

$$\vartheta_0(s) \equiv \vartheta(0; is) = \sum_{n=-\infty}^{+\infty} \mathrm{e}^{-\pi s n^2}. \qquad (19.34)$$

It satisfies a Jacobi identity (the proof uses Poisson's formula),

$$\vartheta_0(s) = (1/s)^{1/2} \vartheta_0(1/s). \qquad (19.35)$$

One can write

$$\sum_{\mathbf{k} \neq 0} \frac{1}{(2\pi\mathbf{k}/L)^2} = \int_{1/\Lambda^2}^{+\infty} \mathrm{d}s \left(\vartheta_0^d(4\pi^2 s/L^2) - 1 \right)$$

$$= \frac{L^2}{4\pi} \int_{4\pi/L^2\Lambda^2}^{+\infty} \mathrm{d}s \left(\vartheta_0^d(s) - 1 \right). \qquad (19.36)$$

The integral in the right-hand side converges exponentially for s large.

For $s \to 0$, equation (19.35) implies

$$\vartheta_0(s) - (1/s)^{1/2} \underset{s \to 0}{\sim} 2s^{-1/2} \, \mathrm{e}^{-\pi/s}, \qquad (19.37)$$

where $s^{-1/2}$ corresponds to the infinite size limit.

It is simple to verify that for $g = O(1)$ they give subleading contributions to the moments.

(ii) After renormalization, the loop corrections can be formally expanded in powers of ρ and φ^2 because the size L provides an IR cut-off, and the zero mode has been removed. Because all contributions are ultraviolet (UV) finite, the coefficients are proportional to powers of L dictated by dimensional analysis. Again dimensional analysis indicates that they give subleading contributions.

The conclusion is that, for $d > 4$, the effective action (19.26) can be simply derived from mean field theory, as in the infinite volume limit, the only modification coming from the last integration over the average field (19.24). In particular, the result (19.42) is, indeed, universal.

19.5.5 Moments: Dimension $d = 4 - \varepsilon$

We now use the RG arguments presented in Section 19.3.3: rather than calculating physical quantities as function of $\{\rho, M, g, L\}$, we can, in the critical domain, set $\Lambda = L = 1$ and $g = g^*$, the IR fixed point value, then replace ρ by $\rho L^{1/\nu}$, M (if one introduces a magnetic field) by $M L^{\beta/\nu}$, and thus φ by $\varphi L^{\beta/\nu}$ in $\Sigma(\varphi)$.

At leading order, one needs g^* at order ε, which is

$$g^* = \frac{48\pi^2 \varepsilon}{N + 8} + O\left(\varepsilon^2\right). \tag{19.47}$$

Then, the moments m_σ at leading order given by the expression (19.40), become

$$m_\sigma(L, \rho) = \frac{L^{-\sigma(d-2+\eta)/2}}{(g^*)^{\sigma/4}} \mu_\sigma\left(\rho L^{1/\nu}(g^*)^{-1/2}\right), \tag{19.48}$$

where μ_σ is defined in equation (19.29). The equation shows that the ε-expansion is not uniform. The method used here, in which the zero mode is treated separately, gives the correct leading order only if ρ is formally assumed to be of order $\varepsilon^{1/2}$ (this condition is realized automatically for $\rho \propto T - T_c = 0$).

Note the appearance of powers of $g^{*1/2}$, which, for ε small, is equivalent to $\varepsilon^{1/2}$. This suggests that physical quantities will have an expansion in powers of $\varepsilon^{1/2}$ rather than ε. The analysis of higher order corrections confirms this observation. Let us exhibit this phenomenon in the one-loop approximation.

One-loop corrections. We then set $L = 1$, $g = g^*$. As we already discussed at the end of Section 19.5.1, at L fixed, all terms in perturbation theory can then be expanded in powers of φ^2 and ρ.

After the change of variables (19.27), φ^2 has a coefficient proportional to $(g^*)^{1/2} \sim \varepsilon^{1/2}$. In the same way, ρ is of order $\varepsilon^{1/2}$. A term contributing to the ℓ-loop order and proportional to $\varphi^{2l} \rho^m$ is of order $\varepsilon^{\ell-1+(l+m)/2}$. The leading two-loop correction comes from the term proportional to φ^2 and thus is of order $\varepsilon^{3/2}$. Therefore, at one-loop order, only the terms proportional to φ^2, $\varphi^2\rho$, φ^4, ρ and ρ^2 have to be taken into account.

After finite renormalizations of ρ and φ, the only correction relevant at one-loop order is related to the coefficient a_2 given in equation (19.39). In terms of renormalized ρ and field φ, the moments at one-loop order can be written as (see also equations (19.32) and (19.46))

$$m_\sigma(L, \rho) = L^{-\sigma(d-2+\eta)/2} \mu_\sigma (\rho L^{1/\nu} + b), \tag{19.49}$$

where the constant b is inferred from equations (19.39) and (19.47),

$$b = a_2 g^{*1/2} = \frac{N+2}{\sqrt{N+8}} \frac{(3\varepsilon)^{1/2}}{6} \int_0^{+\infty} ds \left[\vartheta_0^4(s) - 1 - 1/s^2\right] + O\left(\varepsilon^{3/2}\right). \tag{19.50}$$

The ratio $\mathcal{R}_4(T = T_c)$. We now apply the results to the dimensionless universal ratio (19.41) at T_c $(\rho = 0)$,

$$\mathcal{R}_4 = m_4/m_2^2.$$

From expression (19.49), setting $\rho = 0$, one infers $\mathcal{R}_4(T = T_c)$ at order $\varepsilon^{1/2}$,

$$\mathcal{R}_4(T_c) = \frac{g_{4+N}(b) g_N(b)}{[g_{2+N}(b)]^2}, \tag{19.51}$$

g_σ being defined by equation (19.30).

Using the value of the integral

$$\int_0^{+\infty} ds \left[\vartheta_0^4(s) - 1 - 1/s^2\right] = -0.561843942\ldots,$$

one obtains in three dimensions for $N = 1$,

$$\mathcal{R}_4(T_c) = 1.800\ldots. \tag{19.52}$$

This result can be compared to the mean field value 2.188 and a Monte Carlo numerical estimate 1.6 [192, 194]. The agreement is comparable to other results at order ε.

19.6 The $(\phi^2)^2$ field theory: Cylindrical geometry

In this section, we study the $(\phi^2)^2$ field theory when space is infinite in one dimension, hereafter called (Euclidean) time, and of finite size L with periodic boundary conditions in the remaining $(d-1)$ space dimensions, generalizing the two-dimensional situation discussed in Section 19.2.

To isolate the zero modes, we express the fields in terms of Fourier components in the $(d-1)$ space dimensions:

$$\phi(\tau, x) = \varphi(\tau) + \chi(\tau, x),$$
$$\chi(\tau, x) = (2\pi/L)^{d-1} \sum_{\mathbf{k} \in \mathbb{Z}^{d-1} \neq 0} e^{i2\pi \mathbf{k} \cdot \mathbf{x}/L} \phi_{\mathbf{k}}(\tau). \tag{19.53}$$

To calculate correlation functions of the space integral

$$\varphi(\tau) = L^{1-d} \int \mathrm{d}^{d-1}x \, \phi(x, \tau),$$

one needs only the effective action $\mathcal{S}_L(\varphi)$ obtained by integrating over χ,

$$\exp[-\mathcal{S}_L(\varphi)] = \mathcal{N}^{-1} \int [\mathrm{d}\chi] \exp[-\mathcal{S}(\varphi + \chi)], \qquad (19.54)$$

where the normalization $\mathcal{N}$ is chosen in such a way that $\mathcal{S}_L(\varphi = 0, \rho = 0) = 0$. The partition function $\mathcal{Z}$ and φ-field correlation functions then are given by simple path integrals of quantum mechanics type, with the effective action $\mathcal{S}_L$.

We illustrate the method with the calculation of the finite size correlation length ξ_L.

19.6.1 Finite size correlation length: Leading order calculation

The leading order approximation is obtained by neglecting all corrections due the integration over the $\mathbf{k} \neq \mathbf{0}$ components of the field. The effective action, defined by equation (19.54), reduces to ($\dot\varphi \equiv \partial_\tau \varphi$)

$$\mathcal{S}_L^{(0)}(\varphi) = \mathcal{S}(\varphi) = L^{d-1} \int \mathrm{d}\tau \left[\tfrac{1}{2} \left(\dot\varphi(\tau)\right)^2 + \tfrac{1}{2}\rho\varphi^2(\tau) + \frac{u}{4!} \left(\varphi^2(\tau)\right)^2 \right]. \qquad (19.55)$$

To the action $\mathcal{S}_L^{(0)}$ corresponds the quantum Hamiltonian

$$\hat{H} = -\frac{1}{2L^{d-1}}\nabla_\varphi^2 + L^{d-1}\left(\tfrac{1}{2}\rho\varphi^2 + \frac{u}{4!}(\varphi^2)^2 \right), \qquad (19.56)$$

where $\nabla_\varphi \equiv (\partial/\partial\varphi_1, \ldots, \partial/\partial\varphi_N)$.

We then rescale φ,

$$\varphi \mapsto u^{-1/6}L^{(1-d)/3}\varphi,$$

and set

$$\hat{H} = u^{1/3}L^{(1-d)/3}\hat{H}_{\mathrm{sc.}}. \qquad (19.57)$$

The rescaled quantum Hamiltonian then reads

$$\widehat{H}_{\mathrm{sc.}} = -\tfrac{1}{2}\nabla_\varphi^2 + \tfrac{1}{2}z\varphi^2 + \frac{1}{4!}\left(\varphi^2\right)^2 \qquad (19.58)$$

with

$$z = u^{-2/3}L^{2(d-1)/3}\rho = g^{-2/3}(\Lambda L)^{2(d-4)/3}L^2\rho. \qquad (19.59)$$

In a cylindrical geometry, a quantity of interest is the correlation length ξ_L in the infinite direction. The finite size correlation length ξ_L is related to the difference between the two lowest eigenvalues of $\widehat{H}$ and thus to the lowest eigenvalues E_0 and E_1 of $\hat{H}_{\mathrm{sc.}}$ rescaled according to equation (19.57). Defining,

$$\Xi(z) = (E_1(z) - E_0(z))^{-1}, \qquad (19.60)$$

one finds

$$\xi_L/L = u^{-1/3}L^{(d-4)/3}\Xi(z) = g^{-1/3}(L\Lambda)^{(d-4/3)}\Xi(z). \qquad (19.61)$$

For $\rho = 0$, $\xi_L/L \propto (L\Lambda)^{(d-4)/3}$ and, for $\rho > 0$,

$$\frac{\xi_L}{L} = \frac{\xi_\infty}{L}\sqrt{z}\,\Xi(z).$$

19.6.2 Finite size correlation length: One-loop corrections

We now investigate how loop corrections, due to the integration over the non-zero modes, modify the leading order results.

The one-loop correction $\mathcal{S}^{(1)}(\varphi)$ to the effective action, analogous to expression (19.31), is given in the initial field φ and τ variable by

$$\mathcal{S}_L^{(1)}(\varphi) = \frac{1}{2} \sum_{\mathbf{k}\neq 0} \left\{ \operatorname{tr} \ln \left[1 + \left(-\partial_\tau^2 + (2\pi\mathbf{k}/L)^2\right)^{-1} \left(\rho + u\varphi^2(\tau)/2\right) \right] \right.$$
$$\left. + (N-1)\operatorname{tr}\ln\left[1 + \left(-\partial_\tau^2 + (2\pi\mathbf{k}/L)^2\right)^{-1}\left(\rho + u\varphi^2(\tau)/6\right)\right]\right\}. \quad (19.62)$$

Again, one faces the problem that the one-loop contribution generates non-local interactions.

However, as we discussed in Section 19.2.5, for $L \to 0$, the expressions have a local expansion.

Critical temperature. For $L \to 0$ the first term in an expansion in powers of $(\rho + u\varphi^2(\tau)/6)$, which is the leading term, is automatically local. At the critical temperature ($\rho = 0$),

$$\mathcal{S}_L^{(1)}(\varphi) \sim \tfrac{1}{12}(N+2)u \int \mathrm{d}\tau\, \varphi^2(\tau) \sum_{\mathbf{k}\neq 0} \operatorname{tr}\left(-\partial_\tau^2 + (2\pi\mathbf{k}/L)^2\right)^{-1},$$

from which the infinite size limit has still to be subtracted (the mass renormalization).

Using again the Schwinger representation and regularization with a cut-off Λ, one can rewrite the coefficient as

$$\operatorname{tr}\left(-\partial_\tau^2 + (2\pi\mathbf{k}/L)^2\right)^{-1} = \sum_{\mathbf{k}\neq 0} \frac{1}{2\pi} \int_{1/\Lambda^2}^{+\infty} \mathrm{d}s \int \mathrm{d}\omega\, \mathrm{e}^{-s(\omega^2 + 4\pi^2\mathbf{k}^2/L^2)}$$

$$= \frac{1}{2\sqrt{\pi}} \int_{1/\Lambda^2} \frac{\mathrm{d}s}{\sqrt{s}} \left[\vartheta_0^{d-1}(4\pi s/L^2) - 1\right]$$

$$= \frac{L}{4\pi} \int_{4\pi/L^2\Lambda^2} \frac{\mathrm{d}s}{\sqrt{s}} \left[\vartheta_0^{d-1}(s) - 1\right].$$

Subtracting the infinite size limit, one finally obtains,

$$\mathcal{S}^{(1)}(\varphi) = \tfrac{1}{2}uLa_2 \int \mathrm{d}\tau\, \varphi^2(\tau), \quad (19.63)$$

with

$$a_2 = \frac{1}{24\pi}(N+2)\int_0 \frac{\mathrm{d}s}{\sqrt{s}}\left[\vartheta_0^{d-1}(s) - 1 - s^{(1-d)/2}\right]. \quad (19.64)$$

The corresponding Hamiltonian (19.56) for $\rho = 0$ then becomes

$$\hat{H} = -\frac{1}{2L^{d-1}}\nabla_\varphi^2 + \frac{1}{2}uLa_2\varphi^2 + \frac{u}{4!}L^{d-1}(\varphi^2)^2.$$

After the rescaling

$$\varphi \mapsto u^{-1/6} L^{(1-d)/3} \varphi \,,$$

and setting (equation (19.57)),

$$\hat{H} = u^{1/3} L^{(1-d)/3} \hat{H}_{\text{sc.}} \,,$$

one obtains the rescaled quantum Hamiltonian,

$$\widehat{H}_{\text{sc.}} = -\tfrac{1}{2} \nabla^2_\varphi + \tfrac{1}{2} z \varphi^2 + \frac{1}{4!} \left(\varphi^2 \right)^2 \,, \tag{19.65}$$

with

$$z = a_2 g^{1/3} (\Lambda L)^{(4-d)/3} \,. \tag{19.66}$$

The behaviour of z then depends on the sign of $(d - 4)$.

19.6.3 *Finite size correlation length: Dimensions $d > 4$*

At leading order, for $d > 4$, the parameter (19.59) of the rescaled Hamiltonian (19.58),

$$z = g^{-2/3} (\Lambda L)^{2(d-4)/3} L^2 \rho \,,$$

is small only very close to T_c for ρ infinitesimal and otherwise very large due to the factor $(\Lambda L)^{2(d-4)/3}$. Therefore, the relevant limits are $z = 0$, and $|z| \to \infty$.

For $\rho = 0$, the correlation length (19.61) behaves like

$$\xi_L / L = g^{-1/3} \Xi(0) (\Lambda L)^{(d-4)/3} \gg 1 \,. \tag{19.67}$$

The expression (19.67) exhibits a violation of the naive extension of the scaling form (19.69), proved for $d < 4$.

For $\rho > 0$, $z \to \infty$, and $\Xi(z) \sim z^{-1/2}$. Then,

$$\xi_L \sim \frac{1}{\sqrt{\rho}} \sim \xi_\infty$$

and ξ_L is insensitive to the finite size effects.

For $\rho < 0$, the behaviour changes drastically depending whether the symmetry is continuous or discrete. For $N > 1$, for $z \to -\infty$, one observes that the lowest eigenvalues of the Hamiltonian (19.58) can be obtained by approximating $\hat{H}$ by the angular moment part, fixing the radial coordinate to $|\varphi| = \sqrt{-6z}$. The corresponding eigenvalues are then $\ell(\ell + N - 2)/(-12z)$. It follows that

$$\Xi(z) \underset{z \to -\infty}{\sim} -\frac{12z}{N - 1} \quad \Rightarrow \quad \frac{\xi_L}{L} \sim -\frac{12}{(N-1)g} (L\Lambda)^{d-4} \rho L^2 \gg 1 \,.$$

By contrast, for $N = 1$, instantons are responsible for the splitting between the two lowest-lying states. A WKB analysis yields

$$\ln \Xi(z) \sim 2(-2z)^{3/2} \quad \Rightarrow \quad \ln \xi_L \sim \frac{2}{g} (-2\rho L^2)^{3/2} (L\Lambda)^{d-4} \gg 1. \tag{19.68}$$

The one-loop correction (19.66) for $d > 4$ is small and thus does not modify the asymptotic result.

19.6.4 Finite size correlation length: Dimensions $d = 4 - \varepsilon$

From equation (19.20), one concludes that the finite size correlation length ξ_L has the form (equation (19.61))

$$\xi_L/L \sim \Xi(\rho L^{1/\nu}). \tag{19.69}$$

In particular, at $\rho = 0$, ξ_L grows linearly with L and $\xi_L/L = \Xi(0)$ is a *universal ratio*. Since ξ_L is related to the ratio of the two largest eigenvalues λ_0 and λ_1 of the transfer matrix, one learns also that

$$\lambda_0/\lambda_1 - 1 \sim 1/\xi_L = 1/L\Xi(0).$$

With this knowledge, it is interesting to return to the analysis of the existence of phase transitions.

Since, for $\rho > 0$, ξ_L goes to a constant for large L, and since, for $\rho < 0$, it grows faster than L, as one can verify, the ratio ξ_L/L can be used to determine the critical temperature in transfer matrix calculations.

For $d = 4 - \varepsilon$, at leading order, one can replace g by its IR fixed point value g^*, which is of order ε. Using the result (19.61) and the RG scaling (19.69), one finds

$$\xi_L/L = (g^*)^{-1/3}\Xi[\rho L^{1/\nu}/g^{*2/3}]. \tag{19.70}$$

An analysis of the leading loop corrections to the leading order result again shows that, up to finite renormalizations, only the one-loop correction plays a role in the sense of an ε-expansion.

One thus finds the finite size correlation length ξ_L at one loop,

$$\xi_L(\rho)/L = (g^*)^{-1/3}\Xi(\rho L^{1/\nu} + b), \tag{19.71}$$

with

$$b = a_2 g^{*1/3} = a_2 \left(\frac{48\pi^2\varepsilon}{N+8}\right)^{1/3} + O\left(\varepsilon^{4/3}\right).$$

As in the case of the hypercubic geometry, one observes that the contribution coming from a_3, which is of the same order as the two-loop contribution ($O(\varepsilon^{4/3})$), is negligible at this order. One thus needs only the coefficient $\tilde{a}_2$ of $g^* \int d\tau\, \varphi^2/2$.

The coefficient of φ^2. The coefficient a_2 of $g^* \int d\tau\, \varphi^2/2$ in the expansion of the expression (19.62) is given by equation (19.64),

$$a_2 = \frac{N+2}{24\pi} \int_0^{+\infty} \frac{ds}{\sqrt{s}} \left[\vartheta_0^3(s) - 1 - s^{-3/2}\right]). \tag{19.72}$$

The finite size correlation length at T_c. Substituting the value of a_2 into equation (19.71) one obtains, in particular, the finite size correlation length at T_c at one-loop order,

$$\frac{\xi_L}{L} = \left(\frac{48\pi^2\varepsilon}{N+8}\right)^{-1/3} \Xi\left[K\pi^{-1/3}\frac{N+2}{12}\left(\frac{6\varepsilon}{N+8}\right)^{1/3}\right](1 + O(\varepsilon)) \tag{19.73}$$

with

$$K = \int_0^{+\infty} \frac{ds}{\sqrt{s}} \left[\vartheta_0^3(s) - 1 - s^{-3/2}\right] = -2.8372974\ldots. \tag{19.74}$$

19.7 Continuous symmetries: Finite size effects at low temperature

In models with *continuous symmetries*, in the low temperature ordered phase, the long distance behaviour is described by effective interactions between Goldstone modes (see Chapter 10). It is also interesting to examine the problem of finite size effects in this context.

In the case of the $O(N)$ symmetric N-vector model, universal physical observables can be derived from the low temperature or low coupling expansion of the non-linear σ-model, whose partition function reads (equations (10.12, 10.15))

$$\mathcal{Z} = \int [\mathrm{d}\Theta(\boldsymbol{\pi})] \exp\left[-\frac{\Lambda^{d-2}}{T} \mathcal{S}(\boldsymbol{\pi}) \right], \tag{19.75}$$

where the measure $\mathrm{d}\Theta(\boldsymbol{\pi})$ is the product over all space points of $(1-\boldsymbol{\pi}(x))^{-1/2}$, and the action reads

$$\mathcal{S}(\boldsymbol{\pi}) = \tfrac{1}{2} \int \mathrm{d}^d x \left[\left(\nabla_x \boldsymbol{\pi}(x)\right)^2 + \left(\nabla_x \sigma(x)\right)^2 \right], \tag{19.76}$$

with

$$\sigma(x) = \sqrt{1 - \boldsymbol{\pi}^2(x)} \,.$$

Here, the parameter T is dimensionless, and Λ is a large momentum cut-off provided, for instance, by a regularization lattice of spacing $a = 1/\Lambda$.

Previous considerations concerning RG equations apply to the RG equations derived for the σ-model. After addition to the action of a symmetry breaking term of the form $-H \int \mathrm{d}^d x \, \sigma(x)$, the vertex functions, in a Fourier representation, satisfy equations of the form,

$$\left[\Lambda \frac{\partial}{\partial \Lambda} + \beta(T) \frac{\partial}{\partial T} - \frac{n}{2} \zeta(T) + \rho(T) H \frac{\partial}{\partial H} \right] \tilde{\Gamma}^{(n)}(p_i; T, H, \Lambda) = 0 \,.$$

The finite sizes do not modify the RG equations but only affect the solutions. General solutions of the RG equations now depend on an additional scaling variable $L/\xi(T)$, where, for $d > 2 \, and \, T < T_c$, the RG invariant length $\xi(T)$ is defined by the equation

$$\xi(T) = \Lambda^{-1} T^{1/(d-2)} \exp\left[\int_0^T \left(\frac{1}{\beta(T')} - \frac{1}{(d-2)T'} \right) \mathrm{d}T' \right]. \tag{19.77}$$

Alternatively, the solutions can be parametrized in terms of a size-dependent deviation T_L from the critical temperature, obtained by solving the equation

$$\lambda \frac{\mathrm{d}}{\mathrm{d}\lambda} T(\lambda) = \beta\left[T(\lambda)\right], \quad T(1) = T \tag{19.78}$$

at scale $\lambda = 1/\Lambda L$,

$$T_L \equiv T(1/\Lambda L). \tag{19.79}$$

Then,

$$\ln(\Lambda L) = \int_{T_L}^{T} \frac{dT'}{\beta(T')}, \tag{19.80}$$

which shows in particular that T_L is a function of T and L only through the expected ratio $L/\xi(T)$.

At one-loop order, the RG $\beta(T)$ function in a cut-off scheme has the expansion

$$\beta(T) = (d-2)T + \beta_2(d)T^2 + O\left(T^3\right), \tag{19.81}$$

with

$$\beta_2(d) = -(N-2)/2\pi + O(d-2).$$

Within a $\varepsilon = d - 2$ expansion, for $d > 2$, the RG β-function has, for $N > 2$, in addition to the IR stable zero $T = 0$, another zero $T = T_c \sim 2\pi\varepsilon/(N-2)$, which is the critical temperature and at which the validity of the low temperature expansion ceases.

For $d > 2$, and $T < T_c$ fixed (and thus the length $\xi(T)$ is of order $1/\Lambda$), when ΛL increases, T_L approaches the IR fixed point $T = 0$:

$$T_L \sim \left(\xi(T)/L\right)^{d-2} \ll 1. \tag{19.82}$$

Therefore, finite size effects can be calculated from the low temperature expansion and RG.

At T_c, and more generally in the critical domain, physical quantities can be calculated in an $\varepsilon = d - 2$ expansion. Since T_c is a RG fixed point, $T_L(T_c) = T_c$.

Finally, solving equation (19.80) perturbatively, one finds

$$\frac{1}{T_L} = \frac{(\Lambda L)^{d-2}}{T} + \frac{\beta_2(d)}{d-2}\left[(\Lambda L)^{d-2} - 1\right] + O(T). \tag{19.83}$$

Two dimensions. In two dimensions, a phase transition with order is impossible. The case $N = 2$ is special since it displays a phase transition without order (the Kosterlitz–Thouless transition), which requires a specific discussion. For $N > 2$, the correlation length remains finite in infinite volume for all T and has the form,

$$\xi(T) \propto \Lambda^{-1} \exp\left[\int^{T} \frac{dT'}{\beta(T')}\right]. \tag{19.84}$$

In a finite volume, perturbative calculations can be performed in *two dimensions* even in zero magnetic field H because L provides an IR cut-off. However, because $T = 0$ is then a UV fixed point, T_L goes to zero *for $L/\xi(T)$ small* as

$$T_L \sim \frac{2\pi}{(N-2)\ln(\xi(T)/L)} \tag{19.85}$$

and this is the limit in which physical quantities can be calculated.

Notice here that a statistical (classical) field theory in two dimensions with a finite size in one of the dimensions and periodic boundary conditions, is also a finite temperature quantum field theory (see Chapter 21). Therefore, some of the considerations in this section are also relevant for such a situation.

20 The weakly interacting Bose gas at the critical temperature

In the absence of interactions, at some temperature, the free Bose gas undergoes a Bose–Einstein condensation (BEC). In the presence of interactions, the condensation temperature becomes a critical temperature of the nature of the Helium superfluid transition. However, when the interaction is weak enough, the transition has additional *universal properties*, some being *non-perturbative* in nature. Their main features can be derived using *renormalization group* (RG) methods [196].

This system, whose large scale physics can be studied with a two-component $(\phi^2)^2$ field theory, offers a striking physical example of a general fine tuning, where the coefficient of the $(\phi^2)^2$ interaction is small, because the two-body repulsive interactions are small, and the coefficients of other possible interactions like $(\phi^2)^3$ are even smaller because higher many-body interactions are negligible.

Therefore, the physics on scales larger than the microscopic scale, can be described by a super-renormalizable field theory consistent on all scales. In particular, in addition to the universal large scale properties of the Helium class, the gas also displays a universal short distance behaviour of the Bose–Einstein type.

As a very instructive exercise, we use the RG to prove that T_c *increases linearly with the strength of the interaction*, parametrized in terms of the s-wave scattering length a [195, 196]. Moreover, when $\Delta T_c/T_c$ is expressed in terms of the dimensionless product $an^{1/3}$ (n is the density), the *coefficient is universal*. However, the *coefficient cannot be obtained from perturbative calculations*.

Since the Hamiltonian of the system is the $N = 2$ example of the general N-vector model, we generalize the problem to arbitrary N. The coefficient of $\Delta T_c/T_c$ can then be expanded in powers of $1/N$ using standard large N techniques [47].

The *leading order result*, which is *independent of N* for non-trivial reasons, though it has to be applied to $N = 2$, is in reasonable agreement with numerical estimates.

20.1 Bose gas: Field integral formulation

The partition function, in the grand canonical formalism, of a non-relativistic quantum gas of identical bosons at temperature T and chemical potential μ, has the form,

$$\mathcal{Z}(T,\mu) = \operatorname{tr} e^{-(\mathbf{H}-\mu\mathbf{N})/T},\tag{20.1}$$

where $\mathbf{H}$ is the quantum Hamiltonian and $\mathbf{N}$ the particle number operator.

The partition function $\mathcal{Z}$ can also be expressed as a field integral over complex fields $\bar{\psi}, \psi$ satisfying *periodic boundary conditions in imaginary time*, associated with boson creation and annihilation operators, as

$$\mathcal{Z} = \int [\mathrm{d}\psi(t,x)\mathrm{d}\bar{\psi}(t,x)]\, e^{-\mathcal{S}(\bar{\psi},\psi)},\tag{20.2}$$

where $\mathcal{S}(\bar{\psi},\psi)$ is the imaginary time action.

From Random Walks to Random Matrices. Jean Zinn-Justin, Oxford University Press (2019).
© Jean Zinn-Justin. DOI: 10.1093/oso/9780198787754.001.0001

20.1.1 Euclidean Bose gas action

We assume that the interactions are short range, that the gas is *sufficiently dilute for the two-body interactions to be weak and that three-body or higher interactions are completely negligible.* Therefore, the euclidean action contains only a $(\bar{\psi}\psi)^2$ type interaction.

We are interested here only in long wavelength phenomena. Therefore, *the two-body potential itself can be replaced by a δ-function* and parametrized in terms of the s-wave scattering length $a > 0$ (because the interaction is assumed to be short range and repulsive).

The effective Euclidean action for a Bose gas of particles of mass m takes then, for the dimension $d = 3$, the *local* form

$$\mathcal{S}(\bar{\psi}, \psi) = \int_0^{1/T} \mathrm{d}t \int \mathrm{d}^3 x \left\{ -\bar{\psi}(t,x) \left(\frac{\partial}{\partial t} + \frac{\hbar^2}{2m} \nabla_x^2 + \mu \right) \psi(t,x) \right.$$
$$\left. + \frac{2\pi \hbar^2 a}{m} \left[\bar{\psi}(t,x)\psi(t,x) \right]^2 \right\}, \tag{20.3}$$

where $\nabla_x \equiv (\partial/\partial x_1, \partial/\partial x_2, \partial/\partial x_3)$ and the fields satisfy periodic boundary conditions in imaginary time,

$$\psi(0,x) = \psi(1/T, x), \quad \bar{\psi}(0,x) = \bar{\psi}(1/T, x). \tag{20.4}$$

The condition that the interaction is weak implies that $a \ll \lambdabar$, where $\lambdabar$ is the thermal wavelength,

$$\lambdabar = \hbar\sqrt{2\pi/mT}\,.$$

The model is expected to undergo a Helium-type superfluid transition, a property that will be verified later.

20.1.2 Equation of state and two-point function

Inserting the action (20.3) into the field integral representation (20.2) of the partition function, one notes that the equation of state (the relation between density, temperature and chemical potential) can be expressed in terms of the $\langle \bar{\psi}\psi \rangle$ correlation function G. The density ρ_ψ, in d space dimensions, assuming translation invariant boundary conditions, is given by (*c.f.,* equations (20.1), (20.2) and (20.3)),

$$\rho_\psi = \frac{T}{\text{space volume}} \frac{\partial \ln \mathcal{Z}}{\partial \mu} = \langle \bar{\psi}(0,0)\psi(0,0) \rangle = \int \frac{\mathrm{d}^3 p}{(2\pi)^3} \sum_{\nu \in \mathbb{Z}} \tilde{G}(p, \omega_\nu; \mu), \tag{20.5}$$

where $\langle \bullet \rangle$ means expectation value with the measure $\mathrm{e}^{-\mathcal{S}}$, $\tilde{G}(p, \omega_\nu; \mu)$ is the two-point function in Fourier representation, and the frequencies,

$$\omega_\nu = 2\pi\nu T\,, \quad \nu \in \mathbb{Z}\,,$$

which are discrete, due to the periodic boundary conditions in time (20.4), are also called *Matsubara frequencies*.

20.2 Independent bosons: Bose–Einstein condensation

In most of the chapter, we now keep the dimension $d \le 4$ arbitrary, even though we are mainly interested in $d = 3$, in order to be able to use dimensional continuation to d real and dimensional regularization.

For vanishing self-interaction (here $a = 0$), the partition function can be calculated explicitly.

It is convenient to confine the gas in a hypercubic box of linear size L with periodic boundary conditions and then study the large L limit (although a harmonic potential would be physically more realistic).

Then, the fields $\psi(t, \mathbf{x}), \bar{\psi}(t, \mathbf{x})$ are also periodic functions of period L of all space variables. The momenta, arguments of the Fourier transform, belong to the lattice,

$$\mathbf{p} = \frac{2\pi\hbar}{L}\mathbf{n}, \quad \mathbf{n} \in \mathbb{Z}^d.$$

We denote by $\mathcal{Z}_0(T, \mu)$ the partition function for free bosons. The free energy, $\mathcal{W}_0(T, \mu) = T \ln \mathcal{Z}_0(T, \mu)$, is

$$\mathcal{W}_0(T, \mu) = -T \sum_{\mathbf{n} \in \mathbb{Z}^d} \ln\left(1 - e^{-(p^2/2m - \mu)/T}\right).$$

For $L \gg \lambdabar$ large, the sum over momentum modes can be replaced by an integral (a *semi-classical approximation*), and the free energy per unit volume, which is the pressure Π, becomes ($d\mathbf{n} = d\mathbf{p}L/2\pi\hbar$)

$$\Pi = L^{-d}\mathcal{W}_0(T, \mu) = -T \int \frac{d^d p}{(2\pi\hbar)^d} \ln\left(1 - e^{-(p^2/2m - \mu)/T}\right).$$

The Bose gas is stable only if the chemical potential μ is non-positive.

The derivative of $\ln \mathcal{Z}_0$ with respect to μ/T (T fixed) yields the average particle number and thus the gas density,

$$\rho_\psi = L^{-d}\langle N \rangle = \frac{T}{L^d} \int dt\, d^d x \left\langle \psi(t, x)\bar{\psi}(t, x)\right\rangle$$

$$= \frac{1}{(2\pi\hbar)^d} \int \frac{d^d p}{e^{(p^2/2m - \mu)/T} - 1}. \qquad (20.6)$$

The equation of state (20.6) exhibits the phenomenon of *Bose–Einstein condensation*.

At fixed temperature T, the density ρ_ψ is an increasing function of μ. Since $\mu \le 0$, for space dimensions $d > 2$, ρ_ψ is bounded by the value $\rho_{\psi,c}$ of the integral calculated for $\mu = 0$:

$$\rho_\psi \le \rho_{\psi,c} = \frac{1}{(2\pi\hbar)^d} \int \frac{d^d p}{e^{p^2/2mT} - 1} = \zeta(d/2)\left(\frac{mT}{2\pi\hbar^2}\right)^{d/2} = \zeta(d/2)\lambdabar^{-d}, \quad \cdot \quad (20.7)$$

($\zeta(s)$ is Riemann's ζ-function).

Conversely, at fixed density, the equation of state has a solution up to the minimal temperature

$$T_c = \frac{2\pi\hbar^2}{m}\left(\frac{\rho_\psi}{\zeta(d/2)}\right)^{2/d}.$$

In particular, in three dimensions,

$$T_c^0(\rho_\psi)\underset{d=3}{\propto}(\hbar^2/m)\rho_\psi^{2/3}.$$

To understand the physics below T_c, one has to return to the *exact form in a finite box*, where the momentum modes are discrete and the integral is replaced by a sum of poles. When μ approaches the pole corresponding to the ground state of the one-body Hamiltonian, the sum diverges. One then verifies that a macroscopic fraction of the free Bose gas condenses into the ground state, which here is the zero momentum mode. This is the phenomenon of Bose–Einstein condensation.

In two dimensions, because $\rho_{\psi,c}$ in expression (20.7) diverges, there is no condensation.

The discussion indicates that, *in the limit of vanishing repulsive interactions*, the phase transition of the interacting model (20.3) reduces to the Bose–Einstein condensation of the free Bose gas.

20.3 The weakly interacting Bose gas and the Helium phase transition

20.3.1 *Phase transition and dimensional reduction*

In the interacting model, for $d > 2$, a $U(1)$ phase transition of superfluid Helium type occurs at a critical chemical potential μ_c where $\tilde{G}^{-1}(p = 0, \omega = 0; \mu_c)$ vanishes and, thus, the *correlation length*, which characterizes the decay of correlations at large distance in the disordered phase, *diverges*.

At $d = 2$, the system exhibits the peculiar *Kosterlitz–Thouless transition*: below T_c the correlation length diverges but without ordering, a phase that we do not discuss here.

The theory of critical phenomena tells us that the *universal properties* near a continuous phase transition in systems with dimension $d \leq 4$ depend primarily on contributions from the small momenta or large distance (or infrared (IR)) region.

The inverse propagator, or inverse of the two-point function at leading order for $a \to 0$, in Fourier representation is

$$\tilde{G}^{-1}(p,\omega;\mu) = p^2/2m - i\omega_\nu - \mu, \quad \omega_\nu = 2\pi\nu T.$$

The momentum pole closest to $|p| = 0$ is obtained for $\omega_\nu = 0$. One finds $|p| = \sqrt{-2m\mu} \equiv \hbar/\xi$, where ξ is the correlation length. As soon as the frequency gap $2\pi T$ satisfies

$$2\pi T \gg |p|^2/2m = \hbar^2/m\xi^2 \Leftrightarrow \xi \gg \lambda,$$

the problem is simplified, since the *large scale properties remain sensitive only to the zero-mode $\nu = 0$ corresponding to the $\omega_\nu = 0$ Matsubara frequency*.

In the mode expansion

$$\psi(t,x) = \sum_{\nu \in \mathbb{Z}} e^{i\omega_\nu t} \, \psi_\nu(x), \ \ \bar\psi(t,x) = \sum_{\nu \in \mathbb{Z}} e^{i\omega_\nu t} \, \bar\psi_\nu(x),$$

one can thus omit, at leading order, the non-zero modes.

The entire calculation can then be cast in terms of an *effective classical statistical field theory in d dimensions*, an example of dimensional reduction.

The integration over the higher frequency modes that are not critical can be done perturbatively. Up to subleading corrections, the effect is a renormalization of the coefficients of the zero-mode action.

Since at leading order these modes are completely omitted, the reduced theory is valid only up to momenta of order $\sqrt{mT}$, and the temperature gap provides a ultraviolet (UV, *i.e.*, large momentum) cut-off $\Lambda \propto \sqrt{mT} \propto \hbar/\lambda$.

20.3.2 *Effective classical statistical field theory*

From now on, we set $\hbar = 1$.

We rescale the field ψ_0, in order to introduce more conventional field theory normalizations, and parametrize it in terms of two real fields ϕ_1, ϕ_2 as,

$$\psi_0 = \sqrt{mT}(\phi_1 + i\phi_2), \quad \bar\psi_0 = \sqrt{mT}(\phi_1 - i\phi_2). \tag{20.8}$$

The action becomes

$$\mathcal{Z} = \int [\mathrm{d}\phi(x)] \exp\left[-\mathcal{S}(\phi)\right]$$

with

$$\mathcal{S}(\phi) = \int \left\{ \tfrac{1}{2} [\nabla_x \phi(x)]^2 + \tfrac{1}{2} r \phi^2(x) + \frac{u}{4!} \left[\phi^2(x)\right]^2 \right\} \mathrm{d}^d x\,, \tag{20.9}$$

where a cut-off $\Lambda \propto \hbar/\lambda$ is implied, $r = -2m\mu$ and, for $d = 3$, $u = 96\pi^2 a/\lambda^2$. A phase transition occurs at a value $r = r_c$ where the correlation length diverges.

The action reduces to the ordinary $O(2)$ symmetric $(\phi^2)^2$ relativistic quantum field theory in imaginary time, which is also a classical statistical field theory and is known to describe the *universal properties of the superfluid Helium transition*.

20.4 RG and universality

To gain some general insight into the universal properties of the phase transition, *RG methods* can be used (for details, see Chapter 6). As an example, we discuss here the properties of the *critical two-point correlation function* or its inverse in Fourier representation, the vertex two-point function, which we denote by $\tilde\Gamma^{(2)}$.

We introduce the dimensionless interaction strength (Λ is the large momentum cut-off)

$$g = \Lambda^{d-4} u \propto (a/\lambda)^{d-2} \ll 1\,.$$

The transition point $r = r_c$ is determined by the condition of divergence of the correlation length, which is equivalent to $\tilde\Gamma^{(2)}(p = 0, g, \Lambda)|_{r=r_c} = 0$.

At the transition point, $\tilde{\Gamma}^{(2)}(p, \Lambda, g)$ satisfies an RG equation of the form

$$\left(\Lambda \frac{\partial}{\partial \Lambda} + \beta(g) \frac{\partial}{\partial g} - \eta(g)\right) \tilde{\Gamma}^{(2)}(p, \Lambda, g) = 0, \qquad (20.10)$$

where $\beta(g)$ and $\eta(g)$ are two RG functions calculable as power series in g.

20.4.1 Solution of the RG equations: The IR fixed point

Equation (20.10) can be solved by the method of characteristics. One introduces a dilatation parameter λ and looks for two functions $g(\lambda)$ and $Z(\lambda)$ such that

$$\lambda \frac{\mathrm{d}}{\mathrm{d}\lambda}\left[Z^{-1}(\lambda)\tilde{\Gamma}^{(2)}(p; g(\lambda), \lambda\Lambda)\right] = 0. \qquad (20.11)$$

The function $g(\lambda)$ is the effective interaction at the scale λ.

Differentiating explicitly with respect to λ and using the chain rule, one finds that equation (20.11) is consistent with equation (20.10), provided that

$$\lambda \frac{\mathrm{d}}{\mathrm{d}\lambda} g(\lambda) = \beta(g(\lambda)), \qquad g(1) = g, \qquad (20.12a)$$

$$\lambda \frac{\mathrm{d}}{\mathrm{d}\lambda} \ln Z(\lambda) = \eta(g(\lambda)), \qquad Z(1) = 1. \qquad (20.12b)$$

Equation (20.11) then implies

$$\tilde{\Gamma}^{(2)}(p; g, \Lambda) = Z^{-1}(\lambda)\tilde{\Gamma}^{(2)}(p; g(\lambda), \lambda\Lambda).$$

Rescaling Λ by a factor $1/\lambda$, and using dimensional analysis, one can write the equation as

$$\tilde{\Gamma}^{(2)}(\lambda p; g, \Lambda) = Z^{-1}(\lambda)\lambda^2 \tilde{\Gamma}^{(2)}(p; g(\lambda), \Lambda). \qquad (20.13)$$

The interpretation of the equation is that it is equivalent to decreasing the momentum p at g fixed or to varying the effective interaction at p fixed.

The integration of the flow equation (20.12a) for the effective interaction yields

$$\int_{g}^{g(\lambda)} \frac{\mathrm{d}g'}{\beta(g')} = \ln \lambda.$$

We are interested in the regime $\lambda \to 0$. For $\beta(g) < 0$, the equation implies $g(\lambda) > g$ and, for $\beta(g) > 0$, it implies $g(\lambda) < g$.

Perturbative calculations yield [29, 83]

$$\beta(g) = -(4-d)g + \frac{5}{24\pi^2}g^2 + \mathcal{O}(g^3), \qquad \eta(g) = \frac{1}{18(8\pi^2)^2}g^2 + \mathcal{O}(g^3).$$

For $d < 4$, the β-function has the zero $g = 0$ with a negative slope, which corresponds to the *IR repulsive Gaussian fixed point* and, for $d > 2$, a non-trivial zero g^* with a positive slope (see Fig. 20.1), which, for $\varepsilon = 4 - d$ small, behaves like $g^* = 24\pi^2\varepsilon/5 + \mathcal{O}(\varepsilon^2)$ and which corresponds to an *IR attractive fixed point*.

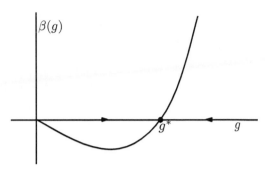

Fig. 20.1 RG β-function and RG flow for $2 < d < 4$.

This fixed point is at the origin of *Wilson–Fisher's* famous ε-expansion [79] (see Section 6.10): all universal quantities like the correlation exponent ν, the scaling equation of state and so on, can be calculated as ε-expansions.

The RG equation (20.10) then implies

$$\tilde{\Gamma}^{(2)}(p, \Lambda, g) \underset{p \to 0}{\propto} p^{2-\eta}, \text{ with } \eta \equiv \eta(g^*) > 0\,.$$

20.4.2 RG equation: Another form of the solution and crossover scale

Combined with dimensional analysis, the solution $\tilde{\Gamma}^{(2)}$ of the RG equation at criticality can be written as

$$\tilde{\Gamma}^{(2)}(p, \Lambda, g) = p^2 \tilde{Z}(g) F\big(p/\Lambda(g)\big) \tag{20.14}$$

with

$$\beta(g)\frac{\partial \ln \tilde{Z}(g)}{\partial g} = \eta(g), \quad \beta(g)\frac{\partial \ln \Lambda(g)}{\partial g} = -1\,. \tag{20.15}$$

Therefore, (with the normalization $\tilde{Z}(0) = 1$),

$$\tilde{Z}(g) = \exp \int_0^g \frac{\eta(g')}{\beta(g')} \mathrm{d}g' = 1 + \mathcal{O}(g^2)\,.$$

On dimensional grounds, the function $\Lambda(g)$ is proportional to Λ. Integrating the second equation (20.15), one then finds

$$\Lambda(g) \propto \Lambda g^{1/(4-d)} \exp\left[-\int_0^g \mathrm{d}g' \left(\frac{1}{\beta(g')} + \frac{1}{(4-d)g'} \right) \right] = \Lambda g^{1/(4-d)}\big(1 + \mathcal{O}(g)\big)\,.$$

A universal behaviour of the correlation function is expected for $|p| \ll \Lambda$.

Here we are interested in interactions $g \ll 1$, which implies $\Lambda(g) \sim g^{1/(4-d)}\Lambda \ll \Lambda$.

The scale $\Lambda(g)$ thus is a *crossover scale* separating a *universal large distance regime*, which is governed by the non-trivial zero $g^* > 0$ of the β-function, from a *universal short distance regime* governed by the Gaussian fixed point, $g = 0$.

In this situation,

$$\tilde{\Gamma}^{(2)}(p) \propto p^{2-\eta}, \text{ for } p \ll \Lambda(g), \quad \tilde{\Gamma}^{(2)}(p) \propto p^2, \text{ for } \Lambda(g) \ll p \ll \Lambda.$$

At a short distance, but one much larger than the microscopic scale, the physical system looks like a Bose–Einstein condensate at the condensation temperature, while, at a large distance, it looks like a critical Helium superfluid system.

This has to be contrasted with the *generic* situation $g = O(1)$. Then, $\Lambda(g) = O(\Lambda)$ and *only the large distance properties are universal.*

20.5 The shift of the critical temperature for weak interaction

As an illustration, we discuss the effect of a weak repulsive two-body interaction on the transition temperature of a dilute Bose gas at fixed density. This question has remained controversial for a long time until it was realized that the *effect is non-perturbative* in nature [195].

RG arguments are required to prove that, for $d = 3$, T_c *increases linearly with the strength of the interaction* parametrized in terms of the s-wave scattering length a [195, 196]. Moreover, when $\Delta T_c/T_c$ is expressed in terms of the dimensionless product $a\rho_\psi^{1/3}$ (ρ_ψ is the density), the *coefficient is universal*. The *coefficient cannot be obtained from perturbative calculations*: however, by generalizing the $O(2)$ invariant field theory to an $O(N)$ theory, it can be calculated as an $1/N$ expansion [196].

20.5.1 The variation of the equation of state

The shift of the critical temperature for weak interaction can be derived from the leading order non-trivial contribution at criticality (in the massless theory) to the equation of state. Taking into account the change of fields (20.8), one infers from equation (20.5),

$$\rho_\psi = \langle \bar{\psi}\psi \rangle = 2mT \langle \phi_1\phi_1 \rangle = 2mT\rho,$$

with

$$\rho = \int^\Lambda \frac{\mathrm{d}^d p}{(2\pi)^d} \frac{1}{\tilde{\Gamma}^{(2)}(p)}, \tag{20.16}$$

where the vertex function $\tilde{\Gamma}^{(2)}$ is the inverse of the two-point function $\langle \phi_1\phi_1 \rangle = \langle \phi_2\phi_2 \rangle$, and the large momentum cut-off Λ is of order $1/\chi$.

Because the interactions are weak, one may imagine calculating the change in the transition temperature by perturbation theory. However, *the perturbative expansion for a critical theory does not exist for any fixed dimension $d < 4$.*

Using equation (20.14) and since $\tilde{Z}(g) = 1 + O(g)$, one finds for the variation of the density (20.16),

$$\delta\rho \underset{g \to 0}{\sim} \int^\Lambda \frac{\mathrm{d}^3 p}{(2\pi)^3} \frac{1}{p^2} \left(\frac{1}{F(p/\Lambda(g))} - 1 \right).$$

Fig. 20.2 Leading large momentum contribution to $\tilde{\Gamma}^{(2)}(p) - p^2$.

From perturbation theory, one infers that, for $d = 3$, the function $F(p)$ behaves for large p as (see Fig. 20.2)

$$F(p) = \tilde{\Gamma}^{(2)}(p)/p^2 = 1 + \mathcal{O}(\ln p/p^2).$$

Therefore, the first correction to the density is convergent at large momentum and independent of the cut-off procedure:

$$\delta\rho = \int \frac{\mathrm{d}^3 p}{(2\pi)^3} \frac{1}{p^2} \left(\frac{1}{F(p/\Lambda(g))} - 1 \right).$$

Similarly, the IR behaviour $\tilde{\Gamma}^{(2)}(p) \propto p^{2-\eta}$ and, thus, $F(p) \propto p^{-\eta}$ implies that the integral is IR convergent. One can thus integrate over $|p|$ from 0 to ∞.

Setting $p = \Lambda(g)k$, one finds that the density variation takes the general form

$$\delta\rho = \Lambda(g) \int \frac{\mathrm{d}^3 k}{(2\pi)^3} \frac{1}{k^2} \left(\frac{1}{F(k)} - 1 \right).$$

The g dependence is entirely contained in $\Lambda(g)$. For g small, since $\Lambda(g) \sim \Lambda g$ and $\rho \propto \Lambda \sim 1/\chi$, one concludes

$$\frac{\delta\rho}{\rho} \propto g \sim (-3/2c_0)\, a\rho_\psi^{1/3},$$

a linear behaviour that, however, is non-perturbative! Moreover, *the amplitude $-3c_0/2$ is universal.*

20.5.2 *The N-vector model: The large N expansion at order $1/N$*

The function $F(p)$ cannot be determined by a non-existing perturbative expansion. Since the Hamiltonian of the system, which also describes the *Helium superfluid transition*, is the $N = 2$ example of the general N-vector model, we generalize the problem to arbitrary N. Indeed, such a generalization provides us with a new tool, the large N-expansion, which makes a calculation at the critical point possible.

The coefficient of $\Delta T_c/T_c$ can then be calculated as an expansion in powers of $1/N$ [196]. Although the application of the result for $N = 2$ is rather tentative, it can be hoped that the $N = 2$ value gives a rough estimate of the true universal value.

The action of the $O(N)$ symmetric generalization of the effective action (20.9) reads

$$\mathcal{S}(\phi) = \int \left\{ \frac{1}{2} [\nabla_x \phi(x)]^2 + \frac{1}{2} r \phi^2(x) + \frac{u}{4!} \left[\phi^2(x) \right]^2 \right\} \mathrm{d}^d x,$$

where $\phi(x)$ is an N-component field.

The corresponding partition function is

$$\mathcal{Z} = \int [\mathrm{d}\phi(x)] \exp\left[-\mathcal{S}(\phi)\right]. \tag{20.17}$$

The basic idea of the large N expansion is the same as mean field theory: for N large, the $O(N)$ invariant quantities self-average and thus have small fluctuations. This observation suggests taking $\phi^2(x)$ instead of $\phi(x)$ as a dynamical field. Technically, introducing an auxiliary field $\lambda(x)$, one uses the identity (equivalent to the *Hubbard transformation*: see, *e.g.*, Ref. [47]) ,

$$\exp\left\{-\int \mathrm{d}^d x \left[\frac{1}{2} r \phi^2(x) + \frac{u}{4!} \left(\phi^2(x)\right)^2\right]\right\}$$
$$\propto \int [\mathrm{d}\lambda(x)] \exp\left[\int \mathrm{d}^d x \left(\frac{3}{2u}\lambda^2(x) - \frac{3r}{u}\lambda(x) - \frac{1}{2}\lambda(x)\phi^2(x)\right)\right],$$

where the λ integration contour is parallel to the imaginary axis.

This identity, introduced into the field integral (20.17), yields

$$\mathcal{Z} = \int [\mathrm{d}\phi(x)\mathrm{d}\lambda(x)] \exp\left[-\mathcal{S}(\phi,\lambda)\right], \tag{20.18}$$

with

$$\mathcal{S}(\phi,\lambda) = \int \left\{\frac{1}{2}\left[\nabla_x \phi(x)\right]^2 - \frac{3}{2u}\lambda^2(x) + \frac{3r}{u}\lambda(x) + \frac{1}{2}\lambda(x)\phi^2(x)\right\}.$$

The new field integral is Gaussian in ϕ, and the integration over the field ϕ can be performed. The dependence of the partition function on N then becomes explicit.

Actually, it is convenient to separate the components of ϕ into one component σ and $(N-1)$ components π, and integrate only over π. This leads to an equivalent expansion in $1/(N-1)$. At leading order, the difference is negligible. For each π-component, the integration is Gaussian and yields,

$$\int [\mathrm{d}\pi(x)] \exp\left\{-\frac{1}{2}\int \mathrm{d}^d x \left[\left(\nabla_x \pi(x)\right)^2 + \lambda(x)\pi^2(x)\right]\right\}$$
$$\propto \det^{-1/2}[-\nabla_x^2 + \lambda(x)] = \exp\left\{-\frac{1}{2}\operatorname{tr}\ln[-\nabla_x^2 + \lambda(x)]\right\},$$

where the general identity $\ln\det = \operatorname{tr}\ln$ has been used.

The partition function then is given by the integral

$$\mathcal{Z} = \int [\mathrm{d}\lambda(x)] [\mathrm{d}\sigma(x)] \exp\left[-\mathcal{S}_N(\lambda,\sigma)\right],$$

with

$$\mathcal{S}_N(\lambda,\sigma) = \int \left[\frac{1}{2}\left(\nabla_x \sigma\right)^2 - \frac{3}{2u}\lambda^2(x) + \frac{3r}{u}\lambda(x) + \frac{1}{2}\lambda(x)\sigma^2(x)\right] \mathrm{d}^d x$$
$$+ \frac{(N-1)}{2} \operatorname{tr}\ln\left[-\nabla_x^2 + \lambda(x)\right].$$

The *large N limit is taken at Nu fixed*. With this condition, $\mathcal{S}_N$ is of order N, and the field integral can be calculated for N large by the *steepest descent method*. The saddle point values correspond to $\sigma^2 = \mathcal{O}(N)$, $\lambda = \mathcal{O}(1)$.

20.5.3 The saddle point equations

One looks for a uniform saddle point $(\sigma(x), \lambda(x)$ space-independent),

$$\sigma(x) = \sigma, \quad \lambda(x) = \lambda,$$

and σ of order $\sqrt{N}$. The action density becomes,

$$\mathcal{S}_N(\lambda, \sigma)/\text{volume} = -\frac{3}{2u}\lambda^2 + \frac{3r}{u}\lambda + \frac{1}{2}\lambda\sigma^2 + \frac{(N-1)}{2}\int^\Lambda \frac{\mathrm{d}^d p}{(2\pi)^d} \ln\left(p^2 + \lambda\right).$$

Differentiating the expression with respect to σ and λ, one obtains the saddle point equations for N large,

$$\lambda\sigma = 0, \quad \frac{\sigma^2}{N} - \frac{6}{Nu}(\lambda - r) + \frac{1}{(2\pi)^d}\int^\Lambda \frac{\mathrm{d}^d p}{p^2 + \lambda} = 0.$$

For $d > 2$, at leading order for N large, in the limit $\lambda = \sigma = 0$, a continuous phase transition is found at the point

$$r = r_c \equiv -\frac{Nu}{6}\frac{1}{(2\pi)^d}\int^\Lambda \frac{\mathrm{d}^d p}{p^2}.$$

However, then $\tilde{\Gamma}^{(2)}(p) = p^2$ at leading order. A next order calculation is required to obtain a non-trivial result for $\delta\rho$.

20.5.4 The two-point function: $1/N$ correction

At order $1/N$, one finds

$$\tilde{\Gamma}^{(2)}(p) = p^2 + \frac{2}{N}\int \frac{\mathrm{d}^d q}{(2\pi)^d}\frac{1}{(6/Nu) + B_d(q)}\left(\frac{1}{(p+q)^2} - \frac{1}{q^2}\right) + \mathcal{O}\left(\frac{1}{N^2}\right),$$

where $B_d(q)$ is the one-loop contribution to the perturbative four-point function (Fig. 20.3),

$$B_d(q) = \int \frac{\mathrm{d}^d k}{(2\pi)^d}\frac{1}{k^2(k+q)^2} \underset{q\to 0}{\sim} b(d)q^{d-4},$$

which is UV finite for $d < 4$. For $d = 3$, $b(1) = 1/8$.

Fig. 20.3 Feynman diagram: the four-point function $B_d(q)$.

The order $1/N$ thus yields the first non-trivial correction to $\tilde{\Gamma}^{(2)}(p)$. One can easily evaluate the quantity

$$\delta\rho = -\frac{2}{N}\int \frac{\mathrm{d}^d p}{(2\pi)^{2d}}\frac{1}{p^4}\frac{\mathrm{d}^d q}{(6/Nu) + b(d)q^{d-4}}\left(\frac{1}{(p+q)^2} - \frac{1}{q^2}\right),$$

by first keeping the dimension d generic, using dimensional regularization and inverting the order of integrations.

After the two integrations, taking the $d = 3$ limit, one obtains $(1/32\pi^2)(Nu/6)$. As expected, $\delta\rho \propto u$:

$$\delta\rho = -u/96\pi^2 = -\frac{a}{\chi^2}.$$

One finally obtains the shift of the transition temperature:

$$\frac{\Delta T_c}{T_c} = \frac{8\pi}{3\zeta(3/2)}\frac{a}{\chi} = c_0 a\rho_\psi^{1/3} \text{ with } c_0 = \frac{8\pi}{3\zeta(3/2)^{4/3}} = 2.33\ldots.$$

Note that, although the final result does not depend on N, the estimate of the universal value c_0 is only valid for N large.

Taking into account, the $1/N$ correction [197], one finds $c_0 = 1.71\ldots$. A number of other methods, such as lattice calculations and the summation of perturbative expansions, suggest $c_0 \approx 1.3$.

Conclusions. Using *RG* arguments, one can show that the *properties of the dilute, weakly interacting Bose gas remain dominated by the UV fixed point up to large length scales*; this is why one can still refer to Bose–Einstein condensation when discussing the phase transition of the interacting Bose gas, although, ultimately, the large scale properties will be governed by the IR fixed point of the universality class of the superfluid Helium transition.

RG arguments enable one to confirm that the relative *shift of the transition temperature at fixed density is proportional to the dimensionless combination* $a\rho_\psi^{1/3}$ (ρ_ψ is the density) for weak interactions. This result is non-perturbative, and the proportionality coefficient, which is *universal*, cannot be obtained from perturbation theory.

Therefore, a *non-perturbative method*, the large N expansion, has been introduced that allows a systematic, analytic, calculation of this coefficient as a power series in $1/N$, where, eventually, one has to set $N = 2$.

The *leading order contribution* is formally of order $1/N$ multiplied by a function of aN, which is kept fixed in the large N limit. *Because, for $d = 3$, the result is linear in a, the $1/N$ factor, somewhat surprisingly, cancels* and the result is independent of N. Adding the 30% $1/N$ correction [197], one finds a value in reasonable agreement with other numerical estimates.

21 Quantum field theory at finite temperature

In simple collider experiments, one generally probes *physics at zero temperature*. Therefore, the study of relativistic quantum field theory (QFT) at finite temperature, was initially motivated by cosmological problems, for example, concerning the restoration of the weak-electromagnetic symmetry in the early universe. However, it has later gained additional attention in connection with high energy heavy ion collisions and speculations about possible phase transitions in quantum chromodynamics (see Chapter 13).

In this chapter, we review the properties at finite temperature of several local, relativistic, translation invariant, QFTs in generic $(1, d)$ dimensions, although $d = 3$ is the most interesting physics application.

The problems we tackle correspond to a highly relativistic and quantum regime and thus we set $\hbar = c = 1$. We use a field integral formalism. Moreover, we discuss QFT at thermal equilibrium and thus the density matrix $\mathrm{e}^{-\beta H}$ rather than the evolution operator. Therefore, time is imaginary and the Euclidean Lagrangian density is $O(d + 1)$ invariant.

In this formalism, relations between thermal QFT in $(1, d)$ dimensions and statistical classical field theory in $(d + 1)$ dimensions become apparent. We especially emphasize that additional physical intuition about QFT at equilibrium at finite temperature can be gained by realizing that QFT can then be considered as an example of a classical statistical field theory in systems with one finite size. Finite temperature T corresponds to a finite size $L = 1/T$ in one direction in the corresponding $(d + 1)$ dimensional classical theory.

In particular, the zero-temperature limit of the quantum theory then corresponds to the usual infinite volume limit of the classical theory.

This identification makes it possible to apply *renormalization group* (RG) ideas and the theory of *finite size effects* of the classical theory to analyse properties of finite temperature QFT (see Chapter 19).

A particular issue is the relevance of quantum effects at finite temperature. More specifically, we discuss the limit of high temperature and the situation of finite temperature phase transitions.

QFT at finite temperature is the relativistic generalization of finite temperature non-relativistic quantum statistical mechanics. In the latter framework, quantum effects are important only at low temperature. A relevant parameter is the ratio between the *thermal wavelength* $\hbar/\sqrt{2\pi m T}$ and a characteristic length scale of the system.

Only when this ratio is large are quantum effects important. Increasing the temperature is at leading order equivalent to decreasing $\hbar$.

Moreover, the transition from quantum to classical behaviour is associated with *dimensional reduction*: the imaginary time disappears.

From Random Walks to Random Matrices. Jean Zinn-Justin, Oxford University Press (2019).
© Jean Zinn-Justin. DOI: 10.1093/oso/9780198787754.001.0001

In a relativistic theory, finite temperature effects dominate in a limit where the temperature T, in energy units, is larger than the temperature-dependent mass m_T (with the speed of light $c = 1$), which characterizes the decay of correlation functions in space directions: $T \gg m_T$.

In this limit, one expects that the statistical properties of finite temperature QFT in $(1, d)$ dimensions can be described by an effective classical statistical field theory in d dimensions, a process called *dimensional reduction*.

We illustrate these ideas with a few simple examples. In the case of the $(\phi^2)^2$ scalar field theory, we construct explicitly the corresponding effective reduced theory at one-loop order, using the technique of mode expansion of fields in the imaginary time variable. We study the non-linear σ-model and the Gross–Neveu model, in the large N limit, where N is the number of field components. Indeed, the large N expansion provides another especially convenient tool to study dimensional reduction.

Finally, we briefly discuss Abelian gauge theories, whose physical realization is quantum electrodynamics (QED), because the relation between the number of vector field components and space-time dimensions affects dimensional reduction.

New technical difficulties arising when the gauge group is non-Abelian will not be examined here (see Chapter 38 in Ref. [18] for details).

21.1 Finite temperature QFT: General considerations

To understand some static properties of finite temperature QFT at thermal equilibrium, we study the partition function $\mathcal{Z} = \operatorname{tr} e^{-H/T}$, where H is the Hamiltonian of the QFT, and T the temperature. For a simple relativistic theory in d space dimensions (and one time dimension) with scalar *boson fields* ϕ and Euclidean (*i.e.*, imaginary time) local action $\mathcal{S}(\phi)$, the quantum partition function is given by a field integral of the form

$$\mathcal{Z} = \int [\mathrm{d}\phi] \exp\left[-\mathcal{S}(\phi)\right], \qquad (21.1)$$

where $\mathcal{S}(\phi)$ is the integral of the Euclidean Lagrangian density,

$$\mathcal{S}(\phi) = \int_0^{1/T} \mathrm{d}\tau \int \mathrm{d}^d x \, \mathcal{L}(\phi),$$

and the field ϕ satisfies *periodic boundary conditions* in the (imaginary) time direction

$$\phi(\tau = 0, x) = \phi(\tau = 1/T, x).$$

By contrast, *fermion fields $\psi(\tau, x)$ satisfy anti-periodic boundary conditions*,

$$\psi(\tau = 0, x) = -\psi(\tau = 1/T, x),$$

a difference very relevant at high temperature.

Regularization. QFT, beyond its formal construction, requires some regularization to be defined as a consequence of unavoidable large momentum or ultraviolet (UV) divergences or, equivalently, short distance singularities, a process called *regularization*. Therefore, we always assume a large momentum cut-off Λ, where Λ^{-1} is a representative of the initial microscopic scale at which the initial theory is defined. For example, in simple theories, like the ϕ^4 field theory, for a scalar propagator we substitute

$$\frac{1}{k^2 + m^2} \mapsto \frac{1}{(k^2 + m^2)(1 + k^2/\Lambda^2)^n},$$

where the integer n is chosen large enough to render all terms in the perturbative expansion finite (but, for non-linear or gauge symmetries, the regularization is more complicated).

The cut-off Λ is much larger than all physical momenta, masses and temperature T, even in the high temperature regime.

For practical calculations, one may also use *dimensional regularization*, a regularization that also performs a partial renormalization.

21.1.1 Classical statistical field theory and RG

The quantum partition function (21.1) can also be interpreted as the partition function of a *classical statistical field theory* in $(d+1)$ dimensions with a finite size $L = 1/T$ in one direction. Correlation functions thus satisfy the RG equations of the corresponding $(d+1)$-dimensional theory.

General results obtained in the study of finite size effects (see also Chapter 19) also apply here. RG equations are only sensitive to short distance singularities and, therefore, finite size effects do not modify RG equations. They affect only the solutions of RG equations, because a new, dimensionless, RG invariant, argument appears, which can be written as the ratio $\xi_L/L \equiv T/m_T$ where the correlation length $\xi_L = 1/m_T$ characterizes the decay of correlation functions in space directions, and m_T is the finite temperature mass.

For L finite ($T > 0$), one expects a crossover from a $(d+1)$-dimensional behaviour when the finite size correlation length ξ_L is small compared to L, here $m_T \gg T$, to a d-dimensional behaviour when ξ_L is large compared to L, here $m_T \ll T$. The latter high temperature regime can be described by an effective d-dimensional theory. Then, the zero-temperature mass m is also small compared to the temperature.

21.1.2 Mode expansion and dimensional reduction

As a consequence of the thermal boundary conditions, fields can be expanded in Fourier modes in the time direction, and the corresponding frequencies are quantized. For boson fields,

$$\phi(\tau, x) = \sum_{\nu \in \mathbb{Z}} e^{i\omega_\nu \tau} \phi_\nu(x), \quad \omega_n = 2\nu\pi T, \tag{21.2}$$

and, for fermion fields,

$$\psi(\tau, x) = \sum_{\nu \in \mathbb{Z}} e^{i\omega_\nu \tau} \psi_\nu(x), \quad \bar{\psi}(\tau, x) = \sum_{\nu \in \mathbb{Z}} e^{-i\omega_\nu \tau} \bar{\psi}_\nu(x), \text{ with } \omega_\nu = (2\nu + 1)\pi T. \tag{21.3}$$

In the limit where the temperature-dependent mass m_T is small compared to the temperature T and for momenta much smaller than the temperature, in the case of bosons, one expects to be able to single out the zero mode while treating all non-zero modes perturbatively: the perturbative integration over the non-zero modes then leads to an *effective field theory* for the zero mode, with a temperature-dependent d-dimensional action, providing an example of *dimensional reduction*.

In this limit, since fermions have no zero mode, one expects to be able to integrate them out.

Mode expansion: A remark. The mode expansions (equations (21.2)and (21.3)) are well-suited to simple situations where the field belongs to a linear space. In the case of non-linear σ-models or gauge theories, the separation of the zero mode is technically more complicated.

21.2 Scalar field theory: Effective theory for the zero mode

In the example of a scalar field theory, we decompose the field ϕ into a sum,

$$\phi(\tau, x) = \varphi(x) + \chi(\tau, x), \tag{21.4}$$

where φ is the zero mode, and χ the sum of all other modes (equation (21.2)),

$$\chi(\tau, x) = \sum_{\nu \in \mathbb{Z}, \neq 0} e^{i\omega_\nu \tau} \phi_\nu(x) \ \Rightarrow \ \int_0^{1/T} d\tau \, \chi(\tau, x) = 0. \tag{21.5}$$

After integration over the χ-field, one can define an effective d-dimensional action $\mathcal{S}_T$ for the zero mode by

$$e^{-\mathcal{S}_T(\varphi)} = \int [d\chi] \exp[-\mathcal{S}(\varphi + \chi)]. \tag{21.6}$$

The reduced action $\mathcal{S}_T$ has the form of a vacuum amplitude in a background field.

Perturbative calculation. The field integral over χ can be evaluated by the steepest descent method. The saddle point can be found be expanding the action $\mathcal{S}(\varphi + \chi)$ in powers of χ:

$$\mathcal{S}(\varphi + \chi) = \mathcal{S}(\varphi) + \int d\tau \, d^d x \, \chi(\tau, x) \, \mathcal{S}^{(1)}(\varphi; \tau, x)$$

$$+ \frac{1}{2!} \int d^d x \, d^d x' d\tau \, d\tau' \, \chi(\tau, x) \chi(\tau', x') \, \mathcal{S}^{(2)}(\varphi; \tau, x; \tau', x') + O(\chi^3), \tag{21.7}$$

with, using the functional differentiation formalism,

$$\mathcal{S}^{(1)}(\varphi; \tau, x) = \left. \frac{\delta \mathcal{S}(\phi)}{\delta \phi(\tau, x)} \right|_{\phi = \varphi},$$

$$\mathcal{S}^{(2)}(\varphi; \tau, x; \tau', x') = \left. \frac{\delta^2 \mathcal{S}(\phi)}{\delta \phi(\tau, x) \delta \phi(\tau', x')} \right|_{\phi = \varphi}. \tag{21.8}$$

Due to locality, after the substitution $\phi = \varphi$, the coefficients of the χ expansion in time reduce to δ-functions (or time derivatives of δ-function). In particular, the contribution linear in χ that contains no zero mode, after time integration vanishes.

As a consequence, the saddle point corresponds to $\chi = 0$.

21.2.1 *Effective reduced action: Leading order*

At leading order, the effective φ action then becomes

$$\mathcal{S}_T(\varphi) = \frac{1}{T} \int \mathrm{d}^d x\, \mathcal{L}(\varphi). \tag{21.9}$$

Here $T = 1/L$ plays, in this leading approximation, the formal role of $\hbar$, and the small T, large L, expansion corresponds to a *loop expansion* of the φ field integral.

The temperature T is small with respect to Λ. If T/Λ is the relevant expansion parameter, which means that the perturbative expansion is dominated by UV contributions, then the effective d-dimensional theory can still be studied with perturbation theory.

This is expected when the number of space dimensions is large enough and field theories are not renormalizable.

However, another dimensionless ratio, m_T/T, can be found, which is small near a phase transition. This may be the relevant expansion parameter for theories which are dominated by small momentum (infrared (IR)) contributions, a problem which arises in low dimensions d, in particular, for the dimension $d = 3$, which is relevant for the theory of fundamental interactions at the microscopic scale. Then, plain perturbation theory is no longer possible or useful.

21.2.2 *One-loop correction to the effective action*

Taking now into account the quadratic term in the expansion (21.7) and integrating over χ, one obtains (using $\mathrm{tr}\,\ln = \ln\det$),

$$\int [\mathrm{d}\chi] \exp\left[-\frac{1}{2!} \int \mathrm{d}^d x\, \mathrm{d}^d x'\mathrm{d}\tau\, \mathrm{d}\tau'\, \chi(\tau, x)\chi(\tau', x')\, \mathcal{S}^{(2)}(\varphi; \tau, x; \tau', x')\right]$$
$$\propto \exp\left[-\tfrac{1}{2} \operatorname{tr}\ln \mathcal{S}^{(2)}(\varphi)\right],$$

where $\mathcal{S}^{(2)}$ is the operator defined by its kernel in equation (21.8), and the trace has to be understood in the sense of operators associated to kernels: for an operator $\mathbf{K}$ to which is associated a kernel $K(s, t)$, $s, t \in \mathbb{R}^{d+1}$, the trace is defined by

$$\operatorname{tr} \mathbf{K} = \int \mathrm{d}^{d+1} s\, K(s, s).$$

The Gaussian integration over χ generates a one-loop correction to the effective action, which becomes

$$\mathcal{S}_T(\varphi) = \frac{1}{T} \int \mathrm{d}^d x\, \mathcal{L}(\varphi) + \tfrac{1}{2} \operatorname{tr}\ln \mathcal{S}^{(2)}(\varphi) + O(\text{two loops}). \tag{21.10}$$

21.2.3 Reduced action at higher orders

More generally, in terms of Feynman diagrams, $\mathcal{S}_T$ is a sum of connected diagrams where internal lines propagate only non-zero modes and external lines contain only zero modes.

Frequency conservation then implies that one-line reducible diagrams vanish. This leads to the identity

$$\mathcal{S}_T(\varphi) = \frac{1}{T}\Gamma(\phi = \varphi), \tag{21.11}$$

where $\Gamma(\phi)$ is the generating functional of vertex or one-particle irreducible functions but calculated with diagrams where the zero mode is omitted. For ϕ constant, it reduces to the thermodynamic potential.

The remaining modes, which appear in propagators, from the d-dimensional field theory viewpoint correspond to massive particles with masses proportional to T, which, thus, acts as an IR cut-off. Therefore, for small fields φ, the effective action $\mathcal{S}_T$ has a *local expansion* in powers of $\varphi(x)$ and its derivatives.

Beyond perturbation theory, the existence of a local expansion relies on the condition $m_T \ll T$

One expects, but this has to be checked carefully, that, in general, loop corrections coming from the integration over massive modes generate contributions that renormalize the terms already present at leading order in the effective action and additional interactions suppressed by powers of T. The resulting effective action contains all possible local interactions consistent with symmetries.

Exceptions are provided by gauge theories where new low dimensional interactions are generated by the breaking of $O(1, d)$ invariance.

21.2.4 Renormalization

If the initial $(1, d)$-dimensional theory has been renormalized, the complete theory is finite in the formal infinite cut-off limit. However, as a consequence of the zero-mode subtraction, cut-off dependent terms may remain in the reduced d-dimensional action. These terms provide the necessary counter-terms which render the perturbative expansion of the effective field theory finite. The effective can thus be written as

$$\mathcal{S}_T(\varphi) = \mathcal{S}_T^{(0)}(\varphi) + \text{ counter-terms}.$$

Correlation functions have finite expressions in terms of the parameters of the effective action, in which the counter-terms have been omitted. The first part $\mathcal{S}_T^{(0)}(\varphi)$ thus satisfies the RG equations of the $(d + 1)$-dimensional theory.

Finally, the local expansion breaks down at momenta of order T. Actually, the temperature T plays the role of an intermediate cut-off. Determining the finite parts may involve some careful calculations.

21.3 The $(\phi^2)^2_{1,d}$ scalar QFT: Phase transitions

We first examine the explicit example of the simple $(\phi^2)^2_{1,d}$ scalar field theory at a finite temperature T. The scalar field ϕ is an N-component vector that satisfies periodic boundary conditions in the Euclidean time direction, $\phi(1/T, x) = \phi(0, x)$. The Euclidean action is $O(N)$ symmetric and reads,

$$\mathcal{S}(\phi) = \int \mathrm{d}\tau \, \mathrm{d}^d x \left[\tfrac{1}{2}\left(\partial_\tau \phi(\tau, x)\right)^2 + \tfrac{1}{2}\left(\nabla_x \phi(\tau, x)\right)^2 \right.$$
$$\left. + \tfrac{1}{2} r \phi^2(\tau, x) + \frac{u}{4!}\left(\phi^2(\tau, x)\right)^2 \right], \tag{21.12}$$

with $0 \le \tau \le 1/T$.

As usual, a *large momentum cut-off* Λ is implied, to render the field theory UV finite. It represents the momentum scale where the QFT as a physical theory breaks down.

At zero temperature, the quantum model is equivalent to a $(d+1)$-dimensional classical statistical field theory, a theory that has been systematically studied since, for $d = 2, 1$, it describes the large scale universal properties of a wide class of continuous macroscopic phase transitions.

Since a QFT is meaningful only if *the physical mass m is much smaller than the cut-off* Λ, this implies that the parameter r must be close to a transition point $r_c(N, u)$, $|r - r_c| \ll \Lambda^2$, where the physical mass vanishes. This is the famous *fine tuning problem*.

At finite temperature, the QFT has the properties of a classical $(d+1)$-dimensional classical field theory with a *finite size* $L = 1/T$ in the time direction.

21.3.1 Phase transitions at zero temperature

Although we are mainly interested in the space dimension $d = 3$, we summarize here the properties of phase transitions at zero temperature, the infinite size limit of the corresponding classical theory in $(d+1)$ dimensions, for several dimensions.

Below dimension $(d+1) = 4$, the large distance behaviour of physical quantities, or their singularities at the transition point r_c, are non-perturbative and their study requires RG methods.

Dimension $d = 1$. For $d = 1$, the field theory for $N = 1$ has an Ising-like phase transition at a critical value $r = r_c$ where the mass vanishes (the correlation length diverges).

For $N = 2$, the field theory has the peculiar Kosterlitz–Thouless phase transition without ordering and, for $N > 2$, no phase transition is possible.

Dimension $d \ge 2$. For all N, the field theory has a phase transition at a special value $r = r_c$ where the mass vanishes. For $N > 1$, $r < r_c$ corresponds to a spontaneously broken symmetry with massless Goldstone bosons. This ordered phase will be examined in the large N limit in Section 21.10, using the more suitable formalism of the non-linear σ-model.

21.3.2 Phase transitions at finite temperature

When the temperature-dependent mass m_T is much smaller than the temperature, or, in the corresponding classical theory, the correlation length $\xi_L \equiv 1/m_T$ is large compared to the finite size, such a theory, from the viewpoint of phase transitions, behaves like a d-dimensional classical theory.

For $d = 1$, no phase transition is possible and the correlation length is bounded.

For $d = 2$, for $N = 1$, an Ising-like phase transition and, for $N = 2$, a *Kosterlitz–Thouless* phase transition without ordering are possible. No phase transition is possible for $N > 2$.

In dimensions $d > 2$, in particular, in the physically relevant dimension $d = 3$, a phase transition is always possible.

In all cases, one expects that such *a finite temperature phase transition can occur only when the system at zero temperature is in the broken (ordered) phase*, since the temperature increases disorder.

More detailed results require an RG analysis with finite size effects and the construction of an effective d-dimensional classical field theory.

21.4 Temperature effects: The temperature-dependent mass

A first and important issue is the behaviour of the temperature-dependent mass m_T (the inverse correlation length).

At leading order (tree approximation), in the symmetric phase, the vertex two-point function (the inverse of the two-point correlation function) in the Fourier representation reads

$$\tilde{\Gamma}^{(2)}(p, \omega_\nu) = p^2 + \omega_\nu^2 + r \,,$$

where the zero-temperature physical mass is $m = \sqrt{r}$. At finite temperature, when the mass m is much smaller than the temperature, the large distance behaviour of the correlation function is governed by the zero-mode contribution. Thus, at this order, $m_T = m$.

However, m_T is renormalized by interactions. If m_T becomes of the order of the temperature, the zero mode is no longer different from other modes. The inverse temperature provides an IR cut-off, eliminating possible IR problems, and one expects to be able to calculate using standard perturbation theory. Actually, one should be able to rearrange the $(d + 1)$ perturbation theory to treat equally all modes.

By contrast, if the temperature-dependent mass remains much smaller than the temperature, perturbation theory may be invalidated by IR contributions. However, one can then integrate over the non-zero modes to generate an effective d-dimensional theory, perform a *local expansion* of the effective action and investigate the large distance properties using RG equations of the d-dimensional theory.

21.5 Phase structure at finite temperature at one loop

To gain some intuition about the phase structure at finite temperature, we first examine the thermodynamic potential density (the one-particle irreducible generating functional for constant fields), the Legendre transform of the free energy and then the two-point function, at one-loop order.

21.5.1 Thermodynamic potential density at one loop

The thermodynamic potential density is given by (see also equation (21.10)),

$$
\begin{aligned}
G(\phi) \equiv \frac{\Gamma(\phi)}{\text{volume}} &= \tfrac{1}{2}r\phi^2 + \frac{1}{4!}u(\phi^2)^2 \\
&\quad + \tfrac{1}{2}\,\mathrm{tr}\ln\left[(-\partial_\tau^2 - \nabla_x^2 + r + \tfrac{1}{6}u\phi^2)\delta_{ij} + \tfrac{1}{3}u\phi_i\phi_j\right] - (\phi = 0) \\
&= \tfrac{1}{2}r\phi^2 + \frac{1}{4!}u(\phi^2)^2 + \tfrac{1}{2}(N-1)\,\mathrm{tr}\ln\left(-\partial_\tau^2 - \nabla_x^2 + r + \tfrac{1}{6}u\phi^2\right) \\
&\quad + \tfrac{1}{2}\,\mathrm{tr}\ln\left(-\partial_\tau^2 - \nabla_x^2 + r + \tfrac{1}{2}u\phi^2\right) - (\phi = 0). \qquad (21.13)
\end{aligned}
$$

The field expectation value is a solution of $\partial G/\partial\phi_i = 0$. This equation factorizes into two equations $\phi = 0$, which yields the symmetric solution, and ($\phi = |\phi|$):

$$
\begin{aligned}
0 &= r + \tfrac{1}{6}u\phi^2 + \tfrac{1}{6}(N-1)uT\int \frac{\mathrm{d}^d k}{(2\pi)^d}\sum_\nu \frac{1}{k^2 + \omega_\nu^2 + r + \tfrac{1}{6}u\phi^2} \\
&\quad + \tfrac{1}{2}uT\int \frac{\mathrm{d}^d k}{(2\pi)^d}\sum_\nu \frac{1}{k^2 + \omega_\nu^2 + r + \tfrac{1}{2}u\phi^2}, \qquad (21.14)
\end{aligned}
$$

with $\omega_\nu = 2\pi\nu T$.

Applying the identity $(A21.12)$ to equation (21.14), one obtains

$$
\begin{aligned}
0 &= r + \tfrac{1}{6}u\phi^2 + \tfrac{1}{6}(N-1)\frac{u}{(2\pi)^d}\int \frac{\mathrm{d}^d k}{\omega_{\mathrm{T}}(k)}\left(\frac{1}{2} + \frac{1}{e^{\omega_{\mathrm{T}}(k)/T} - 1}\right) \\
&\quad + \tfrac{1}{2}\frac{u}{(2\pi)^d}\int \frac{\mathrm{d}^d k}{\omega_{\mathrm{L}}(k)}\left(\frac{1}{2} + \frac{1}{e^{\omega_{\mathrm{L}}(k)/T} - 1}\right), \qquad (21.15)
\end{aligned}
$$

with

$$
\omega_{\mathrm{T}}(k) = \sqrt{k^2 + r + \tfrac{1}{6}u\phi^2}, \qquad \omega_{\mathrm{L}}(k) = \sqrt{k^2 + r + \tfrac{1}{2}u\phi^2}.
$$

The one-loop correction is explicitly the sum of a quantum and a thermal correction. At zero temperature, the expectation value vanishes at the transition point $r = r_c$ given by

$$
r_c = -\tfrac{1}{6}(N+2)u\frac{1}{(2\pi)^d}\int \frac{\mathrm{d}^d k}{2|k|} + O(u^2),
$$

where the integral is implicitly regularized by a large momentum cut-off Λ. We set

$$
r = r_c + \rho, \qquad (21.16)
$$

where the quantity r_c has technically the form of a mass renormalization, and the deviation ρ must satisfy $|\rho| \ll \Lambda^2$.

One can then rewrite equation (21.15) as

$$0 = \rho + \tfrac{1}{6}u\phi^2 + \tfrac{1}{6}(N-1)u \int \frac{d^d k}{(2\pi)^d} \left[\frac{1}{\omega_T(k)} \left(\frac{1}{2} + \frac{1}{e^{\omega_T(k)/T} - 1} \right) - \frac{1}{2|k|} \right]$$

$$+ \tfrac{1}{2}u \int \frac{d^d k}{(2\pi)^d} \left[\frac{1}{\omega_L(k)} \left(\frac{1}{2} + \frac{1}{e^{\omega_L(k)/T} - 1} \right) - \frac{1}{2|k|} \right], \tag{21.17}$$

with

$$\omega_T(k) = \sqrt{k^2 + \rho + \tfrac{1}{6}u\phi^2}, \quad \omega_L(k) = \sqrt{k^2 + \rho + \tfrac{1}{2}u\phi^2}.$$

21.5.2 Critical temperature

A transition is possible at a finite temperature T_c if the equation

$$\rho = -\tfrac{1}{6}(N+2)u \int \frac{d^d k}{(2\pi)^d} \frac{1}{|k|} \frac{1}{e^{|k|/T_c} - 1} + O(u^2) \tag{21.18}$$

has a solution (we have neglected ρ in the right-hand side because $\rho = O(u)$).

This implies $\rho < 0$ and, thus, at zero temperature, the system must be in the broken phase. The integral is convergent at $k = 0$ only for $d > 2$, which we assume below. Since the integral is convergent at large momenta, one can rescale $k \mapsto Tk$ and the relation becomes

$$\tfrac{1}{6}(N+2)uT_c^{d-1} \int \frac{d^d k}{(2\pi)^d} \frac{1}{|k|} \frac{1}{e^{|k|} - 1} = -\rho + O(u^2).$$

If the interaction strength is of order 1 at cut-off scale, we can set

$$u = g\Lambda^{3-d}, \tag{21.19}$$

where $g = O(1)$ is dimensionless. The equation can be rewritten as

$$\tfrac{1}{6}(N+2)g(T_c/\Lambda)^{d-1} \int \frac{d^d k}{(2\pi)^d} \frac{1}{|k|} \frac{1}{e^{|k|} - 1} = (-\rho/\Lambda^2) + O(u^2),$$

which implies the scaling relation

$$T_c/\Lambda \propto (-\rho/\Lambda^2)^{1/(d-1)}.$$

This leading order scaling relation is expected to be modified by logarithms for $d = 3$.

For $d < 3$, the perturbative expansion for the massless theory is IR divergent and the result is non-perturbative.

21.5.3 Two-point function: One-loop calculation in the symmetric phase

The inverse two-point function at one loop in the Fourier representation reads

$$\tilde{\Gamma}^{(2)}(p,\omega_\nu) = p^2 + \omega_\nu^2 + r + \tfrac{1}{6}(N+2)Tu \int \frac{\mathrm{d}^d k}{(2\pi)^d} \sum_\nu \frac{1}{k^2 + \omega_\nu^2 + r} + O(u^2),$$

where a momentum cut-off Λ is implied.

We introduce the function (defined by equation ($A21.16$))

$$\mathcal{T}(T,m) \equiv \mathcal{T}_-(T,m) = \frac{1}{(2\pi)^d} \int \frac{\mathrm{d}^d k}{\omega(k)} \frac{1}{\mathrm{e}^{\omega(k)/T} - 1}, \qquad (21.20)$$

with

$$\omega(k) = \sqrt{k^2 + m^2}.$$

It satisfies,

$$\mathcal{T}(T,m) = m^{(d-1)} \mathcal{T}(T/m, 1), \qquad (21.21)$$

since the integral in equation (21.20) is UV finite.

After summation over the integer ν, the one-loop term can then be rewritten as

$$\tilde{\Gamma}^{(2)}(p,\omega_\nu) = p^2 + \omega_\nu^2 + r + \tfrac{1}{6}(N+2)u \left[\Omega_{d+1}(\sqrt{r}) + \mathcal{T}(T, \sqrt{r}) \right] + O(u^2),$$

with (Fig. 21.1)

$$\Omega_d(m) = \frac{1}{(2\pi)^d} \int \frac{\mathrm{d}^d k}{k^2 + m^2}, \qquad (21.22)$$

the divergent k-integral being regularized at large momenta, for example, by a momentum cut-off Λ.

One again recognizes the sum of the zero-temperature quantum correction and the thermal contribution.

We introduce the shift (21.16). In the symmetric phase $\rho \geq 0$, the two-point function becomes

$$\tilde{\Gamma}^{(2)}(p,\omega_\nu) = p^2 + \omega_\nu^2 + \rho$$
$$+ \tfrac{1}{6}(N+2)u \left[\Omega_{d+1}(\sqrt{\rho}) - \Omega_{d+1}(0) + \mathcal{T}(T, \sqrt{\rho}) \right] + O(u^2). \quad (21.23)$$

The physical mass at zero temperature is

$$m^2 = \rho + \tfrac{1}{12}(N+2)u \left[\Omega_{d+1}(\sqrt{\rho}) - \Omega_{d+1}(0) \right] + O(u^2).$$

The smallest temperature-dependent mass is obtained for $\nu = 0$ and given by

$$m_T^2 = m^2 + \tfrac{1}{6}(N+2)u\mathcal{T}(T,m) + O(u^2),$$

where here we have substituted $r \mapsto m^2$.

Fig. 21.1 One-loop contribution to the two-point function.

As expected, for u small at least, the temperature-dependent mass increases with the temperature. Using the scaling relation (21.21), one can rewrite the expression as

$$\frac{m_T^2}{T^2} = \frac{m^2}{T^2} + \frac{1}{6}(N+2)u\frac{m^{d-1}}{T^2}\mathcal{T}(T/m,1) + O(u^2). \qquad (21.24)$$

Since, generically, $u \propto \Lambda^{3-d}$, where Λ is the momentum cut-off, the correction is of order $(m/\Lambda)^{d-3}(m^2/T^2)$. For $d > 3$, (a situation also relevant to a $(1+4)$ theory with a compact fifth dimension), the correction is small and $m_T \ll T$ is equivalent to $m \ll T$. For $d \le 3$, the issue is non-perturbative and an RG analysis is required.

Symmetric phase at finite temperature. The expression (21.23) is only suited for the phase already symmetric at zero temperature. To describe the situation with symmetry breaking at zero temperature but symmetry restoration at higher temperature, it is necessary to introduce a temperature-dependent shift of ρ of the form (21.18). We thus set

$$\rho = \rho_T - \tfrac{1}{6}(N+2)u\mathcal{T}(T,0).$$

The two-point function can then be written as

$$\tilde{\Gamma}^{(2)}(p,\omega_\nu) = p^2 + \omega_\nu^2 + \rho_T + \tfrac{1}{6}(N+2)u\left[\Omega_{d+1}(\sqrt{\rho_T}) - \Omega_{d+1}(0)\right.$$
$$\left. + \mathcal{T}(T,\sqrt{\rho_T}) - \mathcal{T}(T,0)\right] + O(u^2). \qquad (21.25)$$

The temperature-dependent mass at one loop follows:

$$m_T^2 = \rho_T + \tfrac{1}{6}(N+2)u\left[\Omega_{d+1}(\sqrt{\rho_T}) - \Omega_{d+1}(0) + \mathcal{T}(T,\sqrt{\rho_T}) - \mathcal{T}(T,0)\right]$$

and the relevant regime for dimensional reduction is $\rho_T \ll T^2$, which implies $T - T_c$ small.

21.6 RG at finite temperature

Quite useful information can be inferred from an RG analysis. One important quantity is the ratio $m_T/T \equiv L/\xi_L$, where ξ_L is the correlation length at finite size $L = 1/T$, and m_T, therefore, is the smallest effective mass at finite temperature.

We assume a generic situation in which the coupling constant is of order 1 at cut-off scale Λ (the scale at which the underlying microscopic theory is defined).

We thus parametrize the coupling constant u as in equation (21.19), $u = g\Lambda^{3-d}$, with $g = O(1)$.

The zero-temperature theory satisfies the RG equations of a $(d+1)$-dimensional field theory in infinite volume, which for the connected n-point correlation function in Fourier representation takes the form

$$\left(\Lambda\frac{\partial}{\partial\Lambda} + \beta(g)\frac{\partial}{\partial g} + \frac{n}{2}\eta(g) - \eta_2(g)\rho\frac{\partial}{\partial\rho}\right)\widetilde{W}^{(n)}(p_i; \rho, g, T, \Lambda) = 0, \qquad (21.26)$$

where β, η, η_2 are RG functions calculable as series expansions in powers of g.

At least in a neighbourhood of dimension 3 (*i.e.*, $d+1 = 4$), one finds

$$\beta(g) = -(3-d)g + \frac{N+8}{48\pi^2}g^2 + O(g^3),$$

$$\eta(g) = O(g^2),$$

$$\eta_2(g) = -\frac{N+2}{48\pi^2}g + O(g^2).$$

The dimensionless ratio $m_T/T = R(\rho/T^2, g, \Lambda/T)$ is an RG invariant. It satisfies

$$\left(\Lambda\frac{\partial}{\partial\Lambda} + \beta(g)\frac{\partial}{\partial g} - \eta_2(g)\rho\frac{\partial}{\partial\rho}\right)R = 0.$$

The solution can be parametrized as (method of characteristics)

$$m_T/T = R(\rho/T^2 g, \Lambda/T) = R(\rho(\lambda)/T^2, g(\lambda), \lambda\Lambda/T), \qquad (21.27)$$

where λ is a scale parameter, and $g(\lambda), \rho(\lambda)$ the effective parameters at scale λ. They satisfy the differential equations

$$\lambda\frac{\mathrm{d}g(\lambda)}{\mathrm{d}\lambda} = \beta(g(\lambda)), \quad \lambda\frac{\mathrm{d}\rho(\lambda)}{\mathrm{d}\lambda} = -\rho(\lambda)\eta_2(g(\lambda)), \quad \text{with} \quad g(1) = 1, \ \rho(1) = \rho.$$

Their integration yields

$$\int_g^{g(\lambda)}\frac{\mathrm{d}g'}{\beta(g')} = \ln\lambda, \quad \rho(\lambda) = \rho\exp\left[-\int_g^{g(\lambda)}\mathrm{d}g'\frac{\eta_2(g')}{\beta(g')}\right].$$

21.6.1 *Dimension $d \leq 3$*

In the special dimension $d = 3$, the $(\phi^2)_4^2$ theory is just renormalizable. The positive sign of the RG β-function for g small implies that, in the range in which the β-function remains positive, the effective coupling constant at the physical scale $\lambda \ll 1$ is logarithmically small: $g(\lambda) = O(1/\ln\lambda)$ for $\lambda \to 0$. The field theory is called *IR free* in the sense that, from the RG viewpoint, $g = 0$ is an IR fixed point: this leads to the so-called *triviality issue*: if one insists taking the infinite cut-off limit $(\Lambda \to \infty)$, the renormalized or effective coupling vanishes.

Otherwise, the theory remains perturbative with, however, logarithmic deviations from plain perturbation theory, which can be derived by RG arguments (RG improved perturbation theory).

For example, if the temperature T is the relevant scale, we choose $\lambda = T/\Lambda \ll 1$ and thus

$$g(T/\Lambda) \sim \frac{48\pi^2}{(N+8)\ln(\Lambda/T)}. \tag{21.28}$$

One also finds

$$\rho(T/\Lambda) \propto \frac{\rho}{(\ln \Lambda/T)^{(N+2)/(N+8)}}. \tag{21.29}$$

Then,

$$m_T/T = R\big(\rho(T/\Lambda)/T^2, g(T/\Lambda), 1\big).$$

By contrast, for $m \ll T$, the relevant scale is m, and equation (21.24) becomes

$$\frac{m_T^2}{m^2} - 1 \sim \tfrac{1}{6}(N+2)g(m/\Lambda)\mathcal{T}(T/m, 1). \tag{21.30}$$

This shows that the leading correction is only logarithmically suppressed, because m/Λ is small while the coefficient grows like T^2/m^2 for $m \ll T$. Therefore, in this limit, dimensional reduction is only marginally useful. For $d = 3$, dimensional reduction is mainly useful near a finite temperature phase transition.

Dimension $d = 2$. The three-dimensional classical theory has an IR fixed point $g^* > 0$. Finite size scaling (equation (21.27)) then predicts, that, in the symmetric phase, generically, $m_T/T = O(1)$, except near a finite temperature continuous phase transition. This occurs for $N = 1$ when the system at zero temperature is in the broken phase. Higher values of N require a specific analysis.

21.7 Effective action: Perturbative calculation

We now assume that the temperature T is close to the critical temperature T_c at which a continuous phase transition occurs and where one expects dimensional reduction to be relevant.

21.7.1 Effective action at leading order

The effective field theory in d dimensions can be derived by expanding the field in eigenmodes in the time direction (equation (21.2)) and integrating perturbatively over all non-zero modes (equation (21.6)).

In the notation (21.12), the result at leading order is

$$\mathcal{S}_T^{(0)}(\varphi) = \mathcal{S}(\varphi) = \frac{1}{T}\int d^d x \left\{ \tfrac{1}{2}[\nabla\varphi(x)]^2 + \tfrac{1}{2}r\varphi^2(x) + \frac{1}{4!}u(\varphi^2(x))^2 \right\}. \tag{21.31}$$

After a field rescaling $\varphi \mapsto \varphi T^{1/2}$, the action becomes

$$\mathcal{S}_T^{(0)}(\varphi) = \int d^d x \left\{ \tfrac{1}{2}[\nabla\varphi(x)]^2 + \tfrac{1}{2}r\varphi^2(x) + \frac{1}{4!}uT(\varphi^2(x))^2 \right\}. \tag{21.32}$$

The action (21.32) generates a perturbation theory. In terms of the dimensionless bare coupling $g = u\Lambda^{d-3}$, a quantity generically of order 1, the expansion parameter is $(\Lambda/m_T)^{3-d}(T/m_T)g$. For $d = 3$, the expansion parameter reduces to gT/m_T. Since dimensional reduction is useful only for m_T/T small, even when the running coupling constant $g(T/\Lambda)$ at scale $T \ll \Lambda$, $g(T/\Lambda) = O(1/\ln(\Lambda/T))$, is small, one expects physics to be non-perturbative, being dominated by IR contributions.

For $d < 3$, IR singularities are always present in both the initial and the reduced theory, and the small coupling regime can never be reached for interesting situations.

21.7.2 Effective action: One-loop correction

At one-loop order, we set $r = r_c + \rho$, with

$$r_c = -u\frac{(N+2)}{6}\Omega_{d+1}(0) \, .$$

This leads to the addition to the one-loop contribution of

$$\frac{r_c}{2T}\int \mathrm{d}^d x \, \varphi^2(x) = -u\frac{(N+2)}{12T}\Omega_{d+1}(0)\int \mathrm{d}^d x \, \varphi^2(x) \, . \tag{21.33}$$

The one-loop contribution to the effective action, which also is the one-loop contribution to the generating functional or vertex (or one-particle-irreducible (1PI)) functions, generated by the integration over non-zero modes, takes the form (using $\ln \det = \operatorname{tr} \ln$)

$$\mathcal{S}_T^{(1)}(\varphi) = \tfrac{1}{2}\operatorname{tr}\ln\left[\left(-\partial_\tau^2 - \nabla_x^2 + r + \tfrac{1}{6}u\varphi^2(x)\right)\delta_{ij}\right.$$
$$\left.+\tfrac{1}{3}u\varphi_i(x)\varphi_j(x)\right] - (\varphi = 0). \tag{21.34}$$

At this order, r_c can be neglected and we substitute $r = \rho$.

The situation of interest here is when the temperature-dependent mass m_T (the inverse of the correlation length ξ_L) is much smaller than the temperature T. In the perturbative framework, this also implies $|\rho| \ll T^2$. For $m_T < T$, a *local expansion* in φ is then justified.

The leading order in the derivative expansion can be obtained by treating $\varphi(x)$ as a constant. The evaluation of the one-loop contribution $\mathcal{S}_T^{(1)}$ to the reduced action, in the Fourier representation at fixed momentum, can then be inferred from the calculation of the partition function of the harmonic oscillator.

In a finite large space volume V_d, one can use the identity

$$\operatorname{tr}\ln(-\partial_\tau^2 - \nabla_x^2 + M^2) - \operatorname{tr}\ln(-\partial_\tau^2)$$
$$= V_d\int\frac{\mathrm{d}^d k}{(2\pi)^d}\sum_{\nu\neq 0}\left[\ln(\omega_\nu^2 + k^2 + M^2) - \ln(\omega_\nu^2)\right]$$
$$= 2V_d\int\frac{\mathrm{d}^d k}{(2\pi)^d}\ln\left[2T\sinh\left(\omega(k)/2T\right)/\omega(k)\right], \tag{21.35}$$

where $\omega(k) = \sqrt{k^2 + M^2}$.

When this applied to the expression (21.34), one finds

$$\mathcal{S}_T^{(1)}(\varphi) = \int \mathrm{d}^d x \int \frac{\mathrm{d}^d k}{(2\pi)^d} \Big\{ (N-1) \ln \left[2T \sinh\big(\omega_\mathrm{T}(k)/2T\big)/\omega_\mathrm{T}(k) \right]$$
$$+ \ln \left[2T \sinh\big(\omega_\mathrm{L}(k)/2T\big)/\omega_\mathrm{L}(k) \right] \Big\} - (\varphi = 0), \qquad (21.36)$$

with

$$\omega_\mathrm{T}(k) = \sqrt{k^2 + \rho + \tfrac{1}{6} u \varphi^2(x)}, \quad \omega_\mathrm{L}(k) = \sqrt{k^2 + \rho + \tfrac{1}{2} u \varphi^2(x)}.$$

21.8 Effective action: φ-Expansion

At this order, an expansion in powers of φ makes sense only if $-\rho/T^2 < 4\pi^2$, a condition that constrains ρ only for $\rho < 0$ and, more generally, involves the dimensionless RG invariant ratio $m_T/T = L/\xi_L$, where m_T is the mass parameter of the ordered phase, and is consistent with our assumptions. Moreover, the expansion around $\varphi = 0$ is meaningful only if the expectation value of φ is small enough.

21.8.1 The φ^2 term

For the quadratic term, the local approximation is not needed because the corresponding one-loop diagram is a constant. One obtains

$$\left[\mathcal{S}_T^{(1)} \right]_2 = \frac{1}{12}(N+2) \mathcal{G}_2(T,\rho) \frac{u}{T} \int \mathrm{d}^d x \, \varphi^2(x),$$

with

$$\mathcal{G}_2(T,\rho) = \Omega_{d+1}(\sqrt{\rho}) - T\Omega_d(\sqrt{\rho}) + \mathcal{T}(T,\sqrt{\rho}). \qquad (21.37)$$

The expression is the sum of the zero-temperature result, the subtracted zero-mode contribution and the thermal fluctuations. One can verify that the expression is indeed regular up to $\rho + 4\pi^2 T^2 = 0$.

At one loop, it is necessary to add the mass renormalization (21.33). This partially renormalizes $\mathcal{G}_2$, and replaces it by

$$\mathcal{G}_{2,\mathrm{r}}(T,\rho) = \Omega_{d+1}(\sqrt{\rho}) - \Omega_{d+1}(0) - T\Omega_d(\sqrt{\rho}) + \mathcal{T}(T,\sqrt{\rho}). \qquad (21.38)$$

The total expression is not singular at $\rho = 0$. For $d \geq 2$, the divergent contribution $-T\Omega_d(\sqrt{\rho})$ provides a renormalization for the one-loop perturbative contribution generated by the action $\mathcal{S}_T^{(0)}$.

Order $(\varphi^2)^2$. The quartic term is proportional to the initial interaction:

$$\left[\mathcal{S}_T^{(1)} \right]_4 = -\frac{1}{144}(N+8) \mathcal{G}_4(T,\rho) \frac{u^2}{T} \int \mathrm{d}^d x \, (\varphi^2(x))^2.$$

It is possible to verify that

$$\mathcal{G}_4(T,\rho) = -\frac{\partial}{\partial \rho} \mathcal{G}_2(T,\rho). \qquad (21.39)$$

The one-loop reduced action. We first keep only the terms already present in the tree approximation. After the rescaling $\varphi \mapsto \varphi T^{1/2}$, the effective action becomes

$$\mathcal{S}_T(\varphi) = \int \mathrm{d}^d x \left\{ \frac{1}{2} [\nabla \varphi(x)]^2 + \frac{1}{2} r_1 \varphi^2(x) + \frac{1}{4!} T u_1 \big(\varphi^2(x) \big)^2 \right\}, \tag{21.40}$$

with

$$r_1 = \rho + \tfrac{1}{6}(N+2)u\mathcal{G}_{2,\mathrm{r}}, \quad u_1 = u - \tfrac{1}{6}(N+8)u^2\mathcal{G}_4.$$

Other interactions. For space dimensions $d < 6$, the coefficients of the other interaction terms are no longer UV divergent. Since the zero-mode contribution has been omitted, no IR divergence is generated, even in the massless limit. In this limit, the coefficients are thus proportional to powers of $1/T$ obtained by dimensional analysis. In the normalization (21.40),

$$[\mathcal{S}_T]_{(2n)} \propto g^n (T/\Lambda)^{n(d-3)} T^{d-n(d-2)} \int \mathrm{d}^d x \, (\varphi^2(x))^n,$$

and, therefore, the interactions are increasingly negligible at high temperature, at least for $d \geq 3$.

The local expansion of the one-loop determinant also generates monomials with derivatives. No term proportional to $(\nabla \varphi)^2$ is generated at one-loop order. All other terms with derivatives are finite for $d < 6$ and, thus, the structure of the coefficients again is determined by dimensional analysis. To $2k$ derivatives corresponds an additional factor, $1/T^{2k}$.

Finally, for $\rho \neq 0$ but $\rho/T^2 \ll 1$, one can expand in powers of ρ, and the previous arguments immediately generalize.

21.9   The $(\phi^2)^2$ field theory at finite temperature in the large N limit

Large N methods in vector field theories are well adapted to studying finite temperature QFT because, unlike perturbation theory, they make it possible to study crossovers between different space dimensions.

In Section 21.10, we discuss the $O(N)$ symmetric non-linear σ-model at finite temperature in the large N limit. We briefly describe here the $(\phi^2)^2$ field theory in the same limit in order to be able to compare a few results in both theories [47]. We focus mainly on the dimension $d = 3$. Other dimensions are studied using the formalism of the non-linear σ-model.

21.9.1   The large N limit

To determine the partition function in the large N limit, we use a method that will be useful again in the example of the GN model in Section 21.11.

We consider the partition function

$$\mathcal{Z} = \int [\mathrm{d}\phi \, \mathrm{d}\lambda] \, \mathrm{e}^{-\mathcal{S}(\phi,\lambda)}, \tag{21.41}$$

where the λ-integration runs parallel to the imaginary axis.

The action that reads ($\mu = 1, \ldots, d$)

$$\mathcal{S}(\boldsymbol{\phi}, \lambda) = \int d\tau \, d^d x \left[\tfrac{1}{2} \sum_{\mu} \left(\partial_\mu \boldsymbol{\phi}(\tau, x) \right)^2 + \tfrac{1}{2} \lambda(\tau, x) \boldsymbol{\phi}^2(\tau, x) - \frac{3}{2u} \left(\lambda(\tau, x) - r \right)^2 \right],$$

is then equivalent to the action (21.12) as it is possible to verify by performing the Gaussian integration over λ. Indeed, the integration amounts to replacing λ by the solution of the λ field equation,

$$\tfrac{1}{2} \boldsymbol{\phi}^2(\tau, x) - \frac{3}{u} \left(\lambda(\tau, x) - r \right) = 0 \,.$$

The ϕ-integration. In the form (21.41), it is simple to take the large N limit because one can integrate over $\boldsymbol{\phi}$ explicitly. Actually, it is useful to integrate only over $(N - 1)$ components of $\boldsymbol{\phi}$, which we denote by $\boldsymbol{\pi}$, denoting by σ the remaining component.

For one ϕ-component, one finds,

$$\int [d\phi] \exp \left\{ -\tfrac{1}{2} \int d\tau \, d^d x \left[\left(\partial_\tau \phi(\tau, x) \right)^2 + \left(\nabla_x \phi(\tau, x) \right)^2 + \lambda(\tau, x) \phi^2(\tau, x) \right] \right\}$$

$$\propto \left[\det \left(-\partial_\tau^2 - \nabla_x^2 + \lambda(\cdot) \right) \right]^{-1/2} = \exp \left[-\tfrac{1}{2} \operatorname{tr} \ln \left(-\partial_\tau^2 - \nabla_x^2 + \lambda(\cdot) \right) \right].$$

The large N action. After integration, one obtains

$$\mathcal{Z} = \int [d\sigma d\lambda] \exp \left[-\mathcal{S}_N(\sigma, \lambda) \right], \tag{21.42}$$

with

$$\mathcal{S}_N(\sigma, \lambda) = \tfrac{1}{2} \int d\tau \, d^d x \left[\tfrac{1}{2} \left(\partial_\tau \sigma(\tau, x) \right)^2 + \tfrac{1}{2} \left(\nabla_x \sigma(\tau, x) \right)^2 + \tfrac{1}{2} \lambda(\tau, x) \sigma^2(\tau, x) \right.$$

$$\left. - \frac{3}{2u} \left(\lambda(\tau, x) - r \right)^2 \right] + \tfrac{1}{2}(N - 1) \operatorname{tr} \ln \left[-\partial_\tau^2 - \nabla_x^2 + \lambda(\cdot) \right]. \tag{21.43}$$

The large N limit is taken at Nu fixed. At leading order for N large, the action is proportional to N and the field integral can be calculated by the steepest descent method (and $(N - 1)$ can be replaced by N).

21.9.2 Zero temperature

The saddle points correspond to constant functions $\sigma(x) \equiv \bar{\sigma}$, $\lambda(x) \equiv m^2$, where m is also, at leading order, the mass of the $\boldsymbol{\pi}$ field. The saddle point equations, obtained by differentiating the action with respect to $\bar{\sigma}$ and m^2, are

$$m^2 \bar{\sigma} = 0 \,, \tag{21.44a}$$

$$\bar{\sigma}^2 = \frac{6}{u}(m^2 - r) - N \, \Omega_{d+1}(m) \,, \tag{21.44b}$$

where Ω_d is the one-loop diagram in Fig. 21.1 (see equation (21.22)).

In the broken phase $\bar{\sigma} \neq 0$ and thus $m = 0$. The $\boldsymbol{\pi}$ field is massless and corresponds to the Goldstone bosons generated by the spontaneously broken symmetry. Equation (21.44*b*) then becomes

$$\bar{\sigma}^2 = -\frac{6r}{u} - N\Omega_{d+1}(0).$$

The critical value r_c of the parameter r is determined by the condition $\bar{\sigma} = 0$:

$$r_c = -\tfrac{1}{6}uN\,\Omega_{d+1}(0).$$

For $r > r_c$, $\bar{\sigma}$ vanishes, the symmetry is no longer broken and the common mass of the σ and $\boldsymbol{\pi}$ fields is the solution of

$$m^2 - r - \tfrac{1}{6}uN\,\Omega_{d+1}(m) = 0\,.$$

Since r is of order Λ^2, the condition $m \ll \Lambda$ is only satisfied for $r - r_c$ small. The equation can be rewritten as

$$m^2 - (r - r_c) - \tfrac{1}{6}uN\left[\Omega_{d+1}(m) - \Omega_{d+1}(0)\right] = 0\,.$$

In the specific dimension $d = 3$, u is dimensionless and the equation becomes

$$m^2 - (r - r_c) + \frac{1}{48\pi^2}uNm^2\ln(\Lambda/m) + O(m^2) = 0\,.$$

The m^2 contribution can be neglected at leading order. The result is

$$m^2 \sim \frac{48\pi^2}{uN\ln(\Lambda/m)}(r - r_c) \sim \frac{96\pi^2}{uN\ln(\Lambda^2/(r - r_c))}(r - r_c), \qquad (21.45)$$

a result reflecting the IR freedom of the field theory.

21.9.3   *Finite temperature*

The saddle point equations now read $m_T^2\bar{\sigma} = 0$ and

$$\bar{\sigma}^2 = \frac{6}{u}(m_T^2 - r) - NT\int\frac{d^dk}{(2\pi)^d}\sum_{\nu}\frac{1}{m_T^2 + k^2 + (2\pi\nu T)^2}\,,$$

$$= \frac{6}{u}(m_T^2 - r) - N\left[\Omega_{d+1}(m_T) + \mathcal{T}(T, m_T)\right]\,,$$

where m_T is the temperature-dependent mass.

For example, it is possible to verify that, for $m = 0$, the symmetry is unbroken at finite temperature and $\bar{\sigma} = 0$. The gap equation then can be written as

$$m_T^2 = \tfrac{1}{6}Nu\left[\Omega_{d+1}(m_T) - \Omega_{d+1}(0) + \mathcal{T}(T, m_T)\right].$$

For $d \geq 3$, m/T small is consistent with m_T/T small. The left-hand side can again be neglected. Using the result ($A21.19$), for $d = 3$, one finds

$$m_T^2\ln(\Lambda/m_T) \sim \tfrac{2}{3}\pi^2T^2 \quad\Rightarrow\quad \left(\frac{m_T}{T}\right)^2 \sim \frac{2\pi^2}{3\ln(\Lambda/T)}\,. \qquad (21.46)$$

21.10   The non-linear σ-model at finite temperature for large N

We now discuss another, related example: the $O(N)$ symmetric non-linear σ-model (see also Section 10.4), which, for $N > 1$, is especially suited for describing the ordered phase of the $(\phi^2)^2$ field theory, because the presence of Goldstone modes introduces some specific elements in the analysis.

We recall that it has been proved, within the framework of the $1/N$ expansion, that the non-linear σ-model with a suitable regularization is equivalent to the $(\phi^2)^2$ field theory, both quantum field theories generating two different perturbative expansions of the same physical model [86, 47].

In perturbation theory, due the non-linear character of the group representation, one is confronted with difficulties that also appear in non-Abelian gauge theories. At finite temperature, the separation of the zero mode is also more complicated.

In the large N limit, the problems related to the non-linear character of the group representation are avoided [86, 198]. Therefore, we discuss the finite temperature properties in this framework.

21.10.1   The $O(N)$ symmetric non-linear σ-model

The $O(N)$ symmetric non-linear σ-model can be described as a field theory for an N-component vector field $\mathbf{S}$ belonging to a sphere and thus satisfying the constraint,

$$\mathbf{S}^2(x) = 1 \,. \tag{21.47}$$

In d space dimensions, the action takes the form

$$\mathcal{S}(\mathbf{S}) = \frac{1}{2G} \int \mathrm{d}^{d+1}x \left(\nabla_x \mathbf{S}(x)\right)^2 . \tag{21.48}$$

The QFT has also the interpretation of the formal continuum limit of a classical spin model in $(d+1)$ dimensions (see Chapter 10)

To study the large N limit, it is convenient to enforce the constraint (21.47) by a functional δ-function in a Fourier representation (Section 10.11). The partition function of the non-linear σ-model can then be written as

$$\mathcal{Z} = \int [\mathrm{d}\mathbf{S}(x)\mathrm{d}\lambda(x)] \exp\left[-\mathcal{S}(\mathbf{S}, \lambda)\right] , \tag{21.49}$$

with

$$\mathcal{S}(\mathbf{S}, \lambda) = \frac{1}{2G} \int \mathrm{d}^{d+1}x \left[\left(\nabla_x \mathbf{S}(x)\right)^2 + \lambda(x)\left(\mathbf{S}^2(x) - 1\right)\right] , \tag{21.50}$$

where the λ-integration runs along the imaginary axis.

The parameter G is the coupling constant of the quantum model as well as the temperature of the corresponding classical theory in $(d+1)$ dimensions. It can be parametrized as

$$G = \Lambda^{1-d} g \,, \tag{21.51}$$

where Λ is the cut-off, representative of the inverse microscopic scale, and g a number generically of order 1.

The model is renormalizable in the $d = 1$ dimension, where G is dimensionless. The sign of the RG β-function for G small and $N > 2$ is negative, implying that the model is *asymptotically free* for large momenta like non-Abelian gauge theories in $(1 + 3)$ dimensions.

21.10.2   The large N limit at zero temperature

We first recall how the σ-model can be solved for N large at zero temperature.

In the representation (21.50), the action is quadratic in the field $\mathbf{S}$ and thus the Gaussian integral over the field can be performed.

For one $\mathbf{S}$ component ($\operatorname{tr} \ln = \ln \det$),

$$
\int [\mathrm{d}S] \exp\left\{-\frac{1}{2G} \int \mathrm{d}^{d+1}x \left[(\nabla_x S(x))^2 + \lambda(x) S^2(x) \right]\right\}
$$

$$
\propto \left[\det\left(-\nabla_x^2 + \lambda(.)\right)\right]^{-1/2} = \exp\left[-\tfrac{1}{2} \operatorname{tr} \ln\left(-\nabla_x^2 + \lambda(\cdot)\right)\right]. \tag{21.52}
$$

Again, it is useful to integrate only over $(N-1)$ components of $\mathbf{S}$, which we denote by $\boldsymbol{\pi}$, denoting by σ the remaining component.

After integration one obtains,

$$
\mathcal{Z} = \int [\mathrm{d}\sigma(x)\mathrm{d}\lambda(x)] \exp\left[-\mathcal{S}_N(\sigma, \lambda)\right], \tag{21.53}
$$

with

$$
\mathcal{S}_N(\sigma, \lambda) = \frac{1}{2G} \int \mathrm{d}^{d+1}x \left[(\nabla_x \sigma(x))^2 + (\sigma^2(x) - 1)\lambda(x) \right]
$$

$$
+ \tfrac{1}{2}(N-1) \operatorname{tr} \ln\left[-\nabla_x^2 + \lambda(\cdot)\right]. \tag{21.54}
$$

The large N limit is taken at NG fixed. At leading order, one can replace $(N-1)$ by N. Therefore, the action is proportional to N and, for N large, the field integral can be calculated by the steepest descent method.

The saddle points correspond to constant functions $\sigma(x) \equiv \bar{\sigma}$, $\lambda(x) \equiv m^2$, where m at leading order is also the $\boldsymbol{\pi}$-field mass. The saddle point equations, obtained by differentiating the action with respect to $\bar{\sigma}$ and m^2, are

$$
m^2 \bar{\sigma} = 0 , \tag{21.55a}
$$

$$
\bar{\sigma}^2 = 1 - NG\,\Omega_{d+1}(m) , \tag{21.55b}
$$

where Ω_{d+1} is the cut-off dependent one-loop diagram defined by equation (21.22).

For $d > 1$ and G small, the field expectation value $\bar{\sigma}$ is different from zero, the $O(N)$ symmetry is spontaneously broken and thus m, the mass of the $\boldsymbol{\pi}$-field, vanishes. The $\boldsymbol{\pi}$-field components are the Goldstone bosons induced by the spontaneous symmetry breaking of the continuous $O(N)$ symmetry.

Equation (21.55b) has a solution until $\bar{\sigma}$ vanishes. This determines a critical coupling G_c given by

$$
\frac{1}{NG_c} = \Omega_{d+1}(0) \quad \Leftrightarrow \quad \frac{1}{Ng_c} = \Lambda^{1-d}\Omega_{d+1}(0). \tag{21.56}
$$

The quantity g_c has a finite, but non-universal, limit for $\Lambda \to \infty$.

The field expectation value can be written as

$$\bar{\sigma}^2 = 1 - G/G_c. \tag{21.57}$$

Above G_c, $\bar{\sigma}$ vanishes, the symmetry is unbroken and m, which is now the common mass of the π- and σ-fields, is given by the gap equation

$$\Omega_{d+1}(m) = \frac{1}{NG} \quad \Rightarrow \quad \Omega_{d+1}(0) - \Omega_{d+1}(m) = \frac{1}{N}\left(\frac{1}{G_c} - \frac{1}{G}\right). \tag{21.58}$$

The physical mass m is much smaller than the cut-off Λ only for $G - G_c$ small (the usual *fine tuning*). Depending on the space dimension, one infers the following,
(i) for $d > 3$,

$$m \propto \Lambda^{(d+1)/2}\sqrt{G - G_c} \Leftrightarrow m \propto \Lambda\sqrt{g - g_c}, \tag{21.59}$$

a mean field result from the viewpoint of phase transitions in the corresponding classical model;
(ii) for $d = 3$,

$$\frac{1}{G_c} - \frac{1}{G} \sim \frac{N}{8\pi^2}m^2\ln(\Lambda/m) \Leftrightarrow \frac{1}{g} - \frac{1}{g_c} \sim \frac{N}{8\pi^2}\frac{m^2}{\Lambda^2}\ln(\Lambda/m), \tag{21.60}$$

a result that is related to the IR freedom of the equivalent $(\phi^2)^2$ field theory (see equation (21.45));
(iii) for $d = 2$,

$$m = \frac{4\pi}{N}\left(\frac{1}{G_c} - \frac{1}{G}\right) = \frac{4\pi}{N}\left(\frac{1}{g_c} - \frac{1}{g}\right)\Lambda; \text{ and} \tag{21.61}$$

(iv) for $d = 1$,

$$m \propto \Lambda\, \mathrm{e}^{-2\pi/NG} \quad \Rightarrow \quad G \sim \frac{2\pi}{N}\frac{1}{\ln(\Lambda/m)}, \tag{21.62}$$

a result consistent with the behaviour of the RG β-function for small coupling:

$$\beta(G) = -\frac{N-2}{2\pi}G^2 + O(G^3).$$

The physical domain then corresponds to G logarithmically small.

21.10.3 The σ two-point function

Differentiating the action (21.54) twice with respect to σ, λ, and using the saddle point equations, one obtains the inverse connected σ, λ two-point function, in the Fourier representation, as a 2×2 matrix,

$$\boldsymbol{\Sigma}(p) \equiv \begin{bmatrix} [\mathcal{S}_N]_{\sigma\sigma}(p) & [\mathcal{S}_N]_{\sigma\lambda}(p) \\ [\mathcal{S}_N]_{\sigma\lambda}(p) & [\mathcal{S}_N]_{\lambda\lambda}(p) \end{bmatrix} = \frac{1}{G}\begin{bmatrix} p^2 + m^2 & \bar{\sigma} \\ \bar{\sigma} & -\frac{1}{2}NGB_{d+1}(p) \end{bmatrix}, \tag{21.63}$$

with (Fig. 21.2)

$$B_{d+1}(p) = \frac{1}{(2\pi)^{d+1}}\int \frac{\mathrm{d}^{d+1}k}{[(k+p)^2 + m^2](k^2 + m^2)}. \tag{21.64}$$

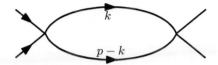

Fig. 21.2 The four-point function: the one-loop diagram $B_d(p)$.

Inverting the matrix $\boldsymbol{\Sigma}$, one obtains the two-point functions or propagators,

$$\begin{bmatrix} \tilde{\Delta}_{\sigma\sigma}(p) & \tilde{\Delta}_{\sigma\lambda}(p) \\ \tilde{\Delta}_{\sigma\lambda}(p) & \tilde{\Delta}_{\lambda\lambda}(p) \end{bmatrix} = \frac{G}{\frac{1}{2}NG(p^2+m^2)B_{d+1}(p)+\bar{\sigma}^2} \begin{bmatrix} \frac{1}{2}NGB_{d+1}(p) & \bar{\sigma} \\ \bar{\sigma} & -p^2-m^2 \end{bmatrix}.$$

In the symmetric phase, $\bar{\sigma} = 0$ and the two fields decouple. The σ propagator is then

$$\tilde{\Delta}_{\sigma\sigma}(p) = \frac{G}{p^2+m^2}.$$

Returning to the expression (21.50), one notes that all field components have the same mass m as expected.

In the broken phase, $m = 0$ and the mass of the π field (the Goldstone boson) vanishes. The possible additional spectrum is given by the zeros of the determinant of $\boldsymbol{\Sigma}$,

$$\det \boldsymbol{\Sigma} = -G\left[\tfrac{1}{2}NGp^2 B_{d+1}(p) + \bar{\sigma}^2\right],$$

where

$$\bar{\sigma}^2 = 1 - NG\Omega_{d+1}(0).$$

For $d = 2$,

$$B_3(p)|_{m=0} = \frac{1}{(2\pi)^3} \int \frac{\mathrm{d}^3 k}{k^2(k+p)^2} = \frac{1}{8|p|}$$

and, for $d = 3$,

$$B_4(p)|_{m=0} \sim \frac{1}{8\pi^2}\ln(\Lambda/|p|).$$

The equation has no solution for $p^2 \geq 0$ because $\det \boldsymbol{\Sigma}$ is strictly negative and, for $p^2 < 0$, because $B_d(p)$ is then complex. For $\bar{\sigma}$ small, one solution corresponds to a complex resonance pole, which, for $\bar{\sigma} \to 0$, approaches the real axis and is the precursor of the additional σ-component.

21.10.4 *The large N limit at finite temperature: The gap equations*

The saddle point equations (21.55a), (21.55b) become

$$m_T^2 \bar{\sigma}_T = 0 \tag{21.65a}$$

$$\bar{\sigma}_T^2 = 1 - NGT \int \frac{\mathrm{d}^d k}{(2\pi)^d} \sum_\nu \frac{1}{k^2 + \omega_\nu^2 + m_T^2}, \quad \text{with } \omega_\nu = 2\pi\nu T, \tag{21.65b}$$

where $\bar{\sigma}_T$ and m_T^2 are the expectation values of the fields σ and λ, respectively.

Equation (21.65*b*) can be rewritten as (equation (*A*21.13))

$$\bar{\sigma}_T^2 = 1 - NG\left[\Omega_{d+1}(m_T) + \mathcal{T}(T, m_T)\right],\tag{21.66}$$

where (see definitions (*A*21.16))

$$\mathcal{T}(T, m_T) \equiv \mathcal{T}_-(T, m_T)$$
$$= \frac{1}{(2\pi)^d} \int \frac{\mathrm{d}^d k}{\omega(k)} \frac{1}{\mathrm{e}^{\omega(k)/T} - 1}, \quad \text{with } \omega(k) = \sqrt{k^2 + m_T^2}, \tag{21.67}$$

is the boson thermal correction.

21.10.5 The symmetric phase

In the symmetric phase, $\bar{\sigma}_T$ vanishes and the saddle point or gap equation (21.65*b*) reduces to

$$\frac{1}{NG} = \Omega_{d+1}(m_T) + \mathcal{T}(T, m_T),\tag{21.68}$$

where m_T is the temperature-dependent mass that governs the decay of correlation functions in space directions.

A phase transition is possible only if the right-hand side remains finite for $m_T \to 0$. As one sees in equation (21.65*b*), when d decreases, IR divergences come first from the contribution of the zero mode: since the k-integral is d-dimensional, a phase transition thus is possible only for $d > 2$. This is an example of *dimensional reduction*, $(d+1) \mapsto d$.

From the point of view of perturbation theory, a crossover between different dimensions leads to technical difficulties, because IR divergences are more severe in lower dimensions. By contrast, as emphasized before, the large N expansion is particularly well-suited for handling the problem, because the large N limit can be derived explicitly in any dimension.

When the system is in a symmetric phase at zero temperature, one can introduce the zero-temperature mass (equation (21.58)) and rewrite equation (21.68) as

$$\mathcal{T}(T, m_T) = \Omega_{d+1}(m) - \Omega_{d+1}(m_T),\tag{21.69}$$

an equation that implies $m_T > m$, is UV finite for $d < 3$ and depends logarithmically on the cut-off for $d = 3$.

Dimension $d = 1$. In dimension $d = 1$, even at zero temperature, a phase transition is impossible: the symmetry remains unbroken and the mass is given by equation (21.62).

For $m_T \ll T$, the integral in equation (21.69) is dominated by the zero-mode contribution. One finds

$$\frac{m_T}{T} \sim \frac{\pi}{\ln(m_T/m)}.\tag{21.70}$$

Since we have assumed $m_T/T \ll 1$, this implies $m_T \gg m$. The logarithmic behaviour at high temperature of the ratio m_T/T corresponds to the UV asymptotic freedom of the classical non-linear σ-model in two dimensions.

Dimensions $d > 1$. In higher dimensions, the system can be in either phase at zero temperature, depending on the value of the coupling constant G. Introducing the critical coupling constant G_c, one can then rewrite the gap equation as

$$\frac{1}{N}\left(\frac{1}{G} - \frac{1}{G_c}\right) = \Omega_{d+1}(m_T) - \Omega_{d+1}(0) + \mathcal{T}(T, m_T). \tag{21.71}$$

21.10.6 Dimension $d = 2$

At finite temperature, the symmetry remains unbroken, because no phase transition is possible in two dimensions. The gap equation is UV finite and, for $G > G_c$, using equation (21.69) and performing the integrals, becomes

$$\frac{m_T - m}{4\pi} = \mathcal{T}(T, m_T).$$

The scaling relation (21.21),

$$\mathcal{T}(T, m_T) = m_T \mathcal{T}(T/m_T, 1),$$

reduces the evaluation of the thermal correction to the calculation of (equation (A21.21))

$$\mathcal{T}(\tau, 1) = -\frac{\tau}{2\pi} \ln(1 - \mathrm{e}^{-1/\tau}).$$

The equation can be rewritten as

$$\frac{m}{2T} = \ln[2\sinh(m_T/2T)] \;\Rightarrow\; \frac{m_T}{2T} = \ln\left[\tfrac{1}{2}\left(\mathrm{e}^{m/2T} + \sqrt{\mathrm{e}^{m/T} + 4}\right)\right].$$

It has the scaling form predicted by finite size RG arguments. For all $1 \gg m/\Lambda \geq 0$ ($1 \gg g - g_c \geq 0$), $m_T/T = O(1)$ and the mass of the zero mode is comparable to the mass of other modes. For example, for $G = G_c$,

$$\frac{m_T}{2T} = \ln[(1 + \sqrt{5})/2].$$

For $d = 2$, the phase with spontaneous breaking at zero temperature corresponds to the analytic continuation $m < 0$ (equation (21.61)). When $|m|/\Lambda$ is non-vanishing but small, which implies a tuning of $G - G_c$, $|m|$ is the *crossover scale* between critical and deep IR behaviours.

For $g - g_c < 0$ fixed, $m \to -\infty$ and thus

$$\frac{m_T}{T} \sim \mathrm{e}^{m/2T} = \exp\left[\frac{2\pi\Lambda}{NT}\left(\frac{1}{g_c} - \frac{1}{g}\right)\right]. \tag{21.72}$$

For $\Lambda \to \infty$, the ratio m_T/T goes to zero and the zero mode has a vanishing mass. Therefore, for $G < G_c$, dimensional reduction is useful.

The domain $G < G_c$ fixed corresponds to the deep IR (perturbative) region where only Goldstone particles propagate. When the coupling constant G goes to zero, the mass m_T has the exponential behaviour characteristic of dimension 2.

21.10.7 Dimension $d \geq 3$

For $d \geq 3$, the situation is different because a phase transition is possible in a d-dimensional classical theory.

Dimension $d = 3$. At $G = G_c$, at leading order, one finds

$$\left(\frac{m_T}{T}\right)^2 \sim \frac{2\pi^2}{3\ln(\Lambda/T)},$$

a logarithmic decrease reflecting the IR freedom of the $(\phi^2)^2$ field theory in $(1+3)$ dimensions. The ratio m_T/T is logarithmically small for large cut-off.

The quantity $\mathcal{T}(T, m_T)$ has a finite limit for $m_T = 0$. The relation (21.71) then determines the critical temperature T_c,

$$\frac{1}{G} - \frac{1}{G_c} = \frac{NT_c^2}{12} \quad \Rightarrow \quad \frac{T_c^2}{\Lambda^2} = \frac{12}{N}\left(\frac{1}{g} - \frac{1}{g_c}\right). \tag{21.73}$$

This relation implies that the system is in the broken phase $G < G_c$ at zero temperature and the parameter g such that $g_c - g = O(T^2/\Lambda^2) \ll 1$, results consistent with the small value of m_T/T for $g = g_c$.

At $g - g_c$ fixed, the critical temperature diverges.

Dimension $d > 3$. In the limit $G = G_c$, one finds

$$(m_T/T)^2 \propto (T/\Lambda)^{d-3} \ll 1. \tag{21.74}$$

The critical temperature behaves like

$$\left(\frac{T_c}{\Lambda}\right)^{d-1} \propto \frac{1}{g} - \frac{1}{g_c},$$

generalizing the form of dimension 3 and again implying that the system at zero temperature is in the broken phase. The parameter g has to be tuned in such way that $T_c \ll \Lambda$. At $g < g_c$ fixed, the critical temperature diverges.

21.10.8 The spontaneously broken phase

This situation occurs only for dimensions $d > 2$. Then, $m_T = 0$, and $\bar{\sigma}_T$ is the solution of (equation (21.65b))

$$\bar{\sigma}_T^2 = 1 - NG\left[\Omega_{d+1}(0) + \mathcal{T}(T, 0)\right] = \bar{\sigma}^2 - NG\mathcal{T}(T, 0),$$

where (equation (A21.19))

$$\mathcal{T}(T, 0) \equiv \mathcal{T}_-(T, 0) = N_d\Gamma(d-1)\zeta(d-1)T^{d-1}.$$

The second form of the equation shows again that temperature increases disorder, since $\bar{\sigma}_T < \bar{\sigma} \equiv \bar{\sigma}_{T=0}$.

The two-point functions. Differentiating the action (21.54) twice with respect to σ, λ, and using the saddle point equations, setting the time component of the momentum to zero, one obtains the inverse connected σ, λ two-point function as a 2×2 matrix, in the Fourier representation,

$$\boldsymbol{\Sigma} = \begin{bmatrix} [\mathcal{S}_N]_{\sigma\sigma}(p) & [\mathcal{S}_N]_{\sigma\lambda}(p) \\ [\mathcal{S}_N]_{\sigma\lambda}(p) & [\mathcal{S}_N]_{\lambda\lambda}(p) \end{bmatrix} = \frac{1}{G} \begin{bmatrix} p^2 & \bar{\sigma}_T \\ \bar{\sigma}_T & -\frac{1}{2}NGB_{d+1}(p) \end{bmatrix}, \tag{21.75}$$

with, here,

$$B_{d+1}(p) = \frac{T}{(2\pi)^d} \sum_\nu \int \frac{\mathrm{d}^d k}{[(k+p)^2 + \omega_\nu^2](k^2 + \omega_\nu^2)}, \quad \omega_\nu = 2\pi\nu T. \tag{21.76}$$

The common denominator of the propagators is the determinant

$$\det \boldsymbol{\Sigma} = -G \left[\bar{\sigma}_T^2 + \tfrac{1}{2}NGp^2 B_{d+1}(p) \right],$$

which is strictly negative for $p^2 \geq 0$ and complex for $p^2 < 0$. Therefore, it does not vanish for p^2 real.

For $d = 3$, for $|p| \to 0$, the integral (21.76) is dominated by the zero mode,

$$B_4(p) = TB_3(p) + \frac{T}{(2\pi)^3} \sum_{\nu \neq 0} \int \frac{\mathrm{d}^3 k}{(k^2 + \omega_\nu^2)^2} + O(p^2)$$

$$= \frac{T}{8|p|} + \frac{1}{4\pi^2} \ln(\Lambda/T) + O(1).$$

Again, $\det \boldsymbol{\Sigma} = 0$ has no solution for $p^2 < 0$, because $B_3(p)$ is complex.

21.11 The GN model at finite temperature for large N

To gain some intuition about the behaviour, when the temperature varies, of fermion systems where a phase transition occurs at zero temperature, we now examine a simple model of self-interacting fermions, the GN model [116]. Since the model is renormalizable only in $(1 + 1)$ dimensions and, even there, the phase structure is non-perturbative, we examine the limit of a *large number N of fermion components*. The results can be generalized to all orders in an $1/N$ expansion.

It can be proven that, within the large component expansion, *the GN model is equivalent to the GNY (Y for Yukawa) model*, at least for generic couplings: the GNY model has the same symmetries, but contains an additional elementary scalar particle coupled to fermions through a Yukawa-like interaction, and is renormalizable in $(1 + 3)$ dimensions. This equivalence provides a simple interpretation for some of the results that are found [118].

21.11.1 The GN model

The GN model is defined by a $U(N)$ symmetric action for a set of N massless Dirac fermions $\{\psi^i, \bar{\psi}^i\}$ of the form

$$\mathcal{S}\left(\bar{\psi}, \psi\right) = -\int \mathrm{d}^{d+1}x \left[\bar{\psi}(x) \cdot \not{\partial}\psi(x) + \tfrac{1}{2}G\left(\bar{\psi}(x) \cdot \psi(x)\right)^2\right]. \qquad (21.77)$$

We assume a coupling constant $G > 0$, which corresponds to an attractive interaction. Since G has a mass dimension $(1 - d)$, we parametrize it also as

$$G = \Lambda^{1-d}g, \qquad (21.78)$$

where Λ is a cut-off and g is dimensionless, assumed to be generically of order 1.

While the $U(N)$ symmetry is here a simple technical device, *the GN model has also in all dimensions a discrete $\mathbb{Z}_2$ symmetry which prevents the addition of a mass term:* with the notation $x_0 \equiv \tau$, $\mu = 0 \ldots d$,

$$\mathbf{x} = \{x_0, \ldots, x_\mu, \ldots, x_d\} \mapsto \tilde{\mathbf{x}} = \{x_0, \ldots, -x_\mu, \ldots, x_d\},$$

together with

$$\begin{cases} \psi(\mathbf{x}) \mapsto \gamma_\mu \psi(\tilde{\mathbf{x}}), \\ \bar{\psi}(\mathbf{x}) \mapsto -\bar{\psi}(\tilde{\mathbf{x}})\gamma_\mu. \end{cases}$$

For $(d+1)$ even, it is equivalent to a *discrete chiral symmetry* and, for $(d+1)$ odd, it corresponds to space reflections. Below, to simplify, as an abuse of language, we will speak about chiral symmetry, irrespective of dimensions.

The perturbative expansion corresponds to a chiral invariant theory with massless fermions. Therefore, the question of a spontaneous chiral symmetry breaking and fermion mass generation is non-perturbative.

For example, in dimension $d = 1$, G is dimensionless and this reflects the property that the GN model is renormalizable. However, in the absence of an explicit symmetry breaking fermion mass term, the perturbative expansion is IR divergent. Therefore, the perturbative prediction of a massless fermion phase at zero temperature cannot be trusted. This is also consistent with the small coupling behaviour of the RG β-function, which shows that the GN model is asymptotically (UV) free but strongly interacting at large distance.

Finite temperature. At the non-zero temperature T, due to the anti-periodic boundary conditions, fermions have an expansion of the form (see equation(21.3)),

$$\psi(\tau, x) = -\psi(\tau + 1/T, x) = \sum_{\omega_\nu = (2\nu+1)\pi T} \mathrm{e}^{i\omega_\nu t}\, \psi_\nu(x),$$

$$\bar{\psi}(\tau, x) = -\bar{\psi}(\tau + 1/T, x) = \sum_{\omega_\nu = (2\nu+1)\pi T} \mathrm{e}^{-i\omega_\nu t}\, \bar{\psi}_\nu(x), \qquad \nu \in \mathbb{Z}.$$

Unlike bosons, due to these anti-periodic conditions, fermions have no zero modes. In the limit where the temperature-dependent mass m_T is much smaller than the temperature, fermions can be completely integrated out.

An equivalent formulation. We introduce an auxiliary scalar field σ and consider the action

$$\mathcal{S}(\bar{\psi},\psi,\sigma) = -\int d\tau\, d^dx \left[\bar{\psi}(\tau,x)\cdot(\not{\partial}+\sigma(\tau,x))\psi(\tau,x) - \frac{1}{2G}\sigma^2(\tau,x)\right]. \quad (21.79)$$

To the chiral transformation of fermions corresponds then the reflection $\sigma \mapsto -\sigma$.

The new action is quadratic in $\sigma(\tau,x)$, and the Gaussian σ integration can be performed. The integration amounts to substituting for σ the solution of the σ field equation,

$$\frac{\delta\mathcal{S}}{\delta\sigma(\tau,x)} = \frac{1}{G}\sigma(\tau,x) - \bar{\psi}(\tau,x)\cdot\psi(\tau,x) = 0.$$

After the substitution, one recovers the GN action (equation (21.77)).

The property that the fermion interaction is attractive and the form (21.79) of the action suggests a new question: does σ represent just a fermion $\bar{\psi}\psi$ pair or can it be considered as a fermion scalar bound state? This again is a non-perturbative question. Quite generally, since limited insight about the physics of the model at *finite temperature* can be gained from perturbation theory, we thus study the GN model for N large, at NG fixed.

Chemical potential. The GN model can be studied in presence of a chemical potential but this situation will not be considered here.

21.11.2 *The GN model at zero temperature for N large: Gap equation and mass spectrum*

To be able to generate a large N expansion, it is necessary to integrate out the fermions to render the N-dependence explicit.

After the introduction of the auxiliary periodic scalar field σ, in the form (21.79), the action becomes quadratic in the fermion fields $\psi, \bar{\psi}$, and the fermion integration can be performed. After integration, the N-dependence becomes explicit.

For $G = O(1/N)$, the resulting, non-local, action for the σ field,

$$\mathcal{S}_N(\sigma) = \frac{1}{2G}\int dt \int d^dx\, \sigma^2(\tau,x) - N\,\mathrm{tr}\ln[\not{\partial}+\sigma(\cdot)]. \quad (21.80)$$

is proportional to N, and the partition function can be evaluated for N large by the steepest descent method.

Saddle point equations. Differentiating the action (21.80), one looks for a saddle point $\sigma(x) = M$ constant, where, in the large N limit, M is also the fermion mass, as the action (21.79) shows. The action density in a large volume V_{d+1} becomes

$$\frac{\mathcal{S}_N(\sigma = M)}{V_{d+1}} \equiv \mathcal{E}(M) = \frac{1}{2G}M^2 - \frac{N}{V_{d+1}}\,\mathrm{tr}\ln(\not{\partial}+M).$$

Differentiating with respect to M, one obtains the saddle point (or gap) equation,

$$\frac{M}{G} = \frac{N}{V_{d+1}}\,\mathrm{tr}\,(\not{\partial}+M)^{-1}.$$

Then,

$$\operatorname{tr}\left(\not{\partial}+M\right)^{-1}=\operatorname{tr}_{\gamma}\int \mathrm{d}^{d}x\,\langle x|\left[\not{\partial}+M^{-1}|x\rangle=V_{d+1}\int \frac{\mathrm{d}^{d+1}k}{(2\pi)^{d+1}}\operatorname{tr}_{\gamma}\frac{1}{i\not{k}+M}\,,$$

where $\operatorname{tr}_{\gamma}$ means the trace is restricted to the space of γ matrices, and we have used the bra-ket quantum notation.

Taking the trace,

$$\operatorname{tr}_{\gamma}\frac{1}{i\not{k}+M}=\frac{M}{k^{2}+M^{2}}\operatorname{tr}_{\gamma}\mathbf{1}\,,$$

one can rewrite the equation as

$$\frac{M}{N'G}-M\Omega_{d+1}(M)=0\,,$$

where $N'=N\operatorname{tr}_{\gamma}\mathbf{1}$ is the total number of fermion degrees of freedom.

The equation factorizes into the equation $M=0$ corresponding to the unbroken (or disordered) phase and the gap equation

$$\frac{1}{N'G}-\Omega_{d+1}(M)=0\,, \tag{21.81}$$

whose solution corresponds to the broken (or ordered) phase. We choose the solution $M\geq 0$.

Solution of the gap equation. For $d=1$, with some definition of the cut-off Λ, the solution of the gap equation is

$$M=\Lambda\,\mathrm{e}^{-(2\pi)/N'G}\,.$$

The chiral symmetry is always broken and the mass M is physical for G small enough such that $M\ll\Lambda$.

For $d>1$, the equation defines a critical value G_{c} where M vanishes,

$$\frac{1}{N'G_{c}}=\Omega_{d+1}(0)\,. \tag{21.82}$$

The gap equation (21.81) has solutions only for $G>G_{c}$. For $G<G_{c}$, the only saddle point is $M=0$ and the chiral symmetry is preserved. At G_{c} a phase transition occurs and for $G>G_{c}$, the symmetry is spontaneously broken and the fermions become massive.

Equation (21.81) can be rewritten as

$$\frac{1}{N'}\left(\frac{1}{G_{c}}-\frac{1}{G}\right)=\Omega_{d+1}(0)-\Omega_{d+1}(M)\,.$$

For $d > 3$, the integral is IR convergent for $M = 0$ and proportional to Λ^{d-3}. Introducing the dimensionless parameter g and, correspondingly, the critical value g_c, one finds

$$\frac{M^2}{\Lambda^2} \propto \frac{1}{g_c} - \frac{1}{g}.$$

The equation shows that M is physical only if $g - g_c$ is small enough (fine tuning).
For $d = 3$, one finds

$$\frac{M^2}{\Lambda^2} \ln(\Lambda/M) \propto \frac{1}{g_c} - \frac{1}{g},$$

a form that reflects the IR freedom of the theory.
For $d = 2$, the integral is UV convergent and, thus,

$$\frac{M}{\Lambda} \propto \frac{1}{g_c} - \frac{1}{g}.$$

In all dimensions $d > 1$, the condition that M is physical implies a fine tuning of $(g - g_c)$.
Finally, calculating the inverse of the σ two-point function $\tilde{\Delta}_\sigma(p)$ at the saddle point,

$$\tilde{\Delta}_\sigma^{-1}(p) = \frac{N'}{2(2\pi)^{d+1}}(p^2 + 4M^2) \int \frac{\mathrm{d}^{d+1}k}{(k^2 + M^2)[(p+k)^2 + M^2]},$$

one observes that, in the broken phase, the σ-field corresponds to a fermion bound state at the $\bar{\psi}\psi$ threshold, since $m_\sigma = 2M$ [199, 118, 200].

21.11.3 *Finite temperature: Gap or saddle point equation*

In the action, the time integration is now limited to $0 \le \tau \le 1/T$, where T is the temperature.

Since, at finite temperature, fermions satisfy *anti-periodic boundary conditions* in the Euclidean time variable, the field $\sigma(\tau, x)$ satisfies *periodic boundary conditions*,

$$\sigma(\tau + 1/T, x) = \sigma(\tau, x),$$

as expected for a boson field.

The saddle point corresponds to a constant field. We thus set $\sigma(\tau, x) = M_T$ and we choose $M_T \ge 0$. Returning to the action (21.79), one notices that, in the large N limit, M_T is no longer the finite temperature fermion mass because, due to the anti-periodic boundary conditions, the decay of the fermion two-point function in space directions is governed by the mass parameter $\sqrt{\pi^2 T^2 + M_T^2}$.

The action density at finite temperature T becomes

$$\mathcal{E}(M_T) = \frac{M_T^2}{2G} - \frac{NT}{V_d} \operatorname{tr} \ln(\not\partial + M_T) - (M_T = 0), \tag{21.83}$$

where V_d is a large d-dimensional volume.

Then,

$$N \operatorname{tr} \ln(\partial\!\!\!/ + M_T) = N' \frac{V_d}{2} \int \frac{\mathrm{d}^d k}{(2\pi)^d} \sum_{\nu \in \mathbb{Z}} \ln(\omega_\nu^2 + k^2 + M_T^2),$$

where $\omega_\nu = (2\nu + 1)\pi T$.

The sum over frequencies yields

$$N \operatorname{tr} \ln(\partial\!\!\!/ + M_T) = N' V_d \int \frac{\mathrm{d}^{d-1} k}{(2\pi)^{d-1}} \ln\left[2\cosh\left(\omega(k)/2T\right)\right], \tag{21.84}$$

with $\omega(k) = \sqrt{k^2 + M_T^2}$.

The gap equation at finite temperature, obtained by differentiating $\mathcal{E}$ with respect to M_T, factorizes into two equations, $M_T = 0$, and the gap equation

$$\frac{1}{N'G} = \Omega_{d+1}(M_T) - \mathcal{T}(T, M_T), \tag{21.85}$$

where $\Omega_{d+1}(M_T)$ is the quantum contribution, and $\mathcal{T}(T, M_T)$ is now the *fermion thermal contribution* given by

$$\mathcal{T}(T, M_T) \equiv \mathcal{T}_+(T, M_T) = \frac{1}{(2\pi)^d} \int \frac{\mathrm{d}^d k}{\omega(k)} \frac{1}{\mathrm{e}^{\omega(k)/T} + 1}. \tag{21.86}$$

The equation (21.85) is analogous to equation (21.68). It differs by the replacement of boson by fermion thermal correction and the sign in front of it.

21.11.4 *Phase transition at finite temperature: The critical temperature*

The dimension $d = 1$ requires a specific discussion: for this see Section 21.11.6.

For $d > 1$, we introduce the critical value G_c (equation (21.82)), where $M \equiv M_{T=0}$ vanishes at zero temperature:

$$\frac{1}{N'}\left(\frac{1}{G_c} - \frac{1}{G}\right) = \Omega_{d+1}(0) - \Omega_{d+1}(M_T) + \mathcal{T}(T, M_T). \tag{21.87}$$

For $G > G_c$, we can also introduce the fermion physical mass $M \equiv m_\psi$, the solution of

$$\frac{1}{N'}\left(\frac{1}{G_c} - \frac{1}{G}\right) = \Omega_{d+1}(0) - \Omega_{d+1}(M), \tag{21.88}$$

and rewrite the gap equation as

$$\Omega_{d+1}(M_T) - \Omega_{d+1}(M) = \mathcal{T}(T, M_T). \tag{21.89}$$

Since $\Omega_d(m)$ is a decreasing function and $\mathcal{T}(T, M_T)$ is positive, the equation shows that $M_T < M$, as expected, since temperature decreases order.

Critical temperature. $\Omega_{d+1}(0) - \Omega_{d+1}(M_T)$ is always positive. Therefore, the gap equation has no solution for $G \leq G_c$, where the chiral symmetry is unbroken at zero temperature. Then, $M_T = 0$ is the minimum and the $\sigma \mapsto -\sigma$ symmetry is not broken at $N = \infty$.

For $d > 1$, $\mathcal{T}(T, 0)$ is IR finite (equation $(A21.20)$),

$$\mathcal{T}(T, 0) = N_d(1 - 2^{2-d})\Gamma(d - 1)\zeta(d - 1)T^{d-1}.$$

and thus, if $G > G_c$, by contrast, one always finds a transition temperature T_c where M_T vanishes, between two Ising-like phases: a low temperature broken phase and a symmetric high temperature phase,

$$T_c \propto \left[\frac{1}{N'}\left(\frac{1}{G_c} - \frac{1}{G}\right)\right]^{1/(d-1)} = [\Omega_{d+1}(0) - \Omega_{d+1}(M)]^{1/(d-1)}. \qquad (21.90)$$

For $d > 3$, this equation implies

$$T_c/\Lambda \propto (M/\Lambda)^{2/(d-1)}$$

and, therefore, $T_c \ll \Lambda$ is implied by $M \ll \Lambda$ but then $M \ll T_c \ll \Lambda$, where M is the zero-temperature fermion mass.

For $d = 3$, the situation is more subtle, since

$$\left(\frac{T_c}{M}\right)^2 \sim \frac{3}{\pi^2} \ln(\Lambda/M),$$

which still implies $M \ll T_c \ll \Lambda$.

By contrast, for $d = 2$, one finds $T_c = M/\ln 4$.

21.11.5 The σ two-point function

Since the mass of the σ zero mode plays an important role, we also calculate the σ two-point correlation function $\tilde{\Delta}_\sigma(p_0 = 0, p)$ by differentiating the action (21.80) twice with respect to $\sigma(\tau, x)$.

Symmetric phase. In the symmetric phase, $M_T = 0$ and, for $d > 1$, in the Fourier representation, the propagator can be written as

$$\tilde{\Delta}_\sigma^{-1}(0, p) = \frac{1}{G} - \frac{1}{G_c} + N'\mathcal{T}(T, 0)$$

$$+ \frac{N'T}{2(2\pi)^d}p^2\sum_\nu \int \frac{\mathrm{d}^d k}{(k^2 + \omega_\nu^2)\left[(p + k)^2 + \omega_\nu^2\right]},$$

where a cut-off regularization is implied, with (equation $(A21.20)$)

$$\mathcal{T}(T, 0) = N_d(1 - 2^{2-d})\Gamma(d - 1)\zeta(d - 1)T^{d-1}.$$

When the σ mass is small compared to T, for example, near the phase transition, for $|p|$ small, the propagator reduces to

$$\tilde{\Delta}_\sigma^{-1}(0, p) = \frac{1}{G} - \frac{1}{G_c} + N'\mathcal{T}(T, 0) + \frac{N'T}{2(2\pi)^d}p^2\sum_\nu \int \frac{\mathrm{d}^d k}{(k^2 + \omega_\nu^2)^2}.$$

The limit $G = G_c$. For $G = G_c$, and $d > 3$, the integral is dominated by large momenta and thus

$$\Lambda^{d-3} m_\sigma^2 \propto T^{d-1},$$

which implies $m_\sigma \ll T$ and justifies the small momentum approximation.

For $d = 3$, one finds

$$m_\sigma^2 \sim \tfrac{2}{3} T^2 / \ln(\Lambda/T)$$

and the approximation is marginally valid.

For $d = 2$, $m_\sigma^2 \sim 16 \ln 2 \, T^2$, the small momentum approximation is no longer valid and the σ and fermion fields have all temperature-dependent masses of order T.

Near the critical temperature. Close enough to the critical temperature T_c, with $T_c \gg T - T_c > 0$, the propagator can be written as

$$\frac{\tilde{\Delta}_\sigma^{-1}(0,p)}{N'} = N_d (1 - 2^{2-d}) \Gamma(d) \zeta(d-1) T_c^{d-2} (T - T_c)$$

$$+ \frac{T}{2(2\pi)^d} p^2 \sum_\nu \int \frac{d^d k}{(k^2 + \omega_\nu^2)^2} \, .$$

For $d > 3$, now

$$m_\sigma^2 / \Lambda^2 \propto T_c^{d-2} (T - T_c) / \Lambda^{d-1} \, .$$

For $d = 3$,

$$m_\sigma^2 \sim \frac{2}{3} \frac{T^2 - T_c^2}{\ln(\Lambda/T)} \underset{T \to T_c}{\sim} \frac{4 T_c (T - T_c)}{3 \ln(\Lambda/T)} \, .$$

For $d = 2$, $m_\sigma^2 \sim 16 M (T - T_c)$.

In all cases, the σ-mass vanishes at T_c like $\sqrt{T - T_c}$, a typical mean field behaviour.

Broken or ordered phase. For $M_T \neq 0$, using the gap equation, one can write the propagator as

$$\tilde{\Delta}_\sigma^{-1}(0,p)$$

$$= \frac{N'T}{2(2\pi)^d} \left(p^2 + 4M_T^2\right) \sum_\nu \int \frac{d^d k}{(k^2 + \omega_\nu^2 + M_T^2)\left[(p+k)^2 + \omega_\nu^2 + M_T^2\right]} \, .$$

Therefore, when the symmetry is broken, the scalar bound state has a mass $2M_T$, generalizing the zero-temperature result. However, M_T is no longer related to the decay of the fermion propagator in space directions, due to the anti-periodic boundary conditions.

Local expansion. When the σ mass or expectation value are small compared to T and because the fermions can be integrated out, one can perform a mode expansion of the field σ, retaining only the zero mode and then a local expansion of the action (21.80), and study it to all orders in the $1/N$ expansion. In the reduced theory, T plays the role of a large momentum cut-off.

Since the resulting reduced action is local and symmetric in $\sigma \mapsto -\sigma$, it describes the physics of an Ising-like transition with short range interactions (unlike what happens at zero temperature).

The leading terms of the effective d-dimensional action have the form

$$\mathcal{S}_d(\sigma) = \int \mathrm{d}^d x \left[\tfrac{1}{2} Z_\sigma (\nabla_x \sigma(x))^2 + \tfrac{1}{2} r \sigma^2(x) + \frac{1}{4!} u \sigma^4(x) \right]. \tag{21.91}$$

The properties of this model are those of the critical ϕ^4 theory and, for $d \le 4$, have been studied by RG improved perturbative techniques.

21.11.6 Dimension $d = 1$

The coupling constant G is dimensionless and the perturbative GN model is renormalizable.

The situation $d = 1$ is doubly special, since, at zero temperature, chiral symmetry is always broken and, at finite temperature, the Ising symmetry is never broken because the effective theory is one dimensional.

Zero temperature. At finite N, the theory has a rich spectrum (see Section 8.5.4). All particles are massive and the masses are proportional to the RG invariant mass scale

$$\Lambda(G) \propto \Lambda \exp\left[-\int^G \frac{\mathrm{d}G'}{\beta(G')} \right],$$

where $\beta(G)$ is the RG β-function (equation (8.22)),

$$\beta(G) = -(N-1)G^2/\pi + O(G^3), \tag{21.92}$$

an expression that shows that the GN model is asymptotically (UV) free and the spectrum non-perturbative.

For $N \to \infty$, since $\Omega_2(0)$ diverges, equation (21.82) shows that G_c vanishes, confirming that chiral symmetry is always spontaneously broken. The mass M is thus the solution of the gap equation (21.81). It satisfies ($N' = 2N$)

$$\frac{1}{2NG} = \Omega_2(M) = \frac{1}{4\pi^2} \int \frac{\mathrm{d}^2 k}{k^2 + M^2} \sim \frac{1}{2\pi} \ln(\Lambda/M) \ \Rightarrow\ M \propto \Lambda\, \mathrm{e}^{-\pi/(NG)}. \tag{21.93}$$

The fermions are massive, with mass M and the boson mass $m_\sigma = 2M$, in agreement with the limit of the finite N result [118]. The explicit form (21.93) implies that, for N large, $\beta(G) = -NG^2/\pi$, a form consistent with expression (21.92).

The mass M is physical only if G is logarithmically small, to compensate for the Λ factor.

Finite temperature and pseudo-critical temperature. The equation (21.89) becomes

$$\Omega_2(M_T) - \Omega_2(M) = \frac{1}{2\pi} \ln(M/M_T) = \mathcal{T}(T/M_T, 1), \tag{21.94}$$

which implies $M > M_T$. The function $\mathcal{T}(z,1)$ is a positive increasing function that varies between 0 and $+\infty$.

This equation has a finite limit for $M_T \to 0$. To evaluate the limit explicitly, it is convenient to consider the generic expressions in d dimensions and take the $d = 1$ limit.

We consider the equation (21.89) taken for $M_T = 0$, which yields the critical temperature, in $d > 1$ continuous dimensions,

$$N_d(1 - 2^{2-d})\Gamma(d-1)\zeta(d-1)T_c^{d-1} = \Omega_{d+1}(0) - \Omega_{d+1}(M)$$
$$= -\frac{N_d}{4\sqrt{\pi}}\Gamma((1-d)/2)\Gamma(d/2)M^{d-1}.$$

Setting $d = 1 + \varepsilon$, using the expansions $(A21.3)$ and $(A21.8)$, $\Gamma(\varepsilon) = 1/\varepsilon - \gamma + O(\varepsilon)$ (γ is Euler's constant), $\zeta(\varepsilon) = -\frac{1}{2}(1 + \varepsilon\ln(2\pi)) + O(\varepsilon^2)$, one finds

$$T_c = M\,\mathrm{e}^\gamma\,/\pi.$$

However, T_c is not a transition temperature. Indeed, for M_T small, dimensional reduction is justified, which here corresponds to one-dimensional quantum mechanics with an action of the form (21.91). A phase transition would imply a degenerate quantum ground state. However, the Hamiltonian of the double-well potential has a unique ground state, due to quantum tunnelling. In the σ-potential, the pseudo-critical temperature instead corresponds to the transition between a a double-well potential for $T < T_c$ to a quartic anharmonic oscillator ($M_T = 0$) for $T > T_c$.

21.12 Abelian gauge theories: The QED example

Notation. In what follows, Greek indices like $\mu = 0, 1, \ldots, d$ refer to space–time components, while Roman indices like $i = 1, \ldots, d$ refer to space components only. Moreover, $\tau \equiv x_0$.

Our goal is to discuss the Abelian gauge field coupled to fermion matter at finite temperature. We discuss the Abelian case because it is much simpler, the mode decomposition being consistent with the gauge structure. Some new technical difficulties arise in non-Abelian gauge theories [201].

Because the gauge field has a number of components which depends on the number of space dimensions, the mode expansion have some new properties and affects gauge transformations.

Here, we consider the simplest non-trivial example of a gauge theory, QED, a theory which is IR free in four dimensions, and therefore, from the RG point of view, has properties similar to the scalar ϕ^4 field theory.

21.12.1   Massive vector field coupled to fermion matter

For quantization purposes, we first consider a *massive* vector field $A_\mu(\tau, x)$ coupled minimally to Dirac fermions $\psi(\tau, x), \bar\psi(\tau, x)$. The action reads

$$\mathcal{S}(A, \bar\psi, \psi) = \int \mathrm{d}\tau\, \mathrm{d}^d x \left[\frac{1}{4e^2} \sum_{\mu,\nu} F_{\mu\nu}^2(\tau, x) + \frac{1}{2e^2} m^2 \sum_\mu A_\mu^2(\tau, x) \right.$$

$$\left. + \bar\psi(\tau, x)(\slashed\partial + i\slashed A(\tau, x) + M)\psi(\tau, x) \right] \tag{21.95}$$

$(\slashed V \equiv \sum_\mu \gamma_\mu V_\mu)$, with

$$F_{\mu\nu}(\tau, x) = \partial_\mu A_\nu(\tau, x) - \partial_\nu A_\mu(\tau, x).$$

With the physical action (21.95), the propagator of the vector field in the Fourier representation has the form,

$$[\tilde\Delta_A]_{\mu\nu}(p) = \frac{\delta_{\mu\nu} + p_\mu p_\nu/m^2}{p^2 + m^2}.$$

At the pole $p^2 = -m^2$, the numerator is a projector, a reflection of the property that a massive vector field has only d components.

For $m = 0$, the field theory is invariant under the *gauge transformations*,

$$A_\mu(\tau, x) \mapsto A_\mu(\tau, x) - \partial_\mu \Omega(\tau, x) \tag{21.96a}$$

$$\psi(\tau, x) \mapsto \mathrm{e}^{i\Omega(\tau, x)} \psi(\tau, x), \quad \bar\psi(\tau, x) \mapsto \mathrm{e}^{-i\Omega(\tau, x)} \bar\psi(\tau, x). \tag{21.96b}$$

In the form (21.95) and for $m \neq 0$, the field theory contains only physical degrees of freedom, but is not renormalizable in $d = 3$ dimensions and has no perturbative $m = 0$ limit. In the field integral representation, after a transformation that does not affect the expectation values of gauge invariant observables, the action (21.95) can be replaced by

$$\mathcal{S}_\xi(A, \psi, \bar\psi) = \mathcal{S}(A, \bar\psi, \psi) + \frac{1}{2\xi} \int \mathrm{d}\tau\, \mathrm{d}^d x \left(\sum_\mu \partial_\mu A_\mu(\tau, x) \right)^2, \tag{21.97}$$

where ξ is a free parameter. The field theory becomes renormalizable in $d = 3$ dimensions and, correspondingly, correlation functions have an $m \to 0$ limit, a property not shared by non-Abelian gauge theories.

21.12.2   Finite temperature

At finite temperature T, the vector field satisfies periodic boundary conditions and can be expanded in a Fourier series of the form

$$A_\mu(\tau, x) = \sum_{n \in \mathbb{Z}} \mathrm{e}^{i\omega_n \tau} A_{\nu,n}(x), \quad \omega_\nu = 2n\pi T. \tag{21.98}$$

Fermion fields satisfy anti-periodic boundary conditions, and the Fourier expansion then takes the form (21.3),

$$\psi(\tau, x) = \sum_{n \in \mathbb{Z}} e^{i\omega_n \tau} \psi_n(x), \ \ \bar{\psi}(\tau, x) = \sum_{n \in \mathbb{Z}} e^{-i\omega_n \tau} \bar{\psi}_n(x), \ \ \ \omega_n = (2n+1)\pi T.$$

High temperature and dimensional reduction. In what follows, because we are interested in high temperature properties, we assume $m, M \ll T$.

At leading order, keeping only the zero modes, one obtains a reduced action in the form of the free theory,

$$\mathcal{S}(A, \bar{\psi}, \psi) = \frac{1}{T} \int d^d x \left\{ \frac{1}{2e^2} \left[(\nabla_x A_0(x))^2 + m^2 A_0^2(x) \right] + \frac{1}{4e^2} \sum_{i,j} F_{ij}^2(x) \right.$$

$$\left. + \frac{1}{2} m^2 \sum_i A_i^2(x) \right\}. \tag{21.99}$$

The $(d+1)$-dimensional massive vector field has been decomposed into a massive scalar field A_0 and a d-dimensional massive vector field A_i. In the massless limit, the scalar field becomes massless and the vector fields become gauge fields. By changing to the gauge (21.97), one ensures a smooth massless limit. It leads to a reduced theory quantized in a generalized Coulomb gauge.

21.12.3 From physical to temporal (Weyl) gauge: zero temperature

To transform the action from the unitary form (21.95) to the temporal gauge, one has to eliminate the A_0 component by a gauge transformation. Therefore, we set (equation (21.96b))

$$\psi(\tau, x) \mapsto e^{i\Omega(\tau, x)} \psi(\tau, x), \ \ \ \bar{\psi}(\tau, x) \mapsto e^{-i\Omega(\tau, x)} \bar{\psi}(\tau, x).$$

After the gauge transformation,

$$\int d\tau \, d^d x \, \bar{\psi}(\tau, x) \slashed{\partial} \psi(\tau, x) \mapsto \int d\tau \, d^d x \, \bar{\psi}(\tau, x) \left[\slashed{\partial} + i \slashed{\partial} \Omega(\tau, x) \right] \psi(\tau, x).$$

At zero temperature, we can choose the Ω solution of

$$\partial_\tau \Omega(\tau, x) = -A_0(\tau, x),$$

to eliminate the A_0 field component. We then perform the gauge transformation on the remaining A_i components,

$$A_i(\tau, x) \mapsto A_i(\tau, x) - \partial_i \Omega(\tau, x).$$

The components of the curvature tensor become,

$$F_{0,i} = \partial_t A_i - \partial_i A_0 \mapsto \partial_t A_i - \partial_i \partial_\tau \Omega - \partial_i A_0 = \partial_t A_i, \ \ \ F_{ij} \mapsto F_{ij}.$$

In the massless limit $m \to 0$, one obtains the action corresponding to the $A_0 = 0$ temporal gauge:

$$S(A, \bar{\psi}, \psi) = \frac{1}{e^2} \int d\tau \, d^d x \left[\tfrac{1}{2} \sum_i (\partial_\tau A_i(\tau, x))^2 + \tfrac{1}{4} \sum_{i,j} F_{ij}^2(\tau, x) \right]$$

$$+ \int d\tau \, d^d x \, \bar{\psi}(\tau, x)(\slashed{\partial} + i \sum_i \gamma_i A_i(\tau, x) + M)\psi(\tau, x). \qquad (21.100)$$

The propagator in the Fourier representation reads

$$[\tilde{\Delta}_A]_{ij}(\omega, p) = \frac{1}{p^2} \left(\delta_{ij} - \frac{p_i p_j}{p^2} \right) + \frac{1}{\omega^2} \frac{p_i p_j}{p^2}, \qquad (21.101)$$

where ω and p correspond to the time and space components, respectively. The pole at $\omega = 0$ leads to several difficulties in perturbation theory.

21.12.4 *From physical to temporal (Weyl) gauge: Finite temperature*

To transform the action from the unitary to the temporal gauge, one again tries to eliminate the A_0 component by a gauge transformation. Therefore, we set

$$\psi(\tau, x) \mapsto e^{i\Omega(\tau, x)} \psi(\tau, x), \quad \bar{\psi}(\tau, x) \mapsto e^{-i\Omega(\tau, x)} \bar{\psi}(\tau, x).$$

At finite temperature, because $\psi(\tau, x), \bar{\psi}(\tau, x)$ are anti-periodic,

$$\psi(\tau + 1/T, x) = -\psi(\tau, x), \quad \bar{\psi}(\tau + 1/T, x) = -\bar{\psi}(\tau, x),$$

$\Omega(\tau, x)$ has to satisfy

$$\Omega(\tau + 1/T, x) = \Omega(\tau, x) \quad (\mathrm{mod}\ 2\pi).$$

It can be decomposed into the sum of a linear function and a periodic function in the form,

$$\Omega(\tau, x) = 2\pi n T \tau + \tilde{\Omega}(\tau, x), \ n \in \mathbb{Z}, \qquad (21.102)$$

with

$$\tilde{\Omega}(\tau + 1/T, x) = \tilde{\Omega}(\tau, x).$$

After the gauge transformation,

$$\int d\tau \, d^d x \, \bar{\psi}(\tau, x) \slashed{\partial} \psi(\tau, x) \mapsto \int d\tau \, d^d x \, \bar{\psi}(\tau, x) \left[\slashed{\partial} + i \slashed{\partial} \Omega(\tau, x) \right] \psi(\tau, x).$$

At finite temperature, we can no longer choose,

$$\partial_\tau \Omega(\tau, x) = -A_0(\tau, x)$$

to eliminate the A_0 field component.

Indeed, $\partial_\tau \Omega$ has no zero mode, unlike A_0. Therefore, we decompose

$$A_0(\tau, x) = B_0(x) + Q_0(\tau, x),$$

where $B_0(x)$ is defined only up to a multiple of $2\pi T$ (equation (21.102)) and

$$Q_0(\tau, x) = \sum_{n \neq 0 \in \mathbb{Z}} \mathrm{e}^{\omega_n \tau} Q_{0,n}(x), \quad \omega_n = 2\pi n T.$$

The equation

$$\partial_\tau \Omega(\tau, x) = -Q_0(\tau, x)$$

has solutions.

We then perform a gauge transformation on the field $A_i(\tau, x)$,

$$A_i(\tau, x) \mapsto A_i(\tau, x) - \partial_i \Omega(\tau, x),$$

which is consistent with the periodicity of the vector field.

We infer

$$F_{0i} = \partial_\tau A_i - \partial_i A_0 \mapsto \partial_\tau A_i - \partial_i Q_0 + \partial_i A_0 = \partial_i B_0.$$

The gauge transformation has eliminated A_0 up to its zero mode. We can then take the zero vector mass limit and obtain the action

$$\mathcal{S}(A_i, B_0, \bar\psi, \psi) = \frac{1}{2Te^2} \int \mathrm{d}^d x \big(\nabla_x B_0(x)\big)^2 + \frac{1}{4e^2} \int \mathrm{d}\tau \, \mathrm{d}^d x \sum_{i,} F_{i,j}^2(\tau, x)$$

$$+ \int \mathrm{d}\tau \, \mathrm{d}^d x \, \bar\psi(\tau, x)(\not\partial + i\gamma_0 B_0(x) + i \sum_i \gamma_i A_i(\tau, x) + M)\psi(\tau, x). \quad (21.103)$$

We still have to quantize the remaining vector components A_i, for example, by adding to the action a term proportional to $(\sum_i \partial_i A_i)^2$.

Note that the usual difficulties that appear in perturbation calculations with the temporal gauge (the gauge field propagator is singular at $\omega = 0$) reduce here to the need to quantize the remaining zero mode, and to the non-explicit renormalizability.

The latter problem can be solved with the help of dimensional regularization, for example (for gauge invariant observables). An alternative possibility is to introduce a renormalizable gauge.

21.12.5　Dimensional reduction

In what follows, we now generalize to N Dirac fermions with $U(N)$ symmetry.

At finite temperature, to generate the effective action for the gauge field zero modes, we have to integrate over all fermion modes (anti-periodic boundary conditions) and over the non-zero modes of the gauge field. We thus set

$$A_i(\tau, x) = B_i(x) + Q_i(\tau, x),$$

with

$$Q_i(\tau, x) = \sum_{n \neq 0} \mathrm{e}^{\omega_n \tau} Q_{i,n}(x).$$

At leading order, one finds a free theory containing a gauge field B_i and a massless scalar field B_0. At one-loop order, only fermion modes contribute. Replacing the gauge field A_μ by its zero mode B_μ and performing the fermion integration explicitly, one obtains the effective action

$$\mathcal{S}_T(B) = \frac{1}{T} \int \mathrm{d}^d x \left[\frac{1}{2e^2} (\nabla_x B_0(x))^2 + \frac{1}{4e^2} \sum_{i,j} B_{ij}^2(x) \right] - N \operatorname{tr} \ln \left(\slashed{\partial} + i \slashed{B} \right) \quad (21.104)$$

with

$$B_{ij}(x) = \partial_i B_j(x) - \partial_j B_i(x).$$

An interesting issue concerns the behaviour of the induced mass of the time component B_0 of the gauge field. We thus calculate the action density for constant B_0.

21.12.6 The action density

The action density as a function of a constant field B_0 is given by

$$\mathcal{E}(B_0) = -\frac{1}{2} N' T \sum_n \int \frac{\mathrm{d}^d k}{(2\pi)^d} \, \ln \left[k^2 + \left(B_0 + (2n+1)\pi T \right)^2 \right],$$

where $N' = N \operatorname{tr} \mathbf{1}$ is the total number of fermion degrees of freedom. One notices that the energy density is a periodic function of B_0 with period $2\pi T$, as expected.

The sum over n can be performed (it can be derived from the general identity $(A21.10)$) and one obtains

$$\mathcal{E}(B_0) = -\frac{1}{2} N' T \int \frac{\mathrm{d}^d k}{(2\pi)^d} \, \ln \big(\cosh(k/T) + \cos(B_0/T) \big). \quad (21.105)$$

We verify that the difference $\mathcal{E}(B_0) - \mathcal{E}(0)$ is UV finite and has a scaling form $T^{d+1} f(B_0/T)$. This is not surprising since, in the zero-temperature limit, no gauge field mass or quartic B_0 potential is generated.

The derivative

$$\mathcal{E}'(B_0) = \frac{1}{2} N' \sinh(B_0/T) \frac{1}{(2\pi)^d} \int \frac{\mathrm{d}^d k}{\cosh(k/T) + \cos(B_0/T)},$$

is negative for $-\pi < B_0/T < 0$ and positive for $0 < B_0/T < \pi$. The action density has a unique minimum at $B_0 = 0$ in the interval $-\pi < B_0/T < \pi$.

A special case is $d = 1$, where one finds

$$\mathcal{E}'(B_0) = \tfrac{1}{2} N' B_0 / \pi, \quad \text{for } |B_0| < \pi, \quad \text{and thus} \quad \mathcal{E}(B_0) = \tfrac{1}{4} N' B_0^2 / \pi.$$

Neglecting all B_0 derivatives, one obtains a local contribution to the action $\mathcal{S}_T$:

$$-N \operatorname{tr} \ln \left(\slashed{\partial} + i \slashed{B} \right) \sim -\frac{1}{2} N' \int \mathrm{d}^d x \int \frac{\mathrm{d}^d k}{(2\pi)^d} \, \ln \big(\cosh(k/T) + \cos(B_0(x)/T) \big).$$

The coefficient $\mathcal{G}_2(d)$ of $\frac{1}{2} \int \mathrm{d}^d x \, B_0^2(x)$, obtained by expanding the expression in powers of B_0, follows:

$$\mathcal{G}_2(d) = N' N_d \Gamma(d) \left(1 - 2^{2-d} \right) \zeta(d-1) T^{d-2}.$$

21.12.7 Discussion

At leading order, one thus obtains a mass m_T that is proportional to $eT^{(d-1)/2}$. If e is generic, that is, of order 1 at the microscopic scale $1/\Lambda$, then $e \propto \Lambda^{(3-d)/2}$ and the scalar mass m_T is proportional to $(\Lambda/T)^{(3-d)/2}T$. It is thus large with respect to the vector masses of the non-zero modes for $d < 3$, and small for $d > 3$.

If one takes into account loop corrections, one finds for $d > 3$ a finite coupling constant renormalization $e \mapsto e_{\mathrm{r}}$, and the conclusion is not changed. The zero mode becomes massive but with a mass small compared to T, justifying mode and local expansions.

For $d = 3$, QED is IR free, since the first coefficient of the RG β-function is

$$\beta_{e^2} = \frac{N}{6\pi^2}e^4 + O(e^6),$$

e_{r} has to be replaced by the effective coupling constant $e(T/\Lambda)$, which is logarithmically small:

$$e^2(T/\Lambda) \sim \frac{6\pi^2}{N\ln(\Lambda/T)},$$

and the scalar mass thus is still small, although only logarithmically,

$$m_{B_0}^2 \propto \frac{T^2}{\ln(\Lambda/T)}.$$

The separation between zero and non-zero modes remains only marginally justified. High temperature QED shares some of the properties of high temperature ϕ^4 field theory and, for the same reason, a perturbative expansion remains meaningful.

Note that, if one is interested in IR physics only, one can in a second step integrate over the massive scalar field B_0.

Finally, for $d < 3$, one finds an IR fixed point and, therefore, one expects that, in massless QED, m_T becomes proportional to T and comparable to all other modes, in particular to all gauge field non-zero modes that become massive vector fields.

Quantization. To quantize the theory, one still has to fix the gauge, using, for instance, a covariant gauge. For what concerns the massive modes, they are here quantized in a unitary non-renormalizable gauge. Therefore, a change of gauge is required for these modes also, for renormalizability. This is not difficult in the Abelian case because we can make an independent gauge transformation on each vector field. Furthermore, we note that the vector masses are not renormalized.

For more details and more systematic QED calculations, we refer to the literature.

A21 Appendix: One-loop contributions

We give here a few mathematical results frequently used in the chapter.

A21.1 Γ and ζ functions

A few identities about Γ and Riemann's ζ functions are useful.

Γ-*function*. The Γ function satisfies

$$\sqrt{\pi}\,\Gamma(2z) = 2^{2z-1}\Gamma(z)\Gamma(z+1/2), \quad \Gamma(z)\Gamma(1-z)\sin(\pi z) = \pi. \qquad (A21.1)$$

These identities imply for the $\psi(z)$ function, $\psi(z) = \Gamma'(z)/\Gamma(z)$,

$$2\psi(2z) = 2\ln 2 + \psi(z) + \psi(z+1/2), \quad \psi(z) - \psi(1-z) + \pi/\tan(\pi z) = 0. \quad (A21.2)$$

We need ($\gamma = -\psi(1)$ is Euler's constant),

$$\Gamma(\varepsilon) = \frac{1}{\varepsilon} - \gamma + O(\varepsilon). \qquad (A21.3)$$

Riemann's ζ function. Riemann's ζ function is defined by

$$\zeta(s) = \sum_{n \geq 1} \frac{1}{n^s}. \qquad (A21.4)$$

It satisfies the reflection formula

$$\zeta(s)\Gamma(s/2) = \pi^{s-1/2}\Gamma\big((1-s)/2\big)\zeta(1-s), \qquad (A21.5)$$

which can be written in different forms using Γ-function relations. Moreover,

$$\sum_{n=1} \frac{(-1)^n}{n^s} = \big(2^{1-s} - 1\big)\zeta(s). \qquad (A21.6)$$

In the chapter, the ζ-function appears in the form

$$\Gamma(s)\zeta(s) = \int_0^\infty \frac{\mathrm{d}\tau\,\tau^{s-1}}{e^s - 1}.$$

Finally,

$$\zeta(1+\varepsilon) = 1/\varepsilon + \gamma + O(\varepsilon), \qquad (A21.7)$$
$$\zeta(\varepsilon) = -\tfrac{1}{2}\big(1 + \varepsilon \ln(2\pi)\big) + O(\varepsilon^2). \qquad (A21.8)$$

A21.2 The one-loop two-point contribution at $T = 0$

The one-loop calculations involve the diagram in Fig. 21.1:

$$\Omega_d(m) = \frac{1}{(2\pi)^d} \int \frac{d^d k}{k^2 + m^2} . \qquad (A21.9)$$

The momentum integral is regularized, that is, cut at a momentum of order Λ, for instance, by a Pauli–Villars regularization,

$$\frac{1}{k^2 + m^2} \mapsto \frac{1}{(k^2 + m^2)(1 + k^2/\Lambda^2)^n},$$

where n is such that $2n + 2 > d$. For $\Lambda \to \infty$, and $d > 2$ it behaves like

$$\Omega_d(m) \propto \Lambda^{d-2},$$

where the proportionality constant depends on the explicit regularization.

The diagram has a finite $m \to 0$ limit for $d > 2$, a property of direct relevance for a number of phase transitions.

In the gap equations, in fact, only differences of the form $\Omega_d(m_1) - \Omega_d(m_2)$ appear. We assume $m_1, m_2 \geq 0$.

For $d = 2$,

$$\Omega_2(m_1) - \Omega_2(m_2) = \frac{1}{2\pi} \ln(m_2/m_1).$$

For $d = 3$,

$$\Omega_3(m_1) - \Omega_3(m_2) = \frac{1}{4\pi}(m_2 - m_1).$$

For $d = 4$,

$$\Omega_4(m_1) - \Omega_4(m_2) = \frac{1}{8\pi^2} \left[-m_1^2 \left(ln(\Lambda/m_1) + c_4 \right) + m_2^2 \left(ln(\Lambda/m_2) + c_4 \right) \right],$$

where c_4 is a numerical, regularization dependent, constant.

A21.3 The thermal corrections at one loop

We first give some details about the evaluation of the gap equation.

A21.3.1 *The gap equation*

The one-loop thermal corrections can be rewritten by using the general identity

$$\sum_{\nu \in \mathbb{Z}} \frac{z}{(\nu + \mu)^2 + z^2} = \frac{\pi \sinh(2\pi z)}{\cosh(2\pi z) - \cos(2\pi \mu)} . \qquad (A21.10)$$

In the boson case, $\mu = 0$ and, thus,

$$\sum_{\nu \in \mathbb{Z}} \frac{z}{\nu^2 + z^2} = \frac{\pi \sinh(2\pi z)}{\cosh(2\pi z) - 1} = 2\pi \left(\frac{1}{2} + \frac{1}{e^{2\pi z} - 1} \right). \qquad (A21.11)$$

Applied to the one-loop correction in $(d+1)$ dimensions, it leads to $(\omega_\nu = 2\pi \nu T)$

$$T \int \frac{\mathrm{d}^d k}{(2\pi)^d} \sum_\nu \frac{1}{k^2 + \omega_\nu^2 + m^2} = \frac{1}{(2\pi)^d} \int \frac{\mathrm{d}^d k}{\omega(k)} \left(\frac{1}{2} + \frac{1}{e^{\omega(k)/T} - 1} \right), \qquad (A21.12)$$

with

$$\omega(k) = \sqrt{k^2 + m^2}\,.$$

The identity can be rewritten as

$$T \int \frac{\mathrm{d}^d k}{(2\pi)^d} \sum_\nu \frac{1}{k^2 + \omega_\nu^2 + m^2} = \Omega_{d+1}(m) + \mathcal{T}_-(T, m), \qquad (A21.13)$$

where we have defined

$$\mathcal{T}_-(T, m) = \frac{1}{(2\pi)^d} \int \frac{\mathrm{d}^d k}{\omega(k)} \frac{1}{e^{\omega(k)/T} - 1}\,.$$

In the fermion case, $\mu = 1/2$ and, thus,

$$\sum_{\nu \in \mathbb{Z}} \frac{z}{(\nu + 1/2)^2 + z^2} = \frac{\pi \sinh(2\pi z)}{\cosh(2\pi z) + 1} = 2\pi \left(\frac{1}{2} - \frac{1}{e^{2\pi z} + 1} \right). \qquad (A21.14)$$

This leads to the identity $(\omega_\nu = (2\nu + 1)\pi T)$

$$T \int \frac{\mathrm{d}^d k}{(2\pi)^d} \sum_\nu \frac{1}{k^2 + \omega_\nu^2 + m^2} = \frac{1}{(2\pi)^d} \int \frac{\mathrm{d}^d k}{\omega(k)} \left(\frac{1}{2} - \frac{1}{e^{\omega(k)/T} + 1} \right).$$

The identity can be rewritten as

$$T \int \frac{\mathrm{d}^d k}{(2\pi)^d} \sum_\nu \frac{1}{k^2 + \omega_\nu^2 + m^2} = \Omega_{d+1}(m) - \mathcal{T}_+(T, m), \qquad (A21.15)$$

with now the combined definitions,

$$\mathcal{T}_\pm(T, m) = \frac{1}{(2\pi)^d} \int \frac{\mathrm{d}^d k}{\omega(k)} \frac{1}{e^{\omega(k)/T} \pm 1}, \qquad (A21.16)$$

with

$$\omega(k) = \sqrt{k^2 + m^2}.$$

The + sign corresponds to fermions and the − sign to bosons.
They lead to the functions (21.20) and (21.86).

Setting $k = mq$, one obtains the scaling relations

$$\mathcal{T}_{\pm}(T,m) = m^{d-1}\frac{1}{(2\pi)^d}\int\frac{d^dq}{\sqrt{q^2+1}}\frac{1}{e^{m/T\sqrt{q^2+1}}\pm 1} = m^{d-1}\mathcal{T}_{\pm}(T/m,1).$$

A general relation. It is simple to verify that

$$\mathcal{T}_{-}(T,m) - \mathcal{T}_{+}(T,m) = 2\frac{1}{(2\pi)^d}\int\frac{d^dk}{\omega(k)}\frac{1}{e^{2\omega(k)/T}-1} = 2\mathcal{T}_{-}(T/2,m), \quad (A21.17)$$

a relation that makes it possible to derive $\mathcal{T}_{+}$ from $\mathcal{T}_{-}$.

Another representation. We first integrate over angles,

$$\mathcal{T}_{-}(\theta,1) = N_d\int\frac{q^{d-1}dq}{\sqrt{q^2+1}}\frac{1}{e^{\sqrt{q^2+1}/\theta}-1},$$

where N_d is the loop factor,

$$N_d = \frac{2}{(4\pi)^{d/2}\Gamma(d/2)}. \quad (A21.18)$$

We then set

$$\sqrt{q^2+1}/\theta = s \;\Rightarrow\; \theta ds = \frac{q\,dq}{\sqrt{q^2+1}}.$$

It follows that

$$\mathcal{T}_{-}(\theta,1) = N_d\theta\int ds\,\theta(\theta^2 s^2 - 1)\left(\theta^2 s^2 - 1\right)^{(d-2)/2}\frac{1}{e^s-1}.$$

A useful limit for $d > 2$ is,

$$\mathcal{T}_{-}(\theta,1)\underset{\theta\to\infty}{\sim} N_d\theta^{d-1}\int_0^\infty ds\frac{s^{d-2}}{e^s-1} = N_d\zeta(d-1)\Gamma(d-1)\theta^{d-1}$$

and, thus,

$$\mathcal{T}_{-}(T,m)\underset{T\gg m}{\sim}\mathcal{T}_{-}(T,0) = N_d\Gamma(d-1)\zeta(d-1)T^{d-1}. \quad (A21.19)$$

It follows,

$$\mathcal{T}_{+}(T,0) = (1 - 2^{2-d})\mathcal{T}_{-}(T,0), \quad (A21.20)$$

an expression that has a finite limit at $d = 2$.

The special dimension $d = 2$. Dimension 2 is special in the sense that $\mathcal{T}_\pm$ can be calculated explicitly. Indeed,

$$\mathcal{T}_-(\theta, 1) = \frac{\theta}{2\pi} \int_{1/\theta}^{\infty} ds \, \frac{1}{e^s - 1} = -\frac{\theta}{2\pi} \ln(1 - e^{-1/\theta}) . \tag{A21.21}$$

Then,

$$\mathcal{T}_+(\theta, 1) = -\frac{\theta}{2\pi} \left[\ln(1 - e^{-1/\theta}) - \ln(1 - e^{-2/\theta}) \right]$$

$$= \frac{\theta}{2\pi} \ln(1 + e^{-1/\theta}) . \tag{A21.22}$$

The special dimension $d = 1$. A useful limit is $\theta \to \infty$. Then,

$$\mathcal{T}_+(\theta, 1) = \frac{1}{\pi} \int_0^\infty \frac{dq}{\sqrt{q^2 + 1}} \frac{1}{e^{\sqrt{q^2+1}/tau} + 1}$$

for $\theta \to \infty$ behaves like

$$\mathcal{T}_+(\theta, 1) = \frac{1}{2\pi} [\ln(\pi\theta) - \gamma] + O(\theta^{-2}) . \tag{A21.23}$$

A21.3.2 The two-point function

We consider the integral

$$\mathcal{I}_d(T) = \frac{T}{(2\pi)^d} \sum_\nu \int \frac{d^d k}{(k^2 + \omega_\nu^2)^2} , \quad \omega_\nu = (2\nu + 1)\pi T .$$

For $d > 3$, the integral is UV dominated and the sum over ν can be replaced by an integral. One finds

$$\mathcal{I}_d(T) \propto \Lambda^{d-3} .$$

For $d = 3$, the large momentum divergence is logarithmic and T acts as an IR cut-off. Thus,

$$\mathcal{I}_3(T) \sim \frac{1}{8\pi^2} \ln(\Lambda/T) .$$

For $d < 3$, $\mathcal{I}_d(T)$ is UV finite. One can first integrate and then sum. One obtains

$$\mathcal{I}_d(T) = 2 \frac{1}{(4\pi)^{d/2}} \Gamma(2 - d/2) \pi^{d-4} \left(1 - 2^{d-4}\right) \zeta(4 - d) T^{d-3} . \tag{A21.24}$$

For $d = 2$, this yields

$$\mathcal{I}_2(T) = \frac{T}{4\pi} \sum_\nu \frac{1}{\omega_\nu^2} = \frac{1}{2\pi^3 T} \sum_{\nu=1}^{\infty} \frac{1}{(2\nu + 1)^2} = \frac{1}{16\pi T} .$$

Finally, for $d = 1$,

$$\mathcal{I}_1(T) = \frac{\zeta(3)}{2\pi^3} \frac{1}{T^2} . \tag{A21.25}$$

22 From random walk to critical dynamics

In this chapter, we discuss Langevin equations, technically, first order in time stochastic differential evolution equations with Gaussian noise, related, in particular, to Brownian motion, random walk, diffusion processes and critical dynamics in classical phase transitions, which is a very important physics application.

For simplicity, we consider only *Langevin equations with gradient driving forces*, mainly time-independent and briefly, time-dependent equations, and *white noise*.

From the Langevin equation, one derives an evolution equation satisfied by the probability distribution of the stochastic variables, the Fokker–Planck (FP) equation. From the FP equation, one can directly infer a few general properties of the time evolution, like *detailed balance*, when the driving force is time independent.

An important physics application is the description of time evolution of correlation functions for phase transitions near the critical temperature, a topic called *critical dynamics* [202]. Indeed, in the framework of critical phenomena, time evolution is generally described by phenomenological Langevin equations.

From a Langevin equation one can also derive a path or field integral representation for expectation values of observables [203]. The corresponding integrands define automatically positive measures. The path integral leads to a simple derivation of *Jarzinsky's relation* when the driving force is time dependent [204].

When the driving force is time independent, the path or field integral formulation exhibits the simplest example of quantum mechanical *supersymmetry* [185].

In the case of Langevin field equations, the field integral representation is especially useful because it makes it possible to use methods of quantum field theory (QFT) to explore the properties of time-dependent correlation functions. In particular, this formalism allows deriving the *renormalization group (RG)* equations that govern critical dynamics [205, 206].

The RG equations satisfied by time-dependent correlation functions generalize the RG equations of the equilibrium theory. Generalized scaling properties follow, which involve a *correlation time* and exhibit the property of slowing down of the dynamics at the critical temperature (*critical slowing down*).

In the theory of critical phenomena, Langevin equations are constrained to lead to a given equilibrium distribution, but an infinite number of Langevin equations can be associated to an equilibrium distribution and, thus, *there are many more dynamic than static universality classes*.

In the case of a ϕ^4-like equilibrium field theory, universal dynamical quantities near the transition can be calculated in the form of a $\varepsilon = 4 - d$-expansion near dimension 4.

In this chapter, we consider only the simplest subclass, *purely dissipative Langevin equations without conservation laws*. In the classification of the review of Halperin and Hohenberg [202], this corresponds to the model A.

From Random Walks to Random Matrices. Jean Zinn-Justin, Oxford University Press (2019).
© Jean Zinn-Justin. DOI: 10.1093/oso/9780198787754.001.0001

22.1 Random walk with gradient driving force

Examples of Langevin equations are provided by first order in time stochastic differential equations of the form ($\dot{q} \equiv dq/dt$)

$$\dot{q}_i(t) = -\tfrac{1}{2}\Omega\frac{\partial\mathcal{U}(\mathbf{q}(t),t)}{\partial q_i} + \nu_i(t), \quad i = 1\ldots d, \tag{22.1}$$

in which $\mathbf{q}(t)$ is the position in $\mathbb{R}^d$ at time t, the driving force is the gradient of a potential $\mathcal{U}(\mathbf{q}, t)$ and a smooth function of $\mathbf{q}$, the constant $\Omega > 0$ is an inverse time scale parameter and $\nu_i(t)$ are stochastic functions called hereafter the *noise*.

Moreover, we assume that the noise has a Gaussian probability distribution $[d\rho(\nu)]$ of the form (*Gaussian white noise*)

$$[d\rho(\nu)] = [d\nu]\exp\left[-\frac{1}{2\Omega}\int dt\sum_i \nu_i^2(t)\right], \tag{22.2}$$

related to Markov's processes. The constant Ω also characterizes the width of the noise distribution. Equivalently, the Gaussian noise can be characterized by its one- and two-point correlation functions:

$$\langle\nu_i(t)\rangle = 0, \quad \langle\nu_i(t)\nu_j(t')\rangle = \Omega\,\delta_{ij}\,\delta(t-t'). \tag{22.3}$$

Notation. In what follows, quite generally, we denote by $\langle\bullet\rangle_\nu$ the average over the noise.

22.1.1 The trajectory probability distribution

Given the position $\mathbf{q}(t) \in \mathbb{R}^d$ at initial time t', $\mathbf{q}(t') = \mathbf{q}'$, the Langevin equation (22.1) with noise distribution (22.2) generates a *time-dependent* probability distribution $P(\mathbf{q}, t; \mathbf{q}', t')$ for the random position $\mathbf{q}(t)$ at time t, which formally can be written as

$$P(\mathbf{q}, t; \mathbf{q}', t') = \left\langle\prod_{i=1}^d \delta\left[q_i(t) - q_i\right]\right\rangle_\nu, \quad t \geq t'. \tag{22.4}$$

Note that the vector $\mathbf{q}$ argument of $P(\mathbf{q}, t; \mathbf{q}', t')$ should not be confused with the function of time $\mathbf{q}(t)$.

As a consequence, if $\mathcal{O}(\mathbf{q})$ is an arbitrary smooth function, the definition (22.4) implies

$$\int P(\mathbf{q}, t; \mathbf{q}', t')\mathcal{O}(\mathbf{q})d^d q = \langle\mathcal{O}(\mathbf{q}(t))\rangle_\nu. \tag{22.5}$$

The distribution $P(\mathbf{q}, t; \mathbf{q}', t')$ plays an important role in what follows and is further discussed in Section 22.3, where it is shown to satisfy an evolution equation, the FP equation [208].

22.1.2 The purely dissipative Langevin equation

When the potential $\mathcal{U}(\mathbf{q})$ is time independent, the Langevin equation

$$\dot{\mathbf{q}}(t) = -\tfrac{1}{2}\Omega\,\nabla_q\mathcal{U}(\mathbf{q}(t)) + \boldsymbol{\nu}(t), \tag{22.6}$$

($\nabla_q \equiv \{\partial/\partial q_1, \ldots, \partial/\partial q_d\}$) is called *purely dissipative*.

If the corresponding function $e^{-\mathcal{U}(q)}$ is normalizable, it is, up to a normalization, the *equilibrium distribution*, that is, the limit of the distribution $P(\mathbf{q}, t; \mathbf{q}', t')$ for $t \to +\infty$.

In the case of the dissipative equation (22.6), time-translation invariance implies that the FP distribution (equation (22.4)) is function only of $t - t'$.

In most of the chapter, in particular, the part devoted to critical dynamics, we deal only with dissipative Langevin equations. However, in Sections 22.2.3, 22.3.4 and 22.4.2, we also briefly discuss a few properties of the dynamics when, in equation (22.6), the force still derives from a potential $\mathcal{U}(\mathbf{q}, t)$, but is explicitly *time dependent*.

22.2 An elementary example: The linear driving force

A Langevin equation for a one-dimensional trajectory with a linear driving force and a Gaussian white noise can be written as

$$\dot{q}(t) = -\tfrac{1}{2}\omega q(t) + \nu(t), \tag{22.7}$$

in which $\nu(t)$ is a Gaussian noise with the distribution (22.2) for $\Omega = 1$:

$$\langle \nu(t) \rangle = 0, \quad \langle \nu(t_1)\nu(t_2) \rangle = \delta(t_1 - t_2).$$

The linear Langevin equation (22.7) provides the simplest example of a purely dissipative Langevin equation [207]. The driving force derives from the potential

$$\mathcal{U}(q) = \tfrac{1}{2}\omega q^2. \tag{22.8}$$

The Langevin equation (22.7) can be solved explicitly. Since $q(t)$ is linearly related to $\nu(t)$, which has a Gaussian distribution, $q(t)$ has also a Gaussian distribution.

We set the boundary condition $q(t') = q'$ at initial time t'.

22.2.1   The Brownian motion: $\omega = 0$

For $\omega = 0$, the solution of the Langevin equation reduces to

$$q(t) = q' + \int_{t'}^{t} \nu(\tau)\mathrm{d}\tau.$$

All correlation functions are determined by the one- and two-point functions. Averaging the equation over the noise, one finds

$$\langle q(t) \rangle_\nu = q'.$$

Moreover, the connected two-point correlation function is

$$\langle q(t_1)q(t_2) \rangle_\nu - q'^2 = \tfrac{1}{2}(t_1 + t_2 - 2t') - \tfrac{1}{2}|t_2 - t_1|.$$

The distribution of $q(t)$ at time t is determined by the one- and two-point functions. Here,

$$\langle q^2(t) \rangle_\nu - \langle q(t) \rangle_\nu^2 = t - t'.$$

Therefore, the probability distribution $P(q, t; q', t')$ (defined in equation (22.4)) is

$$P(q, t; q', t') = \frac{1}{\sqrt{2\pi(t - t')}}\, e^{-(q-q')^2/2(t-t')}.$$

One recognizes the distribution of the simple *random walk* or Brownian motion (see Section 1.3).

22.2.2 Case $\omega \neq 0$

The solution of equation (22.7) is

$$q(t) = q' \, e^{-\omega(t-t')/2} + \int_{t'}^{t} e^{-\omega(t-\tau)/2} \, \nu(\tau) d\tau \, . \tag{22.9}$$

Averaging equation (22.9) over the noise, one finds

$$\langle q(t) \rangle_\nu = q' \, e^{-\omega(t-t')/2} \, . \tag{22.10}$$

From equation (22.9), one infers the connected two-point correlation function:

$$\begin{aligned}
\langle q(t_1)q(t_2) \rangle_\nu^{\text{conn.}} &\equiv \langle q(t_1)q(t_2) \rangle_\nu - \langle q(t_1) \rangle_\nu \, \langle q(t_2) \rangle_\nu \\
&= \int_{t'}^{\inf(t_1,t_2)} dt \, e^{-\omega(t_1+t_2-2t)/2} \\
&= \frac{1}{\omega} \left[e^{-\omega|t_1-t_2|/2} - e^{\omega(2t'-t_1-t_2)/2} \right] .
\end{aligned} \tag{22.11}$$

In particular, the second cumulant of the distribution of $q(t)$ is

$$\left\langle \left(q(t) - \langle q(t) \rangle_\nu \right)^2 \right\rangle_\nu = \frac{1}{\omega} \left[1 - e^{-\omega(t-t')} \right] .$$

Probability distribution. Again, the distribution $P(q,t;q',t')$ is Gaussian and is thus determined by its two first moments $\langle q(t) \rangle_\nu$ and $\langle q^2(t) \rangle_\nu$. Thus,

$$P(q,t;q',t') = \sqrt{\frac{\omega}{2\pi(1 - e^{-\omega(t-t')})}} \, e^{-S(q,q')}, \tag{22.12}$$

with

$$S(q,q') = \frac{\omega}{2} \frac{\left(q - q' \, e^{-\omega(t-t')/2} \right)^2}{1 - e^{-\omega(t-t')}} .$$

FP equation. It is simple to verify that the probability distribution (22.12) satisfies the equation

$$\frac{\partial}{\partial t} P(q,t;q',t') = \frac{1}{2} \frac{\partial}{\partial q} \left(\frac{\partial}{\partial q} + \omega q \right) P(q,t;q',t'),$$

which is called the FP equation (see equation (22.30)).

Detailed balance. We show in Section 22.3 that, in the case of a general dissipative Langevin equation, the distribution $P(q,t;q',t')$ satisfies the *detailed balance condition*, which relates a process and its time-reversed form:

$$P(q,t;q',t') = e^{-\mathcal{U}(q)+\mathcal{U}(q')} \, P(q',t;q,t').$$

Here, it reduces to

$$S(q,q') = \tfrac{1}{2}\omega(q^2 - q'^2) + S(q',q).$$

Equilibrium. For $\omega > 0$, for $t \to +\infty$, the distribution converges towards the *equilibrium distribution*, independent of the initial point q',

$$P_{\text{eq.}}(q) = \sqrt{\omega/2\pi}\, \mathrm{e}^{-\omega q^2/2}, \qquad (22.13)$$

where $\omega q^2/2 \equiv \mathcal{U}(q)$ is the potential (22.8).

By contrast, for $\omega < 0$, for $t \to \infty$, one finds

$$P(q, t; q', t') \sim \sqrt{|\omega|/2\pi}\, \mathrm{e}^{\omega(t-t')/2}\, \mathrm{e}^{\omega q'^2/2}, \qquad (22.14)$$

which is not an equilibrium distribution, since it converges to zero everywhere.

Still for $\omega > 0$, the one- and two-point functions (22.11) for $t_1, t_2 \gg t'$ reduce to

$$\langle q(t_1) \rangle = 0, \quad \langle q(t_1)q(t_2) \rangle_\nu = \frac{1}{\omega}\, \mathrm{e}^{-\omega|t_1 - t_2|/2} = \frac{1}{2\pi} \int \frac{\mathrm{d}k}{k^2 + \omega^2/4}.$$

The parameter $\tau = 2/\omega$ is the *correlation time*, since it characterizes the exponential decay of time correlations. It is also the *relaxation time*, since it characterizes the approach to equilibrium. In what follows, very often we take the $t' \to -\infty$ limit in such a way that the system is at equilibrium at any finite time.

22.2.3 Addition of a time-dependent linear potential: Jarzinsky's relation

The addition of a constant force h in equation (22.7), which then becomes

$$\dot{q}(t) = -\tfrac{1}{2}\omega q(t) + \tfrac{1}{2}h + \nu(t),$$

is equivalent to a translation of $q(t)$ by $-h/\omega$ and all previous results are modified accordingly. For example, for $\omega > 0$, the equilibrium distribution simply becomes

$$P_{\text{eq.}}(q) = \sqrt{\omega/2\pi}\, \mathrm{e}^{-\omega(q-h/\omega)^2/2}. \qquad (22.15)$$

By contrast, we now add a time-dependent force term such that the Langevin equation (22.7) becomes

$$\dot{q}(t) = -\tfrac{1}{2}\omega q(t) + \tfrac{1}{2}h(t) + \nu(t), \qquad (22.16)$$

where $h(t)$ is continuous, vanishes for $t \le t'$ and is differentiable for $t > t'$. Moreover, we assume that, at initial time t', $q(t') = q'$.

The force derives from the potential,

$$\mathcal{U}(q, t) = \tfrac{1}{2}\omega q^2 - h(t)q.$$

The solution of the equation can be written as

$$q(t) = \mathrm{e}^{-\omega(t-t')/2}\, q' + \frac{1}{2} \int_{t'}^{t} \mathrm{d}\tau\, \mathrm{e}^{-\omega(t-\tau)/2}\, h(\tau) + r(t), \qquad (22.17)$$

where $r(t)$ satisfies equation (22.16) with $h(t) \equiv 0$ and $r(t') = 0$:

$$r(t) = \int_{t'}^{t} \mathrm{d}\tau\, \mathrm{e}^{-\omega(t-\tau)/2}\, \nu(\tau). \qquad (22.18)$$

In particular,

$$\langle r(t) \rangle_{\nu} = 0, \quad \rho(t) \equiv \langle r^2(t) \rangle_{\nu} = \frac{1}{\omega}\left(1 - e^{-\omega(t-t')}\right).$$

We also need the connected two-point function:

$$\langle r(t_1)r(t_2) \rangle_{\nu}^{\text{conn.}} = \frac{1}{\omega}\left[e^{-\omega|t_1-t_2|/2} - e^{\omega t' - \omega(t_1+t_2)/2}\right]. \tag{22.19}$$

Jarzinsky's relation [204]. We now assume that the distribution of the initial position $q(t') = q'$ is the equilibrium distribution in zero field (equation (22.13)).

For reasons that will be explained in Sections 22.3.4 and 22.4.2, we want to evaluate the expectation value of the quantity

$$\mathcal{J} = \exp\left[-\int_{t'}^{t''} dt\, \frac{\partial \mathcal{U}}{\partial t}\right] = \exp\left[\int_{t'}^{t''} dt\, \dot{h}(t)q(t)\right].$$

Then,

$$\langle \mathcal{J} \rangle_{\nu} = \exp\left[\int_{t'}^{t''} dt\, \dot{h}(t)\left(e^{-\omega(t-t')/2}\, q' + \tfrac{1}{2}\int_{t'}^{t} d\tau\, e^{-\omega(t-\tau)/2}\, h(\tau)\right)\right]$$

$$\times \left\langle \exp\int_{t'}^{t''} dt\, \dot{h}(t)r(t) \right\rangle_r.$$

First, the q'-dependent factor has to be averaged with the equilibrium distribution (22.12):

$$F_1 \equiv \sqrt{\omega/2\pi}\int dq'\, \exp\left[-\omega q'^2/2 + \int dt\, q'\dot{h}(t)\, e^{-\omega(t-t')/2}\right]$$

$$= \exp\left[\frac{1}{2\omega}\int dt_1\, dt_2\, \dot{h}(t_1)\dot{h}(t_2)\, e^{-\omega(t_1+t_2-2t')/2}\right]. \tag{22.20}$$

A second contribution comes from the linear shift in h. It can be rewritten, after a few transformations, as ($\theta(t)$ is the Heaviside step function, $\theta(t) = 0$ for $t < 0$, and $\theta(t) = 1$ for $t > 0$)

$$\ln F_2 = \tfrac{1}{2}\int_{t'}^{t''} dt\, \dot{h}(t)\int_{t'}^{t} d\tau\, e^{-\omega(t-\tau)/2}\, h(\tau)$$

$$= \tfrac{1}{2}\int_{t'}^{t} dt\, dt_1\, dt_2\, \theta(t_2-t)\theta(t-t_1)\, e^{-\omega(t_2-t)/2}\, \dot{h}(t_1)\dot{h}(t_2)$$

$$= \frac{1}{2\omega}h^2(t'') - \frac{1}{\omega}\int_{t'}^{t} dt_1\, dt_2\, \theta(t_2-t_1)\, e^{-\omega(t_2-t_1)/2}\, \dot{h}(t_1)\dot{h}(t_2). \tag{22.21}$$

A third contribution comes from averaging over $r(t)$ and involves the connected $r(t)$ two-point function:

$$F_3 = \exp\left\{\frac{1}{\omega}\int_{t'}^{t''} dt_1\, dt_2\, \theta(t_2 - t_1)\left[e^{-\omega(t_2-t_1)/2}\right.\right.$$
$$\left.\left. - e^{\omega t' - \omega(t_1+t_2)/2}\right]\dot{h}(t_1)\dot{h}(t_2)\right\}. \tag{22.22}$$

Thus,

$$F_1 F_3 = \exp\left[\frac{1}{\omega}\int_{t'}^{t''} dt_1\, dt_2\, \theta(t_2 - t_1)\,e^{-\omega(t_2-t_1)/2}\,\dot{h}(t_1)\dot{h}(t_2)\right]. \tag{22.23}$$

The second term in F_2 cancels the product $F_1 F_3$, and the product of the three factors then yields

$$\langle \mathcal{J}\rangle_\nu = e^{h^2(t'')/2\omega}.$$

The result has a remarkable property: it depends only on the value of $h(t)$ at t'' and not on the values for $t < t''$ [204]. It is a simple example of the general relation (22.56), since

$$\int dq\, e^{-\mathcal{U}(q,t)}\,/\mathcal{Z}' = \int dq\, e^{-\omega q^2/2 + h(t)q}\,/\mathcal{Z}' = e^{h^2(t)/2\omega},$$

where $\mathcal{Z}'$ is the normalization at time t':

$$\mathcal{Z}' = \int dq\, e^{-\mathcal{U}(q,t')} = \frac{1}{\sqrt{2\pi/\omega}}.$$

22.3   The Fokker–Planck formalism

We have defined by equation (22.4) the *time-dependent* probability distribution $P(\mathbf{q}, t; \mathbf{q}', t')$ for the stochastic position $\mathbf{q}(t)$, which formally can be written as

$$P(\mathbf{q}, t; \mathbf{q}', t') = \left\langle \prod_{i=1}^{d} \delta\left[q_i(t) - q_i\right]\right\rangle_\nu, \quad t \geq t'. \tag{22.24}$$

Because the Langevin equation is local in time, and the values of the noise at different times are uncorrelated, the distribution P satisfies a Markov property,

$$P(\mathbf{q}_3, t_3; \mathbf{q}_1, t_1) = \int d^d q_2\, P(\mathbf{q}_3, t_3; \mathbf{q}_2, t_2) P(\mathbf{q}_2, t_2; \mathbf{q}_1, t_1), \quad t_1 \leq t_2 \leq t_3. \tag{22.25}$$

Therefore, the distribution P is completely determined from its knowledge for small time intervals. It is convenient to introduce the quantum of bra-ket notation (not to be confused with the notation $\langle \bullet \rangle_\nu$ meaning average over the noise ν):

$$P(\mathbf{q}, t; \mathbf{q}', t') \equiv \langle \mathbf{q}|\, \mathbf{P}(t, t')\, |\mathbf{q}'\rangle. \tag{22.26}$$

Then, equation (22.25) can be rewritten in operator form as

$$\mathbf{P}(t_3, t_1) = \mathbf{P}(t_3, t_2)\mathbf{P}(t_2, t_1). \tag{22.27}$$

22.3.1 The FP equation

By differentiating equation (22.27) with respect to $t_3 \equiv t$ and taking the $t_2 = t_3 = t$ limit, one derives the differential equation

$$\frac{\partial \mathbf{P}(t, t')}{\partial t} = -\mathbf{H}(t)\mathbf{P}(t, t'), \text{ with } \mathbf{H}(t) = -\lim_{t' \to t} \frac{\partial \mathbf{P}(t, t')}{\partial t}. \tag{22.28}$$

The equation has the formal structure of the equation relating the Hamiltonian and the evolution operator in quantum mechanics, in imaginary time. The operator $\mathbf{H}$ plays a role formally analogous to a quantum Hamiltonian and thus can be called the FP Hamiltonian.

Similarly, differentiating the equation with respect to t_1 and taking the $t_2 = t_1$ limit, one obtains

$$\frac{\partial \mathbf{P}(t, t')}{\partial t'} = \mathbf{P}(t, t')\mathbf{H}(t'). \tag{22.29}$$

On the other hand, starting from the Langevin equation (22.1) and noise (22.2), one can prove directly (but, the proof has some subtle points: see Section 22.7.1) that equations (22.1) and (22.2) imply that the probability distribution $P(\mathbf{q}, t; \mathbf{q}', t')$ (equation (22.24)) at finite time $t > t'$ satisfies the partial differential evolution equation,

$$\frac{\partial}{\partial t} P(\mathbf{q}, t; \mathbf{q}', t') = \tfrac{1}{2}\Omega \nabla_q \cdot \left[\nabla_q + \nabla_q \mathcal{U}(\mathbf{q}, t)\right] P(\mathbf{q}, t; \mathbf{q}', t'). \tag{22.30}$$

The equation can be rewritten in operator notation as

$$\frac{\partial \mathbf{P}(t, t')}{\partial t} = -\mathbf{H}(t)\mathbf{P}(t, t')$$

with (in the operator notation of quantum mechanics with $\hbar = 1$),

$$\mathbf{H}(t) = \tfrac{1}{2}\Omega \left(\hat{\mathbf{p}}^2 - i\hat{\mathbf{p}} \cdot \nabla_q \mathcal{U}(\hat{\mathbf{q}}, t)\right), \tag{22.31}$$

where $\hat{\mathbf{q}}$ and $i\hat{\mathbf{p}} \equiv \nabla_q$ are operators acting on functions of $\mathbf{q}$ by multiplication and differentiation, respectively.

The FP equation establishes a formal correspondence between stochastic differential equations and statistical quantum mechanics. All observables that can be calculated from the Langevin equation by averaging over the noise, can be recovered by methods of quantum mechanics, using the FP Hamiltonian $\mathbf{H}$.

An integration over space shows that the form of the FP equation automatically ensures that probability is conserved:

$$\frac{\partial}{\partial t} \int \mathrm{d}^d q \, P(\mathbf{q}, t; \mathbf{q}', t') = 0.$$

The conservation equation follows from a property that can formally be expressed by noting that the Hamiltonian $\mathbf{H}$ has a constant as a left eigenvector:

$$\langle 0 | \, \mathbf{H}(t) = 0, \text{ with } \langle 0 | \mathbf{q} \rangle = 1. \tag{22.32}$$

Notation. In what follows, we will sometimes omit the dependence on the initial data and use for $P(\mathbf{q}, t; \mathbf{q}', t')$ the simplified notation $P(\mathbf{q}, t)$ when a reference to the initial boundary condition is not important.

22.3.2 *Dissipative Langevin equation*

In the case of the dissipative Langevin equation (equation (22.6)), $\mathcal{U}$ is time independent and the stochastic process is time-translation invariant. Equation (22.28) can then be integrated to yield

$$\mathbf{P}(t,t') = e^{-(t-t')\mathbf{H}}. \tag{22.33}$$

Moreover, after the transformation

$$\widetilde{\mathbf{H}} = e^{\mathcal{U}(\hat{\mathbf{q}})/2}\,\mathbf{H}\,e^{-\mathcal{U}(\hat{\mathbf{q}})/2},$$

one obtains

$$\widetilde{\mathbf{H}} = \tfrac{1}{2}\Omega\left(-\nabla_q + \tfrac{1}{2}\nabla_q\mathcal{U}\right)\left(\nabla_q + \tfrac{1}{2}\nabla_q\mathcal{U}\right) = \tfrac{1}{2}\Omega\mathbf{A}^\dagger\cdot\mathbf{A}, \tag{22.34}$$

where we have defined

$$\mathbf{A}(q,\nabla_q) \equiv \nabla_q + \tfrac{1}{2}\nabla_q\mathcal{U} \equiv i\hat{\mathbf{p}} + \tfrac{1}{2}\nabla_q\mathcal{U}. \tag{22.35}$$

The FP Hamiltonian (22.31) is thus equivalent to a Hermitian quantum Hamiltonian.

Similarly, if we introduce the operator (with the definition (22.26) for the operator **P**)

$$\tilde{\mathbf{P}}(t,t') = e^{\mathcal{U}(\hat{\mathbf{q}})/2}\,\mathbf{P}(t,t')\,e^{-\mathcal{U}(\hat{\mathbf{q}})/2}, \tag{22.36}$$

or, in matrix element form,

$$\langle \mathbf{q}\,|\tilde{\mathbf{P}}(t,t')|\,\mathbf{q}'\rangle = e^{\mathcal{U}(\mathbf{q})/2}\,\langle\mathbf{q}|\,\mathbf{P}(t,t')\,|\mathbf{q}'\rangle\,e^{-\mathcal{U}(\mathbf{q}')/2}, \tag{22.37}$$

the operator $\tilde{\mathbf{P}}(t,t') = e^{-(t-t')\widetilde{\mathbf{H}}}$ is Hermitian and the kernel $\langle\mathbf{q}|\tilde{\mathbf{P}}(t,t')|\mathbf{q}'\rangle$ satisfies

$$\partial_t\,\langle \mathbf{q}\,|\tilde{\mathbf{P}}(t,t')|\,\mathbf{q}'\rangle = -\tfrac{1}{2}\Omega\Big[-\Delta_q + \tfrac{1}{4}\big(\nabla_q\mathcal{U}(\mathbf{q})\big)^2$$
$$- \tfrac{1}{2}\Delta_q\mathcal{U}(\mathbf{q})\Big]\,\langle\mathbf{q}\,|\tilde{\mathbf{P}}(t,t')|\,\mathbf{q}'\rangle. \tag{22.38}$$

In the form (22.34), one notes that the Hamiltonian $\widetilde{\mathbf{H}}$ is positive. Moreover, if the function $e^{-\mathcal{U}(\mathbf{q})/2}$ is normalizable,

$$\langle\mathbf{q}|0\rangle = e^{-\mathcal{U}(\mathbf{q})/2}, \quad \langle 0|0\rangle = \int d^d q\; e^{-\mathcal{U}(\mathbf{q})} < \infty,$$

it is the eigenfunction corresponding to the eigenvalue 0:

$$\mathbf{A}|0\rangle = 0 \quad\Rightarrow\quad \widetilde{\mathbf{H}}|0\rangle = 0,$$

and thus is the ground state of $\widetilde{\mathbf{H}}$. At large times, the operator $e^{-(t-t')\widetilde{\mathbf{H}}}$ projects onto its ground state. In terms of the probability distribution $P(\mathbf{q},t)$, this implies,

$$\lim_{t\to\infty} P(\mathbf{q},t) = e^{-\mathcal{U}(\mathbf{q})/2}\,e^{\mathcal{U}(\mathbf{q}')/2}\langle\mathbf{q}|0\rangle\langle 0|\mathbf{q}'\rangle = e^{-\mathcal{U}(\mathbf{q})}.$$

At large times, the distribution $P(\mathbf{q},t)$ converges towards the equilibrium distribution $e^{-\mathcal{U}(\mathbf{q})}$.

Moreover, if $\epsilon_1 > 0$ is the second eigenvalue of **H**, then the leading correction to the equilibrium distribution is proportional to $e^{-\epsilon_1 t}$ and, thus, $\tau = 1/\epsilon_1$ is the relaxation and *correlation time*.

By contrast, if the wave function $e^{-\mathcal{U}(\mathbf{q})/2}$ is not normalizable, there exists no equilibrium distribution; the Langevin equation (22.6) has only runaway solutions.

22.3.3 Detailed balance

From (22.36), one infers

$$\langle \mathbf{q}| \, \mathbf{P}^{\dagger}(t,t') \, |\mathbf{q}'\rangle = e^{\mathcal{U}(\mathbf{q})/2} \, \langle \mathbf{q} \, |\tilde{\mathbf{P}}(t,t')| \, \mathbf{q}'\rangle \, e^{-\mathcal{U}(\mathbf{q}')/2},$$

and, since the operator $\mathbf{P}$ is real,

$$\langle \mathbf{q}'| \, \mathbf{P}(t,t') \, |\mathbf{q}\rangle = e^{\mathcal{U}(\mathbf{q})-\mathcal{U}(\mathbf{q}')} \, \langle \mathbf{q}| \, \mathbf{P}(t,t') \, |\mathbf{q}'\rangle, \qquad (22.39)$$

which relates a process and its time reversed form and is the *detailed balance condition*.

The purely dissipative Langevin equation (22.6) thus leads to an evolution that satisfies detailed balance.

In the example of the Langevin equation (22.7), detailed balance follows directly from the explicit expression (22.12) for the probability distribution.

Detailed balance and equilibrium. Detailed balance then determines the equilibrium distribution by a simple argument. We rewrite equation (22.39) as

$$\langle \mathbf{q}'| \, \mathbf{P}(t,t') \, |\mathbf{q}\rangle \, e^{-\mathcal{U}(\mathbf{q})} = e^{-\mathcal{U}(\mathbf{q}')} \, \langle \mathbf{q}| \, \mathbf{P}(t,t') \, |\mathbf{q}'\rangle \, .$$

Integrating over $\mathbf{q}$ and using probability conservation in the right-hand side, one finds

$$\int \mathrm{d}^d q \, \langle \mathbf{q}'| \, \mathbf{P}(t,t') \, |\mathbf{q}\rangle \, e^{-\mathcal{U}(\mathbf{q})} = e^{-\mathcal{U}(\mathbf{q}')} \, .$$

Considering the properties of the operator $\mathbf{P}$, this equation implies that $e^{-\mathcal{U}(\mathbf{q})}$ is the equilibrium distribution, provided it is normalizable.

In particular, we recover that, if the system is initially at equilibrium, it remains in the same state at any later time, as follows also directly from the FP equation (22.30).

Example: The Langevin equation with linear driving force. After the transformation (22.36), the FP Hamiltonian associated with equation (22.7) takes the form

$$\tilde{\mathbf{H}} = \tfrac{1}{2}\hat{p}^2 + \tfrac{1}{8}\omega^2 \hat{q}^2 - \tfrac{1}{4}\omega \, , \qquad (22.40)$$

where one recognizes the Hamiltonian of a shifted quantum harmonic oscillator. The eigenvalues ϵ_n of $\mathbf{H}$ are

$$\epsilon_n = \tfrac{1}{2} \left(n + \tfrac{1}{2} \right) |\omega| - \tfrac{1}{4}\omega \, , \quad n \geq 0 \, . \qquad (22.41)$$

We note that if ω is positive, ϵ_0 vanishes. Correspondingly, $e^{-\omega q^2/2}$ is normalizable and is thus the equilibrium distribution. The second eigenvalue then is $\epsilon_1 = \omega/2$ and we recover the relaxation time.

By contrast, if ω is negative, the lowest eigenvalue is $|\omega|/2 > 0$, there is no equilibrium distribution and one recovers the asymptotic form (22.14).

22.3.4 Gradient time-dependent force and Jarzinsky's relation

When the potential in equation (22.1) is explicitly time dependent, the situation is modified. The FP equation (22.30) can be written as

$$\frac{\partial P(\mathbf{q}, t; \mathbf{q}', t')}{\partial t} = \tfrac{1}{2} \Omega \nabla_q \cdot [\nabla_q + \nabla_q \mathcal{U}(\mathbf{q}, t)] \, P(\mathbf{q}, t; \mathbf{q}', t')$$

$$\equiv \nabla_q \cdot \mathrm{D}_q(t) P(\mathbf{q}, t; \mathbf{q}', t'), \tag{22.42}$$

with

$$\mathrm{D}_q(t) = \nabla_q + \nabla_q \mathcal{U}(\mathbf{q}, t).$$

As a consequence, detailed balance is replaced by a more complicated relation. The transformation (22.37),

$$\langle \mathbf{q} \, | \tilde{\mathbf{P}}(t, t') | \, \mathbf{q}' \rangle = e^{\mathcal{U}(\mathbf{q}, t)/2} \, \langle \mathbf{q} | \, \mathbf{P}(t, t') \, | \mathbf{q}' \rangle \, e^{-\mathcal{U}(\mathbf{q}', t')/2}, \tag{22.43}$$

leads to the equation

$$\partial_t \, \langle \mathbf{q} \, | \tilde{\mathbf{P}}(t, t') | \, \mathbf{q}' \rangle$$

$$= \tfrac{1}{2} \left\{ \Omega \left[\Delta_q - \tfrac{1}{4} (\nabla_q \mathcal{U}(\mathbf{q}, t))^2 + \tfrac{1}{2} \Delta_q \mathcal{U}(\mathbf{q}, t) \right] + \partial_t \mathcal{U}(\mathbf{q}, t) \right\} \langle \mathbf{q} \, | \tilde{\mathbf{P}}(t, t') | \, \mathbf{q}' \rangle . \tag{22.44}$$

Taking the Hermitian conjugate of equation (22.29), one obtains

$$\frac{\partial \mathbf{P}^\dagger(t, t')}{\partial t'} = \mathbf{H}^\dagger(t') \mathbf{P}^\dagger(t, t').$$

The arguments of Section 22.3.2 can then be generalized to derive

$$\mathbf{H}^\dagger(t') = e^{\mathcal{U}(\hat{\mathbf{q}}, t')} \, \mathbf{H}(t') \, e^{-\mathcal{U}(\hat{\mathbf{q}}, t')} .$$

Defining the operator

$$\mathcal{O}(t', t) = e^{-\mathcal{U}(\hat{\mathbf{q}}, t')} \, \mathbf{P}^\dagger(t, t') \, e^{\mathcal{U}(\hat{\mathbf{q}}, t)}, \tag{22.45}$$

one finds

$$\left[\frac{\partial}{\partial t'} + \frac{\partial \mathcal{U}(\hat{\mathbf{q}}, t')}{\partial t'} \right] \mathcal{O}(t', t) = \mathbf{H}(t') \mathcal{O}(t', t),$$

with the boundary condition $\mathcal{O}(t, t) = \mathbf{1}$.

When $\mathcal{U}$ is time independent, this relation again leads to detailed balance. More generally, taking matrix elements, one finds

$$\langle \mathbf{q} | \, \mathcal{O}(t', t) \, | \mathbf{q}' \rangle = e^{-\mathcal{U}(\mathbf{q}, t')} \, \langle \mathbf{q}' | \, \mathbf{P}(t, t') \, | \mathbf{q} \rangle \, e^{\mathcal{U}(\mathbf{q}', t)}, \tag{22.46}$$

where

$$\left[\frac{\partial}{\partial t'} + \frac{\partial \mathcal{U}(\mathbf{q}, t')}{\partial t'} \right] \langle \mathbf{q} | \, \mathcal{O}(t', t) \, | \mathbf{q}' \rangle = -\nabla_q \cdot \mathrm{D}_q(t') \, \langle \mathbf{q} | \, \mathcal{O}(t', t) \, | \mathbf{q}' \rangle .$$

An interpretation of the matrix elements of the operator (22.45) will be provided in Section 22.4.1. Jarzinsky's relation then follows from equation (22.46).

22.4 Path integral representation

We have shown that the probability distribution corresponding to the evolution governed by the Langevin equation (22.1) satisfies a partial differential equation. Correspondingly, it has also a path integral representation, which we now describe.

22.4.1 Dissipative Langevin equation

In the case of the dissipative Langevin equation, equation (22.38), which has exactly the form of a Schrödinger equation in imaginary time, implies for the matrix elements (22.37) the path integral representation,

$$\tilde{P}(\mathbf{q}'',t'';\mathbf{q}',t') \equiv \langle \mathbf{q} \,|\tilde{\mathbf{P}}(t,t')|\, \mathbf{q}' \rangle = \int [\mathrm{d}q(\tau)]\, \mathrm{e}^{-\tilde{\mathcal{S}}(\mathbf{q})/\Omega},$$

with

$$\tilde{\mathcal{S}}(\mathbf{q}) = \tfrac{1}{2}\int_{t'}^{t''}\mathrm{d}\tau\,\left[\dot{\mathbf{q}}^2(\tau) + \tfrac{1}{4}\Omega^2\left(\nabla_q \mathcal{U}(\mathbf{q}(\tau))\right)^2 - \tfrac{1}{2}\Omega^2\Delta_q\mathcal{U}(\mathbf{q}(\tau))\right]. \qquad (22.47)$$

and the boundary conditions $\mathbf{q}(t') = \mathbf{q}'$, $\mathbf{q}(t'') = \mathbf{q}''$. This implies for the probability distribution $P(\mathbf{q}'',t'';\mathbf{q}',t')$ (equation (22.37)),

$$P(\mathbf{q}'',t'';\mathbf{q}',t') = \int [\mathrm{d}q(\tau)]\, \mathrm{e}^{-\mathcal{S}(\mathbf{q})/\Omega},$$

with

$$\mathcal{S}(\mathbf{q}) = \tilde{\mathcal{S}}(\mathbf{q}) + \tfrac{1}{2}\Omega\big(\mathcal{U}(\mathbf{q}'') - \mathcal{U}(\mathbf{q}')\big). \qquad (22.48)$$

We then note that

$$\tfrac{1}{2}\Omega\big(\mathcal{U}(\mathbf{q}'') - \mathcal{U}(\mathbf{q}')\big) = \tfrac{1}{2}\Omega\int_{t'}^{t''}\mathrm{d}\tau\,\dot{\mathbf{q}}(\tau)\cdot\nabla_q\mathcal{U}(\mathbf{q}(\tau)). \qquad (22.49)$$

Therefore, the *dynamic action* $\mathcal{S}(\mathbf{q})$ can be rewritten as

$$\mathcal{S}(\mathbf{q}) = \tfrac{1}{2}\int_{t'}^{t''}\mathrm{d}\tau\,\left\{\left[\dot{\mathbf{q}}(\tau) + \tfrac{1}{2}\Omega\nabla_q\mathcal{U}(\mathbf{q}(\tau))\right]^2 - \tfrac{1}{2}\Omega^2\Delta_q\mathcal{U}(\mathbf{q}(\tau))\right\}. \qquad (22.50)$$

Integration within path integrals. While the relation (22.49) outside a path integral is obvious, this is no longer the case within a path integral, due to the non-commutation between integration over time and path integration. This non-commutation is related to the non-commutation of quantum operators in expression (22.31). From the path integral viewpoint, the reason is that generic paths contributing to the path integral are not differentiable. While the path integral with the action (22.47) is defined, because the action (22.50) contains a linear term in the time derivative, a perturbative calculation of the path integral then involves the undefined quantity sgn(0) (sgn is the sign function). Consistency with the definition (22.48) requires sgn(0) = 0.

22.4.2 Detailed balance and path integral

We assume the boundary conditions, $\mathbf{q}(t') = \mathbf{q}'$, and $\mathbf{q}(t'') = \mathbf{q}''$. We then substitute

$$\int_{t'}^{t''} d\tau \left[\dot{\mathbf{q}} + \tfrac{1}{2}\Omega\nabla_q \mathcal{U}(q)\right]^2 = \int_{t'}^{t''} d\tau \left[\dot{\mathbf{q}} - \tfrac{1}{2}\Omega\nabla_q \mathcal{U}(q)\right]^2 + 2\Omega\left[\mathcal{U}(\mathbf{q}'') - \mathcal{U}(\mathbf{q}')\right],$$

inside the action. The path integral becomes

$$P(\mathbf{q}'', t''; \mathbf{q}', t') = \int [dq(t)]\, e^{-\mathcal{S}(q)/\Omega - \mathcal{U}(\mathbf{q}'') + \mathcal{U}(\mathbf{q}')},$$

with now

$$\mathcal{S}(q) = \tfrac{1}{2}\int_{t'}^{t''} d\tau \left[\left(\dot{\mathbf{q}} - \tfrac{1}{2}\Omega\nabla_q \mathcal{U}(\mathbf{q})\right)^2 - \tfrac{1}{2}\Omega^2\Delta_q\mathcal{U}\right].$$

We then change variables $\mathbf{q}(t) \mapsto \mathbf{q}(t' + t'' - t)$ in the path integral and then $t' + t'' - t \mapsto t$. This only has the effect of changing the sign of $\dot{q}(t)$ in the action. One recovers the initial action,

$$\mathcal{S}(q) = \tfrac{1}{2}\int_{t'}^{t''} d\tau \left[\left(\dot{\mathbf{q}} + \tfrac{1}{2}\Omega\nabla_q \mathcal{U}(\mathbf{q})\right)^2 - \tfrac{1}{2}\Omega^2\Delta_q\mathcal{U}\right],$$

but with the different boundary conditions $\mathbf{q}(t'') = \mathbf{q}'$ and $\mathbf{q}(t') = \mathbf{q}''$. One concludes

$$P(\mathbf{q}'', t''; \mathbf{q}', t') = e^{-\mathcal{U}(\mathbf{q}'') + \mathcal{U}(\mathbf{q}')}\, P(\mathbf{q}', t''; \mathbf{q}'', t'), \tag{22.51}$$

which is the detailed balance condition, a result that can also be easily derived from the FP equation (see Section 22.3.4) and, equivalently, from the form (22.47) of the action.

22.4.3 Time-dependent force deriving from a potential

Deriving the path integral representation in the time-dependent case follows the same lines but with a slight modification. Again, it is safer to start from the matrix elements of $\tilde{\mathbf{P}}(t, t')$, because equation (22.44) does not contain a space differentiation with a space-dependent coefficient.

The corresponding action is then

$$\tilde{\mathcal{S}}(\mathbf{q}) = \int_{t'}^{t''} d\tau \left[\tfrac{1}{2}\dot{\mathbf{q}}^2(\tau) + \mathcal{V}(\mathbf{q}(\tau), \tau)\right] \tag{22.52}$$

with

$$\mathcal{V}(\mathbf{q}(\tau), \tau) = \tfrac{1}{2}\Omega\left[\tfrac{1}{8}\left(\nabla_q\mathcal{U}(\mathbf{q}(\tau), \tau)\right)^2 - \tfrac{1}{2}\Omega\Delta_q\mathcal{U}(\mathbf{q}(\tau), \tau) - \partial_\tau\mathcal{U}(\mathbf{q}(\tau), \tau)\right].$$

The action corresponding to the probability distribution P is then

$$\mathcal{S}(\mathbf{q}) = \tilde{\mathcal{S}}(\mathbf{q}) + \tfrac{1}{2}\Omega\left[\mathcal{U}(\mathbf{q}'', t'') - \mathcal{U}(\mathbf{q}', t')\right]. \tag{22.53}$$

We note that

$$
\mathcal{U}(\mathbf{q}'', t'') - \mathcal{U}(\mathbf{q}', t') = \int_{t'}^{t''} d\tau \frac{d}{d\tau} \mathcal{U}(\mathbf{q}(\tau), \tau)
$$

$$
= \int_{t'}^{t''} d\tau \left[\dot{\mathbf{q}}(\tau) \cdot \nabla_q \mathcal{U}(\mathbf{q}(\tau), \tau) + \partial_\tau \mathcal{U}(\mathbf{q}(\tau), \tau) \right].
$$

After substitution of the second expression in $\mathcal{S}(\mathbf{q})$, the terms with the partial τ-derivative cancel and the expression can again be written as

$$
\mathcal{S}(\mathbf{q}) = \tfrac{1}{2} \int_{t'}^{t''} d\tau \left[\left(\dot{\mathbf{q}}(\tau) + \tfrac{1}{2}\Omega\nabla_q \mathcal{U}(\mathbf{q}(\tau), \tau) \right)^2 - \tfrac{1}{2}\Omega^2 \Delta_q \mathcal{U}(\mathbf{q}(\tau), \tau) \right]. \quad (22.54)
$$

22.4.4 Time-dependent force and Jarzinsky's relation

We assume that $\mathcal{U}(\mathbf{q}, t)$ is explicitly time dependent. However, note that $\mathcal{U}(\mathbf{q}, t)$ is determined by the Langevin equation only up to a function time independent of $\mathbf{q}$. Then, the cross-term in equation (22.54) cannot be integrated and instead yields

$$
\int_{t'}^{t''} d\tau \, \dot{\mathbf{q}} \cdot \nabla_q \mathcal{U}(q(\tau), \tau) = \left[\mathcal{U}(\mathbf{q}(t''), t'') - \mathcal{U}(\mathbf{q}(t'), t') \right]
$$

$$
- \int_{t'}^{t''} d\tau \frac{\partial \mathcal{U}(q(\tau), \tau)}{\partial \tau}. \quad (22.55)
$$

It follows that

$$
\tfrac{1}{2} \int_{t'}^{t''} d\tau \left(\dot{\mathbf{q}} + \tfrac{1}{2}\Omega\nabla_q \mathcal{U}(\mathbf{q}, \tau) \right)^2 + \Omega \int_{t'}^{t''} d\tau \frac{\partial \mathcal{U}}{\partial \tau}
$$

$$
= \tfrac{1}{2} \int_{t'}^{t''} d\tau \left(\dot{\mathbf{q}} - \tfrac{1}{2}\Omega\nabla_q \mathcal{U}(\mathbf{q}, \tau) \right)^2 + \Omega\left(\mathcal{U}(\mathbf{q}(t''), t'') - \mathcal{U}(\mathbf{q}(t'), t') \right).
$$

We then use the same arguments as in the proof of detailed balance and obtain

$$
\left\langle \exp\left[-\int_{t'}^{t''} d\tau \frac{\partial \mathcal{U}}{\partial \tau} \right] \right\rangle_\nu = e^{-\mathcal{U}(\mathbf{q}'', t'') + \mathcal{U}(\mathbf{q}', t')} P_*(\mathbf{q}', t''; \mathbf{q}'', t'),
$$

where the average in the left-hand side is taken at $\mathbf{q}'$ and $\mathbf{q}''$ fixed, and P_* corresponds to the process with $\mathcal{U}(\mathbf{q}, -t)$.

We now assume that $\mathcal{U}(\mathbf{q}, t)$ is constant in time for $t \leq t'$ and the system is at equilibrium at $t = t'$. Therefore, we can integrate the left-hand side over all $\mathbf{q}'$ with the normalized weight $e^{-\mathcal{U}(\mathbf{q}', t')}/\mathcal{Z}'$, where

$$
\mathcal{Z}' = \int d^d q' \, e^{-\mathcal{U}(\mathbf{q}', t')}.
$$

Then,

$$
\left\langle \exp\left[-\int_{t'}^{t''} d\tau \frac{\partial \mathcal{U}}{\partial \tau} \right] \right\rangle_{\nu, \mathbf{q}'} = e^{-\mathcal{U}(\mathbf{q}'', t'')}/\mathcal{Z}'.
$$

Finally, after an integration over all $\mathbf{q}''$, one obtains Jarzinsky's relation [204],

$$\left\langle \exp\left[-\int_{t'}^{t''} \mathrm{d}\tau \, \frac{\partial \mathcal{U}}{\partial \tau}\right]\right\rangle_{\nu,\mathbf{q}',\mathbf{q}''} = \mathcal{Z}''/\mathcal{Z}' = \mathrm{e}^{\mathcal{W}''-\mathcal{W}'}, \qquad (22.56)$$

where $\mathcal{W}'$ and $\mathcal{W}'$ are the initial and final free energies, respectively.

Remarks.

(i) Since, in terms of the Langevin equation, $\mathcal{U}(\mathbf{q},t)$ is defined only up to a function of time, the free energy variation depends on a more precise definition of $\mathcal{U}(\mathbf{q},t)$.

(ii) Relation (22.56) can be modified by adding any total time derivative to $\partial \mathcal{U}/\partial t$.

22.5   The dissipative Langevin equation: Supersymmetric formulation

The path integral associated to the Langevin equation in the special case (22.6),

$$\dot{q}_i(t) = -\frac{\Omega}{2}\frac{\partial \mathcal{U}}{\partial q_i} + \nu_i(t), \qquad (22.57)$$

with the noise (22.2), has a remarkable representation. At the price of introducing a few auxiliary paths, it can be shown to exhibit the simplest form of supersymmetry [185].

22.5.1   *Grassmann coordinates and algebraic properties*

To discuss supersymmetry, it is convenient to add to time and space two *Grassmann coordinates*, θ and $\bar{\theta}$, generators of a Grassmann algebra $\mathfrak{A}$:

$$\theta^2 = \bar{\theta}^2 = 0, \quad \theta\bar{\theta} = -\bar{\theta}\theta.$$

Grassmann derivatives. We also define two linear operations acting on $\mathfrak{A}$:

$$\frac{\partial}{\partial \theta}, \quad \frac{\partial}{\partial \bar{\theta}},$$

such that (with $\theta \mapsto \theta_1, \bar{\theta} \mapsto \theta_2$) with the θ_i they form a representation of fermion creation and annihilation operators:

$$\frac{\partial}{\partial \theta_i}\frac{\partial}{\partial \theta_j} + \frac{\partial}{\partial \theta_j}\frac{\partial}{\partial \theta_i} = 0,$$

$$\frac{\partial}{\partial \theta_i}\theta_j + \theta_j\frac{\partial}{\partial \theta_i} = \delta_{ij},$$

$$\theta_i\theta_j + \theta_i\theta_j = 0.$$

Integrals. We also define definite integrals on $\mathfrak{A}$ by

$$\int \mathrm{d}\theta \equiv \frac{\partial}{\partial \theta}, \quad \int \mathrm{d}\bar{\theta} \equiv \frac{\partial}{\partial \bar{\theta}}.$$

Then, for example,

$$\int \mathrm{d}\theta \int \mathrm{d}\bar{\theta} \, \mathrm{e}^{\mu\bar{\theta}\theta} = \int \mathrm{d}\theta \int \mathrm{d}\bar{\theta} \, (1 + \mu\bar{\theta}\theta) = \mu.$$

Grassmann parity. We define an algebra automorphism on $\mathfrak{A}$ by

$$P(\theta) = -\theta \,, \quad P(\bar{\theta}) = -\bar{\theta} \,.$$

Finally, when the number of generators is even, one can define a 'complex' conjugation in $\mathfrak{A}$ (with the formal properties of a Hermitian conjugation) and a scalar product [6].

22.5.2 Superpaths and covariant derivatives

We introduce a superpath,

$$\phi_i(t; \bar{\theta}, \theta) = q_i(t) + \theta \bar{c}_i(t) + c_i(t) \bar{\theta} + \theta \bar{\theta} \bar{q}_i(t),$$

where $\bar{q}_i(t)$ is an auxiliary path and $\theta, \bar{\theta}, \bar{c}_i(t), c_i(t)$ are independent generators of an infinite-dimensional Grassmann algebra (see also Section 2.9.2),

$$c_i(t)c_j(t') + c_j(t')c_i(t) = 0 \,, \quad \theta c_i(t) + c_i(t)\theta = 0 \,, \quad \bar{\theta} c_i(t) + c_i(t)\bar{\theta} = 0 \,.$$

The dynamic action corresponding to the Langevin equation (22.57) with the noise Gaussian distribution (22.2),

$$[\mathrm{d}\rho(\nu)] = [\mathrm{d}\nu] \exp\left[-\int \mathrm{d}t\, \nu^2(t)/2\Omega\right],$$

and the boundary conditions $\mathbf{q}(t'') = \mathbf{q}''$, $\mathbf{q}(t') = \mathbf{q}'$, can be then rewritten as

$$\mathcal{S}(\phi) = \int_{t'}^{t''} \mathrm{d}t\, \mathrm{d}\bar{\theta}\, \mathrm{d}\theta \left(\frac{2}{\Omega} \bar{\mathrm{D}}\phi \cdot \mathrm{D}\phi + \mathcal{U}(\phi)\right) \tag{22.58}$$

with the definitions of the *covariant derivatives*,

$$\bar{\mathrm{D}} = \frac{\partial}{\partial \bar{\theta}}, \quad \mathrm{D} = \frac{\partial}{\partial \theta} - \bar{\theta}\frac{\partial}{\partial t}. \tag{22.59}$$

For convenience, we have rescaled the action (22.50) by a factor Ω ($\mathcal{S}/\Omega \mapsto \mathcal{S}$).

The equivalence between dynamic actions can be explicitly verified by expanding the superpath. Thus,

$$\bar{\mathrm{D}}\phi = -\mathbf{c} - \theta\bar{\mathbf{q}}, \quad \mathrm{D}\phi = \bar{\mathbf{c}} + \bar{\theta}\bar{\mathbf{q}} - \bar{\theta}\left(\dot{\mathbf{q}} + \theta\dot{\bar{\mathbf{c}}}\right).$$

Also,

$$\mathcal{U}(\phi) = \mathcal{U}(\mathbf{q}) + \nabla_q \mathcal{U}(\mathbf{q}) \cdot \left(\theta\bar{\mathbf{c}} + \mathbf{c}\bar{\theta} + \theta\bar{\theta}\bar{\mathbf{q}}\right) + \tfrac{1}{2}\theta\bar{\theta}\sum_{i,j} \partial^2_{q_i,q_j}\, \mathcal{U}(\mathbf{q})\left(\bar{c}_i c_j + \bar{c}_j c_i\right).$$

One then identifies the coefficient of $\theta\bar{\theta}$ and uses

$$\int \mathrm{d}\bar{\theta}\, \mathrm{d}\theta\, \theta\bar{\theta} = 1 \,.$$

Therefore,

$$S(\phi) = \int_{t'}^{t''} dt \left[\frac{2}{\Omega} \left(\bar{\mathbf{c}} \cdot \mathbf{c} + \bar{\mathbf{q}} \cdot \dot{\mathbf{q}} - \bar{\mathbf{q}}^2 \right) + \bar{\mathbf{q}} \cdot \nabla_q \mathcal{U}(\mathbf{q}) + \partial^2_{q_i,q_j} \, \mathcal{U}(\mathbf{q}) \bar{c}_i c_j \right].$$

Since the action is quadratic in $\bar{\mathbf{q}}$, the integral over $\bar{\mathbf{q}}$ can be performed. This amounts to replacing $\bar{\mathbf{q}}$ by the solution of the $\bar{\mathbf{q}}$ variational equation,

$$\bar{\mathbf{q}} = \tfrac{1}{2}\dot{\mathbf{q}} + \tfrac{1}{4}\Omega\nabla_q\mathcal{U}(\mathbf{q}).$$

The integral over Grassmann yields the determinant ($\theta(t)$ is the Heaviside step function)

$$\det\left(\frac{\partial}{\partial t}\delta_{ij} - \tfrac{1}{2}\Omega\frac{\partial^2\mathcal{U}}{\partial q_i\partial q_j} \right) \propto \det\left(\delta(t-t')\delta_{ij} - \tfrac{1}{2}\Omega\theta(t-t')\frac{\partial^2\mathcal{U}(t)}{\partial q_i\partial q_j} \right),$$

where the choice of $\theta(t)$ as the inverse of $\partial/\partial t$ corresponds to the causality of the Langevin equation.

Using the identity $\ln\det = \mathrm{tr}\ln$ and expanding the logarithm in powers of Ω one notices that all terms vanish due to the $\theta(t)$ functions except the first one, which is proportional to $\theta(0)$ and, thus, undefined. However, a careful analysis indicates that the proper choice is $\theta(0) = 1/2$. The final result is

$$S(\mathbf{q}) = \int dt \left[\frac{1}{2\Omega} \left(\dot{\mathbf{q}} + \tfrac{1}{2}\Omega\nabla_q\mathcal{U}(\mathbf{q}) \right)^2 - \tfrac{1}{4}\Omega\Delta\mathcal{U}(\mathbf{q}) \right],$$

in agreement with expression (22.50).

22.5.3 Supersymmetry

The covariant derivatives $\bar{D}$ and D satisfy the anticommutation relations

$$D^2 = \bar{D}^2 = 0, \quad D\bar{D} + \bar{D}D = -\frac{\partial}{\partial t}. \tag{22.60}$$

We then introduce two other Grassmann-type differential operators,

$$Q = \frac{\partial}{\partial\theta}, \quad \bar{Q} = \frac{\partial}{\partial\bar{\theta}} + \theta\frac{\partial}{\partial t}, \tag{22.61}$$

Both *anticommute* with D and $\bar{D}$ and satisfy

$$Q^2 = \bar{Q}^2 = 0, \quad Q\bar{Q} + \bar{Q}Q = \frac{\partial}{\partial t}. \tag{22.62}$$

The two pairs $D, \bar{D}$ and $Q, \bar{Q}$ provide the simplest example of generators of supersymmetry.

We then observe that Q generates a translation of θ, and the action (22.58) is translation invariant. The operator $\bar{Q}$ is the generator of an additional symmetry. Indeed, if we perform a variation of ϕ of the form

$$\delta\phi_i(t,\theta,\bar{\theta}) = \bar{\varepsilon}\bar{Q}\phi_i\,, \tag{22.63}$$

which in component form reads

$$
\begin{aligned}
\delta q_i &= c_i\bar{\varepsilon}\,, & \delta c_i &= 0\,, \\
\delta\bar{c}_i &= (\bar{q}_i - \dot{q}_i)\,\bar{\varepsilon}\,, & \delta\bar{q}_i &= \dot{c}_i\bar{\varepsilon}\,,
\end{aligned}
\tag{22.64}
$$

we observe that the variation of the action density is a *total derivative*. This is clear for $\mathcal{U}$ because it does not explicitly depend on t and $\bar{\theta}$. For the remaining term, the additional property that $\bar{Q}$ anticommutes with D and $\bar{D}$ has to be used:

$$\delta\left[\bar{D}\phi D\phi\right] = \bar{D}\left[\bar{\varepsilon}\bar{Q}\phi\right]D\phi + \bar{D}\phi D\left[\bar{\varepsilon}\bar{Q}\phi\right] = \bar{\varepsilon}\bar{Q}\left[\bar{D}\phi D\phi\right],$$

which is a total derivative. The action is thus supersymmetric. The operators D and $\bar{D}$ are *covariant derivatives with respect to the supersymmetry*.

This supersymmetry is directly related to the property that the corresponding FP Hamiltonian (22.31) is then equivalent to a positive Hamiltonian of the form (22.34).

A few properties.

(i) The anticommutator (22.62) of $\bar{Q}$ and Q generates time translations. Thus, supersymmetry implies time-translation invariance.

(ii) $\bar{Q}$ and Q seem to have a different structure. But, if one takes the translated time $t + \theta\bar{\theta}$ as a new variable, then $\bar{Q}$ reduces to a $\bar{\theta}$ translation. To emphasize the symmetry between $\bar{Q}$ and Q, or $\bar{\theta}$ and θ, one can also substitute $t \mapsto t + \frac{1}{2}\theta\bar{\theta}$. We find it more convenient to remain with the original variables but this is mainly a matter of taste.

(iii) Considering the fermions as additional dynamic variables, one can define a Hamiltonian associated with the supersymmetric action in boson–fermion space. One finds that the functional integral describes both the Langevin equation and *its time-reversed form*.

22.5.4 *Supersymmetry and detailed balance*

We have already shown that the dissipative Langevin equation leads to detailed balance. As an exercise, we verify this property with the supersymmetric formalism. For this purpose, in the path integral, we change path variables, setting

$$\phi(t;\bar{\theta},\theta) = \psi(t + \theta\bar{\theta};\bar{\theta},\theta). \tag{22.65}$$

Acting on ψ, the differentiation operators become

$$\bar{D} = \frac{\partial}{\partial\bar{\theta}} - \theta\frac{\partial}{\partial t}, \quad D = \frac{\partial}{\partial\theta}. \tag{22.66}$$

The two contributions to the action become

$$\int_{t'}^{t''} dt\, d\bar\theta\, d\theta\, \mathcal{U}\big(\psi(t+\theta\bar\theta;\bar\theta,\theta)\big) = \int_{t'}^{t''} dt\, d\bar\theta\, d\theta\, \mathcal{U}\big(\psi(t;\bar\theta,\theta)\big) + \int_{t'}^{t''} dt\, d\bar\theta\, d\theta\, \theta\bar\theta\frac{\partial\mathcal{U}}{\partial t}$$

$$= \int_{t'}^{t''} dt\, d\bar\theta\, d\theta\, \mathcal{U}\big(\psi(t;\bar\theta,\theta)\big) + \mathcal{U}(\mathbf{q}'') - \mathcal{U}(\mathbf{q}').$$

For the derivative term, we change $t \mapsto t' + t'' - t$, and $\theta \mapsto -\bar\theta$, $\bar\theta \mapsto \theta$. We find

$$\bar D = \frac{\partial}{\partial\theta} - \bar\theta\frac{\partial}{\partial t}, \quad D = -\frac{\partial}{\partial\bar\theta}.$$

But then $\bar D\psi$ and $D\psi$ anticommute and, after the commutation and the change $\psi \mapsto \phi$, we recover the initial expression. This proves detailed balance within the supersymmetric framework.

22.6   Critical dynamics: The Langevin equation in field theory

We now consider Langevin equations for fields as tools to discussing critical dynamics [202]. A number of results obtained in the case of a finite number of degrees of freedom have straightforward extensions to field theory and are not derived again.

For simplicity, we restrict here the discussion to a dissipative Langevin equation for a one-component scalar field $\varphi(t, x)$, t being the time, and $x \in \mathbb{R}^d$. It has the general form

$$\dot\varphi(t, x) = -\tfrac{1}{2}\Omega\frac{\delta\mathcal{H}}{\delta\varphi(t, x)} + \nu(t, x), \tag{22.67}$$

where the constant Ω^{-1} provides a time scale, $\nu(t, x)$ is a stochastic field (the noise) and $\delta/\delta\varphi(t, x)$ denotes the functional derivative, which is defined by the standard algebraic rules of differentiation and, moreover,

$$\frac{\delta}{\delta\varphi(t, x)}\varphi(u, y) = \delta^{(d)}(x - y)\delta(t - u).$$

The functional $\mathcal{H}(\varphi)$ is a time-independent, local Landau–Ginzburg Hamiltonian (or configuration energy), analogous to an imaginary time QFT action. An example is

$$\mathcal{H}(\varphi) = \int d^d x \left[\tfrac{1}{2}\big(\nabla_x\varphi(x)\big)^2 + V\big(\varphi(x)\big)\right]. \tag{22.68}$$

We assume that the stochastic field $\nu(t, x)$ has the Gaussian local distribution (Gaussian white noise)

$$[d\rho(\nu)] = [d\nu]\exp\left[-\int dt\, d^d x\, \nu^2(t, x)/2\Omega\right]. \tag{22.69}$$

Alternatively, it can be characterized by its one- and two-point correlation functions,

$$\langle\nu(t, x)\rangle = 0, \quad \langle\nu(t, x)\nu(t', x')\rangle = \Omega\,\delta(t - t')\delta^{(d)}(x - x').$$

22.6.1 The associated FP equation

Given the noise distribution (22.69) and some initial distribution $\varphi'(x)$ for the field $\varphi(t,x)$ at time t', the Langevin equation generates a time-dependent field distribution, which formally can be written as

$$P\big(\varphi(x),t;\varphi'(x),t'\big) = \left\langle \prod_x \delta\left(\varphi(t,x) - \varphi(x)\right) \right\rangle_\nu .$$

From the Langevin equation, one derives the FP evolution equation

$$\dot{P}(\varphi,t;\varphi',t') = -\Omega \mathbf{H}_{\mathrm{FP}} P(\varphi,t;\varphi',t'), \qquad (22.70)$$

where the operator $\mathbf{H}_{\mathrm{FP}}$, the FP Hamiltonian (not to be confused with the Hamiltonian or classical configuration energy (22.68)), is given by (see equation (22.31) and Section 22.7.1)

$$\mathbf{H}_{\mathrm{FP}}\left(\varphi, \frac{\delta}{\delta\varphi}\right) = -\frac{1}{2}\int \mathrm{d}^d x \frac{\delta}{\delta\varphi(x)}\left[\frac{\delta}{\delta\varphi(x)} + \frac{\delta\mathcal{H}}{\delta\varphi(x)}\right].$$

It has the form of a generalized quantum Hamiltonian but is *not Hermitian*. Its specific form ensures *conservation of probabilities*, since the integral over all fields of the right-hand side vanishes.

The change $P = \mathrm{e}^{-\mathcal{H}/2}\tilde{P}$ transforms $\mathbf{H}_{\mathrm{FP}}$ into

$$\tilde{\mathbf{H}}_{\mathrm{FP}}\left(\varphi, \frac{\delta}{\delta\varphi}\right) = \frac{1}{2}\mathbf{A}^\dagger \mathbf{A} \quad \text{with} \quad \mathbf{A} = \frac{\delta}{\delta\varphi(x)} + \frac{1}{2}\frac{\delta\mathcal{H}}{\delta\varphi(x)}.$$

Therefore, the FP Hamiltonian is equivalent to a Hermitian positive operator. If the distribution $\mathrm{e}^{-\mathcal{H}(\varphi)}$ is normalizable, it is the ground state of the FP Hamiltonian $\mathbf{H}_{\mathrm{FP}}$ with eigenvalue 0.

From the FP equation, one then infers that the Langevin equation (22.67), together with the noise distribution (22.69), generates a dynamics that converges towards the *equilibrium distribution* $\mathrm{e}^{-\mathcal{H}(\varphi)}$.

One also proves that the *purely dissipative Langevin equation* is associated to time evolution with *detailed balance*: as a consequence of the real Hermiticity of $\tilde{\mathbf{H}}_{\mathrm{FP}}$, one can write the condition of detailed balance as

$$P(\varphi',t;\varphi,t') = \mathrm{e}^{-\mathcal{H}(\varphi)+\mathcal{H}(\varphi')}\, P(\varphi,t;\varphi',t').$$

The problem of locality. A serious problem arises when the Hamiltonian is local, which is the situation of short range interactions one would like to investigate. Indeed, more explicitly, $\tilde{\mathbf{H}}_{\mathrm{FP}}$ reads

$$\tilde{\mathbf{H}}_{\mathrm{FP}}\left(\varphi, \frac{\delta}{\delta\varphi}\right) = -\frac{1}{2}\int \mathrm{d}^d x \left[\left(\frac{\delta}{\delta\varphi(x)}\right)^2 - \frac{1}{4}\left(\frac{\delta\mathcal{H}}{\delta\varphi(x)}\right)^2 + \frac{1}{2}\frac{\delta^2\mathcal{H}}{(\delta\varphi(x))^2}\right].$$

In the example of the Hamiltonian (22.68),

$$\frac{\delta \mathcal{H}}{\delta \varphi(x)} = -\nabla_x^2 \varphi(x) + V'(\phi(x)),$$

$$\frac{\delta^2 \mathcal{H}}{\delta \varphi(x) \delta \varphi(y)} = \left[-\nabla_x^2 + V''(\phi(x)) \right] \delta^{(d)}(x - y).$$

The second derivative in the limit $y \to x$ is undefined, being formally proportional to $\delta^{(d)}(0)$. The dynamic field theory at this level already requires a regularization. It is possible to introduce a small non-locality by substituting the kernel $\Omega(x-y)$ for the constant Ω. A lattice regularization is also possible. Alternatively, dimensional regularization cancels this contribution completely but the geometric properties are no longer apparent.

22.6.2   *The linear Langevin equation*

If $\mathcal{H}(\varphi)$ corresponds to a field theory for a free scalar particle of mass m,

$$\mathcal{H}(\varphi) = \tfrac{1}{2} \int \mathrm{d}^d x \left[\left(\nabla_x \varphi(x) \right)^2 + m^2 \varphi^2(x) \right], \tag{22.71}$$

and the equilibrium distribution thus is Gaussian, the Langevin equation reads

$$\dot{\varphi}(t, x) = -\tfrac{1}{2} \Omega \left(-\nabla_x^2 + m^2 \right) \varphi(t, x) + \nu(t, x). \tag{22.72}$$

The equation can be solved by Fourier transformation, setting

$$\varphi(t, x) = \int \mathrm{d}^d k \, \mathrm{e}^{ikx} \, \tilde{\varphi}(t, k).$$

Assuming, for simplicity, a boundary condition at $t = -\infty$ which leads to equilibrium at any finite time, then

$$\tilde{\varphi}(t, k) = \int_{-\infty}^{t} \mathrm{d}u \, \mathrm{e}^{-\Omega(k^2 + m^2)(t-u)/2} \frac{1}{(2\pi)^d} \int \mathrm{d}^d x \, \mathrm{e}^{-ikx} \, \nu(u, x).$$

The expectation value $\langle \tilde{\varphi}(t, k) \rangle_\nu$ vanishes and

$$\langle \tilde{\varphi}(t_1, k_1) \tilde{\varphi}(t_2, k_2) \rangle_\nu = \frac{1}{(2\pi)^d} \frac{\delta^{(d)}(k_1 + k_2)}{k_1^2 + m^2} \mathrm{e}^{-\Omega(k_1^2 + m^2)|t_1 - t_2|/2}. \tag{22.73}$$

The different k modes have different correlation times $\tau(k) = 2/\Omega(k^2 + m^2)$, the slowest mode being $k = 0$ and, thus, $\tau = 2/\Omega m^2 = 2\xi^2/\Omega$, where ξ is the correlation length. In the critical limit $\xi \to \infty$, the correlation time diverges, an example of *critical slowing down*. Defining the dynamic exponent z by

$$\tau \propto \xi^z, \tag{22.74}$$

one finds in the Gaussian model the value $z = 2$.

Finally, for $t_1 = t_2$, the equal-time correlation function is simply the two-point function associated with the static equilibrium Hamiltonian (22.71).

22.7 Time-dependent correlation functions and dynamic action

Beyond the Gaussian theory, the derivation of general properties of a dynamic theory based on the Langevin equation requires more elaborated techniques.

In particular, within the theory of phase transitions, important issues concern the properties of the time evolution near the critical temperature and the generalization to dynamics of the universal static behaviour.

The question can be answered completely in the case of the purely dissipative Langevin equation. The solution to the problem requires constructing an RG for the dynamics as weell. For this purpose, one has first to understand how the Langevin equation for fields renormalizes.

To discuss renormalization, it is convenient to set up a formalism more directly amenable to the ordinary methods of QFT [203]. This can be done by constructing a field integral representation (a generalization of the path integral of Section 22.4.1) of the time-dependent φ-field correlation functions in terms of an associated local action, which, in this framework, it is natural to call *dynamic action*.

When the static Hamiltonian $\mathcal{H}(\varphi)$ is renormalizable (and in the case of white Gaussian noise), one can then prove that the renormalizations of the static theory, together with a *time scale renormalization*, render the Langevin equation finite.

Dynamic RG equations, which generalize the static RG equations, follow. In particular, they imply that, near T_c, the dynamics is affected by a phenomenon of *critical slowing down*, that is, by the *divergence of the correlation or relaxation time*, which is governed by a *new, dynamic, critical exponent*.

22.7.1 Dynamic action

The generating functional $\mathcal{Z}(J)$ of dynamic correlation functions of the field $\varphi(t, x)$ solution of equation (22.67), is given by the noise expectation value

$$\mathcal{Z}(J) = \left\langle \exp\left[\int d^d x\, dt\, J(t, x)\varphi(t, x)\right]\right\rangle_\nu$$
$$\equiv \int [d\nu] \exp\left[-\int d^d x\, dt\left(\frac{1}{2\Omega}\nu^2(t, x) - J(t, x)\varphi(t, x)\right)\right],$$

as can be verified by expanding in powers of the external field J.

We write Langevin equation (22.67) as

$$\mathcal{L}(\varphi; t, x) = \nu(t, x), \text{ with } \mathcal{L}(\varphi; t, x) \equiv \dot{\varphi}(t, x) + \tfrac{1}{2}\Omega\frac{\delta\mathcal{H}}{\delta\varphi(t, x)}.$$

To impose the Langevin equation, we insert into the field integral the identity

$$\int [d\varphi]\det \mathbf{M}\prod_{t,x}\delta\big(\mathcal{L}(\varphi; t, x) - \nu(t, x)\big) = 1,$$

where $\mathbf{M}$ is the differential operator,

$$\mathbf{M} = \frac{\delta\,\mathcal{L}(\varphi; t, x)}{\delta\varphi(t', x')} = \frac{\partial}{\partial t}\delta(t - t')\delta^{(d)}(x - x') + \frac{\Omega}{2}\frac{\delta^2\mathcal{H}}{\delta\varphi(t', x')\delta\varphi(t, x)}.$$

The δ-function can immediately be used to integrate over the noise ν:

$$\mathcal{Z}(J) = \int [\mathrm{d}\varphi] \det \mathbf{M} \exp\left\{ -\int \mathrm{d}^d x \, \mathrm{d}t \left[\frac{1}{2\Omega} \left(\mathcal{L}(\varphi; t, x) \right)^2 - J(t, x)\varphi(t, x) \right] \right\}.$$

For a system with a discrete set of degrees of freedom (a $d = 0$ dimensional or a lattice regularized field theory), the determinant can be calculated, using the identity

$$\det \mathbf{M} \propto \exp \mathrm{tr} \ln \left[1 + \left(\frac{\partial}{\partial t} \right)^{-1} \frac{\Omega}{2} \frac{\delta \mathcal{H}}{\delta \varphi \delta \varphi} \right].$$

As a consequence of the *causality of the Langevin equation*, the inverse of the operator $(\partial/\partial t)\delta(t - t')$ is the kernel $\theta(t - t')$ ($\theta(t)$ is the Heaviside step function). In an expansion in powers of Ω, all terms thus vanish when one takes the trace, except the first one which yields

$$\det M \propto \exp \left\{ \theta(0) \frac{\Omega}{2} \int \mathrm{d}t \, \mathrm{d}^d x \left. \frac{\delta^2 \mathcal{H}}{\delta \varphi(t, x)\delta \varphi(t, x')} \right|_{x'=x} \right\}.$$

For the undefined quantity $\theta(0)$, we choose $\theta(0) = 1/2$, a choice symmetric in time. The final expression then formally reads

$$\mathcal{Z}(J) = \int [\mathrm{d}\varphi] \exp \left[-\mathcal{S}(\varphi) + \int \mathrm{d}^d x \, \mathrm{d}t \, J(t, x)\varphi(t, x) \right],$$

$$\mathcal{S}(\varphi) = \frac{1}{2\Omega} \int \mathrm{d}^d x \, \mathrm{d}t \left[\left(\dot{\varphi}(t, x) \right)^2 + \tfrac{1}{4}\Omega^2 \left(\frac{\delta \mathcal{H}(\varphi)}{\delta \varphi(t, x)} \right)^2 \right]$$

$$- \frac{\Omega}{4} \int \mathrm{d}t \, \mathrm{d}^d x \left. \frac{\delta^2 \mathcal{H}(\varphi)}{\delta \varphi(x, t)\delta \varphi(x', t)} \right|_{x'=x}, \tag{22.75}$$

where we have expanded the square and used the identity,

$$\int_{t'}^{t''} \mathrm{d}t \int \mathrm{d}^d x \, \dot{\varphi}(t, x) \frac{\delta \mathcal{H}(\varphi)}{\delta \varphi(t, x)} = \mathcal{H}(\varphi(t'')) - \mathcal{H}(\varphi(t')),$$

an equation valid inside the field integral only for $\theta(0) = 1/2$. In addition, we assume that the boundary fields at time $t = \pm\infty$ are constant fields, minima of the action.

Finally, the form of the dynamic action is consistent with the solution $P(\varphi, t; \varphi', t')$ of the FP equation (22.70) expressed as a field integral.

22.7.2   *The divergent determinant*

In dimension $d > 0$, the dynamic action is undefined because the Hamiltonian $\mathcal{H}(\varphi)$ is local. Indeed, the contribution of the determinant is formally proportional to the undefined quantity $\delta^{(d)}(0)$, since

$$\ln \det \mathbf{M} \propto \int \mathrm{d}t \, \mathrm{d}^d x \left. \frac{\delta^2 \mathcal{H}}{\delta \varphi(t, x)\delta \varphi(x', t)} \right|_{x'=x} \propto \, {}^{\backprime}\delta^{(d)}(0){}'.$$

The determinant thus has to be regularized. Three possible regularization methods are:

(i) regularize by introducing temporarily a noise distribution slightly non-local in space corresponding to the substitution

$$\frac{1}{\Omega} \int dt\, d^d x\, \nu^2(t,x) \mapsto \int dt\, d^d x\, d^d y\, \Omega^{-1}(x-y)\nu(t,x)\nu(t,y),$$

where Ω^{-1} has to be understood in the sense of operators and, similarly, replace in the Langevin equation (22.67),

$$\Omega \frac{\delta \mathcal{H}}{\delta\varphi(t,x)} \mapsto \int d^d y\, \Omega(x-y) \frac{\delta \mathcal{H}}{\delta\varphi(t,y)};$$

(ii) dimensional regularization, where terms like $\delta^{(d)}(0)$ vanish and, therefore, the determinant can be completely omitted; with this convention, the expression (22.75) can be used in practical perturbative calculations and

(iii) lattice regularization that makes it possible to keep this divergent term in some regularized form in order to preserve the geometric structure of the dynamic action; the geometric properties then determine the form of the renormalization.

22.8 The dissipative Langevin equation and supersymmetry

We have explained how to associate a dynamic action to Langevin or FP equations. Quite generally, the dynamic action has a BRST symmetry (see Section 14.6). This symmetry and its consequences in the form of Ward–Takahashi (WT) identities determine, under some general condition, that the structure of the Langevin equation is stable under renormalization.

As we have shown in Section 22.5, in the particular example of purely dissipative equations with Gaussian noise, the dynamic action has an additional Grassmann symmetry which, combined with the first one, provides the simplest example of supersymmetry: *quantum mechanics supersymmetry*. To exhibit supersymmetry, an alternative formalism has been introduced, based on Grassmann coordinates.

22.8.1 Supersymmetry

We consider again the Langevin equation (22.67),

$$\dot{\varphi}(t,x) = -\tfrac{1}{2}\Omega \frac{\delta \mathcal{H}}{\delta\varphi(t,x)} + \nu(t,x),$$

with the noise Gaussian distribution

$$[d\rho(\nu)] = [d\nu]\exp\left[-\int dt\, d^d x\, \nu^2(t,x)/2\Omega\right].$$

We introduce a superfield in the form,

$$\phi(t,x;\bar{\theta},\theta) = \varphi(t,x) + \theta\bar{C}(t,x) + C(t,x)\bar{\theta} + \theta\bar{\theta}\bar{\varphi}(t,x),$$

where $\theta, \bar{\theta}, \bar{C}(t,x), C(t,x)$ are generators of a Grassmann algebra, and $\bar{\varphi}(t,x)$ an auxiliary scalar field.

Rescaling, for convenience, the Langevin equation by a factor $2/\Omega$, one can rewrite the dynamic action in a supersymmetric form as

$$S(\phi) = \int d\bar{\theta}\, d\theta\, dt \left[\frac{2}{\Omega} \int d^d x\, \bar{D}\phi D\phi + \mathcal{H}(\phi) \right],\qquad (22.76)$$

where $D, \bar{D}$ are the covariant derivatives (22.59).

22.8.2 *WT identities*

One symmetry simply implies that correlation functions are invariant under a translation of the coordinate θ. The second transformation has a slightly more complicated form. It implies that the generating functional $\mathcal{W}(J)$ of connected correlation functions satisfies [206]:

$$\int d^d x\, dt\, d\bar{\theta}\, d\theta\, \bar{Q}\, J(t,x,\theta,\bar{\theta}) \frac{\delta \mathcal{W}}{\delta J(t,x,\theta,\bar{\theta})} = 0\,.$$

It follows that connected correlation functions $W^{(n)}(t_i, x_i, \theta_i, \bar{\theta}_i)$ and vertex functions $\Gamma^{(n)}(t_i, x_i, \theta_i, \bar{\theta}_i)$ satisfy the WT identities,

$$\bar{Q}\, W^{(n)}(t_i, x_i, \theta_i, \bar{\theta}_i) = 0\,, \quad \bar{Q}\, \Gamma^{(n)}(t_i, x_i, \theta_i, \bar{\theta}_i) = 0\,, \qquad (22.77)$$

with

$$\bar{Q} \equiv \sum_{k=1}^{n} \left(\frac{\partial}{\partial \bar{\theta}_k} + \theta_k \frac{\partial}{\partial t_k} \right).$$

Example: A two-point function. The WT identity and causality determine the form of the two-point function. As the relations (22.61) and (22.62) show, supersymmetry implies translation invariance with respect to time and to θ. Therefore, any two-point function $W^{(2)}$ can be written as

$$W^{(2)} = A(t_1 - t_2) + (\theta_1 - \theta_2)\left[(\bar{\theta}_1 + \bar{\theta}_2)B(t_1 - t_2) + (\bar{\theta}_1 - \bar{\theta}_2)C(t_1 - t_2)\right].$$

The WT identity (22.77) then implies

$$2B(t) = \frac{\partial A}{\partial t}.$$

The WT identity does not determine the function C. An additional constraint is provided by causality. For the two-point function, it implies that the coefficient of $\theta_1 \bar{\theta}_2$ vanishes for $t_1 < t_2$, and the coefficient of $\theta_2 \bar{\theta}_1$ vanishes for $t_2 < t_1$.

The last function is thus determined, up to a possible distribution localized at $t_1 = t_2$. One finds

$$2C(t) = -\operatorname{sgn}(t)\frac{\partial A}{\partial t}, \qquad (22.78)$$

where $\operatorname{sgn}(t)$ is the sign of t, and, therefore,

$$W^{(2)} = \left\{ 1 + \tfrac{1}{2}\,(\theta_1 - \theta_2)\left[\bar{\theta}_1 + \bar{\theta}_2 - (\bar{\theta}_1 - \bar{\theta}_2)\operatorname{sgn}(t_1 - t_2)\right]\frac{\partial}{\partial t_1} \right\} A(t_1 - t_2). \quad (22.79)$$

22.9 Renormalization of the dissipative Langevin equation

To be more specific, we now assume that $\mathcal{H}(\varphi)$ has the form

$$\mathcal{H}(\varphi) = \tfrac{1}{2} \int \mathrm{d}^d x \left[(\nabla_x \varphi)^2 + \mathcal{V}(\varphi(x)) \right], \quad \mathcal{V}(\varphi) = \tfrac{1}{2} m^2 \varphi^2 + O(\varphi^3).$$

Then the propagator in the dynamic theory, in the Fourier representation, reads $(\delta(\theta) = \theta)$

$$\tilde{\Delta}(\omega, \mathbf{k}, \theta', \boldsymbol{\theta}) = \frac{\Omega \left[1 + \tfrac{1}{2} i \omega \left(\theta' - \theta \right) \left(\bar{\theta} + \bar{\theta}' \right) + \tfrac{1}{4} \Omega \left(k^2 + m^2 \right) \delta^2 (\bar{\theta}' - \bar{\theta}) \right]}{\omega^2 + \frac{\Omega^2}{4} \left(k^2 + m^2 \right)^2}.$$

Dimensions in the sense of power counting follow:

$$[k] = 1, \ [\omega] = 2, \ [\theta] = [\bar{\theta}] = -1.$$

Similarly (since integration and differentiation over anticommuting variables are identical operations, the dimension of $\mathrm{d}\theta$ is $-[\theta]$),

$$[x] = -1, \ [t] = -2, \ [\mathrm{d}\theta] = [\mathrm{d}\bar{\theta}] = 1 \ \Rightarrow [\mathrm{d}t] + [\mathrm{d}\theta] + [\mathrm{d}\bar{\theta}] = 0.$$

Therefore, the term proportional to $\mathcal{H}(\phi)$ in the action has the same dimension, in the sense of power counting, as in the static case: the static and the dynamic theories have the same power counting and both theories are renormalizable in the same space dimension d.

The supersymmetry transformation is linearly represented on the fields and can be preserved by a suitable regularization. Then, the renormalized action remains supersymmetric. In superfield notation, the most general form of the renormalized action $\mathcal{S}_r$ consistent with power counting and supersymmetry takes the form

$$\mathcal{S}_r(\phi) = \int \mathrm{d}\bar{\theta} \, \mathrm{d}\theta \, \mathrm{d}t \left[\frac{2 Z_\omega}{\Omega} \int \mathrm{d}^d x \, \bar{\mathrm{D}} \phi \mathrm{D} \phi + \mathcal{H}_r(\phi) \right]. \tag{22.80}$$

Only one new renormalization constant, a time scale renormalization, Z_ω, is generated. After renormalization, the drift force in the Langevin equation is thus proportional to the functional derivative of the renormalized Hamiltonian.

22.10 Dissipative Langevin equation: RG equations in $4 - \varepsilon$ dimensions

We have explained in Section 22.9 how the dynamic action renormalizes. We can now derive RG equations for the dynamic theory. Note that we write RG equations for the *renormalized theory, using the subscript* 0 *for the initial (or bare) parameters* [205].

The N-vector model near four dimensions. We consider an N-component field $\boldsymbol{\varphi}$ in $d = 4 - \varepsilon$ space dimensions satisfying the dissipative Langevin equation,

$$\dot{\boldsymbol{\varphi}}(t,x) = -\frac{\Omega_0}{2}\frac{\delta \mathcal{H}(\boldsymbol{\varphi})}{\delta \boldsymbol{\varphi}(t,x)} + \boldsymbol{\nu}(t,x) \tag{22.81}$$

with

$$\mathcal{H}(\boldsymbol{\varphi}) = \int \mathrm{d}^d x \left[\frac{1}{2}\left(\nabla_x \boldsymbol{\varphi}(x)\right)^2 + \frac{1}{2}r_0\boldsymbol{\varphi}^2(x) + g_0\frac{\Lambda^\varepsilon}{4!}\left(\boldsymbol{\varphi}^2(x)\right)^2 \right],$$

where r_0 plays the role of the temperature near T_c, and the coupling constant $g_0 > 0$ is dimensionless.

The noise field has the Gaussian distribution

$$[\mathrm{d}\rho(\boldsymbol{\nu})] = [\mathrm{d}\boldsymbol{\nu}]\exp\left[-\int \mathrm{d}t\,\mathrm{d}^d x\,\boldsymbol{\nu}^2(t,x)/2\Omega_0 \right],$$

in such a way that $\mathrm{e}^{-\mathcal{H}}$ is the $O(N)$ invariant equilibrium distribution.

In terms of the superfield

$$\boldsymbol{\phi} = \boldsymbol{\varphi} + \theta\bar{\mathbf{C}} + \mathbf{C}\bar{\theta} + \theta\bar{\theta}\bar{\boldsymbol{\varphi}},$$

the corresponding dynamic action $\mathcal{S}(\boldsymbol{\phi})$ takes the supersymmetric form

$$\mathcal{S}(\boldsymbol{\phi}) = \int \mathrm{d}t\,\mathrm{d}\bar{\theta}\,\mathrm{d}\theta \left[\int \mathrm{d}^d x\,\frac{2}{\Omega_0}\bar{\mathrm{D}}\boldsymbol{\phi}\mathrm{D}\boldsymbol{\phi} + \mathcal{H}(\boldsymbol{\phi}) \right],$$

with

$$\bar{\mathrm{D}} = \frac{\partial}{\partial\bar{\theta}}, \quad \mathrm{D} = \frac{\partial}{\partial\theta} - \bar{\theta}\frac{\partial}{\partial t}.$$

22.10.1 RG equations at and above T_c

We have shown that static and supersymmetric dynamic theories have the same upper-critical dimension. Therefore, fluctuations are only relevant for dimensions $d \leq 4$. We have also shown that the renormalized dynamic action $\mathcal{S}_\mathrm{r}(\boldsymbol{\phi})$ then takes the form

$$\mathcal{S}_\mathrm{r}(\boldsymbol{\phi}) = \int \mathrm{d}\bar{\theta}\,\mathrm{d}\theta\,\mathrm{d}t \left[\int \mathrm{d}^d x\,\frac{2}{\Omega}Z_\omega\bar{\mathrm{D}}\boldsymbol{\phi}\mathrm{D}\boldsymbol{\phi} + \mathcal{H}_\mathrm{r}(\boldsymbol{\phi}) \right],$$

in which $\boldsymbol{\phi}$ is now the renormalized field, and $\mathcal{H}_\mathrm{r}(\boldsymbol{\phi})$ is the static renormalized Hamiltonian.

To renormalize the dynamic action, we thus need, in addition to the static renormalization constants, only a renormalization of the parameter Ω_0:

$$\Omega_0 = \Omega Z/Z_\omega,$$

where Z is the field renormalization constant.

Introducing the *renormalization scale* μ, we define the RG differential operator,

$$\mathrm{D}_{\mathrm{RG}} = \mu\frac{\partial}{\partial\mu} + \beta(g)\frac{\partial}{\partial g} + \eta_\omega(g)\Omega\frac{\partial}{\partial\Omega} - \frac{n}{2}\eta(g),$$

where the new independent RG function $\eta_\omega(g)$ is given by

$$\eta_\omega(g) = \mu\frac{\mathrm{d}}{\mathrm{d}\mu}\bigg|_{g_0,\Omega_0}\ln\Omega. \tag{22.82}$$

The RG equations satisfied by the vertex functions of the critical theory $(T = T_c)$, in the Fourier representation, then read

$$\mathrm{D}_{\mathrm{RG}}\tilde{\Gamma}^{(n)}(p_i,\omega_i,\boldsymbol{\theta},\mu,\Omega,g) = 0\,,$$

where we have used the vector notation $\boldsymbol{\theta} \equiv (\bar{\theta},\theta)$.

22.10.2 The infrared fixed point

From the study of the ϕ^4 equilibrium field theory, we know that there exists, at least within the ε-expansion, an infrared (IR) fixed point.

At a fixed point g^*, the RG equations reduce to (omitting the $O(N)$ group indices)

$$\left(\mu\frac{\partial}{\partial\mu} + \eta_\omega(g^*)\Omega\frac{\partial}{\partial\Omega} - \frac{n}{2}\eta(g^*)\right)\tilde{\Gamma}^{(n)}(p_i,\omega_i,\boldsymbol{\theta},\mu,\Omega) = 0\,.$$

We then introduce the dynamic exponent $z = 2 + \eta_\omega(g^*)$.

From a dimensional analysis, we infer

$$\tilde{\Gamma}^{(n)}\left(\lambda p_i,\kappa\omega_i,\frac{\boldsymbol{\theta}}{\sqrt{\kappa}},\lambda\mu,\frac{\kappa\Omega}{\lambda^2}\right) = \lambda^{d-n(d-2)/2}\kappa^{1-n}\tilde{\Gamma}^{(n)}(p_i,\omega_i,\boldsymbol{\theta},\mu,\Omega).$$

In the dimensional relation, we choose $\kappa = \Omega\lambda^z\mu^{-\eta_\omega}$. Then, combining the solution of the RG equation with dimensional analysis, we find the dynamic scaling relations

$$\tilde{\Gamma}^{(n)}(\lambda p_i,\omega_i,\boldsymbol{\theta},\mu = \Omega = 1) = \lambda^{d-nd_\varphi-z(n-1)}F^{(n)}(p_i,\lambda^{-z}\omega_i,\boldsymbol{\theta}\lambda^{z/2}),$$

where $d_\varphi = \frac{1}{2}(d-2+\eta)$ is the field dimension.

After some algebra, we obtain the corresponding relations for connected correlation functions:

$$\tilde{W}^{(n)}(\lambda p_i,\omega_i,\boldsymbol{\theta},\mu = \Omega = 1) = \lambda^{(d+z)(1-n)+nd_\varphi}G^{(n)}(p_i,\omega_i\lambda^{-z},\boldsymbol{\theta}\lambda^{z/2}).$$

For $n = 2$, and $\boldsymbol{\theta} = 0$, one obtains the φ-field two-point correlation function that satisfies the scaling relation,

$$\tilde{W}^{(2)}(p,\omega,\boldsymbol{\theta} = 0) \sim p^{-2+\eta-z}G^{(2)}(\omega/p^z).$$

The equal-time correlation function is obtained by integrating over ω. The result is consistent with the static scaling.

The dynamic critical two-point function thus depends on a frequency scale that vanishes at small momentum like p^z or a time scale that diverges like p^{-z}, a sign of critical slowing down.

The RG function η_ω at one-loop order for this model is

$$\eta_\omega(g) = N_d^2 \frac{N+2}{72} \left[6\ln(4/3) - 1\right] g^2 + O\left(\tilde{g}^3\right), \quad N_d = \frac{2}{\Gamma(d/2)(4\pi)^{d/2}}. \quad (22.83)$$

Inserting the fixed point value $g^* = 48\pi^2\varepsilon/(N+8) + O(\varepsilon^2)$, one infers the dynamic critical exponent [209],

$$z = 2 + \frac{N+2}{2(N+8)^2}\left[6\ln(4/3) - 1\right]\varepsilon^2 + 0\left(\varepsilon^3\right).$$

22.10.3 Correlation functions above T_c in the critical domain

We write the RG equations only at the IR fixed point:

$$\left(\mu\frac{\partial}{\partial\mu} + \eta_\omega\Omega\frac{\partial}{\partial\Omega} - \eta_2\rho\frac{\partial}{\partial\rho} - \frac{n}{2}\eta\right)\tilde{\Gamma}^{(n)}(p_i, \omega_i, \boldsymbol{\theta}, \rho, \mu, \Omega) = 0, \quad (22.84)$$

in which ρ is a measure of the deviation from the critical temperature:

$$\rho \propto T - T_c.$$

Dimensional analysis yields

$$\tilde{\Gamma}^{(n)}(p_i, \omega_i, \boldsymbol{\theta}, \rho, \mu, \Omega) = \lambda^{d-n(d-2)/2}\kappa^{1-n}\tilde{\Gamma}^{(n)}\left(\frac{p_i}{\lambda}, \frac{\omega_i}{\kappa}, \boldsymbol{\theta}\sqrt{\kappa}, \frac{\rho}{\lambda^2}, \frac{\mu}{\lambda}, \frac{\Omega\lambda^2}{\kappa}\right).$$

Finally, combining this equation with the RG equation (22.84) and choosing

$$\lambda = \rho^\nu\mu^{\nu\eta^2} \sim \xi^{-1}, \quad \kappa = \Omega\mu^{-\eta_\omega}\lambda^z \sim \xi^{-z},$$

in which ξ is the correlation length, one obtains the scaling relation

$$\tilde{\Gamma}^{(n)}(p_i, \omega_i, \boldsymbol{\theta}, \rho, \mu = 1, \Omega = 1) \sim \xi^{-d+n(d-2+\eta)/2+z(n-1)}$$
$$\times F^{(n)}(p_i\xi, \omega_i\xi^z, \boldsymbol{\theta}\xi^{-z/2}).$$

Only the combination $\omega_i\xi^z$ appears in the right-hand side and this implies, after Fourier transformation, that all times are measured in units of a *correlation time* τ that diverges at the critical temperature as

$$\tau \propto \xi^z, \quad (22.85)$$

an effect called *critical slowing down*.

23 Field theory: Perturbative expansion and summation methods

The determination of the behaviour of perturbative expansions at large orders in quantum field theory, based on the instanton calculus, has confirmed that perturbative expansions are always *divergent*.

When the expansion parameter is not small, the straightforward sum of the terms of the series is useless. However, when divergent series are *Borel summable*, *summation methods* can be found that transform divergent series into convergent sequences. This is, in particular, the situation that is met in the application of quantum field theory methods to critical phenomena and the calculations of critical exponents and other universal quantities (see Sections 5.6 and 6.12).

In this respect, the determination of the large order behaviour is very useful and has led to a development of specific summation methods that can incorporate this information efficiently.

23.1 Divergent series in quantum field theory

Dyson's intuitive argument. We first recall Dyson's intuitive argument, as it has initially been formulated, but apply it to boson theories because it has to be modified in the example of fermion theories like quantum electrodynamics.

The energy of a system of n *bosons* with two-body interactions of strength g has the schematic form

$$E = an + \tfrac{1}{2}bgn^2 \quad a, b > 0\,.$$

For $g > 0$, which corresponds to a repulsive interaction, the minimum of the energy $E = 0$ corresponds to the vacuum $n = 0$. The boson system thus is stable. By contrast, for $g < 0$ and above a critical value n_c given by,

$$\frac{\mathrm{d}E}{\mathrm{d}n} = 0 \implies n_c = -a/bg\,,$$

corresponding to the maximal energy $E_c = -a^2/2bg$, the energy decreases without lower bound. If a system is initially in the now metastable state or vacuum $n = 0$, because, in a non-relativistic quantum statistical theory, in the second quantized formulation, or in quantum field theory, the boson number fluctuates, the system is unstable under quantum (barrier penetration) or statistical fluctuations. Therefore, physical observables are singular at $g = 0$.

Moreover, E_c is the height of the barrier that has to be overcome. A WKB type analysis then indicates that, for g small, the lifetime of the metastable vacuum state will be of order $\mathrm{e}^{c/g}$, $c > 0$, and one can argue that the terms of the perturbative expansion in powers of g increase like $k!$ at large order k.

Semi-classical evaluation of functional integrals (*instanton calculus* [26–28]) have confirmed this intuition and led to a precise evaluation of the behaviour of perturbative expansions at large orders.

From Random Walks to Random Matrices. Jean Zinn-Justin, Oxford University Press (2019).
© Jean Zinn-Justin. DOI: 10.1093/oso/9780198787754.001.0001

23.1.1 A special class of divergent series

We focus here on a class of divergent series [210] one meets commonly in quantum mechanics or quantum field theory (an exception is provided by theories involving fermions and without boson self-interactions).

Let $f(z)$ be an analytic function singular at $z = 0$ and holomorphic in a sector of the complex plane $S \equiv \{0 < |z| < |z_0|, |\operatorname{Arg} z| < \alpha/2\}$, in which it has a formal Taylor series expansion

$$f(z) \sim \sum_{k=0}^{\infty} f_k z^k. \qquad (23.1)$$

Moreover, we assume that one can find two positive constants A and M such that

$$\left| f(z) - \sum_{k=0}^{K-1} f_k z^k \right| \le M A^{-K} K! \, |z|^K \quad \forall K > 0 . \qquad (23.2)$$

Unlike convergent series, divergent series do not, in general, define unique analytic functions. The smallest uncertainty can be estimated by calculating the minimum of the bound when K varies at z fixed:

$$\min_{K} M A^{-K} K! \, |z|^K = M A^{-K} K! \, |z|^K \Big|_{K=A/|z|} \underset{|z| \to 0}{\sim} M \sqrt{2\pi A/|z|} \, \mathrm{e}^{-A/|z|} . \qquad (23.3)$$

This result confirms that these series are numerically useful only if the argument $|z|$ is small enough. Moreover, in general, they define only functions up to contributions of order $\mathrm{e}^{-A'/z}$, where, for real analytic functions, $A' \ge A/\cos(\alpha/2)$.

23.1.2 Borel summable series. Borel transformation

A general theorem of complex analysis states that no analytic function can satisfy a bound of the form (23.3) in a sector with an angle α larger than π. Therefore, for $\alpha > \pi$, the series defines a unique function.

The unique function can be determined by introducing its Borel transform (Watson theorem),

$$B_f(z) = \sum_{k=0}^{\infty} \frac{f_k}{k!} z^k. \qquad (23.4)$$

Formally, in the sense of power series,

$$f(z) = \int_0^{\infty} \mathrm{d}s \, \mathrm{e}^{-s} \, B_f(sz). \qquad (23.5)$$

From the bound (23.2), one infers that $B_f(z)$ is holomorphic in a disc of radius A. Moreover, it can be proved that $B_f(z)$ is also analytic in a sector

$$|\operatorname{Arg} z| \in [0, \tfrac{1}{2}(\alpha - \pi)[, \qquad (23.6)$$

and does not increase faster than an exponential in the sector, so that integral (23.5) converges for $|z|$ small enough and inside the sector

$$|\operatorname{Arg} z| < \alpha/2 .$$

In addition, it can be shown that the integral (23.5) satisfies a bound of type (23.2). Hence, this integral representation yields the unique function which has the asymptotic expansion (23.1) in the sector S, and the series is called *Borel summable*.

We consider in this chapter only series that are known or believed to be Borel summable and thus define unique functions. Other situations are much more involved and require a specific analysis (see Chapter 24).

Series summation. When a series is Borel summable, it remains the practical problem to sum it, that is, to extract from the series a sequence of approximations converging to the expanded function.

For this purpose, a number of practical methods have been developed in the context of non-relativistic quantum mechanics and field theory, most notably Padé approximants, Borel–Padé summation and Borel transformation with mapping, which we describe in more detail because it has led to a very important application: a precise and reliable determination of critical exponents.

We also explain the less known order-dependent mapping (ODM) summation method. We recall the basis of the method. For a class of series, we give intuitive arguments to explain its convergence and illustrate its properties by a few examples.

23.2 An example: The perturbative $(\phi^2)^2$ field theory

In the early 1970s, following Wilson [80], it was shown that universal properties of a whole class of macroscopic physical systems near a continuous phase transitions can be derived from an $O(N)$ symmetric $(\phi^2)^2$ field theory. In particular, this led to renormalization group (RG)-based methods for calculating critical exponents and other universal quantities.

The field theory is defined by the Euclidean (or imaginary time) effective action, a local functional of the N-component field $\phi(x)$, $x \in \mathbb{R}^d$,

$$\mathcal{S}(\phi) = \int \mathrm{d}^d x \left[\tfrac{1}{2} \left[\nabla \phi(x)\right]^2 + \tfrac{1}{2} r \phi^2(x) + \frac{g}{4!} \left(\phi^2(x)\right)^2 \right], \tag{23.7}$$

where r and $g > 0$ are two parameters (for $d \geq 2$, a large momentum cut-off Λ is implied). To this action is associated a functional measure $\mathrm{e}^{-\mathcal{S}(\phi)}/\mathcal{Z}$, where $\mathcal{Z}$ is the partition function given by the field integral

$$\mathcal{Z} = \int [\mathrm{d}\phi] \, \mathrm{e}^{-\mathcal{S}(\phi)}. \tag{23.8}$$

Observables can then be derived from correlation functions defined by (for $N = 1$)

$$\langle \phi(x_1) \ldots \phi(x_n) \rangle = \mathcal{Z}^{-1} \int [\mathrm{d}\phi] \phi(x_1) \ldots \phi(x_n) \, \mathrm{e}^{-\mathcal{S}(\phi)}. \tag{23.9}$$

For $d = 0$, the integral is reduced to a N-dimensional integral, and, for $d = 1$, to the path integral representation of the non-relativistic, $O(N)$ symmetric, quartic anharmonic oscillator. Dimensions $d > 1$ correspond to quantum field theories, and the expression (23.7) is then somewhat symbolic, since the theory has to be modified at a short distance or large momentum Λ to regularize ultraviolet (UV) divergences, and renormalized to cancel them.

For $2 \leq d < 4$, the most interesting physics applications concern critical phenomena [18]. Physical quantities are then critical exponents, ratio of critical amplitudes, equations of state and other universal observables.

Finally, $d = 4$ is relevant to the theory of fundamental interactions at the microscopic scale.

The main analytic tool for calculating physical quantities in quantum field theory is the perturbative expansion. For the $(\phi^2)^2$ field theory with Euclidean action (23.7), the perturbative expansion is an expansion in powers of the positive parameter g. The value $g = 0$ corresponds to a singularity, since the field integral (23.8) is not defined for $g < 0$. The *perturbative expansion is divergent for all $g > 0$.*

For $d < 4$, the perturbative expansion is known to be Borel summable [211] while, for $d = 4$, due to the presence of renormalons [212] and the triviality issue (infrared (IR) freedom), the renormalized perturbative expansion is most unlikely to be Borel summable.

23.2.1 *The perturbative expansion: Large order behaviour*

For all $d < 4$, for any physical observable f, in agreement with Dyson's intuition, a semi-classical calculation confirms that the large order behaviour of the coefficients f_k of the term of order g^k has the general structure

$$f_k \underset{k \to \infty}{\propto} (-1)^k k^b a^k k! \,, \tag{23.10}$$

where a depends only on d and b is a half-integer that depends on N and the physical observable. The corresponding values of the parameter $A = 1/a$ are:

$$d = 0 : \ A = 3/2 \,,$$
$$d = 1 : \ A = 8 \quad [222] \,,$$
$$d = 2 : \ A = 35.10268957367896(1) \quad \text{(Zinn-Justin, unpublished)},$$
$$d = 3 : \ A = 113.38350781527714(1) \quad \text{(Zinn-Justin, unpublished)}. \tag{23.11}$$

For the quartic anharmonic oscillator ($d = 1$), the result was first derived using the Schrödinger equation [222]. For any $d < 4$, it has been inferred from a steepest descent calculation of the field integral (23.8) for $g < 0$ [26, 223].

For $d = 4$, one finds $A = 16\pi^2$ [26] but, to the semi-classical contribution coming from the steepest descent calculation, a contribution due to the large momentum singularities of Feynman diagrams has, in general, to be added (*renormalons*).

Similar results have been obtained for a number of quantum field theories. When the formal expansion parameter is Planck's constant, a divergence of the form (23.10) is in general found (except for theories with fermions and without boson self-interactions, like QED), but the parameter a may be complex. For an early review, see Ref. [213].

Conclusion. It follows from the large order behaviour analysis that, when the expansion parameter is not small, a summation of the perturbative expansion is required.

23.3 Renormalized perturbation theory: Callan–Symanzik equations

From now on, we restrict the presentation to the example $N = 1$ of action (23.7) (Ising-like or liquid–vapour transitions), but the generalization to arbitrary N is straightforward.

Since the most precise field theory estimates of critical exponents in three dimensions are based on the renormalized perturbation theory and the corresponding RG equations in Callan–Symanzik (CS) form [87] (following an initial suggestion of Parisi, 1973), we briefly recall the CS formalism (see also Section 15.2.1).

For $d < 4$, the perturbative expansion for the critical (*i.e.*, massless) theory is IR divergent and, thus, only a massive (non-critical) theory can be defined. CS equations are useful, inhomogeneous, variants of the RG equations in the framework of the so-called renormalized field theory, which apply to the massive theory.

23.3.1 CS equations

Renormalized vertex (or one-particle-irreducible (1PI)) functions in the Fourier representation $\tilde{\Gamma}_{\mathrm{r}}^{(n)}$, functions of a renormalized mass m and a dimensionless coupling g_{r}, are defined by tuning the parameters of the initial action (23.7) and the field renormalization Z as functions of the cut-off Λ in such a way that

$$\tilde{\Gamma}_{\mathrm{r}}^{(n)}(p, g_{\mathrm{r}}, m)\Big|_{g_{\mathrm{r}}, m \text{ fixed}} = \lim_{\Lambda \to \infty} Z^{n/2}(g, \Lambda/m) \tilde{\Gamma}^{(n)}(p_i, g, r, \Lambda), \qquad (23.12)$$

has a limit when the large momentum cut-off Λ goes to infinity.

Equation (23.12) defines Z, m, g_{r} only up to finite renormalizations. Therefore, in $d \leq 4$ dimensions, a standard choice is to completely define the renormalized vertex functions by the conditions

$$\tilde{\Gamma}_{\mathrm{r}}^{(2)}(p, m, g_r) = m^2 + p^2 + O(p^4), \quad \tilde{\Gamma}_{\mathrm{r}}^{(4)}(0,0,0,0) = m^{4-d} g_{\mathrm{r}}, \qquad (23.13)$$

where $m \ll \Lambda$ is proportional to the physical mass or the inverse of the correlation length, and g_{r} is dimensionless.

In addition, one needs the vertex functions $\Gamma^{(1,n)}$, derived from the generalization of the correlation functions (23.9) to the expectation values

$$\left\langle \tfrac{1}{2}\phi^2(y)\phi(x_1)\cdots\phi(x_n) \right\rangle.$$

Their renormalization is determined by the condition

$$\tilde{\Gamma}_{\mathrm{r}}^{(1,2)}(0; 0, 0; g_{\mathrm{r}}, m) = 1.$$

One then proves the RG equations in the CS form,

$$\left(m\frac{\partial}{\partial m} + \beta(g_{\mathrm{r}})\frac{\partial}{\partial g_{\mathrm{r}}} - \frac{n}{2}\eta(g_{\mathrm{r}}) \right) \tilde{\Gamma}_{\mathrm{r}}^{(n)}(p_i; g_{\mathrm{r}}, m)$$
$$= \left(2 - \eta(g_{\mathrm{r}})\right) m^2 \tilde{\Gamma}_{\mathrm{r}}^{(1,n)}(0; p_i; g_{\mathrm{r}}, m), \qquad (23.14)$$

where the right-hand side refers to the vertex function related to the expectation value of n fields ϕ and $\frac{1}{2}\int \mathrm{d}^d x\,\phi^2(x)$.

CS equations for general correlation functions with $\phi^2(x)$ insertions involve another independent RG function, $\eta_2(g_r)$.

The advantage of the CS equations is that they can be derived also *at fixed dimension* $d < 4$ ($d = 2, 3$ for $N = 1$, and $d = 3$ for $N > 1$).

However, they are predictive only if the right-hand side becomes negligible for large momenta $|p_i| \gg m$ (but still $|p_i| \ll \Lambda$), a property that can only be explicitly verified in four dimensions or, after dimensional continuation, within the framework of the $\varepsilon = 4 - d$-expansion.

23.3.2 RG functions in three dimensions in the CS formalism

Since the number of Feynman diagrams has a factorial increase and, for $d > 1$, the number of momentum integrations is proportional to the number of loops, the difficulty of evaluating the successive perturbative terms increases very rapidly. Moreover, questions like regularization and renormalization arise.

Therefore, Nickel's calculation of RG functions, from which critical exponents can be derived, in the $d = 3$, $O(N)$ symmetric $(\phi^2)^2$ field theory, *up to order* g^7 [89], has been a *remarkable achievement*. For $N = 1$, he has extended the calculations to the equation of state.

As a physically relevant example of the use of summation methods, we thus consider the calculation of critical exponents in three dimensions, for $N = 1$, based on the expansion of the RG functions at order g^7.

To evaluate critical exponents, the first important function is the RG β-function $\beta(g_r)$ whose zeros determine the IR fixed points. For example, for $N = 1$,

$$\tilde{\beta}(\tilde{g}) = -\tilde{g} + \tilde{g}^2 - \tfrac{308}{729}\tilde{g}^3 + 0.3510695977\tilde{g}^4 - 0.3765268283\tilde{g}^5$$
$$+ 0.49554751\tilde{g}^6 - 0.749689\tilde{g}^7 + O\left(\tilde{g}^8\right), \tag{23.15}$$

where $\tilde{g} = 3g_r/(16\pi)$, and g_r is the renormalized interaction strength defined implicitly by the conditions (23.13). Then,

$$\beta(g_r) = \frac{16\pi}{3}\tilde{\beta}(\tilde{g}). \tag{23.16}$$

The RG fixed point is given by the non-vanishing zero $\tilde{g}^*$ of the RG β-function (23.15). Critical exponents are then obtained from the two RG functions [89],

$$\gamma^{-1}(\tilde{g}) = 1 - \tfrac{1}{6}\,\tilde{g} + \tfrac{1}{27}\,\tilde{g}^2 - 0.0230696212\,\tilde{g}^3 + 0.0198868202\,\tilde{g}^4$$
$$- 0.0224595215\,\tilde{g}^5 + 0.0303679053\,\tilde{g}^6 - 0.046877951\,g^7 + O(g^8),$$
$$\eta(\tilde{g}) = 0.0109739368\,\tilde{g}^2 + 0.0009142222\,\tilde{g}^3 + 0.0017962228 g^4$$
$$- 0.0006537035\,\tilde{g}^5 + 0.0012749100\,\tilde{g}^6 - 0.001697694\,\tilde{g}^7 + O(g^8), \tag{23.17}$$

by setting $\tilde{g} = \tilde{g}^*$.

23.4 Summation methods and critical exponents

The number of possible summation methods is essentially unlimited. We describe here only a few, which have been applied to physics problems. However, we emphasize that the *optimal summation methods* for a given series, are the methods that makes the most efficient use of all the known properties of the series.

To calculate critical exponents, one must first determine the non-trivial zero $\tilde{g}^*$ of the β-function (23.15) and then various physical quantities like critical exponents for $\tilde{g} = \tilde{g}^*$. The first terms of the expansion (23.15) indicate that $\tilde{g}^*$ is a number of order 1 and, thus, a numerical determination of $\tilde{g}^*$ from the series (23.15) clearly requires a summation method.

We briefly describe here three methods: Padé approximation, Padé–Borel method and Borel transformation with conformal mapping.

Another method based only on the analytic properties of the expanded function, the ODM method has also been investigated, which we explain more thoroughly in Section 23.5.

Apparent errors. Numerical results relevant for experimental physics cannot be reported without error estimates. In the absence of mathematical bounds, this is the most delicate and time-consuming task, and the quoted error can only be considered as an educated guess.

23.4.1 Padé approximants

In particle physics, in the theoretical framework of hadron physics, in the 1960s, it was realized that, in the case of the strong nuclear force, unlike QED, expansion parameters (like the nucleon–nucleon–pion coupling) were large and, therefore, perturbation theory was useless, leading many physicists to even reject quantum field theory as a framework to describe such phenomena.

However, before the large order behaviour was even determined, it was proposed [172] to sum the perturbative expansion using Padé approximants [214], and the idea was applied to the ϕ^4 field theory in $d = 4$ dimensions, which was considered to be a phenomenological model for pion physics.

Since only two or three terms could be calculated, the possible convergence of the Padé summation could hardly be checked. Nevertheless, the results obtained in this way made much better physical sense than those provided by plain perturbation theory (for a review, see [215]), even though we have since realized that they could make sense only at very low energy, due to the triviality issue.

Padé approximants: Definition. We consider an analytic function f that has a Taylor series expansion of the form

$$f(z) = \sum_{k=0}^{\infty} f_k z^k. \tag{23.18}$$

(The = sign has to understood here and later in the sense of series expansion.)

From the series, one derives Padé approximants, which are rational functions of the form $P_M(z)/Q_N(z)$, where P_M and Q_N are polynomials of degrees M and N, respectively, satisfying [214]

$$Q_N(z)f(z) = P_M(z) + O\left(z^{N+M+1}\right), \quad Q_N(0) = 1. \tag{23.19}$$

With $(K+1)$ successive terms of the series, one can construct all $[M, N]$ Padé approximants with $N + M \leq K$.

The Padé approximation can be used generally when the location and nature of the singularities of a function f in the complex plane are unknown, but the method is especially adapted to meromorphic functions and to the special class of Stieltjes functions.

The main shortcoming of the Padé method is that, even for a rather general class of functions where Padé approximants are known to converge, they converge only in measure. For example, spurious poles may occasionally appear close to the region of interest (see the value of γ for $k = 5$ in Table 23.1). This property may be the source of instabilities in the results and, therefore, makes an empirical determination of errors, especially for short series, often quite difficult.

From the $[n+1, n]$ and $[n, n]$ approximants (related to the continued fraction) to the RG β-function (23.15), more precisely to $\tilde{\beta}(\tilde{g})/\tilde{g}$, one infers for the zeros of β-function the successive values reported in Table 23.1.

Table 23.1

Series summed by Padé approximants: zero $\tilde{g}^*$ of the $\beta(g)$ function and exponents γ and ν for $\tilde{g}^* = 1.411$ in the ϕ_3^4 field theory. The $k = 7$ values for $\tilde{g}^* = 1.42$ are $\nu = 0.6313$, $\gamma = 1.2421$.

k	2	3	4	5	6	7
$\tilde{g}^*$	1.	1.73159	1.37138	1.45709	1.41593	1.42317
ν	0.61394	0.61706	0.62629	0.63727	0.62890	0.63027
γ	1.21807	1.22964	1.23668	1.27493	1.23932	1.24039

Continued fractions. Padé approximants are related to the expansion in continued fractions.

For example, the continued fraction defined by

$$f_n(z) = f_n(0) + \frac{z}{f_{n+1}(z)}, \quad f_0(z) \equiv f(z),$$

truncated at $n = N$ (*i.e.*, taking the limit $f_{N+1} \to \infty$), generates $[[N/2], [N/2]]$ and $[[N/2] + 1, [N/2]]$ approximants ($[N/2]$ is the integer part of $N/2$).

23.4.2 Methods based on Borel transformation

Several summation methods are based on the Borel transformation. We consider an analytic function $f(z)$ which has an expansion of the form (23.18). One defines its Borel transform $B_f(z)$ by equation (23.4),

$$B_f(z) = \sum_{k=0}^{\infty} \frac{f_k}{k!} z^k.$$

We assume that the coefficients f_k are such, for example, that $f_{k+1}/f_k \sim ak$ for k large. The expansion of B_f then converges in a disc of radius $1/|a|$. In the sense of formal series, one recovers $f(z)$ by expanding in powers of z the integral representation (23.5),

$$f(z) = \int_0^{\infty} \mathrm{d}s \, \mathrm{e}^{-s} B_f(sz).$$

However, to determine $f(z)$ for finite z, an analytic continuation of the Borel transform B_f outside the circle of convergence in a neighbourhood of the real positive axis is required.

Padé approximants [89] have then been used for this purpose, leading to the so-called Borel–Padé method.

With a more precise knowledge of the large order behaviour, a more efficient method has been developed, combining a Borel transformation (actually Borel–Leroy) and a conformal mapping [218, 31]. The results obtained in this way have been proved to be quite reliable and precise enough for most experiments.

23.4.3 Borel transformation and conformal mapping

For the RG β-function in three dimensions, which has the expansion

$$\tilde{\beta}(\tilde{g}) = \sum_k \tilde{\beta}_k \, \tilde{g}^k,$$

which is known to be *Borel summable*, the large order behaviour has the form (23.10),

$$\tilde{\beta}_k \underset{k \to \infty}{\sim} C(-a)^k k^{7/2} k! \,,$$

with $a = 16\pi/3A = 0.147774232\ldots$, where A is given by equation (23.11).

In this example, it is useful to introduce a generalized Borel–Laplace transformation (here, Borel–Leroy):

$$B_\sigma(\tilde{g}) = \sum_k \frac{\beta_k}{\Gamma(k + \sigma + 1)} \tilde{g}^k,$$

where σ is a free parameter that can be tuned to improve convergence. Then,

$$\beta(\tilde{g}) \stackrel{\cdot}{=} \int_0^{+\infty} t^\sigma \, \mathrm{e}^{-t} B_\sigma(\tilde{g}t)\mathrm{d}t \,.$$

From the large order behaviour, one infers that the function $B_\sigma(z)$ is analytic in a disc of radius $1/a$, where it is defined by the perturbative expansion. It is then necessary to perform an analytic continuation in a neighbourhood of the real positive semi-axis.

For the $(\phi^2)^2$ theory and critical exponents, since only a small number of perturbative coefficients is available, a precise continuation requires a domain of analyticity larger than what is rigorously established. Le Guillou and Zinn-Justin [218] assumed maximal analyticity, that is, analyticity in a cut-plane. The continuation was then be achieved by a using a *conformal mapping $z \mapsto u$ of the cut-plane onto a disc* (Fig. 23.1),

$$z = \frac{4u/a}{(1-u)^2}.$$

Finally, additional modifications were introduced to optimize the summation method by dealing with the singularity at $u = 1$ (for details, see Ref. [218]).

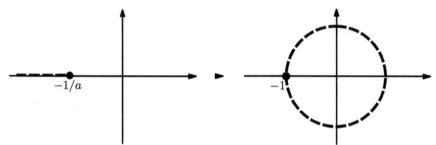

Fig. 23.1 Mapping a cut-plane onto a disc.

Later, further optimizations of the summation technique such as introducing the homographic transformation $\tilde{g} = g'/(1 + \alpha g')$, where $\alpha > 0$ is variational parameter, to displace possible singularities, as well as using some additional seven-loop contributions, have led to slightly improved estimates of critical exponents [31]. As a consequence, the summation method, ultimately, depended on three parameters. The criteria of best convergence and minimal sensitivity were then used to determine these parameters. The sensitivity of the results when the parameters are varied played a role in the determination of the apparent errors.

Other critical exponents are obtained from the two RG functions (23.17) [89] by setting $\tilde{g} = \tilde{g}^*$. The series for $\gamma^{-1}(\tilde{g})$ and $\eta(\tilde{g})/\tilde{g}^2$, and independently the series for the function $\nu^{-1}(\tilde{g})$, have also been summed, although they are related to the γ and η by $\gamma(\tilde{g}) = \nu(\tilde{g})(2 - \eta(\tilde{g}))$. A verification of this relation after summation gives some information about the errors.

To give an idea of the apparent convergence, a few typical results are displayed in Table 23.2 (for details, see Section 5.7).

Table 23.2

Series summed by the method based on Borel transformation and mapping for the zero $\tilde{g}^*$ of the $\beta(g)$ function and the exponents γ and ν in the ϕ_3^4 field theory.

k	2	3	4	5	6	7
$\tilde{g}^*$	1.8774	1.5135	1.4149	1.4107	1.4103	1.4105
ν	0.6338	0.6328	0.62966	0.6302	0.6302	0.6302
γ	1.2257	1.2370	1.2386	1.2398	1.2398	1.2398

23.5 ODM summation method

We discuss now in more detail the ODM summation method, because it is not widely known.

The ODM summation method [216, 217] is based on some knowledge of the analytic properties of the function that is expanded. It applies both to convergent series and to divergent series, although it is mainly useful in the latter case. Among several, three applications that we do not discuss here are the hydrogen atom in a very strong magnetic field [218], the equation of state of the universality class of the Ising model [99], and the imaginary cubic potential in quantum mechanics [219].

After the method has been proposed, some rigorous convergence proofs have been given (where the method is called δ-expansion) but they are not always optimal [220, 221].

23.5.1 The general method

Again, let $f(z)$ be an analytic function that has the Taylor series expansion

$$f(z) = \sum_{\ell=0} f_\ell z^\ell.$$

When the Taylor series has a finite radius of convergence, to continue the function in the whole domain of analyticity, one can map the domain onto a circle, while preserving the origin.

Divergent series: An intuitive idea. In the case of a divergent series, one adds to the domain of analyticity a disc $|z| < r$ of variable radius r and applies a similar mapping. Of course, the transformed series is still divergent. Then, one recalls the empirical rule that, for a divergent series, one is instructed to truncate the series at the term of minimal modulus, the last term giving an order of magnitude of the error (see Section 23.1.1). One adjusts then the radius r order by order, in such a way that the last added term corresponds always to the minimum.

In what follows, we consider only functions analytic in a sector (as in the example of Fig. 23.2) and mappings $z \mapsto \lambda$ of the form

$$z = \rho\zeta(\lambda), \quad \zeta(\lambda) = \lambda + O\left(\lambda^2\right),$$

where $\zeta(\lambda)$ is an explicit analytic function, and ρ *an adjustable parameter.*

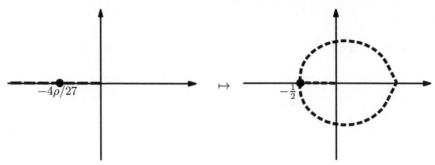

Fig. 23.2 Example: mapping ($\alpha = 3$) for a function analytic in a cut-plane.

Although the transformed series is still divergent at ρ fixed, we shall verify on a few examples that, by adjusting ρ order by order (here, we limit ourselves to Borel summable examples), *one can construct a convergent algorithm.*

After the transformation, f is given by a Taylor series in λ of the form

$$f\big(z(\lambda)\big) = \sum_{k=0}^{\infty} P_k(\rho)\lambda^k,$$

where the coefficients $P_k(\rho)$ are polynomials of degree k in ρ. Since the result is formally independent of the parameter ρ, the parameter can be chosen freely.

The k-th approximant $f^{(k)}(z)$ is constructed in the following way: one truncates the expansion at order k and chooses ρ so as to cancel the last term. Since $P_k(\rho)$ has k roots (real or complex), one chooses for ρ the largest possible root (in modulus) ρ_k for which $P'_k(\rho)$ is small. This leads to a sequence of approximants

$$f^{(k)}(z) = \sum_{\ell=0}^{k} P_\ell(\rho_k)\lambda^\ell(\rho_k, z), \quad \text{with} \quad P_k(\rho_k) = 0.$$

In the case of convergent series, it is expected that ρ_k has a non-vanishing limit for $k \to \infty$. By contrast, for divergent series, it is expected that ρ_k goes to zero for large k as

$$\rho_k = O\left(f_k^{-1/k}\right).$$

The intuitive idea here is that ρ_k corresponds to a 'local' radius of convergence.

Since ρ_k goes to zero, the function $\zeta(\lambda)$ must diverge for a finite value of λ. Below, we choose $\lambda = 1$ by convention.

Other choices. In the case of real functions, when the relevant zeros are complex, it is often convenient to choose minima of the polynomials P_k, solutions of

$$P'_k(\rho_k) = 0,$$

selecting, in general, the largest zero for which P_k is small. Other mixed criteria involving a combination of P_k and P'_k can also be used. Indeed, the *approximant is not very sensitive to the precise value of ρ_k, within errors.* Finally, $P_{k+1}(\rho_k)$ gives an order of magnitude of the error.

23.5.2 Functions analytic in a cut-plane: Heuristic convergence analysis

Although some rigorous convergence results have been established [220], these are not optimal. Therefore, we give here heuristic but quantitative arguments [217] that show the nature of the convergence of the ODM method.

Large order behaviour. Following the analysis in Ref. [216], to simplify, we consider a real function analytic in a cut-plane with a cut along the real negative axis (Fig. 23.2) and a Cauchy representation of the form

$$E(g) = \frac{1}{\pi} \int^{0_-} dg' \frac{\Delta(g')}{g' - g},$$

but the generalization is simple.

Moreover, we assume that

$$\Delta(g) \underset{g \to 0_-}{\propto} g^b \, e^{A/g}, \; A > 0. \tag{23.20}$$

The function $E(g)$ can be expanded in powers of g:

$$E(g) = \sum_k E_k g^k, \; \text{with } E_k = \frac{1}{\pi} \int^{0_-} \frac{dg}{g^{k+1}} \Delta(g).$$

The assumption (23.20) then implies the large order behaviour

$$E_k \underset{k \to \infty}{\propto} (-A)^{-k} \Gamma(k - b) \sim (-A)^{-k} k^{-b-1} k!,$$

exactly of the form (23.10).

A class of mappings. We introduce the mapping

$$g = \rho \frac{\lambda}{(1 - \lambda)^\alpha}, \quad \alpha > 1. \tag{23.21}$$

The Cauchy representation then can be written as

$$E\big(g(\lambda)\big) = \frac{1}{\pi} \int^{0_-} d\lambda' \frac{\Delta\big(g(\lambda')\big)}{\lambda - \lambda'} + R(\lambda),$$

where $R(\lambda)$ is a sum of contributions from cuts at finite distance from the origin. We expand

$$E\big(g(\lambda)\big) = \sum_k P_k(\rho)[\lambda(g)]^k, \tag{23.22}$$

with

$$P_k(\rho) = \frac{1}{\pi} \int^{0_-} d\lambda \, \Delta\big(g(\lambda)\big) \lambda^{-k-1} + \text{finite distance contributions.} \tag{23.23}$$

For $k \to \infty$, the factor λ^{-k} favours small values of λ but, for too small values of λ, the exponential decay of $\Delta(g(\lambda)$ takes over. Thus, $P_k(\rho_k)$ can be evaluated by the *steepest descent method*.

With the ansatz that, at the saddle point, $\lambda < 0$ is independent of k and

$$\rho_k \sim R/k, \ R > 0,$$

which implies $g(\lambda) \to 0$, $\Delta(g)$ can be replaced by its asymptotic form (23.20) for $g \to 0_-$. At leading order, the saddle point equation reduces to

$$\frac{\mathrm{d}}{\mathrm{d}\lambda} \left(\frac{A}{R\lambda}(1 - \lambda)^\alpha - \ln|\lambda| \right) = 0. \tag{23.24}$$

Since the equation depends only on the ratio R/A, in what follows we set $R/A = \mu$.

The equation can be rewritten as

$$\mu + \frac{1}{\lambda}(1 - \lambda)^{\alpha-1}\big((\alpha - 1)\lambda + 1\big) = 0.$$

We now exclude the exceptional case where the mapping (23.21) removes all singularities, except at the origin, (and this excludes the integral of Section 23.6).

Then $P_{k+1}(\rho_k)$, whose behaviour is given by the saddle point,

$$P_{k+1}(\rho_k) = O(\mathrm{e}^{\sigma k}), \text{ with } \sigma = \frac{1}{\lambda\mu}(1 - \lambda)^\alpha - \ln \lambda, \tag{23.25}$$

cannot decrease exponentially with k. This implies another equation,

$$\frac{1}{\lambda\mu}(1 - \lambda)^\alpha - \ln|\lambda| = 0. \tag{23.26}$$

This is indeed the region where the contribution coming from the cut at the origin, and that from the other finite distance singularities, are comparable, and where the zeros of $P_k(\rho)$ can lie.

Note that, quite generally, as a function of μ,

$$\frac{\mathrm{d}\sigma}{\mathrm{d}\mu} = -\frac{(1 - \lambda)^\alpha}{\lambda\mu^2} > 0,$$

since λ at the saddle point is negative.

Returning to the expansion (23.22), at g fixed, from the behaviour of ρ_k, one infers

$$1 - \lambda \sim (R/kg)^{1/\alpha} \ \Rightarrow \ \lambda^k \sim \mathrm{e}^{-k^{1-1/\alpha}(R/g)^{1/\alpha}}.$$

In a generic situation, one then expects $P_k(\rho_k)$ to behave like

$$P_k(\rho_k) = O(\mathrm{e}^{Ck^{1-1/\alpha}})$$

and the domain of convergence depends on the sign of the constant C.

Table 23.3

Saddle points.

α	3/2	2	5/2	3	4
μ	4.031233504	4.466846120	4.895690188	5.3168634291	6.1359656420
$-\lambda$	0.2429640300	0.2136524524	0.1896450439	0.1699396648	0.14003129119

For $C > 0$, the domain of convergence is

$$|g| < RC^{-\alpha}[\cos(\operatorname{Arg} g/\alpha)]^{\alpha}.$$

For $\alpha > 2$, this domain extends beyond the first Riemann sheet and requires analyticity of the function $E(g)$ in the corresponding domain.
For $C < 0$, the domain of convergence is the union of the sector $|\operatorname{Arg} g| < \pi\alpha/2$ and the domain

$$|g| > R|C|^{-\alpha}[-\cos(\operatorname{Arg} g/\alpha)]^{\alpha}.$$

Again, for $\alpha > 2$, this domain extends beyond the first Riemann sheet.

23.5.3 Examples
For $\alpha = 3/2$, combining equations (23.24) and (23.26), one finds

$$\mu = 4.031233504, \quad \lambda = -0.2429640300.$$

For $\alpha = 2$, equation (23.24) becomes

$$\lambda^2 - \mu\lambda - 1 = 0 \;\Rightarrow\; \lambda = \tfrac{1}{2}(\mu - \sqrt{\mu^2 + 4}).$$

For $\mu = 3.017759126\ldots$ one recovers the exponential rate of convergence (23.27).
In the case of additional singularities, with the additional equation (23.26), one obtains

$$\mu = 4.466846120\ldots, \quad \lambda = -0.2136524524\ldots.$$

To give a few other examples, again combining equations (23.24) and (23.26), one obtains the results displayed in Table 23.3.

23.6 Application: The simple integral $d = 0$

For $r = 1$, the integral (23.8) in the case $d = 0$ reduces to the simple integral

$$Z(g) = \frac{1}{\sqrt{2\pi}} \int \mathrm{d}x\, \mathrm{e}^{-x^2/2 - gx^4/4!},$$

and the convergence of the ODM method can be studied analytically.

23.6.1 The optimal mapping

Analytic properties suggest that the optimal mapping is given by setting

$$g = \frac{\rho\lambda}{(1-\lambda)^2} \quad \text{and} \quad Z(g) = (1-\lambda)^{1/2}f(\lambda).$$

Then, f has an expansion of the form

$$f(\lambda) = \sum_k P_k(\rho)[\lambda(g)]^k .$$

Convergence can be studied analytically. First, f has the representation

$$f(\lambda) = \frac{1}{\sqrt{2\pi}} \int ds \ e^{-s^2/2+\lambda(s^2/2-\rho s^4/24)} = P_k(\rho)[\lambda(g)]^k,$$

with

$$P_k(\rho) = \frac{1}{k!}\frac{1}{\sqrt{2\pi}} \int ds \ e^{-s^2/2}(s^2/2 - \rho s^4/24)^k.$$

Setting

$$s^2/2 = kt, \quad \rho = R/k,$$

one can rewrite the expression as

$$P_k(\rho) = \frac{k^{k+1/2}}{k!} \sqrt{\frac{1}{\pi}} \int \frac{dt}{\sqrt{t}} \ e^{-kt}(t - Rt^2/6)^k.$$

For $k \to \infty$, the integral can be evaluated by the steepest descent method. The saddle point equation is

$$Rt^2/6 - (1 + R/3)t + 1 = 0$$

and, thus,

$$t = \frac{3}{R}\left(1 + R/3 \pm \sqrt{1 + R^2/9}\right).$$

For k odd, the zero corresponds to a cancellation between the two saddle points. This yields the equation

$$e^{2\sqrt{R^2+9}/R} = \frac{\sqrt{R^2+9}+R}{\sqrt{R^2+9}-R} \quad \Leftrightarrow \quad e^{\sqrt{R^2+9}/R} = \tfrac{1}{3}\left(\sqrt{R^2+9}+R\right).$$

As expected, one finds

$$\rho_k \sim \frac{R}{k}, \quad \text{with } R = 4.526638689\ldots \quad (\mu = R/A = 3.017759126\ldots).$$

The minimum is given by one of the saddle points

$$P_k(\rho) \propto e^{-3k/R} = (0.5154353381\ldots)^k. \tag{23.27}$$

At g fixed, λ converges to 1. More precisely,

$$\lambda = 1 - \sqrt{R/kg} + O(1/k) \quad \Rightarrow \quad \lambda^k \sim e^{-\sqrt{Rk/g}}.$$

The approximants converge geometrically on the entire Riemann surface, a situation possible only because the function $Z(g)$ has no other singularity at finite distance, and an algebraic singularity at infinity.

23.6.2 Numerical application

The slope is found to be $1/k\rho_k \approx 0.2209$, in agreement with the prediction $1/R = 0.2209$ (once even–odd order oscillations are taken into account). The logarithm of the error has a slope $0.696/0.685$, to be compared with the prediction $3/R = 0.66$ (see Table 23.4).

Table 23.4

ODM method for the integral $d = 0$ for $g \to \infty$: $Z(g) \sim g^{-1/4}(1/2) * 24^{1/4} * \sqrt{\pi}/\Gamma(3/4)$. We define $[Z(g)g^{1/4}]_{\text{exact}} - [Z(g)g^{1/4}]_{\text{ODM}} = \delta$.

k	5	10	15	20	25	30
$1/\rho$	1.131726	2.35036	3.34050	4.5594	5.5495	6.8614
$-\delta$	$5.7 * 10^{-3}$	$2.5 * 10^{-5}$	$3.7 * 10^{-6}$	$2.2 * 10^{-8}$	$3.4 * 10^{-9}$	$5.1 * 10^{-11}$
$\ln\|\delta\|$	-5.1578	-10.5921	-12.5008	-17.5923	-19.4855	-23.6818
k	35	40	45	50	55	60
$1/\rho$	7.7586	8.9778	9.9678	11.1869	12.1769	13.3958
$-\delta$	$3.5 * 10^{-12}$	$2.5 * 10^{-14}$	$3.9 * 10^{-15}$	$2.9 * 10^{-17}$	$4.5 * 10^{-18}$	$3.4 * 10^{-20}$
$\ln\|\delta\|$	-26.3535	-31.2859	-33.1625	-38.0643	-39.9364	-44.8208

23.6.3 Alternative mapping

Another mapping that also regularizes the point at infinity is

$$g = \rho \frac{\lambda}{(1-\lambda)^4}.$$

Numerical results for $g = 5$ are displayed in Table 23.5.

Table 23.5

ODM method for the integral $d = 0$ for $g = 5$: we define $[Z(g)]_{\text{exact}} - [Z(g)]_{\text{ODM}} = \delta$.

k	5	10	15	20	25	30
$1/\rho$	0.5918	1.0297	1.5627	2.0779	2.5865	3.1376
$\|\delta\|$	$1.1 * 10^{-3}$	$3.7 * 10^{-5}$	$1.7 * 10^{-6}$	$9.2 * 10^{-7}$	$1.1 * 10^{-7}$	$3.5 * 10^{-9}$
$\ln\|\delta\|$	-6.7454	-10.2069	-13.2837	-13.8898	-16.0103	-19.4614
k	35	40	45	50	55	60
$1/\rho$	3.6167	4.1877	4.6557	5.3021	5.6959	6.2458
$-\delta$	$3.4 * 10^{-9}$	$8.0 * 10^{-10}$	$1.0 * 10^{-10}$	$2.9 * 10^{-11}$	$1.2 * 10^{-11}$	$2.4 * 10^{-12}$
$\ln\|\delta\|$	-19.4706	-20.9453	-22.9796	-24.2450	-25.0907	-26.7433

Finally, for $k = 65, 70$,

$$1/\rho_k = 6.7479\,, \ 7.5724\,, \quad \ln|\delta| = -30.3657\,, \ -28.9505\,.$$

On the average, between $k = 5$ and $k = 70$,

$$\rho_k \sim R/k\,, \text{ with } R = 9.75\,.$$

The results displayed in Table 23.3 lead to the prediction

$$R = 9.2039\,.$$

The error is about

$$\delta = -1.59 \frac{k^{3/4}}{g^{1/4}}$$

and 1.59 has to be compared with the expected asymptotic value (Section 23.5.2) 1.74 if one assumes convergence for all $g > 0$.

23.7 The quartic anharmonic oscillator: $d = 1$

The path integral corresponds to the quantum Hamiltonian

$$H = -\tfrac{1}{2}\left(\frac{\mathrm{d}}{\mathrm{d}x}\right)^2 + \tfrac{1}{2}x^2 + \frac{g}{4!}x^4\,.$$

As an example, we consider the perturbative expansion of the ground state energy. Variational and scaling arguments suggest the mapping

$$g = \frac{\rho\lambda}{(1-\lambda)^{3/2}}\,, \qquad E = \frac{\epsilon}{(1-\lambda)^{1/2}}\,. \tag{23.28}$$

Then,

$$\epsilon = \sum_k P_k(\rho)[\lambda(g)]^k\,.$$

Numerically, one finds that. for k large, the limit of $k\rho_k$ agrees with the result of Table 23.3. A reasonable fit of the data is

$$\frac{A}{k\rho_k} = 0.248063\cdots + \frac{1.7}{k^{2/3}}\,.$$

Again, λ converges to 1:

$$\lambda = 1 - \left(\frac{R}{kg}\right)^{2/3} + O(k^{-4/3}) \ \Rightarrow \ \lambda^k \sim \mathrm{e}^{-Ck^{1/3}/g^{2/3}}\,, \text{ with } C = R^{2/3} = 10.1317\ldots.$$

Finally, a numerical fit yields

$$P_k(\rho_k) \propto \mathrm{e}^{-9.7k^{1/3}}\,.$$

The error at order k is thus of order $\mathrm{e}^{-k^{1/3}(9.7 + Cg^{-2/3})}$. One finds convergence in the first Riemann sheet and a part of the second.

23.8 ϕ^4 **field theory in** $d = 3$ **dimensions**

In Ref. [216], the ODM method was applied to functions of the initial parameter g of the action (23.7), instead of the renormalized parameter g_r introduced in Section 23.4.3. Then, the point of physical interest is $g \to \infty$, which corresponds to the zero $\tilde{g}^*$ of the β-function (23.15).

Due to UV divergences, a needed regularization and renormalization, scaling arguments are no longer applicable to determine an appropriate mapping. The relation (23.16) between initial and renormalized parameter shows that, for $g \to \infty$, physical observables have an expansion in powers of $g^{-\omega}$, where the exponent $\omega = \tilde{\beta}'(\tilde{g}^*)$.

This then suggests the mapping

$$g = \rho \frac{\lambda}{(1 - \lambda)^{1/\omega}},$$

but the difficulty is that ω has to be inferred from the series (23.15) itself. The results obtained in this way [216] are consistent with those obtained in [31] ($\omega = 0.80(1)$ from Borel transformation and mapping), but empirical errors are more difficult to assess. Also, the expected rate of convergence is of order $e^{-\text{const.}k^{1-\omega}} = e^{-\text{const.}k^{0.2}}$, which is rather slow (see Table 23.6) (see Ref. [216] for details). Finally, the information about the large order behaviour cannot easily be incorporated.

Table 23.6

Series in the ϕ_3^4 field theory for the correction exponent ω summed by ODM with $\omega_i = 0.79$ and $d\omega_{\text{cal.}}/d\omega_i = -0.6$. Adjusting ω_i, one finds $\omega = 0.77$.

k	2	3	4	5	6
$\omega_{\text{cal. } k}$	0.552	0.754	0.711	0.767	0.759

Here, to illustrate the flexibility of the method, we work directly with functions of $\tilde{g}$. We also take into account the covariance of the β-functions under a change of parametrization:

$$\beta_1(g_1) = \frac{dg_1}{dg_2} \beta_2(g_2). \tag{23.29}$$

This transformation rule is such that the derivative of the β-function at a zero (a fixed point), which is a physical observable, remains unchanged.

Equation (23.29) suggests a mapping of the form

$$\tilde{g} = \rho \left(\frac{1}{(1 - \lambda)^\alpha} - 1 \right),$$

with a suitable choice of the parameter α, since, unlike a mapping of the form (23.21), it introduces no new singularity.

We thus set

$$\beta_\lambda(\lambda) = \frac{(1-\lambda)^{\alpha+1}}{\alpha\rho}\tilde{\beta}\big(\tilde{g}(\lambda)\big).$$

A few trials, without trying to optimize, suggest the value $\alpha = 3/2$ and this is the value we have adopted. The results for the zero $\tilde{g}^*$, for comparison with the method outlined in Section 23.4.3, and the exponent $\omega = \beta'_\lambda(\lambda^*)$, are given in Table 23.7. The order of magnitude of the errors can only be estimated by the sensitivity to the precise choice of the parameter ρ (*e.g.*, the zero of last term or its derivative).

Table 23.7

Results for the zero $\tilde{g}^*$ of the β-function and the exponent ω summed by the ODM method in the ϕ_3^4 field theory. With the method from Section 23.4.3 ([31]), one finds $\tilde{g}^* = 1.411 \pm 0.004$ and $\omega = 0.799 \pm 0.011$.

k	3	4	5	6	7
$\tilde{g}^*$	1.09871	1.39330	1.41771	1.41737	1.41744
ω	1	0.7984	0.7804	0.7806	0.7807

The indications are $\Delta\tilde{g}^* \approx \Delta\omega \approx 0.006$. In the case of complex zeros, we have given only the real part in the tables.

Other critical exponents are obtained from the two RG functions (23.17) by setting $\tilde{g} = \tilde{g}^*$. The results, displayed in Table 23.8, can be compared with the results of Ref. [31]:

$$\gamma = 1.2396 \pm 0.0013, \quad \nu = 0.6304 \pm 0.0013, \quad \eta = 0.0335 \pm 0.0025.$$

One observes a satisfactory agreement.

Table 23.8

Critical exponents $\gamma = \gamma(\tilde{g}^*)$, $\nu = \nu(\tilde{g}^*)$ and $\eta = \eta(\tilde{g}^*)$, for $\tilde{g}^* = 1.411$.

k	3	4	5	6	7
γ	1.23717	1.23486	1.23845	1.23820	1.23923
ν	0.62521	0.62486	0.62746	0.62771	0.62865
η		0.0290	0.0289	0.0297	0.0306

24 Hyper-asymptotic expansions and instantons

In quantum field theory, the main *analytic* tool for determining physical quantities is perturbation theory. However, simple arguments indicate that the perturbative expansion is always divergent. As a consequence, *when the interaction is not weak enough, the plain expansion is useless.*

Unlike convergent series, *divergent series do not determine, in general, a unique analytic function.* The question is then, *to which extent does the perturbative expansion determine a field theory?*

An exception is provided by the class of divergent *Borel summable series* to which correspond a unique function.

The questions are then, *which field theories generate Borel summable expansions?* and *what kind of additional information should one derive from a field theory* to complement the expansion, *when it is not Borel summable,* in order to determine completely physical quantities from the expansion?

Some information about the first question is provided by the study of the large order behaviour of perturbation theory (formal expansions in powers of $\hbar$, or loop expansions in terms of Feynman diagrams).

The behaviour of the perturbative expansion at large orders was first derived in quantum mechanics for simple Hamiltonians with polynomial potentials from a study of the Schrödinger equation [222].

In the late 1970s, the large order behaviour has been determined more systematically, following Lipatov [26], in quantum mechanics and quantum field theory by using functional integral representations [223, 28, 27]. It was realized that the large order behaviour in quantum mechanics with analytic potentials and in super-renormalizable quantum field theories is related to barrier penetration effects in the semi-classical limit, *generalized to include formal barrier penetration for physical and non-physical values of the parameters in the action* [28].

In the framework of functional integration, barrier penetration amplitudes are obtained by the steepest descent method and correspond to non-trivial saddle points of the form of finite action solutions of the imaginary time classical equations of motion called *instantons*.

However, in renormalizable quantum field theories, like $\phi^4_{d=4}$ or quantum chromodynamics (QCD), *renormalons* [212] generated by ultraviolet or infrared singularities, yield additional contributions to the large order behaviour.

Although the large order behaviour only indicates a possible Borel summability, it can show that a series is *not Borel summable.* In particular, this happens when *physical barrier penetration* occurs [28].

Examples of the latter situation are metastable perturbative vacua decaying by barrier penetration or potential with degenerate minima connected by tunnelling.

In this chapter, we consider only the situation of degenerate classical minima.

From Random Walks to Random Matrices. Jean Zinn-Justin, Oxford University Press (2019).
© Jean Zinn-Justin. DOI: 10.1093/oso/9780198787754.001.0001

Moreover, since examples in quantum field theory correspond to renormalizable theories like QCD or $CP(N-1)$ models, which are difficult to work out, we focus on non-relativistic one-dimensional quantum mechanics with analytic potentials [213], mainly *polynomials* [38] and, on a case by case study, to a few other analytic potentials like the periodic *cosine potential.*

We consider specifically expansions in powers of $\hbar$ or any equivalent parameter, and energy eigenvalues of order $\hbar$. In terms of Feynman diagrams, this corresponds to an expansion in the number of loops.

A question then arises: is it possible to define in some way the sum of the perturbative expansion and then to add to it some non-perturbative corrections in order to recover the exact functions?

In Section 24.1, we first briefly recall the notion of Borel summability (see also Chapter 23) and discuss the ground state energy of the quartic anharmonic oscillator, a simple example of Borel summable series.

Then, we discuss a few explicit non-Borel summable examples, the spectra of potentials with *degenerate minima*, like the quartic double-well or the periodic cosine potentials. The structure of the *exact hyper-asymptotic semi-classical expansions* of low-lying energy levels for a few *analytic potentials* is described [39]. The expansions involve an *infinite number of perturbative series*, but all of them can be generated by expanding generalized Bohr–Sommerfeld quantization formulae that can be parametrized in terms of a *few spectral functions*, two in the simplest examples [39]. Later, it was pointed out that the two are actually related [224].

The form of the hyper-asymptotic expansions was initially conjectured from a generalization of the results obtained by semi-classical evaluations of partition functions based on the path integral formalism.

The *infinite number of quasi-saddle points, also called multi-instantons*, generated by the steepest descent method yields contributions that can be *summed exactly at leading order* [39]. The same strategy could still be useful in problems where our present understanding is more limited, in particular, in quantum field theory, where the Schrödinger equation is no longer useful.

Finally, these properties have a direct interpretation within the framework of the *complex WKB expansion* of the solutions of the Schrödinger equation [225].

24.1 Divergent series and Borel summability

Let $f(z)$ be an analytic function holomorphic in a sector S of the complex plane,

$$S \equiv \{|\text{Arg}\, z| \le \alpha/2, \quad |z| \le |z_0|\}, \tag{24.1}$$

which has in S the Taylor series expansion

$$f(z) = \sum_{k=0}^{\infty} f_k z^k, \tag{24.2}$$

where the series (24.2) diverges for all $z \ne 0$, and satisfies the bound

$$\left| f(z) - \sum_{k=0}^{K-1} f_k z^k \right| \le M A^{-K} K! |z|^K \quad \forall K. \tag{24.3}$$

Moreover, we assume that the domain of interest is z real positive.

The best estimate of the function is obtained by minimizing, at z fixed, the error in (24.3) as a function of the order K. For $|z|$ small, on finds in S an error of order

$$\varepsilon(z) \propto e^{-A/|z|},\tag{24.4}$$

which shows that, in general, an asymptotic series does not define a unique function but only an infinite class of functions.

However, if the angle α satisfies,

$$\alpha > \pi,\tag{24.5}$$

the function f is unique and can be recovered by a Borel transformation. The series is then called *Borel summable*.

Borel transformation. The Borel transform $B_f(z)$ of $f(z)$ is defined by

$$B_f(z) = \sum_0^\infty \frac{f_k}{k!} z^k.\tag{24.6}$$

The bound (24.3) implies the bound on the coefficients of B_f,

$$|f_k/k!| < M A^{-k}.\tag{24.7}$$

Thus, $B_f(z)$ is analytic at least in a disc of radius A and defined by the series. Furthermore, it can be proved that the integral

$$f(z) = \int_0^\infty e^{-t} B_f(zt) \mathrm{d}t,\tag{24.8}$$

exists. Indeed, as a consequence of the inequality (24.5), $B_f(z)$ is also analytic in a sector

$$|\mathrm{Arg}\, z| \in [0, \tfrac{1}{2}(\alpha - \pi)[,\tag{24.9}$$

and does not increase faster than an exponential in the sector, so that integral (24.8) converges for $|z|$ small enough and inside the sector

$$|\mathrm{Arg}\, z| < \alpha/2.$$

In addition, it can be shown that the right-hand side of equation (24.8) satisfies a bound of type (24.3). Hence, this integral representation yields the unique function which has the asymptotic expansion (24.2) in the domain (24.1).

Example of non-Borel summable series. If the relevant values of the argument of the series are $z \geq 0$ and the coefficients f_k behave for $k \to \infty$ as

$$f_k \propto A^{-k} k!, \quad \text{with } A > 0,$$

the Borel transform has a singularity on the real positive axis at $z = A$, and the integral (24.8) does not exist. The series cannot be Borel summable. This is a situation one meets in some quantum theories.

24.2 Perturbative expansion and path integral

Partition function and path integral. We consider one-dimensional non-relativistic quantum Hamiltonians of the form,

$$H = \frac{g}{2} \left(\frac{\mathrm{d}}{\mathrm{d}q} \right)^2 + \frac{1}{g} V(q), \quad \text{with } V(q) = \tfrac{1}{2}q^2 + O(q^3) \ge 0, \tag{24.10}$$

where $V(q)$ is an analytic function such that $V(q) \to +\infty$ for $|q| \to \infty$.

The quantum partition function $\mathcal{Z}(\beta) = \mathrm{tr}\, \mathrm{e}^{-\beta H}$ (β is the inverse temperature) has a simple path integral representation. It is given by the sum over closed paths,

$$\mathcal{Z}(\beta) \propto \int_{q(-\beta/2)=q(\beta/2)} [\mathrm{d}q(t)] \exp\left[-\frac{1}{g} \int_{-\beta/2}^{\beta/2} \left[\tfrac{1}{2}\dot{q}^2(t) + V\big(q(t)\big) \right] \mathrm{d}t \right].$$

Partition function and Fredholm determinant. Since the Hamiltonian (24.10) has a discrete spectrum, the partition function has the large β expansion

$$\mathcal{Z}(\beta) \equiv \mathrm{tr}\, \mathrm{e}^{-\beta H} = \sum_{N \ge 0} \mathrm{e}^{-\beta E_N}, \quad \text{with } E_N > E_{N-1}. \tag{24.11}$$

From the partition function, on infers the trace $G(E)$ of the resolvent of H (after analytic continuation and possible subtraction),

$$G(E) = \mathrm{tr}\, \frac{1}{H-E} = \int_0^\infty \mathrm{d}\beta\, \mathrm{e}^{\beta E}\, \mathcal{Z}(\beta). \tag{24.12}$$

The poles of $G(E)$ yield the spectrum of the Hamiltonian H. The Fredholm determinant $\mathcal{D}(E) = \det(H-E)$, which vanishes on the spectrum, is then given by

$$\frac{\partial}{\partial E} \ln \mathcal{D}(E) = -G(E). \tag{24.13}$$

Perturbative expansion. For $g \to 0$, the path integral can be evaluated by the *steepest descent method*. Saddle points are solutions $q_c(t)$ to the Euclidean equations of motion.

When the potential has a unique minimum located at $q = 0$, the leading saddle point is $q_c(t) \equiv 0$. A systematic expansion around the saddle point then yields the perturbative expansion of the partition function. Eigenvalues of the Hamiltonian are recovered from a subsequent large β expansion of the partition function or by calculating the zeros of the determinant $\mathcal{D}(E)$.

In the case of a *potential with degenerate minima*, all minima are the starting points of a perturbative expansion, and the partition is the sum of the corresponding contributions.

24.3 The quartic anharmonic oscillator: A Borel summable example

Before discussing some more complicated non-Borel summable examples, let us consider the energy eigenvalues of a quantum Hamiltonian of the form (24.10), the quartic anharmonic oscillator,

$$H = -\tfrac{1}{2}\left(\mathrm{d}/\mathrm{d}q\right)^2 + \tfrac{1}{2}q^2 + \tfrac{1}{4}gq^4, \; g > 0\,. \tag{24.14}$$

Since the parameter g plays the role of $\hbar$, we set $\hbar = 1$.

The partition function has the path integral representation,

$$\mathrm{tr}\,\mathrm{e}^{-\beta H} = \int_{q(-\beta/2)=q(\beta/2)} [\mathrm{d}q(t)]\exp\left[-\mathcal{S}(q)\right], \tag{24.15}$$

where $\mathcal{S}(q)$ is the Euclidean (or imaginary time) action

$$\mathcal{S}(q) = \int_{-\beta/2}^{\beta/2}\left[\tfrac{1}{2}\dot{q}^2(t) + \tfrac{1}{2}q^2(t) + \tfrac{1}{4}gq^4(t)\right]\mathrm{d}t \tag{24.16}$$

and one integrates over paths satisfying periodic boundary conditions.

Generalization of arguments applicable to finite dimensional integrals indicates that the path integral (24.15) defines a function of g analytic in the cut-plane with a cut along the entire negative axis.

In this domain, for g small, the integral is dominated by the saddle point $q(t) \equiv 0$. Therefore, it can be calculated by expanding the integrand in powers of g and integrating term by term.

This leads to the perturbative expansion of the partition function and thus, expanding it for β large (equation (24.11)), of the energy eigenvalues $E_N(g)$ of the Hamiltonian. It takes the form

$$E_N(g) = N + \tfrac{1}{2} + \sum_{k=1}^{\infty} E_{N,k}g^k. \tag{24.17}$$

The eigenvalues are solutions of a spectral equation (related to $\mathcal{D}(E)$) of the form

$$N + \tfrac{1}{2} = B(E, g). \tag{24.18}$$

For example, at order g^3, the function B has the expansion

$$B = E - g\left(\tfrac{3}{8}E^2 + \tfrac{3}{32}\right) + g^2\left(\tfrac{35}{64}E^3 + \tfrac{85}{256}E\right)$$
$$- \tfrac{1}{64}g^3\left(\tfrac{1155}{16}E^4 + \tfrac{2625}{32}E^2 + \tfrac{1995}{256}\right) + O(g^4).$$

Inverting equation (24.18) and expanding the solution as power series in g, one obtains the expansions (24.17) of the eigenvalues $E_N(g)$.

24.3.1 Cauchy representation and barrier penetration

Since $E_N(g)$ is analytic in the cut-plane and can be shown to behave like $g^{1/3}$ for g large, it has a Cauchy representation of the form [226]

$$E_N(g) = \frac{1}{2} + \frac{g}{\pi} \int_{-\infty}^{0} \frac{\operatorname{Im} E_N(g')\, dg'}{g'(g' - g)}. \qquad (24.19)$$

Expanding both sides in powers of g, one obtains the integral representation

$$E_{N,k} = \frac{1}{\pi} \int_{-\infty}^{0} \frac{\operatorname{Im} E_N(g)\, dg}{g^{k+1}}, \quad \text{for } k > 0. \qquad (24.20)$$

For $g < 0$, the Hamiltonian (24.14) is not bounded from below. The function $\operatorname{Im} E_N(g)$ is then proportional to the decay rate per unit time by tunnelling of a quantum particle initially located near the metastable minimum $q = 0$ at level N.

For $g \to 0_-$, the barrier penetration amplitude $\operatorname{Im} E_N(g)$ can be derived from a semi-classical approximation.

24.3.2 Barrier penetration and instantons: The ground state $N = 0$

In a path integral representation, barrier penetration effects can be calculated by the steepest descent method and correspond to non-constant solutions of the imaginary time equations of motion, with finite action for $\beta \to \infty$, called *instantons*.

The action (24.16) is directly the imaginary time (or Euclidean) action. The equation of motion reads

$$-\ddot{q}(t) + q(t) + g q^3(t) = 0.$$

For $\beta \to \infty$, the finite action solution corresponds to a trajectory that leaves $q = 0$ at time $t = -\infty$, is reflected at $q = \pm\sqrt{-2/g}$ and returns at $t = +\infty$ at $q = 0$. The general solution is

$$q_c(t) = \pm \left(-\frac{2}{g}\right)^{1/2} \frac{1}{\cosh(t - t_0)}, \qquad (24.21)$$

where t_0 is an integration constant, and the corresponding classical action is

$$\mathcal{S}(q_c) = -\frac{4}{3g}. \qquad (24.22)$$

One finds two one-parameter families of saddle points. To complete the calculation of the saddle point contribution at leading order, one has to sum over all saddle points. This involves factorizing the measure in the path integral into an integration measure over the *collective coordinate* t_0 and over the orthogonal modes (this generates a non-trivial Jacobian) [227]. The integration over t_0 is exact, and the integration over the orthogonal modes is performed in the Gaussian approximation. One then obtains, at leading order for g small and negative,

$$\operatorname{Im} E_0(g) \underset{g \to 0_-}{\sim} \left(\frac{8}{\pi}\right)^{1/2} \frac{1}{\sqrt{-g}} e^{4/3g}. \qquad (24.23)$$

For $k \to \infty$, due to the factor g^{-k}, the Cauchy integral (24.20) is dominated by the small negative g values. One infers [222, 223],

$$
E_{0,k} \underset{k \to \infty}{\sim} \frac{1}{\pi} \int^{0-} \left(\frac{8}{\pi} \right)^{1/2} \frac{1}{\sqrt{-g}} \frac{e^{4/3g}}{g^{k+1}} \left[1 + O(g) \right] dg
$$

$$
= - \left(\frac{6}{\pi^3} \right)^{1/2} \left(-\frac{3}{4} \right)^k \Gamma(k + 1/2) \left[1 + O(1/k) \right]. \qquad (24.24)
$$

This result confirms that the perturbative series is divergent and determines the nature of the divergence. Successive corrections to the semi-classical result yield a series in powers of g which, when integrated, generates a systematic expansion in powers of $1/k$. The corrections can be obtained by substituting in the first equation (24.24) $e^{-A(E,g)}$ for $e^{4/3g}$, with

$$
A(E, g) = -\frac{4}{3g} - g \left(\tfrac{17}{16} E^2 + \tfrac{67}{192} \right) + g^2 \left(\tfrac{227}{128} E^3 + \tfrac{671}{512} E \right) + O(g^3)
$$

[228–230], and using a solution of the spectral equation (24.18) for E. Note that A and B are related. They satisfy,

$$
\frac{\partial E}{\partial B} = -\frac{2}{a} \left(Bg + g^2 \frac{\partial A}{\partial g} \right), \qquad (24.25)
$$

where a is the instanton action, here $a = -4/3$, and A and E are considered functions of (B, g). The relation is a variant of equation (24.37), first proved for the double-well and cosine potentials [224]. It reduces the determination of both functions to the determination of the simpler perturbative spectral function $B(E, g)$.

The Borel transform $B_{E_0}(t)$ then has as a singularity closest to the origin at the point $t = -4/3$. Globally, one proves that the Borel transform is analytic in a cut-plane and that the perturbative expansion is *Borel summable*.

More generally, a necessary condition for Borel summability is that quantum tunnelling occurs only for non-physical (including complex) values of parameters in the action (here, $g < 0$). By contrast, physical tunnelling leads to positive actions and, thus, to singularities of the Borel transform on the real positive axis, and non-Borel summability.

The instanton method generalize to quantum field theories. For example, the ϕ^4 field theory, the d-dimensional generalization of the quartic anharmonic oscillator, has analogous perturbative expansions for $d < 4$, and these are also Borel summable.

The $O(n)$ symmetric quartic anharmonic oscillator. In the example of the $O(n)$ symmetric anharmonic oscillator in n dimensions,

$$
H = -\tfrac{1}{2} \nabla_q^2 + \tfrac{1}{2} \mathbf{q}^2 + \tfrac{1}{4} g \left(\mathbf{q}^2 \right)^2 , \qquad (24.26)
$$

the expression (24.24) generalizes in the form [223]

$$
E_{0,k} \underset{k \to \infty}{\sim} -\frac{6^{n/2}}{\pi \Gamma(n/2)} \left(-\frac{3}{4} \right)^k \Gamma(k + n/2). \qquad (24.27)
$$

The argument $(k + n/2)$ of the Γ function reflects the additional $(n - 1)$ collective coordinates parametrizing the instanton solution, since $q_c(t)$ is replaced by $\hat{\mathbf{n}} q_c(t)$, with $\hat{\mathbf{n}}^2 = 1$.

24.4 The double-well potential: Generalized Bohr–Sommerfeld quantization formulae

Barrier penetration and non-Borel summability. For analytic potentials with degenerate classical minima, like the double-well potential, expansions in powers of $\hbar$ are *non-Borel summable*, because quantum tunnelling occurs. Then, *additional contributions* of order $\exp(-\text{const.}/\hbar)$, generated by *quantum tunnelling*, have to be added to the perturbative expansion. Therefore, the exact determination of eigenvalues starting from their expansion for $\hbar$ small becomes a non-trivial problem.

Hyper-asymptotic expansions. In this situation, we describe conjectures that, at least in two examples, the quartic double-well and the cosine potentials, give systematic procedures for calculating energy eigenvalues, for $\hbar$ *finite*, from expansions that are shown to contain powers of $\hbar$, $\ln \hbar$ and $\exp(-\text{const.}/\hbar)$ [39].

Generalized Bohr–Sommerfeld quantization formulae. Generalized Bohr–Sommerfeld quantization formulae make it possible to derive the *infinite number of series* that appear in such formal expansions from a few *WKB expansions* [39, 229, 230].

This relation with the WKB expansion is not completely trivial. Indeed, the perturbative expansion corresponds from the viewpoint of the WKB approximation to a situation with confluent singularities and thus, for example, the usual WKB expressions for barrier penetration are singular when the energy goes to zero. It requires a complex generalization of the WKB method [225].

Finally, these conjectures have found a natural interpretation in the framework of Ecalle's theory [232] of *resurgent functions*, as has been shown by Pham's collaborators [233].

24.4.1 The quartic double-well potential: Perturbative expansion

The simplest example of a potential with degenerate minima is the *quartic double-well potential*. The corresponding Hamiltonian can be written as

$$H = -\frac{g}{2}\left(\frac{\mathrm{d}}{\mathrm{d}q}\right)^2 + \frac{1}{g}V(q), \quad V(q) = \frac{1}{2}q^2(1-q)^2,$$

where the symbol g plays the role of $\hbar$ and energy eigenvalues are measured in units of $\hbar$, a normalization adapted to perturbative expansions.

We describe below the hyper-asymptotic structure of the perturbative expansion of its energy eigenvalues.

Symmetry and perturbative expansion. The potential is symmetric in $q \leftrightarrow (1-q)$ and thus the Hamiltonian commutes with the corresponding reflection operator,

$$P\psi(q) = \psi(1-q) \;\Rightarrow\; [H,P] = 0\,.$$

The common eigenfunctions of H and P satisfy

$$H\psi_{\epsilon,N}(q) = E_{\epsilon,N}(g)\psi_{\epsilon,N}(q), \quad P\psi_{\epsilon,N}(q) = \epsilon\psi_{\epsilon,N}(q), \tag{24.28}$$

where $\epsilon = \pm 1$, and $E_{\epsilon,N}(g) = N + 1/2 + O(g)$.

One can separate eigenvalues according to the parity of eigenfunctions by considering the two partition functions

$$\mathcal{Z}_{\pm}(\beta) = \text{tr}\left[\tfrac{1}{2}(1 \pm P)\, e^{-\beta H}\right] = \sum_{N=0} e^{-\beta E_{\pm,N}}. \tag{24.29}$$

The eigenvalues $E_{\epsilon,N}$ ($\epsilon = \pm 1$) are then poles of

$$G_{\epsilon}(E) = \int_0^{\infty} d\beta\, e^{\beta E}\, \mathcal{Z}_{\epsilon}(\beta).$$

Correspondingly, the Fredholm determinant (24.13) factorizes:

$$\mathcal{D}(E) = \mathcal{D}_+(E)\mathcal{D}_-(E). \tag{24.30}$$

Perturbative expansion. The potential has two symmetric degenerate classical minima located at $q = 0, 1$. The expansions of the potential near each minimum,

$$V(q) = \tfrac{1}{2}q^2 + O(q^3), \text{ or } V(q) = \tfrac{1}{2}(1-q)^2 + O((1-q)^3).,$$

generate two perturbative expansions. However, due to the symmetry, the expansions are identical, and the spectrum thus is twice degenerate to all orders in an expansion in powers of g: $\mathcal{D}_+(E) = \mathcal{D}_-(E)$. For example, for the ground state energy,

$$E_{+,0}(g) = E_{-,0}(g) = \sum_{\ell=0}^{\infty} E_{0,\ell}^{(0)} g^{\ell}. \tag{24.31}$$

24.4.2 Quantum tunnelling and energy splitting

The degeneracy is lifted by quantum tunnelling, which, in the path integral formalism, is obtained by calculating the contribution of the neighbourhood of a nontrivial classical (instanton) solution relating the two minima (see Sections 24.3.2 and 24.5.2). For the two lowest-lying levels, the full calculation of the path integral in the saddle point approximation then yields,

$$E_{-,0}(g) - E_{+,0}(g) \underset{g\to 0}{\sim} 2\frac{e^{-1/6g}}{\sqrt{\pi g}}. \tag{24.32}$$

While the instanton contribution dominates the energy difference, one may wonder whether it makes sense to add such a small contribution directly to the divergent perturbative expansion. The answer depends on the rate of the divergence of the series. However, the answer is known here without further calculation, for accidental reasons. A relation in the sense of *formal power series* between the $O(2)$ anharmonic oscillator and the double-well potential (24.26) [228, 231] has been discovered such that

$$[E_0(g)]_{O(2)}(-4g) = 2[E_0(g)]_{\text{double-well}}. \tag{24.33}$$

Therefore, the large order behaviour for the double-well potential can be inferred from expression (24.27). One obtains

$$E_{0,k} \underset{k\to\infty}{\sim} -\frac{1}{\pi}3^{k+1}\Gamma(k+1). \tag{24.34}$$

All terms have the same sign and this confirms that the series is not Borel summable (unlike the $O(2)$ series). According to the analysis of Section 23.1.1 and equation (23.3), the perturbative expansion of the double-well defines E_0 up to order $\mathrm{e}^{-1/3g} \ll \mathrm{e}^{-1/6g}$ and, therefore, it makes sense to add the energy splitting to the perturbative expansion.

An intriguing observation is that the larger order behaviour has a factor $\Gamma(k+1)$. For the $O(2)$ quartic anharmonic oscillator, the origin is simple: the instanton solution depends on two parameters (*cf.*, the comment after equation (24.27)). By contrast, any finite action solution to the double-well equation of motion can depend only on one time integration constant. The solution to the apparent paradox will be explained in Section 24.5.3.

24.4.3 The hyper-asymptotic expansion

In Ref. [39], we have conjectured that the eigenvalues $E_{\epsilon,N}(g)$ have an exact semi-classical expansion (an hyper-asymptotic or trans-series expansion) of the form

$$
E_{\epsilon,N}(g) = \sum_0^\infty E_{N,l}^{(0)} g^l
$$

$$
+ \sum_{n=1}^\infty \left(\frac{2}{g}\right)^{Nn} \left(-\epsilon \frac{\mathrm{e}^{-1/6g}}{\sqrt{\pi g}}\right)^n \sum_{k=0}^{n-1} (\ln(-2/g))^k \sum_{l=0}^\infty \mathcal{E}_{N,nkl} g^l. \qquad (24.35)
$$

The power series $\sum e_{N,nkl} g^l$ are *not Borel summable* for $g > 0$, and the expansion thus requires an interpretation.

By contrast, for g negative, for example, the series are Borel summable and $\ln(-g)$ is also real. Starting from $g < 0$, one then proceeds by *analytic continuation* to $g = |g| \pm i0$ consistently for the series and $\ln(-g)$.

In the analytic continuation, the Borel sums become complex, with the imaginary parts being exponentially smaller by about a factor of $\mathrm{e}^{-1/3g}$ than the real parts. These imaginary contributions are cancelled by the perturbative imaginary parts coming from the function $\ln(-2/g)$, a phenomenon that finds a natural explanation in the framework of resurgence theory [232].

24.4.4 Generalized Bohr–Sommerfeld quantization formula

We have also conjectured [39] that all the series in equation (24.35) are generated by an expansion for g small of two *spectral equations or generalized Bohr–Sommerfeld quantization formulae*, which, in the case of the double-well potential, read ($\epsilon = \pm 1$)

$$
\mathcal{D}_\epsilon(E) = \frac{1}{\Gamma\left(\frac{1}{2} - B(E,g)\right)} + \frac{\epsilon i}{\sqrt{2\pi}} \left(-\frac{2}{g}\right)^{B(E,g)} \mathrm{e}^{-A(E,g)/2} = 0, \qquad (24.36)
$$

with

$$
B(E,g) = -B(-E,-g) = E + \sum_{k=1} g^k b_{k+1}(E),
$$

$$
A(E,g) = -A(-E,-g) = \frac{1}{3g} + \sum_{k=1} g^k a_{k+1}(E).
$$

The coefficients $b_k(E)$, $a_k(E)$ are odd or even polynomials in E of degree k. The four first orders, for example, are

$$B = E + g\left(3E^2 + \tfrac{1}{4}\right) + g^2\left(35E^3 + \tfrac{25}{4}E\right) + g^3\left(\tfrac{1155}{2}E^4 + \tfrac{735}{4}E^2 + \tfrac{175}{32}\right)$$
$$+ g^4\left(\tfrac{45045}{4}E^5 + \tfrac{45045}{8}E^3 + \tfrac{31185}{64}E\right) + O\left(g^5\right),$$

$$A = \frac{1}{3g} + g\left(17E^2 + \tfrac{19}{12}\right) + g^2\left(227E^3 + \tfrac{187}{4}E\right) + g^3\left(\tfrac{47431}{12}E^4 + \tfrac{34121}{24}E^2 + \tfrac{28829}{576}\right)$$
$$+ O\left(g^4\right).$$

The function $B(E,g)$ can be inferred from the complex WKB perturbative expansion. The function $A(E,g)$ has initially been determined at this order by a combination of analytic and numerical calculations.

However, more recently, for the double-well and cosine potentials, the following remarkable relation was proved [224], using differential equation techniques,

$$\frac{\partial E}{\partial B} = -\frac{1}{a}\left(2Bg + g^2\frac{\partial A}{\partial g}\right), \qquad (24.37)$$

where a is the instanton action (here, $a = 1/3$), and A and E are considered functions of (B,g) (see also equation (24.25), for a variant).

24.4.5 Multi-instanton contributions at leading order

Replacing the functions A and B by their leading terms, one obtains the spectral equations

$$\frac{e^{-1/6g}}{\sqrt{2\pi}}\left(-\frac{2}{g}\right)^E = -\frac{\epsilon i}{\Gamma(\tfrac{1}{2} - E)}$$

$$\Leftrightarrow \quad \frac{\cos \pi E}{\pi} = \epsilon i \frac{e^{-1/6g}}{\sqrt{2\pi}}\left(-\frac{2}{g}\right)^E \frac{1}{\Gamma(\tfrac{1}{2} + E)}. \qquad (24.38)$$

Expanding then the equation in powers of $e^{-1/6g}$, one obtains contributions of order $e^{-n/6g}$ for $g \to 0$ that, from the point of view of the path integral representation, correspond to successive multi-instanton contributions at leading order (see Section 24.5.3).

For example, the term proportional to $e^{-1/6g}$, which can be identified with the *one-instanton contribution at leading order*, is

$$E_N^{(1)}(g) = -\frac{\epsilon}{N!}\left(\frac{2}{g}\right)^{N+1/2}\frac{e^{-1/6g}}{\sqrt{2\pi}}(1 + O(g)).$$

The next term, the *two-instanton contribution*, is ($\psi = (\ln \Gamma)'$)

$$E_N^{(2)}(g) = \frac{1}{(N!)^2}\left(\frac{2}{g}\right)^{2N+1}\frac{e^{-1/3g}}{2\pi}\left[\ln(-2/g) - \psi(N+1) + O\left(g\ln g\right)\right].$$

More generally, the n-th power, which can be identified with the n-instanton contribution at leading order, has the form

$$E_N^{(n)}(g)$$

$$= (-1)^n \left(\frac{2}{g}\right)^{n(N+1/2)} \left(\frac{\mathrm{e}^{-1/6g}}{\sqrt{2\pi}}\right)^n \left[P_n^{(N)}\left(\ln(-2/g)\right) + O\left(g\left(\ln g\right)^{n-1}\right)\right],$$

in which $P_n^N(\sigma)$ is a polynomial of degree $(n-1)$. For example, for $N = 0$, one finds (γ is Euler's constant)

$$P_1^{(0)}(\sigma) = 1\,, \quad P_2^{(0)}(\sigma) = \sigma + \gamma\,, \quad P_3^{(0)}(\sigma) = \frac{3}{2}\,(\sigma + \gamma)^2 + \frac{\pi^2}{12}\,.$$

24.4.6 Large order behaviour of perturbative series

After an analytic continuation from g negative to g positive, the Borel sums become complex, with the imaginary part being exponentially smaller by about a factor $\mathrm{e}^{-1/3g}$ than the real part.

Consistently, the function $\ln(-2/g)$ also becomes complex, with the imaginary part $\pm i\pi$. Since the eigenvalues are real, the sum of all contributions must be real and the imaginary parts must cancel.

For example, the *non-perturbative imaginary part of the Borel sum of the perturbation series* cancels the *perturbative imaginary part of the two-instanton contribution*. For the ground state,

$$\mathrm{Im}\, E^{(0)}(g) \underset{g\to 0}{\sim} \frac{1}{\pi g}\, \mathrm{e}^{-1/3g}\, \mathrm{Im}\left[P_2^{(0)}\left(\ln(-2/g)\right)\right] = -\frac{1}{g}\, \mathrm{e}^{-1/3g}\,.$$

The coefficients of the perturbative expansion

$$E^{(0)}(g) = \sum_k E_k^{(0)} g^k$$

of the ground state energy, are related to the imaginary part by a Cauchy integral ($k > 1$):

$$E_k^{(0)} = \frac{1}{\pi} \int_0^\infty \mathrm{Im}\left[E^{(0)}(g)\right] \frac{\mathrm{d}g}{g^{k+1}}\,.$$

For $k \to \infty$, the integral is dominated by small g values. Thus,

$$E_k^{(0)} \underset{k\to\infty}{\sim} -\frac{1}{\pi} \int_0^\infty \frac{\mathrm{e}^{-1/3g}}{g^{k+2}}\,\mathrm{d}g = -\frac{1}{\pi}\, 3^{k+1} k!\,,$$

a result in agreement with expression (24.34). Similarly, since $\mathrm{Im}\, E^{(1)}(g)$ and $\mathrm{Im}\, E^{(3)}(g)$ cancel at leading order,

$$\mathrm{Im}\, E^{(1)}(g) \sim 3\pi \left(\frac{\mathrm{e}^{-1/6g}}{\sqrt{\pi g}}\right)^3 \left[\ln(2/g) + \gamma + O(g\ln(g))\right].$$

The coefficients of the expansion

$$E^{(1)}(g) = -\frac{1}{\sqrt{\pi g}}\, e^{-1/6g}\left(1 + \sum_{k}^{\infty} E_k^{(1)} g^k\right)$$

are given by the Cauchy integral

$$E_k^{(1)} = -\frac{1}{\pi}\int_0^\infty \left\{\mathrm{Im}\left[E^{(1)}(g)\right]\sqrt{\pi g}\, e^{1/6g}\right\}\frac{dg}{g^{k+1}}.$$

Combining both equations, one finds

$$E_k^{(1)} \sim -\frac{3}{\pi}\int_0^\infty \left(\ln\frac{2}{g}+\gamma\right)e^{-1/3g}\,\frac{dg}{g^{k+2}} \sim -\frac{3^{k+2}}{\pi}k!\,(\ln 6k + \gamma).$$

The result have been verified numerically by calculating many terms of the corresponding series.

The real part of the two-instanton contribution. Another test is provided by the evaluation of the ratio dominated for $g \ll 1$ by the two-instanton contribution (Fig. 24.1),

$$\Delta(g) = 4\frac{\left\{\frac{1}{2}\left(E_{0,+}+E_{0,-}\right)-\mathrm{Re}\left[\text{Borel sum } E^{(0)}(g)\right]\right\}}{\left(E_{0,+}-E_{0,-}\right)^2\left(\ln 2g^{-1}+\gamma\right)} = 1 + 3g + \cdots.$$

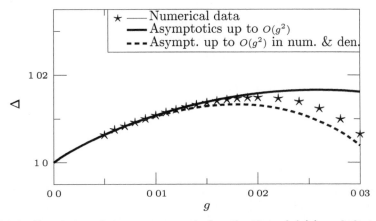

Fig. 24.1 Comparison between a numerical evaluation of $\Delta(g)$ and the asymptotic expansion for g small.

24.5 Instantons and multi-instantons

The conjecture, and its generalization to other analytic potentials, were initially motivated by a summation of leading order *multi-instanton* contributions. The method may be worth recalling, since it could be useful for other, less understood, problems.

24.5.1 Partition function and symmetries

For a symmetric double-well potential, one can separate eigenvalues according to the parity of eigenfunctions by considering the two partition functions (equation (24.29))

$$\mathcal{Z}_\pm(\beta) = \mathrm{tr}\left[\tfrac{1}{2}(1 \pm P)\,\mathrm{e}^{-\beta H}\right] = \sum_{N=0} \mathrm{e}^{-\beta E_{\pm,N}},$$

where P is the reflection operator.

The projected partition functions can then be rewritten as

$$\mathcal{Z}_\pm(\beta) = \tfrac{1}{2}\big(\mathcal{Z}(\beta) \pm \mathcal{Z}_\mathrm{a}(\beta)\big),$$

where

$$\mathcal{Z}_\mathrm{a}(\beta) \equiv \mathrm{tr}\left(P\,\mathrm{e}^{-\beta H}\right) \tag{24.39}$$

has also a path integral representation but with twisted boundary conditions.

This strategy generalizes to other symmetries of the Hamiltonian and leads to path integrals where the boundary conditions are twisted by symmetry transformations.

24.5.2 Potentials with symmetric degenerate minima

As noticed at the end of Section 24.2, in the case of potentials with degenerate minima, one must sum over the perturbative contributions generated by all minima (constant saddle points): to each minimum corresponds a set of eigenvalues, and thus several perturbative expansions are generated.

For a *symmetric double-well potential*, the two expansions are identical. Therefore, all eigenvalues are twice degenerate to all orders in perturbation theory:

$$E_{\pm,N}(g) = E_N^{(0)}(g) \equiv \sum_{n=0}^{\infty} E_{N,n}^{(0)} g^n.$$

In the case of a symmetric potential with two degenerate minima, it is also useful to consider the quantity (24.39). It is given by a path integral with the same integrand but with the twisted boundary conditions $q(-\beta/2) = P(q(\beta/2))$:

$$\mathcal{Z}_\mathrm{a}(\beta) \propto \int_{q(-\beta/2)=P(q(\beta/2))} [\mathrm{d}q(t)]\exp\left[-\frac{1}{g}\int_{-\beta/2}^{\beta/2}\left[\tfrac{1}{2}\dot{q}^2(t) + V\big(q(t)\big)\right]\mathrm{d}t\right]. \tag{24.40}$$

Instantons. In the case of potentials corresponding to metastable perturbative vacua, or those with degenerate minima, additional contributions to the perturbative eigenvalues are found, corresponding to barrier penetration effects.

For the perturbative ground state, which is obtained from the infinite β limit, the non-perturbative contributions to the path integral come from saddle points that take the form of non-constant *finite action solutions of the Euclidean (or imaginary time) equations of motion* called *instantons* (see also Section 24.3.2).

In the case of the symmetric double-well potential, the path integral representation of $\mathcal{Z}(\beta)$ is dominated by the constant saddle points $q(t) \equiv 0, 1$. By contrast, *constant saddle points do not contribute* to the path integral representation of $\mathcal{Z}_{\mathrm{a}}(\beta)$, because they do not satisfy the boundary conditions. The leading contributions correspond to finite action solutions (*instantons*) connecting the two minima of the potential (see Fig. 24.2) and are is non-perturbative.

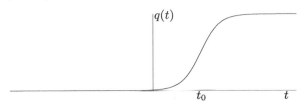

Fig. 24.2 Instanton solution.

In the example of the quartic double-well potential, the solutions are

$$q_{\mathrm{c}}(t) = \left(1 + \mathrm{e}^{\pm(t-t_0)} \right)^{-1} \;\Rightarrow\; \mathcal{S}(q_{\mathrm{c}}) = 1/6 \,.$$

Since the two solutions depend on an integration constant t_0 (the instanton position), one finds two one-parameter families of degenerated saddle points.

The corresponding contribution to the path integral is proportional, at leading order for $g \to 0$ and for $\beta \to \infty$, to $\mathrm{e}^{-1/(6g)}$ and thus is *non-perturbative*, as expected.

The complete calculation involves integrating exactly over the time t_0 (the collective coordinate), which, for β finite, varies in $[0, \beta]$, and over the remaining fluctuations in the Gaussian approximation. The two lowest eigenvalues are given by ($\epsilon = \pm 1$)

$$E_{\epsilon,0}(g) = \lim_{\beta \to \infty} -\frac{1}{\beta} \ln \mathcal{Z}_\epsilon(\beta) \underset{g \to 0, \beta \to \infty}{=} E_0^{(0)}(g) - \epsilon E_0^{(1)}(g), \quad \text{with}$$

$$E_0^{(1)}(g) = \frac{1}{\sqrt{\pi g}} \, \mathrm{e}^{-1/6g} \left(1 + O(g) \right).$$

Ambiguities coming from the perturbative expansion appear only at order $\mathrm{e}^{-1/3g}$ (see equations (24.4) and (24.34)).

24.5.3   Multi-instantons

For β finite, one finds subleading saddle points, which correspond to additional oscillations in the well of the potential $-V(q)$. For $\beta \to \infty$, the limit of the action of the solutions with n oscillations is $n \times 1/6$.

However, the *Gaussian integral at the saddle point diverges for $\beta \to \infty$.* Indeed, in this limit, the classical solutions decompose into a *succession of largely separated instantons*, and fluctuations that change the distances between instantons induce only infinitesimal variations of the action.

Moreover, the saddle points have a direction of instability. We will give an intuitive solution to both problems, which, although not totally satisfactory, yields the correct result.

To solve the first problem, one *sums over all configurations of largely separated instantons*, which are connected in a smooth way, and which become solutions of the equation of motion only asymptotically, for infinite separation.

They depend on n *collective coordinates*, the distance between instantons. The action then has a dependence on the collective coordinates, called *instanton interaction*.

The two-instanton configuration. We first examine a two-instanton (more precisely, an instanton–anti-instanton) configuration. In the infinite β limit, the one-instanton configuration can be written as

$$q_{\pm}(t) = f\big(\mp(t - t_0)\big), \ f(t) \equiv 1/\left(1 + \mathrm{e}^t\right) = 1 - f(-t),$$

where the constant t_0 characterizes the instanton position.

It is simple to verify that a configuration $q_{\mathrm{c}}(t)$ that is the sum of instantons separated by a distance θ, up to an additive constant adjusted in such a way as to satisfy the boundary conditions (Fig. 24.3),

$$q_{\mathrm{c}}(t) = f(t - t_0 - \theta/2) + f(-t + t_0 - \theta/2) - 1 = f(t - t_0 - \theta/2) - f(t - t_0 + \theta/2),$$

has the required properties: it is differentiable and, for θ large, but fixed, it minimizes the variation of the action. The corresponding action is

$$S(q_c) = \tfrac{1}{3} - 2\,\mathrm{e}^{-\theta} + O\left(\mathrm{e}^{-2\theta}\right).$$

It will become apparent later that contributions to the classical action of order $\mathrm{e}^{-2\theta}$ give only corrections of order g.

For β large, but finite, symmetry between θ and $(\beta - \theta)$ implies

$$S(q_c) = \tfrac{1}{3} - 2\,\mathrm{e}^{-\theta} - 2\,\mathrm{e}^{-(\beta-\theta)} + \text{subleading contributions}.$$

Note that a two-instanton configuration depends on two collective coordinates. Moreover, it yields the leading instanton contribution to the partition function $\mathcal{Z}(\beta)$. Thus, it governs the large order behaviour of perturbation theory, and this explains the form of the large order behaviour (24.34).

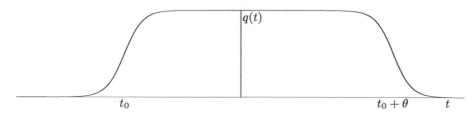

Fig. 24.3 A two-instanton configuration.

24.5.4 *The general multi-instanton action*

For a succession of n instantons (more precisely, alternatively instantons and anti-instantons) separated by times θ_i with

$$\sum_{i=1}^{n} \theta_i = \beta \,,$$

the classical action $\mathcal{S}_c(\theta_i)$ is then

$$\mathcal{S}_c(\theta_i) = \frac{n}{6} - 2\sum_{i=1}^{n} e^{-\theta_i} + O\left(e^{-(\theta_i + \theta_j)}\right). \tag{24.41}$$

For n even, the n-instanton configurations contribute to $\operatorname{tr} e^{-\beta H}$, while, for n odd, they contribute to $\operatorname{tr}\left(P e^{-\beta H}\right)$ (P is the reflection operator).

24.5.5 *The multi-instanton contribution*

The evaluation of the contribution to the path integral of the neighbourhood of the n-instanton configuration, at leading order, is straightforward but slightly technical. One finds that the n-instanton contribution to the combination

$$\mathcal{Z}_\epsilon(\beta) = \tfrac{1}{2} \operatorname{tr}\left[(1 + \epsilon P) e^{-\beta H}\right],$$

($\epsilon = \pm 1$) can be written as

$$\mathcal{Z}_\epsilon^{(n)}(\beta) = e^{-\beta/2} \frac{\beta}{n} \left(\epsilon \frac{e^{-1/6g}}{\sqrt{\pi g}}\right)^n \int_{\substack{\theta_i \geq 0 \\ \sum \theta_i = \beta}} \prod_i d\theta_i \, \exp\left(\frac{2}{g}\sum_{i=1}^{n} e^{-\theta_i}\right). \tag{24.42}$$

Statistical physics interpretation. The expression can be interpreted as the classical partition function of a gas of n particles separated by distances θ_i on an axis of length β with nearest-neighbour exponentially decreasing interactions $e^{-\theta}$. The sum over n yields the partition function in the grand canonical formulation, with the fugacity $e^{-1/6g}/\sqrt{\pi g}$.

Neglecting the instanton interactions (the free gas approximation) and summing over n, one recovers the one-instanton approximation to the energy eigenvalues.

Beyond the one-instanton approximation: The instability problem. If one examines the classical action (24.41), one discovers that the *interaction between instantons is attractive.* Therefore, for g small, the *dominant contributions to the integral come from configurations in which the instantons are close.* For such configurations, the concept of instanton is no longer meaningful, since the configurations cannot be distinguished from fluctuations around the constant or the one-instanton solution.

Such a difficulty should have been anticipated. In the case of potentials with degenerate minima, the perturbative expansion is not Borel summable and the *series determines eigenvalues only up to exponentially decreasing contributions that are of the order of two-instanton contributions.* But if the perturbative expansion is ambiguous at the two-instanton order, n-instanton contributions with $n \geq 2$ are not defined. To proceed any further, *it is necessary to first define the sum of the perturbative expansion.*

In the example of the quartic double-well potential, *the perturbation series is Borel summable for g negative* because it is directly related to the $O(2)$ anharmonic oscillator (see relation (24.33)). Therefore, we *define* the sum of the perturbation series as the analytic continuation of this Borel sum from g negative to $g = |g| \pm i0$. In the Borel transformation , this can be shown to be equivalent to integrating above or below the real positive axis. Simultaneously, for *g negative, the interaction between instantons becomes repulsive and the multi-instanton contributions become meaningful.*

Therefore, we first calculate, for g small and negative, both the sum of the perturbation series and the multi-instanton contributions, and then perform an analytic continuation to g positive of all quantities consistently.

24.5.6 The sum of leading order instanton contributions

We assume that, initially, g is negative and calculate the sum of leading n-instanton contributions to the trace of the resolvents,

$$G_\epsilon(E) = \sum_{n=1}^\infty \int_0^\infty \mathrm{d}\beta \; \mathrm{e}^{\beta E} \, \mathcal{Z}_\epsilon^{(n)}(\beta),$$

where $\mathcal{Z}_\epsilon^{(n)}(\beta)$ is given by equation (24.42). The integration over β is immediate, and the integrals over the θ_i's then factorize. Evaluating the unique integral for $g \to 0_-$, summing over n and adding the resolvent of the harmonic oscillator, one finds for the resolvent $G_\epsilon(E)$ a result consistent with the expression (24.38),

$$G_\epsilon(E) = -\frac{\partial}{\partial E} \ln \mathcal{D}_\epsilon(E) \;\Rightarrow\; \mathcal{D}_\epsilon(E) = \frac{1}{\Gamma(\frac{1}{2} - E)} + \epsilon i \left(-\frac{2}{g}\right)^E \frac{\mathrm{e}^{-1/6g}}{\sqrt{2\pi}}.$$

24.6 Perturbative and exact WKB expansions

The conjectures, motivated by semi-classical evaluation of path integrals (instanton calculus), have been confirmed by considerations based on the Schrödinger equation,

$$[H\psi](q) \equiv -\frac{g}{2}\psi''(q) + \frac{1}{g}V(q)\psi(q) = E\psi(q),$$

where the *potential V is an entire function*. This property makes it possible to *extend the Schrödinger equation and its solutions to the complex q-plane.*

24.6.1 Riccati equation and complex Bohr–Sommerfeld quantization formula
Setting
$$S(q) = -g\psi'(q)/\psi(q),$$
one transforms the Schrödinger equation into the Riccati equation,

$$gS'(q) - S^2(q) + 2V(q) - 2gE = 0. \tag{24.43}$$

One decomposes

$$S(q) = S_+(q) + S_-(q),$$

where in the sense of formal series expansions,

$$S_\pm(q; g, E) = \pm S_\pm(q; -g, -E).$$

Equation (24.43) is equivalent to the two equations

$$gS'_- - S_+^2 - S_-^2 + 2V(q) - 2gE = 0, \quad gS'_+ - 2S_+S_- = 0. \tag{24.44}$$

The quantization condition (or spectral equation) can then be written as

$$-\frac{1}{2i\pi g} \oint_C dz\, S_+(z, E) = N + \tfrac{1}{2},$$

where N is also the number of real zeros of the eigenfunction, and C is a contour in the complex plane that encloses them.

The equation can also be written as

$$\exp\left[-\frac{1}{g} \oint_C dz\, S_+(z)\right] + 1 = 0. \tag{24.45}$$

This elegant formulation [225], restricted, however, to one dimension and analytic potentials, *bypasses the difficulties generally associated with turning points.*

It allows a *smooth transition between a WKB expansion* (an expansion for $g \to 0$ at gE fixed), in our normalization, and *a perturbative expansion* (expanding for $g \to 0$ at E fixed), which can be obtained by expanding the WKB expansion at E fixed.

24.6.2 Complex WKB expansion

At leading order in the WKB limit, the function S_+ reduces to

$$S_+(q) = S(q) = S_0(q), \text{ with } S_0(q) = \sqrt{2V(q) - 2gE}$$

and the quantization condition becomes

$$N + \tfrac{1}{2} = B(E, g) = -\frac{1}{2i\pi g} \oint_C dz\, S_0(z, E),$$

where the contour C, in the case of a unique classical minimum, encloses the cut of $S_0(q)$ that joins the turning points.

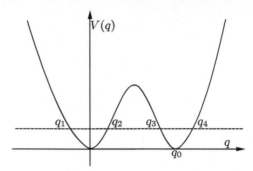

Fig. 24.4 Degenerate minima: the four turning points.

By contrast, if the *potential has two degenerate, not necessarily symmetric, minima*, for E small enough, the function $S_0(q)$ has the four branch points $q_1 < q_2 < q_3 < q_4$ on the real axis (Fig. 24.4).

One can then define two functions $B_1(E, g)$ and $B_2(E, g)$ which, at leading order, correspond to contours enclosing the cuts $[q_1, q_2]$ and $[q_3, q_4]$ and generate two perturbative expansions.

One can also organize the cuts differently in such a way that a contour can enclose the cut $[q_2, q_3]$. Comparing with the conjecture, one infers the decomposition

$$\frac{1}{g} \oint_{C[q_2, q_3]} \mathrm{d}z \, S_+(z) = A(E, g) + \ln(2\pi) - \sum_{i=1}^{2} \ln \Gamma\left(\tfrac{1}{2} - B_i(E, g)\right)$$
$$+ B_i(E, g) \ln(-g/2C_i),$$

where, at leading order in the WKB expansion, the contour encloses the cut $[q_2, q_3]$ and the constants C_i are chosen such that $A(E, g)$ has no term of order g^0.

24.7 Other analytic potentials: A few examples

Other analytic potentials have been considered [229]. We give here a few examples.

24.7.1 Asymmetric wells

For a potential with two asymmetric wells,

$$V(q) = \tfrac{1}{2}\omega_1^2 q^2 + O(q^3), \quad V(q) = \tfrac{1}{2}\omega_2^2(q - q_0)^2 + O\left((q - q_0)^3\right),$$

the spectral equation takes the form

$$\frac{2\pi}{\Gamma\left(\tfrac{1}{2} - B_1(E, g)\right) \Gamma\left(\tfrac{1}{2} - B_2(E, g)\right)} + \left(-\frac{2C_1}{g}\right)^{B_1(E, g)} \left(-\frac{2C_2}{g}\right)^{B_2(E, g)} \mathrm{e}^{-A(g, E)} = 0,$$

where $B_1(E, g)$ and $B_2(E, g)$ are determined by the *perturbative expansions around the two minima of the potential*,

$$B_1(E, g) = E/\omega_1 + O(g), \quad B_1(E, g) = E/\omega_2 + O(g).$$

The constants C_1 and C_2 are adjusted in such a way that

$$A(E,g) - a/g = O(g), \quad a = 2 \int_0^{q_0} \mathrm{d}q \sqrt{2V(q)}.$$

From the poles of Γ-functions for $g \to 0$, one sees that the spectral equation yields two sets of energy eigenvalues,

$$E_N = \left(N + \tfrac{1}{2}\right)\omega_1 + O(g), \quad E_N = \left(N + \tfrac{1}{2}\right)\omega_2 + O(g).$$

The same expression *contains the instanton contributions to the two different sets of eigenvalues.*

It is possible to verify that, while multi-instanton contributions are singular in the limit $\omega_1 = \omega_2$, the spectral equation is regular in this limit.

One-instanton contribution and large order behaviour. The spectral equation can again be used to infer the large order behaviour of perturbation theory from the imaginary part of the leading instanton contribution by writing a dispersion integral. Setting $\omega_1 = 1$, $\omega_2 = \omega$, one infers that the coefficients $E_{N,k}$ of the perturbative expansion of the energy eigenvalue $E_N(g) = N + \tfrac{1}{2} + O(g)$ behave, for order $k \to \infty$, like

$$E_{N,k} \underset{k \to \infty}{=} K_N \frac{\Gamma(k + (N+1/2)(1+1/\omega))}{a^{k+(N+1/2)(1+1/\omega)}}\left(1 + O\left(k^{-1}\right)\right).$$

24.7.2 The periodic cosine potential

We consider the Hamiltonian

$$H = -\frac{g}{2}\left(\frac{\mathrm{d}}{\mathrm{d}q}\right)^2 + \frac{1}{16g}(1 - \cos 4q).$$

The cosine potential is still an *entire function but no longer a polynomial.* On the other hand, the periodicity of the potential simplifies the analysis, because eigenfunctions can be classified according to their behaviour under a translation of one period,

$$\psi_\varphi(q + \pi/2) = \mathrm{e}^{i\varphi}\,\psi_\varphi(q), \quad 0 \le \varphi < 2\pi .$$

The generalized Bohr–Sommerfeld quantization formula then takes the form

$$\left(\frac{2}{g}\right)^{-B(E,g)} \frac{\mathrm{e}^{A(E,g)/2}}{\Gamma(\tfrac{1}{2} - B(E,g))} + \left(\frac{-2}{g}\right)^{B(E,g)} \frac{\mathrm{e}^{-A(E,g)/2}}{\Gamma(\tfrac{1}{2} + B(E,g))} = \frac{2\cos\varphi}{\sqrt{2\pi}},$$

with

$$B = E + g\left(E^2 + \tfrac{1}{4}\right) + g^2\left(3E^3 + \tfrac{5}{4}E\right) + g^3\left(\tfrac{25}{2}E^4 + \tfrac{35}{4}E^2 + \tfrac{17}{32}\right)$$
$$+ g^4\left(\tfrac{245}{4}E^5 + \tfrac{525}{8}E^3 + \tfrac{721}{64}E\right) + O(g^5),$$

$$A = \frac{1}{g} + g\left(3E^2 + \tfrac{3}{4}\right) + g^2\left(11E^3 + \tfrac{23}{4}E\right) + g^3\left(\tfrac{199}{4}E^4 + \tfrac{341}{8}E^2 + \tfrac{215}{64}\right) + O(g^4).$$

Here the relation (24.37) is again satisfied with $a = 1$ and reads

$$\frac{\partial E}{\partial B} = -g\left(2B + g\frac{\partial A}{\partial g}\right). \tag{24.46}$$

25 Renormalization group approach to matrix models

The study of the statistical properties of *random matrices of large size* has a long history, dating back to Wigner [234], who suggested using *Gaussian ensembles* to give a statistical description of the spectrum of complex Hamiltonians and derived the famous semicircle distribution, followed by contributions from Mehta, Gaudin and Dyson, among others. Later, 't Hooft [235] noticed that, in $SU(N)$ non-Abelian gauge theories, *tessalated surfaces can be associated to Feynman diagrams* and that the *large N expansion corresponds to an expansion in successive topologies.*

This observation has not yet led to a solution of non-Abelian gauge theories in four dimensions, even for N large, but, based on this observation, in the mid 1980s it was realized that some ensembles of random matrices in *the large size* and so-called *double scaling limit* could be used as toy models for *two-dimensional quantum gravity coupled to conformal matter and string theory* or as examples of *critical statistical models on some type of random surfaces* [236].

A tremendous development of random matrix theory followed. With the use of increasingly sophisticated mathematical methods, a number of matrix models have been solved exactly [237]. However, the somewhat paradoxical situation is that either models can be solved exactly or little can be said about them. In particular, one has a limited knowledge about path integrals where paths are matrices, and even less is known about integrals over fields that belong to large matrices, although it is to these integrals that the important example of non-Abelian gauge theories belongs.

Therefore, since the solved models exhibit *critical points* and *universal properties*, it is tempting to use *renormalization group* (RG) ideas to determine universal properties, without solving models explicitly. In this spirit, equations can be written that correspond to a recursive integration over components of matrices. However, to solve these equations, even in the large N limit, an approximation scheme is required [240].

To understand some of the difficulties encountered, similar ideas have also been applied to $O(N)$ symmetric vector models [241], which are models that can more generally be solved explicitly in the large N limit [47].

We describe here an approximate RG implementation. Improving the leading order RG equation has been achieved to a limited extent [241] but the interpretation of the results is not straightforward. This example illustrates again the point that RG is an idea but not a method: it requires a suitable implementation, which is, generally, not easy to guess.

From Random Walks to Random Matrices. Jean Zinn-Justin, Oxford University Press (2019).
© Jean Zinn-Justin. DOI: 10.1093/oso/9780198787754.001.0001

25.1 One-Hermitian matrix models and random surfaces: A summary

We consider an integral over $N \times N$ *Hermitian matrices* M of the form

$$e^{Z(N)} = \int dM \; e^{-N \operatorname{tr} V(M)},$$

where V is a polynomial, for example,

$$V(M) = \tfrac{1}{2}M^2 + \tfrac{1}{3}gM^3. \tag{25.1}$$

The formal expansion of $Z(N, g)$ in powers of g generates a sum of *connected fat* Feynman diagrams (Fig. 25.1), whose duals in the example (25.1) are *connected triangulated surfaces* (Fig. 25.2). If all triangles have a unit area, *the term of order g_3^n is the sum of surfaces of area n.*

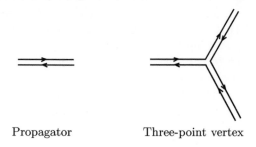

Propagator   Three-point vertex

Fig. 25.1 Matrix integrals with cubic potential: faithful Feynman rules.

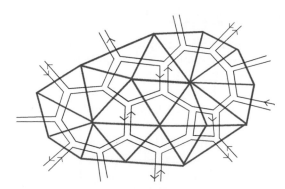

Fig. 25.2 Matrix integrals with cubic potential: Feynman diagram and dual triangulated surface.

Moreover, $Z(N, V)$ has a large N topological expansion,

$$Z(N, V) = N^2 Z_0(V) + Z_1(V) + \frac{1}{N^2} Z_2(V) + \cdots = \sum_{h=0} N^{2-2h} Z_h(V),$$

where $Z_h(V)$ is the sum of all *surfaces of genus h.*

25.2 Continuum and double scaling limits

By tuning one parameter in the polynomial $V(M)$, one can, in general, reach a continuum limit of the discretized surface and, more generally, a double scaling limit.

25.2.1 Continuum limit: The cubic example

The function $Z_0(g = g_3^2)$ can be calculated, for example, by diagonalizing the Hermitian matrices,

$$\mathbf{M} = \mathbf{U}^\dagger \mathbf{\Lambda} \mathbf{U},$$

integrating over unitary transformations $\mathbf{U}$ and evaluating the remaining integral over eigenvalues $\mathbf{\Lambda}$ by the *steepest descent method* [238].

The contribution to the partition function of the singularity at $g = 0$ is exponentially suppressed for $N \to \infty$. Instead, for the cubic example, one finds that the partition function has a singularity at a critical value $g = g_c = 1/12\sqrt{3}$, where the singular contribution to $Z_0(g)$ has the form,

$$Z_{0,\text{sing.}}(g) \propto (g - g_c)^{2-\alpha}.$$

The specific heat exponent α (called the *string susceptibility exponent* and denoted γ in the string terminology) is here $\alpha = -1/2$. At g_c the *average surface area*,

$$A = \langle n \rangle = \frac{\partial \ln Z_{0,\text{sing.}}(g)}{\partial g} \propto \frac{2 - \alpha}{g - g_c},$$

diverges, making it possible to define a *continuum limit* by shrinking the size of the triangles to zero.

25.2.2 The double scaling limit

It can be shown that, at the same critical value, $Z_h(g)$ is singular, with the behaviour

$$Z_h(g) \propto (g_c - g)^{(2-\alpha)(1-h)}.$$

Setting

$$\tau^{-1} = N(g - g_c)^{(2-\alpha)/2},$$

one defines a *double scaling limit* in which $N \to \infty$ and $g \to g_c$ at τ fixed.

In the double scaling limit (relevant for string theory), one finds a sum over all continuum surfaces with increasing genus:

$$Z(N, g) \sim Z(\tau) = \sum_{h=0} \tau^{2-2h} f_h.$$

The double scaling limit can be determined by introducing, after reducing the matrix integral to an integral over eigenvalues, a set of orthogonal polynomials [239].

25.2.3 Generalizations

Critical and multicritical points. For a general polynomial V, at fixed genus, by tuning only one parameter of V, one can generate only a singularity with $\alpha = -1/2$ corresponding to two-dimensional pure quantum gravity (central charge $c = 0$).

By choosing for V a polynomial of degree high enough and *tuning a larger number of parameters*, it is possible to reach *multicritical points* with values of the (string susceptibility) exponent $\alpha = 3/2 - m$, $m > 0$.

The Liouville field theory makes it possible to relate α to the central charge c by

$$\alpha = \tfrac{1}{12}\left(c - 1 - \sqrt{(1-c)(25-c)}\right).$$

Integrals over several matrices. A limited class of *integrals over several matrices* can also be evaluated in the double scaling limit.

Critical points can then be interpreted as *conformal matter on some type of random surfaces* or *two-dimensional quantum gravity coupled to conformal matter*.

The simplest example is

$$e^{Z(N)} = \int dM_1\, dM_2\; e^{-N\,\mathrm{tr}\,V(M_1,M_2)},$$

where

$$V(M_1, M_2) = \tfrac{1}{2}M_1^2 + \tfrac{1}{2}M_2^2 - cM_1M_2 + \tfrac{1}{4}g\left(M_1^4 + M_2^4\right),$$

which has a $\mathbb{Z}_2$ symmetry (the exchange $M_1 \leftrightarrow M_2$) and can be interpreted as an Ising model on a random surface.

By tuning the two parameters, one describes a critical Ising model on a continuum random surface.

25.3 The RG approach

The central result of matrix model theory is the *existence of a double scaling limit*: a continuum limit with critical exponents that depend on how the coupling constants of the theory are tuned.

The properties of the double scaling limit are reminiscent of the theory of continuous phase transitions. There, the critical behaviour and, more generally, the universal critical properties can be determined by studying infrared fixed points of an *RG*. This suggests trying to implement an RG strategy in matrix models as well.

We have to construct some kind of RG flow. Indeed, *from such RG flows, scaling laws and exponents, which characterize the double scaling limit, may follow automatically.*

We have to understand how the coupling constants of the theory, the size N of the matrices and the matrix parameters, evolve under a rescaling of the regularization length introduced by the discretization of the surface (the 'world sheet', in the string terminology).

The strategy will be to determine how, for N large, a variation $N \mapsto N + \delta N$ of the *size of the matrices* can be compensated by a variation of the *matrix parameters* [240].

However, as in the *Wilson's scheme of partial integration in a momentum shell* [100, 101], for matrix models, the RG flow also generates an infinite number of coupling constants.

Explicit calculations then require some approximation scheme (for critical phenomena, one introduces the $4 - \varepsilon$, $2 + \varepsilon$, or $1/N$ expansions), and we describe two below.

25.3.1   *The matrix RG flow*

We consider an integral over $N \times N$ Hermitian matrices $\boldsymbol{\phi}_N$ of an *even* (for simplicity) analytic function S_N of the form,

$$\mathrm{e}^{\mathcal{Z}(N,\mathcal{V})} = \int \mathrm{d}\boldsymbol{\phi}_N \, \exp\left[-S_N\left(\boldsymbol{\phi}_N, \mathcal{V}\right)\right], \qquad (25.2)$$

with

$$S_N(\boldsymbol{\phi}_N, \mathcal{V}) \equiv N \operatorname{tr} \mathcal{V}(\boldsymbol{\phi}_N) \text{ and } \mathcal{V}(\phi) = \tfrac{1}{2}\phi^2 + \sum_2^\infty \frac{g_k}{2k}\phi^{2k}. \qquad (25.3)$$

In what follows, we set $g_2 = g$.

Flow equations. Flow equations are then obtained by *integrating out one column and one row* of $\boldsymbol{\phi}_{N+1}$. We set

$$\boldsymbol{\phi}_{N+1} = \begin{pmatrix} \boldsymbol{\varphi}_N & \mathbf{v} \\ \mathbf{v}^\dagger & \omega \end{pmatrix}, \qquad (25.4)$$

where $\boldsymbol{\varphi}_N$ is an $N \times N$ matrix, $\mathbf{v}$ is an N-component complex vector, and ω is a real number.

Noticing that contributions involving ω are of relative order $1/N$, as a first order approximation, we neglect them.

We define

$$\exp\left[-\tilde{S}_N(\boldsymbol{\varphi}_N, \tilde{\mathcal{V}})\right] \sim \mathcal{N}_N \int \mathrm{d}^N\mathbf{v}\, \mathrm{d}^N\mathbf{v}^* \, \exp\left[-S_{N+1}\left(\boldsymbol{\phi}_{N+1}, \mathcal{V}\right)\right],$$

with a normalization $\mathcal{N}_N$ such that $\tilde{S}_N(0, \tilde{\mathcal{V}}) = 0$.

Gaussian approximation. The expansion of $\operatorname{tr} \mathcal{V}(\boldsymbol{\phi}_{N+1})$ in powers of $\mathbf{v}$ *truncated at quadratic order* is

$$\operatorname{tr} \mathcal{V}(\boldsymbol{\phi}_{N+1})|_{\mathrm{quad.}} = \operatorname{tr} \mathcal{V}(\boldsymbol{\varphi}_N) + \mathbf{v}^* \cdot \mathcal{V}'(\boldsymbol{\varphi}_N)\boldsymbol{\varphi}_N^{-1}\mathbf{v}. \qquad (25.5)$$

With this approximation, the integration over $\mathbf{v}$ is Gaussian and yields

$$\tilde{S}_N(\boldsymbol{\varphi}_N) = (N + 1) \operatorname{tr} \mathcal{V}(\boldsymbol{\varphi}_N) + \operatorname{tr} \ln\left[\mathcal{V}'(\boldsymbol{\varphi}_N)\boldsymbol{\varphi}_N^{-1}\right]. \qquad (25.6)$$

The corrections of order $|\mathbf{v}|^4$ generate *products of traces* and are large only if $\mathcal{V}'(\mu)/\mu$ is small.

We renormalize $\varphi_N = \zeta \phi_N$ with ζ chosen such that

$$\tilde{S}_N = \left[\tfrac{1}{2}(N+1) + g\right] \operatorname{tr} \varphi_N^2 + O(\varphi_N^4) = \tfrac{1}{2} N \operatorname{tr} \phi_N^2 + O(\phi_N^4).$$

It follows that

$$\zeta^2 = 1 - (2g+1)/N + O(1/N^2). \tag{25.7}$$

After the matrix renormalization, the variation $\delta\mathcal{V}$ of the function $\mathcal{V}$ leads to the flow equation

$$N\delta\mathcal{V}(\mu) \equiv \mathcal{B}(\mathcal{V}) = N \frac{\partial}{\partial N}\mathcal{V}(\mu) \tag{25.8}$$

(μ real), with

$$\mathcal{B}(\mathcal{V}) = \mathcal{V}(\mu) - \left(g + \tfrac{1}{2}\right)\mu\mathcal{V}'(\mu) + \ln\left(\mathcal{V}'(\mu)/\mu\right). \tag{25.9}$$

The fixed point equation is $\mathcal{B}(\mathcal{V}) = 0$. We set $\rho = \tfrac{1}{2}\mu^2$, and $V(\rho) = \mathcal{V}(\mu)$. The equation becomes

$$\mathcal{B}(V) = V(\rho) - (2g+1)\,\rho V'(\rho) + \ln\left(V'(\rho)\right). \tag{25.10}$$

25.3.2 Linear perturbative approximation

In an approximation of small deviation from the Gaussian matrix integral, that is, $V(\rho) - \rho$, and thus $V'(\rho) - 1$ small, the determinant contribution $\ln(V'(\rho))$ can be replaced by $V'(\rho) - 1$. The fixed point equation reduces to the linear differential equation

$$\mathcal{B}(V) = V(\rho) - (2g+1)\rho V'(\rho) + V'(\rho) - 1 = 0.$$

Integrating, one finds

$$V^*(\rho) = 1 - [1 - (2g+1)\rho]^{1/(1+2g)}.$$

Since V^* must be a regular function (and assuming $g + \tfrac{1}{2} > 0$), one infers

$$\frac{1}{1+2g_c} = m, \quad m \text{ positive integer} \;\Rightarrow\; g_c = \frac{1}{2m} - \frac{1}{2}.$$

Thus,

$$V^*(\rho) = 1 - (1 - \rho/m)^m.$$

One notices that, as expected, the approximation is mainly valid for ρ small and very poor for $\rho = m$, where $V'(\rho)$ vanishes.

One finds [240]:

for $m = 1$, the Gaussian fixed point;

for $m = 2$, $\rho - \tfrac{1}{4}\rho^2$ instead of the exact *minimal* potential $\rho - \tfrac{1}{12}\rho^2$;

for $m = 3$, $\rho - \tfrac{1}{3}\rho^2 + \tfrac{1}{27}\rho^3$ instead of $\rho - \tfrac{1}{9}\rho^2 + \tfrac{2}{405}\rho^3$.

More generally, it is possible to verify the intriguing relation, independent of m,

$$V_{\text{exact}}^*(\rho) = \int_0^1 \frac{\mathrm{d}s}{s\sqrt{1-s}} V^*(\rho s/2).$$

In particular,

$$\frac{\text{coeff. } \rho^k \text{ exact}}{\text{coeff. } \rho^k} = 2^k \frac{\Gamma(k+1/2)}{\Gamma(k)\sqrt{\pi}},$$

For successive values of k, the ratio is 1, 3, 15, ...

25.3.3 Stability analysis

Stability is governed by the eigenvalues of the operator $\delta\mathcal{B}/\delta V$. For the m-th multicritical point, the eigenvalue equation reads

$$\frac{\delta\mathcal{B}}{\delta V}h = \Omega h - 2\delta g(h)\rho(1-\rho/m)^{m-1} = \kappa h\,,$$

with

$$\Omega = 1 + \left(1 - \frac{\rho}{m}\right)\frac{\mathrm{d}}{\mathrm{d}\rho}\,,$$

where h and κ are the eigenvector and the eigenvalue, respectively, and $\delta g = \frac{1}{2}h''(0)$.
   The solution of the equation such that $h(0) = 0$ is

$$h(\rho) = 2m\delta g\left[(1-\rho/m)^{m-1}\frac{1-\rho(1-m\kappa)}{\kappa(1-m\kappa)} - \frac{(1-\rho/m)^{-m(\kappa-1)}}{\kappa(1-m\kappa)}\right]. \qquad (25.11)$$

The function h must be regular for $\rho = m$ and thus

$$\kappa_p = 1 - \frac{p}{m}\,, \quad p \text{ integer}\,.$$

Equation (25.11) shows that the values $\kappa = 0$ and $\kappa = 1/m$ are excluded and thus $p \neq m, m-1$.
   For $m = 1$, $p > 1$ and thus $\kappa < 0$, and the *Gaussian fixed point is thus stable*.
   For $m = 2$, the only positive value is $\kappa = 1$ and one recovers *one direction of instability*.
   In general, one finds $(m-1)$ eigenvalues that correspond to relevant perturbations (or direction of instabilities). By tuning $(m-2)$ parameters, one can select the eigenvector corresponding to the smallest eigenvalue $\kappa = 2/m$ $(p = m-2)$.
   The solution of the flow equation near the fixed point takes the form

$$g(N) - g_c \propto N^\kappa\,.$$

Since Z_0, the leading term in the topological expansion, is proportional to N^2, one infers

$$Z_{0,\text{sing.}} \propto (g - g_c)^{2/\kappa}.$$

Therefore,

$$2 - \alpha = 2/\kappa \ \Rightarrow \ \alpha = 2 - 2/(2/m) = 2 - m \,,$$

in good qualitative and even semi-quantitative agreement with the exact value

$$\alpha = \tfrac{3}{2} - m \,.$$

Full equation. The exact flow equations as given by the Gaussian integration with

$$\mathcal{B}(V) = V(\rho) - (2g + 1)\rho V'(\rho) + \ln\big(V'(\rho)\big) = 0 \,, \tag{25.12}$$

can be studied and generalized to path integrals with matrices. However, the interpretation of the results requires additional study [241].

Vector models. The exact flow equation as given by the Gaussian integration are essentially identical to those found for the corresponding vector (matrices replaced by vectors) models. This remains true for path integrals [241].

Bibliography

[1] J. Zinn-Justin, *Transitions de phase et groupe de renormalisation*, EDP Sciences and CNRS Editions (Les Ulis 2005); English version *Phase Transitions and Renormalization Group*, Oxford Univ. Press (Oxford 2007).

[2] N. Wiener, *Differential space*, J. of Math. and Phys. 2 (1923) 131–174, reprinted in *Selected Papers of N. Wiener*, p. 55, Y. M. Lee, Norman Levinson, W. T. Martin eds., MIT Press (Cambridge 1965). See also N. Wiener, *The Average value of a Functional*, Proc. London Math. Soc. 22 (1924) 454–467.

[3] J. Zinn-Justin, *Path integral*, Scholarpedia 4 (2009) 8674.

[4] R.P. Feynman, *Space-time approach to non-relativistic quantum mechanics*, Rev. Mod. Phys. 20 (1948) 367–387.

[5] R.P. Feynman, A.R. Hibbs, *Quantum mechanics and path integrals*, Mc Graw-Hill (New York, 1965).

[6] For a modern presentation and more references, see, for example, J. Zinn-Justin, *Intégrale de chemin en mécanique quantique: introduction*, EDP Sciences and CNRS Editions (Les Ulis 2003); English version, *Path Integrals in Quantum Mechanics*, Oxford Univ. Press (Oxford 2005); Russian translation, Fizmatlit (Moscow 2006).

[7] G. Wentzel, *Zur Quantenoptik*, Z. Physik 22 (1924) 193–199.

[8] P.A.M. Dirac, *The Lagrangian in quantum mechanics*, Physik. Z. Sowjetunion 3 (1933) 64, reprinted in *Selected Papers on Quantum Electrodynamics*, J. Schwinger, ed., Dover (New York 1958).

[9] G.C. Wick, *The evaluation of the collision matrix*, Phys. Rev. 80 (1950) 268–272.

[10] Feynman's contributions to the basic rules of QED are presented in, R. Feynman, *The theory of positrons*, Phys. Rev. 76 (1949) 749–759; *Space-time approach to quantum electrodynamics*, Phys. Rev. 76 (1949) 769–789.

[11] A.M. Polyakov, *Compact gauge fields and the infrared catastrophe*, Phys. Lett. 59B (1975) 82–84.

[12] G. 't Hooft, *Symmetry breaking through Bell–Jackiw anomalies*, Phys. Rev. Lett. 37 (1976) 8–11; *Computation of the quantum effects due to a four-dimensional pseudoparticle*; Phys. Rev. D 14 (1976) 3432–3450.

[13] L.D. Faddeev, *The Feynman integral for singular Lagrangians*, Theor. Math. Phys. 1 (1969) 1–13.

[14] V. Bargmann, *On a Hilbert space of analytic functions and an associated integral transform part I*, Commun. Pure and Appl. Math. 14 (1961) 187–214.

[15] J.R. Klauder, *The action option and a Feynman quantization of spinor fields in terms of ordinary c-numbers*, Ann. Phys. 11 (1960) 123–126, an article where integrals over complex phase space variables are introduced;
      In S.S. Schweber, *On Feynman quantization*, J. Math. Phys. 3 (1962) 831, a general formulation is found and the relation with the holomorphic formalism introduced in Ref. [14] is shown explicitly.

[16] Functional methods to deal with boson and fermion systems are described in F.A. Berezin, *The Method of Second Quantization* (Academic Press, New York 1966), Elsevier reprint 2012.

[17] P.T. Matthews, A. Salam, *Propagators of quantized field*, Nuovo Cim. 2 (1955) 120–134.

[18] J. Zinn-Justin, *Quantum Field Theory and Critical Phenomena*, (Oxford Univ. Press, 1989), fourth edition: Int. Ser. Monogr. Phys., 113 (2002).

[19] J.L. Lagrange, *Mécanique analytique*, Chez La Veuve Desaint, Libraire (Paris 1788), re-edited in *Oeuvres de Lagrange*, J-A Serret ed., Gauthier–Villars, Paris (1867–1892).

[20] W.R. Hamilton, *On a general method in dynamics*, Philosophical Transaction of the Royal Society Part II (1834) 247–308; *ibidem, Second essay on a general method in dynamics*, Part I (1835) 95–144.

[21] See for example, T.W.B. Kibble, *Classical Mechanics* (second Edition), European Physics Series, Mc Graw Hill (London 1973).

[22] J.C. Maxwell, *On physical lines of force*, Philosophical Magazine (1861).
      An article accompanied an 8 December 1864 presentation by Maxwell to the Royal Society of Maxwell's equations, *A dynamical theory of the electromagnetic Field*, Philosophical Transactions of the Royal Society of London 155, 459–512 (1865) followed later by *A Treatise on Electricity and Magnetism*, Oxford Clarendon Press 1873.

[23] The theory of General Relativity is presented in A. Einstein, *Die Feldgleichungen der Gravitation*, Sitzungsberichte der Preussischen Akademie der Wissenschaften zu Berlin: 844–847, (1915); *Die Grundlage der allgemeinen Relativitätstheorie*, Annalen der Physik 354 (1916) 769–822.

[24] K.G. Wilson, *Renormalization group and critical phenomena. II. Phase-space cell analysis of critical behavior*, Phys. Rev. B4 (1971) 3184–3205.

[25] P.G. de Gennes, *Exponents for the excluded volume problem as derived by the Wilson method*, Phys. Lett. A38 (1972) 339–340.

[26] L.N. Lipatov, *Divergence of the perturbation-theory series and pseudoparticles*, JETP Lett. **25** (1977) 104–107; *Divergence of the perturbation-theory series and the quasi-classical theory*, Sov. Phys. JETP **45** (1977) 216–223.

[27] A number of relevant articles are reprinted in *Large Order Behaviour of Pertur-bation Theory*, Current Physics Sources and Comments, Vol. 7, J.C. Le Guillou, J. Zinn-Justin (eds.), North-Holland, Elsevier Amsterdam (1990).

[28] E. Brézin, J. C. Le Guillou, J. Zinn-Justin, *Perturbation theory at large order. II - Role of the vacuum instability*, Phys. Rev. D15 (1977) 1558–1564.

[29] See for example, E. Brézin, J.C. Le Guillou, J. Zinn-Justin, *Wilson's theory of critical phenomena and Callan–Symanzik equations in 4-ε dimensions*, Phys. Rev. D8 (1973) 434–440; *Approach to scaling in renormalized perturbation theory*, Phys. Rev. D8 (1973) 2418–2430.

[30] J.C. Le Guillou, J. Zinn-Justin, *Critical exponents for the n-vector model in three dimensions from field theory*, Phys. Rev. Lett. 39 (1977) 95–98; *Critical exponents from field theory*, Phys. Rev. B21 (1980) 3976–3998.

[31] R. Guida, J. Zinn-Justin, *Critical exponents of the N-vector model*, J. Phys. A 31 (1998) 8103–8121.

[32] J.C. Le Guillou, J Zinn-Justin, *Accurate critical exponents from the ϵ-expansion*, J. Phys. Lett. (Paris) 46 (1985) 137–141.

[33] M. Campostrini, A. Pelissetto, P. Rossi, E. Vicari, *25th-order high-temperature expansion results for three-dimensional Ising-like systems on the simple-cubic lattice*, Phys. Rev. E 65 (2002) 066127.

[34] M. Hasenbusch, *Finite size scaling study of lattice models in the three-dimensional Ising universality class*, Phys. Rev. B 82 (2010) 174433.

[35] S. El-Showk *et al*, *Solving the 3D Ising model with the conformal bootstrap*, Phys. Rev. D86 (2012) 025022 and *II. c-minimization and precise critical exponents*, J. Stat. Phys. 157 (2014) 869–914;
H. Shimada, S. Hikami, *Fractal dimensions of self-avoiding walks and Ising high-temperature graphs in 3D conformal bootstrap*, J. Stat. Phys. 165 (2016) 1006–1035.

[36] J.A. Lipa *et al.*, *Heat capacity and thermal relaxation of bulk Helium very near the lambda point*, Phys. Rev. Lett. 76 (1996) 944–947.

[37] M. Campostrini, M. Hasenbusch, A. Pelissetto, E. Vicari, *Theoretical estimates of the critical exponents of the superfluid transition in ^4He by lattice methods*, Phys. Rev. B 74 (2006) 144506.

[38] E. Brézin, G. Parisi, J. Zinn-Justin, *Perturbation theory at large orders for a potential with degenerate minima*, Phys. Rev. D16 (1977) 408–412.

[39] J. Zinn-Justin, *Multi-instanton contributions in quantum mechanics*, Nucl. Phys. B192 (1981) 125–140; *ibidem, II*, Nucl. Phys. B218 (1983) 333–348; *Expansion around instantons in quantum mechanics*, J. Math. Phys. 22 (1981) 511–520.

[40] L.D. Faddeev, V.N. Popov, *Feynman diagrams for the Yang–Mills field*, Phys. Lett. 25B (1967) 29–30; *Perturbation theory for gauge-invariant fields*, Kiev report No. ITP 67–36.

[41] C.N. Yang, R.L. Mills, *Conservation of isotopic spin and isotopic gauge invariance*, Phys. Rev. 96 (1954) 191–195.

[42] R.P. Feynman, *Quantum theory of gravitation*, Acta Phys. Polon. 24 (1963) 697–722.

[43] C. Becchi, A. Rouet, R. Stora, *The Abelian Higgs–Kibble model. Unitarity of the S operator*, Phys. Lett 52B (1974) 344; *Renormalization of the Abelian Higgs–Kibble model*, Comm. Math. Phys. 42 (1975) 127; *Renormalization of gauge theories*, Ann. Phys. (NY) 98 (1976) 287–321; I. Tyutin, Preprint of Lebedev Physical Institute, 39 (1975).

[44] K.G. Wilson, *Confinement of quarks*, Phys. Rev. D 10 (1974) 2445–2459.

[45] The quantization was discussed in I.S. Gerstein, R. Jackiw, B.W. Lee, S. Weinberg, *Chiral loops*, Phys. Rev. D3 (1971) 2486–2492.

[46] J. Honerkamp, K. Meetz, *Chiral-invariant perturbation theory*, Phys. Rev. D3 (1971) 1996–1998; J. Honerkamp, *Chiral loops*, Nucl. Phys. B36 (1972) 130–140; A.A. Slavnov, L.D. Faddeev, *Invariant perturbation theory for nonlinear chiral Lagrangians*, Theor. Math. Phys. 8 (1971) 843–850.

[47] For a review of large N techniques see M. Moshe, J. Zinn-Justin, *Quantum field theory in the large N limit: A review*, Phys. Rep. 385 (2003) 69–228.

[48] W. Heisenberg, *Über quantentheoretische Umdeutung kinematischer und mechanischer Beziehungen*, Z. Physik 33 (1925) 879–893; *Über den anschaulichen Inhalt der quantentheoretischen Kinematik und Mechanik*, Z. Physik 43 (1927) 172–198.

[49] E. Schrödinger, *An undulatory theory of the mechanics of atoms and molecules*, Phys. Rev. 28 (1926) 1049–1070.

[50] P.A.M. Dirac, *The quantum theory of the electron*, Proceedings of the Royal Society A 117 (1928) 610–624.

[51] W. Heisenberg, W. Pauli, *Zur Quantendynamik der Wellenfelder*, Z. Physik 56 (1929) 1–61; *ibidem, II*, Z. Physik 59 (1930) 168–190.

[52] I. Waller, *Bemerkungen über die Rolle der Eigenenergie des Elektrons in der Quantentheorie der Strahlung*, Z. Physik 62 (1930) 673–676;
R. Oppenheimer, *Note on the theory of the interaction of field and matter*, Phys. Rev. 35 (1930) 461–477.

[53] V. Weisskopf, *Über die Selbstenergie des Elektrons*, Z. Physik 89 (1934) 27; *Berichtigung zu der Arbeit: Über die Selbstenergie des Elektrons*, Z. Physik 90 (1934) 817.

[54] L.D. Landau, *Zur Theorie der Phasenumwandlungen I*, Physikalische Zeitschrift der Sowjetunion 11 (1937) 26; *On the theory of phase transitions*, Zh. Eksp. Teor. Fiz. 7 (1937) 9–32, reprinted in Ref. [64], p. 193–216.

[55] L. Onsager, *Crystal statistics. I. A two-dimensional model with an order–disorder transition*, Phys. Rev. 65 (1944) 117–149.

[56] W.E. Lamb, Jr., R.C. Retherford, *Fine structure of the hydrogen atom by a microwave method*, Phys. Rev. 72 (1947) 241–243.

[57] H.A. Bethe, *The electromagnetic shift of energy levels*, Phys. Rev. 72 (1947) 339–341.

[58] J. Schwinger, *On quantum-electrodynamics and the magnetic moment of the electron*, Phys. Rev. 73 (1948) 416–417.

[59] S. Tomonaga, *On a relativistically invariant formulation of the quantum theory of wave fields*, Prog. Theoret. Phys. I (1946) 27–42;
S. Tomonaga, J. R. Oppenheimer, *On infinite field reactions in quantum field theory*, Phys. Rev. 74 (1948) 224–225.

[60] J. Schwinger, *Quantum electrodynamics. I. A covariant formulation*, Phys. Rev. 74 (1948) 1439–1461.

[61] F.J. Dyson, *The radiation theories of Tomonaga, Schwinger, and Feynman*, Phys. Rev. 75 (1949) 486–502.

[62] J. Schwinger ed., *Selected Papers on Quantum Electrodynamics*, Dover Publications, (New York 1958).

[63] V.L. Ginzburg, L.D. Landau, *On the theory of superconductivity*, Zh. Eksp. Teor. Fiz. 20 (1950) 1064; English translation in [64] p. 546–568.

[64] L. D. Landau, *Collected Papers*, D. Ter-Haar ed. (Oxford: Pergamon Press, 1965).

[65] E.C.G. Stueckelberg, A. Peterman, *Normalization of constants in the quanta theory*, Helv. Phys. Acta 26 (1953) 499–520.

[66] M. Gell-Mann, F.E. Low, *Quantum electrodynamics at small distances*, Phys. Rev. 95 (1954) 1300–1312.

[67] N.N. Bogoliubov, D.V. Shirkov, *Introduction to the Theory of Quantized Fields* (1959), third edition John Wiley, (New York 1980).

[68] S. Weinberg, *A model of leptons*, Phys. Rev. Lett. 19 (1967) 1264–1266.

[69] S. L. Glashow, J. Iliopoulos, L. Maiani, *Weak interactions with lepton-hadron symmetry*, Phys. Rev. D 2 (1970) 1285–1292.

[70] F. Englert, R. Brout, *Broken symmetry and the mass of vector bosons*, Phys. Rev. Lett. 13 (1964) 321–323.

[71] P.W. Higgs, *Broken symmetries and the masses of gauge bosons*, Phys. Rev. Lett. 13 (1964) 508–509.

[72]   G.S. Guralnik, C.R. Hagen, T.W.B. Kibble, *Conservation laws and massless particles*, Phys. Rev. Lett. 13 (1964) 585–587.

[73]   B.S. DeWitt, *Quantum theory of gravity II, III*, Phys. Rev. 162 (1967) 1195–1239; *ibidem*, 1239–1256.

[74]   G. 't Hooft, *Renormalization of massless Yang-Mills fields*, Nucl. Phys. B33 (1971) 173–199; *Renormalizable Lagrangians for massive Yang-Mills fields*, B35 (1971) 167–188.

[75]   G. 't Hooft, M. Veltman, *Combinatorics of gauge fields*, Nucl. Phys. B50 (1972) 318–353.

[76]   B.W. Lee, J. Zinn-Justin, *Spontaneously broken gauge symmetries, I, II, III*, Phys. Rev. D5 (1972) 3121–3137, 3137–3155, 3155–3160; *ibidem, IV General gauge formulation*, Phys. Rev. D7 (1973) 1049–1056.

[77]   L.P. Kadanoff, *Scaling laws for Ising models near T_c*, Physics 2 (1966) 263–273.

[78]   K.G. Wilson, *Renormalization group and critical phenomena. I. Renormalization group and the Kadanoff scaling picture*, Phys. Rev. B 4 (1971) 3174–3183.

[79]   K.G. Wilson, M.E. Fisher, *Critical Exponents in 3.99 dimensions*, Phys. Rev. Lett. 28 (1972) 240–243.

[80]   K.G. Wilson, *Feynman-graph expansion for critical exponents*, Phys. Rev. Lett., 28 (1972) 548–551.

[81]   E. Brézin, D. J. Wallace, K. G. Wilson, *Feynman-graph expansion for the equation of state near the critical point*, Phys. Rev. B7 (1973) 232–239.

[82]   F.J. Wegner, *Corrections to scaling laws*, Phys. Rev. B5 (1972) 4529–4536.

[83]   See Ref. [29] and:
       C. Di Castro, *The multiplicative renormalization group and the critical behaviour in $d = 4 - \varepsilon$ dimensions*, Lett. Nuovo Cim. 5 (1972) 69–74;
       G. Mack, *Conformal invariance and short distance behavior in quantum field theory* in Strong Interaction Physics, Lecture Notes in Physics 17, p. 300–334, W. Rühl, A. Vancura eds. (Springer Berlin 1973);
       P.K. Mitter, *Callan–Symanzik equations and ε expansions*, Phys. Rev. D7 (1973) 2927–2942;
       B. Schroer, *Theory of critical phenomena based on the normal-product formalism*, Phys. Rev. B8 (1973) 4200–4208.

[84]   The RG β-function for gauge theories at one-loop order is reported in
       H.D. Politzer, *Reliable perturbative results for strong interactions?* Phys. Rev. Lett. 30 (1973) 1346–1349;
       D.J. Gross, F. Wilczek, *Ultraviolet behavior of non-Abelian gauge theories*, Phys. Rev. Lett. 30 (1973) 1343–1346.

[85] A.M. Polyakov, *Interaction of Goldstone particles in two dimensions. Applications to ferromagnets and massive Yang-Mills fields*, Phys. Lett. 59B (1975) 79–81.

[86] E. Brézin, J. Zinn-Justin, *Renormalization of the non-linear σ model in $2 + \varepsilon$ dimensions—Application to the Heisenberg ferromagnets*, Phys. Rev. Lett. 36 (1976) 691–694; *Spontaneous breakdown of continuous symmetries near two dimensions*, Phys. Rev. B14 (1976) 3110–3120.

[87] C.G. Callan, *Broken scale invariance in scalar field theory*, Phys. Rev. D2 (1970) 1541–1547;
K. Symanzik, *Small distance behaviour in field theory and power counting*, Comm. Math. Phys. 18 (1970) 227–246.

[88] G. Parisi, lectures given at the Cargèse summer school 1973, later incorporated into the article *Field-theoretic approach to second-order phase transitions in two- and three-dimensional systems*, J. Stat. Phys. 23 (1980) 49–82.

[89] G.A. Baker, B.G. Nickel, M.S. Green, D.I. Meiron, *Ising-model critical indices in three dimensions from the Callan–Symanzik equation*, Phys. Rev. Lett. 36 (1976) 1351–1354.

[90] C.D. Anderson, *The positive electron*, Phys. Rev. 43 (1933) 491–494.

[91] V.F. Weisskopf, *On the self-energy and the electromagnetic field of the electron*, Phys. Rev. 56 (1939) 72.

[92] V.F. Weisskopf, *Über die Elektrodynamik des Vakuums auf Grund des Quanten-Theorie des Elektrons*, Dan. Mat. Fys. Medd. 14, 6 (1936) 1–39.

[93] K. Hepp, *Proof of the Bogoliubov–Parasiuk theorem on renormalization*, Comm. Math. Phys. 2 (1966) 301–326.

[94] W. Zimmermann, *Convergence of Bogoliubov's method of renormalization in momentum space*, Comm. Math. Phys. 15 (1969) 208–234.

[95] H. Epstein, V. Glaser, *The role of locality in perturbation theory*, Annales IHP A 19 (1973) 211–295.

[96] L.D. Landau, I.I. Pomeranchuk, *On point interactions in quantum electrodynamics*, Dokl. Akad. Nauk SSSR 102 (1955) 489 reprinted in [64] p. 654–658.

[97] S. Weinberg, *Ultraviolet divergences in quantum theories of gravitation*, General Relativity: An Einstein centenary survey, S.W. Hawking, W. Israel, eds., p.790–831, Cambridge University Press, (1979).

[98] L.P. Kadanoff *et al*, *Static phenomena near critical points: Theory and experiment*, Rev. Mod. Phys. 39 (1967) 395–431.

[99] R. Guida, J. Zinn-Justin, *3D Ising model: the scaling equation of state*, Nucl. Phys. B 489 (1997) 626–652.

[100] K.G. Wilson, J.B. Kogut, *The renormalization group and the ε expansion*, Phys. Rep. 12 (1974) 75–199.

[101] F.J. Wegner, F.J. Houghton, *Renormalization group equation for critical phenomena*, Phys. Rev. A 8 (1973) 401–412.

[102] Dimensional regularization has been introduced by
J. Ashmore, *A method of gauge-invariant regularization*, Lett. Nuovo Cim. 4 (1972) 289–290;
G. 't Hooft, M. Veltman, *Regularization and renormalization of gauge fields*, Nucl. Phys. B44 (1972) 189–213;
C.G. Bollini, J.J. Giambiagi, *Lowest order "divergent" graphs in v-dimensional space*, Phys. Lett. B40 (1972) 566–568; *Dimensional renormalization: The number of dimensions as a regularizing parameter*, Nuovo Cim. B12 (1972) 20–26.

[103] RG equations satisfied by unrenormalized (bare) correlation functions in renormalizable field theories have first been derived in J. Zinn-Justin, *Wilson's theory of critical phenomena in renormalized perturbation theory*, lectures given at the Cargèse summer school 1973, Saclay preprint T73/049 unpublished, later incorporated into Ref. [104].

[104] E. Brézin, J.C. Le Guillou, J. Zinn-Justin, *Field theoretical approach to critical phenomena* in *Phase Transitions and Critical Phenomena*, Vol. 6, C. Domb, M.S. Green eds., Academic Press (London 1976), p. 127–247.

[105] This chapter is an edited version of J. Zinn-Justin, *Critical phenomena: field theoretical approach*, Scholarpedia 5 (2010) 8346.

[106] E. Brézin, J.C. Le Guillou, J. Zinn-Justin, B.G. Nickel, *Higher order contributions to critical exponents*, Phys. Lett. A44 (1973) 227–228;
A.A. Vladimirov, D.I. Kazakov, O.V. Tarasov, *Calculation of critical exponents by quantum field theory methods*, Zh. Eksp. Teor. Fiz. 77 (1979) 1035–1045;
K.G. Chetyrkin, F.V. Tkachov, *Integration by parts: The algorithm to calculate β-functions in 4 loops*, Nucl. Phys. B 192 (1981) 159–204;
K.G. Chetyrkin, S.G. Gorishny, S.A. Larin, F.V. Tkachov, *Five-loop renormalization group calculations in the $g\varphi^4$ theory*, Phys. Lett. 132B (1983) 351–354;
S.G. Gorishny, S.A. Larin, F.V. Tkachov, *ε-Expansion for critical exponents: The $O(\epsilon^5)$ approximation*, Phys. Lett. 101A (1984) 120–123;
H. Kleinert, J. Neu, V. Schulte-Frohlinde, K.G. Chetyrkin, S.A. Larin, *Five-loop renormalization group functions of O(n)-symmetric φ^4-theory and ε-expansions of critical exponents up to ϵ^5*, Phys. Lett. B 272 (1991) 39–44; *Erratum-ibid.*, Phys. Lett. B 319 (1993) 545.

[107] E. Vicari, J. Zinn-Justin, *Fixed point stability and decay of correlations*, New J. Phys. 8 (2006) 321.

[108] D.J. Wallace, *Critical behaviour of anisotropic cubic systems*, J. Phys. C6 (1973) 1380–1404;
A. Aharony, *Universal critical behaviour* in *Phase Transitions and Critical Phenomena*, Vol. 6, C. Domb, M.S. Green eds., Academic Press (London 1976), p. 394–402.

[109] D.J. Wallace, R.K.P. Zia, *Gradient flow and the renormalization group*, Phys. Lett. 48A (1974) 325–326; *Gradient properties of the renormalisation group equations in multicomponent systems*, Ann. Phys. 92 (1975) 142–163.

[110] L. Michel, *Renormalization group fixed points of general n-vector models*, Phys. Rev. B29 (1984) 2777–2783.

[111] H. Osborn, A. Stergiou, *Seeking fixed points in multiple coupling scalar theories in the ε-expansion*, JHEP (2018) 2018:51.

[112] See, for example, Chapter 27 in Ref. [18].

[113] S. Weinberg, *Effective gauge theories*, Phys. Lett. B91 (1980) 51–55.

[114] Y. Nambu, G. Jona-Lasinio, *Dynamical model of elementary particles based on an analogy with superconductivity. I, II*, Phys. Rev. 122 (1961) 345–358; 124 (1961) 246–254.

[115] K.G. Wilson, *Quantum field theory models in less than 4 dimensions*, Phys. Rev. D7 (1973) 2911–2926.

[116] D.J. Gross, A. Neveu, *Dynamical symmetry breaking in asymptotically free field theories*, Phys. Rev. D10 (1974) 3235–3253.

[117] D.J. Gross, *Applications of the renormalization group in high energy physics* in *Methods in Field Theory*, Proc. of Les Houches physics school 1975, R. Balian, J. Zinn-Justin eds. (North-Holland, Amsterdam 1976).

[118] J. Zinn-Justin, *Four fermion interaction near four dimensions*, Nucl. Phys. B367 (1991) 105–122.

[119] Early articles are
Y. Nambu, *Axial vector current conservation in weak interactions*, Phys. Rev. Lett. 4 (1960) 380–382;
J. Goldstone, *Field theories with "Superconductor" solutions*, Nuovo Cim. 19 (1961) 154–164;
J. Goldstone, A. Salam, S. Weinberg, *Broken symmetries*, Phys. Rev. 127 (1962) 965–970.

[120] See for example, Chapter 31 of Ref. [18].

[121] E. Brézin, J. Zinn-Justin, J.C. Le Guillou, *Renormalization of the non linear σ model in 2+ε dimensions*, Phys. Rev. D14 (1976) 2615–2621.

[122] N.D. Mermin, H. Wagner, *Absence of ferromagnetism or antiferromagnetism in one- or two-dimensional isotropic Heisenberg models*, Phys. Rev. Lett. 17 (1966) 1133–1136;
S. Coleman, *There are no Goldstone bosons in two dimensions*, Comm. Math. Phys. 31 (1973) 259–264.

[123] J. M. Kosterlitz, D. J. Thouless, *Ordering, metastability and phase transitions in two-dimensional systems*, J. Phys. C6 (1973) 1181–1203.

[124] J. Polchinski, *Renormalization and effective Lagrangians*, Nucl. Phys. B231 (1984) 269–295.

[125] A detailed derivation along the same lines is found in chapter 16 of Ref. [1].

[126] D. Bessis, J. Zinn-Justin, *One-loop renormalization of the nonlinear σ model*, Phys. Rev. D 5 (1972) 1313–1323.

[127] F. David, *Cancellations of infrared divergences in the two-dimensional nonlinear sigma models*, Comm. Math. Phys. 81 (1981) 149.

[128] This chapter is adapted from R. Guida, J Zinn-Justin, *Gauge invariance*, (2008), Scholarpedia, 3 :8287.

[129] G. Kirchhoff, *Über die Bewegung der Elektricität in Leitern*, Annalen der Physik und Chemie 102 (1857) 529–544; reprinted in *Gesammelte Abhandlungen von G. Kirchhoff*, p. 154–168 (J. A. Barth, Leipzig 1882).

[130] W.Weber, *I. Elektrodynamische Maassbestimmungen*, Annalen der Physik und Chemie, 73 (1848) 193–240, shortened version of the 1846 paper published in the Abhandlungen der Königlichen Sächsischen Gesellschaft der Wissenschaften, Leipzig.

[131] J.C. Maxwell, *On Faraday's Lines of Force*, Transactions of the Cambridge Philosophical Society, X(I) (1855).

[132] H. Helmholtz, *Ueber die Bewegungsgleichungen der Elektrizität für ruhende leitende Körper*, Journal für die reine und angewandte Mathematik 72 (1870) 57–129.

[133] H. Weyl, *Gravitation und Elektrizität*, Sitzungsber. Preuss. Akad. Wiss. (1918) 465–480; *Reine Infinitesimalgeometrie*, Math. Z. 2 (1918) 384–411; *Eine neue Erweiterung der Relativitätstheorie*, Ann. Phys. 59 (1919) 101–133.

[134] E. Schrödinger, *Über eine bemerkenswerte Eigenschaft der Quantenbahnen eines einzelnen Elektrons*, Z. Phys. 12 (1922) 13–23.

[135] V. Fock, *Über die invariante Form der Wellen und der Bewegungsgleichungen für einen geladenen Massenpunkt*, Z. Phys. 39 (1927) 226–232.

[136] F. London, *Quantenmechanische Deutung der Theorie von Weyl*, Z. Phys. 42 (1927) 375–389.

[137] H. Weyl, *Gruppentheorie und Quantenmechanik*, Hirzel, (Lepzig 1928).

[138] E. Noether, *Invariante Variationsprobleme*, Nachr. d. König. Gesellsch. d. Wiss.
zu Göttingen, Math-Phys. Klasse, (1918) 235–257. English translation *Invariant variation problems* by M.A. Tavel in Transport Theory and Statistical
Physics. 1(3) (1971) 183–207; arxiv.org/abs/physics/0503066.

[139] O. Darrigol, *Electrodynamics from Ampère to Einstein*, Oxford Univ. Press
(Oxford 1999).

[140] J.C. Taylor ed., *Gauge Theories in the Twentieth Century*, Imperial College
Press (London 2001).

[141] J.D. Jackson, L.B. Okun, *Historical roots of gauge invariance*, Rev. Mod. Phys.
73 (2001) 663–680.

[142] V.N. Gribov, *Quantization of non-Abelian gauge theories*, Nucl. Phys. B139
(1978) 1–19.

[143] B.W. Lee, *Renormalization of the σ-model*, Nucl. Phys. B9 (1969) 649–672.

[144] K. Symanzik, *Renormalizable models with simple symmetry breaking*, Comm.
Math. Phys. 16 (1970) 48–80.

[145] J. Schwinger, *Gauge invariance and mass*, Phys. Rev. 125 (1962) 397–398;
ibidem, II, Phys. Rev. 128 (1962) 2425–2429.

[146] P.W. Anderson, *Plasmons, gauge pnvariance and mass*, Phys. Rev. 130 (1963)
439–442.

[147] A. Klein, B.W. Lee, *Does spontaneous breakdown of symmetry imply zero-mass
particles?* Phys. Rev. Lett. 12 (1964) 266–268.

[148] W. Gilbert, *Broken symmetries and massless particles*, Phys. Rev. Lett. 12
(1964) 713–714.

[149] P.W. Higgs, *Broken symmetries, massless particles and gauge fields*, Phys. Lett.
12 (1964) 132–133.

[150] P.W. Higgs, *Spontaneous symmetry breakdown without massless bosons*, Phys.
Rev. 145 (1966) 1156–1163.

[151] T.W.B. Kibble, *Symmetry breaking in non-Abelian gauge theories*, Phys. Rev.
155 (1967) 1554–1561.

[152] R. P. Feynman, *Quantum theory of gravitation*, Acta Physica Polonica 24 (1963)
697–722.

[153] B.W. Lee, *Renormalizable massive vector-meson theory-Perturbation theory of
the Higgs phenomenon*, Phys. Rev. D5 (1972) 823–835.

[154] A.A. Slavnov, *Ward identities in gauge theories*, Theor. Math. Phys. 10 (1972)
99–107;
J.C. Taylor, *Ward identities*, Nucl. Phys. B33 (1971) 436–444.

[155] See Ref. [18], chapter 16.

[156] J. Zinn-Justin (1975), *Renormalization of gauge theories*, Bonn lectures 1974, published in *Trends in Elementary Particle Physics*, Lecture Notes in Physics 37 p. 1–39, H. Rollnik, K. Dietz eds., Springer Verlag (Berlin 1975).

[157] J. Zinn-Justin, *Renormalization problems in gauge theories*, in Proc. of the 12th School of Theoretical Physics, Karpacz (1975) *Functional and probabilistic methods in quantum field theory*, Acta Universitatis Wratislaviensis 368 (1976) 435–453, Saclay preprint t76/048.

[158] M. Lüscher, P. Weisz, *Scaling laws and triviality bounds in the lattice ϕ^4 theory (III). n-component model*, Nucl. Phys. B318 (1989) 705–741.

[159] J. Elias-Miro *et al*, *Higgs mass implications on the stability of the electroweak vacuum*, Phys. Lett. B 709 (2012) 222–228;
G. Degrassi*et al*, *Higgs mass and vacuum stability in the Standard Model at NNLO*, JHEP 08 (2012) 2012:98.

[160] H. Kluberg-Stern, J-B. Zuber, *Ward identities and some clues to the renormalization of gauge invariant operators*, Phys. Rev. D12 (1975) 467–481.

[161] A.A. Slavnov, *Invariant regularization of gauge theories*, Theor. Math. Phys. 13 (1972) 1064–1066.

[162] S. Coleman, D.J. Gross, *Price of asymptotic freedom*, Phys. Rev. Lett. 31 (1973) 851–854.

[163] R. D. Peccei, Helen R. Quinn, *CP conservation in the presence of pseudoparticles*, Phys. Rev. Lett. 38 (1977) 1440–1443; *Constraints imposed by CP conservation in the presence of pseudoparticles*, Phys. Rev. D 16 (1977) 1791–1797;
S. Weinberg, *A new light boson?* Phys. Rev. Lett. 40 (1978) 223–226;
F. Wilczek, *Problem of strong P and T invariance in the presence of instantons*, Phys. Rev. Lett. 40 (1978) 279–282.

[164] J. Kogut, L. Susskind, *Hamiltonian formulation of Wilson's lattice gauge theories*, Phys. Rev. D 11 (1975) 395–408.

[165] The overlap Dirac operator was found to provide solutions to the Ginsparg–Wilson relation, P.H. Ginsparg, K.G. Wilson, *A remnant of chiral symmetry on the lattice* Phys. Rev. D25 (1982) 2649–2657.

[166] H. Neuberger, *Exactly massless quarks on the lattice*, Phys. Lett. B417 (1998) 141–144; *More about exactly massless quarks on the lattice*, Phys. Lett. B427 (1998) 353–355.

[167] D.B. Kaplan, *A method for simulating chiral fermions on the lattice*, Phys. Lett. B288 (1992) 342–347.

[168] M. Creutz, *Monte Carlo study of quantized SU(2) gauge theory*, Phys. Rev. D 21 (1980) 2308–2315.

[169] S. Dürr *et al*, *Ab initio determination of light hadron masses*, Science 322 (2008) 1224–1227.

[170] This chapter is adapted from J. Zinn-Justin, *Zinn-Justin equation*, Scholarpedia, 4 (2009) 7120.

[171] G. Barnich, M. Henneaux, *Renormalization of gauge invariant operators and anomalies in Yang-Mills theory*, Phys. Rev. Lett. 72 (1994) 1588–1591;
G. Barnich, F. Brandt, M. Henneaux, *Local BRST cohomology in the antifield formalism: I. General theorems*, Comm. Math. Phys. 174 (1995) 57–91.

[172] D. Bessis, M. Pusterla, *Unitary Padé approximants in strong coupling field theory and application to the calculation of the ρ- and f0-meson Regge trajectories*, Nuovo Cim. A54 (1968) 243–294;
J.L. Basdevant, D Bessis, J Zinn-Justin, *Padé approximants in strong interactions. Two-body pion and kaon systems*, Nuovo Cim. A60 (1969) 185–238.

[173] S.L. Adler, *Axial-vector vertex in spinor electrodynamics*, Phys. Rev. 177 (1969) 2426–2438.

[174] S.L. Adler, W.A. Bardeen, *Absence of higher-order corrections in the anomalous axial-vector divergence equation*, Phys. Rev. 182 (1969) 1517–1536.

[175] W.A. Bardeen, *Anomalous Ward identities in spinor field theories*, Phys. Rev. 184 (1969) 1848–1859.

[176] A short selection of articles and lectures further discussing chiral anomalies is
D.J. Gross, R. Jackiw, *Effect of anomalies on quasi-renormalizable theories*, Phys. Rev. D6 (1972) 477–493;
H. Georgi, S.L. Glashow, *Gauge theories without anomalies*, Phys. Rev. D6 (1972) 429–431;
C. Bouchiat, J. Iliopoulos, Ph. Meyer, *An anomaly-free version of Weinberg's model*, Phys. Lett. 38B (1972) 519–523;
M. E. Peskin, in *Recent Advances in Field Theory and Statistical Mechanics*, Les Houches summer school 1982, R. Stora, J.-B. Zuber eds., Les Houches summer school proc. 39 (North-Holland, Amsterdam 1984);
L. Alvarez-Gaumé, *Fundamental problems of gauge theory*, Erice 1985 G. Velo, A.S. Wightman eds. (Plenum Press, New York 1986).

[177] W. Pauli, F. Villars, *On the invariant regularization in relativistic quantum theory*, Rev. Mod. Phys. 21 (1949) 434–444.

[178] The use of dimensional regularization in problems with chiral anomalies has been proposed in D.A. Akyeampong, R. Delbourgo, *Dimensional regularization, abnormal amplitudes and anomalies*, Nuovo Cim. A17 (1973) 578–586.

[179] M. Lüscher, *Exact chiral symmetry on the lattice and the Ginsparg–Wilson relation*, Phys. Lett. B428 (1998) 342–345.

[180] K. Fujikawa, *Comment on chiral and conformal anomalies*, Phys. Rev. Lett. 44 (1980) 1733–1736; *Path integral for gauge theories with fermions*, Phys. Rev. D21 (1980) 2848–2858, Erratum, Phys. Rev. D22 (1980) 1499.

[181] This chapter is directly inspired from J. Zinn-Justin, *Chiral Anomalies and Topology*, School on Topology and Geometry in Physics, Rot an der Rot, Germany, September 24–28 2001, in Lecture Notes in Physics 659 (2005) 167–236.

[182] J.S. Bell, R. Jackiw, *A PCAC puzzle: $\pi_0 \mapsto \gamma\gamma$ in the σ-model*, Nuovo Cim. A60 (1969) 47–61.

[183] J. Steinberger, *On the use of subtraction fields and the lifetimes of some types of meson decay*, Phys. Rev. 76 (1949) 1180–1186.

[184] The index of the Dirac operator in a gauge background is related to Atiyah–Singer's theorem, M. Atiyah, R. Bott, V. Patodi, *On the heat equation and the index theorem*, Invent. Math. 19 (1973) 279–330.

[185] E. Witten, *Dynamical breaking of supersymmetry*, Nuclear Phyics B 188 (1981) 513–554.

[186] J. Wess, B. Zumino, *Consequences of anomalous ward identities*, Phys. Lett. B37 (1971) 95–97.

[187] M. Lüscher, *The secret long range force in quantum field theories with instantons*, Phys. Lett. 78B (1978) 465–467;
A. D'Adda, P. Di Vecchia, M. Lüscher, *A $1/n$ expandable series of non-linear σ models with instantons*, Nucl. Phys. B146 (1978) 63–76; *Confinement and chiral symmetry breaking in CP^{n-1} models with quarks*, B152 (1979) 125–144;
G. Münster, *A study of CP^{n-1} models on the sphere within the $1/n$ expansion*, Nucl. Phys. B218 (1983) 1–31;
P. Di Vecchia *et al*, *The transition from the lattice to the continuum: CP^{N-1} models at large N*, Nucl. Phys. B235 (1984) 478–520;
M. Campostrini, P. Rossi, *$1/N$ expansion of the topological susceptibility in the CP^{N-1} models*, Phys. Lett. B272 (1991) 305–312; *CP^{N-1} models in the $1/N$ expansion*, Phys. Rev. D45 (1992) 618–638; Erratum-*ibid.*, D46 (1992) 2741–2742;
M. Campostrini, P. Rossi, E Vicari, *Monte Carlo simulation of CP^{N-1} models*, Phys. Rev. D46 (1992) 2647–2662.

[188] G. V. Dunne, M. Ünsal, *Continuity and resurgence: Towards a continuum definition of the $CP(N-1)$ model*, Phys. Rev. D 87 (2013) 025015.

[189] E.B. Bogomolnyi, *The stability of classical solutions* Sov. J. Nucl. Phys. 24 (1976) 449–454;
M.K. Prasad, C.M. Sommerfeld, *Exact classical solution for the 't Hooft monopole and the Julia–Zee dyon*, Phys. Rev. Lett. 35 (1975) 760–762.

[190] Early articles and reviews about finite size effects are

M.E. Fisher in *Critical Phenomena, Proceedings of the International School of Physics Enrico Fermi*, Varenna 1971, M.S. Green ed. (Academic Press, New York 1972);

Y. Imry, D. Bergman, *Critical points and scaling laws for finite systems*, Phys. Rev. A3 (1971) 1416–1418;

M.E. Fisher, M.N. Barber, *Scaling theory for finite-size effects in the critical region*, Phys. Rev. Lett. 28 (1972) 1516–1519.

[191] See also M.N. Barber in *Phase Transitions and Critical Phenomena* vol. 8, C. Domb, J. Lebowitz eds. (Academic Press, New York 1983), and the papers collected in *Finite Size Scaling*, J.L. Cardy ed. (North-Holland, Amsterdam 1988).

[192] A presentation of finite size effects and Monte-Carlo simulations are found in K. Binder, *Finite size scaling analysis of Ising model block distribution functions*, Z. Phys. B43 (1981) 119–140.

[193] The field theoretical analysis has been performed in

E. Brézin, *An investigation of finite size scaling*, J. Phys. (Paris) 43 (1982) 15–22;

E. Brézin, J. Zinn-Justin, *Finite size effects in phase transitions*, Nucl. Phys. B257 [FS14] (1985) 867–893;

J. Rudnick, H. Guo, D. Jasnow, *Finite-size scaling and the renormalization group*, J. Stat. Phys. 41 (1985) 353–373;

M. Lüscher, *A new method to compute the spectrum of low-lying states in massless asymptotically free field theories*, Phys. Lett. 118B (1982) 391–394; *Some analytic results concerning the mass spectrum of Yang-Mills gauge theories on a torus*, Nucl. Phys. B219 (1983) 233–261.

[194] A collection of review articles can also be found in V. Privman ed., *Finite Size Scaling and Numerical Simulation of Statistical Systems*, World Publishing (Singapore 1990).

[195] G. Baym *et al*, *The transition temperature of the dilute interacting Bose gas*, Phys. Rev. Lett. 83 (1999) 1703–1706.

[196] G. Baym, J.-P. Blaizot, J. Zinn-Justin, *The transition temperature of the dilute interacting Bose gas for N internal states*, Euro. Phys. Lett. 49 (2000) 150–155.

[197] P. Arnold, B. Tomasik, T_c *for dilute Bose gases: Beyond leading order in 1/N*, Phys. Rev. A 62 (2000) 063604.

[198] W.A. Bardeen, B.W. Lee, R.E. Shrock, *Phase transition in the nonlinear σ model in a (2 + ε)-dimensional continuum*, Phys. Rev. D14 (1976) 985–1005.

[199] Among the authors that have advocated the Nambu–Jona-Lasinio mechanism to generate a composite Higgs particle, see, for example:

Y. Nambu, in *New Theories in Physics*, Kazimierz 1988, Z. Ajduk, S. Pokorski, A. Trautman eds. (World Scientific, Singapore 1989);

A. Miransky, M. Tanabashi, K. Yamawaki, *Is the t quark responsible for the mass of W and Z bosons?* Mod. Phys. Lett. A4 (1989) 1043–1053; *Dynamical electroweak symmetry breaking with large anomalous dimension and t quark condensate*, Phys. Lett. B221 (1989) 177–183.

[200] A. Hasenfratz, P. Hasenfratz, K. Jansen, J. Kuti, Y. Shen, *The equivalence of the top quark condensate and the elementary Higgs field*, Nucl. Phys. B365 (1991) 79–97.

[201] See, for example, J. Zinn-Justin, *Quantum field theory at finite temperature: An introduction*, SPhT lectures unpublished (2000) arXiv: hep-ph/0005272.

[202] P.C. Hohenberg, B.I. Halperin, *Theory of dynamic critical phenomena*, Rev. Mod. Phys. 49 (1977) 435–479.

[203] P.C. Martin, E.D. Siggia, H.A. Rose, *Statistical dynamics of classical systems*, Phys. Rev. A8 (1973) 423–437.

[204] C. Jarzinsky, *Nonequilibrium equality for free energy differences*, Phys. Rev. Lett. 78 (1997) 2690–2693.

[205] C. De Dominicis, L. Peliti, *Field-theory renormalization and critical dynamics above Tc: Helium, antiferromagnets, and liquid-gas systems*, Phys. Rev. B18 (1978) 353–376 and references therein.

[206] See also Chapter 18 in Ref. [18].

[207] P. Langevin, *Sur la théorie du mouvement brownien*, CR Acad. Sci. Paris. 146 (1908) 530–533.

[208] A.D. Fokker, *Die mittlere Energie rotierender elektrischer Dipole im Strahlungsfeld*, Ann. der Physik 348 (1914) 810–820;

M. Planck, *Über einen Satz der statistischen Dynamik und seine Erweiterung in der Quantentheorie*, Sitzungberichte der preuss. Akademie der Wissenschaften 24 (1917) 324–341;

For more early references see S. Chandrasekhar, *Stochastic problems in physics and astronomy*, Rev. Mod. Phys. 15 (1943) 1–89.

[209] N.V. Antonov, A.N. Vasil'ev, *Critical dynamics as a field theory*, Theor. Math. Phys. 60 (1984) 671–679.

[210] E. Borel, *Mémoire sur les séries divergentes*, Ann. Sci. École Norm. Sup., 16 (1899) 9–131.

[211] J.-P. Eckmann, J. Magnen, R. Sénéor, *Decay properties and Borel summability for the Schwinger functions in $P(\Phi)_2$ theories*, Comm. Math. Phys., 39 (1975) 251–271; J. Magnen, R. Sénéor, *Phase space cell expansion and Borel summability for the Euclidean ϕ_3^4 theory*, Comm. Math. Phys., 56 (1977) 237–276.

[212] G. Parisi, *Singularities of the Borel transform in renormalizable theories*, Phys. Lett. 76B (1978) 65–66;

for a review see M. Beneke, *Renormalons*, Phys. Rep. 317 (1999) 1–142.

[213] J. Zinn-Justin, *Perturbation series at large orders in quantum mechanics and field theories: application to the problem of resummation*, Phys. Rep. 70 (1981) 109–167.

[214] G.A. Baker, P. Graves-Morris, *Padé Approximants*, Encyclopedia of Mathematics and its applications 59, Cambridge Univ. Press (Cambridge 1996).

[215] For a review see

J. Zinn-Justin, *Strong interactions dynamics with Padé approximants*, Phys. Rep. 1 (1971) 55–102.

[216] For the order-dependent mapping the original article is:

R. Seznec, J. Zinn-Justin, *Summation of divergent series by order dependent mappings: Application to the anharmonic oscillator and critical exponents in field theory* , J. Math. Phys. 20 (1979) 1398—1408.

[217] J. Zinn-Justin, U.D. Jentschura, *Order-dependent mappings: Strong-coupling behavior from weak-coupling expansions in non-Hermitian theories*, J. Math. Phys. 51 (2010) 072106.

[218] J.C. Le Guillou, J. Zinn-Justin, *The hydrogen atom in strong magnetic fields: summation of the weak field series expansion*, Ann. Phys. (N.Y.) 147 (1983) 57–84.

[219] U.D. Jentschura, J. Zinn-Justin, *Imaginary cubic perturbation: numerical and analytic study*, J. Phys. A 43 (2010) 425301.

[220] R. Guida, K. Konishi, H. Suzuki, *Convergence of scaled delta expansion: Anharmonic oscillator*, Ann. Phys. (NY) 41 (1995) 152–184; *Improved Convergence Proof of the Delta Expansion and order dependent mappings*, Ann. Phys. (NY) 249 (1996)109–145.

[221] A. Duncan, H. F. Jones, *Convergence proof for optimized δ-expansion: Anharmonic oscillator*, Phys. Rev. D47 (1993), 2560–2572;

see also, C. M. Bender, A. Duncan, H. F. Jones, *Convergence of the optimized δ expansion for the connected vacuum amplitude: Zero dimensions*, Phys. Rev. D49 (1994), 4219–4225.

[222] C.M. Bender, T.T. Wu, *Large order of perturbation theory*, Phys. Rev. Lett. 27 (1971) 461–465.

[223] E. Brézin, J. C. Le Guillou, J. Zinn-Justin, *Perturbation theory at large order. I. The φ^{2N} interaction*, Phys. Rev. D 15 (1977) 1544–1577.

[224] G. V. Dunne, M. Unsal, *Generating energy eigenvalue trans-series from perturbation theory*, Phys. Rev. D 89 (2014) 041701(R).

[225] A. Voros, *The return of the quartic oscillator. The complex WKB method*, Annales de l'IHP A39 (1983) 211–338.

[226] The analytic continuation in the g-plane is discussed in
G. Parisi, *Asymptotic estimates in perturbation theory*, Phys. Lett. 66B (1977) 167–169;
E.B. Bogolmony, *Calculation of the Green functions by the coupling constant dispersion relations*, Phys. Lett. 67B (1977) 193–194.

[227] The method of collective coordinates was first described in this context by
J. Zittartz, J.S. Langer, *Theory of bound states in a random potential*, Phys. Rev. 148 (1966) 741–747;
J.S. Langer, *Theory of the condensation point*, Ann. Phys. (NY) 41 (1967) 108–157.

[228] J. Zinn-Justin, *Expansion around instantons in quantum mechanics*, J. Math. Phys. 22 (1981) 511–520.

[229] U.D. Jentschura, J. Zinn-Justin, *Higher-order corrections to instantons*, J. Phys. A34 (2001) L1–L6; *Multi-instantons and exact results I, II: conjectures, WKB expansions, and instanton interactions*, Ann. Phys. 313 (2004) 197; *ibidem*, (2004) 269–325.

[230] U.D. Jentschura, A. Surzhykov, J. Zinn-Justin, *Multi-instantons and exact results III: Unification of even and odd anharmonic oscillators*, Ann. Phys. 325 (2010) 1135–1172.

[231] A.A Andrianov, *The large N expansion as a local perturbation theory*, Ann. Phys. 140 (1982) 82–100.

[232] J. Ecalle, *Les Fonctions Résurgentes*, Vols. I - III, (Publ. Math. Orsay, 1981).

[233] E. Delabaere, H. Dillinger, F. Pham, *Exact semiclassical expansions for one-dimensional quantum oscillators*, J. Math. Phys. 38, 6126 (1997);
E. Delabaere, F. Pham, *Resurgent methods in semi-classical asymptotics*, Annales de l'IHP, A 71, 1 (1999);
C. Mitschi, D. Sauzin, *Divergent Series, Summability and Resurgence I*, Lecture notes in mathematics, Springer (2016).

[234] E.P. Wigner, *Characteristic vectors of bordered matrices with infinite dimension*, Ann. Math. 62 (1955) 548–564; *ibidem, II*, Ann. Math. 65 (1957) 203–207; *On the distribution of the roots of certain symmetric matrices*, Ann. Math. 67 (1958) 325–327.

[235] G. 't Hooft *A planar diagram theory for strong interactions*, Nucl. Phys. B72 (1974) 461–473.

[236] F. David, *Planar diagrams, two-dimensional lattice gravity and surface models*, Nucl. Phys. B 257 (1985) 45–58;

J. Ambjorn, B. Durhuus,, J. Fröhlich, *Diseases of triangulated random surface models, and possible cures*, Nucl. Phys. B257 (1985) 433–449;

V.A. Kazakov, I.K. Kostov, A.A. Migdal, *Critical properties of randomly triangulated planar random surfaces*, Phys. Lett. B 157 (1985) 295–300.

[237] For a review of early developments see P. Di Francesco, P. Ginsparg, J. Zinn-Justin, *2D gravity and random matrices*, Phys. Rep. 254 (1995) 1–133.

[238] E. Brézin, C. Itzykson, G. Parisi, J.B. Zuber, *Planar diagrams*, Comm. Math. Phys. 59 (1978) 35–51.

[239] D. Bessis, C. Itzykson, J.B. Zuber, *Quantum field theory techniques in graphical enumeration*, Advances in Applied Math. 1 (1980) 109–157.

[240] E. Brézin, J. Zinn-Justin, *Renormalization group approach to matrix models*, Phys. Lett. B 288 (1992) 54–58.

[241] J. Zinn-Justin, *Random vector and matrix theories: A renormalization group approach*, J. Stat. Phys. 157 (2014) 990–1016.

Index

Abelian anomaly 284

Abelian gauge theory 273
  finite temperature 408

Abelian Landau–Ginzburg–Higgs
  model 200

action
  dimensional analysis 116
  effective 114

adjoint representation 238

anharmonic oscillator 475

anomaly 222, 268
  Abelian 284
  calculation 287
  cancellation 305

anti-periodic boundary conditions 374, 400

area law 232

asymmetric wells 490

asymptotic freedom 128, 145, 209, 221,
  229, 260, 262, 263, 297, 393, 400

asymptotic safety 61, 259, 260, 262, 263

asymptotic series 472

auxiliary field 370

auxiliary regulator fields 267

axial anomaly 223, 298

axial current 285

axion 224, 333

β-function 77, 94
  non-linear σ-model 165

band structure 319, 320

bare charge 58

bare mass 58

barrier penetration 27, 320, 471, 476

Bogomolnyi's inequality 324, 329

Bohr–Sommerfeld quantization formula
  generalized 472, 480, 491

Borel
  summability 452, 472, 477
  summable 471, 473, 477
  summation 452
  transformation 79, 452, 459, 473
    and mapping 460

Bose–Einstein condensation 361, 363,
  364

boson thermal correction 396

bosonization 291

boundary conditions
  twisted 347

Brillouin zone 150

Brownian motion 3, 17, 423

Brownian paths 17

BRST
  closed 218, 242, 252
  cohomology 219
    operator 218, 247
  exact 218, 242, 252
  invariance 193
  invariant polynomials 243
  symmetry 50, 218, 237, 239, 241,
    251,444
    and group elements 252
    the origin 250

Callan–Symanzik equations 455

canonical commutation relations 182,
  266, 278

canonical transformations 249

central charge 496

chemical potential 361

chiral
anomaly 288, 291, 293, 322
gauge transformations 285
symmetry 125, 197, 205, 273, 333, 400
spontaneous breaking 291, 297
transformations on the lattice 308
non-Abelian 311

classical gauge field equations 214

classical statistical field theory 43

cluster property 86

coherence length 200

cohomology operator 242, 252

collective coordinate 476

colour group 209

colour quantum number 220, 226, 305

complex WKB expansion 472

confinement property 221, 226, 229, 291, 297

conformal mapping 79

conformal matter 496

connected correlation functions 86

connection 191, 211
1-form 192

continued fractions 458

continuous phase transition 371

continuous symmetries 358

continuum limit 4, 16, 495

correction exponent 79

correlation functions 16, 24
scaling form 168

correlation length 61, 88, 336
exponent ν 77, 88, 169

correlation time 425, 429

Coulomb gauge 29

covariant derivative 183, 189, 191, 211, 238, 273, 436

covariant quantization 50

covariant time derivative 184

CP violation 209

CP^1 model 327

CP^{N-1} models 322

critical
domain 73, 88, 262
dynamics 439
exponents 79, 456
order-dependent mapping 470
Ising model 496
slowing down 441, 442, 449
temperature 261, 382

crossover 375
scale 146, 257, 262, 367, 397

cubic anisotropy 103

cubic fixed point 104

curvature of the connection 239

curvature tensor 191, 212, 227

curvature 2-form 192

cut-off 115

cylindrical geometry 340, 353

decay of critical correlations 110

deconfinement transition 226, 236

degenerate minima 471

density matrix 18, 41, 373
at thermal equilibrium 19, 31, 41

detailed balance 424, 430, 433, 438

diffusion equation 19

dimension of field 169

dimensional
analysis 89
continuation 91, 92, 146, 343
reduction 337, 365, 373, 374, 376, 386, 410, 412
regularization 92, 157, 219, 267, 278, 371, 375

Dirac fermion 270

Dirac matrices 189

Dirac operator 299

Dirac operator
index 299

discrete Z_2 chiral symmetry 205

dissipative Langevin equation 429

divergence of the correlation length 365

divergences
 infrared 153
divergent series 452, 471
domain wall fermions 267, 316
 lattice 318
double scaling limit 493, 495
dynamic
 action 432, 442, 443, 447
 critical exponent 449
 critical two-point function 449
 exponent 441, 442, 448
 principle 183
ε-expansion 99, 261
effective
 charge 60, 71
 field theory 33, 66, 67, 70, 85, 111, 386
 Gross-Neveu model 125
 integral 336, 337
 interaction 366
 large distance theories 62
 local quantum field theory 255
electric field 179
electromagnetic π_0 decay 222, 290
electromagnetic tensor 37
electron self-energy 56
emergent symmetry 104
equation of state 362
 scaling form 168
equilibrium distribution 425, 429
Euler–Lagrange equations 35, 178
exterior algebra 252

Faddeev–Popov
 determinant 239
 ghost fields 50, 193, 199, 202, 217, 237, 240
fermion doubling 273
fermion mass generation 125, 205
fermion propagator in a gauge field 274
Feynman diagram 57
 fat 494
Feynman's

gauge 219
 parametrization 123
 path integral 26
field anomalous dimension η 77
field correlation functions 86
field integral 32
field theory
 renormalizable 76
fields
 partial integration 142
fine tuning 58, 67, 91, 114, 125, 205, 208, 256, 262, 263, 394, 403
finite size 373
 correlation length 354, 357
 effects 343, 375
 hypercubic geometry 347
 scaling 335, 345
finite temperature phase transition 380
finite temperature 336, 373
fixed point 6, 63
 attractive 73
 equation 7
 local stability 85
fixed points
 line of 9
flavour quantum number 220, 333
flow equations 497
flux tube 235
Fokker–Planck
 equation 422, 428, 440
 Hamiltonian 428, 440
Fredholm determinant 474
functional δ-function 392
functional differentiation 86
functional integration 13
gap equation 401
gauge
 action 213
 condition 48
 covariant 216

gauge (continued)
  field 37, 47, 191, 211, 238
    topological properties 301
  fixing 29, 48, 213, 237, 245
    locality 194
  independence 189, 218, 243
  invariance 47, 177, 178, 180
    a dynamic principle 180
    obstruction 303
  invariant
    action 213
    observables 216
    operators 192
  theory
    lattice action 282
    semi-classical vacuum 223
  transformation 36, 37, 179, 182, 210, 238, 409
    infinitesimal 211
Gaussian
  critical theory 87
  dimension 89
  distribution 15
    universality 3
  expectation values 24
  field theory 87
  fixed point 65, 75, 111, 141, 152
  path integral 23
  renormalization 90, 115
  white noise 422, 439
Gauss's law 179, 215
General Relativity 194
generalized conjugate momentum 181
ghost field 219, 277
  action 277
ghost number conservation 237
Ginsparg–Wilson relation 307
global group transformations 183
global linear symmetries 271
gluonium 226
gluons 209, 220

Goldstone boson 145, 198, 289, 333, 391, 393
Goldstone modes 260
gradient flow 107
Grassmann
  algebra 31
  differentiation and integration 31
  path integral 31
Gribov copies 194
Gribov's ambiguity 217
Gross–Neveu model 124, 126, 262
Gross–Neveu–Yukawa model 124, 204, 205
Hamiltonian 179
  formalism 214
harmonic oscillator 23
harmonic well 338
heat equation 4
Helium superfluid transition 361
Helium-type superfluid transition 362
Higgs boson 58, 195
Hilbert–Einstein action 37
Hlder condition 17
holomorphic formalism and bosons 30
holomorphic path integral 30
homotopy classes 223, 324
hyper-asymptotic expansion 472, 480
hypercubic symmetry 11
imaginary time Schrödinger equation 22
infrared
  divergences 91, 153
  fixed point 366, 448
  freedom 206, 391, 398, 414
  regularization 154
  regulator 219
instanton 27, 45, 223, 319, 321, 324, 471, 476, 479
  calculus 79
  interaction 486
  $SU(N)$ model 222

irrelevant 112
  operators 85
  perturbation 9, 76

Jarzinsky's relation 426, 435

Kadanoff's renormalization group 83

Landau–Ginzburg Hamiltonian 439
Landau's gauge 48, 216, 240
Landau's theory of critical phenomena
  61, 83
Langevin equation 422
  linear 441
  purely dissipative 423
large N
  action 390
  expansion 369
  field components 52
  limit 370, 389
  $1/N$ correction 371
  techniques 52
large order behaviour 45, 79, 454, 479,
  482, 491
  perturbation theory 471
  quartic anharmonic oscillator 476
lattice
  chiral transformations 308
  Dirac operator in a gauge background
    309
  gauge theory 50
  $O(N)$ vector model 145
  regularization 219, 267, 280
    numerical simulations 50
line of fixed points 176
linear driving force 423
link variables 226
local
  action 362
  expansion 113, 122, 341, 342, 355
  field theory 32
  group transformations 183
  Markov process 15
  stability 104

$U(1)$ transformations 183
locality 2, 32, 56, 57, 71, 86, 114
loop expansion 471
low temperature expansion 149
low temperature limit 147

magnetic field 179
many component field 52
marginal 112
  operators 85
  perturbation 9
Markov property 1, 20
Markovian process 1
mass renormalization 76, 91, 97
matrix γ_5 279
matrix renormalization 498
Matsubara frequencies 362
Maxwell's equations 36
mean field theory 61
Mermin–Wagner–Coleman theorem
  145
method of characteristics 96, 366
metric tensor on the sphere 155
minimal sensitivity 460
minimal subtraction scheme 95
mode expansion 375
mode regularization 299
momentum quantization 346
momentum shell 64, 84
Monte Carlo methods
  simulations 235
multi-instanton 472, 481, 485
  action 486
  contribution 487

N-vector models 98
Nambu–Goldstone bosons 197
neutral pion 290
non-Abelian
  anomaly 301
  gauge
    fields 209, 298
    theories 190, 198, 237, 276, 374

transformations 238
non-decoupling of scales 71
non-linear σ-model 51, 130, 14, 146, 154, 260
  critical temperature 167
  field renormalization 160
  on the lattice 147
  stability of the Gaussian fixed point 152
  temperature renormalization 160
  the problem of the measure 156
non-linear representation 149
non-renormalizable interactions 117
non-renormalizable quantum field theory 112

$O(3)$ non-linear σ-model 327
$O(N)$ symmetric action 369
one-loop correction 377
one-loop Feynman diagrams 271
one-loop reduced action 389
operator ordering 267
order-dependent mapping method 461
  convergence analysis 463
Ornstein–Zernicke form 88
overlap fermions 267, 311

Padé approximants 457
Padé–Borel method 459
parallel transport 190, 192, 210, 225
partial field integration
  differential form 144
partition function 18, 22, 41
path integral 12, 16, 38, 185, 432
  barrier penetration 27
  classical statistical physics 21
  perturbative calculation 25
  phase space variables 39
  quantum evolution 26
  stationary phase method 26
path-ordered integral 212
Pauli matrices 327
Pauli–Villars's regularization 219

penetration length 200
perimeter law 232
periodic boundary conditions 42, 335, 346, 361, 374
periodic cosine potential 319, 491
periodic hypercube 347
perturbation theory 90
perturbative
  renormalization group 92
    equations 65, 118
  renormalization theory 33
  universality 93
phase space coordinates 179
phase space path integral 29
phase space variables 39
pions 333
plaquette action 51, 227
power counting 116
power-law decay 87
propagator
  $O(N)$ model 151

quantization
  gauge theories 47
  non-linear σ-model 51
  quantum electrodynamics 48
quantum anomalies 283
quantum chromodynamics 209, 226, 263
quantum electrodynamics 54, 55, 273
  action 56
  inconsistency 61
quantum field theory 53, 55
  in imaginary time 76
  lattice approximation 50
quantum gas 361
quantum gravity
  two-dimensional 496
quantum mechanics 181
quantum partition function 42, 474
quantum time evolution 26
quantum tunnelling 27

quark 209, 333
quartic anharmonic oscillator 25, 468
quartic double-well potential 478
quenched approximation 236

random matrices 493
random variable
  renormalization 7
random walk 1, 14, 423
  lattice 11
reduced action 377
redundant operators 85
redundant perturbation 10
regularization 33, 58, 71, 76, 82, 88, 112,
  114, 237, 267, 375, 441
  dimensional 157
  field theory 33
  higher derivatives 268
regularized theory 82
regulator fields 271, 273
relaxation time 425, 429
relevant 112
  operators 85
  perturbation 75
  perturbation 9
renormalizability 66
renormalizable field theory 65, 117
renormalizable quantum field theory 59,
  71, 112, 120
renormalization 7, 33, 82, 120, 237,
  246, 378
  constants 93
  local monomials 117
  process 112
  scale 83, 205
  strategy 59
  theorem 92, 93
  tuning 33
renormalization group 6, 53, 60, 62, 64,
  72, 83, 373, 375, 442, 493, 496
  β-function 166, 206, 260
    ϕ_3^4 field theory 456

equations 71, 77, 83, 94, 97, 118,
  128, 344, 366, 385, 446, 448
  in the critical domain 94
  non-linear σ-model 165, 166
  solution 366, 367
  exact 64
  fixed point 62, 456
  fixed trajectories 207
  functional equations 140
  functions 456
  general 137
  linearized flow 104
  matrix flow 497
  methods 365
  transformation 7
renormalized
  action 205, 220
  charge 60
  field theory 120
  parameters 59
  renormalization group equations 95
  theory 82
renormalons 454, 471
Riccati equation 488
roughening transition 234

σ-model
  non-linear 51
saddle point equation 371
scalar potential 177, 180
scale decoupling 61, 62, 70
scaling
  dimension 4, 7
    of the field 77, 96
  property 4
  relations 77, 79, 99, 448, 449
Schrödinger equation 181
Schrödinger picture 278
Schwinger model 200, 291
Schwinger's representation 276
self-adjoint quantum operators 182
self-duality equations 223

semi-classical vacuum 223
series summation 453
short distance cut-off 82
short distance insensitivity 60, 63
short distance singularities 88
short range interactions 61
Slavnov–Taylor identities 245
$SO(2)$ model 175
specific heat exponent 495
spectral equation 475, 480
spin connection 194
spin waves 153
spinors 189
spontaneous chiral symmetry breaking
    204, 205
spontaneous symmetry breaking 147, 196
spontaneously broken symmetry 391
stability analysis 73, 499
stable fixed point
    uniqueness 109
Standard Model 195, 333
    particle physics 305
static magnetic field 28
statistical classical field theory 373
statistical field theory 75
steepest descent method 27, 45, 52, 370,
    393, 464, 471, 476, 495
string susceptibility 495
string tension 235
strong coupling expansion 232
strong CP violation 209, 224, 332
structure constants of the Lie algebra 214
summation methods 457
superconductor 199
superfield 444, 447
superpath 436
super-renormalizable
    field theory 339, 361
    quantum field theory 117
supersymmetric quantum mechanics 283
supersymmetry 271, 314, 437, 444

symmetries and regularization 267
symmetry restoration 153
θ-vacuum 224
Taylor series expansion 472
temperature-dependent exponent 176
temporal gauge 49, 215, 229, 410
    perturbative expansion 216
tessalated surfaces 493
thermal equilibrium 18, 22, 41, 373, 374
thermal wavelength 362, 373
thermodynamic potential 381
time-dependent force 433
time scale renormalization 442, 446
top quark 204
topological
    character 283
    charge 223, 321, 324, 325
    expansion 494, 499
topology and anomalies 289
transition temperature
    shift 372
        universal value 372
translational invariant theories 86
trans-series 480
tricritical behaviour 117
triviality issue 119, 206, 254, 259, 385
twisted boundary conditions 347
two-point function 383
    perturbative calculation 164
$U(1)$ gauge group 48
$U(1)$ problem 222, 224, 334
ultraviolet divergences 56, 266
ultraviolet fixed point 261
unitarity 57
unitary evolution 39, 181
unitary gauge 201
universal properties 6, 343, 365, 493
universal ratio 353, 357
universality 9, 61, 63, 81, 82, 111
    at T_c 351
    class 63, 69, 73, 83, 111

variational principle 35, 178, 180

vector potential 36, 177, 180

vertex functions 97

Ward–Takahashi identities 220, 277, 334, 445

wave functions 183

weak-electromagnetic interactions 304

Wess–Zumino consistency conditions 305

Weyl's or temporal gauge 193, 215

Wick's theorem 24, 25

Widom's scaling form 99

Wiener measure 16

Wilson loop 225

Wilson–Fisher's ε-expansion 78, 81, 367

Wilson–Fisher's fixed point 98

Wilson's fermions 281

winding number 223, 324, 327

WKB complex method 478

WKB expansion 489

Yang–Mills field 238

zero mode 335, 336, 346, 364, 376, 412